Proteomics of Human Body Fluids

Proteomics of Human Body Fluids

Principles, Methods, and Applications

Edited by

Visith Thongboonkerd, MD, FRCPT

Medical Molecular Biology Unit
Office for Research and Development
Faculty of Medicine Siriraj Hospital, Mahidol University
Bangkok, Thailand

HUMANA PRESS ✱ TOTOWA, NEW JERSEY

999 Riverview Drive, Suite 208
Totowa, New Jersey 07512

This publication is printed on acid-free paper. ∞
ANSI Z39.48-1984 (American Standards Institute)

Permanence of Paper for Printed Library Materials.

Cover design by Karen Schulz

For additional copies, pricing for bulk purchases, and/or information about other Humana titles, contact Humana at the above address or at any of the following numbers: Tel.: 973-256-1699; Fax: 973-256-8341; E-mail: orders@humanapr.com; or visit our Website: www.humanapress.com

Printed in the United States of America. 10 9 8 7 6 5 4 3 2 1

eISBN 978-1-59745-432-2

Library of Congress Cataloging in Publication Data

Proteomics of human body fluids : principles, methods, and applications
/ edited by Visith Thongboonkerd.
p. ; cm.
Includes bibliographical references and index.
ISBN: 978-1-58829-657-3 (alk. paper)
1. Body fluids—Analysis. 2. Proteins—Analysis. 3. Proteomics.
I. Thongboonkerd, Visith.
[DNLM: 1. Body Fluids—chemistry. 2. Body Fluids—secretion.
3. Proteome—analysis. 4. Proteomics—methods. QU 105 P967 2007]
RB52.P765 2007
616.07'56—dc22 2006025679

Preface

Rapid growth of the "proteomics" field during the past twelve years has contributed to significant advances in science. To date, proteomic technologies have been widely applied to examining various kinds of biological materials. Clinical proteomics is the concept of using proteomic techniques to evaluate the proteomes in clinical samples for biomarker discovery and for better understanding of normal physiology and pathogenic mechanisms of human diseases. Translating the proteomic information to clinical practice may lead to the ultimate goals of earlier diagnosis, improved therapeutic outcome and successful prevention of diseases.

Human body fluids are produced, secreted, and/or excreted from various tissues or organs. Major compositions of body fluids are water, organic substances and inorganic compounds. These compositions vary in each body fluid, making one's function different from the others. Analyses of protein components in individual fluids would increase current knowledge on the biology and physiology of various organ systems, and on the pathophysiology of diseases, which cause alterations in protein production, secretion, and/or excretion from the affected tissues or organs into the body fluids. Additionally, human body fluids are the main targets and valuable sources for biomarker discovery. As the high-throughput capability and applicability of proteomics for analyzing proteins in the body fluids have been already proven, these desires are most likely achievable by using proteomic technologies.

Proteomics of Human Body Fluids: Principles, Methods, and Applications is the first and nearly complete collection of applications of proteomics to analyze various human body fluids, including plasma, serum, urine, cerebrospinal fluid, pleural effusion, bronchoalveolar lavage fluid, sputum, nasal lavage fluid, saliva, pancreatic juice, bile, amniotic fluid, milk, nipple aspirate fluid, seminal fluid, vitreous fluid, dialysate, and ultrafiltrate yielded during renal replacement therapy. The book has been divided into two main parts. The first part provides basic principles and strategies for proteomic analysis of human body fluids, written by the leading experts in the proteomics field. The second part offers more details regarding methodologies, recent findings and clinical applications of proteomic analysis of each specific type of human body fluids, written by the authorities in their respective fields. Perspectives and future directions of each subject are also discussed. This book, therefore, covers almost everything one needs to know about proteomics of human body fluids.

I would like to thank all the authors who have contributed to this book, which would not be possible without their willingness to give the valuable time from their tight schedule and to share their knowledge and experience. I hope that the book will prove to be a useful source of references for all who are interested in this rapidly growing field of science and that the information inside this book will facilitate the progress of current and future studies on the human body fluid proteomes.

***Visith Thongboonkerd,** MD, FRCPT*

Contents

Preface v
Contributors ix

Part I: Principles of Proteomics Applied to Human Body Fluids

1 Proteomic Strategies for Analyzing Body Fluids
Sung-Min Ahn and Richard J. Simpson ***3***

2 Sample Preparation of Body Fluids for Proteomics Analysis
Natalia Govorukhina and Rainer Bischoff ***31***

3 Multiplexed Immunoassays for Protein Profiling in Human Body Fluids
Silke Wittemann, Dominic P. Eisinger, Laurie L. Stephen, and Thomas O. Joos ***71***

4 Deciphering the Hieroglyphics of Functional Proteomics Using Small Molecule Probes
Wayne F. Patton ***83***

5 Modification-Specific Proteomic Analysis of Glycoproteins in Human Body Fluids by Mass Spectrometry
Jakob Bunkenborg, Per Hägglund, and Ole Nørregaard Jensen ***107***

6 Plasma Proteome Database
Malabika Sarker, G. Hanumanthu, and Akhilesh Pandey ***129***

7 2D PAGE Databases for Proteins in Human Body Fluids
Christine Hoogland, Khaled Mostaguir, Jean-Charles Sanchez, Denis F. Hochstrasser, and Ron D. Appel ***137***

8 Bioinformatics and Experimental Design for Biomarker Discovery
Marc R. Wilkins and Sybille M. N. Hunt ***147***

9 Integrative Omics, Pharmacoproteomics, and Human Body Fluids
K. K. Jain ***175***

Part II: Proteomic Analysis of Specific Types of Human Body Fluids: *Methods, Findings, Applications, Perspectives, and Future Directions*

10 The Human Plasma and Serum Proteome
Gilbert S. Omenn, Rajasree Menon, Marcin Adamski, Thomas Blackwell, Brian B. Haab, Weimin Gao, and David J. States ***195***

11 Proteomics of Human Urine
Visith Thongboonkerd, Pedro R. Cutillas, Robert J. Unwin, Stefan Schaub, Peter Nickerson, Marion Haubitz, Harald Mischak, Dobrin Nedelkov, Urban A. Kiernan, and Randall W. Nelson **225**

12 Proteomics of Human Cerebrospinal Fluid
Margareta Ramström and Jonas Bergquist **269**

13 Proteomics of Pleural Effusion
Joost Hegmans, Annabrita Hemmes, and Bart Lambrecht **285**

14 Proteomics of Bronchoalveolar Lavage Fluid and Sputum
Ruddy Wattiez, Olivier Michel, and Paul Falmagne **309**

15 Proteomics of Sinusitis Nasal Lavage Fluid
Begona Casado, Simona Viglio, and James N. Baraniuk **327**

16 Proteomics of Human Saliva
Francisco M. L. Amado, Rui M. P. Vitorino, Maria J. C. Lobo, and Pedro M. D. N. Domingues **347**

17 Proteomics of Human Pancreatic Juice
Mads Grønborg, Anirban Maitra, and Akhilesh Pandey **377**

18 Proteomics of Human Bile
Troels Z. Kristiansen, Anirban Maitra, and Akhilesh Pandey **399**

19 Proteomics of Amniotic Fluid
David Crettaz, Lynne Thadikkaran, Denis Gallot, Pierre-Alain Queloz, Vincent Sapin, Joël S. Rossier, Patrick Hohlfeld, and Jean-Daniel Tissot **415**

20 Proteomics of Human Milk
Amedeo Conti, Maria Gabriella Giuffrida, and Maria Cavaletto **437**

21 Proteomics of Nipple Aspirate Fluid in Nonlactating Women
Edward R. Sauter **453**

22 Proteomics of Seminal Fluid
Benjamin Solomon and Mark W. Duncan **467**

23 Proteomics of Vitreous Fluid
Atsushi Minamoto, Ken Yamane, and Tomoko Yokoyama **495**

24 Proteomics of Human Dialysate and Ultrafiltrate Fluids Yielded by Renal Replacement Therapy
Michael Walden, Stefan Wittke, Harald Mischak, and Raymond C. Vanholder , for the European Uremic Toxin Work Group (EUTox) **509**

Index **521**

Contributors

MARCIN ADAMSKI • *Michigan Proteomics Alliance for Cancer Research, Center for Computational Medicine and Biology, University of Michigan Medical School, Ann Arbor, MI. Present address: Faculty of Biological and Chemical Sciences, Integrative Biology School, The University of Queensland, Brisbane, Australia*

SUNG-MIN AHN • *Joint ProteomicS Laboratory, Ludwig Institute for Cancer Research & the Walter and Eliza Hall Institute of Medical Research, P.O. Box 2008, Royal Melbourne Hospital Parkville, Victoria 3050, Australia*

FRANCISCO M. L. AMADO • *Department of Chemistry, University of Aveiro, 3810-193 Aveiro, Portugal*

RON D. APPEL • *Proteome Informatics Group, Swiss Institute of Bioinformatics and Computer Science Department, University of Geneva, Geneva, Switzerland*

JAMES N. BARANIUK • *Proteomics Laboratory, Georgetown University, Washington, DC*

JONAS BERGQUIST • *Department of Chemistry, Analytical Chemistry, Uppsala University, Uppsala, Sweden*

RAINER BISCHOFF • *University of Groningen, Center of Pharmacy, Department of Analytical Biochemistry, Antonius Deusinglaan 1, 9713 AV Groningen, The Netherlands*

THOMAS BLACKWELL • *Michigan Proteomics Alliance for Cancer Research, Center for Computational Medicine and Biology, University of Michigan Medical School, Ann Arbor, MI*

JAKOB BUNKENBORG • *Department of Biochemistry and Molecular Biology, University of Southern Denmark, Campusvej 55, DK-5230 Odense M, Denmark*

BEGONA CASADO • *Proteomics Laboratory, Georgetown University, Washington, DC; and Department of Biochemistry "A Castellani", University of Pavia, Italy*

MARIA CAVALETTO • *Department of Enviroment and Life Sciences, University of Piemonte Orientale, Alessandria, Italy*

AMEDEO CONTI • *National Council of Research—Institute of Science of Food Production, Section of Torino, Italy*

DAVID CRETTAZ • *Service Régional Vaudois de Transfusion Sanguine, Lausanne, Switzerland*

PEDRO R. CUTILLAS • *Ludwig Institute for Cancer Research, University College London Branch; and Department of Biochemistry and Molecular Biology, University College London, London, UK*
PEDRO M. D. N. DOMINGUES • *Department of Chemistry, University of Aveiro, 3810-193 Aveiro, Portugal*
MARK W. DUNCAN • *Division of Endocrinology, Metabolism & Diabetes, Department of Pediatrics, University of Colorado Cancer Center Proteomics Core, University of Colorado Health Sciences Center, Denver, CO*
DOMINIC P. EISINGER • *Upstate USA, Lake Placid, NY*
PAUL FALMAGNE • *Department of Proteomics and Protein Biochemistry, 6 Avenue du Champs de Mars, University of Mons-Hainaut, B-7000, Belgium*
DENIS GALLOT • *Unité de Médecine Materno-Fœtale, Hôtel-Dieu, CHU, Clermont-Ferrand, France*
WEIMIN GAO • *Michigan Proteomics Alliance for Cancer Research. Present address: The Institute of Environmental and Human Health, Department of Environmental Toxicology, Texas Tech University, Lubbock, TX*
MARIA GABRIELLA GIUFFRIDA • *National Council of Research—Institute of Science of Food Production, Section of Torino, Italy*
NATALIA GOVORUKHINA • *University of Groningen, Center of Pharmacy, Department of Analytical Biochemistry, Antonius Deusinglaan 1, 9713 AV Groningen, The Netherlands*
MADS GRØNBORG • *McKusick-Nathans Institute of Genetic Medicine and Department of Biological Chemistry, Johns Hopkins University, Baltimore, MD; and Department of Biochemistry and Molecular Biology, University of Southern Denmark, Campusvej 55, Odense M, Denmark*
BRIAN B. HAAB • *Michigan Proteomics Alliance for Cancer Research, Van Andel Research Institute, Grand Rapids, MI*
PER HÄGGLUND • *Biochemistry and Nutrition Group, Biocentrum DTU 224-124, Technical University of Denmark, DK-2800 Kgs, Lyngby, Denmark*
G. HANUMANTHU • *Institute of Bioinformatics, Bangalore, India*
MARION HAUBITZ • *Department of Nephrology, Medizinische Hochschule Hannover, D-30623 Hannover, Germany*
JOOST HEGMANS • *Department of Pulmonary Medicine, Erasmus MC, Rotterdam, The Netherlands*
ANNABRITA HEMMES • *Department of Pulmonary Medicine, Erasmus MC, Rotterdam, The Netherlands*
DENIS F. HOCHSTRASSER • *Clinical Chemistry Laboratory, Geneva University Hospital, Geneva, Switzerland*
PATRICK HOHLFELD • *Département de Gynécologie et Obstrétrique, CHUV, Lausanne, Switzerland*

CHRISTINE HOOGLAND • *Proteome Informatics Group, Swiss Institute of Bioinformatics, Geneva, Switzerland*
SYBILLE M. N. HUNT • *Proteome Systems, Locked Bag 2073, North Ryde 1670, Sydney, Australia*
K. K. JAIN • *Jain PharmaBiotech, Blaesiring 7, CH-4057 Basel, Switzerland*
OLE NØRREGAARD JENSEN • *Department of Biochemistry and Molecular Biology, University of Southern Denmark, Campusvej 55, DK-5230 Odense M, Denmark*
THOMAS O. JOOS • *NMI Natural and Medical Sciences Institute at the University of Tuebingen, Reutlingen, Germany*
URBAN A. KIERNAN • *Intrinsic Bioprobes, Inc., 625 S. Smith Road, Suite #22, Tempe, AZ*
TROELS Z. KRISTIANSEN • *McKusick-Nathans Institute of Genetic Medicine and Department of Biological Chemistry, Johns Hopkins University, Baltimore, MD; and Department of Biochemistry and Molecular Biology, University of Southern Denmark, Campusvej 55, Odense M, Denmark*
BART LAMBRECHT • *Department of Pulmonary Medicine, Erasmus MC, Rotterdam, The Netherlands*
MARIA J. C. LOBO • *High Institute of Health Sciences-North, Department of Dental Sciences, 4585-116 Gandra PRD, Portugal*
ANIRBAN MAITRA • *McKusick-Nathans Institute of Genetic Medicine, The Sol Goldman Pancreatic Cancer Research Center, and Departments of Biological Chemistry, Pathology and Oncology, Johns Hopkins University, Baltimore, MD*
RAJASREE MENON • *Michigan Proteomics Alliance for Cancer Research, Center for Computational Medicine and Biology, University of Michigan Medical School, Ann Arbor, MI*
OLIVIER MICHEL • *Clinical of Allergology and Respiratory Diseases, CHU Saint-Pierre- ULB. Brussels, B-1000, Belgium*
ATSUSHI MINAMOTO • *Department of Ophthalmology and Visual Science, Division of Frontier Medical Science, Graduate School of Biomedical Sciences, Hiroshima University, Japan*
HARALD MISCHAK • *Mosaiques Diagnostics and Therapeutics AG, Mellendorfer Str. 7, D-30625 Hannover; and Department of Nephrology, Medizinische Hochschule Hannover, D-30623 Hannover, Germany*
KHALED MOSTAGUIR • *Proteome Informatics Group, Swiss Institute of Bioinformatics, Geneva, Switzerland*
DOBRIN NEDELKOV • *Intrinsic Bioprobes, Inc., 625 S. Smith Road, Suite #22, Tempe, AZ*
RANDALL W. NELSON • *Intrinsic Bioprobes, Inc., 625 S. Smith Road, Suite #22, Tempe, AZ*

PETER NICKERSON • *Manitoba Centre for Proteomics; and Faculty of Medicine, University of Manitoba, Winnipeg, MB, Canada*
GILBERT S. OMENN • *Michigan Proteomics Alliance for Cancer Research, Center for Computational Medicine and Biology, Departments of Internal Medicine and Human Genetics, University of Michigan Medical School, Ann Arbor, MI*
AKHILESH PANDEY • *McKusick-Nathans Institute of Genetic Medicine, and Departments of Biological Chemistry, Oncology and Pathology, Johns Hopkins University, Baltimore, MD*
WAYNE F. PATTON • *Biochemistry Department, PerkinElmer Life and Analytical Sciences, 549 Albany Street, Boston, MA*
PIERRE-ALAIN QUELOZ • *Service Régional Vaudois de Transfusion Sanguine, Lausanne, Switzerland*
MARGARETA RAMSTRÖM • *Department of Chemistry, Analytical Chemistry, Uppsala University, P.O. Box 599, SE-751 24 Uppsala, Sweden*
JOËL S. ROSSIER • *DiagnoSwiss SA, Monthey, Switzerland*
JEAN-CHARLES SANCHEZ • *Biomedical Proteomics Research Group, Geneva University Hospital, Geneva, Switzerland*
VINCENT SAPIN • *Laboratoire de Biochimie Médicale, INSERM U.384, CHU, Clermont-Ferrand, France*
MALABIKA SARKER • *Institute of Bioinformatics, Bangalore, India*
EDWARD R. SAUTER • *Department of Surgery, University of Missouri, Columbia, MO*
STEFAN SCHAUB • *Manitoba Centre for Proteomics, Winnipeg, MB, Canada; and Department for Transplant Immunology and Nephrology, University Hospital Basel, Petersgraben 4, 4031 Basel, Switzerland*
RICHARD J. SIMPSON • *Joint ProteomicS Laboratory, Ludwig Institute for Cancer Research & the Walter and Eliza Hall Institute of Medical Research, P.O. Box 2008, Royal Melbourne Hospital Parkville, Victoria 3050, Australia*
BENJAMIN SOLOMON • *Division of Endocrinology, Metabolism & Diabetes, Department of Pediatrics, University of Colorado Cancer Center Proteomics Core, University of Colorado Health Sciences Center, Denver, CO*
DAVID J. STATES • *Michigan Proteomics Alliance for Cancer Research, Center for Computational Medicine and Biology, Department of Human Genetics, University of Michigan Medical School, Ann Arbor, MI*
LAURIE L. STEPHEN • *Upstate USA, Lake Placid, NY*
LYNNE THADIKKARAN • *Service Régional Vaudois de Transfusion Sanguine, Lausanne, Switzerland*
VISITH THONGBOONKERD • *Medical Molecular Biology Unit, Office for Research and Development, Faculty of Medicine Siriraj Hospital, Mahidol University, Bangkok 10700, Thailand*

JEAN-DANIEL TISSOT • *Service Régional Vaudois de Transfusion Sanguine, Lausanne, Switzerland*
ROBERT J. UNWIN • *Centre for Nephrology and Department of Physiology, Royal Free and University College Medical School, University College London, London, UK*
RAYMOND C. VANHOLDER • *Nephrology Section, Department of Internal Medicine, University Hospital, Gent, Belgium*
SIMONA VIGLIO • *Department of Biochemistry "A Castellani", University of Pavia, Italy*
RUI M. P. VITORINO • *Department of Chemistry, University of Aveiro, 3810-193 Aveiro, Portugal*
MICHAEL WALDEN • *Mosaiques Diagnostics and Therapeutics AG, Mellendorfer Str. 7, D-30625 Hannover, Germany*
RUDDY WATTIEZ • *Department of Proteomics and Protein Biochemistry, 6 Avenue du Champs de Mars, University of Mons-Hainaut, B-7000, Belgium*
MARC R. WILKINS • *School of Biotechnology and Biomolecular Sciences, The University of New South Wales, Sydney, NSW 2052, Australia*
SILKE WITTEMANN • *NMI Natural and Medical Sciences Institute at the University of Tuebingen, Reutlingen, Germany*
STEFAN WITTKE • *Mosaiques Diagnostics and Therapeutics AG, Mellendorfer Str. 7, D-30625 Hannover, Germany*
KEN YAMANE • *Department of Ophthalmology and Visual Science, Division of Frontier Medical Science, Graduate School of Biomedical Sciences, Hiroshima University, Japan*
TOMOKO YOKOYAMA • *Department of Ophthalmology and Visual Science, Division of Frontier Medical Science, Graduate School of Biomedical Sciences, Hiroshima University, Japan*

I

Principles of Proteomics Applied to Human Body Fluids

1

Proteomic Strategies for Analyzing Body Fluids

Sung-Min Ahn and Richard J. Simpson

Summary

The rapid development of molecular and cell biology in the latter part of the last century has led us to the understanding that many diseases, including cancer, are caused by perturbations of cellular networks, which are triggered by genetic changes and/or environmental challenges. These perturbations manifest by changing cellular protein profiles, which, in turn, alter the quantitative relationship of tissue-specific proteins shed into the tissue/organ microenvironment. Such altered protein expression profiles in body fluids constitute molecular signatures or fingerprints that reflect the original perturbation of cellular networks. The exciting challenge of modern proteomics is to identify such signatures for various disease states—then the body fluids will become windows into disease and potential biospecimen sources for biomarkers of disease. *(1)*.

Key Words: Body fluids; biomarkers; proteome; expression proteomics; targeted proteomics; sample collection; enrichment; prefractionation.

1. Introduction

1.1. Historical Perspective and Biological Context

Claude Bernard, the 19th century physiologist, introduced the concept of *milieu intérieur* or the internal environment, defining it as the circulating organic liquid that surrounds and bathes all tissue elements *(2)*. Bernard regarded extracellular fluid as the internal environment of the body and emphasized the importance of maintaining the constancy of that environment. This notion still holds true today, especially in the era of proteomics. Indeed, the detection of protein perturbations in the internal environment is one of the major goals of the fledgling field of proteomics.

Approximately 60% of the adult human body is fluid, which is mainly distributed between two compartments: the extracellular space and the intracellular

From: *Proteomics of Human Body Fluids: Principles, Methods, and Applications*
Edited by: V. Thongboonkerd © Humana Press Inc., Totowa, NJ

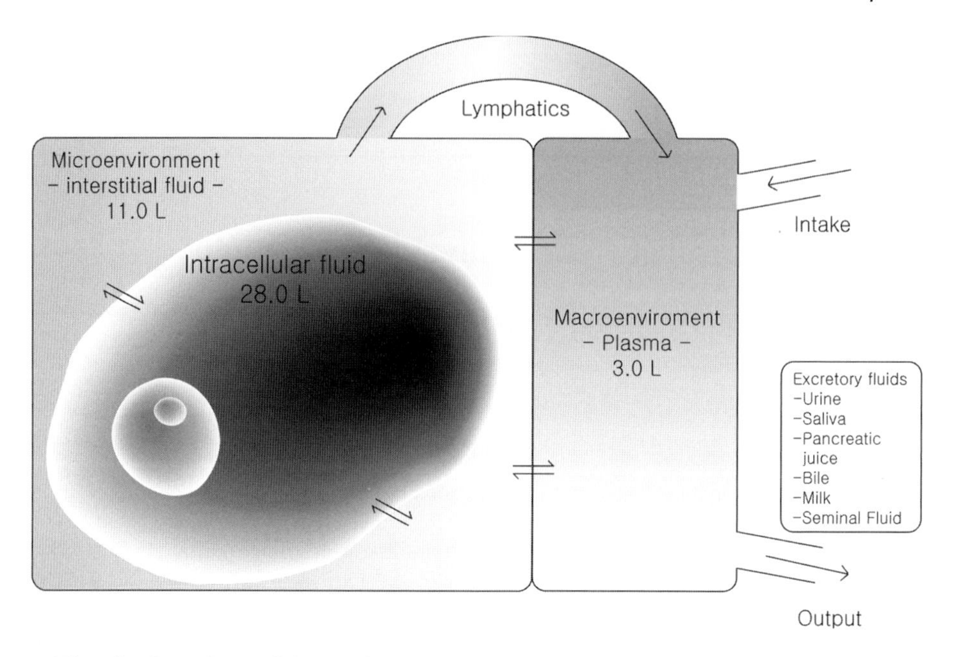

Fig. 1. Overview of body fluids and the internal environment. In an average 70-kg person, the total body water is about 60% of the body weight (approx 42.0 L). The intracellular fluid constitutes about two-thirds of the total body water (approx 28.0 L), whereas the extracellular fluid constitutes the rest (approx 14.0 L). The interstitial fluid comprises more than 75% of the extracellular fluid (approx 11.0 L), whereas plasma comprises the remaining 25% (approx 3.0 L) *(3)*.

space *(3)*. The extracellular fluid is broadly divided into the interstitial fluid and the blood plasma, which can be referred to as the microenvironment and the macroenvironment, respectively **(Fig. 1)**. Tissues consist of cellular elements (parenchymal and stromal cells) and extracellular elements (extracellular matrix and tissue interstitial fluid). In the literature, the term *tissue microenvironment* usually refers to both cellular and extracellular elements *(4)*. In this chapter, however, microenvironments are limited to tissue interstitial fluid (TIF) only, which surrounds and bathes tissues. Since parenchyma, stroma, and blood all contribute to the microenvironment, their individual secreted or shed protein profiles are reflected together in the overall protein profile of the microenvironment. The microenvironment, the interstitial fluid, is in direct contact with cells, exchanging molecules with the intracellular fluid, whereas the macroenvironment, the plasma, continuously communicates with all microenvironments throughout the body, delivering nutrients and signals and receiving feedback directly or indirectly via the lymphatics.

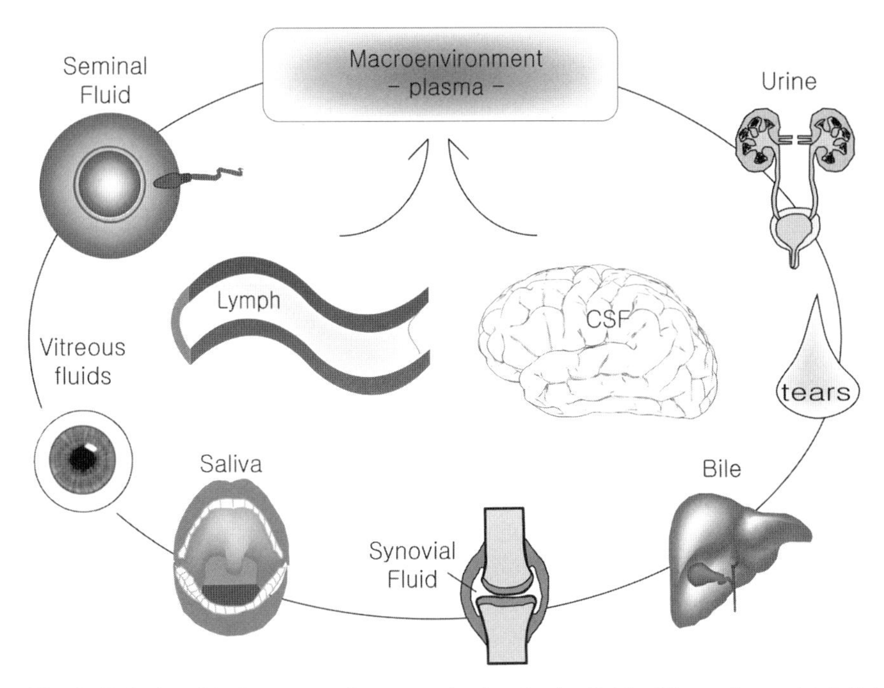

Fig. 2. Relationship between plasma and other body fluids. Tissue interstitial fluid (TIF) enters terminal lymphatics and becomes lymph. (Thus, the composition of lymph in the terminal lympatics is almost identical to that of TIF.) Hence, all lymph in the body empties into the venous system at the junctures of internal jugular veins and subclavian veins *(94)*. CSF is secreted from the choroid plexus at a rate of about 500 mL/d, which is approximately three to four times as much as its total volume (about 150 mL). Excess CSF is absorbed by the arachnoidal villi, which have vesicular holes large enough to allow the free flow of CSF, proteins, and even particles as large as red and white blood cells into the venous system *(95)*. Since lymph and CSF drain into the blood, theoretically all proteins in lymph and CSF are present in the blood. For this reason, lymph and CSF proteomes are considered subproteomes of the plasma. Other body fluids such as urine and tears represent plasma to varying extents while having unique characteristics of their own.

Plasma is important in any proteomic analysis of human body fluids, not only because every cell in the body leaves a record of its physiological state in the products it sheds into the blood *(5)*, but also because it influences most other body fluids. Therefore, it is important to understand the proteomes of various body fluids in the context of plasma. Theoretically, lymph and cerebrospinal fluid (CSF) proteomes are subproteomes of plasma since these body fluids eventually drain into plasma. Other body fluids represent plasma to varying extents while having unique characteristics of their own (**Fig. 2**).

1.2. Differential Enrichment of Biomarkers in Body Fluids

One of the main challenges of proteomics is to find molecular signatures or biomarkers of disease. In plasma, high-abundance proteins such as albumin and transferrin constitute approx 99% of the total protein and the remaining 1% is assumed to include many potential biomarkers that are typically of low abundance *(6)*. Therefore, removal of high-abundance proteins has become a common practice to enrich for low-abundance proteins in plasma. (This issue will be discussed in more details later in this chapter, as well as in other chapters.) However, before trying to remove high-abundance proteins from plasma, the concept of differential enrichment of biomarkers in various body fluids needs to be considered. **Figure 3** illustrates a simplified relationship of the concentration of secreted or membrane-shed cellular proteins in TIF, lymph, and blood. For example, if there is a cancer in the sigmoid region of the colon, cancer cells will secrete or shed cancer-specific proteins into the microenvironment. Such proteins traffic from the TIF to the lymph, being diluted during the process. Lymph fluids from various regions of the body merge and eventually drain into the circulatory system. Approximately 2.5 L of lymph drains into the systemic circulation per day, whereas about 3 L of plasma (approx 5 L of blood) is ejected from the heart every minute. Therefore, the dilution factor is at least 1.5×10^3. (Lymph fluids from different tissues have different tissue-specific proteins. This additional consideration is not included here). Given that only a 10-fold enrichment can be achieved by removing the top six most abundant proteins in plasma, the advantage of using TIF *(7)* or lymph *(8)* rather than plasma seems considerable in discovering biomarkers. For example, the study of Sedlaczek and colleagues *(9)* highlights the differential enrichment of CA125, an ovarian cancer marker, in different body fluids from patients with ovarian carcinoma. **Table 1** summarizes their comparative analysis of CA125 in sera, cyst fluids, and ascites. According to this study, the median value of CA125 is approx 64-fold higher in cyst fluid than in serum.

Malignant ascites is another example of differential enrichment of secreted or membrane-shed proteins. Some cancers such as colorectal and ovarian cancers can be seeded onto peritoneal cavity and cause malignant ascites via various mechanisms. According to Trape and colleagues *(10)*, carcinoembryonic antigen (CEA) levels in malignant ascites are in the range of 33,540 ng/mL maximum, which is more than 5×10^3-fold higher than the normal plasma level of CEA (<5 ng/mL). Although the availability of clinical specimens often becomes the bottleneck of body fluid research owing to a paucity of clinical specimens and ethical considerations, understanding and utilizing the differential enrichment of biomarkers may open a new window of opportunity for discovering otherwise undetectable low-abundance biomarkers.

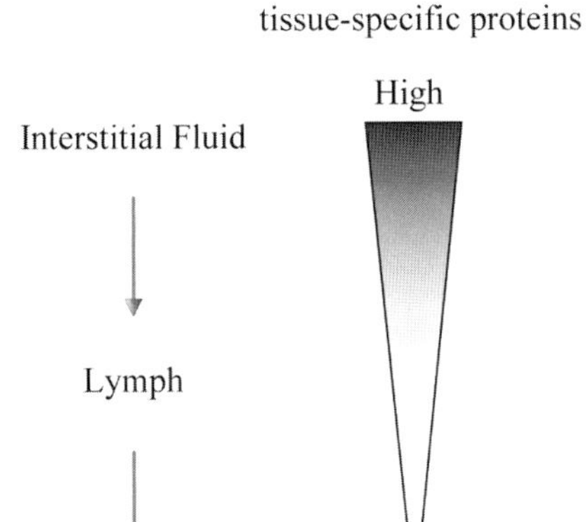

Fig. 3. Dilution of tissue-specific proteins in TIF, lymph, and blood.

Table 1
Median Levels of Ca125 in Serum, Cyst Fluid, and Ascites From Patients With Ovarian Neoplasms

	CA125 (U/mL)		
Histologic type	Serum	Ascites	Cyst fluid
Serous carcinoma	696.0	18,563.0	44,850.0
Endometroid carcinoma	661.0	14,415.5	32,150.0
Mucinous adenocarcinoma	67.0	3521.5	3930.5
Undifferentiated carcinoma	860.7	3909.5	—
Serous cystadenoma	7.1	—	42150.0
Serous cyst	4.8	—	6851.5
Mucinous adenoma	10.8	—	5691.5

From Sedlaczek et al. *(9)*, with permission.

2. Proteomic Approaches for Studying Human Body Fluids

Proteomics, a newly emerging postgenomic technology that allows one to unravel the biological complexity encoded by the genome at the protein level, is built on technologies that allow one to analyze large numbers of proteins in a single experiment. Broadly, there are two main facets of proteomics research:

1. *Expression proteomics,* which aims to catalog the proteome, i.e., the full complement of proteins expressed by the genome in any given cell, tissue, or body fluid at a given time.

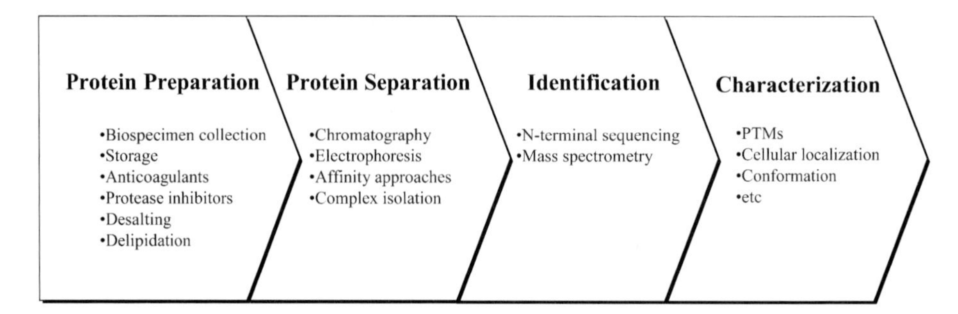

Fig. 4. Pipeline of technologies used in the field of proteomics. The overall success of qualitative and quantitative proteomics relies on the success of the individual technologies involved in a proteomic pipeline. PTMs, posttranslational modifications.

2. *Targeted proteomics,* which strives to determine the cellular functions of genes directly at the protein level (e.g., protein-protein interactions, posttranslational modifications, protein localizations within cells) ***(11)***.

Currently, the major focus of proteomics of body fluids is expression proteomics, especially the quantitative differences in protein profiles between physiological and pathological states.

Figure 4 summarizes the pipeline of technologies that comprise the field of proteomics strategies. Each step in the pipeline involves defined technologies, each of which is technically challenging and of equal importance. Needless to say, the overall success of any proteomics research depends on the success of the individual step in the proteomic pipeline. For solving specific biological questions, the combination of various options in each step provides much flexibility in experimental design. In this chapter, we address current issues and technologies involved in each step of proteomics with an emphasis on their application to body fluid research.

2.1. Biospecimen Collection and Storage

Success in proteomics very much depends on careful biospecimen preparation. In clinical chemistry, many factors are known to cause variations in biospecimen precollection, collection, and postcollection stages ***(12)***. Therefore, a standardized protocol for sample collection, processing, and storage is essential for reproducible experiments within a given laboratory and, especially, from one laboratory to another. (When we compare a large set of data from different laboratories, it is important that we are actually comparing "apples with apples!"). This was one of the major technical issues addressed in the pilot phase of the Human Plasma Proteome Project, the first systematic international effort devoted to analyzing a body fluid ***(13)***.

Among the various body fluids, blood is the most sampled and studied, yet its optimal sample preparation is still problematic. For proteomic analysis, blood can be collected as serum or plasma. When blood is removed from the body and allowed to clot, it separates into a solid clot containing blood cells and fibrin, and a liquid phase termed serum. If an anticoagulant such as heparin or EDTA is added, the liquid phase is termed plasma ***(12)***. From a clinical chemistry perspective, serum differs from plasma only in that it lacks fibrinogen. From a proteomics perspective, however, the differences between serum and plasma can be considerable. The physiological and biochemical difference between serum and plasma is demarcated by the activation of the coagulation cascade, which involves the sequential activation of proteases ***(14)***. The activated proteases during this process will in turn have proteolytic effects on other proteins. According to a recent report of the Human Plasma Proteome Organisation (HUPO), a significant number of peptides differed between serum and plasma specimens (especially intracellular, coagulation-dependent, and enzymatic activity-derived peptides) ***(15)***. The issue of coagulation can also be applied to other body fluids. Extravascular coagulations are observed in lymph ***(16)*** and synovial fluid ***(17)***, and it is likely that most of the internal body fluids have coagulation factors from blood to some extent.

Hulmes and colleagues ***(18)*** have addressed questions regarding plasma collection, stabilization, and storage procedures for proteomic analysis of clinical samples. According to their research, addition of a protease inhibitor cocktail directly to plasma collection tubes prior to phlebotomy, centrifugation within 1 h of blood draw, snap-freezing aliquots immediately in a dry ice/alcohol bath, and storing frozen aliquots in a −70°C freezer can improve sample qualities for proteomic analysis. This recommendation is supported in the report of the HUPO Plasma Proteome Project on specimen collection and handling ***(15)***.

There are a number of anticoagulants that prevent the coagulation of blood. In clinical chemistry, EDTA, heparin, and citrate are the most widely used, and the choice of anticoagulants is important since the manner in which they behave differs. Unlike EDTA and heparin, citrate is used as a concentrated solution in a ratio of 1 part to 9 parts of blood ***(19)***, which itself introduces dilution effects and variation. Heparin is a highly charged molecule, thus being able to prevent binding of molecules to charged surfaces ***(15)***. Although EDTA can interfere with assays when divalent cations are necessary, it does not have dilution effects nor does it interfere with charged molecules. Therefore, EDTA seems to be the anticoagulant of choice for proteomic analysis of body fluids when the primary aim is to catalog and quantitate proteins. However, the choice of anticoagulants may also depend on the specific aim or protein targets of experiments since anticoagulants can affect the stability of some proteins, if not all (e.g., osteocalcin) ***(20,21)***. Heparin, citrate, and EDTA have been reported to yield no obvious *m/z* (mass per charge) peaks in typical proteome analysis, yet some types of blood collection tubes designed to reduce protein degradation contain aprotinin or other protease inhibitors that will appear as *m/z* peaks and pose a potential problem with the interpretation of mass spectra if they are not recognized as exogenous additives to a specimen ***(22)***.

With regard to the use of protease inhibitors, all data from the HUPO Plasma Proteome Project on specimen collection and handling are consistent with the benefits of blocking protease activity and, perhaps more importantly, of blocking this activity immediately, during sample acquisition ***(15)***.

Finally, the limitations of current storage methods using −70 to −80°C freezers are worth mentioning. It has been reported that some degree of degradation occurs over time in coagulation factors of stored plasma samples, presumably owing to renewal of enzymatic activity, albeit minimal, even at −80°C ***(23)***. In this context, Rouy and colleagues ***(24)*** reported that the plasma level of metalloproteinase-9 (MMP-9) decreased by 90% after 2 yr of storage at −80°C, whereas those of MMP-2 remained constant. It is surprising that two enzymes, which share many properties, behave in different manners under the same storage condition. Therefore, careful validation and interpretation are essential when we analyze a large set of body fluid samples stored in tissue banks over a period of time since at least some proteins may show different levels of stability. Topics relating to specimen collection and storage of other body fluids are dealt with in other chapters of this book.

2.2. Sample Loading: How and What to Compare

Typically, when cell or tissue lysates are subjected to proteomic analysis, equal amounts of protein are compared (e.g., 100 μg protein from each sample for 2D electrophoresis [2-DE]). In body fluid research, however, the analysis of samples based on equal protein load may cause serious problems because even the normal interval of total protein levels is very wide (e.g., it ranges from 68.0 to 86.0 mg/mL in plasma). To illustrate this potential problem, let us consider these two hypothetical patients.

Patient A

Total plasma protein	86 mg/mL
CEA	4.9 ng/mL

Patient B

Total plasma protein	68 mg/mL
CEA	4.9 ng/mL

(CEA is a tumor marker for colorectal cancer; the normal range is <5 ng/mL.)

In current medical practice, total plasma protein levels are not considered when we interpret individual protein levels (i.e., they are treated as independent variables). Therefore, CEA levels of both patients will be regarded as normal.

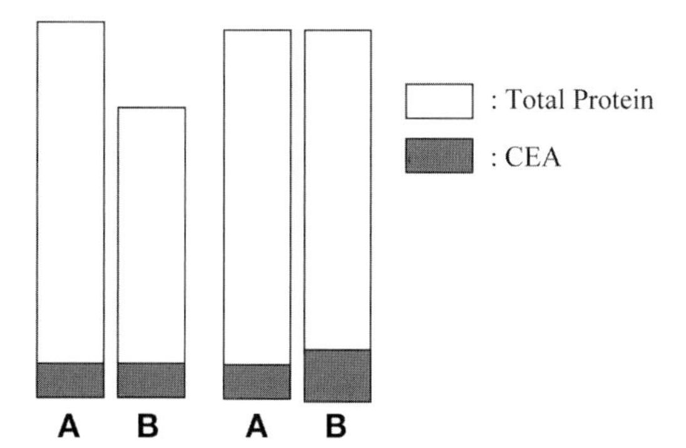

Fig. 5. Volume-based vs protein quantity-based comparison. Left panel illustrates the comparison of the same volumes of plasma from patients A and B. Total protein levels differ, yet carcinoembroyonic antigen (CEA) levels are the same. If we were to analyze the same quantities of proteins instead (i.e., 86 µg of protein for each patient), patterns will look like the right panel. In this case, total protein levels are the same (since we loaded the same quantities), yet CEA levels differ. When the same quantities of samples are compared, it is assumed that the total protein levels in patients' plasma are more or less identical. This assumption does not hold true for most body fluids.

Now, let us assume that we perform 2-DE using plasma from these two patients and that the dynamic range of detection is approx 10^{12}. (In reality, it is about 10^4.) Then, to load 86 µg of each sample to immobilized pH gradient (IPG) strips, we load 1 µL of plasma for patient A, which contains 4.9 pg of CEA, and 1.26 µL of plasma for patient B, which contains 6.2 pg of CEA. After 2-DE, the CEA spots will be selected as differentially expressed spots. (Differential gel electrophoresis [DIGE] can detect quantitative changes as low as 10% [25]). In this approach, over 20% of variation is introduced because the interpretation of CEA levels is dependent on total protein levels. In other words, CEA levels or other biomarker levels, which we try to detect, can vary according to the total protein concentrations **(Fig. 5)**.

When we deal with cell or tissue lysate, the situation is totally different. For example, if we compare radiation-treated with nontreated CaCo2 cell lines using 2-DE, we are trying to detect changes in essentially identical systems, and loading the same amount of samples (e.g., 100 µg protein from each sample) will be a reasonable way to guarantee equal comparison.

There are two ways of solving this problem. The first is by simply loading the same volume of body fluids. Although this method is perfectly compatible with the current practice of clinical laboratories, it may not be an ideal solution for expression proteomics of body fluids wherein prefractionation, such as depletion of high-abundance proteins, is commonly required. When depletion is used, for example, volumetric information is difficult to preserve and invariably lost. If we use multiple affinity columns for depleting high-abundance proteins, volumetric information pertaining to original samples would be lost, first, by dilution during chromatographic separation and, second, by a desalting and concentrating step.

The second solution is to normalize data based on the total protein concentrations. This method provides information about relative concentration of proteins, and data can be normalized even after extensive prefractionation since quantitative information is easier to preserve.

Let us go back to the example of CEA above. In that example, more than 20% of variation was introduced just because we loaded the same quantities of protein from each sample assuming, willingly or unwillingly, that total protein levels of two samples were identical. This systematic variation can be easily corrected by calculating a normalization factor from total protein levels of each sample and applying it to the data.

If we select the sample from patient A as a baseline, the normalization factor will be 68 mg/mL divided by 86 mg/mL. Then, the CEA level of patient B will be corrected to be 4.9 pg by multiplying the compensation factor to the original data $[(68/86) \times 6.2 = 4.9]$.

If we keep track of quantitative information in each prefractionation step, it is possible to calculate proper normalization factors. For example, after depleting high-abundance proteins using affinity chromatography, we can get information about how much protein is depleted from the total proteome (e.g., 85% is depleted), which can be used for normalizing data.

2.3. Prefractionation and Fractionation

The development of proteomics technologies has enabled us to analyze a large number of proteins simultaneously. 2-DE, arguably the most widely used separation technique in proteomics ***(26,27)***, can resolve more than 5000 proteins in one gel and detect less than 1 ng of proteins per spot ***(28)***. Nevertheless, at least two technical challenges need to be overcome before proteomics can realize its full potential for protein expression profiling of body fluids. First, body fluids contain an enormous number of proteins. For example, it is claimed that plasma alone contains more than 1 million protein forms ***(6)***. Another important consideration is the problem of the dynamic range of protein abundances ***(29)***. In plasma, the dynamic range of protein abundances can extend up to 12 orders of magnitude ***(30)***, which far exceeds the current dynamic range of 2-DE (approx 10^4) ***(31)***. To circumvent these problems, good separation strategies are essential. The essence of prefractionation is the enrichment of the target population of proteins (e.g., removal of high-abundance proteins and/or the isolation of subpopulation of proteins—e.g., glycoproteins, phosphoproteins, glycosylphosphatidylinisotol (GPI)-anchored proteins, cysteine-containing proteins) whereas the essence of fractionation is the maximal separation of a complex protein mixture into its individual components (e.g., 2-DE and multidimensional chromatography). Prefractionation and fractionation technologies are too complicated for a single review. For more detailed reviews, *see* Simpson ***(32)*** and Righetti et al. ***(33)***.

2.3.1. Reduction of Dynamic Range of Protein Abundances

Various human body fluids including plasma, CSF, ascites, and lymph are characterized by the presence of high-abundance proteins, which preclude effective analysis of low-abundance proteins (akin to searching for "needles" in a haystack). For example, 22 high-abundance proteins represent about 99% of the total proteins in plasma ***(34)***. Therefore, in any proteomic strategy for analyzing body fluids, the reduction of the dynamic range of protein abundances must be addressed in order to "drill down" to the low-abundance proteins for analysis. Two opposite approaches will be briefly introduced here. One approach reduces the dynamic range by depleting high-abundance proteins; the other achieves the goal by increasing the relative copy numbers of low-abundance proteins. These two may be called the *yin* and *yang* approaches to reducing the dynamic range of protein abundances.

2.3.1.1. Yin Approach of Reducing Dynamic Range of Protein Abundances

2.3.1.1.1. Depletion of High-Abundance Proteins. Depletion of high-abundance proteins is probably the most commonly used prefractionation technique for body fluid research. (Govorukhina and Bischoff discuss it in more detail in Chapter 2.) As just mentioned, this approach aims to reduce the dynamic range of protein abundances by removing high-abundance proteins, and it has been successfully adopted for body fluid research. For example, Pieper and colleagues ***(35)*** have shown that immunoaffinity subtraction chromatography can improve the resolution of low-abundance proteins in plasma. Although this approach is very useful in body fluid research, there are two issues that require careful consideration. The first is the limitation of this approach, which becomes evident if we take albumin, a major high-abundance protein in various body fluids, with about 50 mg/mL in plasma, as an example ***(36)***. If any depletion strategy were able to remove 99.9% of albumin from the plasma (according to a recent report, the efficiency ranges from 96.0 to 99.4% ***[37]***), the remaining (contaminating) concentration of albumin would still be approx 50 μg/mL. This concentration is 1×10^4 fold higher than CEA levels (approx 5 ng/mL) and 5×10^6 fold higher than levels of interleukin-6 (IL-6; approx 10 pg/mL). Considering that the current dynamic range of 2-DE is approx 10^4 ***(31)***, this simple comparison shows that in addition to the depletion of high-abundance proteins, technologies for the efficient separation and enrichment of low-abundance proteins to detectable levels have to be further utilized and developed ***(38)***.

The second issue is the possibility that depletion of high-abundance proteins may diminish the chances of finding low-abundance proteins bound to and carried by high-abundance carrier proteins such as albumin ***(39)***. Although this concept is still controversial and has not proved its importance, we should be careful not to lose extra information when subtracting a portion of proteome before initial analysis.

2.3.1.2. Yang Approach of Reducing Dynamic Range of Protein Abundances

2.3.1.2.1. Reduction of Dynamic Range With a Peptide Library. This approach (enrichment of the general population of proteins to the same degree) involves constructing a large peptide library via combinatorial chemistry. Using just 20 natural amino acids and making six reaction steps, 20^6 peptide ligands can be made; owing to this enormous diversity, there is theoretically a ligand for every protein, antibody, and peptide. When a complex protein mixture such as plasma is incubated with this library under large overloading conditions, high-abundance proteins saturate their specific affinity ligands and excess is removed during the washing step, whereas low-abundance proteins continue to concentrate on their specific affinity ligands. After processing, the eluate has all the representatives of the original mixture, but with much reduced dynamic range since high-abundance proteins are significantly diluted and low-abundance proteins are concentrated ***(33)***. Although the efficiency and efficacy of this approach are not yet clear in this early stage, it is free from potential problems associated with depletion and may work as a complementary method to depletion strategy.

2.3.1.2.2. Reduction of Dynamic Range With Selective Capture Methods. This approach involves enrichment of the selective population of proteins. The peptide library mentioned in **Subheading 2.3.1.2.1.** reduces the dynamic range of protein abundances by enriching the general population of proteins to the same degree. There are, however, other methods by which we can enrich selective populations of proteins using the unique characteristics of each group. Here, we briefly introduce two examples, which are important for analyzing body fluids.

Immunoprecipitation is the most classical example of enriching a selective population of proteins. This technique is based on the immunoaffinity between antibodies and their target proteins. Immunoprecipitation has been successfully applied to the analysis of protein isoforms, phosphorylated proteins, and protein-protein complexes ***(33)***.

Glycoprotein capture is another good example. This method, specifically targeting glycoproteins, is highly relevant to body fluid research, since protein glycosylation is prevalent in extracellular proteins, and many clinical biomarkers in body fluids are also glycoproteins ***(40)***. Currently, there are two main approaches for capturing glycoproteins. Kaji and colleagues ***(41)*** combined the classical lectin chromatography with isotope-coded tagging and mass spectrometry (MS). In this method, glycopeptides, generated by tryptic digestion of protein mixture, are captured by the lectin column. Then captured glycopeptides are isotope-tagged with ^{18}O and identified by multidimensional liquid chromatography (LC) MS. In the other approach, Zhang and colleagues ***(40)*** used hydrazide chemistry to capture glycoproteins through conjugation, which is followed by isotope labeling and identification/quantitation by tandem mass spectrometry (MS/MS).

2.3.2. Electrophoresis and Liquid Chromatography

Electrophoresis and liquid chromatography are two main streams of separation technology. In this section, only a limited number of topics will be discussed, with an emphasis on their application to body fluids. For more detailed reviews, *see* Westermeier and Grona ***(42)***, Simpson ***(43)***, and Mant and Hodges ***(44)***.

2.3.2.1. Isoelectric Focusing

Proteins are amphoteric molecules that carry a positive, negative, or zero net charge, depending on the pH of their surroundings. Therefore, when placed in a pH gradient within an electric field, proteins will migrate to the pH where they have no net charge. Isoelectric focusing (IEF) takes advantage of this phenomenon ***(45)***.

In addition to its well-known application as the first dimension analysis of 2-DE, IEF can also be used as a prefractionation technique or in combination with non-gel-based technologies such as liquid chromatography. Here we will briefly introduce three applications of IEF: prefractionation IEF for narrow-range IPGs, free-flow electrophoresis (FFE), and chromatofocusing.

2.3.2.1.1. Prefractionation-IEF for Narrow-Range Immobilized ph Gradients. The use of multiple narrow overlapping IPGs is the best remedy for increasing the resolution of 2-DE to avoid multiple proteins in a single spot for unambiguous protein identification and to facilitate the application of higher protein amounts for the detection of minor components ***(46)***.

When narrow-range IPGs are loaded with a body fluid (e.g., plasma), a massive disturbance of the focusing process ensues, stemming from two main problems. The first problem is that it is usually not possible to achieve high loads of protein, actually focused, on narrow pH gradients since most of the loaded proteins have p*I*s outside the pH range of the IPG. The second problem is the severe disturbance caused by extraneous proteins, which migrate to the ends of the strip, where they collect in highly concentrated zones in charged states. Therefore, it is essential to prefractionate body fluids into isoelectric fractions that correspond to the pH ranges of the IPGs ***(47)***.

To achieve this goal, various liquid-phase IEF apparati such as the Rotofor™ ***(48)***, the multicompartment electrolyzer ***(49)***, and the zoom fractionator ***(50)*** have been developed. In general, these apparati, except the Rotofor, have multiple compartments separated by separation barriers with a defined pH, and the pH range of a fraction in a compartment is determined by the pH of separation barriers at both ends. After IEF, each fraction can be loaded to corresponding narrow-pH IPG strips. For a more detailed discussion of each technique, see Righetti et al. ***(33)*** and Zuo and Speicher ***(51)***. Recently, Tang and colleagues ***(52)*** reported on four-dimensional analysis, which combines the depletion of high-abundance proteins, liquid-phase IEF, and 1-DE, followed by nanocapillary reversed phase high-performance liquid chromatography (RP-HPLC) tryptic peptide separation prior to MS/MS analysis to detect low-abundance proteins in human plasma and serum.

Görg and colleagues ***(53)*** developed a solid-phase prefractionation IEF using granulated gels. In brief, a Sephadex slurry is made with Sephadex G-200 superfine and a solution containing urea, 3-[(3-cholamidopropyl)dimethylamino]-1-propanesulfonate (CHAPS), dithiothreitol (DTT), and carrier ampholytes. This slurry is mixed with the sample solution, and the mixture is pipeted into the trough of the template inserted into the IPG DryStrip kit for IEF. After IEF, individual Sephadex fractions are removed with a spatula and applied onto rehydrated, narrow-pH-range IPG strips. When IPG-IEF is performed, prefractionated proteins in the Sephadex fraction are electrophoretically transferred to IPG strips and focused. This method does not require special equipment and is relatively free from protein dilution and loss, which may occur in liquid-phase IEF.

2.3.2.1.2. Free-Flow Electrophoresis and Chromatofocusing. Although the prefractionation-IEF methods introduced above are specifically devised for 2-DE, FFE, and chromatofocusing, they provide better results in combination with liquid chromatography; they can also be used with gel-based technologies. Both FFE and chromatofocusing are liquid-phase IEF, and, as their names indicate, the basic principles of FFE and chromatofocusing are based on electrophoresis and chromatography, respectively.

In FFE, samples are continuously injected into a carrier ampholyte solution flowing as a thin film between two parallel plates. By introducing an electric field perpendicular to the flow direction, proteins are separated by IEF according to their different p*I* values and collected ***(29,54)***. Key advantages of this method are improved sample recovery (owing to the absence of solid membrane supports) and sample loading capacity. (Sample loading is continuous during FFE and hence not rate limiting *[29]*).

Although FFE can be coupled off-line to sodium dodecyl sulfate polyacrylamide gel electrophoresis (SDS-PAGE) ***(54)***, the restricted separation capacity of SDS-PAGE presents limitations in resolution, recovery of low-M_r proteins, and sample loadability. In contrast, when FFE is coupled off-line to RP-HPLC, the high resolving power produced in the first-dimension IEF step, in which very narrow-range pH gradients can easily be generated, coupled to the high resolution of modern RP-HPLC stationary phases, extends the resolving power of this 2D protein separation system over other previously described 2D systems based solely on coupled HPLC columns ***(29)***. For the fundamental principles and experimental protocols of FFE, including the introduction of commercial instrumentation, *see* Krivankova and Bocek ***(55)*** and Weber et al. ***(56)***. For a detailed protocol for the application of FFE for proteins and peptides, *see* Moritz and Simpson ***(57)***.

In chromatofocusing, usually a weak anion exchanger is used as the matrix in which the functional groups are amines, and the eluent is a buffer containing a large number of buffering species, which together give a uniform buffering capacity over a broad pH range. Unlike ion-exchange chromatography, in which a pH gradient is normally formed using a gradient mixer, chromatofocusing takes advantage of the buffering action of the charged functional groups on the matrix, and the pH gradient is formed automatically as the eluting buffer titrates the functional groups on the matrix. As elution progresses, the pH at each point in the column is gradually lowered, and proteins with different p*I* values will migrate at different distances on the column before binding. In this way, proteins elute in the order of their p*I*s ***(58)***. Yan and colleagues ***(59)*** used chromatofocusing coupled to nonporous (NPS) RP-HPLC for fractionating and comparing protein expression using a drug-treated cell line vs the same untreated cell line. This method provides a 2D map based on p*I* values and hydrophobicity and has been shown to be highly reproducible for quantitative differential expression analysis. Soldi and colleagues ***(60)*** used a commercial platform combining chromatofocusing and NPS RP-HPLC for protein profiling of human urine and showed that this method could be a complementary system to 2-DE in body fluid research.

2.3.2.2. 2D Electrophoresis

2-DE has been the most commonly used technique in the field of proteomics since its development in 1975 by O'Farrell ***(26)*** and Klose ***(27)***. This technique couples IEF in the first dimension with SDS-PAGE in the second dimension and allows the separation of complex mixtures of proteins according to their respective p*I* and M_r values. Depending on the gel size and pH gradient used, 2-DE can resolve more than 5000 proteins simultaneously (more than 2000 proteins routinely) and can detect less than 1 ng of protein per spot ***(46)***. Since 2-DE has suffered from problems such as reproducibility, resolution, proteins with extremes of p*I*, and recovery of hydrophobic proteins ***(61)***, it is now seriously challenged by other non-gel-based approaches. However, as Rabilloud ***(31)*** pointed out, if the goal of the proteomic experiment is to look for quantitative changes, 2-DE will remain unrivalled for some time.

Body fluid research is closely related to finding disease biomarkers, and therefore, quantitative analysis of differentially expressed proteins in normal and disease groups is important. Although 2-DE is a good quantitative tool, its ability has been hampered by important limitations. First, the predominant protein staining methods are either not sensitive enough (Coomassie brilliant blue) or have a limited linearity (silver staining) ***(62)***. The application of radioactive labeling or fluorescent stains can alleviate these problems, but only partially. Second, the intrinsic gel-to-gel variation of 2-DE masks the biological difference between the samples and compromises any quantitative comparison of protein expression levels ***(63)***. DIGE ***(64)*** circumvents many of the issues associated with traditional 2-DE, such as reproducibility and limited dynamic range, and allows for more accurate and sensitive quantitative analysis ***(65)***.

In DIGE, two samples are labeled in vitro using two different fluorescent cyanine dyes (CyDyes; Amersham Biosciences) differing in their excitation and emission wavelengths, then mixed before IEF, and separated on a single 2D gel. After consecutive excitation with both wavelengths, the images are overlaid and subtracted (normalized), whereby especially differences (e.g., up- or down-regulated, and/or posttranslationally modified proteins) between the two samples can be visualized ***(46)***. This multiplex approach instead of the "one gel one sample" approach solves most of the problems associated with gel-to-gel variation, spot matching, and normalization. In addition, CyDye has a detection limit of 150 to 500 pg for a single protein with a linear response in protein concentration over 5 orders of magnitude, whereas silver staining has a protein detection limit of approx 1 ng with a dynamic range of less than 2 orders of magnitude ***(65)***. At least five replicate gels should be run per sample for quantitative analysis in traditional 2-DE, and owing to the high variability from sample comparisons run in different gels, the threshold for accepting a meaningful variation is set at a factor of 2.0 (100% variation) ***(66)***. In contrast, DIGE can detect quantitative changes as low as 10% with 95% confidence, and the use of an internal standard helps to minimize false positives and negatives ***(25)***.

For biomarker research using body fluids, protein profiling of a large set of samples is essential. In this case, the primary benefit of sample multiplexing is that a pooled standard can be included on each gel, which comprises equal amounts of each sample and represents the average of all the samples being analyzed. The pooled standard approach is used to normalize protein abundance

Table 2
Commonly Used Liquid Chromatography Methods in Proteomics

Principle of separation	Type of chromatography
Size and shape	Size-exclusion chromatography (gel-filtration or gel-permeation chromatography)
Net charge	Ion-exchange chromatography
Hydrophobicity	Hydrophobic interaction chromatography
	Reversed-phase high-performance liquid chromatography
Antigen-antibody interaction	Immunoaffinity chromatography
Isoelectric point (p*I*)	Chromatofocusing
Metal binding	Immobilized metal ion affinity chromatography

measurements across multiple gels in an experiment, making it possible to compare more than two samples accurately ***(67)***. A more detailed review of multiplexed dye technologies is presented by Patton in chapter 4.

2.3.2.3. Liquid Chromatography

Chromatography is a widely used technique for separating the components of a mixture by allowing the sample (the analyte) to distribute between the stationary and mobile phases. Stationary phases, the key elements of LC, are made of the support matrix chemically coated with a bonded phase containing functional groups that provide the desired specific binding interaction ***(68)***. **Table 2** summarizes commonly used chromatographic methods and their principles of separation. Chromatography can be used for the enrichment of low-abundance proteins as well as for multidimensional analysis of body fluids. For a more detailed review of chromatographic methods for separating proteins and peptides, *see* Simpson ***(32)*** and Mant and Hodges (***44***).

2.3.2.3.1. Chromatographic Prefractionation for 2-DE. As mentioned above, immunoaffinity chromatography is the most commonly used prefractionation tool in body fluid research. However, other chromatographic methods such as ion-exchange chromatography and RP-HPLC can also be considered for prefractionation. Combining a chromatographic step with 2-DE provides a third orthogonal dimension for protein separation. For example, if we use RP-HPLC before 2-DE, we are separating proteins based on a combination of their hydrophobicity, p*I*, and M_r. For an overview of chromatographic prefractionation prior to 2-DE, *see* Lescuyer et al. ***(69)***.

2.3.2.3.2. Multidimensional Analysis Using Chromatography. As we discussed in chromatofocusing, multiple chromatographic methods can be coupled for separating the components of a mixture (e.g., combination of chromatofocusing

Table 3
MDLC of Intact Proteins versus MudPIT

	MDLC	MudPIT
Peptide fragment correlation	Good correlation between peptides and their original protein	Poor correlation owing to digestion prior to fractionation
Mixture complexity	Still complex	About 50-fold more complex
Computational requirements	Less challenging; MS/MS data search can be supported by elution profiles and protein properties.	One of the biggest challenges
Solubility	Problematic	A significant advantage over MDLC

Abbreviations: MDLC, multidimensional high-performance liquid chromatography; MS, mass spectrometry; MudPIT, multidimensional protein identification Technology.

and RP-HPLC). This approach, termed multidimensional HPLC (MDLC), can be fully automated to join the various separation steps into a single seamless procedure and can also interface protein and peptide separations directly to mass spectrometers. In MDLC of intact proteins, the protein complex is fractionated and digested to peptides for subsequent mass spectrometric analysis. In an alternative MDLC, termed multidimensional protein identification technology (MudPIT), complex protein samples are enzymatically digested to produce extremely complex peptide mixtures, which are then subjected to multidimensional chromatographic separations and mass spectrometric analysis. For detailed reviews of MDLC of intact proteins and MudPIT, see Apffel *(70)* and Wolters et al. *(71)*. **Table 3** presents a summary of comparisons between MDLC of intact proteins and MudPIT. This table is based on Apffel's review *(70)*.

2.4. Mass Spectrometry

MS has become the method of choice for the identification and characterization of proteins in complex mixtures, largely as a result of the development of soft ionization methods for proteins and the availability of gene and genome sequence databases *(11)*. In expression proteomics of body fluids, the main applications of MS are determination of primary structure of peptides, quantitative analysis, and characterization of posttranslational modifications. For a review of MS-based proteomics, see Aebersold and Mann *(72)*, and for a "hands-on" description of current MS-based proteomics methods, see the proteomics laboratory manual of Simpson *(43)*. Glycosylation, an important posttranslational modification in body fluids, will be discussed by Bunkenborg and colleagues in chapter 5.

Table 4
Current Approaches to Protein Identification Using Mass Spectrometry (MS)

MS approach	Features
Top-down	Analysis of the fragmentation pathway of intact proteins Complete sequence coverage (useful for examining site-specific mutations and posttranslational modifications) ***(96)*** Requirement of specialized equipment Early stage of development
Bottom-up	Digestion of proteins to peptides prior to MS Limited sequence coverage, but improved sequencing properties and detection efficiencies of peptide ***(97)***
PMF	Comparison of peptide mass fingerprints with virtual fingerprints obtained by theoretical cleavage of protein sequences in databases
MS/MS	Peptide sequencing by analyzing the fragmentation patterns of peptides More sensitive and specific than PMF

Abbreviations: PMF, peptide mass fingerprinting; MS/MS, tandem mass spectrometry

2.4.1. Protein Identification

2.4.1.1. Protein Identification Using Mass Spectrometry

Currently, the bottom-up approach using MS/MS is most widely used for protein identification. In MS/MS, peptide ions are isolated, fragmented, and analyzed to produce MS/MS spectra. Then these experimental MS/MS spectra are compared with the theoretical MS/MS spectra generated from protein sequence databases using search algorithms, which assign scores indicating the degree of similarity between the experimental and theoretical MS/MS spectra *(73)* (**Table 4**).

However, it has been increasingly realized that the protein inference problem, i.e., the task of assembling the sequences of identified peptides to infer the proteins of their origin, is far from being trivial and requires special attention. Protein digestion makes peptides, not proteins, the currency of MS/MS, and the connectivity between peptides and proteins is lost at the digestion stage, which complicates computational analysis and biological interpretation of the data, especially in the case of higher eukaryotic organisms in which the same peptide sequence can be present in multiple different proteins *(74)*.

2.4.1.2. MS/MS Search Algorithms for Protein Identification

Currently MS/MS search algorithms scoring functions can essentially be classified into two categories: heuristic and probabilistic algorithms. Heuristic algorithms, such as SEQUEST, Spectrum Mill, X!Tandem, and Sonar, correlate the experimental MS/MS spectrum with a theoretical spectrum and calculate a score based on the similarity between the two. On the other hand, probabilistic algorithms, such as MASCOT, model to some extent the peptide fragmentation process (e.g., ladders of sequence ions) and calculate the probability that a particular peptide sequence produced the observed spectrum by chance. Important considerations when one is carrying out MS/MS database searches are the specified search parameters (i.e., mass tolerance, which is dependent on the instrument and calibration), search strategy (i.e., semitryptic vs tryptic), chosen protein sequence database to query (i.e., IPI vs NCBI NR, which is dependent on the particular experiment), and chosen search engine. In addition, it is recommended to use an algorithm that demonstrates high sensitivity in conjunction with an algorithm that demonstrates high specificity ***(75)***. For more detailed discussion, see Kapp et al. ***(75)*** and Sadygov et al. ***(76)***. For publication guidelines for peptide and protein identification data, *see* Carr et al. ***(77)***.

2.4.2. Quantitative Analysis Using Mass Spectrometry

MS has been used successfully to identify and characterize proteins in complex mixtures (especially suborganellar proteomes such as mitochondria, phagosomes, Golgi bodies, and so on; for reviews, *see* Taylor et al. ***[78]*** and Brunet et al. ***[79]***). MS research so far has been mainly qualitative, yet the recent advent of new methodologies provides the opportunity to obtain quantitative proteomics data sets. There are two approaches in MS-based quantitative proteomics: *stable isotope labeling*, which permits direct comparison of two proteome states in the same analysis, and *ion current-based quantitation* (label-free methods), which compares the ion currents of the same peptides in different experiments ***(80)***.

Two main approaches, based on stable isotope methods, are currently used for relative quantitation using MS: metabolic labeling and chemical tagging. In metabolic labeling, stable isotope-labeled atoms are metabolically incorporated into newly synthesized proteins in vivo. These labeled cells (or their lysates) are then added as an internal standard to cells grown in material with natural abundance isotopes at the beginning of the experiment to account for errors accrued during sample preparation and measurement ***(81)***. This approach can be used for model organisms, as well as cell lines in culture. In stable isotope labeling in cell culture (SILAC), a prototype approach of metabolic labeling in cell culture, mammalian cell lines are grown in a defined medium containing isotope-labeled amino acids. Samples grown in the presence of the natural and heavy isotopes can be pooled and analyzed together. Then, the signal intensities of the light and heavy versions of the same peptides are measured, which allows their relative quantitation ***(82)***. In chemical tagging, protein samples are labeled with chemical tags of light and heavy formats. After labeling, samples are pooled and analyzed together for the same purpose just explained for metabolic labeling. In isotope-coded affinity tag (ICAT), a prototype approach of chemical tagging, cysteine residues of proteins samples are labeled with biotinylated tags of light and heavy formats. After labeling, samples are pooled, digested, and analyzed. As in SILAC, the signal intensities of the light and heavy versions of the same peptide are measured for relative quantitation ***(83)***. Although metabolic labeling has only a limited value in human body fluid research, chemical tagging is fully applicable to it. There are other derivatives of chemical tagging methods such as iTRAQ™, or ^{18}O labeling ***(84,85)***. For a more detailed review of stable isotope methods and metabolic labeling of proteins, *see* Schneider and Hall ***(86)*** and Beynon and Pratt ***(87)***.

Recently, Pan and colleagues ***(88)*** reported a new approach for the detection and quantification of targeted proteins in a complex mixture. In this method, proteotypic peptides that uniquely represent proteins are selected from databases and used as reference peptides. The reference peptides are generated by chemical synthesis and contain heavy stable isotopes. Protein samples are digested and combined with a mixture of defined amounts of isotopically labeled reference peptides. Then the peptide mixture is separated by capillary RP-HPLC and deposited on a matrix-assisted laser desorption ionization (MALDI) plate to be analyzed using a MALDI tandem mass spectrometer. The identification and quantification of targeted proteins is based on searching and identifying the corresponding signature peptide pairs directly ***(88)***. This method holds the promise that it can improve the throughput and confidence of protein identification as well as allowing absolute quantitation. For a more detailed discussion of the proteotypic peptide approach, *see* Kuster et al. ***(89)***.

The second approach of MS-based quantitation is comparing the ion currents of peptides. Quantitation of small molecules by integration of LC-MS-extracted ion currents (XIC) has a long history in analytical chemistry, and similar quantitation techniques have been applied to proteolytic digests of protein mixtures ***(90)***. Obvious advantages of XIC-based quantification are that no labeling is used and it can be performed with any type of sample, whereas clear disadvantages are the multiple occasions for quantitation error to occur during sample processing and LC-MS analysis, as well as the presence of interfering substances in one of the states to be compared (i.e., extremely reproducible conditions are required) ***(80)***. According to Higgs and colleagues ***(90)***, it appears that relatively small (20%) changes in protein relative levels between different biological sample sets are discoverable with a fully automated sample processing and analysis system, which is implemented using a high-throughput computational environment.

3. Future Perspectives

As previously mentioned, one of the main goals of body fluid proteomics is to find protein fingerprints or biomarkers, which may reflect various disease states, and it is becoming more evident that a single biomarker cannot represent all the complex mechanisms behind diseases including cancer. According to Diamandis ***(91)***, the prevailing view in cancer biomarker research is that the most powerful single cancer biomarkers may have already been discovered. Probably, we are now bound to discover biomarkers that could be used in panels with improved sensitivity and specificity (i.e., a multivariate approach). In other words, we may need to construct more detailed patterns to detect a certain phenotype. Then the next challenge will be to understand the relationships between the components comprising disease-specific patterns. Although this approach, referred to as *integrative systems biology*, has been increasingly applied to the study of animal models or single cells in which informative pathway information can be gained at an early stage of analysis, its application has been limited in body fluids in which the relationships between the components may not be revealed without further study ***(92)***. Recently, Davidov and colleagues ***(93)*** reported methods for the differential integrative analysis of plasma. Solely based on a body fluid analysis, their effort represents the first attempt to explain the relationships between molecular phenotypic fingerprints by combining quantitative proteomic and metabolomic data. In the future, it will be possible to enhance this approach by including the genomic component in the form of differential transcription analysis of multiple tissues and make it truly global with respect to understanding pleiotropic effects of gene perturbation on body fluids ***(93)***.

Acknowledgments

This work was supported by Australia-Asia Awards (to S.M.A.) and a National Health and Medical Research Council program grant (no. 280912) (to R.J.S.).

References

1. Skandarajah A, Moritz RL, Tjandra JJ, Simpson RJ. Proteomic analysis of colorectal cancer: discovering novel biomarkers. *Expert Rev Proteomics* 2005;2: 681–692.
2. Robin ED. Limits of the internal environment. In: Robin ED, ed. *Claude Bernard and the Internal Environment: A Memorial Symposium*. New York: M. Dekker, 1979:257–267.
3. Guyton AC, Hall JE. The body fluid compartments: extracellular and intracellular fluids, interstitiatl fluid and edema. In: Guyton AC, Hall JE, eds. *Textbook of Medical Physiology*, 10th ed. Philadelphia: W.B. Saunders, 2000:264–294.
4. Liotta LA, Kohn EC. The microenvironment of the tumour-host interface. *Nature* 2001;411:375–379.
5. Liotta LA, Ferrari M, Petricoin E. Clinical proteomics: written in blood. *Nature* 2003;425:905.
6. Anderson NL, Anderson NG. The human plasma proteome: history, character, and diagnostic prospects. *Mol Cell Proteomics* 2002;1:845–867.

7. Celis JE, Gromov P, Cabezon T, et al. Proteomic characterization of the interstitial fluid perfusing the breast tumor microenvironment: a novel resource for biomarker and therapeutic target discovery. *Mol Cell Proteomics* 2004;3: 327–344.
8. Leak LV, Liotta LA, Krutzsch H, et al. Proteomic analysis of lymph. *Proteomics* 2004;4:753–765.
9. Sedlaczek P, Frydecka I, Gabrys M, Van Dalen A, Einarsson R, Harlozinska A. Comparative analysis of CA125, tissue polypeptide specific antigen, and soluble interleukin-2 receptor alpha levels in sera, cyst, and ascitic fluids from patients with ovarian carcinoma. *Cancer* 2002;95:1886–1893.
10. Trape J, Molina R, Sant F. Clinical evaluation of the simultaneous determination of tumor markers in fluid and serum and their ratio in the differential diagnosis of serous effusions. *Tumour Biol* 2004;25:276–281.
11. Simpson RJ. Role of separation science in proteomics. In: Simpson RJ, ed. *Purifying Proteins for Proteomics: A Laboratory Manual*. Cold Spring Harbor, NY: Cold Spring Harbor Laboratory Press, 2004:1–15.
12. Dufour R. Sources and control of preanalytical variation. In: Kaplan LA, Pesce AJ, Kazmierczak SC, eds. *Clinical Chemistry: Theory, Analysis, Correlation*, 4th ed. St. Louis, MO; Mosby, 2003:64–82.
13. Omenn GS. The Human Proteome Organization Plasma Proteome Project pilot phase: reference specimens, technology platform comparisons, and standardized data submissions and analyses. *Proteomics* 2004;4:1235–1240.
14. Rand MJ, Murray RK. Plasma proteins, immunoglobulins, and blood coagulation. In: Murray RK, Granner DK, Mayes PA, Rodwell VW, eds. *Harper's Biochemistry,* 25th ed. New York: McGraw-Hill, 2000:737–762.
15. Rai AJ, Gelfand CA, Haywood BC, et al. HUPO Plasma Proteome Project specimen collection and handling: towards the standardization of parameters for plasma proteome samples. *Proteomics* 2005;5:3262–3277.
16. Muller N, Danckworth HP. [Coagulation properties of the extravascular fluid. I. Coagulation factors in thoracic-duct lymph]. *Z Lymphol* 1980;4:11–17.
17. Chang P, Aronson DL, Borenstein DG, Kessler CM. Coagulant proteins and thrombin generation in synovial fluid: a model for extravascular coagulation. *Am J Hematol* 1995;50:79–83.
18. Hulmes JD, Bethea B, Ho K, et al. An investigation of plasma collection, stabilization, and storage procedures for proteomic analysis of clinical samples. *Clin Proteomics* 2004;1:17–32.
19. Young DS, Bermes EW. Specimen collection and other preanalytical variables. In: Burtis CA, Ashwood ER, eds. *Tietz Fundamentals of Clinical Chemistry*, 5th ed. Philadelphia: Saunders, 2001:30–54.
20. Durham BH, Robinson J, Fraser WD. Differences in the stability of intact osteocalcin in serum, lithium heparin plasma and EDTA plasma. *Ann Clin Biochem* 1995;32:422–423.
21. Chan BY, Buckley KA, Durham BH, Gallagher JA, Fraser WD. Effect of anticoagulants and storage temperature on the stability of receptor activator for nuclear

factor-kappa B ligand and osteoprotegerin in plasma and serum. *Clin Chem* 2003;49:2083–2085.

22. Drake SK, Bowen RA, Remaley AT, Hortin GL. Potential interferences from blood collection tubes in mass spectrometric analyses of serum polypeptides. *Clin Chem* 2004;50:2398–2401.
23. Lewis MR, Callas PW, Jenny NS, Tracy RP. Longitudinal stability of coagulation, fibrinolysis, and inflammation factors in stored plasma samples. *Thromb Haemost* 2001;86:1495–1500.
24. Rouy D, Ernens I, Jeanty C, Wagner DR. Plasma storage at −80 degrees C does not protect matrix metalloproteinase-9 from degradation. *Anal Biochem* 2005; 338:294–298.
25. Knowles MR, Cervino S, Skynner HA, et al. Multiplex proteomic analysis by two-dimensional differential in-gel electrophoresis. *Proteomics* 2003;3:1162–1171.
26. O'Farrell PH. High resolution two-dimensional electrophoresis of proteins. *J Biol Chem* 1975;250:4007–4021.
27. Klose J. Protein mapping by combined isoelectric focusing and electrophoresis of mouse tissues. A novel approach to testing for induced point mutations in mammals. *Humangenetik* 1975;26:231–243.
28. Gorg A, Drews O, Weiss W. Separation of proteins using two-dimensional gel electrophoresis. In: Simpson RJ, ed. *Purifying Proteins for Proteomics: A Laboratory Manual.* Cold Spring Harbor, NY: Cold Spring Harbor Laboratory Press, 2004:391–430.
29. Moritz RL, Ji H, Schutz F, et al. A proteome strategy for fractionating proteins and peptides using continuous free-flow electrophoresis coupled off-line to reversed-phase high-performance liquid chromatography. *Anal Chem* 2004;76: 4811–4824.
30. Corthals GL, Wasinger VC, Hochstrasser DF, Sanchez JC. The dynamic range of protein expression: a challenge for proteomic research. *Electrophoresis* 2000;21: 1104–1115.
31. Rabilloud T. Two-dimensional gel electrophoresis in proteomics: old, old fashioned, but it still climbs up the mountains. *Proteomics* 2002;2:3–10.
32. Simpson RJ. *Purifying Proteins for Proteomics: A Laboratory Manual.* Cold Spring Harbor, NY: Cold Spring Harbor Laboratory Press, 2004.
33. Righetti PG, Castagna A, Antonioli P, Boschetti E. Prefractionation techniques in proteome analysis: the mining tools of the third millennium. *Electrophoresis* 2005;26:297–319.
34. Tirumalai RS, Chan KC, Prieto DA, Issaq HJ, Conrads TP, Veenstra TD. Characterization of the low molecular weight human serum proteome. *Mol Cell Proteomics* 2003;2:1096–1103.
35. Pieper R, Su Q, Gatlin CL, Huang ST, Anderson NL, Steiner S. Multi-component immunoaffinity subtraction chromatography: an innovative step towards a comprehensive survey of the human plasma proteome. *Proteomics* 2003;3:422–432.
36. Zolg JW, Langen H. How industry is approaching the search for new diagnostic markers and biomarkers. *Mol Cell Proteomics* 2004;3:345–354.

37. Bjorhall K, Miliotis T, Davidsson P. Comparison of different depletion strategies for improved resolution in proteomic analysis of human serum samples. *Proteomics* 2005;5:307–317.
38. Fountoulakis M, Juranville JF, Jiang L, et al. Depletion of the high-abundance plasma proteins. *Amino Acids* 2004;27:249–259.
39. Mehta AI, Ross S, Lowenthal MS, et al. Biomarker amplification by serum carrier protein binding. *Dis Markers* 2003;19:1–10.
40. Zhang H, Li XJ, Martin DB, Aebersold R. Identification and quantification of N-linked glycoproteins using hydrazide chemistry, stable isotope labeling and mass spectrometry. *Nat Biotechnol* 2003;21:660–666.
41. Kaji H, Saito H, Yamauchi Y, et al. Lectin affinity capture, isotope-coded tagging and mass spectrometry to identify N-linked glycoproteins. *Nat Biotechnol* 2003; 21:667–672.
42. Westermeier R, Gronau S. *Electrophoresis in Practice: A Guide to Methods and Applications of DNA and Protein Separations*, 4th ed. Weinheim: Wiley-VCH, 2005.
43. Simpson RJ. *Proteins and Proteomics: A Laboratory Manual*. Cold Spring Harbor, NY: Cold Spring Harbor Laboratory Press, 2003.
44. Mant CT, Hodges RS. *High-Performance Liquid Chromatography of Peptides and Proteins: Separation, Analysis, and Conformation*. Boca Raton; CRC Press, 1991.
45. Stochaj W, Berkelman T, Laird N. Preparative two-dimensional gel electrophoresis with immobilized pH gradients. In: Simpson RJ, ed. *Proteins and Proteomics: A Laboratory Manual*. Cold Spring Harbor, NY: Cold Spring Harbor Laboratory Press, 2003:143–218.
46. Gorg A, Weiss W, Dunn MJ. Current two-dimensional electrophoresis technology for proteomics. *Proteomics* 2005;5:826–827.
47. Herbert BR, Righetti PG, McCarthy J, et al. Sample preparation for high-resolution two-dimensional electrophoresis by isoelectric fractionation in an MCE. In: Simpson RJ, ed. *Purifying Proteins for Proteomics: A Laboratory Manual*. Cold Spring Harbor, NY: Cold Spring Harbor Laboratory Press, 2004:431–442.
48. Bier M. Recycling isoelectric focusing and isotachophoresis. *Electrophoresis* 1998;19:1057–1063.
49. Herbert BR, Righetti PG, McCarthy J, et al. Sample preparation for high-resolution two-dimensional electrophoresis by isoelectric fractionation in an MCE. In: Simpson RJ, ed. *Purifying Proteins for Proteomics: A Laboratory Manual*. Cold Spring Harbor, N.Y.: Cold Spring Harbor Laboratory Press, 2004:431–442.
50. Zuo X, Speicher DW. A method for global analysis of complex proteomes using sample prefractionation by solution isoelectrofocusing prior to two-dimensional electrophoresis. *Anal Biochem* 2000;284:266–278.
51. Zuo X, Speicher DW. Microscale solution isoelectrofocusing: a sample prefractionation method for comprehensive proteome analysis. *Methods Mol Biol* 2004;244:361–375.
52. Tang HY, Ali-Khan N, Echan LA, Levenkova N, Rux JJ, Speicher DW. A novel four-dimensional strategy combining protein and peptide separation methods

enables detection of low-abundance proteins in human plasma and serum proteomes. *Proteomics* 2005;5:3329–3342.

53. Gorg A, Boguth G, Kopf A, Reil G, Parlar H, Weiss W. Sample prefractionation with Sephadex isoelectric focusing prior to narrow pH range two-dimensional gels. *Proteomics* 2002;2:1652–1657.
54. Hoffmann P, Ji H, Moritz RL, et al. Continuous free-flow electrophoresis separation of cytosolic proteins from the human colon carcinoma cell line LIM 1215: a non two-dimensional gel electrophoresis-based proteome analysis strategy. *Proteomics* 2001;1:807–818.
55. Krivankova L, Bocek P. Continuous free-flow electrophoresis. *Electrophoresis* 1998;19:1064–1074.
56. Weber PJA, Weber G, Eckerskorn C. Protein purification using free-flow electrophoresis. In: Simpson RJ, ed. *Purifying Proteins for Proteomics: A Laboratory Manual.* Cold Spring Harbor, NY: Cold Spring Harbor Laboratory Press, 2004:463–478.
57. Moritz RL, Simpson RJ. Liquid-based free-flow electrophoresis-reversed-phase HPLC: a proteomic tool. *Nat Methods* 2005;2:863–873.
58. Mohammad J. Chromatofocusing. In: Simpson RJ, ed. *Purifying Proteins for Proteomics: A Laboratory Manual.* Cold Spring Harbor, NY: Cold Spring Harbor Laboratory Press, 2004:355–379.
59. Yan F, Subramanian B, Nakeff A, Barder TJ, Parus SJ, Lubman DM. A comparison of drug-treated and untreated HCT-116 human colon adenocarcinoma cells using a 2-D liquid separation mapping method based upon chromatofocusing PI fractionation. *Anal Chem* 2003;75:2299–2308.
60. Soldi M, Sarto C, Valsecchi C, et al. Proteome profile of human urine with two-dimensional liquid phase fractionation. *Proteomics* 2005;5:2641–2647.
61. Lilley KS, Razzaq A, Dupree P. Two-dimensional gel electrophoresis: recent advances in sample preparation, detection and quantitation. *Curr Opin Chem Biol* 2002;6:46–50.
62. Westermeier R, Marouga R. Protein detection methods in proteomics research. *Biosci Rep* 2005;25:19–32.
63. Van den Bergh G, Arckens L. Fluorescent two-dimensional difference gel electrophoresis unveils the potential of gel-based proteomics. *Curr Opin Biotechnol* 2004;15:38–43.
64. Unlu M, Morgan ME, Minden JS. Difference gel electrophoresis: a single gel method for detecting changes in protein extracts. *Electrophoresis* 1997;18: 2071–2077.
65. Lilley KS, Friedman DB. All about DIGE: quantification technology for differential-display 2D-gel proteomics. *Expert Rev Proteomics* 2004;1:401–409.
66. Marengo E, Robotti E, Antonucci F, Cecconi D, Campostrini N, Righetti PG. Numerical approaches for quantitative analysis of two-dimensional maps: a review of commercial software and home-made systems. *Proteomics* 2005;5: 654–666.

67. Alban A, David SO, Bjorkesten L, et al. A novel experimental design for comparative two-dimensional gel analysis: two-dimensional difference gel electrophoresis incorporating a pooled internal standard. *Proteomics* 2003;3:36–44.
68. Simpson RJ. Introduction to chromatographic methods for protein and peptide purification. In: Simpson RJ, ed. *Purifying Proteins for Proteomics: A Laboratory Manual.* Cold Spring Harbor, NY: Cold Spring Harbor Laboratory Press, 2004: 41–74.
69. Lescuyer P, Hochstrasser DF, Sanchez JC. Comprehensive proteome analysis by chromatographic protein prefractionation. *Electrophoresis* 2004;25:1125–1135.
70. Apffel A. Multidimensional chromatography of intact proteins. In: Simpson RJ, ed. *Purifying Proteins for Proteomics: A Laboratory Manual.* Cold Spring Harbor, NY: Cold Spring Harbor Laboratory Press, 2004:75–100.
71. Wolters DA, Washburn MP, Yates JR 3rd. An automated multidimensional protein identification technology for shotgun proteomics. *Anal Chem* 2001;73:5683–5690.
72. Aebersold R, Mann M. Mass spectrometry-based proteomics. *Nature* 2003;422: 198–207.
73. Weatherly DB, Atwood JA 3rd, Minning TA, Cavola C, Tarleton RL, Orlando R. A heuristic method for assigning a false discovery rate for protein identifications from mascot database search results. *Mol Cell Proteomics* 2005.
74. Nesvizhskii AI, Aebersold R. Interpretation of shotgun proteomics data: The protein inference problem. *Mol Cell Proteomics* 2005;4:1419–1440.
75. Kapp EA, Schutz F, Connolly LM, et al. An evaluation, comparison, and accurate benchmarking of several publicly available MS/MS search algorithms: sensitivity and specificity analysis. *Proteomics* 2005;5:3475–3490.
76. Sadygov RG, Cociorva D, Yates JR 3rd. Large-scale database searching using tandem mass spectra: looking up the answer in the back of the book. *Nat Methods* 2004;1:195–202.
77. Carr S, Aebersold R, Baldwin M, Burlingame A, Clauser K, Nesvizhskii A. The need for guidelines in publication of peptide and protein identification data: Working Group on Publication Guidelines for Peptide and Protein Identification Data. *Mol Cell Proteomics* 2004;3:531–533.
78. Taylor SW, Fahy E, Ghosh SS. Global organellar proteomics. *Trends Biotechnol* 2003;21:82–88.
79. Brunet S, Thibault P, Gagnon E, Kearney P, Bergeron JJ, Desjardins M. Organelle proteomics: looking at less to see more. *Trends Cell Biol* 2003;13:629–638.
80. Ong SE, Mann M. Mass-spectrometry-based proteomics turns quantitative. *Nat Chem Biol* 2005;1:252–262.
81. Wu CC, MacCoss MJ, Howell KE, Matthews DE, Yates JR 3rd. Metabolic labeling of mammalian organisms with stable isotopes for quantitative proteomic analysis. *Anal Chem* 2004;76:4951–4959.
82. Ong SE, Blagoev B, Kratchmarova I, et al. Stable isotope labeling by amino acids in cell culture, SILAC, as a simple and accurate approach to expression proteomics. *Mol Cell Proteomics* 2002;1:376–386.

83. Gygi SP, Rist B, Gerber SA, Turecek F, Gelb MH, Aebersold R. Quantitative analysis of complex protein mixtures using isotope-coded affinity tags. *Nat Biotechnol* 1999;17:994–999.
84. Ross PL, Huang YN, Marchese JN, et al. Multiplexed protein quantitation in *Saccharomyces cerevisiae* using amine-reactive isobaric tagging reagents. *Mol Cell Proteomics* 2004;3:1154–1169.
85. Stewart II, Thomson T, Figeys D. 18O labeling: a tool for proteomics. *Rapid Commun Mass Spectrom* 2001;15:2456–2465.
86. Schneider LV, Hall MP. Stable isotope methods for high-precision proteomics. *Drug Discov Today* 2005;10:353–363.
87. Beynon RJ, Pratt JM. Metabolic labeling of proteins for proteomics. *Mol Cell Proteomics* 2005;4:857–872.
88. Pan S, Zhang H, Rush J, et al. High throughput proteome screening for biomarker detection. *Mol Cell Proteomics* 2005;4:182–190.
89. Kuster B, Schirle M, Mallick P, Aebersold R. Scoring proteomes with proteotypic peptide probes. *Nat Rev Mol Cell Biol* 2005;6:577–583.
90. Higgs RE, Knierman MD, Gelfanova V, Butler JP, Hale JE. Comprehensive label-free method for the relative quantification of proteins from biological samples. *J Proteome Res* 2005;4:1442–1450.
91. Diamandis EP. Mass spectrometry as a diagnostic and a cancer biomarker discovery tool: opportunities and potential limitations. *Mol Cell Proteomics* 2004;3:367–378.
92. van der Greef J, Stroobant P, van der Heijden R. The role of analytical sciences in medical systems biology. *Curr Opin Chem Biol* 2004;8:559–565.
93. Davidov E, Clish CB, Oresic M, et al. Methods for the differential integrative omic analysis of plasma from a transgenic disease animal model. *Omics* 2004;8:267–288.
94. Guyton AC, Hall JE. The microcirculation and the lymphatic system: capillary fluid exchange, interstitial fluid, and lymph flow. In: Guyton AC, Hall JE, eds. *Textbook of Medical Physiology*, 10th ed. Philadelphia: WB Saunders, 2000:162–174.
95. Guyton AC, Hall JE. Cerebral blood flow; the cerebrospinal fluid; and brain metabolism. In: Guyton AC, Hall JE, eds. *Textbook of Medical Physiology*, 10th ed. Philadelphia: WB Saunders, 2000:709–715.
96. Bogdanov B, Smith RD. Proteomics by FTICR mass spectrometry: top down and bottom up. *Mass Spectrom Rev* 2005;24:168–200.
97. Steen H, Mann M. The ABC's (and XYZ's) of peptide sequencing. *Nat Rev Mol Cell Biol* 2004;5:699–711.

2

Sample Preparation of Body Fluids for Proteomics Analysis

Natalia Govorukhina and Rainer Bischoff

Summary

This chapter gives an overview of various approaches to sample preparation of body fluids with special emphasis on serum. The methodology is presented in sections covering protein depletion, protein enrichment using affinity ligands, protein chip technology in conjunction with mass spectrometry, and the use of magnetic beads. Since many of these methods are also relevant for other body fluids like urine, applications are highlighted briefly. The last part of the chapter deals with the link between sample preparation and downstream separation methods such as 2D electrophoresis or HPLC. A distinction is made between the analysis of high-molecular-weight proteins (the proteome) and the lower molecular weight part (the peptidome).

Key Words: Proteomics; sample preparation; plasma; serum; HPLC; electrophoresis; biomarker; clinical chemistry; urine; mass spectrometry.

1. Introduction

1.1. Proteomics of Human Body Fluids

The analysis of human body fluids constitutes one of the most important approaches to the diagnosis of disease and the following of therapeutic interventions. Human body fluids carry information about the status of the organism that may help in the recognition of physiological misbalances when overt pathological symptoms are not yet present. Analyzing the constituents of body fluids presents a number of challenges, the most difficult being the discrimination between variability in composition caused by an ongoing disease process and natural variability. The composition of body fluids varies due to endogenous, possibly pathological, processes and many environmental influences such as diet and life style and the way the organism deals with them (e.g., metabolism and detoxification). This

From: *Proteomics of Human Body Fluids: Principles, Methods, and Applications*
Edited by: V. Thongboonkerd © Humana Press Inc., Totowa, NJ

variability is most obvious when one is analyzing samples from different persons (cross-sectional studies) but is also present, albeit to a lesser extent, when one is analyzing samples from the same person over a given period (longitudinal studies). Variability cannot be avoided but may be reduced by careful selection of the study population. At any rate, the discovery of disease-related changes in the composition of body fluids requires the study of a significant number of samples from patients and controls and a careful statistical interpretation of the results.

From an analytical chemistry point of view, body fluids constitute highly complex biological samples containing cells, proteins, peptides, and many metabolites. Thus preparation of body fluids is unavoidable prior to determining the concentration or amount of a given set of constituents. Sample pretreatment and all further downstream steps will affect the ultimate analytical result and must therefore be carefully controlled and validated. It is not easy to give a general overview of sample pretreatments for body fluids, since each of them requires an adapted protocol, which in turn needs to be tailored to certain groups of analytes. More detailed procedures are given in Part II of this book, which deals with various kinds of body fluids. In this chapter we focus on sample pretreatments for the analysis of proteins and peptides in serum. Although serum is just one example of a body fluid, albeit an important one, we will use it to highlight general principles of sample pretreatments that have a bearing on other kinds of body fluids, as touched upon in **Subheading 2.6**.

The first step after taking a blood sample from a patient is to treat it in a way that makes it suitable for storage and subsequent analysis. A common initial step is to separate blood cells from soluble components, for example, by low-speed centrifugation. During sampling and centrifugation, it is pivotal to avoid disruption or activation of cells, notably hemolysis of red blood cells (which is shown by an orange to red color of the supernatant) and activation of platelets. The remaining supernatant, the blood plasma, may be stored as such in the case that anticoagulants were added during collection to prevent blood clotting. Alternatively, blood clotting may be allowed or induced by leaving plasma at room temperature for a few hours. Deciding whether to store plasma or serum for subsequent analyses is important. Although plasma is easier to prepare, it requires the presence of efficient anticoagulants for long-term stability. The components of the coagulation, fibrinolytic, and complement systems are all sensitive to contact with unnatural surfaces, such as plastic containers, glass, or injection needles, and there is a risk of activating these systems during processing steps (e.g., during chromatography or solid-phase extraction). The preparation of serum requires coagulation of the plasma, which is a complex biochemical process that may be difficult to control. In most hospital or laboratory settings, coagulation is effectuated at room temperature for 1 to 4 h. During this time the endogenous coagulation system is activated, leading to a cascade of proteolytic events that results in the formation of a fibrin-containing blood clot, which is usually removed by centrifugation.

It is obvious that activating a proteolytic system can have serious consequences for subsequent proteomic analyses, and some authors have noted that the coagulation time affects the resulting serum ***(1,2)***. However, proteolytic events associated with coagulation are highly controlled due to the sequence specificity of the major proteases (thrombin, factor Xa) and their well-defined location in the coagulation pathway (factor VIIIa, factor XIa) (**Fig. 1**). It is thus not clear whether the coagulation time affects the final composition of the proteome significantly, but there are indications that the lower molecular weight part, the so-called peptidome, is altered (Schulz-Knappe, personal communication).

In our own studies, which applied tryptic digestion prior to LC-MS analysis (the shotgun approach), we have not observed major changes in the resulting profiles (**Fig. 2**). It is, however, important to validate this sample processing step carefully within the context of the overall analytical scheme (e.g., the complete protein vs the shotgun and peptidomics approach), because coagulation time is not well controlled in most laboratory or hospital settings and experience shows that it is hard to impose strict rules on hospital personnel with respect to this parameter. Finally, for retrospective studies on already acquired and stored serum samples, it is not possible to influence this step; thus the decision here is whether to include these samples in the analysis or not.

The way samples are initially treated determines in part what kind of analytes can be detected and quantified. Although this is true for any kind of analyte, it is particularly critical for the analysis of proteins and peptides, which are susceptible to degradation, precipitation, chemical modification (e.g., oxidation), adsorption to the walls of containers, and so on. Establishing a well-controlled and reproducible sampling procedure is therefore critical for any study involving human body fluids.

The proteomes in body fluids differ significantly from intracellular or tissue-derived proteomes, which are the subject of most proteomics studies. Systemic body fluids, like blood, sample the whole organism and give an average picture of the physiological state of that organism at a given point in time. Notably, blood contains a few high-abundance proteins that are to a large extent produced and secreted by the liver. In contrast, urine is a much more dilute body fluid that samples the metabolic end products from blood. Its composition is greatly influenced by the status of the kidneys. Although every body fluid presents particular challenges with respect to sample pretreatment, it is fair to say that blood is one of the most difficult body fluids to analyze.

In the following, we will highlight a number of options for sample pretreatment prior to proteomics analysis. Our focus will be directed at serum, but the principles are applicable to other body fluids. We will try to emphasize that

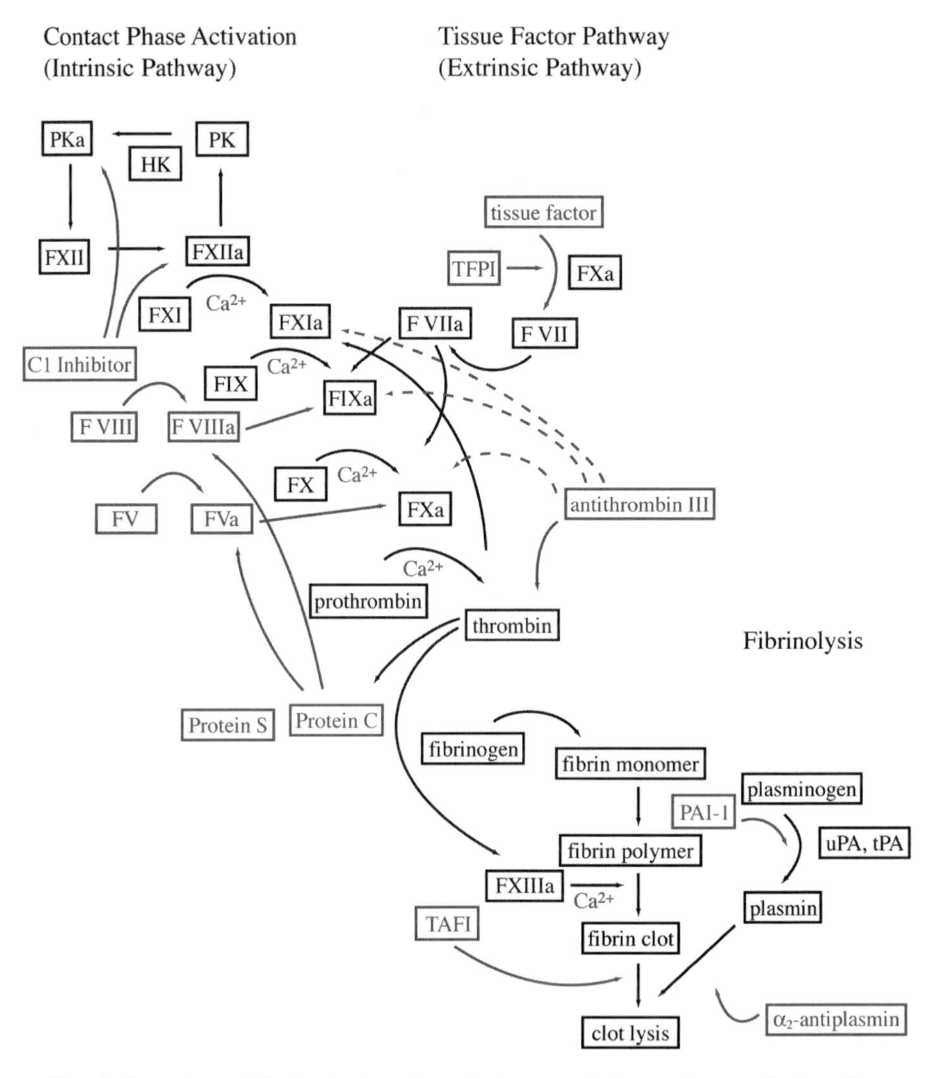

Fig. 1. Overview of the intrinsic and extrinsic coagulation pathways. Both pathways are activated during preparation of serum from blood plasma. Proteolytic activity associated with coagulation may affect the profile of low-molecular-weight proteins and peptides used for peptidomics analysis. F, factor; PK, protein kinase; PKa, protein kinase A; TAFI, tissue angiogenesis factor inhibitor; TFPI, tissue factor pathway inhibitor; TPA, tissue plasminogen activator; UPA, urokinase plasminogen activator. Reproduced with permission from Tapper H, Herwald H. Modulation of hemostatic mechanisms in bacterial infectious diseases. *Blood* 2000;96:2329–2337.

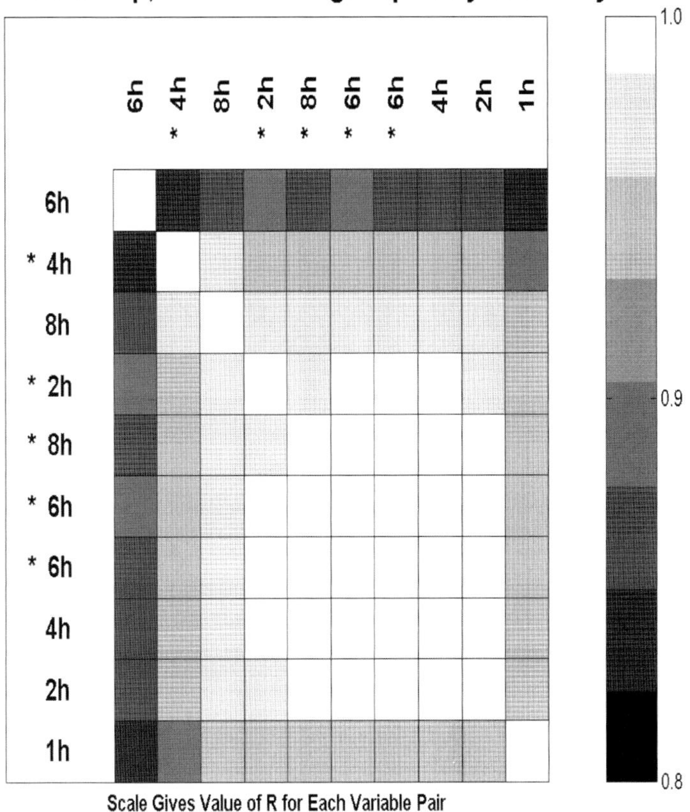

Fig. 2. Correlation map of LC-MS data sets obtained from the analysis of serum (male) after depletion and tryptic digestion. Coagulation at room temperature was allowed to proceed for 1, 2, 4, 6, or 8 h. As indicated in this plot, high correlation of all data sets was found, with correlation coefficients above 0.8 in each case.

there are strategic choices to be made early on in the analytical procedure that will determine the final result.

1.2. Methodological Overview

There is no single approach to proteomics in body fluids. It is likely that the comprehensive analysis of the proteome of any given body fluid is still beyond our reach despite great methodological advances in recent years. A major challenge is the concentration range of proteins in most body fluids, which spans about 12 orders of magnitude *(3)* (**Fig. 3**). Furthermore, it is difficult to predict the number of proteins in body fluids owing to processing events (like the

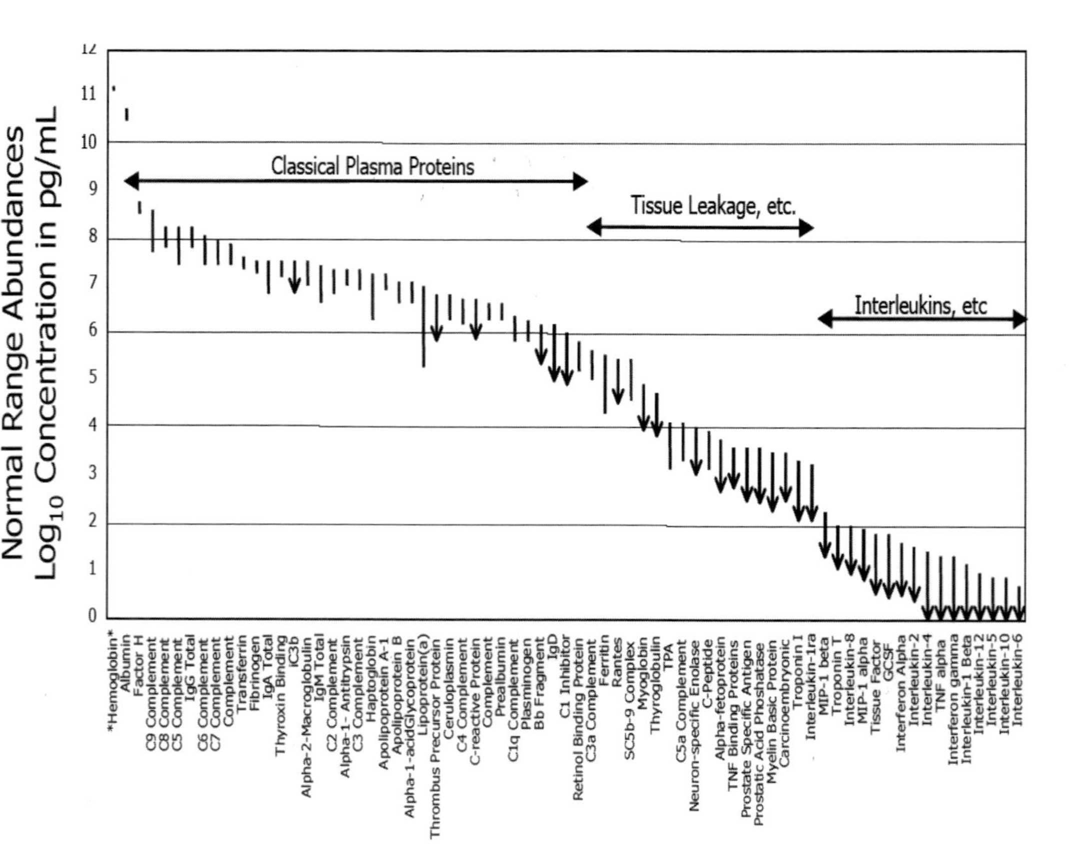

Fig. 3. Concentration range of proteins found in human plasma. It is noteworthy that there is a difference of more than 11 orders of magnitude between the most concentrated and the very low-abundance proteins. Reproduced with permission from Anderson NL, Anderson NG. The Human Plasma Proteome: history, character, and diagnostic prospects. *Mol Cell Proteomics* 2002;1:845–867.

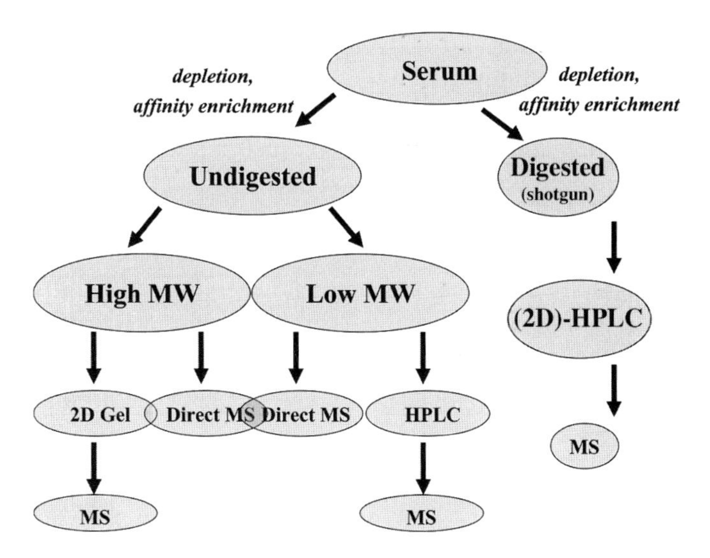

Fig. 4. Schematic overview of different approaches of sample preparation. The main initial difference is whether undigested or trypsin-digested proteins are being analyzed. A second level of differentiation comes into play when one is deciding whether to analyze the high- or the low-molecular-weight (MW) fraction of the proteome. High-molecular-weight proteins are best analyzed by 2D electrophoresis, whereas peptides are more amenable to high-performance liquid chromatography (HPLC) mass spectrometry (MS).

generation of smaller proteins and peptides from larger precursors, posttranslational modifications, and the fact that proteins can enter body fluids by well-defined pathways like secretion) and also cell and tissue turnover as a result of necrosis or apoptosis. From a methodological point of view, the proteome of a body fluid may be roughly divided into high- and low-molecular-weight compartments **(Fig. 4)**. This discrimination is rather arbitrary and is mostly defined by the size-dependent separation method used for prefractionation.

Body fluids have been prefractionated by ultrafiltration with membranes of various cutoff values *(4–6)*. Although ultrafiltration appears to be an easy separation methodology with an apparently clear-cut separation mechanism, its application to complex biological samples shows that discrimination between proteins above and below the nominative cutoff of the membrane is never complete *(6)*. Effects such as adsorption of proteins to the membrane, the generation of a polarization layer close to the membrane surface, and deformation of the pores in relation to the *g* force can all affect filtration. Ultrafiltration is also performed on a large scale in patients with renal insufficiency, and this material has been the source for many studies of bioactive proteins and peptides below approximately 20 kDa *(7–9)*. Although the kidney itself is performing much more complex tasks than just ultrafiltration, urine may be considered an ultrafiltrate of blood and thus has a considerably lower concentration of high-molecular-weight proteins.

An elegant combination of ultrafiltration and chromatography is based on restricted access materials (RAMs), which have an adsorbing internal pore surface and a nonadsorbing external surface ***(10,11)***. The pore diameter in most RAMs is about 6 nm, which corresponds roughly to a cutoff value of 20 kDa. RAM chromatography has been integrated into analytical systems with the goal of analyzing the low-molecular-weight part of the proteome of blood diafiltrates ***(12,13)***. Although a clear enrichment of the low-molecular-weight fraction was observed, a considerable amount of albumin was still present even after RAM chromatography.

The decision of whether to work with undigested proteins or to digest proteins with trypsin prior to further analyses is of principal importance in an analytical scheme (**Fig. 4**). Performing separations of very complex mixtures of proteins is difficult owing to the wide range of physicochemical properties and the possibility that proteins will denature, aggregate, or even precipitate under separation conditions. The most universally applicable separation method for proteins is 2D polyacrylamide gel electrophoresis (2D-PAGE), whereby all proteins are denatured from the beginning and kept in a denatured state throughout separation. This reduces the risk of aggregation and precipitation with subsequent loss of proteins as well as that of proteolysis. There is no comparable universal chromatographic method, and it is thus necessary to develop an appropriate fractionation scheme for groups of proteins or individual proteins. However, as 2D-PAGE has limitations with respect to low-molecular-weight proteins of 10 to 20 kDa and hydrophobic or basic proteins, alternatives are being developed.

One approach is based on the so-called shotgun method, whereby the complete protein mixture is digested with trypsin (other proteases are conceivable as well for this purpose but are not widely used) and the generated peptides are separated by 2D high-performance liquid chromatography (HPLC) ***(14–16)***. Shotgun proteomics has the advantage of overcoming many of the difficulties related to very hydrophobic or otherwise intractable proteins at the expense of rendering the separation problem quite daunting. Assuming that serum contains about 10^5 different protein forms each of which generates 50 tryptic peptides, one has to deal with a mixture of about 5×10^6 peptides to be separated. Fortunately, not all peptides need to be separated into single peaks, and not all peptides of each protein need to be identified by mass spectrometry to trace them back to the protein of origin. A drawback of the shotgun method is that not all regions of a protein are covered by the analysis, which may mean that some possibly relevant posttranslational modifications or processed forms are missing. Nevertheless, the excellent separation capacity of HPLC for peptides compared with complete proteins and the much easier identification of peptides by tandem mass spectrometry have accelerated the use of shotgun proteomics in the biomarker discovery area. The daunting separation problem posed by this approach has also driven recent new developments in HPLC stationary phase chemistry and technology that increase separation efficiency and reduce analysis time ***(17–20)***.

The presence of a few high-abundance proteins in body fluids such as albumin has driven developments to deplete these proteins specifically and thus to increase the loading capacity by a factor of 5 to 10 ***(21–27)***. In addition to increasing the protein or peptide load, depletion also augments the capacity to detect peptides derived from lower abundance proteins ***(28)***. Recent research has shown, however, that depletion of high-abundance proteins does not proceed without the loss of some low-molecular-weight proteins and peptides ***(29)***. It is thus important to decide whether to deplete or rather try to design a fractionation strategy that deals with high-abundance proteins such as albumin or immunoglobulins ***(30)***.

Arguably, very low-abundance proteins in the ng/mL to pg/mL range cannot be detected in complex protein mixtures such as serum even after depletion. Many regulatory proteins such as cytokines or some of the known tumor-specific markers reside in this concentration range and are presently measured by immunological methods. To reach into this lower concentration range, it is often necessary to enrich a given set of proteins by affinity chromatography using highly selective antibodies or group-specific ligands like lectins. The use of protein-specific antibodies limits the scope of the analysis to those proteins that are recognized, and we cannot make new discoveries unless they are related to the targeted protein. Group-specific affinity ligands such as lectins or antibodies directed at a common structural element such as phosphotyrosine represent a compromise between the comprehensive proteomics approach, which often fails to detect low-abundance proteins, and the highly specific methods. For example, lectins have been specifically applied to the discovery of tumor-specific glycoprotein markers, since tumor cells often produce proteins with aberrant glycosylation patterns ***(31,32)***. Lectins have also been used to enrich glycoproteins from complex protein mixtures or glycopeptides from tryptic digests of such mixtures ***(33–36)***. In general, targeted approaches require a hypothesis concerning the role that different kinds of proteins may play in a given disease in order to chose appropriate affinity ligands for enrichment.

2. Sample Preparation

2.1. Preparation of Plasma and Serum

Between two fundamentally different compartments of the blood, namely, blood cells and the actual fluid, most clinical analyses are done on derivatives of the fluid, like plasma or serum. Discussion continues on whether serum or plasma should be used, but this may also depend on the general practice of the hospital that provides samples for analysis, notably, whether samples are analyzed from existing collections. Preparation of plasma requires addition of anticoagulants, such as EDTA, citrate, and/or heparin, whereas serum contains no extra additives. However, serum lacks components of the coagulation system, such as thrombin and fibrinogen, since they are part of or become entrapped in the blood clot and are thus removed from the serum. In addition, other proteins or peptides that have some binding affinity to the clot may also be partially depleted. Being a proteolytic process, coagulation generates peptide fragments from larger proteins that may especially affect the composition of the peptidome. Interestingly, comparison of plasma made with EDTA, citrate, or heparin also shows variation in protein composition ***(37)***.

Sampling blood for plasma or serum preparation is routine in most hospital laboratories and a reasonably standardized procedure is in place using commercial reagents and materials. However, most laboratory technicians and nurses are not aware of the specific requirements of proteomics and thus need to be informed. Very restrictive standard operating procedures (SOPs), notably with respect to the coagulation time and conditions, are often difficult to follow in routine hospital operations. Biomarkers discovered thus far therefore need to be robust enough to be useful in a routine clinical laboratory, and very unstable proteins or peptides are probably not of interest in the long run.

Recently, Schulte et al. ***(2)*** reported that a considerable number of peptides were found in serum but not in human plasma. The authors suggest that these peptides appeared as a result of a clotting-related proteolytic activity. This might be indicative of artifacts generated as a result of the clotting reaction, which is disturbing with respect to peptidomics. The authors therefore propose to use human plasma for this purpose.

An example of the preparation of plasma for biomarker discovery by Peptidomics® (Schulz-Knappe, personal communication) involves taking a blood sample from a superficial vein of the cubital region. The blood sampling procedure should not take longer than 1 min, and EDTA is used as the anticoagulant. Prior to collection, the first sample (approximately 2.5 mL) is discarded. To remove platelets, the sample is centrifuged at 2,000*g* for 10 min. The final plasma sample (approximately 1.5 mL) should be frozen within 30 min after being taken and stored at –80°C.

Serum is made by letting a fresh blood sample coagulate (with or without thrombin as the activator) and either filtering it through a gel or collecting the liquid fraction after centrifugation. As discussed just before, there are possible disadvantages to preparing serum or plasma, but this will also depend on the analytical question (e.g., low- vs high-molecular-weight proteins).

Although their hypothesis has not been proved, Sorace and Zhan *(38)* suggest that variations in blood coagulation might be a significant factor in obscuring clinical proteomics data sets. The source of variation can be both technical and natural. Schulte et al. *(2)* found that a naturally occurring Val-34 to Leu mutation in the activation peptide of factor XIII (FXIIIA) not only affected the process of blood clotting but also correlated with a lower incidence of myocardial infarction and ischemic stroke and an increased risk of hemorrhagic stroke. According to our results, different clotting times ranging from 1 to 8 h in the preparation of serum samples resulted in highly correlated liquid chromatography-mass spectrometry (LC-MS) data sets when we analyzed serum proteins after depletion and trypsin digestion. Correlation coefficients above 0.8 were found for all samples after we selected the 37 top information-rich *m/z* traces using the CODA component detection algorithm *(39)* (**Fig. 2**).

2.2. Removal of High-Abundance Proteins

As the presence of abundant proteins in most biofluids used for diagnostic purposes decreases the capacity of analytical methods to detect low-abundance proteins or peptides, a range of approaches has been developed to reduce the total amount of protein. Blood serum is a complex mixture of thousands of proteins and peptides. However, few of the serum proteins are present in extremely high amounts compared with the rest of the serum components. (Human serum albumin [HSA] constitutes 57–71% and γ-globulins 8–26% of the total of all human serum proteins). The 10 most abundant proteins account for 97% of all the protein content in plasma *(3)*. In a recent publication, the authors claim that the search for specific markers occurs in a fraction of less than 1% of all plasma proteins *(40)*.

Removal of these high-abundance proteins increases the loading capacity of the analytical system by a factor of 5 to 10 and thus improves the detection of low-abundance proteins. Several affinity columns are presently on the market based on dye ligands or antibodies for albumin removal and protein A or G for the removal of immunoglobulins *(41)*. Technically simple approaches that allow processing of multiple samples in parallel based on HSA- and IgG-binding spin columns or filters have been developed *(42)*. For HSA binding, two types of stationary phases are generally used: (1) those based on dye ligands such as Cibacron-Blue and derivatives thereof *(43)*, and (2) those based on specific antibodies against human serum albumin *(42)* raised in mammals (IgG) but also in chickens (IgY), as recently described *(23)*. HSA was also successfully removed by affinity capture on immobilized phage-derived peptides *(44)*. Recently, a synthetic peptide derived from protein G was used for HSA affinity chromatography and depletion of HSA from human plasma *(45)*. The column could easily be regenerated with alkaline treatment owing to the stability of the

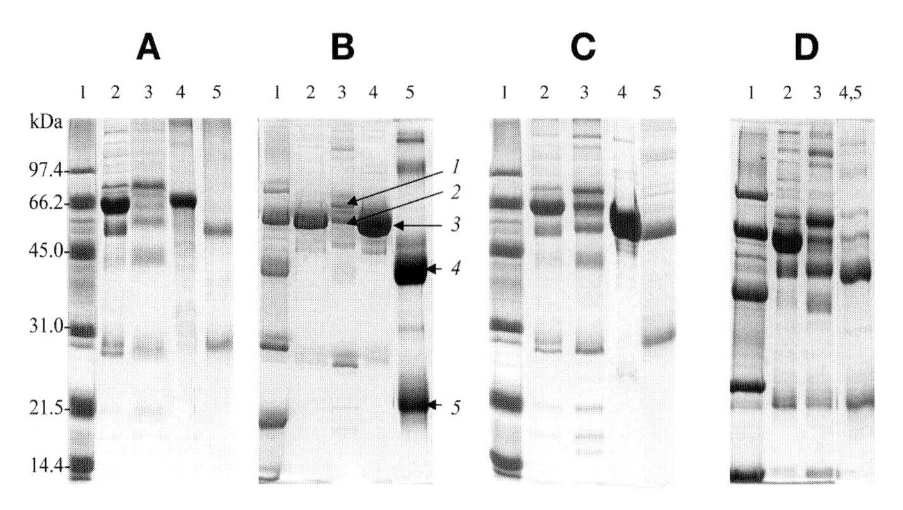

Fig. 5. Depletion of albumin and γ-globulins from human serum. In each lane 8 to 10 μg of protein were loaded, and gels were stained with Coomassie Blue G-250. **(A)** POROS Anti-HSA and POROS Protein G columns. **(B)** HiTrap Blue and HiTrap Protein G columns. **(C)** Merck Albumin Removal column and HiTrap Protein G column. **(D)** Aurum Serum Protein column. Lanes: 1, standards; 2, crude serum; 3, depleted serum; 4, bound protein eluted from albumin-depleting columns; and 5, bound γ-globulins eluted from columns. Labeled protein bands: 1, serotransferrin; 2, α_1-antitrypsin; 3, albumin; 4, 5, γ-globulins, heavy and light chains, respectively. Reproduced with permission from Govorukhina NI, Keizer-Gunnink A, van der Zee AGJ, de Jong S, de Bruijn HWA, Bischoff R. Sample preparation of human serum for the analysis of tumor markers: comparison of different approaches for albumin and [gamma]-globulin depletion. *J Chromatogr A* 2003;1009:171–178.

peptide, and its specificity and capacity were quite high. However, the column is presently not commercially available. Removal of IgG is exclusively done by well-established methods based on immobilized protein A, protein G, or protein L, owing to their high affinity and selectivity ***(46–51)***. Comparative studies of HSA- and IgG-binding columns based on Poros® polystyrene-divinylbenzene beads (Applied Biosystems) ***(24)*** as well as on Mimetic Blue (ProMetic BioSciences) and HiTrap Blue (Amersham Biosciences) for HSA removal have been recently performed ***(25)***.

We tested several approaches specifically to reduce the level of high-abundance proteins in serum based on either specific antibodies, dye ligands, (for albumin), or protein A or G (for γ-globulins) ***(22)***. Analysis by sodium dodecyl sulfate (SDS)-PAGE **(Fig. 5)** and LC-MS after tryptic digestion of the remaining proteins **(Fig. 6)**, showed that reduction with albumin-directed antibodies was most effective, albeit not complete ***(28)***. A more recently introduced multiple affinity removal column, which depletes certain high-abundance

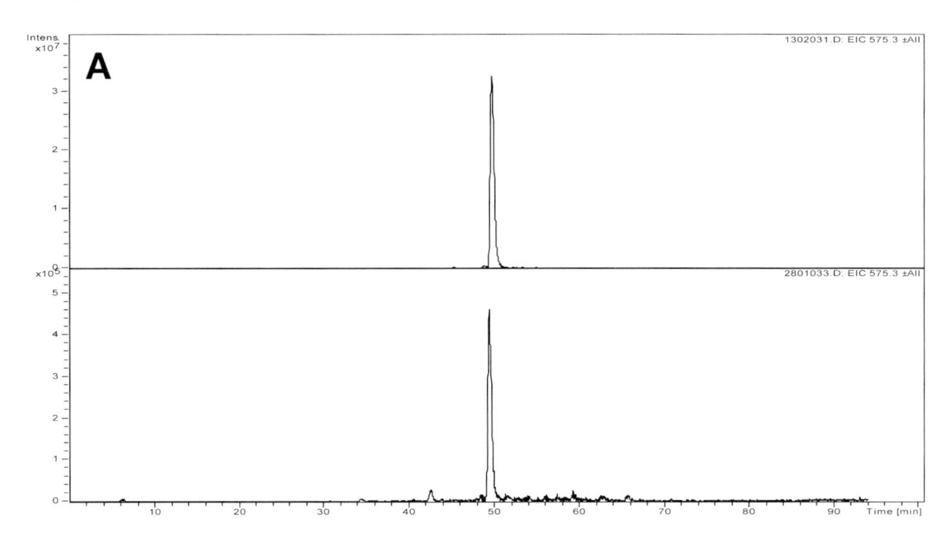

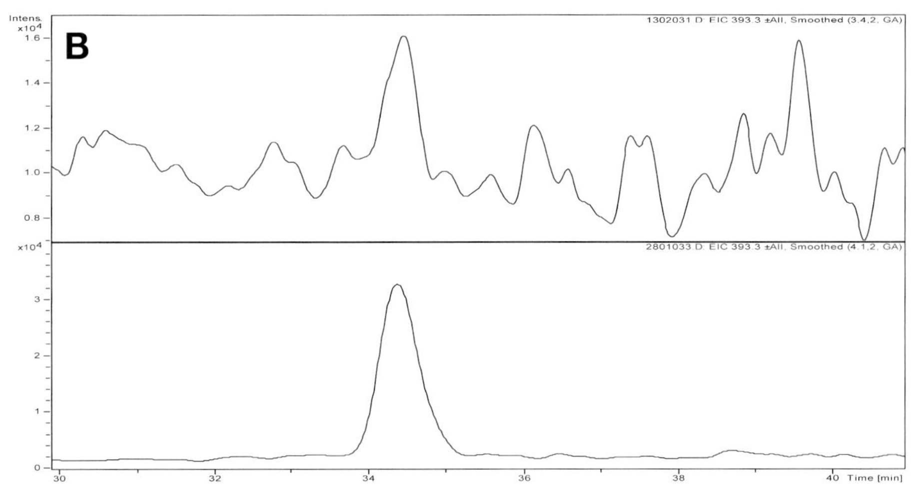

Fig. 6. Efficiency **(A)** and selectivity **(B)** of albumin removal from human serum using an antialbumin immunoaffinity column. **(A)** Extracted ion chromatogram of m/z = 575.3 (doubly charged molecular ion of peptide LVNEVTEFAK; positions 41–50 in human serum albumin) of tryptic digests of human serum (upper trace; peak height 3.2×10^7) or of human serum after depletion with an antialbumin immunoaffinity column (lower trace; peak height 4.6×10^5). **(B)** Extracted ion chromatogram of m/z = 393.3 (doubly charged molecular ion of peptide IVDLVK; positions 193–198 in human α_1-antitrypsin) of tryptic digests of human serum (upper trace; peak height 16,052) or of human serum after depletion with an antialbumin immunoaffinity column (lower trace; peak height 32,607). Note the much cleaner detection of this peptide fragment after depletion and the increased overall peak height. Reproduced with permission from Bischoff R, Luider TM. Methodological advances in the discovery of protein and peptide disease markers. *J Chromatogr B* 2004;803:27–40.

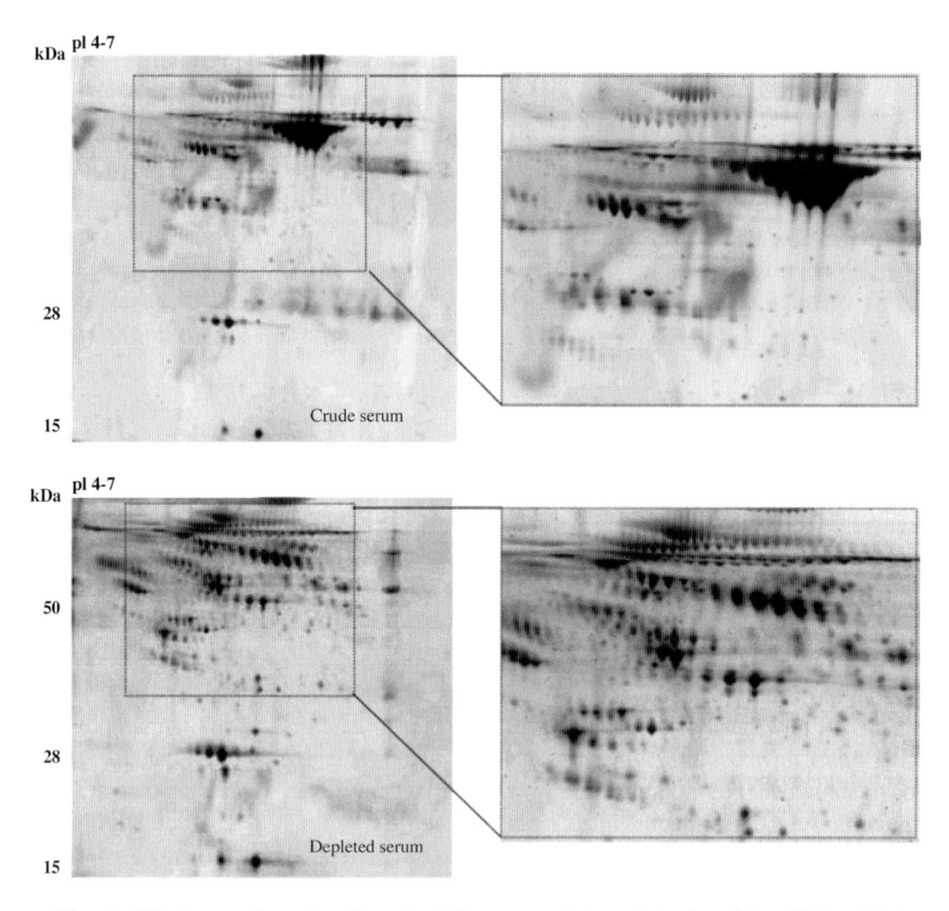

Fig. 7. 2D electrophoresis of crude (70 μg protein) and depleted (multiple affinity removal column, 100 μg protein) serum samples. On the right side is a zoom view of the area containing albumin. Reproduced with permission from Bjorhall K, Miliotis T, Davidsson P. Comparison of different depletion strategies for improved resolution in proteomic analysis of human serum samples. *Proteomics* 2005;5:307–317.

proteins (albumin, IgG, IgA, transferrin, haptoglobin, and α_1-antitrypsin ***[41]***) proved to be most effective in our hands and provided more reproducible results during LC-MS analysis regarding retention times and peak areas than previously evaluated methods (unpublished data). Similar results were recently published for 2D gel electrophoresis **(Fig. 7)** ***(41)***.

Removal of high-abundance proteins by ultrafiltration through cellulose filters with a cutoff of 30 kDa proved to be less successful ***(6)***. Many known "landmark" proteins of low molecular weight (<30 kDa) were missing upon 2D gel electrophoresis. Ultrafiltration has, however, the advantage of allowing one to concentrate the low-molecular-weight fraction of the proteome and was found to be useful for analysis of low-molecular-weight proteins (LMWs) by LC-MS after prefractionation by strong cation-exchange HPLC ***(5)***. To prevent binding of LMWs to serum carriers, particularly albumin, 20% acetonitrile was used. In the resulting fraction, the authors could identify 314 unique proteins including cytokines, growth factors, and transcription factors, which are proteins of low abundance that are very difficult to detect by other methods without serum depletion. This method of sample preparation at a cutoff of 10 kDa was recently used to analyze the LMW fraction of pooled serum from patients with ovarian cancer by nanoLC-electrospray ionization-fourier transform ion cyclotron resonance (ESI-FT-ICR)-MS and analyzed statistically ***(4)***.

A quite different set of methods uses electrophoretic approaches to fractionate complex samples and to separate high-abundance proteins from those of low abundance. The basic principle is based on preparative isoelectric focusing and/or free-flow electrophoresis in solution, whereby the crude sample is prefractionated in a specially designed chamber owing to different electrophoretic mobilities or isoelectric points of the proteins. The HSA-rich fraction was discarded, and other fractions were pooled or analyzed separately. The method was originally reported more than 10 yr ago ***(52)*** and is still in limited use ***(53–55)***. Some commercially available systems can be used for electrophoretic prefractionation (e.g., the Zoom IEF fractionator® [Invitrogen] or the system produced by Weber).

There is one particular problem associated with the removal of serum albumin and globulins. These proteins appear to fulfill the function of carriers for less abundant proteins ***(29,56)***. This is especially critical for LMWs, since they can escape kidney clearance only when bound to high-molecular-weight carrier proteins. Many of these LMWs are found to be associated with the development of cancer and could therefore be extremely important biomarkers (*see* **ref. *29***). Binding of LMWs to high-abundance, high-molecular-weight proteins may be used advantageously based on a two-step procedure, whereby abundant carrier proteins are first specifically bound to the corresponding affinity column followed by elution of the bound LMWs using a gradient ***(29,56)***.

2.3. Targeted Enrichment of Individual Proteins or Protein Families

Since many disease-specific biomarkers are likely of low to very low abundance in body fluids, it is a major challenge to detect them using profiling methods. Based on a given hypothesis about the disease mechanism, it is therefore often advisable to use targeted, affinity-based methods for enrichment prior to analysis. A combination of proteomics technology with targeted enrichment that does not require a very "sharp" hypothesis is based on group-specific ligands like lectins ***(57,58)*** in case of glycoproteins or activity-based probes (ABPs) in the case of proteases or other enzymes ***(59–65)***.

2.3.1. Lectins

Glycosylation of proteins is a posttranslational modification that is easily affected by cellular growth conditions. Modifications of glycosylation patterns are therefore often observed in fast growing cancerous cells compared with their quiescent counterparts ***(31,66)***. Analysis of the carbohydrate portion of proteins is a rather complex task, since the glycosyl moiety is usually a branched chain polymer with an enormous variety in length, composition, and complexity. Studies of glycoproteins can be divided into two types, first identification of the proteins and their glycosylation sites and second the more demanding analysis of the structure of the attached glycosyl residues themselves.

Glycosyl residues can be linked to the protein core via asparagine (*N*-linked glycans) or bound via serine or threonine (*O*-linked glycans). For *N*-linked glycans, *N*-acetylglucosamine (GlcNAc) is the first monosaccharide in the chain, whereas *N*-acetylgalactosamine (GalNAc) is most often found for *O*-linked glycans. In addition to being potentially interesting as biomarkers, failure of proper glycosylation can cause severe abnormalities ***(67)***.

Although the exact structure of glycosylated proteins varies considerably, probably all known glycoproteins can be enriched by lectin affinity chromatography ***(68)***. There are several commercially available lectin affinity columns, which differ in specificity and are used widely in early stages of the isolation of glycoproteins ***(69)***. The specificity of many lectins is known (**Table 1**), allowing us to design complementary enrichment schemes rationally.

Recently five lectins, concanavalin A (Con A), wheat germ agglutinin (WGA), jacalin, lentil lectin (LCA), and peanut lectin (PNA), were tested for capturing glycoproteins from human serum ***(70)***. At first, the authors depleted the high-abundance proteins from human serum with a multiple affinity depletion column (*see* **Subheading 2.2.**) followed by enrichment on single- or multiple-lectin affinity columns. The enriched proteins were eluted with buffers containing specially selected sugars. The resulting fractions were analyzed by LC-MS after digestion with trypsin. **Figure 8** gives an example of how the analysis of proteins in serum can be focused to a subset containing a fucose residue by prior enrichment on a column containing the fucose-specific Lotus tetragonolobus agglutinin (LTA) ***(34)***.

2.3.2. Activity-Based Profiling of Proteases

Standard proteomics techniques give information about the relative abundance of proteins and possibly posttranslational modifications. In most cases, however, these techniques do not provide information about biological activity. In recent years another branch of proteome analysis has developed to tackle this problem with the development of affinity-labeling techniques, generally called

Table 1
Selected Lectins With Their Specificities

Lectin	Specificity
Concanavalin A (ConA)	Glucosyl and mannosyl residues of *N*-linked oligosaccharides
Wheat germ agglutinin (WGA)	Chitobiose core (di-*N*-acetylglucosamine) and *N*-acetylneuraminic acid
Peanut agglutinin (PNA)	T-antigen (Galb1-3GalNAc) found in *O*-glycans of mucin-type proteins
Aleuria aurantia (AAL)	L-fucose-containing oligosaccharides
Galectines	N-acetyllactoseamine (LacNAc)-containing glycans found in both *N*- and *O*-glycans

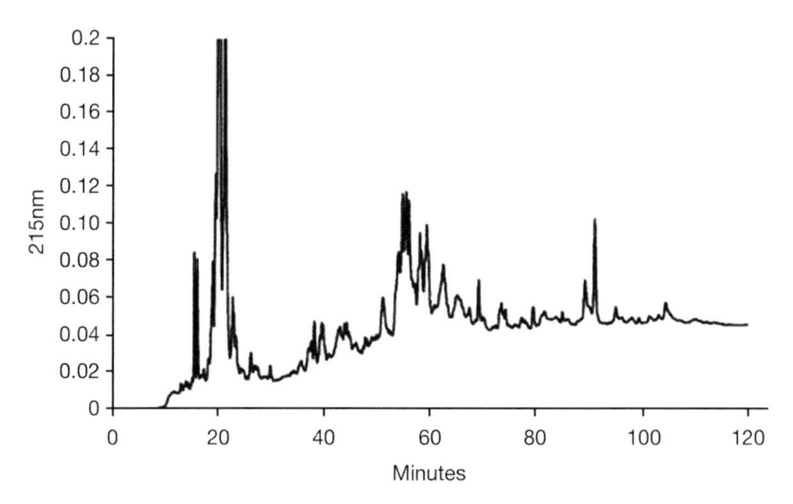

Fig. 8. Enrichment of fucose-containing peptides in serum proteins after tryptic digestion. An LTA lectin affinity column was used to select the peptides followed by deglycosylation and reversed-phase HPLC. Reproduced with permission from Xiong L, Andrews D, Regnier F. 2003. Comparative proteomics of glycoproteins based on lectin selection and isotope coding. *J Proteome Res* 2003;2:618–625.

activity-based protein profiling (ABPP) ***(71–74)***. This line of research focuses on profiling the activity of families of enzymes like the various types of proteases. A derivative of this work is to use affinity ligands, like protease inhibitors, to enrich classes of proteins based on their activity ***(64)***. Arguably, it is the activity of enzymes that is involved in disease development and that may therefore serve as biomarkers rather than the abundance, since most enzymes are present as inactive proforms that are activated upon appropriate (or inappropriate) stimuli.

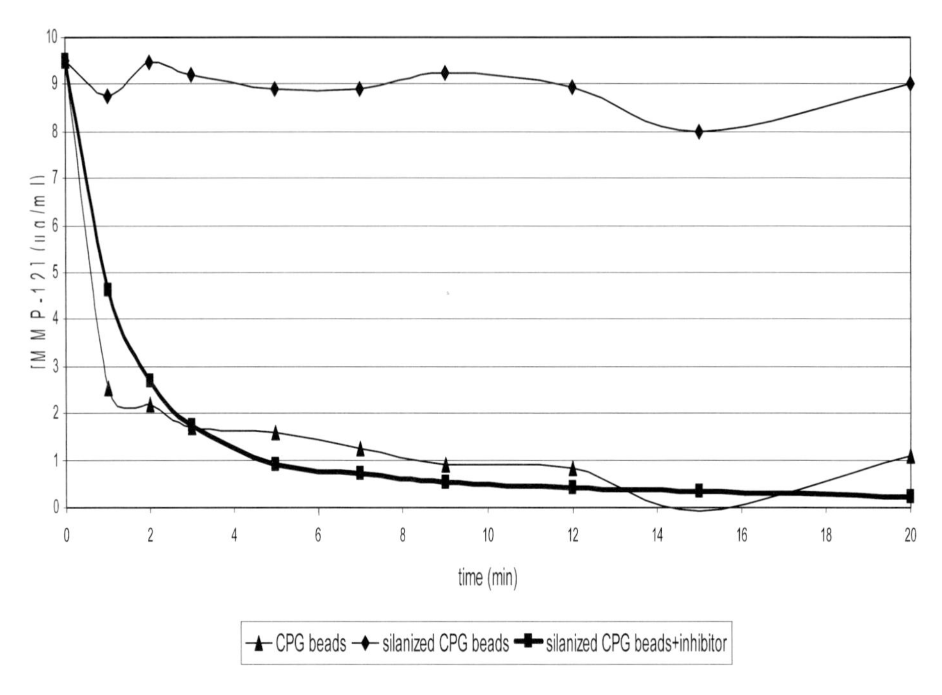

Fig. 9. Binding of the catalytic domain of matrix metalloprotease-12 (MMP-12) to unmodified controlled porosity glass beads (▲), silanized with a diol layer (u) or silanized and then coupled to a reversible MMP inhibitor (n).

ABPs have been described for cysteine proteases ***(73,75–77)***, serine hydrolases, including serine proteases ***(61,78)***, and also metalloproteases ***(59)***. In most cases the labels contain biotin, which allows one not only to visualize but also to isolate the labeled proteins. Even in vivo labeling, for example, in tissue biopsies or cells in culture is feasible ***(73)***. By employing this strategy, sample depletion for abundant proteins can be bypassed as long as the inhibitors or other affinity ligands are sufficiently specific and nonspecific binding to the support materials can be minimized. **Figure 9** gives an example of how strongly some proteins may bind to materials used for the immobilization of affinity ligands and how nonspecific binding may be overcome by chemical derivatization of the surface.

Nonspecific binding to affinity ligands or the surfaces on which they have been immobilized makes stringent controls necessary. For example, preparing nonfunctionalized "control" materials and competing with the affinity interaction by adding an excess of ligand to the binding buffer are common ways of assessing specificity. Some authors have also denatured the proteins by heat treatment prior to binding as a control. As an example, a complex mixture of extracellular serine hydrolases was successfully identified by MS after the hydrolases were enriched by affinity chromatography. In addition, resolution of protein bands on SDS-PAGE was improved upon deglycosylation of the enriched enzymes ***(79,80)***.

Successful affinity-based profiling requires suitable affinity ligands. To address a wider range of proteins based on their activity, combinatorial chemistry approaches have been developed ***(81)***. A new promising concept of in vivo click chemistry protein labeling utilizing the copper(I)-catalyzed azide-alkyne cycloaddition reaction was recently introduced ***(82)***. Present protein enrichment approaches are mostly limited to catalytically active proteins, such as enzymes. A great deal of protein-protein interactions, however, remain largely unexplored.

2.4. Protein Chip Technology (SELDI-TOF-MS)

Since natural body fluids are too complex to analyze directly, investigators are in constant search of new techniques of subfractionation prior to MS. Most often prefractionation is done by LC or electrophoresis, but simple adsorption/washing/desorption methods are finding more widespread use, as they are rather fast and can be automated more easily. In classical matrix-assisted laser desorption/ionization mass spectrometry (MALDI-MS), prefractionated protein samples are digested with trypsin prior to analysis (peptide mass fingerprinting). In this version all peptides are indiscriminately deposited on the MALDI target plate and entrapped in the light-absorbing matrix.

The central idea of surface-enhanced laser desorption/ionization (SELDI) is to use adsorptive surfaces, mainly based on well-known chromatographic principles, to bind a subfraction of proteins from a sample and to analyze the bound proteins or peptides by MALDI-MS. By varying the adsorptive surface, different groups of proteins can be bound and analyzed. This technology has more recently been further developed and commercialized under the trade name ProteinChip® (Ciphergen Biosystems, Palo Alto, CA) and has found widespread application, notably in the medical and clinical research community ***(83)***. The original mass spectrometer was a simple linear MALDI-time of flight (TOF) system, but interfaces have now been developed that allow coupling of the ProteinChip technology to tandem mass spectrometers of the quadrupole-TOF hybrid type.

The key components of this technology are Protein Chip Arrays and the Protein Chip Reader. The array comprises a set of different surfaces, such as a hydrophobic, hydrophilic, or metal chelate to which the biological samples are added. The unbound proteins are washed away, and the bound fraction is subjected to MS analysis. Optionally, it is possible to digest the bound proteins with trypsin to facilitate their identification. However, since most chips bind many diverse proteins, interpretation of the results after trypsin digestion is not always obvious because it is not straightforward to link the observed peptides back to the proteins that gave rise to an increased or decreased peak in the original spectrum. Alternatively, the bound fraction can be subjected to MS analysis without trypsin digestion. As a recent example of this approach, cystatin C was suggested as a biomarker in the diagnosis of Creutzfeld-Jakob disease ***(84)***. Direct fragmentation of the high-molecular-weight ions to obtain sequence information for identification would be most advantageous, but requires high-end mass spectrometers such as FT-ICR-MS instruments ***(85–87)***.

As with any mass spectrometric method dealing with highly complex mixtures, there is a competition between different molecules to ionize (also known as *ion suppression*). It is thus unlikely that the mass spectrum obtained from a ProteinChip will give a true representation of the proteins or peptides adsorbed on the chip. Most applications of SELDI-TOF-MS to body fluids therefore generate rather simple mass spectra, which can be easily analyzed. A number of applications of SELDI-TOF to biological samples, notably plasma, serum, or urine, have shown that samples taken from patients differ significantly from those from healthy controls or from patients with other kinds of disease, opening the possibility of discriminating patient groups and performing early diagnosis of, for example, ovarian or breast cancer ***(88,89)***. However, recent efforts to reproduce these results have met with limited success, and the jury is still out on whether this fairly straightforward approach to sample preparation will lead to clinically relevant results ***(90,91)***.

Probably one of the most impressive studies using SELDI-TOF in recent years was the detection of an antiviral factor secreted by CD8 T-cells upon infection with HIV-1 in immunologically stable patients, which was identified as a member of the α-defensin family by subsequent isolation and protein sequencing ***(92)***. This factor had been known since 1986 but had eluded identification for 15 yr ***(93)***.

2.5. Automated Sample Preparation Using Magnetic Beads

The automation of sample preparation in light of increasing sample throughput and reproducibility is an important aspect of clinical proteomics. In analogy to the previously described ProteinChips, it is possible to prepare samples by adsorption/washing/desorption on magnetic beads (or other kinds of beads). Magnetic beads are an effective tool for fast concentration of diluted samples and for the crude separation of proteins and peptides prior to MS analysis. Magnetic beads are widely used in automated immunoassays, cell purification, and more recently the detection of bacterial pathogens. Mostly, the assay is targeted at individual proteins, like prostate-specific antigen (PSA), which is captured with a biotinylated anti-PSA antibody (anti-F-PSA-M30-IgG) and subsequently bound to streptavidin-coated magnetic beads ***(94)***. This latter approach, however, is targeted to a specific protein and is not applicable to proteomic studies in a broad context. In another example, magnetic nanoparticles modified with vancomycin were used to trap Gram-positive bacteria ***(95)***. The method was able to detect *Staphylococcus aureus* in a 3-mL urine sample at a concentration of 7×10^4 CFU/mL (colony-forming units) by MALDI-TOF-MS. A combination of affinity trapping with MS has also been successful in detecting bacterial and viral infections based on immobilized lectins ***(96,97)***.

Application of magnetic beads to clinical proteomics has emerged only recently mainly based on adapted liquid handling systems ***(98)***. Serum was precipitated with ethanol to remove larger proteins, and the remaining polypeptides in the supernatant were bound to reversed-phase super-paramagnetic silica beads. The washed and eluted peptides were profiled by MALDI-TOF/TOF-MS with the possibility of performing partial sequencing and identification by MS/MS. Four hundred polypeptides were detected in 50 μL serum (range 0.8–15 kDa), and discrimination between samples from brain tumor patients and healthy controls was 96.4% based on a learning algorithm.

2.6. Analysis of Other Body Fluids

Sample preparation is equally important for proteomics and peptidomics in other body fluids, and many of the methods and considerations developed for serum are suitable. Urine is probably the second most relevant body fluid after blood for general proteome studies, owing to its availability. Urine is in fact filtered blood plasma, so it might be representative of the protein spectrum of blood, but with lower protein concentrations. In normal conditions the kidney restricts passage of plasma proteins above approx 40 kDa during filtration in the glomeruli. Proximal renal tubules reabsorb filtered proteins and degrade them. Total amounts of secreted protein per voiding vary from 1 to 10 mg, whereas in pathology, the protein concentrations can dramatically increase ***(99,100)***. Urine collects the metabolic end products of the organism destined for excretion, and its composition is therefore more variable than that of serum or plasma. In particular, the composition and concentration of proteins and peptides in urine are strongly affected by nutrition, the day/night cycle, and the health status of the kidneys. It is thus important to try to control and document these parameters as well as possible.

Although protein concentration in urine is much lower than in serum (approximately 1000-fold) and filtration takes place in the kidneys, albumin is still the major protein. Proteome maps of human urinary proteins were recently constructed after LC-MS analysis of trypsin-digested unfractionated urine ***(101,102)***, by 2D electrophoresis after acetone precipitation ***(103)***, and by depletion of high-abundance proteins (albumin and IgG) followed by ultrafiltration and 2D electrophoresis ***(104)***. An equivalent of urine, human hemofiltrate, was also analyzed by restricted access chromatography to select the peptidome followed by 2D HPLC and MS ***(12,13)***. Urine has furthermore been analyzed by capillary electrophoresis coupled to electrospray ionization MS ***(105)***, establishing a "normal" urinary protein profile. Combined with new analysis software, this analytical method is presently under further investigation.

Normalization of the data obtained to an internal standard that takes biological variation into account is critical for urine. This has been realized in clinical chemistry for a long time, and creatinine is widely used for this purpose. However, whether creatinine is also a suitable normalization standard for proteomics and peptidomics studies in urine is questionable. Normalization based on the total protein content or the area under the curve of the HPLC-UV trace may be preferable. It remains to be seen whether the urinary spectrum of proteins and peptides can be successfully used to detect human diseases short of those related to the kidney or general inflammation.

Usually urine samples should be collected under sterile conditions, cooled down, and treated with protease inhibitors. The next steps of sample pretreatment are variable from one publication to the other. For example, proteins can be concentrated by precipitation with trifluoroacetic acid followed by centrifugation ***(106)***. The resulting sample can be further applied to 1D or 2D electrophoresis, or subjected to solid-phase extraction and trypsin digestion followed by LC-MS analysis. Pieper et al. ***(104)*** compared urinary proteomes of healthy and renal cell carcinoma patients. Initially, cooled samples with added protease inhibitors were cleared by centrifugation and concentrated by membrane filtration. Samples were further desalted and fractionated by gel filtration on Superdex G-75. The resulting sample of more than 30 kDa proteins was passed through a depleting column specific for albumin, IgG, and α-1-acid glycoprotein. The final comparative analysis was done by 2D electrophoresis and mass spectrometry.

Urine samples were recently used for comparative studies of normal and lung cancer patients ***(107)***. The collected urine samples were first desalted by gel filtration (PD-10 columns) followed by lyophilization. The pellet was resuspended in phosphate buffer, extracted with methanol/ chloroform, and precipitated with trichloroacetic acid/acetone to remove organic acids and lipids. Finally, the sample was fractionated with HPLC and 1D or 2D electrophoresis followed by MALDI-MS and MS-MS. The images obtained of the gels demonstrated a quite impressive number of protein spots, but albumin and IgG were still quite abundant.

Whereas easily obtainable body fluids such as blood, urine, saliva, or tears are samples of first choice for human proteomic studies, more specialized samples are frequently used to evaluate the condition of a given organ system. It is implied that a sample taken closer to the diseased organ will show changes in protein composition that are more closely related to the disease than those occurring in blood or urine.

Recently, proteomics of bronchoalveolar lavage fluid (BALF) was reviewed ***(108,109)***. Bronchoalveolar lavage samples the epithelial lining of the lung and is frequently analyzed in cases of severe respiratory diseases (chronic obstructive pulmonary disease, severe asthma, pulmonary fibrosis, and others). BALF consists of a soluble part often used for biomarker analysis and cells derived from the lung tissue or the blood (alveolar macrophages, lymphocytes, neutrophils, and eosinophils). It is noteworthy that most proteins found in BALF correspond to abundant plasma proteins, indicating "plasma leakage" into the alveolar space owing to the lavage procedure. A map of the BALF proteome showed up to 1400 different proteins on a 2D gel ***(108)***, with some proteins at higher concentrations than in serum or plasma. These proteins are likely directly derived from the lung. Removing albumin as the most abundant protein in BALF by RAM chromatography allows one to process larger volumes and thus to detect lower abundance components **(Fig. 10)**.

3. The Linkage to Separation Methods and Mass Spectrometry

The analysis of complex proteomes requires that dedicated and effective sample preparation be followed by high-resolution separation to reduce complexity to a level that can be handled by MS in terms of protein or peptide ionization, identification, and quantification. A wide range of separation methods has been applied to proteins and peptides, and it is beyond the scope of this chapter to review them all. The main purpose of the ensuing sections is to highlight how sample preparation of body fluids affects the downstream separation procedures and MS. To this end two of the major separation methods will be highlighted, notably 2D electrophoresis for whole proteins and HPLC for the low-molecular-weight fraction of proteins and peptides or protein digests.

3.1. 2D Gel Electrophoresis of Proteins in Body Fluids

The presence of a few high-abundance proteins in most body fluids poses a problem for 2D electrophoresis. 2D gels have a limited loading capacity of some hundred micrograms of total protein, which makes the detection of medium- to low-abundance proteins difficult if not impossible unless the sample is prefractionated. Considering that albumin in serum represents 40 mg/mL of the 80 mg/mL total protein concentration and that established tumor markers circulate at concentrations of a few ng/mL or even less, it is clear that applying, for example, 500 μg of total protein to a gel (corresponding to approximately 6 μL of serum) will yield only about 10 pg (approximately 0.12 fmol for a protein of 50 kDa) of a given tumor marker in the original sample. Even assuming a recovery of 100%, this is clearly below the detection level of any protein staining technique and definitely an amount that cannot be identified by in-gel digestion and MS. Without

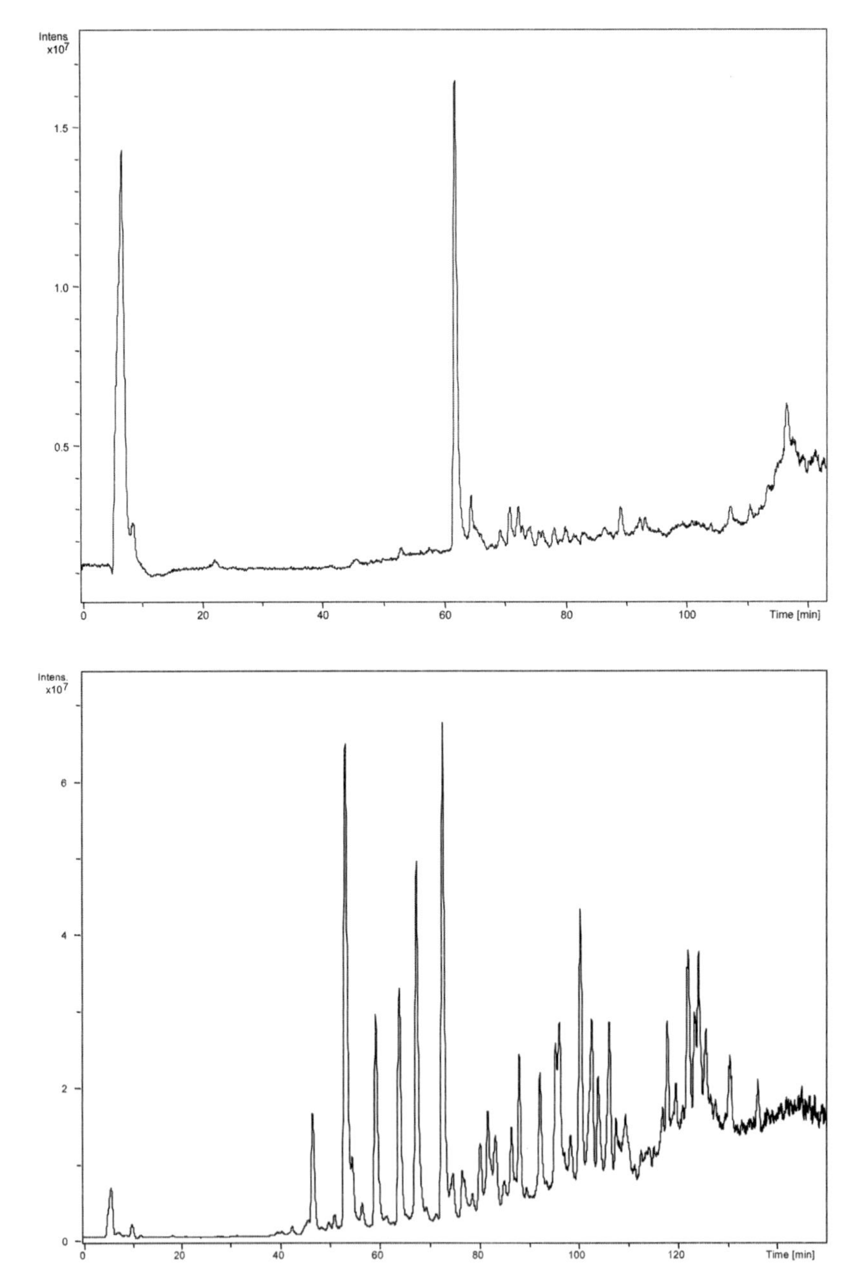

Fig. 10. Sample preparation of bronchoalveolar lavage fluid (BALF) by restricted access material (RAM) chromatography. The upper panel shows the reversed-phase HPLC analysis of 10 μL BALF (major peak is albumin), and the lower panel shows the analysis of 1 mL BALF after sample preparation.

any enrichment or prefractionation, 2D electrophoresis will not be able to reveal proteins at concentrations much below the μg/mL range, an area that is largely occupied by well-known plasma or serum proteins that are likely not relevant as disease-specific biomarkers **(Fig. 3)** *(3)*.

One way to enhance the capacity of 2D electrophoresis to detect proteins at lower concentrations is to remove high-abundance proteins selectively, as described in **Subheading 2.2.** An affinity column developed to deplete the six most abundant proteins from serum resulted in a reduction in total protein by about a factor of 10 ***(41)***. The effect of this removal step on 2D electrophoresis can be appreciated in **Fig. 7**, which shows that after depletion a range of proteins becomes visible. However, increasing the loading capacity by a factor of 10 is not sufficient to reach into the ng/mL concentration range for complex body fluids like serum.

Another strategy to cover more of the low-abundance proteins is to enrich some of them specifically. This is naturally at the expense of losing the overview over the proteome and thus potentially missing relevant markers. Enrichment depends critically on the selection of appropriate ligands (*see* **Subheading 2.3.**) in combination with stationary phases of low nonspecific protein binding. Both requirements are not easy to meet, but it is often the elimination of nonspecific binding that poses the greatest problems. As an example of the effect of nonspecific binding, **Fig. 9** shows the binding kinetics of a metalloprotease to controlled porosity glass beads (with or without silanization to render the surface more hydrophilic) and to the silanized beads containing an immobilized metalloprotease inhibitor.

Prefractionation of the sample is another option to reduce complexity prior to 2D electrophoresis, at the expense, however, of having to run multiple gels for a single sample. This is often not a viable option owing to the work-intensive nature of 2D gels. An alternative is to select narrow pH ranges to visualize only that part of the proteome that does not coincide with the high-abundance proteins. Unfortunately, most serum proteins have similar isoelectric points between pH 5 and 6 (*see* **Fig. 7**), making fractionation difficult. More recently, prefractionation by preparative in-solution isoelectric focusing has emerged as a first step in body fluid analysis prior to 2D electrophoresis and also chromatography ***(110)***. This approach has the advantage that proteins are fractionated based on a clear-cut physicochemical parameter, their isoelectric point, but multiple fractions still need to be analyzed, meaning that an efficient, preferably automated method should be used downstream.

3.2. LC-MS of Proteins and Peptides in Body Fluids

Based on the above discussion, it is obvious that 2D electrophoresis is not the method of choice for analyzing large series of clinical samples in quest of new disease-specific markers. Consequently, other methods have been sought to reduce the workload of 2D gels, methods that make use of automated equipment. In addition to the already described direct combination of sample preparation on protein chips or magnetic beads with MS, as outlined in **Subheadings 2.4**. and **2.5.**, there is increasing interest in the combination of online sample preparation with LC (LC-MS). In the following two examples, a focus on the low-molecular-weight part of the body fluid proteome (also referred to as the *peptidome*) and the shotgun proteomics approach requiring proteolytic digestion will be highlighted. Indeed, there are many possibilities of integrating the sample preparation step with the ensuing separation, but these two approaches may serve as examples.

3.2.1. Peptidomics

Dividing the proteome of body fluids into a high- and a low-molecular-weight fraction (the so-called peptidome *[9,111–114]*) is an approach to detect lower abundance small proteins and peptides. Although it is restricted to a certain molecular weight range, the peptidome contains extensive information about processes in the organism that may be relevant for diagnosis and follow-up of therapy. This is a deliberate choice of sample pretreatment, eliminating most of the high-abundance serum or plasma proteins. An additional advantage of focusing on the molecular weight region below 15 to 20 kDa is that these molecules are more easily separated and recovered by reversed-phase HPLC (RP-HPLC), which is the preferred method for coupling to MS.

There are a number of techniques that allow elimination of the fraction of the proteome above approximately 20 kDa, such as ultrafiltration (*see* **Subheading 3.2.**), precipitation with acids or organic solvents (*see* also **Subheading 3.2.**), or the combination of ultrafiltration with adsorption chromatography, e.g., RAM chromatography ***(10,11)***. Full integration of all analytical steps in an automated system is often desirable for biofluid analysis in a clinical or biomedical environment to increase throughput, reduce the need for skilled personnel, and increase reproducibility. Furthermore, documentation is often facilitated by using an integrated, fully automated analytical system. Combining the "unit operations" of sample pretreatment, separation, and detection in the case of peptidomics was achieved in a system described by Wagner et al. ***(12)*** and further developed by Machtejevas et al. ***(13)***. **Figure 11** shows the instrumental setup combining selection of the peptidome from human hemofiltrate by RAM chromatography followed by prefractionation on a strong cation exchanger and finally separation by RP-HPLC. Although this setup was not coupled online to a mass spectrometer, analysis of selected fractions after RP-HPLC by MALDI-TOF-MS showed that complexity of the original hemofiltrate had been reduced to such a level that most of the fractions contained one major peptide or small protein **(Figs. 12** and **13)**.

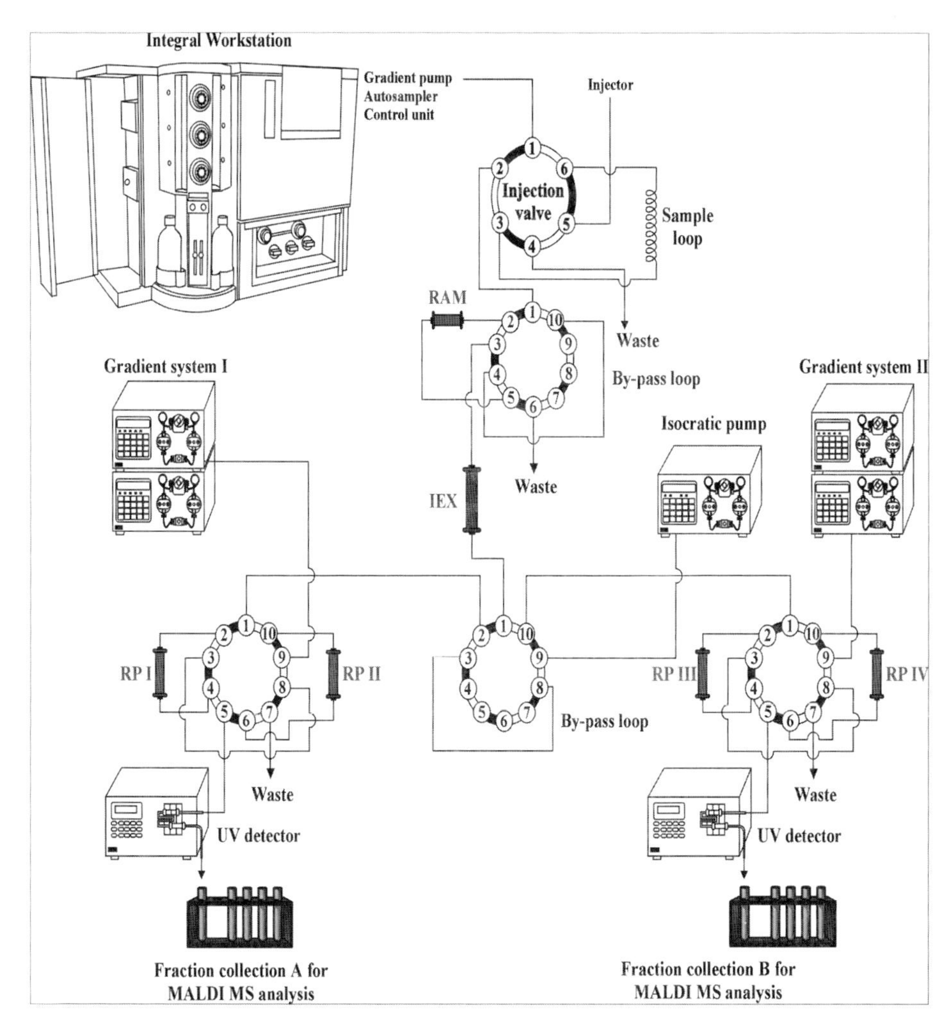

Fig. 11. Schematic diagram of an online comprehensive 2D HPLC system including an integrated sample preparation step by restricted access chromatography. Strong cation-exchange HPLC is used in the first dimension (IEX) followed by rapid reversed-phase (RP) HPLC on four nonporous particle-packed columns working in parallel. Reproduced with permission from Machtejevas E, John H, Wagner K, et al. Automated multi-dimensional liquid chromatography: sample preparation and identification of peptides from human blood filtrate. *J Chromatogr B* 2004;803:121–130.

3.2.2. Shotgun Proteomics

Applying HPLC separation to the high-molecular-weight region of the proteome requires prior proteolytic digestion, since most complex proteins are not stable under the denaturing conditions of RP-HPLC and are thus not quantitatively

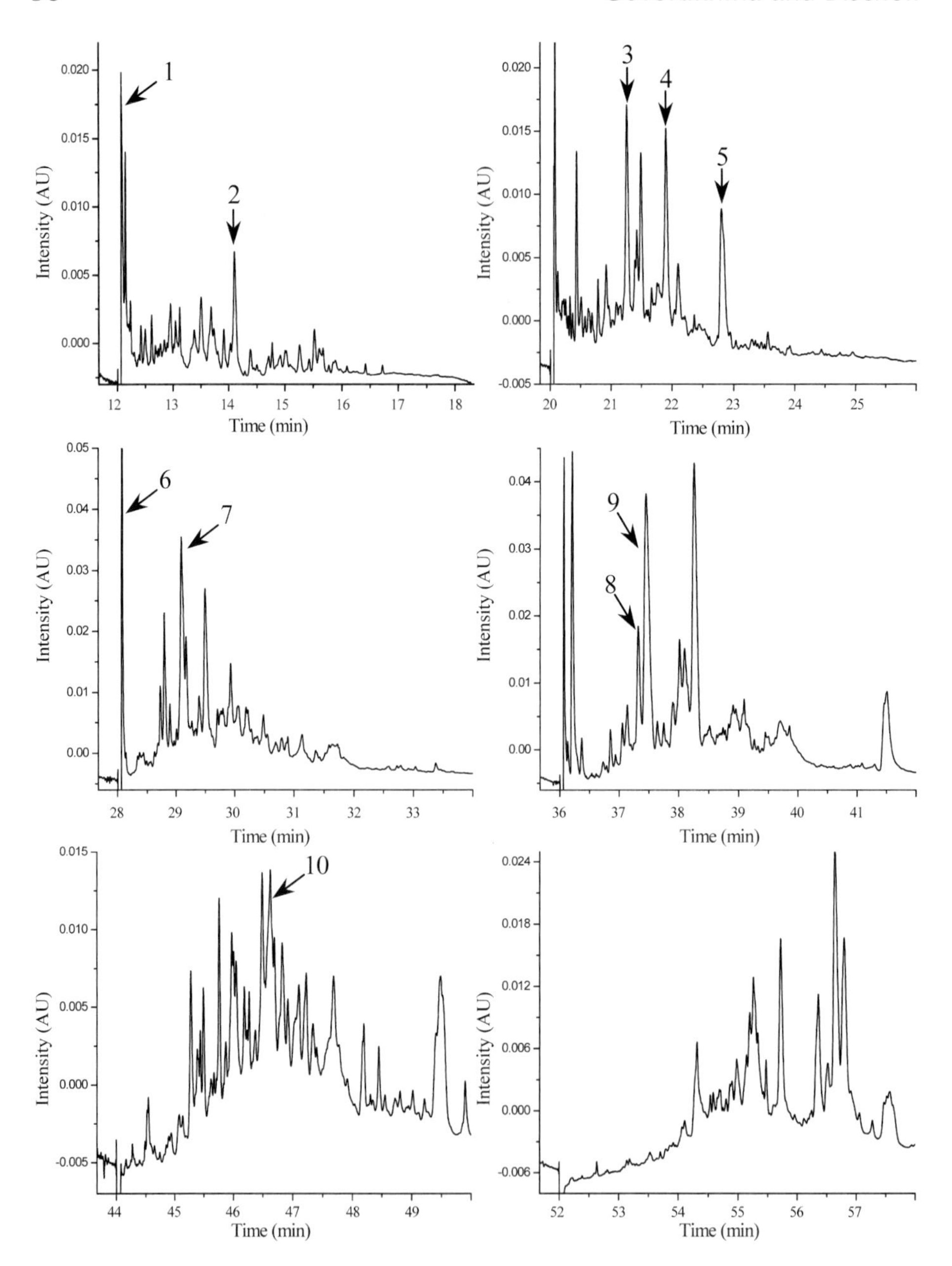

Fig. 12. Selected reversed-phase chromatograms from human hemofiltrate processed through the integrated, multidimensional chromatography system shown in **Fig. 11.** Numbered and marked peak fractions 1 to 10 were selected for MS and sequence analysis. Reproduced with permission from Machtejevas E, John H, Wagner K, et al. Automated multi-dimensional liquid chromatography: sample preparation and identification of peptides from human blood filtrate. *J Chromatogr B* 2004;803:121–130.

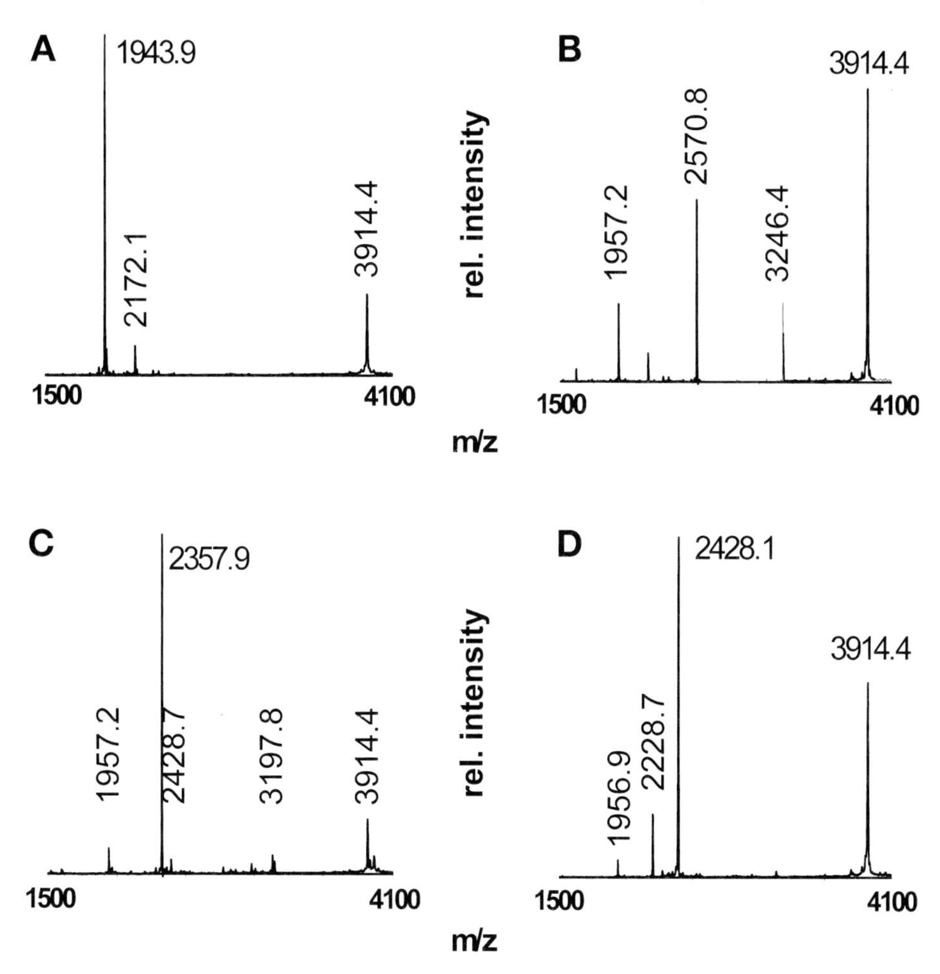

Fig. 13. MALDI-TOF mass spectra of selected peaks from the reversed-phase HPLC fractions shown in **Fig. 12.** Spectra **(A)**, **(B)**, **(C)**, and **(D)** correspond to peak fractions 4, 7, 8, and 9, respectively. Measurement was performed in the linear positive operation mode using a matrix of α-cyano-4-hydroxycinnamic acid mixed with fucose. The peak at *m/z* = 3914.4 is an internal standard. Reproduced with permission from Machtejevas E, John H, Wagner K, et al. Automated multi-dimensional liquid chromatography: sample preparation and identification of peptides from human blood filtrate. *J Chromatogr B* 2004;803:121–130.

recovered. In addition, it is not possible to identify proteins based on their molecular weight only, and fragmentation of large proteins is impossible in most commonly employed mass spectrometers.

This has led to the shotgun proteomics approach, whereby the entire proteome is first digested with trypsin followed by one or multidimensional chromatographic separations of the peptides *(14,16)*. Most of the observations about high-abundance proteins masking those of lower abundance that were made with respect to 2D electrophoresis also apply to the shotgun method. This is partially because of the limited loading capacity of the chromatographic columns, especially when nanoLC-MS (loadability in the ng–μg range) is the final analytical step. Removing high-abundance proteins by affinity chromatography prior to digestion is one way of increasing the loading of medium- and low-abundance proteins. The effect of depletion of albumin on the detection of a tryptic peptide derived from α_1-antitrypsin, whose concentration in serum is about 20-fold lower, shows the clear improvement even for a protein that is still considered to belong to the high-abundance class **(Fig. 6)**. Depleting high-abundance proteins and notably albumin can also have drawbacks, however, as albumin is known to bind and carry numerous compounds including small proteins and peptides that may be codepleted. Zhou et al. *(29)* showed, for example, that albumin binds some 210 proteins and peptides, which could be identified in the albumin-containing fraction after depletion. In some of our own studies, we observed that a small protein added as an internal standard (horse heart cytochrome C) at pmol/μL concentrations was codepleted to about 25% of the original amount when the six most abundant proteins were removed by affinity chromatography *(125)*. It is important to note that although depletion removes some other proteins and peptides, this seems to be rather reproducible, at least when judged from the results with horse heart cytochrome C. (The relative standard deviation of peak areas without normalization is 10–30%.)

The enormously complex mixture generated by shotgun proteomics usually requires multiple separation dimensions prior to MS. This has inspired researchers to develop novel ways of performing "coupled column" HPLC such as the multidimensional protein identification technology (MudPIT) *(15)* or the integrated setup outlined in **Fig. 11.** Because analysis of individual fractions from earlier dimensions in such a multidimensional HPLC approach requires 12 to 24 h per sample, more efficient and rapid separation methods are under development. One of them is based on reducing the particle diameter of the chromatographic materials to 1 μm, which results in very high backpressures that can only be delivered by special HPLC equipment *(115–118)*; another approach is based on monolithic materials that support very high linear flow rates at pressures amenable to common HPLC equipment *(17,119–123)*. These developments show that in addition to the major advances in MS, which have made proteomics as we know it today possible, there is also considerable activity in the fields of sample preparation and separation methodology. It is only through integration of these unit operations into an analytical strategy that the challenges of body fluid analysis can be tackled and possibly the entire dynamic range covered. Much remains to be done.

4. Conclusions

Proteomics of body fluids is a rapidly expanding field driven by the search for better biomarkers for disease diagnosis, follow-up on therapy, and evaluation of the response of patients to newly developed pharmaceuticals. The analysis of body fluids has a long tradition in clinical chemistry and serves to support decision making by clinicians in many respects. Because of recent methodological developments in separation science, MS, and bioinformatics, there has been a surge of efforts to apply them to biomarker discovery, often focusing on biomarker patterns rather than individual molecules. Sample preparation, the indispensable and very critical first step in an analytical method, has attracted less attention, and its relevance is often underestimated. As outlined in this chapter, the approach to sample preparation is an important decision of strategic relevance for the ensuing analyses. It is therefore pivotal to weigh the pros and cons of each approach in light of the final goal. We hope that the overview given in this chapter will guide the reader in this complex methodological field.

Acknowledgments

We thank Dr. Theo Reijmers (Groningen Bioinformatics Institute) for calculations related to **Fig. 2** and Dr. Begona Barroso (Department of Analytical Biochemistry) for contributing the data for **Fig. 10**. We are also very grateful to Dr. Peter Schulz-Knappe (BioVision, Hannover, Germany) for communicating unpublished information for use in this chapter.

References

1. Pusch W, Flocco MT, Leung SM, Thiele H, Kostrzewa M. Mass spectrometry-based clinical proteomics. *Pharmacogenomics* 2003;4:463–476.
2. Schulte I, Tammen H, Schulz-Knappe P, Selle H. Peptides in body fluids and tissues as markers of disease. *Exp Rev Mol Diagn* 2005;5:145–157.
3. Anderson NL, Anderson NG. The Human Plasma Proteome: history, character, and diagnostic prospects. *Mol Cell Proteomics* 2002;1:845–867.
4. Johnson KL, Mason CJ, Muddiman DC, Eckel JE. Analysis of the low molecular weight fraction of serum by LC-dual ESI-FT-ICR mass spectrometry: precision of retention time, mass, and ion abundance. *Anal Chem* 2004;76:5097–5103.
5. Tirumalai RS, Chan KC, Prieto DA, Issaq HJ, Conrads TP, Veenstra TD. Characterization of the low molecular weight human serum proteome. *Mol Cell Proteomics* 2003;2:1096–1103.
6. Georgiou HM, Rice GE, Baker MS. Proteomic analysis of human plasma: failure of centrifugal ultrafiltration to remove albumin and other high molecular weight proteins. *Proteomics* 2001;1:1503–1506.
7. Schulz Knappe P, Schrader M, Standker L, et al. Peptide bank generated by large-scale preparation of circulating human peptides. *J Chromatogr A* 1997;776: 125–132.

8. Schulz Knappe P, Raida M, Meyer M, Quellhorst EA, Forssmann WG. Systematic isolation of circulating human peptides: the concept of peptide trapping. *Eur J Med Res* 1996;1:223–236.
9. Raida M, Schulz-Knappe P, Heine G, Forssmann WG. Liquid chromatography and electrospray mass spectrometric mapping of peptides from human plasma filtrate. *J Am Soc Mass Spectrom* 1999;10:45–54.
10. Racaityte K, Lutz ESM, Unger KK, Lubda D, Boos KS. Analysis of neuropeptide Y and its metabolites by high-performance liquid chromatography-electrospray ionization mass spectrometry and integrated sample clean-up with a novel restricted-access sulphonic acid cation exchanger. *J Chromatogr A* 2000; 890:135–144.
11. Boos KS, Grimm CH. High-performance liquid chromatography integrated solid-phase extraction in bioanalysis using restricted access precolumn packings. *TrAC Trends Anal Chem* 1999;18:175–180.
12. Wagner K, Miliotis T, Marko-Varga G, Bischoff R, Unger KK. An automated on-line multidimensional HPLC system for protein and peptide mapping with integrated sample preparation. *Anal Chem* 2002;74:809–820.
13. Machtejevas E, John H, Wagner K, et al. Automated multi-dimensional liquid chromatography: sample preparation and identification of peptides from human blood filtrate. *J Chromatogr B* 2004;803:121–130.
14. Wolters DA, Washburn MP, Yates JR, III. An automated multidimensional protein identification technology for shotgun proteomics. *Anal Chem* 2001;73:5683–5690.
15. Washburn MP, Wolters D, Yates JR. Large-scale analysis of the yeast proteome by multidimensional protein identification technology. *Nat Biotechnol* 2001;19: 242–247.
16. McDonald WH, Yates JR, III. Shotgun proteomics and biomarker discovery. *Dis Markers* 2002;18:99–105.
17. Barroso B, Lubda D, Bischoff R. Applications of monolithic silica capillary columns in proteomics. *J Proteome Res* 2003;2:633–642.
18. Strittmatter EF, Ferguson PL, Tang K, Smith RD. Proteome analyses using accurate mass and elution time peptide tags with capillary LC time-of-flight mass spectrometry. *J Am Soc Mass Spectrom* 2003;14:980–991.
19. Shen Y, Tolic N, Masselon C, et al. Ultrasensitive proteomics using high-efficiency on-line micro-SPE-nanoLC-nanoESI MS and MS/MS. *Anal Chem* 2004;76:144–154.
20. Adkins JN, Varnum SM, Auberry KJ, et al. Toward a human blood serum proteome: analysis by multidimensional separation coupled with mass spectrometry. *Mol Cell Proteomics* 2002;1:947–955.
21. Li C, Lee KH. Affinity depletion of albumin from human cerebrospinal fluid using Cibacron-blue-3G-A-derivatized photopatterned copolymer in a microfluidic device. *Anal Biochem* 2004;333:381–388.
22. Govorukhina NI, Keizer-Gunnink A, van der Zee AGJ, de Jong S, de Bruijn HWA, Bischoff R. Sample preparation of human serum for the analysis of tumor

markers: comparison of different approaches for albumin and [gamma]-globulin depletion. *J Chromatogr A* 2003;1009:171–178.

23. Hinerfeld D, Innamorati D, Pirro J, Tam SW. Serum/plasma depletion with chicken immunoglobulin Y antibodies for proteomic analysis from multiple mammalian species. *J Biomol Tech* 2004;15:184–190.
24. Greenough C, Jenkins RE, Kitteringham NR, Pirmohamed M, Park BK, Pennington SR. A method for the rapid depletion of albumin and immunoglobulin from human plasma. *Proteomics* 2004;4:3107–3111.
25. Fountoulakis M, Juranville JF, Jiang L, et al. Depletion of the high-abundance plasma proteins. *Amino Acids* 2004;27:249–259.
26. Chromy BA, Gonzales AD, Perkins J, et al. Proteomic analysis of human serum by two-dimensional differential gel electrophoresis after depletion of high-abundant proteins. *J Proteome Res* 2004;3:1120–1127.
27. Bjorhall K, Miliotis T, Davidsson P. Comparison of different depletion strategies for improved resolution in proteomic analysis of human serum samples. *Proteomics* 2005;5:307–317.
28. Bischoff R, Luider TM. Methodological advances in the discovery of protein and peptide disease markers. *J Chromatogr B* 2004;803:27–40.
29. Zhou M, Lucas DA, Chan KC, et al. An investigation into the human serum "interactome". *Electrophoresis* 2004;25:1289–1298.
30. Solassol J, Marin P, Demettre E, et al. Proteomic detection of prostate-specific antigen using a serum fractionation procedure: potential implication for new low-abundance cancer biomarkers detection. *Anal Biochem* 2005;338:26–31.
31. Troyer DA, Mubiru J, Leach RJ, Naylor SL. Promise and challenge: markers of prostate cancer detection, diagnosis and prognosis. *Dis Markers* 2004;20:117–128.
32. Baldus SE, Engelmann K, Hanisch FG. MUC1 and the MUCs: a family of human mucins with impact in cancer biology. *Crit Rev Clin Lab Sci* 2004;41:189–231.
33. Xiong L, Regnier FE. Use of a lectin affinity selector in the search for unusual glycosylation in proteomics. *J Chromatogr B* 2002;782:405–418.
34. Xiong L, Andrews D, Regnier F. Comparative proteomics of glycoproteins based on lectin selection and isotope coding. *J Proteome Res* 2003;2:618–625.
35. Schulenberg B, Beechem JM, Patton WF. Mapping glycosylation changes related to cancer using the Multiplexed Proteomics technology: a protein differential display approach. *J Chromatogr B Analyt Technol Biomed Life Sci* 2003;793:127–139.
36. Geng M, Zhang X, Bina M, Regnier F. Proteomics of glycoproteins based on affinity selection of glycopeptides from tryptic digests. *J Chromatogr B Biomed Sci Appl* 2001;752:293–306.
37. Drake R, Cazares L, Corica A, et al. Quality control, preparation and protein stability issues for blood serum and plasma used in biomarker discovery and proteomic profiling assays. *Bioprocessing J* 2004;July/August 43–49.
38. Sorace JM, Zhan M. A data review and re-assessment of ovarian cancer serum proteomic profiling. *BMC Bioinformatics* 2003;4:24.

39. Windig W, Phalp JM, Payne AW. A Noise and background reduction method for component detection in liquid chromatography/mass spectrometry. *Anal Chem* 1996;68:3602–3606.
40. Zolg JW, Langen H. How industry is approaching the search for new diagnostic markers and biomarkers. *Mol Cell Proteomics* 2004;3:345–354.
41. Bjorhall K, Miliotis T, Davidsson P. Comparison of different depletion strategies for improved resolution in proteomic analysis of human serum samples. *Proteomics* 2005;5:307–317.
42. Wang YY, Chan DW, Wang YY, Cheng P. A simple affinity spin tube filter method for removing high-abundant common proteins or enriching low-abundant biomarkers for serum proteomic analysis. *Proteomics* 2003;3:243–248.
43. Gianazza E, Arnaud P. Chromatography of plasma proteins on immobilized Cibacron Blue F3-GA. Mechanism of the molecular interaction. *Biochem J* 1982;203:637–641.
44. Sato AK, Sexton DJ, Morganelli LA, et al. Development of mammalian serum albumin affinity purification media by peptide phage display. *Biotechnol Prog* 2002;18:182–192.
45. Baussant T, Bougueleret L, Johnson A, et al. Effective depletion of albumin using a new peptide-based affinity medium. *Proteomics* 2005;5:973–977.
46. Bjorck L, Kronvall G. Purification and some properties of streptococcal protein G, a novel IgG-binding reagent. *J Immunol* 1984;133:969–974.
47. Akerstrom B, Bjorck L. A physicochemical study of protein G, a molecule with unique immunoglobulin G-binding properties. *J Biol Chem* 1986;261: 10,240–10,247.
48. Akerstrom B, Brodin T, Reis K, Bjorck L. Protein G: a powerful tool for binding and detection of monoclonal and polyclonal antibodies. *J Immunol* 1985;135: 2589–2592.
49. Guss B, Eliasson M, Olsson A, et al. Structure of the IgG-binding regions of streptococcal protein G. *EMBO J* 1986;5:1567–1575.
50. Fahnestock SR. Cloned streptococcal protein G genes. *Trends Biotechnol* 1987; 5:79–83.
51. Roque AC, Taipa MA, Lowe CR. An artificial protein L for the purification of immunoglobulins and Fab fragments by affinity chromatography. *J Chromatogr A* 2005;1064:157–167.
52. Horvath ZS, Corthals GL, Wrigley CW, Margolis J. Multifunctional apparatus for electrokinetic processing of proteins. *Electrophoresis* 1994;15:968–971.
53. Pang L, Fryksdale BG, Chow N, Wong DL, Gaertner AL, Miller BS. Impact of prefractionation using Gradiflow on two-dimensional gel electrophoresis and protein identification by matrix assisted laser desorption/ionization-time of flight-mass spectrometry. *Electrophoresis* 2003;24:3484–3492.
54. Rothemund DL, Locke VL, Liew A, Thomas TM, Rylatt DB, Wasinger V. Depletion of the highly abundant protein albumin from human plasma using the Gradiflow. *Proteomics* 2003;3:279–287.

55. Weber G, Bocek P. Recent developments in preparative free flow isoelectric focusing. *Electrophoresis* 1998;19:1649–1653.
56. Mehta AI, Mehta AI, Ross S, et al. Biomarker amplification by serum carrier protein binding. *Dis Markers* 2004;19:1–10.
57. Steel LF, Mattu TS, Mehta A, et al. A proteomic approach for the discovery of early detection markers of hepatocellular carcinoma. *Dis Markers* 2001;17: 179–189.
58. Block TM, Comunale MA, Lowman M, et al. Use of targeted glycoproteomics to identify serum glycoproteins that correlate with liver cancer in woodchucks and humans. *Proc Natl Acad Sci U S A* 2005;102:779–784.
59. Saghatelian A, Jessani N, Joseph A, Humphrey M, Cravatt BF. Activity-based probes for the proteomic profiling of metalloproteases. *Proc Natl Acad Sci U S A* 2004;101:10,000–10,005.
60. Ovaa H, Kessler BM, Rolen U, Galardy PJ, Ploegh HL, Masucci MG. Activity-based ubiquitin-specific protease (USP) profiling of virus-infected and malignant human cells. *Proc Natl Acad Sci U S A* 2004;101:2253–2258.
61. Liu YS, Patricelli MP, Cravatt BF. Activity-based protein profiling: the serine hydrolases. *Proc Natl Acad Sci U S A* 1999;96:14694–14699.
62. Kumar S, Zhou B, Liang F, Wang WQ, Huang Z, Zhang ZY. Activity-based probes for protein tyrosine phosphatases. *Proc Natl Acad Sci U S A* 2004;101: 7943–7948.
63. Jessani N, Cravatt BF. The development and application of methods for activity-based protein profiling. *Curr Opin Chem Biol* 2004;8:54–59.
64. Freije JR, Bischoff R. Activity-based enrichment of matrix metalloproteinases using reversible inhibitors as affinity ligands. *J Chromatogr A* 2003;1009:155–169.
65. Falgueyret JP, Black WC, Cromlish W, et al. An activity-based probe for the determination of cysteine cathepsin protease activities in whole cells. *Anal Biochem* 2004;335:218–227.
66. Xiong L, Regnier FE. Use of a lectin affinity selector in the search for unusual glycosylation in proteomics. *J Chromatogr B* 2002;782:405–418.
67. Grunewald S, Matthijs G, Jaeken J. Congenital disorders of glycosylation: a review. *Pediatr Res* 2002;52:618–624.
68. Rudiger H, Gabius HJ. Plant lectins: Occurrence, biochemistry, functions and applications. *Glycoconjugate J* 2001;18:589–613.
69. Wiener MC, Van Hoek AN, Wiener MC. A lectin screening method for membrane glycoproteins: application to the human CHIP28 water channel (AQP-1). *Anal Biochem* 1996;241:267–268.
70. Yang Z, Hancock WS. Approach to the comprehensive analysis of glycoproteins isolated from human serum using a multi-lectin affinity column. *J Chromatogr A* 2004;1053:79–88.
71. Nazif T, Bogyo M. Global analysis of proteasomal substrate specificity using positional-scanning libraries of covalent inhibitors. *Proc Natl Acad Sci U S A* 2001;98:2967–2972.

72. Jeffery DA, Bogyo M. Chemical proteomics and its application to drug discovery. *Curr Opin Biotechnol* 2003;14:87–95.
73. Greenbaum D, Baruch A, Hayrapetian L, et al. Chemical approaches for functionally probing the proteome. *Mol Cell Proteomics* 2002;1:60–68.
74. Adam GC, Cravatt BF, Sorensen EJ. Profiling the specific reactivity of the proteome with non-directed activity-based probes. *Chem Biol* 2001;8:81–95.
75. Bogyo M, Verhelst S, Bellingard-Dubouchaud V, Toba S, Greenbaum D. Selective targeting of lysosomal cysteine proteases with radiolabeled electrophilic substrate analogs. *Chem Biol* 2000;7:27–38.
76. Borodovsky A, Ovaa H, Kolli N, et al. Chemistry-based functional proteomics reveals novel members of the deubiquitinating enzyme family. *Chem Biol* 2002;9:1149–1159.
77. Faleiro L, Kobayashi R, Fearnhead H, Lazebnik Y, Faleiro L. Multiple species of CPP32 and Mch2 are the major active caspases present in apoptotic cells. *EMBO J* 1997;16:2271–2281.
78. Kidd D, Liu Y, Cravatt BF. Profiling serine hydrolase activities in complex proteomes. *Biochemistry* 2001;40:4005–4015.
79. Jessani N, Cravatt BF. The development and application of methods for activity-based protein profiling. *Curr Opin Chem Biol* 2004;8:54–59.
80. Jessani N, Liu Y, Humphrey M, Cravatt BF. Enzyme activity profiles of the secreted and membrane proteome that depict cancer cell invasiveness. *Proc Natl Acad Sci U S A* 2002;99:10,335–10,340.
81. Speers AE, Adam GC, Cravatt BF. Activity-based protein profiling in vivo using a copper(i)-catalyzed azide-alkyne [3 + 2] cycloaddition. *J Am Chem Soc* 2003;125:4686–4687.
82. Speers AE, Cravatt BF. Profiling enzyme activities in vivo using click chemistry methods. *Chem Biol* 2004;11:535–546.
83. Petricoin EF, Zoon KC, Kohn EC, Barrett JC, Liotta LA. Clinical proteomics: translating benchside promise into bedside reality. *Nat Rev Drug Discov* 2002;1:683–695.
84. Sanchez JC, Guillaume E, Lescuyer P, et al. Cystatin C as a potential cerebrospinal fluid marker for the diagnosis of Creutzfeldt-Jakob disease. *Proteomics* 2004;4:2229–2233.
85. Patrie SM, Charlebois JP, Whipple D, et al. Construction of a hybrid quadrupole/fourier transform ion cyclotron resonance mass spectrometer for versatile MS/MS above 10 kDa. *J Am Soc Mass Spectrom* 2004;15:1099–1108.
86. Horn DM, Zubarev RA, McLafferty FW. Automated de novo sequencing of proteins by tandem high- resolution mass spectrometry. *Proc Natl Acad Sci U S A* 2000;97:10,313–10,317.
87. Ge Y, ElNaggar M, Sze SK, et al. Top down characterization of secreted proteins from *Mycobacterium tuberculosis* by electron capture dissociation mass spectrometry. *J Am Soc Mass Spectrom* 2003;14:253–261.
88. Vlahou A, Gregory B, Wright J, et al. A novel approach toward development of a rapid blood test for breast cancer. *Clin Breast Cancer* 2003;4:203–209.

89. Petricoin EF, Ardekani AM, Hitt BA, et al. Use of proteomic patterns in serum to identify ovarian cancer. *Lancet* 2002;359:572–577.
90. Diamandis EP. Mass spectrometry as a diagnostic and a cancer biomarker discovery tool: opportunities and potential limitations. *Mol Cell Proteomics* 2004;3: 367–378.
91. Diamandis EP. OvaCheck: doubts voiced soon after publication. *Nature* 2004; 430:611.
92. Zhang L, Yu W, He T, et al. Contribution of human alpha-defensin 1, 2, and 3 to the anti-HIV-1 activity of CD8 antiviral factor. *Science* 2002;298:995–1000.
93. Walker CM, Moody DJ, Stites DP, Levy JA. CD8+ lymphocytes can control HIV infection in vitro by suppressing virus replication. *Science* 1986;234:1563–1566.
94. Peter J, Unverzagt C, Krogh TN, Vorm O, Hoesel W. Identification of precursor forms of free prostate-specific antigen in serum of prostate cancer patients by immunosorption and mass spectrometry. *Cancer Res* 2001;61:957–962.
95. Lin YS, Weng MF, Chen YC, Tsai PJ. Affinity capture using vancomycin-bound magnetic nanoparticles for the MALDI-MS analysis of bacteria. *Anal Chem* 2005;77:1753–1760.
96. Bundy JL, Fenselau C. Lectin and carbohydrate affinity capture surfaces for mass spectrometric analysis of microorganisms. *Anal Chem* 2001;73:751–757.
97. Bundy J, Fenselau C. Lectin-based affinity capture for MALDI-MS analysis of bacteria. *Anal Chem* 1999;71:1460–1463.
98. Villanueva J, Philip J, Entenberg D, et al. Serum peptide profiling by magnetic particle-assisted, automated sample processing and MALDI-TOF mass spectrometry. *Anal Chem* 2004;76:1560–1570.
99. Celis JE, Rasmussen HH, Vorum H, et al. Bladder squamous cell carcinomas express psoriasin and externalize it to the urine. *J Urol* 1996;155:2105–2112.
100. Marshall T, Williams KM, Marshall T. Clinical analysis of human urinary proteins using high resolution electrophoresis methods. *Electrophoresis* 1998;19: 1752–1770.
101. Spahr CS, Davis MT, McGinley MD, et al. Towards defining the urinary proteome using liquid chromatography-tandem mass spectrometry. I. Profiling an unfractionated tryptic digest. *Proteomics* 2001;1:93–107.
102. Davis MT, Spahr CS, McGinley MD, et al. Towards defining the urinary proteome using liquid chromatography-tandem mass spectrometry. II. Limitations of complex mixture analyses. *Proteomics* 2001;1:108–117.
103. Thongboonkerd V, McLeish KR, Arthur JM, Klein JB. Proteomic analysis of normal human urinary proteins isolated by acetone precipitation or ultracentrifugation. *Kidney Int* 2002;62:1461–1469.
104. Pieper R, Gatlin CL, McGrath AM, et al. Characterization of the human urinary proteome: a method for high-resolution display of urinary proteins on two-dimensional electrophoresis gels with a yield of nearly 1400 distinct protein spots. *Proteomics* 2004;4:1159–1174.
105. Wittke S, Fliser D, Haubitz M, et al. Determination of peptides and proteins in human urine with capillary electrophoresis-mass spectrometry, a suitable tool for

the establishment of new diagnostic markers. *J Chromatogr A* 2003;1013: 173–181.
106. Pang JX, Ginanni N, Dongre AR, Hefta SA, Opitek GJ. Biomarker discovery in urine by proteomics. *J Proteome Res* 2002;1:161–169.
107. Tantipaiboonwong P, Sinchaikul S, Sriyam S, Phutrakul S, Chen ST. Different techniques for urinary protein analysis of normal and lung cancer patients. *Proteomics* 2005;5:1140–1149.
108. Wattiez R, Falmagne P. Proteomics of bronchoalveolar lavage fluid. *J Chromatogr B* 2005;815:169–178.
109. Noel-Georis I, Bernard A, Falmagne P, Wattiez R. Database of bronchoalveolar lavage fluid proteins. *J Chromatogr B* 2002;771:221–236.
110. Righetti PG, Castagna A, Antonioli P, Boschetti E. Prefractionation techniques in proteome analysis: the mining tools of the third millennium. *Electrophoresis* 2005;26:297–319.
111. Schulz-Knappe P, Zucht HD, Heine G, Jurgens M, Hess R, Schrader M. Peptidomics: the comprehensive analysis of peptides in complex biological mixtures. *Comb Chem High Throughput Screen* 2001;4:207–217.
112. Schulz-Knappe P, Schrader M. Peptidomics in biomarker and drug discovery. *Curr Drug Discov* 2003;21–24.
113. Schrader M, Schulz-Knappe P. Peptidomics technologies for human body fluids. *Trends Biotechnol* 2001;19(10 Suppl).
114. Heine G, Zucht HD, Jürgens M, et al. High-resolution peptide mapping of cerebrospinal fluid: a novel concept for diagnosis and research in central nervous system diseases. *J Chromatogr B* 2002;782:353–361.
115. MacNair JE, Patel KD, Jorgenson JW. Ultrahigh-pressure reversed-phase capillary liquid chromatography: isocratic and gradient elution using columns packed with 1.0-micron particles. *Anal Chem* 1999;71:700–708.
116. MacNair JE, Lewis KC, Jorgenson JW. Ultrahigh-pressure reversed-phase liquid chromatography in packed capillary columns. *Anal Chem* 1997;69: 983–989.
117. MacNair JE, Opiteck GJ, Jorgenson JW, Moseley MA, III. Rapid separation and characterization of protein and peptide mixtures using 1.5 microns diameter nonporous silica in packed capillary liquid chromatography/mass spectrometry. *Rapid Commun Mass Spectrom* 1997;11:1279–1285.
118. Mellors JS, Jorgenson JW. Use of 1.5-micron porous ethyl-bridged hybrid particles as a stationary-phase support for reversed-phase ultrahigh-pressure liquid chromatography. *Anal Chem* 2004;76:5441–5450.
119. Xiong L, Zhang R, Regnier FE. Potential of silica monolithic columns in peptide separations. *J Chromatogr A* 2004;1030:187–194.
120. Xie S, Allington RW, Svec F, Frechet JMJ. Rapid reversed-phase separation of proteins and peptides using optimized 'moulded' monolithic poly(styrene-co-divinylbenzene) columns. *J Chromatogr A* 1999;865:169–174.
121. Walcher W, Toll H, Ingendoh A, Huber CG. Operational variables in high-performance liquid chromatography-electrospray ionization mass spectrometry

of peptides and proteins using poly(styrene-divinylbenzene) monoliths. *J Chromatogr A* 2004;1053:107–117.

122. Kimura H, Tanigawa T, Morisaka H, et al. Simple 2D-HPLC using a monolithic silica column for peptide separation. *J Separation Sci* 2004;27:897–904.
123. Hennessy TP, Boysen RI, Huber MI, Unger KK, Hearn MTW. Peptide mapping by reversed-phase high-performance liquid chromatography employing silica rod monoliths. *J Chromatogr A* 2003;1009:15–28.
124. Kemperman RF, Horvatovich PL, Hoekman B, et al. Comparative urine analysis by liquid chromatography-mass spectrometry and multivariate statistics: method development, evaluation, and application to proteinuria. *J Proteome Res* 2007;6:194–206.
125. Govorukhina NI, Reijmers TH, Nyangoma SO, et al. Analysis of human serum by LC-MS: improved sample preparation and data analysis. *J Chromatogr A* 2006;1120:142–150.

3

Multiplexed Immunoassays for Protein Profiling in Human Body Fluids

Silke Wittemann, Dominic P. Eisinger, Laurie L. Stephen, and Thomas O. Joos

Summary

This chapter describes the development and use of suspension antibody microarrays for protein profiling in several human body fluids. In suspension microarrays, which allow the simultaneous analysis of a variety of analytes within a single experiment, capture antibodies are coupled onto color-coded microspheres. Applications are described, with emphasis on analyses of proteins present in different types of body fluids, like serum or plasma, tears, cerebrospinal, pleural, and synovial fluids, and cell culture supernatants. The chapter is divided into the generation of suspension microarrays, sample preparation, processing, and validation of analytical performance.

Key Words: Suspension microarray; microspheres; immunoassay; protein profiling; biological fluids; serum; pleura; cell culture supernatants; tears; cerebrospinal fluid; synovial fluid.

1. Introduction

Protein microarray technology allows the simultaneous determination of a large variety of analytes from a minute amount of sample within a single experiment. Assay systems based on this technology are currently applied for the identification and quantitation of proteins. Protein microarray technology is of major interest for proteomic research in basic and applied biology as well as for diagnostic applications. Miniaturized and parallelized assay systems have reached adequate sensitivity and, hence, have the potential to replace single-plex analysis systems.

The well-known planar microarray-based systems are perfectly suited for screening a large number of target proteins; suspension assays are an interesting alternative, especially when the number of parameters of interest is comparably low.

From: *Proteomics of Human Body Fluids: Principles, Methods, and Applications*
Edited by: V. Thongboonkerd © Humana Press Inc., Totowa, NJ

Suspension assay systems employ different color-coded or size-coded microspheres as the solid support for the capture molecules. A flow cytometer, which is able to identify each individual type of bead and quantify the amount of captured targets on each individual bead, is used as a readout system. In a first step, antigen-specific capture antibodies are immobilized on the individual bead types. Different bead types are combined and incubated with the sample of interest. A labeled secondary antibody detects captured analytes and is visualized with a fluorescent reporter system. Sensitivity, reliability, and accuracy are similar to those observed with standard microtiter ELISA procedures ***(1)***. Color-coded microspheres can be used to perform up to 100 different assay types simultaneously. The flow cytometer identifies several thousand microspheres in a second and simultaneously quantitates the amount of captured analytes ***(2–6)***. Today, suspension microarrays are currently advanced within the field of miniaturized, multiplexed ligand-binding assays with respect to automation and throughput. Appropriate sensitivity, precision, and reliability must be demonstrated for the miniaturized, parallelized assay systems, before they can be applied for screening or diagnostic purposes.

This chapter describes the development and use of suspension antibody microarrays for protein profiling in several human body fluids. Standard methodology guidance is described to validate immunoassays *(7)* and to determine the sensitivity, precision, and accuracy of the multiplexed analysis.

2. Materials

2.1. Equipment and Materials

1. Centrifuge: 5415D (Eppendorf).
2. Vortex mixer (Neolab).
3. Ultrasonic bath.
4. Thermomixer (Eppendorf).
5. Luminex100 instrument (Luminex).
6. Vacuum manifold (Millipore).
7. Filterplates (Millipore 96-well plate, cat. no. MAB1250).
8. Microcentrifuge tubes (Starlab 1.5 mL, cat. no. I1415-2500).
9. Carboxylated beads (Qiagen, cat. no. 922400 or Luminex).

2.2. Common Reagents

1. Bovine serum albumin (BSA; Roth, cat. no. T844.2).
2. Phosphate-buffered saline (PBS; Fischer Scientific, cat. no. 9472615).
3. 1-Ethyl-3-(3-dimethylaminopropyl)-carbodiimide HCl (EDC; Pierce).
4. Sulfo-NHS (*N*-hydroxysulfosuccinimide; Pierce).
5. Detection reagent: streptavidin-phycoerythrin (streptavidin-PE) stock solution (1 mg/mL) in 100 m*M* NaCl, 100 m*M* sodium phosphate, pH 7.5, containing 2 m*M* sodium azide (Molecular Probes, cat. no. S21388).

2.3. Buffers

1. Activation buffer: 100 m*M* sodium phosphate (Na_2HPO_4), pH 6.2.
2. Coupling buffer: 50 m*M* MES, pH 5.0.
3. Washing buffer: 0.05% Tween-20 in PBS, pH 7.4.
4. Blocking/storage (B/S) buffer: 1% BSA fraction IV (Roth, cat. no. T844.2) in 1X PBS.
5. Assay buffer: formulation—1% BSA fraction IV in 1X PBS.
6. Wash buffer: 1X PBS.

3. Methods

3.1. Principle

The principle of suspension antibody microarrays is based on sandwich immunoassays, as represented in **Fig. 1**. First, capture antibodies are coupled to carboxylated microspheres and then samples are incubated with coupled microspheres. Bound analytes are detected with biotinylated antibodies. PE-labeled streptavidin is used for signal detection. Finally, microspheres are identified by a flow cytometer, allowing quantitation of the amount of captured analytes.

3.2. Production of Suspension Microarrays: Antibody Coupling to Carboxylated Microspheres (see Note 1)

Using proven carbodiimide coupling chemistry, antibodies are covalently immobilized on carboxylated beads via the amine groups in lysine side chains. Before the coupling procedure, the beads are first activated using EDC/Sulfo-NHS. Antibodies should not contain foreign protein, azide, glycine, Tris, or any other reagent containing primary amine groups. Otherwise, antibodies must be purified by gel-filtration chromatography or dialysis before use.

3.2.1. Bead Activation

1. Sonicate the carboxylated bead stock suspension for 15 to 20 s to yield a homogeneous bead suspension. Vortex bead stock suspension thoroughly for at least 10 s. Take 2.5×10^6 beads per coupling reaction.
2. Transfer the bead stock suspension to a Starlab microcentrifuge tube.
3. Briefly centrifuge bead suspension (a quick spin up to 3000*g* is sufficient), and discard the supernatant.
4. Wash beads with 80 µL activation buffer. Briefly vortex and centrifuge at 10,000*g* for 2 min. Discard the supernatant and repeat the washing step.
5. Resuspend beads in 80 µL of activation buffer. Sonicate for 15 to 20 s to yield a homogeneous bead suspension.
6. Freshly prepare EDC solution (50 mg/mL) and Sulfo-NHS solution (50 mg/mL) (*see* **Notes 2** and **3**).

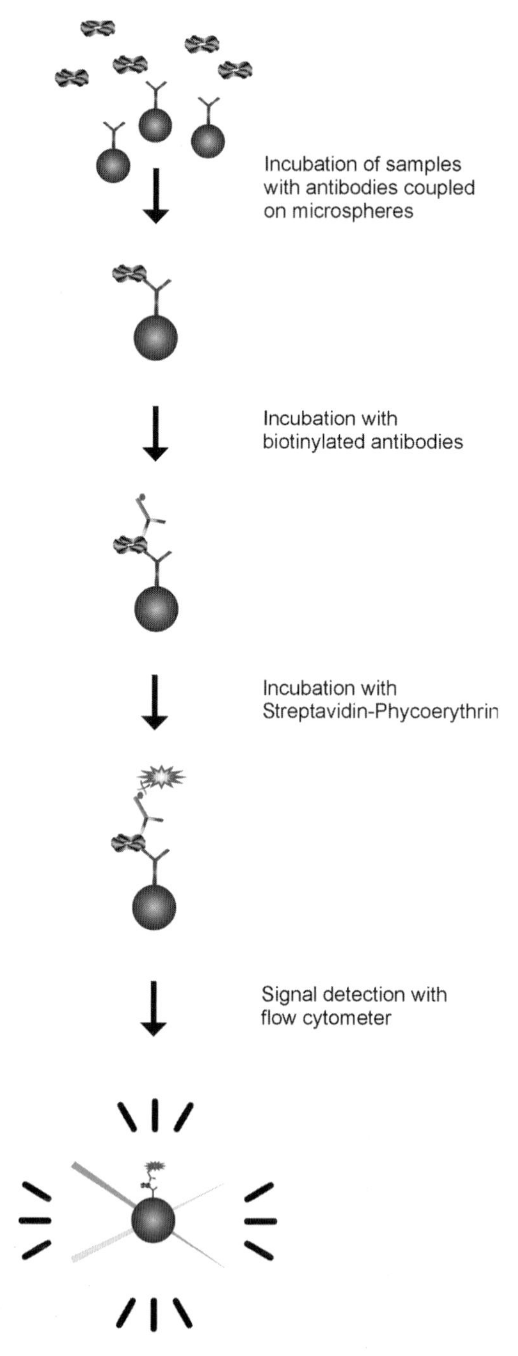

Fig. 1. Processing of suspension microarrays. Schematic representation of the steps required for performing a suspension microarray immunoassay.

7. Add 10 μL EDC solution and 10 μL Sulfo-NHS solution to the bead suspension. Incubate for 20 min at room temperature (15–25°C) in the dark.

3.2.2. Coupling of Antibodies to Activated, Carboxylated Beads

1. Dilute your protein stock solution with coupling buffer to a concentration of 100 μg/mL in a volume of 500 μL.
2. Centrifuge beads at 10,000*g* for 2 min and discard the supernatant.
3. Wash beads with 500 μL of coupling buffer. Briefly vortex and centrifuge at 10,000*g* for 2 min. Discard the supernatant and repeat the washing step.
4. Add the diluted antibody solution (500 μL) from **step 1**.
5. Wrap tube in aluminum foil to avoid light exposure. Agitate the tube with activated beads and antibody solution on a plate shaker at 900 rpm at room temperature (15–25°C) for 2 h.

3.2.3. Washing and Storage of Coupled, Carboxylated Beads

1. Centrifuge beads at 10,000*g* for 2 min and carefully remove and discard the supernatant.
2. Wash beads with 500 μL of washing buffer. Briefly vortex and centrifuge at 10,000*g* for 2 min. Discard the supernatant and repeat the washing step.
3. Resuspend the bead pellet in 1 mL Blocking/Storage (B/S) buffer including 0.05% azide.
4. Determine the bead concentration of the suspension using a cell-counting chamber.

3.2.4. Counting Beads Using a Cell-Counting Chamber

1. Add 5 μL of beads to 45 μL of PBS and mix.
2. Fill the hemacytometer with 10 μL bead suspension by placing the pipet tip against the loading V of the hemacytometer at a 45° angle. Slowly release the sample between the slide and the cover slip until the counting chamber is loaded. It is important to fill both sides of the chamber and wait 2 to 3 min to allow the beads to settle.
3. Count the cells in two opposite corners of the scored chamber and take an average. Each of the nine squares on the grid has an area of 1 square mm, and the coverglass rests 0.1 mm above the floor of the chamber. Thus, the volume over the central counting area is 0.1 mm^3 or 0.1 mL. Multiply the average number of beads each central counting area by 10,000 to obtain the number of beads per mL *of diluted sample*. Multiply by the dilution factor of 10 to get beads/mL.
4. Store beads at 25X, typically 5×10^6 beads/mL.

3.3. Processing of Bead-Based Multiplex Assays

3.3.1. Sample Preparation

Here the preparation of proteins from either clinical specimens or cell culture for use in the multiplexed assay is described.

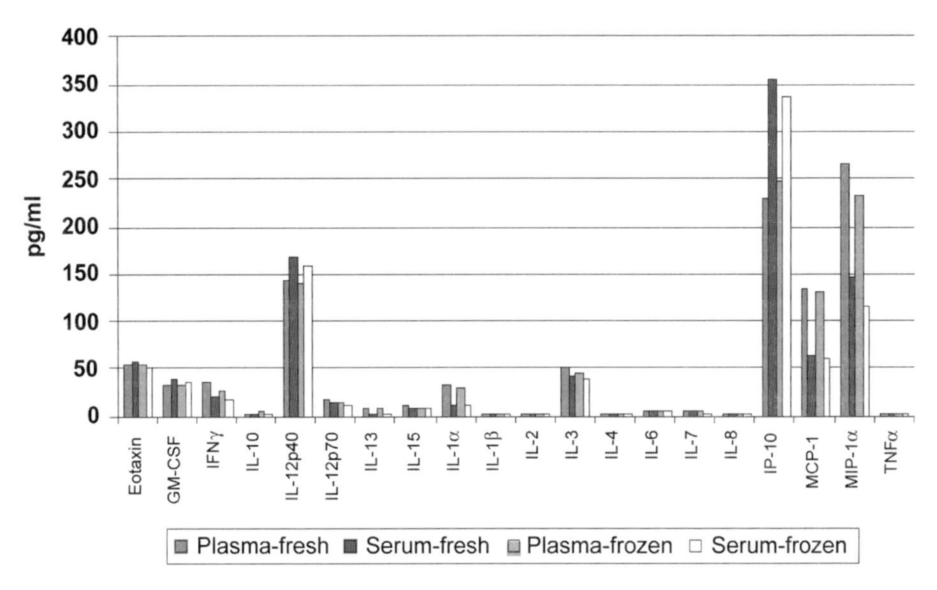

Fig. 2. Serum and plasma were drawn from eight individuals. Half of the sample was subjected to a freeze-thaw cycle. Cytokines were analyzed using cytokine suspension arrays. There were no differences in cytokine levels attributed to the freeze-thaw cycle. Differences between plasma and serum occurred for interleukin-1α (IL-1α), interferon-γ-inducible protein-10 (IP-10), and monocyte chemotactic protein-1 (MCP-1). GM-CSF, granulocyte/macrophage colony-stimulating factor; IFN-γ, interferon-γ; MIP-1α, macrophage inflammatory protein-1α; TNF-α, tumor necrosis factor-α.

3.3.1.1. Serum or Plasma Samples

Serum and plasma samples should be spun down (8000*g*) prior to the assay to remove particulate and lipid layers. This will prevent blocking of the wash plate, as well as the sample needle. **Caution:** samples should be handled as biohazards because they might carry infectious agents. As freezing-thawing cycles might result in a measurable breakdown in some proteins (e.g., cytokines) samples should be aliquoted for any experiment. Storage of aliquoted samples at –80°C is recommended. When we analyzed eight matched serum and plasma samples on the Luminex platform, no differences were seen between fresh samples and those that had undergone a freeze-thaw for levels of tumor necrosis factor-α (TNF-α), eotaxin, interleukin-13 (IL-13), monocyte chemotactic protein-1 (MCP-1), interferon-γ (IFN-γ), IL-12p70, macrophage inflammatory protein-1α (MIP-1α), interferon-γ-inducible protein (IP-10), or granulocyte/macrophage colony-stimulating factor (GM-CSF) (**Fig. 2**). There was, however, a significant increase in IL-1α after the freeze-thaw, suggesting that this process may liberate IL-1α from insoluble receptors. IL-1α and MCP-1 levels were significantly higher in plasma compared with the matched serum sample. In contrast, IP-10 was higher in serum (**Fig. 2**). Another important consideration in analyzing serum or plasma samples is the need for an appropriate buffer (described in **Subheading 3.3.2.**).

3.3.1.2. Cell Culture Samples

Before use, cell culture supernatants should be centrifuged at 14,000*g* to remove any particulates. Cell culture supernatants can be diluted in their corresponding culture medium. As for serum samples, cell culture supernatants should be aliquoted and frozen at –80°C for any experiment.

3.3.1.3. Tears

Tears can be collected and stored for future analysis. In many disease states, levels of IgA increase and may require blocking with animal serum or removal by filtration. As the volume of tear sample is minimal, modifications can be made to the assay to allow for lower volume incubations, without loss of sensitivity. The primary incubation of 5 μL of either tears or standard can be performed in microcentrifuge tubes with beads that have been concentrated fivefold. This will act to increase the effective concentration of the analytes. After the primary incubation, buffer can be added and the mixture transferred to a 96-well wash plate to continue the assay according to the protocol. Experiments performed with 10-plex cytokine assays on the Luminex platform have demonstrated that assay sensitivity was maintained using the low-volume protocol.

3.3.1.4. Cerebrospinal, Synovial, and Pleural Fluids

Precious samples with limited volume such as cerebrospinal fluid (CSF) and synovial fluid are ideal candidates for multiplex analysis. For synovial fluid, animal serum should be added to prevent the binding of heterophilic antibodies and rheumatoid factor (RF) binding, which can cause false positives. For cytokine assays, samples can be filtered with a 50-kDa filter to remove interfering antibodies. Another recently described method to remove RF from serum uses protein L *(8)*. The low-volume protocol described in **Subheading 3.3.1.3.** may be suitable if sample volume is below 50 μL. When CSF samples were analyzed for 22 cytokines using the Luminex platform *(9)*, 11 cytokines were detected. The authors performed spike recovery experiments and describe recoveries as good.

3.3.2. Diluent

It is important that the diluent selected for reconstitution and for dilution of the standards reflect the environment of the samples being measured. Diluents for specific sample types have to be validated prior to use. For analyzing cell culture samples, standards and samples are diluted in the respective cell culture medium. It is important to use the same lot of fetal bovine serum as there are significant differences between lots in compounds that may interfere with the assay. Another factor to ensure is sample pH, which will affect antibody binding. For assaying serum samples, each laboratory should develop and validate an appropriate diluent. We suggest starting with PBS supplemented with 10 to 50% animal serum (e.g., fetal calf serum, horse serum, goat serum, or depleted human serum). The goal is to mimic the serum matrix to ensure similar binding kinetics in both the serum and standard samples. The serum samples may also require dilution with a small amount of serum to prevent false positives, as some human antibodies may show reactivity toward the mouse captures. Generally, 1 to 2% of each species of antibodies is sufficient. The serum diluent must not be used to dilute the detection antibody or the streptavidin-PE.

3.3.3. Detection Antibody

The concentration of detection antibody used can be varied to create an immunoassay with a different sensitivity and dynamic range. The authors typically use the detection antibody at a concentration between 0.5 μg/mL and 1.0 μg/mL. The quantitative range of the assay can be shifted by changing the antibody concentration. A dilution of the detection antibody shifts the standard curve to the lower concentration range, whereas an increased concentration shifts the curve to the higher concentration range.

3.3.4. General Protocol for Processing Bead-Based Multiplex Assays for the Determination of Proteins in Human Body Fluids

1. Centrifuge the sample at 14,000*g* to precipitate any particulates before diluting into appropriate diluent. Dilution factors will vary depending on sample type and concentration of analyte.
2. Resuspend the standard into the appropriate diluent and prepare an eight-point standard curve using twofold serial dilutions.
3. Wet the filter plate with 100 μL Assay buffer.
4. Plate fitting: add 50 μL standard or sample to each well.
5. Sonicate coupled beads for 15 to 20 s to yield a homogeneous suspension. Vortex beads thoroughly for at least 10 s.
6. Dilute beads to 1500 beads per well and add 25 μL diluted bead suspension to each well.
7. Incubate in the dark at room temperature for 2 h (*see* **Note 4**).
8. Washing step: apply vacuum manifold to the bottom of the filter plate to remove liquid. Wash by adding 100 μL assay buffer. Repeat washing step twice. Resuspend beads in 75 μL assay buffer.
9. Add 25 μL detection antibody solution to each well.
10. Incubate in the dark at room temperature for 1.5 h.

11. Washing step: apply vacuum manifold to the bottom of the filter plate to remove liquid. Wash by adding 100 µL assay buffer. Repeat washing step twice. Resuspend beads in 75 µL assay buffer.
12. Add 25 µL streptavidin-PE solution to each well.
13. Incubate in the dark at room temperature for 0.5 h.
14. Washing step: apply vacuum manifold to the bottom of the filter plate to remove liquid. Wash by adding 100 µL assay buffer. Repeat washing step twice. Resuspend beads in 125 µL assay buffer.
15. Incubate on a plate shaker for 1 min.
16. Read results on Luminex 100 instrument.
17. Data evaluation: we recommend extrapolating sample concentrations from a 4-PL or 5-PL curve.

3.4. Validation of Analytical Performance of Miniaturized Multiplexed Protein Assays

3.4.1. Accuracy

Accuracy is expressed by the closeness of the measured value to the true value. Accuracy should be assessed using a minimum of five determinations over a minimum of three concentrations across the expected range of the assay. A deviation of 15% of the measured value from the true value is acceptable. Several methods for estimating accuracy are possible:

1. Accuracy may be determined by comparison of the measured analyte values with those of reference data.
2. Accuracy may be determined by adding known quantities of the analyte into an appropriate test matrix (e.g., serum, plasma). Then the recovery is expressed as the measured analyte concentration relative to added analyte concentration. The Recovery (%) is calculated as follows:

$$\text{Recovery (\%)} = \frac{\text{Measured analyte concentration}}{\text{Background analyte concentration in test matix + added analyte concentration}} \times 100$$

3.4.2. Selectivity

Selectivity can be assessed by performing cross-reactivity experiments, whereby the multiplex assay is performed with each of the standards assayed separately. This will ensure that your capture antibody is selective for its respective analyte only in the assay.

3.4.3. Specificity

Specificity is defined by the ability of an assay to measure unequivocally the amount of an analyte in the presence of interfering substances. Nonspecificity might be derived from cross-reactivity of the antibody used in the assay with other proteins or antibodies present in the sample.

3.4.4. Precision

Precision is expressed by the closeness of agreement between a series of repeated measurements. Precision should be assessed using a minimum of five determinations over a minimum of three concentrations across the expected range of the assay. The mean value should be within 15% of the coefficient of variation (CV).

3.4.4.1. Repeatability

Also termed *intraassay precision*, repeatibility expresses the precision under constant conditions. Measurements are performed within 1 d by the same analyst using identical reagents and the same instrument.

3.4.4.2. Reproducibility

Also known as *interassay precision*, reproducibility expresses the precision by changing measurement conditions, which may involve different analysts, reagents, instruments, and laboratories.

3.4.5. Limits of Detection and Quantitation (see **Note 5**)

3.4.5.1. Detection Limit

The limit of detection (LOD) is the lowest amount of analyte in a sample that can be detected but not quantitated as an exact value. According to the International Union of Pure and Applied Chemistry (IUPAC) definition *(2)* the LOD is estimated as the mean of the zero standard signal plus 3 times the standard deviation (SD) obtained on the zero standard signal:

$$\text{LOD} = \text{Mean}_{\text{zero standard}} + 3 \times \text{SD}_{\text{zero standard}}$$

3.4.5.2. Quantitation Limit

The limit of quantitation (LOQ) is the lowest amount of analyte in a sample that can be quantitated with an acceptable statistical significance. According to the IUPAC definition, the LOQ is estimated as the mean of the zero standard signal plus 10 times the SD obtained on the zero standard signal:

$$\text{LOQ} = \text{Mean}_{\text{zero standard}} + 10 \times \text{SD}_{\text{zero standard}}$$

3.4.6. Linearity

The linearity is defined as the ability of an analytical procedure to produce signals, which are directly proportional to the analyte concentration of the sample.

3.4.7. Range

The range of an analytical procedure is defined by the interval between the upper and lower amount of analyte that can be detected with a suitable level of accuracy, precision, and linearity.

3.4.8. Robustness

Robustness expresses the extent to which measured values remain unaffected by small variations in method parameters (e.g., temperature, reagent concentration) or instrumental parameters. The robustness indicates the reliability of an analytical procedure during normal usage.

4. Notes

1. This method can also be adapted for coupling reactions of antigens, receptors, or other proteins.
2. Minimize the exposure of EDC and Sulfo-NHS to air, and close containers tightly. Use fresh aliquots for each coupling reaction and discard after use.
3. Sulfo-NHS solution (50 mg/mL) can be prepared and stored at –20°C.
4. The incubation time can be varied. The authors typically incubate between 30 min and 2 h. The primary incubation of bead and sample can be performed overnight at 4°C for greater low-end sensitivity.
5. The detection limit depends primarily on the quality of antibodies used. Additionally, the detection limit is influenced by detection conditions (e.g., antibody concentration, incubation time) and the complexity of the multiplex assay and matrix proteins.

References

1. Morgan E, Varro R, Sepulveda H, et al. Cytometric bead array: a multiplexed assay platform with applications in various areas of biology. *Clin Immunol* 2004; 110:252–266.
2. Dasso J, Lee J, Bach H, Mage RG. A comparison of ELISA and flow microsphere-based assays for quantification of immunoglobulins. *J Immunol Methods* 2002; 263:23–33.
3. Carson RT, Vignali DA. Simultaneous quantitation of 15 cytokines using a multiplexed flow cytometric assay. *J Immunol Methods* 1999;227:41–52.
4. Dunbar SA, Vander Zee CA, Oliver KG, Karem KL, Jacobson JW. Quantitative, multiplexed detection of bacterial pathogens: DNA and protein applications of the Luminex LabMAP system. *J Microbiol Methods* 2003;53:245–252.
5. Joos TO, Stoll D, Templin MF. Miniaturised multiplexed immunoassays. *Curr Opin Chem Biol* 2002;6:76–80.
6. Prabhakar U, Eirikis E, Davis HM. Simultaneous quantification of proinflammatory cytokines in human plasma using the LabMAP assay. *J Immunol Methods* 2002;260:207–218.

7. Findlay JW, Smith WC, Lee JW, et al. Validation of immunoassays for bioanalysis: a pharmaceutical industry perspective. *J Pharm Biomed Anal* 2000;21:1249–1273.
8. de Jager W, Prakken BJ, Bijlsma JW, Kuis W, Rijkers GT. Improved multiplex immunoassay performance in human plasma and synovial fluid following removal of interfering heterophilic antibodies. *J Immunol Methods* 2005;300:124–135.
9. Natelson BH, Weaver SA, Tseng CL, Ottenweller JE. Spinal fluid abnormalities in patients with chronic fatigue syndrome. *Clin Diagn Lab Immunol* 2005;12:52–55.

4

Deciphering the Hieroglyphics of Functional Proteomics Using Small Molecule Probes

Wayne F. Patton

Summary

Fluorescence-based total protein detection is generally acknowledged to provide superior capabilities relative to the classical staining methods. Simple measurement of differences in protein expression, however, does not directly measure changes in protein activity and often fails to detect key posttranslational modifications. A number of fluorescence-based strategies tailored to the analysis of protein posttranslational modifications, functional domains, and enzymatic activities have been introduced recently. The new fluorescent staining methods have spurred the rapid development of sophisticated imaging devices that provide the highest possible detection capabilities. A limitation of the newly devised functional proteomics stains is that the abundances of the target proteins are often quite low, and as a consequence, protein prefractionation has increasingly become an integral part of proteomics analysis. It is expected that fluorescence-based approaches, originally devised for gels, will ultimately be adapted to liquid chromatography-mass spectrometry, protein microarrays, and cell-based assays. Additionally, the powerful capabilities of differential protein expression profiling will be engineered into future probes, allowing for multiplexed analysis of posttranslational modifications and enzyme activities. Integrating protein prefractionation and small molecule probe-based detection with multiwavelength imaging significantly expands the power of gel electrophoresis for the large-scale elucidation of the functional properties of proteins.

Key Words: Review; proteomics; posttranslational modifications; protein prefractionation; fusion tags; activity-based probes.

1. Introduction

Although the characterization of genes is undeniably central to the discipline of modern biology, to unravel fully the complex molecular basis of life, a comprehensive understanding of the functional aspects of proteins themselves must be obtained as well. Proteins are the primary mediators of almost all physiological

From: *Proteomics of Human Body Fluids: Principles, Methods, and Applications*
Edited by: V. Thongboonkerd © Humana Press Inc., Totowa, NJ

and pathophysiological processes, and most drugs developed by the pharmaceutical industry specifically target proteins. In contrast to genomics, the essential feature of proteomics, large-scale and all-inclusive analysis of proteins, is relatively difficult to implement fully because of the heterogeneous physicochemical properties displayed by proteins, an inability to amplify proteins by any polymerase chain reaction (PCR)-like process, and the lack of predefined complementary binding partners for specific proteins. Nonetheless, global strategies for the comprehensive analysis of protein posttranslational modifications, functional domains, and enzymatic activities are being devised and increasingly implemented in the field of proteomics. The objective of this review article is to communicate the unique capabilities of gel-based proteomics with respect to the analysis of functional properties of proteins.

2. Detecting Total Protein Profiles

Global quantification of protein expression levels has served as a fundamental component of most proteomics investigations to date. However, the simple measurement of differences in protein expression levels alone does not provide a direct determination of changes in protein activity and often fails to detect key posttranslational modifications that regulate protein activity. The accurate determination of total protein levels does play a fundamental role in evaluating functional alterations in proteins, since such measurements provide a normalizing factor for verifying that different samples are present in equivalent amounts on the gel and assist in distinguishing between changes in the activity levels or degree of posttranslational modification of a protein compared with simple changes in protein expression levels without concomitant changes in the assayed functional attribute.

A variety of options are available for detecting and quantifying proteins in polyacrylamide gels *(1)*. Some representative fluorescence-based molecular probes for the detection of protein expression levels in polyacrylamide gels are summarized in **Fig. 1**. Fluorescence-based total protein detection strategies have gained prominence in recent years and are generally acknowledged to provide superior capabilities relative to the classic staining technologies based on Coomassie Blue and silver stains. The detection capability of a representative fluorescence-based stain, SYPRO® Ruby protein gel stain, is compared with colloidal Coomassie Blue stain and silver stain in **Fig. 2**. Fluorescence-based stains often provide better capabilities than their colorimetric counterparts including superior linear dynamic range of quantification, excellent limits of detection, and outstanding compatibility with mass spectrometry-based protein characterization and identification techniques *(2–4)*.

An increasingly wide range of molecular probes are currently available for the fluorescence-based detection of protein expression levels in sodium dodecyl sulfate (SDS)-polyacrylamide gels, as summarized in **Fig. 1**. The fluorophore

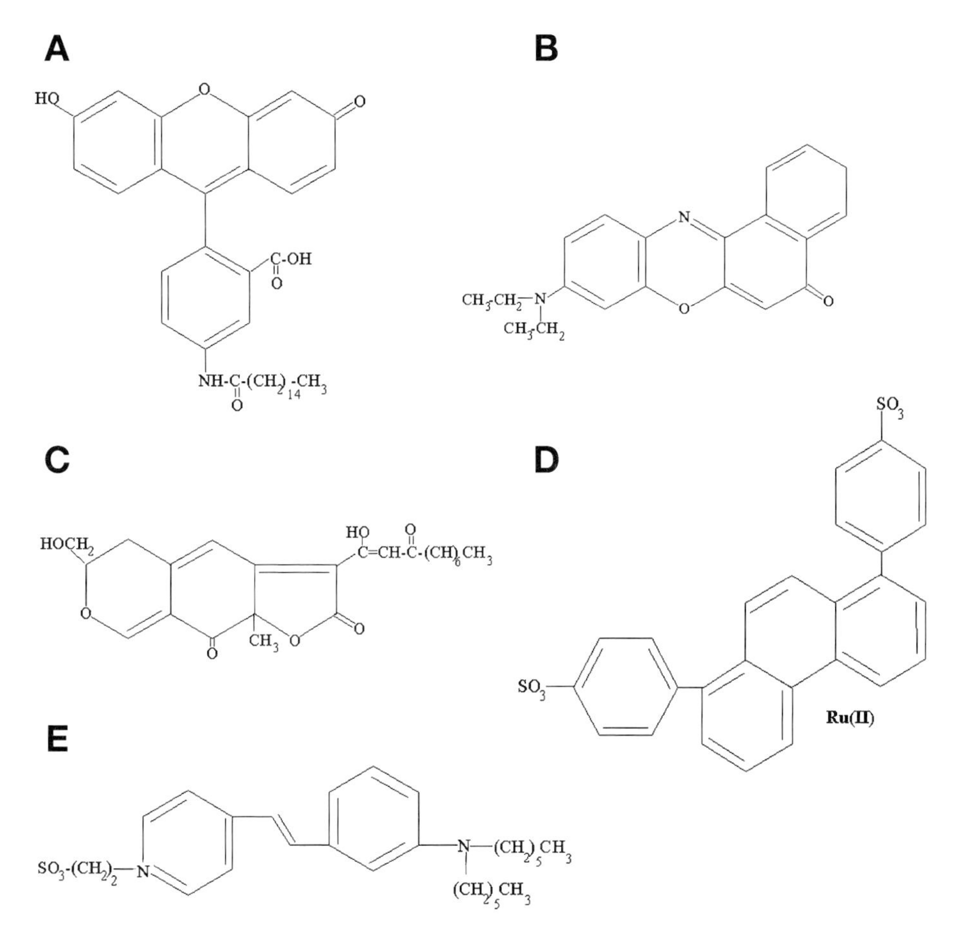

Fig. 1. Structures of some fluorophores commonly utilized for detecting and quantifying proteins in polyacrylamide gels. **(A)** 5-Hexadecanoylaminofluorescein *(16)*. **(B)** Nile Red dye *(10)*. **(C)** Deep Purple dye *(15)*. **(D)** Ruthenium Tris (bathophenanthroline disulfonate) (RuBPSA) *(7)*. Only one of the three bathophenanthroline disulfonate ligands surrounding the ruthenium ion is depicted in the illustration. This fluorophore is structurally similar to SYPRO Ruby protein gel stain. **(E)** Structural analog of OGT MP17 *(17)*. The specific structure of the dye was not revealed, but a class of dyes was referenced. SYPRO Orange and Red dyes also belong to this general class of organic fluorophores.

represents but one ingredient necessary for a successful detection method, and its actual formulation into a stain ultimately determines performance characteristics. For example, ruthenium tris (bathophenanthroline disulfonate) (RuBPSA) **(Fig. 1D)** is widely regarded to be structurally quite similar to SYPRO Ruby protein gel stain (Molecular Probes, Eugene, OR), although simple mass measurement of the two dyes on a matrix-assisted laser desorption ionization-time of

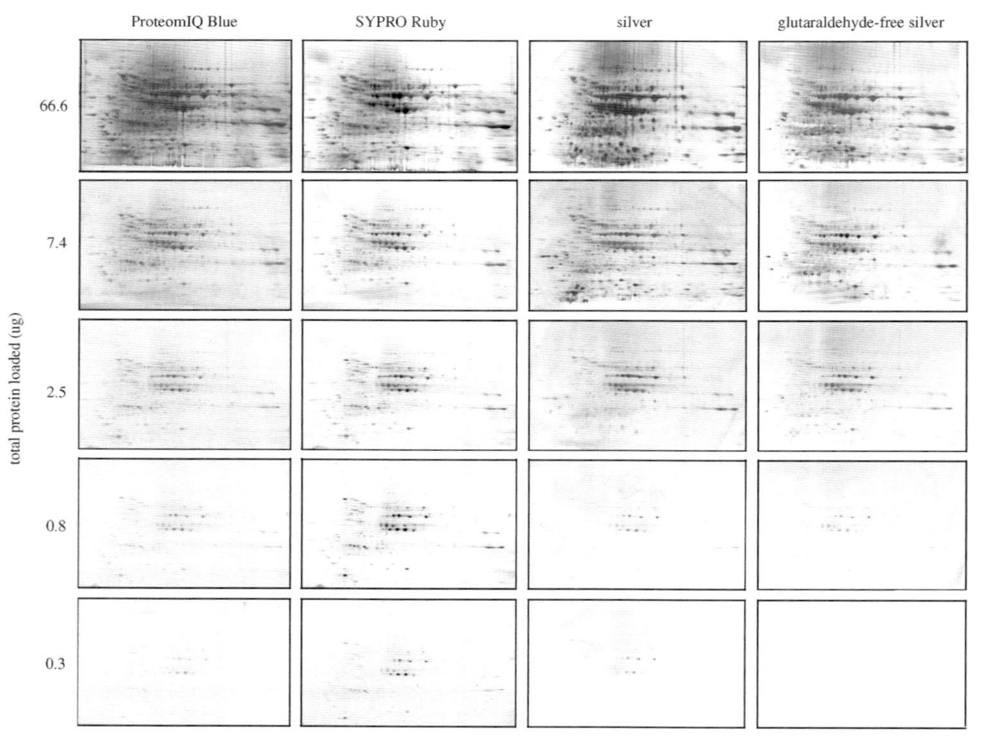

Fig. 2. Comparison of the sensitivity of several different fluorescent and colored stains commonly used in gel-based proteomics. Yeast cells were lysed by sonication in ProteomeIQ Resuspension reagent C7BzO. Samples were fractionated by 2DGE using pH 3-10 immobilized pH gradient gels and 6 to 15% linear gradient SDS-polyacrylamide gels. Gels were stained using colloidal Coomassie Blue stain, SYPRO Ruby stain, a standard silver stain method, and the same silver stain method with glutaraldehyde omitted to improve downstream compatibility of the stain with subsequent protein identification methods involving MALDI-TOF mass spectrometry. (Courtesy of Alexander Lazarev, Myra Robinson, and Gary B. Smejkal, Proteome Systems, Woburn, MA.)

flight (MALDI-TOF) mass spectrometer demonstrates that they are not identical. Whereas SYPRO Ruby stain was created based on principles gleaned from the colloidal Coomassie Blue stain literature, the RuBPSA stain was created using the principles of silver staining ***(5–9)***. Consequently, whereas SYPRO Ruby stain provides background-free end-point staining using a minimal number of procedural steps in as little as 3 h, the RuBPSA procedure is a relatively complex eight-step procedure requiring harsh fixatives and about 26 h to complete.

The fluorophores shown in **Fig. 1** may be classified into three fundamental categories based on their mechanisms of targeting to the proteins. The categories include dyes that bind indirectly to proteins through detergent intercalation, dyes that bind directly but noncovalently with proteins, and dyes that bind covalently with proteins. Gel staining procedures originally devised for the SDS-intercalating Nile Red dye form the basis of formulations incorporating a range of other fluorophores and offering better staining characteristics than the original dyes, including SYPRO Orange dye, SYPRO Red dye, SYPRO Tangerine dye, OGT MP 17 dye, 5-hexadecanoylamino-fluorescein, and Deep Purple® dye (Epicocconone, Belijan Red) ***(10–18)***. The principle limitation of Nile Red dye has been its very low solubility in aqueous media, resulting in a tendency for precipitation during staining. The improved dyes belonging to the family of SDS-intercalating stains all share common structural features. All contain at least one aliphatic tail that promotes association with the SDS micelle, an aromatic chromophore for signal generation and, importantly, hydrophilic functionalities (sulfonate or hydroxyl residues) to promote aqueous solubility. In addition to the structural features cited, Deep Purple dye also forms a reversible covalent linkage with protein primary amines during staining, resulting in a shift in fluorescence to a longer wavelength.

SYPRO Ruby protein gel stain and the structurally related RuBPSA are prominent members of the family of fluorescent dyes that bind noncovalently with proteins ***(4–9)***. Other members include the less widely used europium tris (bathophenanthroline disulfonate) and SYPRO Rose Plus dyes ***(19–20)***. The transition metal ion in these stains is responsible for their luminescent properties, and, unlike most conventional organic fluorophores, the metal ions exhibit a relatively large Stoke's shift, which facilitates detection using simple optical filtering methods. Ruthenium-based stains have proved to be more suitable than previously devised europium-based stains for gel-based applications because they are excitable at either 300 or 450 nm, allowing ultraviolet transilluminator, visible laser, or xenon-arc lamp illumination. This permits the stains to be visualized on almost any gel imaging device available to the investigator. The stains are thought to bind to proteins primarily through interaction of dye sulfonate groups with basic amino acid residues, although some hydrophobic interaction through the bathophenanthroline group itself is also likely. A recent independent comparison of Deep Purple dye, SYPRO Ruby dye, 5-hexadecanoylamino-fluorescein, silver stain, and Coomassie Blue stain, using Progenesis gel analysis software (Perkin-Elmer), found SYPRO Ruby dye to be superior with respect to the capability of detecting the greatest number of proteins with the highest reproducibility (lowest coefficient of variance) ***(18)***. SYPRO Ruby dye has also been shown to be considerably more photostable than alternative dyes, such as Deep Purple dye ***(21)***. A head-to-head comparison of the performance of SYPRO Ruby and Deep Purple stains is shown in **Fig. 3**. The two-color overlay indicates that each dye favors different subsets of proteins, with certain highly acidic proteins being especially well detected by SYPRO Ruby dye.

With the advent of advanced infrared-based gel imagers, such as the Odyssey® Infrared Imaging System (Li-Cor Biosciences, Lincoln, NE), there have been rumors circulating at various scientific conferences of Coomassie Blue stain once again becoming the preeminent fluorescent stain for gel-based proteomics ***(22)***. Careful analysis of the limits of detection and linear dynamic range associated with this reincarnation of an old staining method will determine whether infrared imaging of Coomassie Blue stain is competitive with the cadre of other validated fluorescent stains already commercially available on the market. Certainly, compatibility with downstream protein identification methods, such as mass spectrometry, is well established, but the requirement of a specialized imager may be a drawback to general acceptance of the approach.

Several independent investigations have confirmed the original manufacturer claims concerning the superior reproducibility, equivalent sensitivity, and better compatibility of SYPRO Ruby stain compared with silver stain ***(23,24)***. Unfortunately, SYPRO Ruby dye is very expensive, and the fluorescent stain providing the most cost-effective performance for routine proteomic analysis comprising high and moderate abundance proteins was determined to be 5-hexadecanoylamino-fluorescein ***(18)***. Popular approaches for making SYPRO Ruby dye itself more cost effective include diluting the stain 50:50 with deionized water and reusing the stain ***(25,26)***.

A number of approaches for the irreversible covalent labeling of proteins with fluorophores have been devised in the past, with the sulfhydryl-reactive monobromobimane and the amine-reactive or sulfhydryl-reactive cyanine dyes being the most commonly employed fluorophores for this purpose ***(27–30)***. Fluorescence-based 2D difference gel electrophoresis (DIGE), employing reactive cyanine dyes, has had a tremendous impact in gel-based proteomics, allowing two to three samples to be run on a single 2D gel ***(31)***. This multiplexing strategy significantly improves the quantitative accuracy and statistical confidence of the 2D gel electrophoresis (2-DGE) method, by minimizing gel-to-gel variation and allowing inclusion of a pooled internal standard.

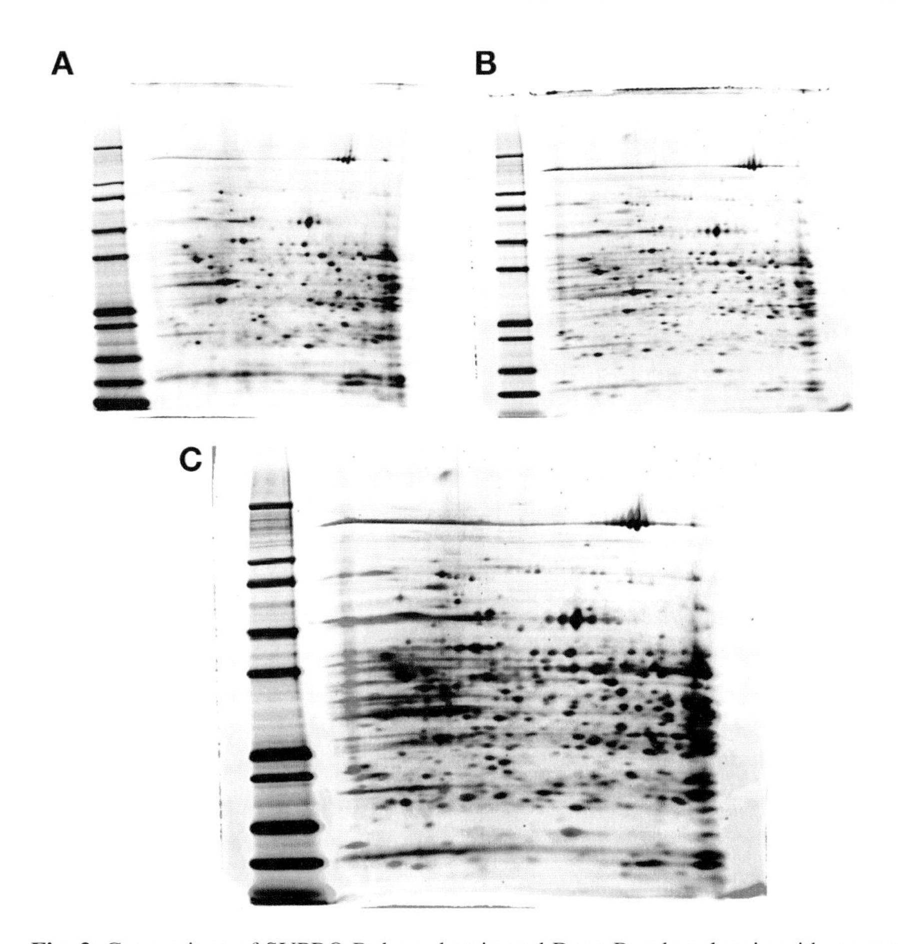

Fig. 3. Comparison of SYPRO Ruby gel stain and Deep Purple gel stain with respect to the different proteins detectable after 2DGE and staining. An extract of rat liver cells (20 μg) was separated by isoelectric point on Zoom pH 4 to 7 gel strips and then by molecular weight on NuPAGE 4 to 12% polyacrylamide, Bis-Tris slab gels (Invitrogen, Carlsbad, CA). Gels were stained in parallel with either SYPRO Ruby gel stain (Invitrogen) or Deep Purple gel stain (GE Healthcare, Chalfont St. Giles, UK), following the most updated staining protocols of each manufacturer. The resultant fluorescent signals from the two stains were visualized using an FX® Pro Plus gel imager (Bio-Rad, Hercules, CA) using the 488-nm excitation line of an argon-ion laser for SYPRO Ruby stain and the 532-nm excitation line of a YAG laser for Deep Purple stain. Fluorescent signals from both stains were detected using a 555-nm long-pass emission filter. **(A)** Yeast proteins visualized with Deep Purple stain. **(B)** Yeast proteins visualized with SYPRO Ruby stain. **(C)** Computer-generated overlay of images from Deep Purple stain (green) and SYPRO Ruby stain (fuchsia), using Z3 software (Compugen, Tel Aviv, Israel). (Courtesy of Courtenay Hart, Molecular Probes/Invitrogen, Eugene.)

3. Detecting Protein Functional Attributes

As discussed, a variety of options for fluorescence-based detection of protein expression levels in polyacrylamide gels are currently available, and it has become somewhat routine work to devise additional ones at this point. The new scientific frontier for biological analysis in gel electrophoresis has certainly become the detection of functional attributes of proteins. Only a relatively few breakthroughs have been reported, and, clearly, there is a room for further advancements in the field, particularly in terms of improving detection sensitivity *(1)*. Fundamentally, the requirements for developing a good gel stain that is selective for a particular functional attribute are much more stringent than that for a nonspecific total protein stain. With total protein stains, nonspecific or secondary binding processes only serve to enhance detection further, whereas with stains targeting particular functional attributes, even specific to nonspecific binding ratios of 100:1 confound interpretation of results, because proteins in the biological sample itself may vary in concentration over a range of four or more orders of magnitude. Consequently, it is more than just a good idea to evaluate any functional proteomics stain in parallel with a total protein stain, so as to differentiate between specific staining of proteins and very weak nonspecific staining of highly abundant components in the sample.

3.1. Phosphorylation

Reversible protein phosphorylation has been intensely studied since the discovery of its central role in regulating enzyme activity in the 1950s. Phosphoproteomics, the systematic parallel analysis of the phosphorylation status of large sets of proteins involved in the regulatory circuitry of cells and tissues, is a relatively new field likely to drive research in the postgenomics era for a number of years to come. Phosphorylation arguably represents the single most critical posttranslational modification in biology, impacting almost every functional aspect of a cell, from mitosis through apoptosis. The modification plays a central role in both the regulation of signal transduction pathways and the modulation of enzyme activity. Phosphoproteins are most often detected by autoradiography after incorporation of ^{32}P or ^{33}P into cultured cells or after incorporation into subcellular fractions using protein kinases. Radiolabeling remains the most sensitive and effective method for detecting protein phosphorylation in gels. However, this approach is somewhat limited with respect to the range of biological materials that can be analyzed, as examination of clinically derived samples requires in vivo labeling, which is not usually feasible. The hazardous nature of radiolabels, as well as the accompanying disposal costs for radioactive waste, also makes the radiolabeling approach somewhat onerous to perform.

Pro-Q® Diamond phosphoprotein stain (Molecular Probes) detects phosphoproteins in SDS-polyacrylamide gels, isoelectric focusing gels, 2D gels, electroblots, and protein microarrays through a mechanism that combines a trivalent transition metal ion and a fluorescent metal ion-indicator dye ***(32–39)***. The stain has also recently been adapted to phosphate quantification of phosphoproteins and phosphopeptides from solution, detection of phosphopeptides by high-performance liquid chromatography, and isolation of phosphopetides using magnetic beads. The staining technique is rapid, simple to perform, readily reversible, and fully compatible with analytical procedures such as MALDI-TOF mass spectrometry. Pro-Q Diamond dye is capable of detecting as little as 1 to 8 ng of a phosphoprotein. The linear range of signal response for the fluorescent dye allows rigorous quantitation of phosphorylation changes over a 500- to 1000-fold concentration range. Using heart tissue, the technology has been employed successfully to identify two novel phosphorylated proteins present in complex I of the mitochondrial respiratory chain ***(32,36)***. A recent study comparing ^{32}P labeling with Pro-Q Diamond dye staining concluded that results from the two approaches do not correlate perfectly ***(39)***. This is understandable, as even in a comparison of 4-h vs 16-h of radiolabeling, only 71% of the spots were found to be correlated with one another ***(39)***. ^{32}P radiolabeling is dependent on the turnover rates of the individual phosphorylation sites, whereas the fluorescent stain measures the actual abundance of the posttranslational modification within the protein.

The only commercialized staining alternative for detecting phosphoproteins in gels requires alkaline hydrolysis of serine or threonine phosphate esters, precipitation of the released inorganic phosphate with calcium, formation of an insoluble phosphomolybdate complex, and then visualization of the complex with methyl green dye, as commercialized in the GelCode phosphoprotein detection kit (Pierce). The staining method is not particularly sensitive, with μg to mg amounts of phosphoprotein required to obtain a discernable signal. In addition, phosphotyrosine residues escape detection with the stain.

3.2. Glycosylation

Protein glycosylation is typically detected after periodate oxidation of glycan moieties, followed by Schiff's base conjugation of a hydrazide or hydrazine tag. A sensitive green-fluorescent glycoprotein-specific stain, Pro-Q Emerald 300 dye (Molecular Probes), detects glycoproteins directly in polyacrylamide gels or on polyvinylidene difluoride membranes ***(40–42)***. As little as 300 pg of α1-acid glycoprotein (40% carbohydrate) may be detected in gels after staining with the dye, although less highly glycosylated proteins are generally detectable at the 1- to 2-ng level. As is typical with fluorescence-based detection approaches, Pro-Q Emerald 300 dye provides a 500- to 1000-fold linear dynamic range of detection. One limitation of the stain is that a UV transilluminator-based imaging system is required for detection. However, using a related stain, Pro-Q Emerald 488 dye, glycoproteins may be detected using visible light excitation sources, albeit with an overall 10-fold poorer detection sensitivity. The GlycoProfile III fluorescent glycoprotein detection kit (Sigma-Aldrich) also detects glycoproteins by a periodate oxidation and Schiff's base conjugation and is suitable for fluorescence-based detection of glycoproteins using UV excitation. The dye is not as sensitive as Pro-Q Emerald 300 dye, detecting roughly 150 ng of a glycoprotein in gels. Finally, it should be noted that the commonly employed acid fuchsin-based colorimetric glycoprotein stain may be used as a fluorescent stain if imaged for fluorescence with a 532-nm excitation source and a 675-nm long-pass filter ***(41)***.

A metabolic approach for the labeling and identification of cell surface glycans has gained prominence in recent years ***(43–47)***. The approach is based on metabolically incorporating an unnatural sugar precursor containing an azide functional group into cell surface glycans using the cell's endogenous enzymes. The azide functional group is then induced to undergo a covalent reaction with an exogenously added tag containing a phosphine residue to form an amide bond, via the Bertozzi-Staudinger ligation reaction. Triarylphosphine can be ester conjugated to a wide range of probes including biotin, various fluorophores, or the FLAG tag in order to facilitate detection of the glycoproteins. Recently, an analogous approach has been devised for the labeling and detection of farnesylated proteins, suggesting that the chemical strategy is broadly applicable to the analysis of a wide range of protein posttranslational modifications ***(48)***.

In another recently described method, covalent modification of *O*-linked β-*N*-acetylglucosamine (*O*-GlcNAc) is achieved using an engineered galactosyltransferase enzyme to label proteins modified with this sugar selectively using a ketone-biotin tag. The biotin tag permits enrichment of low-abundance *O*-GlcNAc-containing proteins from complex cell lysates and subsequent detection of the modification by standard streptavidin-based Western blotting techniques ***(49,50)***.

3.3. Oxidative Posttranslational Modifications

The biological effects of nitric oxide are largely mediated through *S*-nitrosylation of cysteine thiol residues on a variety of proteins, ranging from ion channels to nuclear regulatory proteins. Protein *S*-nitrosylation is becoming recognized as an important mechanism for the reversible posttranslational regulation of protein activity and consequently cellular function ***(51–54)***. Until relatively recently, the detection of protein *S*-nitrosylation in cells and tissues was hampered by the unavailability of suitable techniques to detect *S*-nitrosylated proteins explicitly. The Biotin-Switch method is the first approach to provide detection of protein *S*-nitosylation after SDS-polyacrylamide gel electrophoresis and electroblotting ***(55,56)***. The basic technique, commercialized as the NitroGlo nitrosylation detection kit (Perkin-Elmer), provides an optimized procedure for detecting protein *S*-nitrosylation that has been validated using known *S*-nitrosylated proteins, such as creatine phosphokinase and H-ras. The detection approach involves chemically blocking free cysteine residues, followed by reduction of *S*-nitrosylated cysteine to free thiol residues. The free thiol residues are then reacted with a biotinylating reagent. Consequently, only the cysteine residues, which formerly possessed nitroso modifications, are tagged with a biotin label. Gel electrophoresis is then used to separate the various proteins in the sample directly, or the labeled proteins are selectively enriched using streptavidin affinity chromatography prior to fractionation. Electroblotting, using chemiluminescence reagents, such as the Western Lightning kit (Perkin-Elmer), and subsequent imaging of the blots, reveals the *S*-nitrosylated proteins in the sample.

Another significant posttranslational modification generated by free radicals is the oxidative conversion of a tyrosine residue into a tyrosyl radical, which can subsequently react with another tyrosyl radical to generate a dityrosyl residue or alternatively with nitrogen dioxide to form nitrotyrosine ***(57,58)***. Formation of dityrosine or nitrotyrosine residues in proteins alters them both structurally and functionally. A fluorescent tyramine-fluorescein conjugate, referred to as TyrFluo, allows detection of protein oxidation events in intact cells ***(57–60)***. The probe facilitates detection of oxidatively modified extracellular proteins; a related cell-permeable probe, acetylTyrFluo, permits detection of modified proteins intracellularly. Although not commercially available, TyrFluo is readily synthesized by conjugating the succinimidyl ester of 6-(fluorescein-5-carboamido)hexanoic acid/6-(fluorescein-6-carboamido)hexanoic acid with the amine group of tyramine. Owing to the relatively low levels of tyrosyl radicals encountered in biological specimens, detection of the TyrFluo-conjugated proteins is typically performed after electroblotting using a horseradish peroxidase-conjugated anti-fluorescein antibody and a chemiluminescent substrate, rather than directly in polyacrylamide gels by fluorescence-based measurement.

Carbonylation has been used extensively as an indicator of protein oxidative modification ***(61–67)***. Typically, the labeling schemes widely used for detection of carbonyl groups are analogous to those used for detection of glycoproteins by periodic acid Schiff's base (PAS) formation. Instead of intentionally oxidizing the proteins with periodic acid, however, the native oxidative products present in the sample are assayed directly. Biotin-hydrazide, 2,4-dinitrophenylhydrazine, and fluorescein hydrazine have been employed to label the carbonyl groups of oxidized proteins. Pro-Q Emerald 300 dye can also be employed in order to detect carbonyl groups of oxidatively modified proteins.

3.4. α-Helical Transmembrane Domains

Integral membrane proteins are considered important drug targets by the pharmaceutical industry, as they play key roles in converting extracellular binding events to intracellular signals. These proteins characteristically consist of one or more hydrophobic, transmembrane domains that interact with the hydrophobic component of lipid bilayer membranes. Unfortunately, proteomic profiles derived from 2-DGE lack highly hydrophobic proteins, particularly integral membrane proteins containing more than one α-helical transmembrane domain. This is primarily owing to the very poor resolution of this class of proteins in the isoelectric focusing component of the procedure, arising from their poor solubilization by nonionic detergents, even in the presence of high concentrations of urea. Detection of proteins containing two or more α-helical transmembrane domains was recently accomplished using a combination of SDS-polyacrylamide gel electrophoresis and fluorescence-based staining methods ***(68)***. With the Pro-Q Amber dye formulation (Molecular Probes), staining in gels is achieved through interaction of a hydrophobic fluorophore with the amphiphilic portion of the protein after removal of SDS by fixation and washing. As little as 5 to 10 ng of a seven-membrane spanning protein, such as bacteriorhodopsin, is detectable using the staining method; proteins containing two α-helical transmembrane domains are generally detectable at the 8- to 10-ng level.

3.5. Fusion Tags

The oligohistidine domain is a transition metal-binding peptide sequence consisting of 4 to 10 consecutive histidine residues. Recombinant proteins are commonly fused to oligohistidine affinity tag sequences in order to assist in their purification or retention by immobilized nickel-ion affinity chromatography. The tag domain may be incorporated into the N-terminal, C-terminal, or internal sites of the protein using standard molecular biology techniques. Detection of oligohistidine-tagged fusion proteins after polyacrylamide gel electrophoresis has been accomplished by Western blotting strategies that involve the transfer of proteins from an SDS-polyacrylamide gel to a nitrocellulose or polyvinylidene difluoride membrane, the blocking of unoccupied sites on the membrane with protein or detergent solutions, incubation with an appropriate oligohistidine-binding reagent (primary antibody or biotin-nitrilotriacetic acid), incubation with a secondary detection reagent (antibody-reporter enzyme conjugate, streptavidin-reporter enzyme conjugate), and subsequent incubation with a visualization reagent (colorimetric, fluorogenic, or chemiluminescent compound; reviewed in **ref. *1***).

Two novel fluorophore-nitrilotriacetic acid conjugates, Pro-Q Sapphire 365 and Pro-Q Sapphire 488 oligohistidine gel stains (Molecular Probes), facilitate direct fluorescence-based detection of oligohistidine-tagged fusion proteins in

SDS-polyacrylamide gels, without immunoblotting, reporter enzymes, or secondary detection reagents ***(69)***. Subsequently, other fluorophore-chelate conjugates were commercialized, such as the Pro-Q Sapphire 532 oligohistidine gel stain (Molecular Probes) and the InVision® His-tag in-gel stain (Invitrogen, Carlsbad, CA) ***(41,70)***. The fluorescence-based approaches are capable of detecting 20 to 65 ng of oligohistidine-tagged fusion protein from samples derived from whole cell extracts.

A method for site-specific fluorescent labeling of recombinant proteins has been developed using a biarsenical fluorescein or rhodamine derivative and tetracysteine epitope-tagged proteins ***(71–74)***. The detection reagent has recently been commercialized as the Lumino® system (Invitrogen). The Lumino reagent is virtually nonfluorescent until it binds to a tetracysteine sequence. The reagent may be applied directly to protein samples before electrophoresis, permitting the visualization of fusion proteins during or immediately after gel electrophoresis using a standard UV transilluminator or any of a variety of gel imaging instruments.

Finally, a small C-terminal tetrapeptide motif containing selenocysteine, referred to as a Sel-tag, shows promise as a fusion tag for proteomics applications ***(75)***. The selenocysteine residue of a selenoprotein is analogous to the cysteine residue except that a selenium atom replaces the sulfur atom. Selenocysteine is cotranslationally incorporated into proteins using a recoded UGA stop codon. The selenocysteine residue has unique biochemical attributes, such as low pKa, high electrophilicity, and higher chemical reactivity than a cysteine residue. These properties make selenocysteine amenable for the attachment of various tags. Proteins containing the Sel-tag may be purified using 4-phenylarsine oxide Sepharose column chromatography. The γ-emitting radionucleotide, ^{75}Se, can be metabolically incorporated into the selenoproteins, provided that excess cysteine is added to prevent nonspecific radiolabeling of cysteine or methionine residues. Electrophilic compounds, such as 5-iodoacetamidefluorescein, can be used to label Sel-tagged proteins fluorescently, provided that labeling is conducted at acidic pH value and in the presence of a reducing agent, such as dithiothritol, using a short reaction time. Similarly, positron-emitting ^{11}C-labeled methyl iodide may be used to label the selenoproteins. The Sel-tag technology is currently limited somewhat by the requirement that the tag be introduced at the C-terminus of proteins and the fact that limited bacterial selenocysteine incorporation efficiency leads to substantial amounts of truncated protein. Additionally, under certain circumstances the selenium atom can be lost from the tagged protein.

3.6. Activity-Based Proteomics

Activity-based protein profiling employs chemical approaches for determining the activity of proteins in complex samples ***(76–87)***. Two strategies have been employed with respect to profiling specific classes of enzymes, directed approaches tailored to specific classes of enzymes and nondirected approaches that profile enzymes from several different classes. The strategies differ from conventional zymography, as exemplified by the in-gel fluorescence-based detection of β-glucuronidase activity, in that detection is achieved by labeling of enzymes in the specimen prior to analysis by gel electrophoresis, rather than through renaturing enzymes in gels and subsequently detecting their activity ***(88)***.

The directed profiling approach has been employed to detect serine hydrolases, cysteine proteases, protein tyrosine phosphatases, glycosidase, and metalloproteases ***(78,83,85)***. Typically, a reactive small molecule inhibitor known to target particular enzymes is coupled to a linker group and a detection and/or isolation tag to create a directed activity-based probe. For example, serine hydrolases, such as thrombin, trypsin, urokinase plasminogen activator, phospholipase A_2, acetylcholinesterase, and fatty acid amide hydrolase, may all be detected directly in gels using a probe consisting of a fluorophosphonate reactive group, an alkyl or polyethylene glycol linker, and a rhodamine or fluorescein reporter group ***(76,87)***. The enzymes are then detected by direct in-gel fluorescence imaging.

For enzyme classes in which covalent active site-directed inhibitors have not yet been identified, the nondirected strategy is most appropriate. With nondirected probes, a combinatorial strategy is employed in which libraries of candidate probes are synthesized and screened against complex samples for activity-dependent protein reactivity. For example, a library of sulfonate ester reactive groups has been coupled to an aliphatic linker and a rhodamine signaling molecule or biotin affinity purification tag for broad-based analysis of enzyme activities, including glutathione *S*-transferase, tissue transglutaminase, aldehyde dehydrogenase, enoyl-CoA hydratase, epoxide hydrolase, thiolase, and platelet-type phosphofructokinase ***(77,80)***.

The directed and nondirected activity-based strategies for proteomics have produced a considerable list of molecular probes that may be employed separately or in combination to characterize enzyme activities on a proteome-wide scale. A recent development in this field is the creation of "tag-free" probes, which are cell-permeable and thus allow in vivo labeling of enzyme activities ***(82)***. In this scheme, the enzymes are labeled with a reactive functionality that also contains a biologically inert coupling agent, and then the reporter group (biotin or fluorophore), equipped with a complementary coupling agent, is subsequently attached. Comprehensive reviews concerning activity-based proteomics have previously been published and should be consulted by those interested in more information on this exciting technology ***(83,84,86,87)***.

4. Advanced Imaging Instrumentation

The recent development of the new fluorescent staining methods cited has spurred the rapid evolution of sophisticated imaging devices capable of providing the highest possible detection capabilities for them, including CCD camera-based multiwavelength imaging platforms and gel scanners with multiple laser excitation sources (reviewed in **refs.** *2,3,* and *89*). The DIGE dyes, SYPRO Ruby dye, and Pro-Q Diamond dye have served as the most popular stains for benchmarking these instruments, and many of the newer machines are equipped with specialized filters and even application software pull-down menus specifically tailored to these popular dyes.

Cooled CCD camera-based imaging systems offer a great deal of flexibility in terms of stains and labels that may be analyzed *(2,3)*. Instruments based on CCD camera technology are suitable for the quantification of absorbance, fluorescence, chemiluminescence, and even radioactivity (employing storage phosphor plates or autoradiography film). CCD cameras capture images by integrating signals over a user-defined period, improving the detection of even weakly fluorescent signals. This important feature often makes protein bands detectable that cannot be observed by eye alone. Gel imaging systems typically use 14- or 16-bit cooled CCD cameras, providing a quantitative capability that extends over as much as 3 to 4 orders of magnitude in intensity. However, the image resolution of most fixed CCD cameras is often relatively poor, providing only 200 μm of resolution across the entire field, when, for example, larger format 2D gels are imaged. In addition, most CCD camera-based imaging systems employ UV illumination to excite fluorescent dyes, which is suboptimal for the multitude of fluorescent dyes and tags that absorb maximally in the visible region of the spectrum.

Alternatively, a number of gel imaging instruments that employ photomultiplier tube (PMT) detectors combined with laser light sources, such as the FLA-5000® fluorescent imager (Fuji, Tokyo, Japan), are commercially available for the analysis of proteins separated by polyacrylamide gel electrophoresis. These instruments serially pass an illumination beam over each point of the sample in a 2D raster pattern format, by optical scanning, mechanical scanning, or a combination of the two approaches *(2,3)*. Modern laser-based gel scanners usually employ multiple PMTs and lasers to allow simultaneous detection of multiple fluorophores in a gel. Laser-based gel scanners routinely achieve 25 to 50 μm spatial resolution but lack the depth of focus to allow convenient analysis of nonstandard objects, such as multiwell plates. Another disadvantage of laser scanners is that they are limited to imaging fluorophores that spectrally match the output of the limited number of laser sources they are equipped with. Thus, certain commonly employed dyes, such as Pro-Q Emerald 300 glycoprotein stain, cannot be imaged on these devices.

A third category of gel imager, the ProXPRESS® 2D Proteomic Imaging System (Perkin-Elmer), is a newly introduced instrument that enables the use of a wide range of fluorescent and colored dyes owing to its multiwavelength emission and excitation capabilities. This highly sensitive multiwavelength imaging platform is equipped with a 16-bit slow-scan CCD camera cooled to –35°C, a UV transilluminator, a Xenon arc light source, and ProSCAN® 4.0 image acquisition software. Spatial resolution is a particularly important parameter for 2-DGE, as it is critical to capture a sufficient number of data points per spot in order to perform Gaussian smoothing and spot segmentation. As alluded to earlier, the spatial resolution provided by conventional fixed CCD camera imaging devices is usually inferior to that of laser-based gel scanners and photographic film. However, the ProXPRESS device mechanically scans the CCD camera over the gel or blot, acquiring multiple images that are subsequently automatically stitched together to form a complete image. As a consequence, the system readily delivers the same 30- to 50-μm spatial resolution obtained with high-end laser scanners. With respect to multiplexing applications, the device is capable of acquiring images using different filter combinations, with the result that as many as four different fluorescent labels are detectable from a single gel.

Figure 4 demonstrates the quantitative capability of this xenon-arc-based imaging system using the fluorescent phosphoprotein-selective stain, Pro-Q Diamond dye, and two model phosphoproteins, ovalbumin and β-casein. Signal-to-noise ratios for the 0.4-ng phosphoprotein bands on the gel were all greater than 3.0, demonstrating that this quantity of phosphoprotein is readily detectable by the ProXPRESS gel imager. This sensitivity of detection compares favorably with imaging of the dye using laser-based gel scanners, which are reported to provide detection limits of 1 to 8 ng for the same model phosphoproteins *(33)*.

5. Protein Fractionation and Detection

Technological innovations and improvements related to fluorescence-based functional proteomics have progressed considerably over the past 6 yr. One limitation of many of the newly devised tools and technologies relates to the abundances of the target proteins possessing the assayable attribute relative to the abundances of the bulk housekeeping proteins in tissues and cells. Enzymes, signal transduction proteins, and plasma membrane proteins often serve catalytic roles in biological systems and are consequently present at levels that may be 1 or more orders of magnitude lower than structural components, such as cytoskeletal or extracellular matrix proteins. Compounding this problem is the fact that many of the fluorescence-based approaches for the detection of protein functional activities are currently about an order of magnitude less sensitive than approaches employed to detect the total protein profile. This is entirely understandable, as the number of functional groups that serve as anchors for dye binding is reduced in the case of the specialized dyes. For example, whereas an

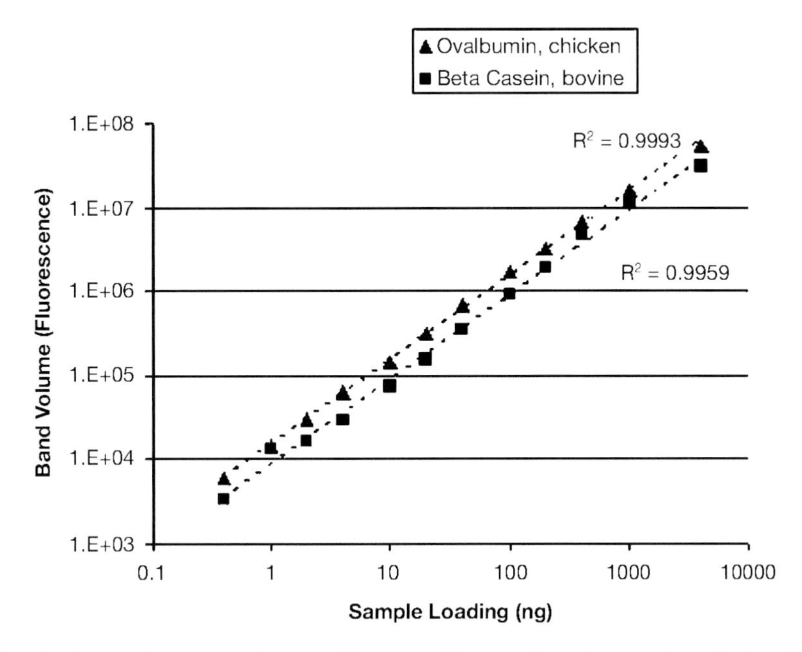

Fig. 4. Dynamic range of quantification of two phosphoproteins obtained from SDS-polyacrylamide gels stained with Pro-Q Diamond dye and imaged using the ProXPRESS 2D proteomic imager. Band volumes were plotted against sample loading for the two phosphoproteins (ovalbumin and β-casein). The proteins were loaded in such a manner as to produce a four order of magnitude dilution series. A double logarithmic scale was employed to facilitate evaluation of instrument performance at both low and high fluorescence signal levels. (Courtesy of Elaine Scrivener and Peter Jackson, PerkinElmer Life and Analytical Sciences, Seer Green, UK.)

average phosphoprotein may possess a dozen or so lysine and arginine residues that can potentially interact with sulfonate residues from a fluorescent total protein stain, it may only contain one or two phosphorylated residues that a phosphate-selective stain can bind to. The vast majority of fluorescent probes used to assay protein functional activity in gels only display micromolar or weaker binding affinities for their targets. Although incremental improvements in binding affinities are likely in the future, only modest improvements in overall detection sensitively are anticipated. Consequently, protein prefractionation has increasingly become an inescapable reality in proteomic analysis.

5.1. Prefractionation Strategies

Many of the detection strategies cited in this review may be used effectively with prefractionated samples that have been separated by SDS-polyacrylamide or 2-DGE, to provide a more comprehensive proteome-wide analysis of protein characteristics. A number of techniques for the fractionation of complex protein samples have been optimized for proteomic applications, including sucrose-gradient fractionation, microscale solution-phase isoelectric focusing, ultrafiltration, and ion-exchange membrane chromatography-based fractionation ***(90–94)***. Of these approaches, chromatographic prefractionation offers certain performance advantages ***(94)***. The chromatographic step is a straightforward approach to applying a third dimension of separation to a sample that is largely independent of the isoelectric point and molecular weight fractionations produced in 2-DGE. Even ion exchange chromatography only bears a modest resemblance to isoelectric focusing in most circumstances, as surface-charge features rather than overall protein isoelectric point are involved in binding to the column packings. Under nondenaturing conditions, separation is also influenced by the existence of protein complexes in the sample. Successful application of membrane-based microscale ion-exchange fractionation to uncover differentially expressed proteins in brain samples from patients with Alzheimer's disease and a control group demonstrated that 13 protein differences could be identified ***(92)***. The differences were not apparent when unfractionated material was evaluated by 2-DGE alone.

5.2. Trifunctional Probes

Typically, two types of chemical probes for assaying functional attributes of proteins have been developed in the past. In the first instance, a detectable moiety, such as a fluorophore, is attached to the targeted functional attribute, but no mechanism for enrichment of the labeled protein is explicitly provided, although antidye antibodies might be used in this capacity. Examples of this type of probe include Pro-Q Diamond phosphoprotein stain and Pro-Q Emerald glycoprotein stain. In the second instance, an affinity tag, such as biotin, is introduced, but subsequent detection of the targeted attribute requires additional steps, such as electroblotting and immunodetection. The biotin-switch approach for identification of *S*-nitrosylated proteins is an example of this type of probe. A few trifunctional probes containing a moiety suitable for binding to the functional modification, a second handle suitable for enrichment of the target, and a third providing a means for detection have been introduced ***(80,95)***. A trifunctional phenyl sulfonate probe, for example, has been devised for the consolidated visualization and affinity purification of certain enzymes, such as 3-β-hydroxysteroid dehydrogenase, platelet-type phosphofructokinase, and type II tissue transglutaminase, by a combination of avidin affinity chromatography and in-gel fluorescence detection ***(80)***. Chemical probes that facilitate both isolation and detection should prove highly useful for the visualization and identification of specific functional attributes on a proteome-wide scale.

6. Future Perspectives

A presentation of the relative merits and shortcomings of gel-based vs gel-free approaches to proteomics goes well beyond the scope of this chapter. Several papers devoted to this topic have already been published in the scientific literature, with the authors of this review historically championing the gel-based techniques ***(1,96–99)***. It is safe to state that the days when the term "proteomics" was synonymous with "2-DGE/mass spectrometry" are now behind us, and alternate approaches, such as isotope-coded affinity tagging (ICAT), protein microarrays, and multidimensional protein identification technology (MudPIT) will increasingly play significant roles in proteomic investigations of the future. The throughput and resolution required for a particular proteomic investigation must be determined prior to deciding which technological approach offers the optimal breadth and depth of analysis needed to assign functional attributes to the proteins under study. Generally speaking, liquid chromatography-mass spectrometry methods provide higher resolution but lower throughput, whereas SDS-polyacrylamide gel electrophoresis provides lower resolution but higher throughput.

Clearly the polyacrylamide gel represents a specialized environment, and technologies optimized for this matrix are not always directly transferable to liquid chromatography-based assay formats or direct analysis in living cells and tissues. The past 6 yr have demonstrated that it is feasible to identify a host of protein functional attributes using fluorescence-based small molecule probes combined with polyacrylamide gel electrophoresis. It is anticipated increasingly that the approaches originally applied to gels will be adapted to other technology platforms, such as liquid chromatography-mass spectrometry, protein microarrays, cell-based assays, and even whole animal imaging. Proceeding forward, specialized small molecule probes will likely each be designed with the ability to serve as a sort of "proteomics Swiss army knife" for a particular protein functional attribute, displaying broad-based capabilities in cell imaging, electrophoretic separations, liquid chromatography, mass spectrometry, and protein microarrays. Additionally, the powerful capabilities of differential protein expression profiling, as already demonstrated by DIGE technology, will become incorporated into the probes, allowing for the multiplexed analysis of a host of protein functional attributes, including posttranslational modifications and enzyme activities. Prototypes for these advanced tools may already be gleaned from the literature, as exemplified by pairs of reactive fluorophores for examining differences in protease activity displayed between two samples, using both light microscopy and gel electrophoresis as analytical readouts ***(100)***.

References

1. Patton WF. *J Chromatogr B Anal Technol Biomed Life Sci* 2002;771:3–31.
2. Patton WF. *Electrophoresis* 2000;21:1123–1144.

3. Patton WF. *Biotechniques* 2000;28:944–957.
4. Lopez MF, Berggren K, Chernokalskaya E, Lazarev A, Robinson M, Patton WF. *Electrophoresis* 2000;21:3673–3683.
5. Berggren K, Chernokalskaya E, Steinberg TH, et al. *Electrophoresis* 2000;21: 2509–2521.
6. Berggren KN, Schulenberg B, Lopez MF, et al. *Proteomics* 2002;2:486–498.
7. Rabilloud T, Strub JM, Luche S, van Dorsselaer A, Lunardi J. *Proteomics* 2001;1: 699–704.
8. Lamanda A, Zahn A, Roder D, Langen H. *Proteomics* 2004;4:599–608.
9. Berger K, Wissmann D, Ihling C, et al. *Mol Cell Endocrinol* 2004;227:21–30.
10. Daban JR, Bartolome S, Samso M. *Anal Biochem* 1991;199:169–174.
11. Alba FJ, Bermudez A, Bartolome S, Daban JR. *Biotechniques* 1996;21:625–626.
12. Steinberg TH, Jones LJ, Haugland RP, Singer VL. *Anal Biochem* 1996;239:223–237.
13. Steinberg TH, Haugland RP, Singer VL. *Anal Biochem* 1996;239:238–245.
14. Steinberg TH, Lauber WM, Berggren K, Kemper C, Yue S, Patton WF. *Electrophoresis* 2000;21:497–508.
15. Mackintosh JA, Choi HY, Bae SH, et al. *Proteomics* 2003;3:2273–2288.
16. Kang C, Kim HJ, Kang D, Jung DY, Suh M. *Electrophoresis* 2003;24:3297–3304.
17. Garcia A, Prabhakar S, Brock CJ, et al. *Proteomics* 2004;4:656–668.
18. Chevalier F, Rofidal V, Vanova P, Bergoin A, Rossignol M. *Phytochemistry* 2004; 65:1499–1506.
19. Lim MJ, Patton WF, Lopez MF, Spofford KH, Shojaee N, Shepro D. *Anal Biochem* 1997;245:184–195.
20. Kemper C, Berggren K, Diwu Z, Patton WF. *Electrophoresis* 2001;22:881–889.
21. Smejkal GB, Robinson MH, Lazarev A. *Electrophoresis* 2004;25:2511–2519.
22. Smejkal GB. *Expert Rev Proteomics* 2004;1:1–7.
23. Nishihara JC, Champion KM. *Electrophoresis* 2002;23:2203–2215.
24. White IR, Pickford R, Wood J, Skehel JM, Gangadharan B, Cutler P. *Electrophoresis* 2004;25:3048–3054.
25. Krieg RC, Paweletz CP, Liotta LA, Petricoin EF 3rd. *Biotechniques* 2003;35: 376–378.
26. Ahnert N, Patton WF, Schulenberg B. *Electrophoresis* 2004;25:2506–2510.
27. Urwin VE, Jackson P. *Anal Biochem* 1993;209:57–62.
28. Berggren KN, Chernokalskaya E, Lopez MF, Beechem JM, Patton WF. *Proteomics* 2001;1:54–65.
29. Unlu M, Morgan ME, Minden JS. *Electrophoresis* 1997;18:2071–2077.
30. Shaw J, Rowlinson R, Nickson J, et al. *Proteomics* 2003;3:1181–1195.
31. Van den Bergh G, Arckens L. *Curr Opin Biotechnol* 2004;15:38–43.
32. Schulenberg B, Aggeler R, Beechem JM, Capaldi RA, Patton WF. Analysis of steady-state protein phosphorylation in mitochondria using a novel fluorescent phosphosensor dye. *J Biol Chem* 2003;278:27,251–27,255.
33. Steinberg TH, Agnew BJ, Gee KR, et al. *Proteomics* 2003;3:1128–1144.
34. Martin K, Steinberg TH, Goodman T, et al. *Comb Chem High Throughput Screen* 2003;6:331–339.

35. Goodman T, Schulenberg B, Steinberg TH, Patton WF. *Electrophoresis* 2004;25: 2533–2538.
36. Schulenberg B, Goodman TN, Aggeler R, Capaldi RA, Patton WF. *Electrophoresis* 2004;25:2526–2532.
37. Murray J, Marusich MF, Capaldi RA, Aggeler R. *Electrophoresis* 2004;25: 2520–2525.
38. Nagata-Ohashi K, Ohta Y, Goto K, et al. *J Cell Biol* 2004;165:465–471.
39. Chen Z, Southwick K, Thulin CD. *J Biomol Tech* 2004;15:249–256.
40. Steinberg TH, Pretty On Top K, Berggren KN, et al. *Proteomics* 2001;1: 841–855.
41. Hart C, Schulenberg B, Steinberg TH, Leung WY, Patton WF. *Electrophoresis* 2003;24:588–598.
42. Schulenberg B, Beechem JM, Patton WF. *J Chromatogr B Anal Technol Biomed Life Sci* 2003;793:127–139.
43. Saxon E, Bertozzi CR. *Science* 2000;287:2007–2010.
44. Hang HC, Yu C, Kato DL, Bertozzi CR. *Proc Natl Acad Sci U S A* 2003;100: 14,846–14,851.
45. Vocadlo DJ, Hang HC, Kim EJ, Hanover JA, Bertozzi CR. *Proc Natl Acad Sci U S A* 2003;100:9116–9121.
46. Prescher JA, Dube DH, Bertozzi CR. *Nature* 2004;430:873–877.
47. Luchansky SJ, Argade S, Hayes BK, Bertozzi CR. *Biochemistry* 2004;43: 12,358–12,366.
48. Kho Y, Kim SC, Jiang C, et al. *Proc Natl Acad Sci U S A* 2004;101:12,479–12,484.
49. Khidekel N, Arndt S, Lamarre-Vincent N, et al. *J Am Chem Soc* 2003;125: 16,162–16,163.
50. Khidekel N, Ficarro SB, Peters EC, Hsieh-Wilson LC. *Proc Natl Acad Sci U S A* 2004:101:13,132–13,137.
51. Ravi K, Brennan LA, Levic S, Ross PA, Black SM. *Proc Natl Acad Sci U S A* 2004;101:2619–2624.
52. Marshall HE, Hess DT, Stamler JS. *Proc Natl Acad Sci U S A* 2004;101: 8841–8842.
53. Reynaert NL, Ckless K, Korn SH, et al. *Proc Natl Acad Sci U S A* 2004;101: 8945–8950.
54. Chung KK, Thomas B, Li X, et al. *Science* 2004;304:1328–1331.
55. Jaffrey SR, Snyder SH. *Sci STKE* 2001;2001:PL1.
56. Jaffrey SR, Erdjument-Bromage H, Ferris CD, Tempst P, Snyder SH. *Nat Cell Biol* 2001;3:193–197.
57. Czapski GA, Avram D, Sakharov DV, Wirtz KW, Strosznajder JB, Pap EH. *Biochem J* 2002;365:897–902.
58. Avram D, Romijn EP, Pap EH, Heck AJ, Wirtz KW. *Proteomics* 2004;4: 2397–2407.
59. van der Vlies D, Wirtz KW, Pap EH. *Biochemistry* 2001;40:7783–7788.
60. Czapski GA, Avram D, Wirtz KW, Pap EH, Strosznajder JB. *Med Sci Monit* 2001;7:606–609.

61. Ahn B, Rhee SG, Stadtman ER. *Anal Biochem* 1987;161:245–257.
62. Talent JM, Kong Y, Gracy RW. *Anal Biochem* 1998;263:31–38.
63. Conrad CC, Talent JM, Malakowsky CA, Gracy RW. *Biol Proced Online* 2000; 2:39–45.
64. Conrad CC, Choi J, Malakowsky CA, et al. *Proteomics* 2001;1:829–834.
65. Reinheckel T, Korn S, Mohring S, Augustin W, Halangk W, Schild L. *Arch Biochem Biophys* 2000;376:59–65.
66. Yoo BS, Regnier FE. *Electrophoresis* 2004;25:1334–1341.
67. Kjaersgard IV, Jessen F. *J Agric Food Chem* 2004;52:7101–7107.
68. Hart C, Schulenberg B, Patton WF. *Electrophoresis* 2004;25:2486–2493.
69. Hart C, Schulenberg B, Diwu Z, Leung WY, Patton WF. *Electrophoresis* 2003;24:599–610.
70. Kapanidis AN, Ebright YW, Ebright RH. *J Am Chem Soc* 2001;123: 12,123–12,125.
71. Adams SR, Campbell RE, Gross LA, et al. *J Am Chem Soc* 2002;124:6063–6076.
72. Griffin BA, Adams SR, Jones J, Tsien RY. *Methods Enzymol* 2000;327:565–578.
73. Griffin BA, Adams SR, Tsien RY. *Science* 1998;281:269–272.
74. Feldman G, Bogoev R, Shevirov J, Sartiel A, Margalit I. *Electrophoresis* 2004; 25:2447–2451.
75. Johansson L, Chen C, Thorell J, et al. *Nature Methods* 2004;1:61–66.
76. Liu Y, Patricelli MP, Cravatt BF. *Proc Natl Acad Sci U S A* 1999;96: 14,694–14,699.
77. Adam GC, Cravatt BF, Sorensen EJ. *Chem Biol* 2001;8:81–95.
78. Adam GC, Sorensen EJ, Cravatt BF. *Mol Cell Proteomics* 2002;1:81–90.
79. Adam GC, Sorensen EJ, Cravatt BF. *Nat Biotechnol* 2002;20:805–809.
80. Adam GC, Sorensen EJ, Cravatt BF. *Mol Cell Proteomics* 2002;1:828–835.
81. Kidd D, Liu Y, Cravatt BF. *Biochemistry* 2001;40:4005–4015.
82. Speers AE, Adam GC, Cravatt BF. *J Am Chem Soc* 2003;125:4686–4687.
83. Speers AE, Cravatt BF. *Chembiochem* 2004;5:41–47.
84. Speers AE, Cravatt BF. *Chem Biol* 2004;11:535–546.
85. Saghatelian A, Jessani N, Joseph A, Humphrey M, Cravatt BF. *Proc Natl Acad Sci U S A* 2004;101:10,000–10,005.
86. Jessani N, Cravatt BF. *Curr Opin Chem Biol* 2004;8:54–59.
87. Patricelli MP, Giang DK, Stamp LM, Burbaum JJ. *Proteomics* 2001;1: 1067–1071.
88. Kemper C, Steinberg TH, Jones L, Patton WF. *Electrophoresis* 2001;22:970–976.
89. Lopez MF, Mikulskis A, Golenko E, et al. *Proteomics* 2003;3:1109–1116.
90. Hanson BJ, Schulenberg B, Patton WF, Capaldi RA. *Electrophoresis* 2001; 22:950–959.
91. Chernokalskaya E, Gutierrez S, Pitt AM, Leonard JT. *Electrophoresis* 2004;25: 2461–2468.
92. Lopez MF, Mikulskis A, Golenko E, et al. *Electrophoresis* 2004;25:2557–2563.
93. Schulenberg B, Patton WF. *Electrophoresis* 2004;25:2539–2544.
94. Lescuyer P, Hochstrasser DF, Sanchez JC. *Electrophoresis* 2004;25:1125–1135.

95. Wilbur DS, Chyan MK, Hamlin DK, et al. *Bioconjug Chem* 2002;13:1079–1092.
96. Gygi SP, Corthals GL, Zhang Y, Rochon Y, Aebersold R. *Proc Natl Acad Sci U S A* 2000;97:9390–9395.
97. Herbert BR, Harry JL, Packer NH, Gooley AA, Pedersen SK, Williams KL. *Trends Biotechnol* 2001;19:S3–S9.
98. Lopez MF, Melov S. *Circ Res* 2002;90:380–389.
99. Patton WF, Schulenberg B, Steinberg TH. *Curr Opin Biotechnol* 2002;13:321–328.
100. Greenbaum D, Baruch A, Hayrapetian L, et al. *Mol Cell Proteomics* 2002;1:60–68.

5

Modification-Specific Proteomic Analysis of Glycoproteins in Human Body Fluids by Mass Spectrometry

Jakob Bunkenborg, Per Hägglund, and Ole Nørregaard Jensen

Summary

Glycosylation of proteins is a very common, diverse, and heterogeneous type of modification, especially for proteins with extracellular destinations. This chapter describes some general strategies for the enrichment of glycoproteins and glycopeptides with an emphasis on proteomic analysis of *N*-glycosylated proteins in body fluids and other complex samples. An approach for identification of *N*-glycosylated proteins and mapping of their glycosylation sites is described. In this approach, glycoproteins are initially selectively purified by lectin chromatography. Following tryptic digestion, glycopeptides are enriched by hydrophilic interaction chromatography (HILIC). Glycan heterogeneity is then reduced by treating the glycopeptides with endoglycosidases. The resulting peptides are then analyzed by matrix-assisted laser desorption/ionization (MALDI) mass spectrometry and nano-flow reversed-phase liquid chromatography tandem mass spectrometry (LC-MS/MS). The analysis allows the identification of *N*-glycosylation sites and is demonstrated on a mixture of standard proteins.

Key Words: Proteomics; posttranslational modifications; mass spectrometry; lectin; HILIC; glycosylation; glycoproteomics; plasma proteins.

1. Introduction

The analysis of glycoproteins in human body fluids is one of the most complex and challenging areas of proteomic research. In human plasma, protein concentrations span over 10 orders of magnitude *(1)*, and heterogeneous processing through glycosylation and other modifications increases the complexity even further, making identification of low-abundance proteins a very difficult task. Glycoproteins are known to be involved in many diseases (e.g., cancer *[2]*), and it may therefore be of particular interest to identify low-abundance glycoprotein

From: *Proteomics of Human Body Fluids: Principles, Methods, and Applications*
Edited by: V. Thongboonkerd © Humana Press Inc., Totowa, NJ

biomarkers leaking into body fluids from diseased cells or tissues. Glycosylation is one of the most common types of posttranslational modifications of both secreted and cell surface proteins. A wide range of glycans may be covalently attached to proteins, and glycosylations are usually categorized depending on the specific amino acid involved (*see* Spiro for a review ***[3]***). In *N*-linked glycosylation, lipid-linked dolichol phosphate oligosaccharide precursors are transferred to asparagine residues located in the sequon N*X*S/T/C (where *X* is any amino acid except proline). *O*-linked glycans are usually attached to the hydroxyl group of serine or threonine, but linkages to tyrosine, hydroxylysine, and hydroxyproline have also been reported. A single mannose residue can be covalently attached to the C-2 atom in the indole ring of tryptophans, and this *C*-mannosylation only takes place on the first tryptophan in the consensus sequence W*XX*W ***(4)***. The *O*- and *N*-linked glycans are often further processed into different glycoforms depending on the protein structure and of the levels of sugar-nucleotide donors, glycosidases, and glycosyltransferases present at the time the protein travels through the endoplasmatic reticulum and Golgi apparatus. Many proteins are associated to cell surfaces by glycosylphosphatidylinositol (GPI) anchors covalently bound to the C-terminus. Finally, glycation is a nonenzymatic modification whereby glucose reacts preferentially with primary amine groups from lysine residues or the N-termini of proteins ***(5)***.

The complexity of the task at hand when one is analyzing protein glycosylation can be illustrated by a very impressive study of the human complement regulatory protein CD59 ***(6)***. CD59 is both *O*- and *N*-glycosylated and has been found in several body fluids including urine, milk, and saliva, although it is normally membrane-bound by a GPI anchor. The pool of glycans attached to the single *N*-glycosylation site contains at least 123 glycan structures after desialylation with the largest structure detected, containing approximately 25 monosaccharide residues. The full proteomic analysis of glycoproteins in human body fluids requires not only identification of the masses of all glycoforms present and their relative abundances, but also determination of the branching, linkage patterns, and configurations of each glycan attached to each glycosylation site in the protein. Currently, this comprehensive information cannot be obtained through the use of a single technique, but the combination of chromatography, enzymatic processing, and soft ionization tandem mass spectrometry (MS/MS) provides very powerful analysis tools. This chapter gives a brief general overview of strategies used for glycoprotein characterization in a proteomics context (**Subheading 1.**) and describes in detail a pragmatic approach for mapping *N*-glycosylation sites in complex mixtures (**Subheadings 2.** and **3.**). This approach is based on glycoprotein enrichment by lectin chromatography, glycopeptide enrichment by hydrophilic liquid interaction chromatography (HILIC), and complexity reduction by enzymatic

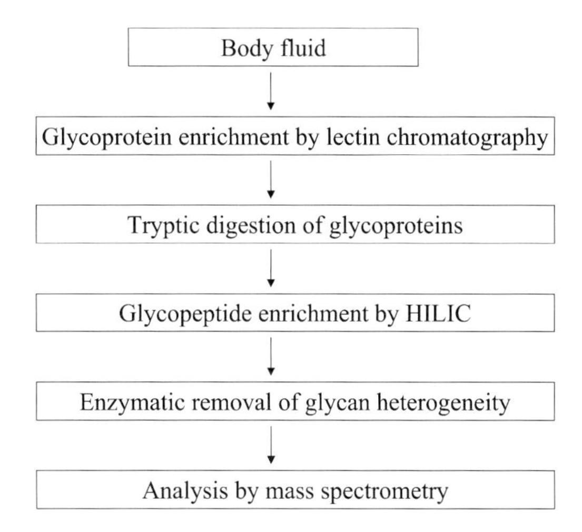

Fig. 1. Strategy for identification of *N*-glycosylation sites in human body fluids. Lectin chromatography is used to enrich for a subset of proteins with particular glycoforms. The proteins can be separated by various means (e.g., SDS-PAGE, size-exclusion, and so on) followed by tryptic digestion. The glycopeptides are enriched by hydrophilic interaction chromatography (HILIC) or a second round of lectin chromatography. The glycosylated peptides are then treated enzymatically to remove the glycan heterogeneity, as outlined in **Fig. 2.** Finally, the resulting pool of peptides is analyzed by liquid chromatography coupled to tandem mass spectrometry.

deglycosylation **(Fig. 1)**. The samples are ultimately analyzed by either matrix-assisted laser desorption/ionization (MALDI) MS or nanoflow liquid chromatography (LC) coupled online to electrospray ionization tandem mass spectrometry (ESI-MS/MS).

1.1. Glycoprotein Enrichment

One of the main strategies in proteomic analysis is to reduce complexity by targeting a subclass of proteins or peptides through their amino acid composition (e.g., cysteines) or posttranslational modifications (e.g., phosphorylation). In the context of body fluid analysis, this step also serves as a way to remove various nonproteinaceous small molecules that may interfere with subsequent enzymatic digestions (e.g., trypsin inhibitors). Glycoproteins and glycopeptides can be enriched by a number of different techniques, which are listed in **Table 1**. One of the most commonly used methods for glycoprotein enrichment is lectin affinity chromatography. Lectins are carbohydrate-binding proteins of nonimmune origin with a wide range of specificities for different glycans. More than 100 different lectins are commercially available, and a representative list of some lectins, which are widely used to isolate and characterize glycoconjugates and

Table 1
Glycosylation-Specific Enrichment Strategies

Enrichment strategy	Principle
Lectin chromatography	Noncovalent interactions between a glycan and a carbohydrate binding protein
Antibody	Noncovalent interactions between a glycan and a carbohydrate binding protein
Hydrophilic interaction chromatography (HILIC)	Hydrophilic oligosaccharides in an apolar mobile phase bind strongly to the polar stationary phase
Boronic acid	Reversible pH-dependent formation of cyclic ester between *cis*-diols and immobilized phenylboronic acid
Porous graphite	Reversed-phase chromatography on an apolar stationary phase
Periodate oxidation hydrazine coupling	The aldehyde groups resulting from periodate oxidation react with hydrazine coupled to a resin or an affinity tag
Chemoenzymatic tagging	Enzymatic incorporation of saccharide analogs that can be derivatized with biotin affinity tags
β-Elimination/Michael addition	*O*-glycans attached to serine and threonine are removed by β-elimination under alkaline conditions, and a suitable affinity tag is introduced by Michael addition

released glycans, is displayed in **Table 2**. More elaborate schemes using serial lectin affinity chromatography (SLAC) have been utilized to separate glycoforms *(7)*. Monoclonal antibodies have been raised against different carbohydrate epitopes and have been used to probe glycosylation patterns *(8)*.

A drawback of lectin and antibody chromatography is that the high specificity for particular oligosaccharide structures only pulls out a subset of the glycans attached to the given glycosylation sites. A less biased strategy utilizes periodate oxidation of glycan hydroxyl groups followed by coupling to either a hydrazide-conjugated biotin group or an immobilized hydrazide column ***(9,10)***. This approach can be combined with peptide-*N*-glycosidase F (PNGase F) digestion to release the peptide moiety of conjugated *N*-glycosylated peptides. However, it should be noted that this method is limited to glycans possessing *cis* vicinal diols. Boronic acid affinity chromatography ***(11)*** is based on the formation of a cyclic ester between *cis*-diols and immobilized phenylboronic acid at basic pH (pH 8.5); this reaction can be reversed at acidic pH. Khidekel et al. ***(12)*** have used a

Table 2
Lectins Commonly Used in Affinity Purification of Body Fluids

Lectin name	Source	Carbohydrate binding	Oligosaccharide preferences[a]
Concanavalin A (Con A)	*Canavalia ensiformis* (jack bean)	α-Man α-Glc α-GlcNAc	Branched mannose
Wheat germ agglutinin (WGA)	*Triticum vulgaris* (wheat germ)	β-GlcNAc Sialic acid	GlcNAc(β1,4)GlcNAc
Jacalin	*Artocarpus integrifolia* (jackfruit)	α-Gal β-GlcNAc	Gal(β1,3)GalNAc
Sambucus nigra agglutinin (SNA)	*Sambucus nigra* (elder)	β-Gal Sialic acid	NANA(α2,6)Gal
Lotus tetragonolobus	*Lotus tetragonolobus* (asparagus pea)	α-Fuc	Fuc(α1,2)Gal Fuc(α1,3)GlcNAc
Dolichos biflorus agglutinin (DBA)	*Dolichos biflorus* (horse gram)	α-GlcNAc	GalNAc(α1,3)GalNAc
Maackia amurensis (MAA)	*Maackia amurensis* (Amur maackia)	Sialic acid	NANA(α2,3)Gal
Ulex europaeus I agglutinin (UEA I)	*Ulex europaeus* (gorse)	α-Fuc	Fuc(α1,2)Gal
Ulex europaeus II agglutenin (UEA II)	*Ulex europaeus* (gorse)	β-GlcNAc	Gal(β1,4)GlcNAc

[a]The lectins react with a specific structural carbohydrate motif, and the affinities depend on the glycan structure and oligosaccharide composition. For example, Con A has a specificity for high-mannose and biantennary complex-type chains.

chemoenzymatic strategy whereby a ketone-containing galactose analog is ligated to the existing glycan by β-1,4-galactosyltransferase and then conjugated to biotin. Glycans can also be enriched by strategies that utilize the hydrophilic properties of the glycans. The term *hydrophilic interaction chromatography* was coined by Alpert ***(13)*** to describe the form of normal phase chromatography (hydrophilic stationary phase) whereby the analytes are loaded in a hydrophobic, predominantly organic mobile phase and eluted with a more polar mobile phase. This technique has been successfully applied to isolate and separate both carbohydrates ***(14,15)*** and glycopeptides ***(16,17)***. Porous graphitic columns ***(18)*** and microcolumns ***(19)*** have also been used to capture very hydrophilic glycopeptide conjugates resulting from proteolytic digestions with nonspecific proteases. Finally, *O*-glycosylated peptides can be labeled and affinity-tagged by modification with various nucleophiles in Michael addition reactions in combination with β-elimination ***(20)***, and providing a handle for enrichment and analysis.

1.2. Deglycosylation Methods

PNGase F and endoglycosidase H (Endo H) are commonly used to reduce the glycan heterogeneity by deglycosylation of *N*-linked glycans **(Fig. 2)**. PNGase F cleaves the bond between asparagine and the innermost *N*-acetylglucosamine (GlcNAc) residue of the *N*-glycan. Since asparagine is converted to aspartic acid in this reaction, glycosylation sites can by identified by a mass increase of 1 Dalton **(Fig. 2A)**. However, this mass shift can also be the result of "true" deamidation, which is a common modification of human proteins in vivo and in vitro ***(21)***. To eliminate this ambiguity, PNGase F may be used in the presence of ^{18}O-labeled water, which leads to a mass increase of 3 Daltons per glycosylation site in deglycosylated peptides ***(22,23)***.

N-linked glycans can also be deglycosylated using Endo H or other endo-β-*N*-acetylglucosaminidases **(Fig. 2B)**. These enzymes hydrolyze the *O*-glycosidic bond between the two innermost GlcNAc residues attached to asparagine. Thus, a deglycosylated peptide will retain a single GlcNAc residue (mass increase 203 Daltons) attached to asparagine. If an α-1-6-linked fucose residue is attached to the innermost GlcNAc residue, this residue will also be retained on the peptide after deglycosylation (mass increase of 203 + 146 Daltons). In MS/MS experiments, the oxonium ion of the retained GlcNAc residue (*m/z* 204) can be used as a diagnostic marker for glycopeptides in, e.g., interpretation of tandem mass spectra and precursor ion scanning ***(24)***.

One drawback of these enzymes in a comprehensive analysis of glycosylation sites is that no single enzyme is active on all the major classes of *N*-glycans, i.e., high mannose, complex, and hybrid types. To circumvent this problem, a combination of enzymes that are active, when used in concert, on all the major classes can be utilized. For example, Endo H is active on high-mannose-type glycosylation, and Endo D is active on complex and hybrid-type glycosylation when used together with a set of exoglycosidases ***(25,26)***.

O-glycan structures can vary considerably, and there is no commercially available *endo*-acting enzyme that is generally applicable for deglycosylation of these. For example, *O*-glycanase (Endo-β-*N*-acetylgalactosaminidase) displays strict substrate specificity and is only active on the disaccharide galactosyl-β-1-3-GalNAc attached to serine or threonine ***(27)***. As an alternative to enzymatic deglycosyation, *O*-glycans can be removed by β-elimation reactions under alkaline conditions ***(20,28)***. In this reaction, previously glycosylated serine and threonine residues are converted to dehydroalanine and dehydrobutyric acid, respectively. This results in mass shifts of 18 Daltons that can be used for identification of glycosylation sites. A drawback with the β-elimination reaction is that phosphorylated serine and threonine residues also may be modified in β-elimination reactions. As previously mentioned, this reaction can be combined with Michael addition reaction for selective

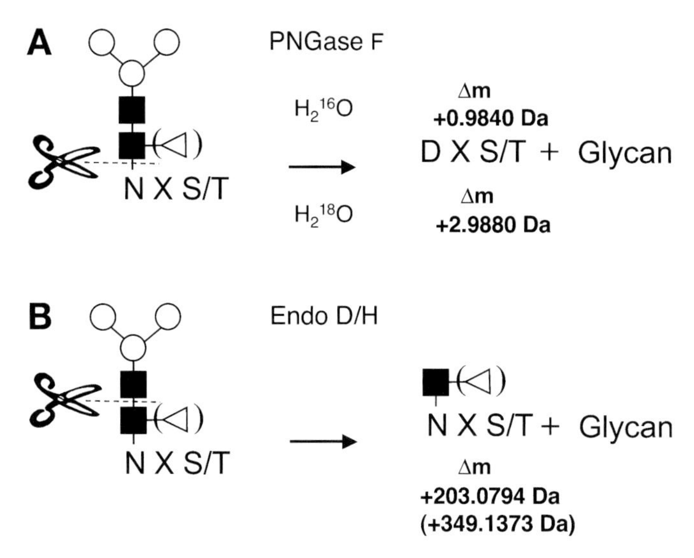

Fig. 2. Deglycosylation schemes. To facilitate the analysis, two different types of enzymes can be used to reduce glycan heterogeneity. The simplest scheme utilizes PNGase F to remove the glycan entirely. In this reaction, the asparagine is converted to aspartic acid with a concomitant mass increase of 1 Dalton. Since deamidation of asparagine is a common modification, it is recommended to perform the PNGase F treatment in ^{18}O water to avoid false positives. The second scheme utilizes a mixture of exoglycosidase D and H and some exoglycosidases to remove the bulk of the glycan structure, leaving a single *N*-acetylglucosamine residue attached to the peptide. The retained glycan can be monitored by, e.g., precursor ion scanning. This method can also be used for detecting core fucosylation, but, if required, the fucose group may be eliminated by using a fucosidase in the enzyme cocktail.

tagging and enrichment of modified peptides or for relative quantitation ***(20,29)***. It is also possible to remove the majority of *O*- and *N*-linked glycans chemically with trifluoromethanesulphonic acid ***(30)***.

GPI anchors can be released from the C-terminus of proteins using phosphatidylinositol phospholipase C (PI-PLC), which hydrolyzes the bond between the lipid and the phosphatidylinositol moiety. This enzyme has been used in combination with mass spectrometry in a proteomics approach to identify GPI-anchored proteins ***(31)***.

1.3. Model Glycoproteins

Human body fluids have very complex compositions, and it is not clear how, e.g., plasma changes with age, gender, health status, genetics, blood type, sample handling, and so on. To develop and demonstrate the current methodologies, we have chosen to work on a small set of four commercially available model proteins (α-1-acid glycoprotein, fetuin, ovalbumin, and ribonuclease B). We have used these model proteins for two reasons: first because it is easier to troubleshoot a simple, well-characterized system and second because, using these readily available proteins, other researchers hopefully should be able to reproduce the results presented here. α-1-Acid glycoprotein is a highly glycosylated plasma protein that contains five *N*-linked glycosylation sequons within its 183-amino-acids-long polypeptide chain—the glycans are mainly of the complex type. Ribonuclease B contains a single *N*-glycosylation site with mainly high-mannose-type oligosaccharides attached. Ovalbumin also displays a single *N*-glycosylation site but contains both high mannose and hybrid glycans. Fetuin has four *O*-glycosylation sites and three *N*-glycosylation sites, with the asparagine-linked glycans being mainly of the triantennary complex type. It should also be noted that even though a mixture of well-described glycoproteins is used in this validation protocol, unexpected proteins and peptides may be identified. This may be owing to variations in the purity of the commercial preparations of the proteins (e.g., ovalbumin samples are often contaminated with ovomucoid) and because proteins degrade over time (nontryptic cleavages, deamidations, and so on) unless precautions are taken. **Table 3** gives an overview of the peptides identified using the outlined protocols.

1.4. Mass Spectrometry

Two MS-based methods for analysis of deglycosylated peptides are outlined. Peptide mass fingerprinting with MALDI-MS is well suited for monitoring the purity of the glycopeptide enrichment and deglycosylation reactions **(Fig. 3)**. For in-depth analysis and for separation of more complex samples, direct nanoflow LC-ESI-MS/MS is a very useful tool. By fragmentation of the peptides in MS/MS, the exact modification site can be determined. The most common method to produce peptide fragmentation is collision-induced dissociation, whereby the ionized peptides are accelerated and made to collide with an inert gas; then the mass-to-charge ratios of the resulting fragment ions are measured.

One problem with glycopeptide conjugates is that the glycan fragments very easily upon collisional activation compared with the peptide backbone, and therefore it is often difficult to identify both the glycan and the peptide. Electron capture dissociation *(32,33)* or electron transfer dissociation *(34)* fragment the glycopeptides with efficient cleavage of the peptide backbone while leaving the glycan structure intact, but these techniques are not commonplace yet. In the methods presented here the analytical problems of collision-induced dissociation are decreased by reducing the size of the glycan to get an efficient cleavage of the peptide backbone. The complete removal of the glycan using PNGase F leads to the simplest peptide MS/MS spectra. However, the partial deglycosylation

Table 3
Glycopeptides Identified From the Set of Test Proteins and the Observed Monoisotopic Molecular Masses After Endo D/H Digestion (+203.08 Daltons per Glycosylation Site) or PNGase F Digestion in ^{18}O Water (+2.99 Daltons per Glycosylation Site)[a]

Protein	MH$^+$ PNGase F ^{18}O	MH+ Endo D/H	Peptide sequence
Fetuin (P12763)			
72–103	3674.76	3874.85	RPTGEVYDIEIDTLETTCHVLDPTPLA**n**CSVR
144–159	1871.93	2072.02	KLCPDCPLLAPL**n**DSR
145–159	1743.83	1943.92	LCPDCPLLAPL**n**DSR
160–187	3019.56	3219.65	VVHAVEVALATFNAES**n**GSYLQLVEISR
α-1-Acid glycoprotein 1 (P02763)			
19–42	2562.37	2762.46	qIPLCANLVPVPIT**n**ATLDQITGK
43–57	1940.91	2141.00	WFYIASAFRNEEY**n**K
58–81	2898.45	3098.54	SVQEIQATFFYFTP**n**KTEDTIFLR
87–101	1918.89	2118.98	QDQCIY**n**TTYLNVQR
87–108	2679.25	3079.43	QDQCIY**n**TTYLNVQRE**n**GTISR
α-1-Acid glycoprotein 2 (P19652)			
19–38	2191.17	2391.26	qIPLCANLVPVPIT**n**ATLDR
26–38	1411.81	1611.90	LVPVPIT**n**ATLDR
87–108	2670.20	3070.38	QNQCFY**n**SSYLNVQRE**n**GTVSR
Ovalbumin (P01012)			
287–322	3813.84	4013.93	MEEKY**n**LTSVLMAMGITDVFSSSANLSGISSAESLK
Ovomucoid (P01005)			
25–38	1612.72	1812.81	AEVDCSRFP**n**ATDK
25–41	1926.88	2126.97	AEVDCSRFP**n**ATDKEGK
81–106	2952.16	3352.34	EHDGECKETVPM**n**CSSYA**n**TTSEDGK
88–106	2096.84	2497.02	ETVPM**n**CSSYA**n**TTSEDGK
189–209	2358.07	2558.16	CNFCNAVVES**n**GTLTLSHFGK

[a]The numbering scheme refers to the SwissProt entry sequence, and the deglycosylated asparagine is denoted with **n**. Cysteines are carbamidomethylated, and q is pyroglutamine.

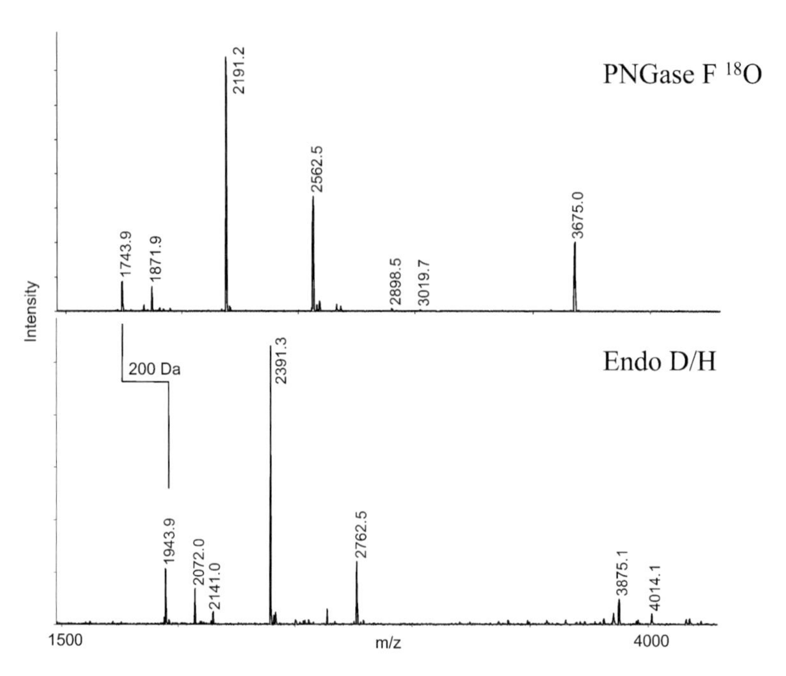

Fig. 3. MALDI spectra of the HILIC-enriched deglycosylated peptides from the reference proteins. The upper panel displays the peptides deglycosylated with PNGase F in ^{18}O water. The lower panel displays the peptides deglycosylated with Endo H/D. There is a 200-Dalton mass difference per glycosylation site between the two deglycosylation methods. Note that the HILIC enrichment of glycopeptides is very efficient, and very few nonglycosylated peptides are observed. The sequences corresponding to the masses can be found in **Table 3**.

strategy using Endo D and Endo H has the benefit that the remaining *N*-acetylglucosamine gives rise to a diagnostic oxonium ion at *m/z* 204.08 in the tandem mass spectra that can aid interpretation. Another advantage is that it is, at least in some cases, possible to detect core fucosylation. Examples of tandem mass spectra of glycopeptides after digestion with Endo D/Endo H and PNGase F are shown in **Figs. 4** and **5**, respectively. In an alternative strategy, glycopeptide conjugates with intact glycan structures but only very short (three to five amino acids) peptide moieties are analyzed by MS/MS. This method can be used to analyze glycan structures in purified proteins but is not suitable for analysis of complex mixtures *(18)*.

2. Materials

The equipment listed below merely reflects the current set-up and should not be taken as an endorsement.

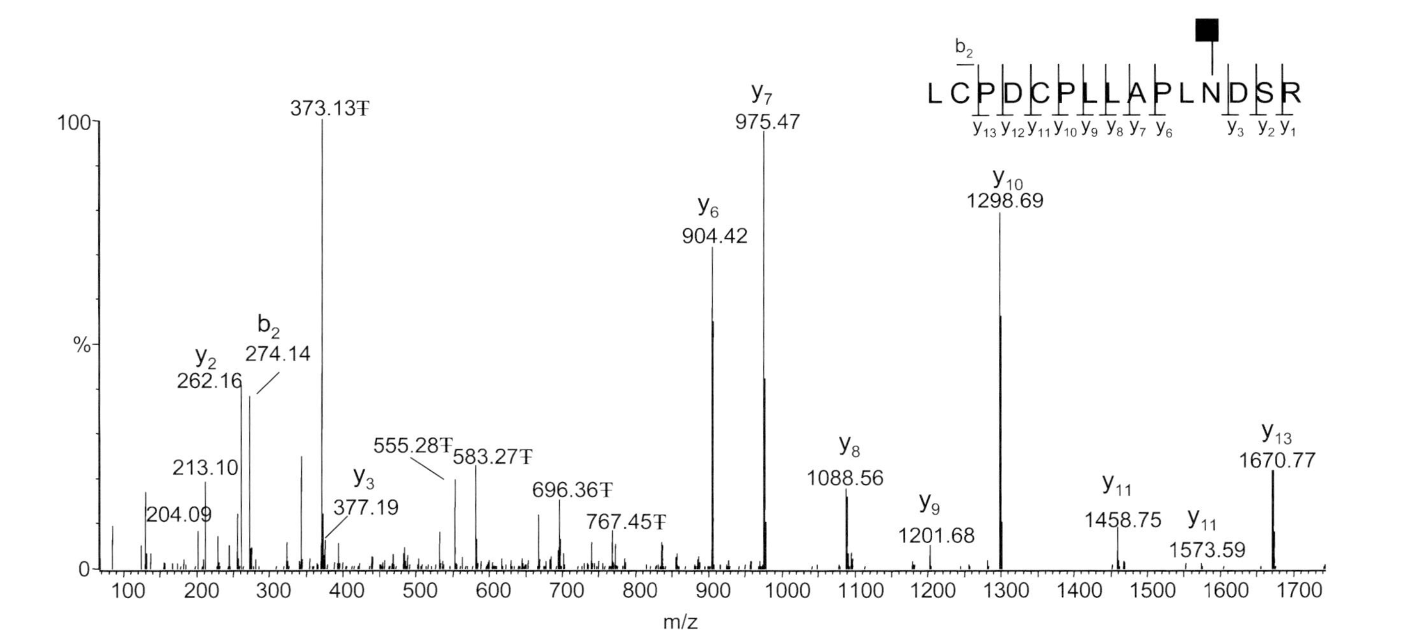

Fig. 4. ESI-QTOF tandem mass spectrum of a glycopeptide that has been subjected to deglycosylation with Endo D and Endo H. The spectrum of the doubly charged ion at *m/z* 972.48 corresponds to the peptide LCPDCPLLAPLNDSR from fetuin carrying a single GlcNAc residue—the oxonium ion of the GlcNAc can be observed at *m/z* 204.1. The masses marked with Ŧ arise from internal fragmentation.

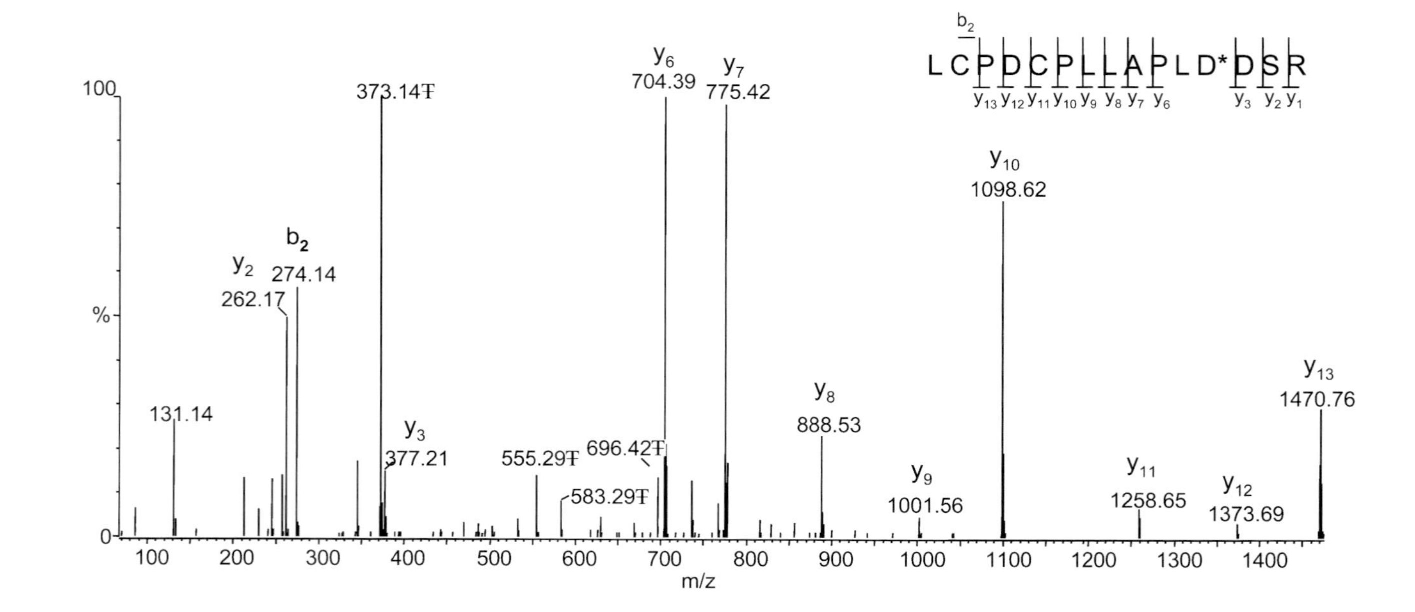

Fig. 5. ESI-QTOF tandem mass spectra of a glycopeptide that has been subjected to deglycosylation with PNGase F in ^{18}O water. The spectrum of the doubly charged ion at *m/z* 872.45 corresponds to the peptide LCPDCPLLAPLNDSR from fetuin with a single asparagine converted into an ^{18}O-containing aspartic acid (D*). The masses marked with Ŧ arise from internal fragmentation.

2.1. Solvents

1. Acetonitrile (ACN), high-performance liquid chromatography (HPLC) grade (Fisher Scientific, Loughborough, UK).
2. Formic acid (FA), Suprapur (Merck, Darmstadt, Germany).
3. Acetic acid (AcOH), Suprapur, (Merck).
4. Isopropanol (Merck).
5. All aqueous solvents were made with UHQ water purified on a Purelab ultra system from ELGA (Stoke, UK).

2.2. Lectin Chromatography

1. Concanavalin A, immobilized on agarose (Vector, Burlingame, CA).
2. Loading buffer: 50 m*M* HEPES (Sigma-Aldrich) pH 7.4, 1 m*M* $MnCl_2$, 1 m*M* $CaCl_2$, 150 m*M* NaCl (all Sigma-Aldrich).
3. Elution buffer: 100 m*M* methyl-α-D-mannopyranoside (Sigma-Aldrich) in 10 m*M* HEPES, pH 7.4.

2.3. Reference Proteins

1. α-1-Acid glycoprotein (orosomucoid), human (Sigma, St. Louis, MO).
2. Fetuin, bovine (Roche, Mannheim, Germany).
3. Ovalbumin, chicken (Sigma-Aldrich, St. Louis, MO).
4. Ribonuclease B, bovine (Sigma-Aldrich).
5. Serum albumin, bovine (BSA) (Sigma-Aldrich).

2.4. In-Solution Digestion

1. 8 *M* Urea (Sigma-Aldrich) in 400 m*M* NH_4HCO_3 (Sigma-Aldrich), pH 7.8.
2. 6 *M* Guanidine hydrochloride (GndHCl; Sigma-Aldrich), 50 m*M* Tris-HCl (Sigma-Aldrich), pH 7.8.
3. DL-Dithiothreitol (DTT), (Sigma-Aldrich).
4. Iodoacetamide (Sigma-Aldrich).
5. Trypsin, sequence grade (Promega, Madison, WI).

2.5. HILIC

1. ZIC-HILIC resin (10 μm, Sequant, Umeå, Sweden).
2. GELoader tips (Eppendorf, Hamburg, Germany).
3. Loading solvent: 80% ACN, 1% FA.
4. Elution solvent: 1% FA.

2.6. PNGase F Deglycosylation

1. Peptide *N*-glycosidase F (Roche, Mannheim, Germany).
2. Pefabloc Sc (AEBSF) (Roche).
3. $H_2{}^{18}O$ 99% (Icon Isotopes, Summit, NJ).

2.7. Endoglycosidase Deglycosylation

1. Endoglycosidase D from *Streptococcus pneumoniae* (ICN Biomedicals, Aurora, OH).
2. Endoglycosidase H from *Streptomyces plicatus* (Roche).
3. Neuraminidase from *Arthrobacter ureafaciens* (Roche).
4. β-galactosidase from *Bos taurus* testes(ProZyme, San Leandro, CA).
5. *N*-acetyl-β-D-glucosaminidase from *Diplococcus pneumoniae* (Roche).
6. 100 m*M* Ammonium acetate buffer, pH 5.5 (Sigma-Aldrich).
7. Pefabloc Sc (AEBSF) (Roche).

2.8. Microcolumn Sample Preparation

1. POROS R2 20 resin (20 μm, Applied Biosystems, Foster City, CA).
2. GELoader tips (Eppendorf, Hamburg, Germany).
3. Loading solvent, 5% FA.
4. Elution solvent, MALDI-matrix solution.

2.9. MALDI

1. Bruker Ultraflex TOF/TOF mass spectrometer (Bruker Daltonik, Bremen, Germany).
2. 2,5-Dihydroxybenzoic acid (DHB) (Sigma-Aldrich).
3. α-Cyano-4-hydroxycinnamic acid (CHCA) (Sigma-Aldrich).
4. 2,6-Dihydroxyacetophenone (Sigma-Aldrich).
5. Diammonium hydrogen citrate (Sigma-Aldrich).

2.10. LC-MS Setup

1. In-house reversed-phase columns:
   a. Kasil (PQ, Berwyn, PA).
   b. Formamide (Sigma-Aldrich).
   c. Fused silica (Composite Metal, Ilkley, UK).
   d. Ceramic scoring wafer (Restek, Bellefonte, PA).
   e. Reprosil-pur C18 reversed-phase material, 3 μm (Dr. Maisch, Ammerbuch-Entringen, Germany).
   f. Pressure bomb, (Proxeon, Odense, Denmark).
   g. Magnetic stirrer (Struers, Ballerup, Denmark).
   h. Microteflon-coated magnetic bar (Sigma-Aldrich).
2. HPLC equipment: Ultimate nanoflow pump (LC Packings, Amsterdam, Holland) with a Famos autosampler (LC Packings).
3. Mass spectrometer: QTOF Micro (Waters/Micromass, Manchester, UK).
4. HPLC mobile phase for directly hyphenated LCMS:
   a. Solvent A: 0.5% acetic acid.
   b. Solvent B: 0.5% acetic acid, 80% acetonitrile.

3. Methods

3.1. Lectin Chromatography of Glycoproteins

Enrichment of glycoproteins from body fluids is often necessary to reduce complexity and to remove nonglycosylated species such as serum albumin. Lectins have been widely applied to human plasma (*see*, e.g., refs. ***16*** and ***35–37*** for more detailed descriptions).

1. 50 μL Concanavalin A agarose beads are placed in an Eppendorf tube and washed three times with loading buffer.
2. 100 μL human plasma is diluted with 400 μL loading buffer and added to the ConA beads.
3. The beads are incubated for 30 min at room temperature.
4. The beads are washed three times with loading buffer.
5. The glycoproteins are eluted with two times 50 μL elution buffer.

3.2. In-Solution Digestion of Proteins

The proteins are dissolved under strongly denaturing conditions, the disulfide bonds are reduced, and the cysteines are alkylated (*see* **Note 1**). The proteins are then digested using trypsin (*see* Højrups *[38]* on digestion for a more complete discussion of conditions for different proteases). For testing and validation purposes, we use a mixture of four glycoproteins: α-1-acid glycoprotein, fetuin, ovalbumin, and ribonuclease B. First 100 μg of each protein was mixed from stock solutions and the mixture was lyophilized.

1. The lyophilized sample is redissolved in 28.5 μL alkylation buffer containing 6 *M* guanidine hydrochloride in 50 m*M* Tris-HCl pH 7.6. (*see* **Note 2**).
2. 1.5 μL of 100 m*M* DTT is added to the solubilized protein and incubated at 50°C for 15 min.
3. The sample is allowed to cool to room temperature, 3.0 μL of 110 m*M* iodoacetamide is added, and the solution is incubated for 15 min at room temperature in the dark.
4. 2.0 μL of 100 m*M* L-cysteine is added to quench the alkylation (*see* **Note 3**).
5. The mixture is diluted by adding 100 μL H_2O.
6. 10 μg of sequence grade trypsin dissolved in 20 μL water is added and incubated at 37°C overnight.
7. The resulting glycopeptide stock solution can be aliquoted and stored in the freezer.

3.3. Glycopeptide Enrichment by HILIC

HILIC has been used successfully for separation of amino acids, peptides, oligonucleotides, glycans, and various small polar compounds. We have used HILIC for enrichment of glycosylated peptides prior to mass spectrometric analysis using HILIC resin microcolumns packed into GELoader tips (Eppendorf). The use of in-house disposable microcolumns has several advantages (relatively cheap and easy to produce, flexible column sizes, avoidance of carryover, small elution volumes), and for these reasons, the format has been widely used with a number of different resins ***(16,39,40)***.

1. A slurry of HILIC material is prepared by placing a small amount of HILIC resin in an Eppendorf tube. To remove contaminants, the resin is washed twice in 100% methanol, twice in 5% FA, twice again in 100% methanol, and then stored in 100% methanol.
2. To retain the beads, the end of a GELoader pipette tip is constricted by squeezing its narrow end and twisting.
3. The HILIC resin slurry is vortexed, and a small amount is placed in the top of the GELoader tip. The liquid is then forced through the tip by applying pressure with a disposable plastic syringe fitted to the GELoader opening by an adapter piece made by cutting both ends of a 200-μL disposable pipette tip.
4. The length of the column is adjusted depending on the amount of sample, but usual column lengths are 0.3 to 1 cm.
5. The HILIC column is equilibrated by passing 20 μL loading solvent (80% ACN, 1% FA) through it.
6. The in-solution tryptic digest is dissolved in 10 to 20 μL loading buffer and applied to the column (*see* **Note 4**).
7. The column is washed two times with 10 μL loading solvent.
8. The glycopeptides are eluted in 1 to 5 μL of 1% FA depending on column size and downstream processing scheme.

3.4. Deglycosylation Using PNGase F

PNGase F is active on most *N*-linked glycans found in mammals. In the process, the asparagine to which the glycan is attached is converted into an aspartic acid, leading to a mass increase of 1 Dalton **(Fig. 2A)**. To avoid false positives from in vivo or in vitro deamidations, which are especially common if the following residue is glycine, we perform the enzymatic reaction in $H_2{}^{18}O$ water, leading to the incorporation of ^{18}O into aspartic acid and a mass increase of 3 Daltons.

1. 1 mU of PNGase F is added to an Eppendorf tube containing 15 μL of 50 m*M* Tris-HCl pH 7.8.
2. The enzyme is lyophilized and redissolved in 15 μL of 99% $H_2{}^{18}O$.
3. 1.0 μL of 100 m*M* Pefabloc is added to the sample (*see* **Note 5**).
4. The HILIC eluent is lyophilized and redissolved with the 15 μL of 99% $H_2{}^{18}O$-PNGase F solution.
5. The sample is incubated at 37°C overnight.

3.5. Deglycosylation Using Endo-β-N-Acetylglucosaminidases

To target a broad range of *N*-glycan structures, a mixture of two endo-β-*N*-acetylglucosaminidases (Endo H and Endo D) is used. Endo H is active on high-mannose-type glycosylation, and Endo D is active on complex and hybrid-type glycosylation when used together with the set of exoglycosidases listed below. When applied to analysis of the reference proteins used in this study, a similar number of glycopeptides was identified using PNGase F and Endo D/Endo H (**Fig. 3** and **Table 3**).

1. 1.0 µL of 100 m*M* Pefabloc is added to the HILIC eluent (*see* **Note 6**).
2. The sample is lyophilized and redissolved in 30.0 µL of 100 m*M* ammonium acetate, pH 5.5.
3. A glycosidase stock solution is prepared with the following activities (*see* **Note 7**).
   a. 1 mU/µL Endo D.
   b. 5 mU/µL Endo H.
   c. 0.5 mU/µL Neuraminidase.
   d. 1 mU/µL β-galactosidase.
   e. 0.5 mU/µL *N*-acetyl-β-D-glucosaminidase.
4. 1 µL of glycosidase stock mixture is added to the sample and incubated at 37°C overnight.

3.6. MALDI-MS

MALDI-MS offers an excellent method for making fast assessments of various steps of the enrichment strategy. We find that the matrix 2,6-dihydroxyacetophenone (DHAP) containing di-ammonium hydrogen citrate (DAHC) *(41)* (*see* **Note 8**) is a very good supplement to DHB and CHCA. Monitoring the HILIC process is very convenient with MALDI-MS after deglycosylation, and an example is shown in **Fig. 3**. The spectra were acquired on a Bruker Ultraflex TOF/TOF mass spectrometer. The monoisotopic molecular mass of the glycopeptides treated with PNGase F in ^{18}O water and Endo D/H glycosidase differs by 200.09 Daltons per glycosylation site (the mass difference between ^{18}O aspartic acid and *N*-acetylglucosamine-asparagine). The spectra are not very complex, and similar signals are observed in both spectra. The most intense ions can be identified as glycopeptides, showing that the HILIC enrichment of glycopeptides is efficient.

3.6.1. Preparation of DHAP/DAHC Matrix

1. 10 mg DHAP is dissolved in 1 mL of 1:1 EtOH/ACN.
2. 100 µL of 1 *M* aqueous DAHC is added.
3. The mixture is vortexed.

3.6.2. Sample Clean-Up and Deposition

1. A POROS R2-20 microcolumn is packed in a GELoader tip, washed with 10 µL solution containing 70% ACN and 5% FA, and equilibrated with 10 µL of 5% FA.
2. 10 µL of 5% FA is put in the top, a suitable volume (1–10 µL) of sample is added to the 10 µL of 5% FA, and the mixture is loaded onto the resin.
3. The resin is washed with 10 µL of 5% FA.
4. The sample is eluted directly on the target using 0.5 to 1 µL matrix.

3.7. LC-MS

For more complex glycopeptide mixtures, it is necessary to separate the peptides prior to analysis, and we have used directly coupled liquid chromatography (LC) MS/MS. We use a setup with a fairly short (1–1.5 cm) in-house precolumn packed with C18 reversed-phase material in a 75-μm ID fused silica capillary with a silicate frit. The sample is loaded onto the precolumn with a flow of 4 μL/min solvent A. After loading, the peptides are eluted at a flow rate of approximately 100 nL/min through a 7 to 10-cm C18 analytical column packed in a 50-μm ID fused silica capillary with a silicate frit. The setup used is the current compromise among chromatographic performance, ruggedness, ease of maintenance, and cost. The resulting tandem mass spectra can be processed and submitted to a search engine with glycosylation as a variable modification. **Table 3** lists the peptides identified by LC-MS and the monoisotopic masses. Some of the expected tryptic glycopeptides from the reference proteins are either too small (e.g., the glycopeptide NLTK from ribonuclease B with a monoisotopic mass of 474.3 Daltons) or too large to yield informative tandem mass spectra in ESI-MS. **Figures 4** and **5** show examples of the resulting tandem mass spectra of a glycosylated peptide from fetuin. The spectra are fairly similar, but the *m/z* 204 ion from the GlcNAc oxonium ion should be noted for the Endo D/H-treated peptides—in many spectra it is more prominent, and often it is the most intense peak.

3.7.1. Preparation of In-House Columns

1. 8 μL of formamide is added to an Eppendorf tube containing 44 μL kasil.
2. The mixture is briefly vortexed, and fused silica capillaries with a suitable inner diameter and length (approx 20 cm) are quickly dipped into the solution.
3. The capillaries are placed in a 110°C oven for 1 h, and the frit length is adjusted by cutting with a ceramic scoring wafer.
4. A slurry of reversed-phase material is suspended in isopropanol and placed in a flat-bottomed vial with a magnetic microstirrer bar.
5. The slurry is placed in the pressure bomb with a magnetic stirrer underneath to prevent the resin from sedimenting. By applying pressure, the liquid is forced through the capillary, and, when the column has reached a suitable length, the pressure is gently removed.
6. The columns are tested using an injection of 20 fmol in-solution digestion of BSA.

3.7.2. LC-MS/MS Procedures

1. An aliquot of 8 to 10 μL sample dissolved in solvent A is loaded onto the precolumn at a flow rate of approx 4 μL/min.
2. The flow path is then altered, and the primary flow is split before the precolumn so that a flow of approx 100 nL/min comes through the analytical column.
3. Peptides are eluted with a linear gradient from 0 to 40% solvent B in 35 min directly into the mass spectrometer followed by a washing and reequilibration step.

4. The mass spectrometer is operated in data-dependent mode whereby multiply charged ions trigger the acquisition of tandem mass spectra of the three most abundant ions.
5. The centroided data are searched against a database with modified asparagine as a variable modification (*see* **Note 9**).

4. Notes

1. The addition of L-cysteine can be omitted, but this often results in a substantial number of peptides having carbamidomethylations of the N-terminal amino group, leading to a mass increase of 57 Daltons. Obviously, the carbamidomethylation of the N-terminus can be accounted for during data analysis by specifying a variable modification in the database search parameters, but we prefer to diminish this side reaction.
2. Digestions in high concentration of urea can often lead to carbamylations of lysine and N-terminal amines—elevated temperatures should be avoided to minimize this side reaction.
3. The addition of cysteine can be skipped, but it tends to reduce the number of peptides that get carbamidomethylated at the N-terminal amine group.
4. When one is redissolving the glycopeptides from a guanidine-HCl in-solution digestion, it is necessary to dilute the chaotrope to concentrations lower than 1 *M* because there is a phase separation at higher concentrations with the HILIC loading solvent.
5. Pefabloc is added to avoid back-exchange of ^{18}O by trypsin whereby the C-terminal carboxylic acid group can incorporate either one or two ^{18}O if there is a slight carryover of trypsin from the HILIC chromatography.
6. Pefabloc is added to avoid digesting the endoglycosidase stock mixture. The glycosidase mixture is often also contaminated with other stabilizing enzymes—we have found both BSA and human tropomyosin in large quantities in the Endo D from ICN. Peptides from these contaminants add unnecessarily to the complexity and can lead to erroneous identifications—we commonly digest and analyze a small aliquot of the enzyme to avoid this.
7. The endoglycosidase stock mixture can be stored in the refrigerator for days (and probably more) and in the freezer for longer periods. Repeated freeze-thaw cycles should be avoided. The core-linked fucose group can be removed by the addition of a fucosidase to the glycosidase mixture. This reduces complexity but at the cost of information on core fucosylation.
8. The DHAP/DHAC matrix is a very cold matrix that reduces in-source fragmentation and metastable decay. The DHAC also serves to suppress cation adduction to sialylated glycopeptides.
9. The composite monoisotopic mass for *N*-linked GlcNAc-asparagine is 317.12 Daltons, for fucosylated *N*-linked GlcNac-asparagine it is 463.18 Daltons, and for ^{18}O-deamidated asparagine it is 117.03 Daltons.

Acknowledgments

The Carlsberg foundation is gratefully acknowledged for supporting J.B. P.H. is supported by a long-term fellowship from the Federation of European Biochemical Societies. O.N.J. is a Lundbeck Foundation Research Professor.

References

1. Anderson NLA. The human plasma proteome. *Mol Cell Proteomics* 2002;1:845–864.
2. Kannagi R, Izawa M, Koike T, Miyazaki K, Kimura N. Carbohydrate-mediated cell adhesion in cancer metastasis and angiogenesis. *Cancer Sci* 2004;95: 377–384.
3. Spiro RG. Protein glycosylation: nature, distribution, enzymatic formation, and disease implications of glycopeptide bonds. *Glycobiology* 2002;12:43R–56R.
4. Hofsteenge J, Muller DR, de Beer T, Loffler A, Richter WJ, Vliegenthart JF. New type of linkage between a carbohydrate and a protein: C-glycosylation of a specific tryptophan residue in human RNase Us. *Biochemistry* 1994;33: 13,524–13,530.
5. Iberg N, Fluckiger R. Nonenzymatic glycosylation of albumin in vivo. Identification of multiple glycosylated sites. *J Biol Chem* 1986;261: 13,542–13,545.
6. Rudd PM, Morgan BP, Wormald MR, et al. The glycosylation of the complement regulatory protein, human erythrocyte CD59. *J Biol Chem* 1997;272:7229–7244.
7. Cummings RD, Kornfeld S. Fractionation of asparagine-linked oligosaccharides by serial lectin-agarose affinity chromatography. A rapid, sensitive, and specific technique. *J Biol Chem* 1982;257:11,235–11,240.
8. Peracaula R, Royle L, Tabares G, et al. Glycosylation of human pancreatic ribonuclease: differences between normal and tumor states. *Glycobiology* 2003; 13:227–244.
9. Zhang H, Li XJ, Martin DB, Aebersold R. Identification and quantification of N-linked glycoproteins using hydrazide chemistry, stable isotope labeling and mass spectrometry. *Nat Biotechnol* 2003;21:660–666.
10. Zhang H, Yi EC, Li XJ, et al. High throughput quantitative analysis of serum proteins using glycopeptide capture and liquid chromatography mass spectrometry. *Mol Cell Proteomics* 2005;4:144–155.
11. Li Y, Larsson EL, Jungvid H, Galaev I, Mattiasson B. Affinity chromatography of neoglycoproteins. *Bioseparation* 2000;9:315–323.
12. Khidekel N, Ficarro SB, Peters EC, Hsieh-Wilson LC. Exploring the O-GlcNAc proteome: direct identification of O-GlcNAc-modified proteins from the brain. *Proc Natl Acad Sci U S A* 2004;101:13,132–13,137.
13. Alpert AJ. Hydrophilic-interaction chromatography for the separation of peptides, nucleic acids and other polar compounds. *J Chromatogr A* 1990;499:177–196.
14. Alpert AJ, Shukla M, Shukla AK, et al. Hydrophilic-interaction chromatography of complex carbohydrates. *J Chromatogr A* 1994;676:191–222.
15. Shimizu Y, Nakata M, Kuroda Y, Tsutsumi F, Kojima N, Mizuochi T. Rapid and simple preparation of N-linked oligosaccharides by cellulose-column chromatography. *Carbohydr Res* 2001;332:381–388.
16. Hagglund P, Bunkenborg J, Elortza F, Jensen ON, Roepstorff P. A new strategy for identification of N-glycosylated proteins and unambiguous assignment of their glycosylation sites using HILIC enrichment and partial deglycosylation. *J Proteome Res* 2004;3:556–566.

17. Wada Y, Tajiri M, Yoshida S. Hydrophilic affinity isolation and MALDI multiple-stage tandem mass spectrometry of glycopeptides for glycoproteomics. *Anal Chem* 2004;76:6560–6565.
18. An HJ, Peavy TR, Hedrick JL, Lebrilla CB. Determination of N-glycosylation sites and site heterogeneity in glycoproteins. *Anal Chem* 2003;75:5628–5637.
19. Larsen MR, Hojrup P, Roepstorff P. Characterization of gel-separated glycoproteins using two-step proteolytic digestion combined with sequential micro-columns and mass spectrometry. *Mol Cell Proteomics* 2005;4:107–119.
20. Wells L, Vosseller K, Cole RN, Cronshaw JM, Matunis MJ, Hart GW. Mapping sites of O-GlcNAc modification using affinity tags for serine and threonine post-translational modifications. *Mol Cell Proteomics* 2002;1:791–804.
21. Robinson NE, Robinson AB. Deamidation of human proteins. *Proc Natl Acad Sci USA* 2001;98:12,409–12,413.
22. Gonzalez J, Takao T, Hori H, et al. A method for determination of N-glycosylation sites in glycoproteins by collision-induced dissociation analysis in fast atom bombardment mass spectrometry: identification of the positions of carbohydrate-linked asparagine in recombinant alpha-amylase by treatment with peptide-N-glycosidase F in ^{18}O-labeled water. *Anal Biochem* 1992;205:151–158.
23. Kuster B, Mann M. ^{18}O-labeling of N-glycosylation sites to improve the identification of gel-separated glycoproteins using peptide mass mapping and database searching. *Anal Chem* 1999;71:1431–1440.
24. Jebanathirajah J, Steen H, Roepstorff P. Using optimized collision energies and high resolution, high accuracy fragment ion selection to improve glycopeptide detection by precursor ion scanning. *J Am Soc Mass Spectrom* 2003;14:777–784.
25. Koide N, Muramatsu T. Endo-beta-N-acetylglucosaminidase acting on carbohydrate moieties of glycoproteins. Purification and properties of the enzyme from Diplococcus pneumoniae. *J Biol Chem* 1974;249:4897–4904.
26. Tarentino AL, Plummer TH Jr, Maley F. The release of intact oligosaccharides from specific glycoproteins by endo-beta-N-acetylglucosaminidase H. *J Biol Chem* 1974;249:818–824.
27. Iwase H, Hotta K. Release of O-linked glycoprotein glycans by endo-alpha-N-acetylgalactosaminidase. *Methods Mol Biol* 1993;14:151–159.
28. Greis KD, Hayes BK, Comer FI, et al. Selective detection and site-analysis of O-GlcNAc-modified glycopeptides by beta-elimination and tandem electrospray mass spectrometry. *Anal Biochem* 1996;234:38–49.
29. Vosseller K, Hansen KC, Chalkley RJ, et al. Quantitative analysis of both protein expression and serine/threonine post-translational modifications through stable isotope labeling with dithiothreitol. *Proteomics* 2005;5:388–398.
30. Edge ASB. Deglycosylation of glycoproteins with trifluoromethanesulphonic acid: elucidation of molecular structure and function. *Biochem J* 2003;376:339–350.
31. Elortza F, Nuhse TS, Foster LJ, Stensballe A, Peck SC, Jensen ON. proteomic analysis of glycosylphosphatidylinositol-anchored membrane proteins. *Mol Cell Proteomics* 2003;2:1261–1270.

32. Mirgorodskaya E, Roepstorff P, Zubarev RA. Localization of O-glycosylation sites in peptides by electron capture dissociation in a Fourier transform mass spectrometer. *Anal Chem* 1999;71:4431–4436.
33. Hakansson K, Cooper HJ, Emmett MR, Costello CE, Marshall AG, Nilsson CL. Electron capture dissociation and infrared multiphoton dissociation MS/MS of an N-glycosylated tryptic peptic to yield complementary sequence information. *Anal Chem* 2001;73:4530–4536.
34. Hogan JM, Pitteri SJ, Chrisman PA, McLuckey SA. Complementary structural information from a tryptic N-linked glycopeptide via electron transfer ion/ion reactions and collision-induced dissociation. *J Proteome Res* 2005;4:628–632.
35. Bunkenborg J, Pilch BJ, Podtelejnikov AV, Wisniewski JR. Screening for N-glycosylated proteins by liquid chromatography mass spectrometry. *Proteomics* 2004;4:454–465.
36. Xiong L, Regnier FE. Use of a lectin affinity selector in the search for unusual glycosylation in proteomics. *J Chromatogr B Analyt Technol Biomed Life Sci* 2002;782:405–418.
37. Yang Z, Hancock WS. Approach to the comprehensive analysis of glycoproteins isolated from human serum using a multi-lectin affinity column. *J Chromatogr A* 2004;1053:79–88.
38. Hojrup P. Proteolytic peptide mapping. *Methods Mol Biol* 2004;251:227–244.
39. Gobom J, Nordhoff E, Mirgorodskaya E, Ekman R, Roepstorff P. Sample purification and preparation technique based on nano-scale reversed-phase columns for the sensitive analysis of complex peptide mixtures by matrix-assisted laser desorption/ionization mass spectrometry. *J Mass Spectrom* 1999;34:105–116.
40. Larsen MR, Cordwell SJ, Roepstorff P. Graphite powder as an alternative or supplement to reversed-phase material for desalting and concentration of peptide mixtures prior to matrix-assisted laser desorption/ionization-mass spectrometry. *Proteomics* 2002;2:1277–1287.
41. Pitt JJ, Gorman JJ. Matrix-assisted laser desorption/ionization time-of-flight mass spectrometry of sialylated glycopeptides and proteins using 2,6-dihydroxyacetophenone as a matrix. *Rapid Commun Mass Spectrom* 1996;10:1786–1788.

6

Plasma Proteome Database

Malabika Sarker, G. Hanumanthu, and Akhilesh Pandey

Summary

The plasma proteome is one of the most important proteomes from a diagnostic standpoint, as quantitative changes between normal and diseased states can be used to discover novel biomarkers for clinical diagnosis and therapeutic monitoring of diseases. HUPO's Plasma Proteome Project has involved characterization of the plasma proteome by an international consortium using technologies ranging from antibody arrays to mass spectrometry and has helped in the construction of a substantial catalog of plasma proteins. Our group has developed the Plasma Proteome Database (http://www.plasmaproteomedatabase.org) as a comprehensive resource of annotated data for posttranslational modifications, single-nucleotide polymorphisms, tissue expression, subcellular localization, and disease involvement of plasma proteins. This database should serve to facilitate further research to help understand the plasma proteome in health and disease.

Key Words: Proteomics; annotation; bioinformatics; database; plasma; mass spectrometry.

1. Introduction

Proteins constitute approximately 7% of the human plasma, which is perhaps the most complex human proteome, as blood circulates to all organs and organ systems in the body. True plasma proteins can be considered as those proteins that carry out their function in the circulation, excluding those that leak into the plasma because of cell or tissue damage *(1)*. Anderson et al. *(2)* have classified plasma proteins as follows: proteins secreted by solid tissues, immunoglobulins, receptor ligands, temporary passengers, tissue leakage products, aberrant secretions, and foreign proteins. The liver secretes most of the classical plasma proteins. Leakage of proteins from tissues can occur as a result of cell death or damage, as is the case of diagnostic markers such as cardiac troponins, creatine kinase, and myoglobin. Aberrant secretions involve cancer markers that are released from tumors and other diseased tissues.

From: *Proteomics of Human Body Fluids: Principles, Methods, and Applications*
Edited by: V. Thongboonkerd © Humana Press Inc., Totowa, NJ

The number of protein products that can be identified in the plasma proteome is much higher than the expected number based on the number of genes because of different sizes (precursor forms, degradation products, and splice variants) and posttranslational modifications (PTMs) (e.g., glycosylation, proteolytic cleavage). The protein profile of plasma is dominated by a subset of abundant proteins including albumin (35–50 mg/mL) and immunoglobulins (5–18 mg/mL). Low-abundance proteins such as tissue leakage products (pg/mL to mg/mL concentration range) and cytokines (pg/mL concentration) are difficult to identify by proteomic methods without prior depletion or fractionation steps. Changes in concentration of plasma proteins can provide information about disease processes. Among the abundant proteins, serum albumin is clinically used as an indication of severe liver disease or malnutrition. Interleukin-6, on the other hand, is a low-abundance protein that is a sensitive indicator of inflammation or infection ***(2)***. A recent study showed the feasibility of using quantitative proteomic approaches for identification of proteins differentially expressed between different states (i.e., normal vs diseased states) ***(3)***.

2. Plasma Proteome in Diagnostics

The human plasma proteome is an excellent body fluid proteome for diagnosis of diseases as well as for therapeutic monitoring. Other body fluids including cerebrospinal fluid and synovial fluid share some of the protein repertoire with plasma but are generally more difficult to obtain and/or process quickly in a clinical setting. **Table 1** shows some of the diseases associated with the plasma proteins (data taken from http://www.plasmaproteomedatabase.org). One major limitation for having proteomic-based diagnostic tools for regular clinical testing has been the absence of an informative proteomic map, similar to that of the human genome. In this regard, Ping et al. ***(4)*** described a community effort to annotate a list of proteins identified by the Human Plasma Proteome Project (HPPP) and classified subproteomes in the human plasma with potential relevance to cardiovascular and liver disease, DNA binding, coagulation, and mononuclear phagocytosis.

3. Plasma Proteome Database as a Web Resource for Plasma Proteins

The Human Plasma Proteome Organization (HUPO; http://www.hupo.org) was established in 2001 to help in defining and prioritizing objectives to enhance our understanding of the human proteome and to organize international collaborations in proteomic research. HUPO's Plasma Proteome Project involved participation by a number of laboratories around the world in characterizing the plasma proteome using technologies ranging from antibody arrays to mass spectrometry ***(5–6)***. As part of this initiative, our group developed the Plasma Proteome Database (PPD; http://www.plasmaproteomedatabase.org) as a comprehensive web-based resource of the annotated data including PTMs, protein isoforms, tissue expression, and

Table 1
Association of Plasma Proteins With Human Diseases

Protein name	Associated inherited diseases
Prion protein	Creutzfeldt-Jakob disease; Gerstmann-Straussler syndrome; fatal familial insomnia
Insulin	Hyperproinsulinemia; diabetes mellitus
Proopiomelanocortin	Adrenal insufficiency
Protein C	Hereditary thrombophilia; venous thrombosis; arterial thromboembolic disease; protein C deficiency
Protein S, α	Protein S deficiency
Recombination activating gene 1	Severe combined immunodeficiency; Omenn syndrome
Retinoblastoma 1	Retinoblastoma
Retinol-binding protein 4	Retinol-binding protein deficiency
Rhodopsin	Retinitis pigmentosa; congenital stationary night blindness
Rhodopsin kinase	Oguchi disease
Ribonuclease L	Prostate cancer
Rod outer segment membrane protein 1	Retinitis pigmentosa
Ryanodine receptor 1	Malignant hyperthermia; central core disease
Ryanodine receptor 2	Arrhythmogenic right ventricular cardiomyopathy type 2 (ARVD2); polymorphic ventricular tachycardia
Arrestin	Oguchi disease
Solute carrier family 6 (neurotransmitter transporter, serotonin), member 4	Long/short promoter polymorphism
Solute carrier family 5, member 1	Glucose/galactose malabsorption
Sodium channel, neuronal type I, α-subunit	Epilepsy; severe myoclonic epilepsy of infancy (SMEI)
Somatostatin receptor 5	Acromegaly
Spectrin, α-erythrocytic 1	Hereditary elliptocytosis; hereditary pyropoikilocytosis; hereditary spherocytosis
Spectrin, β I	Hereditary spherocytosis
Matrix metalloproteinase 3	Coronary atherosclerosis
Succinate dehydrogenase complex, subunit B, iron sulfur protein	Pheochromocytoma; paraganglioma
Thyroglobulin	Simple goiter
Transforming growth factor-1β	Camurati-Engelmann disease
Triosephosphate isomerase 1	Triosephosphate isomerase deficiency
Platelet-derived growth factor-β	Meningioma; dermatofibrosarcoma protuberans
CCL5	Human immunodeficiency virus type 1 disease progression

Table 2
Summary of Entries in the Plasma Proteome Database (PPD)

Entry	No.
Unique genes whose products are annotated in PPD	3778
Isoforms	7614
Proteins showing isoform specific localization	100
Proteins with signal sequences	1081
Proteins with isoform-specific expression	222
Posttranslational modifications	5234
Type	
Phosphorylation	2429
Glycosylation	1026
Proteolytic cleavage	709

disease involvement of plasma proteins *(7)*. **Table 2** gives overall statistics of the protein entries in the PPD. An important aspect of this database is the annotation of all protein isoforms from the literature and clustering them on the basis of the gene of origin. The data from the HPPP were deposited into the PRIDE database (http://www.ebi.ac.uk/pride), which is a central repository for peptide data *(8)*. Another source of peptide data is the Human Plasma Peptide Atlas from 28 liquid chromatography-tandem mass spectrometry (LC-MS/MS) experiments that can be browsed using the EnsEMBL genome browser *(9)*.

Proteins in the PPD can be accessed using the query or browse options. The query page allows searching based on database accession numbers in addition to other features of proteins such as size or disease involvement. The browse page provides access to proteins based on their function, protein domain architecture, motifs, or PTMs. The entire dataset can be downloaded as an XML file.

4. Annotation of Plasma Proteins in the PPD

The PPD includes proteins identified on the basis of two or more peptides by the HPPP *(4,5)* as well as those that have been described in the literature to be present in plasma *(6–7,10–15)*. One of the most important aspects of the PPD is the annotation of protein isoforms. We have considered two different classes of isoforms: those arising from alternative splicing or from sequence changes (e.g., mutations). One example of a protein isoform is the EYA4 gene (http://plasmaproteomedatabase.org/protein/HPRD_04648/molecule_page), in which the deletion of 4846 bp causes the frame shift after 193 amino acids and introduces 29 new residues and a premature stop codon. This particular deletion causes cardiac myopathy in humans *(16)*. Although we were able to obtain some isoforms from sequence databases, we had to resort to the primary literature to reconstruct a total of 107 protein isoforms derived from 60 unique genes.

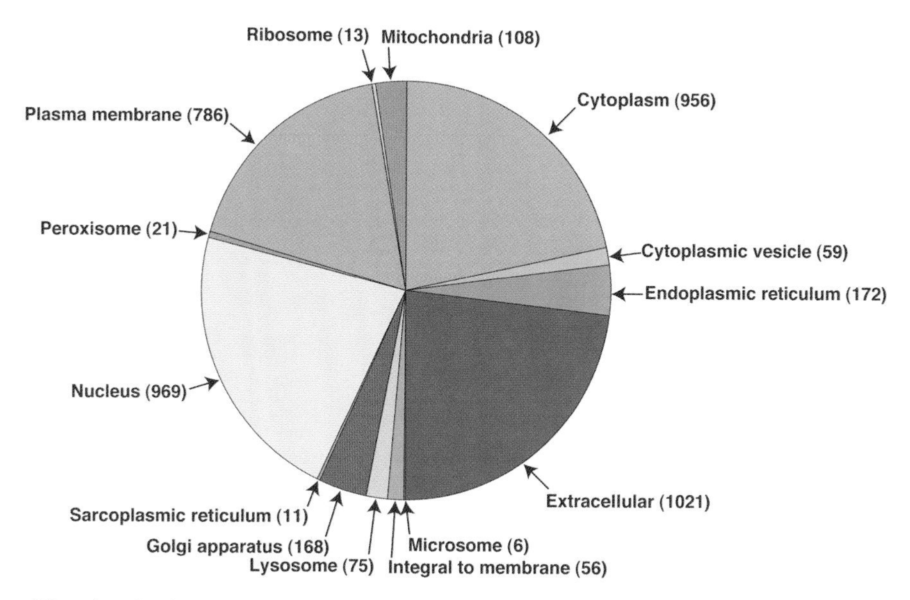

Fig. 1. Distribution of subcellular localization of plasma proteins in the Plasma Proteome Database.

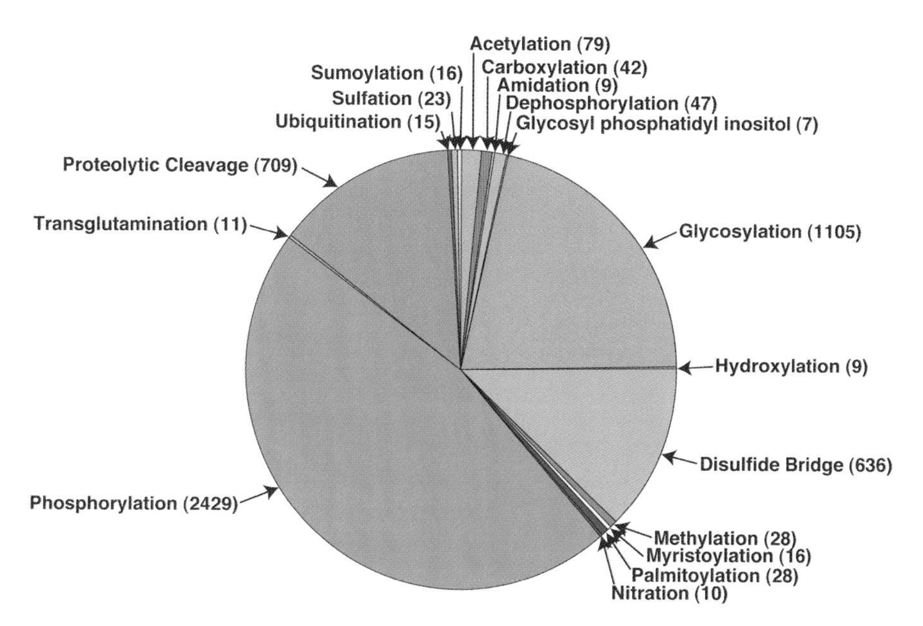

Fig. 2. Distribution of posttranslational modifications of plasma proteins in the Plasma Proteome Database.

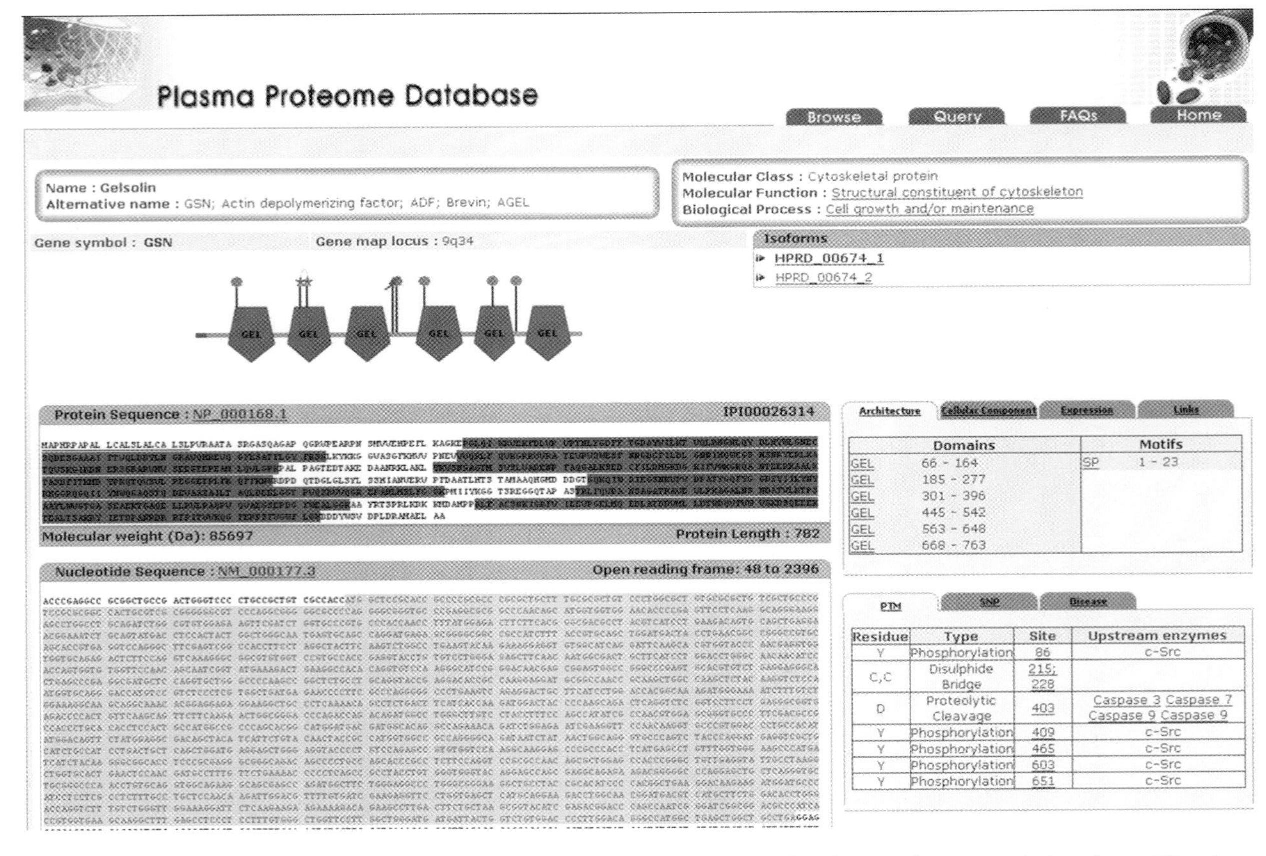

Fig. 3. A screenshot of the protein gelsolin, highlighting isoforms, localization, and posttranslational modifications.

Each isoform was annotated with single-nucleotide polymorphisms (SNPs), PTMs, and tissue expression data. Subcellular localization, function, and process were annotated according to gene ontology terms. **Figures 1** and **2** show the distribution of different subcellular localization and PTMs of plasma proteins in the PPD, respectively. Some of the isoforms of the proteins specifically localize to particular subcellular organelles. For example, gelsolin protein (http://www.plasmaproteomedatabase.org/protein/HPRD_00674/molecule_page) has two isoforms that differ in their N-termini, having an extra 21 amino acids in the first isoform (which codes for a signal peptide and localizes it as an extracellular protein), whereas the other isoform lacks the signal peptide and remains in the cytoplasm **(Fig. 3)**. Gelsolin is an actin-binding protein that severs and caps actin filaments and binds actin monomers to nucleate new filaments in a calcium-dependent manner. Expression of plasma gelsolin in an intracellular reducing compartment, such as the cytoplasm, causes the plasma gelsolin to lack the disulfide bond, which induces a conformation such that the severing rate is several folds slower than that of oxidized gelsolin ***(17)***. Isoform-specific subcellular localization is annotated for 100 proteins in the PPD. In addition, this database also contains information on isoform-specific expression for more than 200 proteins.

All PTMs in PPD were annotated from the literature. Protein architecture was annotated by giving first preference to the literature, and then to SMART (http://smart.embl-heidelberg.de/) and other prediction tools such as PSORT II (http://psort.nibb.ac.jp/form2.html) and TMHMM (http://www.cbs.dtu.dk/services/TMHMM/). Coding SNPs were mapped onto the nucleotide and protein sequences.

Overall, we anticipate that the PPD will become a resource for information about plasma proteins for researchers and clinicians alike.

Acknowledgments

The authors thank all the researchers at the Institute of Bioinformatics for their active participation in the Plasma Proteome Database project.

References

1. Putnam FW. In: Putnam FW, ed. *The Plasma Proteins: Structure, Function, and Genetic Control*. New York, Academic Press, 1975–1987:1–55.
2. Anderson NL, Anderson NG. The human plasma proteome: history, character, and diagnostic prospects. *Mol Cell Proteomics* 2002;11:845–867.
3. Qian WJ, Jacobs JM, Camp DG 2nd, et al. Comparative proteome analyses of human plasma following in vivo lipopolysaccharide administration using multidimensional separations coupled with tandem mass spectrometry. *Proteomics* 2005;5:572–584.
4. Ping P, Vondriska TM, Creighton CJ, et al. A functional annotation of subproteomes in human plasma. *Proteomics* 2005;5:3506–3519.

5. Rai AJ, Gelfand CA, Haywood BC, et al. HUPO Plasma Proteome Project specimen collection and handling: towards the standardization of parameters for plasma proteome samples. *Proteomics* 2005;5:3262–3277.
6. Omenn GS, States DJ, Adamski M, et al. Overview of the HUPO Plasma Proteome Project: results from the pilot phase with 35 collaborating laboratories and multiple analytical groups, generating a core dataset of 3020 proteins and a publicly-available database. *Proteomics* 2005;5:3226–3245.
7. Muthusamy B, Hanumanthu G, Suresh S, et al. Plasma Proteome Database as a resource for proteomics research. *Proteomics* 2005;5:3531–3536.
8. Martens L, Hermjakob H, Jones P, et al. PRIDE: the proteomics identifications database. *Proteomics* 2005;5:3537–3545.
9. Deutsch EW, Eng JK, Zhang H, et al. Human Plasma PeptideAtlas. *Proteomics* 2005;5:3497–3500.
10. Chan KC, Lucas DA, Hise D, et al. Analysis of the human serum proteome. *Clin Proteomics* 2004;1:101–225.
11. Anderson NL, Polanski M, Pieper R, et al. the human plasma proteome: a nonredundant list developed by combination of four separate sources. *Mol Cell Proteomics* 2004;3:311–326.
12. Shen Y, Jacobs JM, Camp DG, et al. Ultra-high-efficiency strong cation exchange LC/RPLC/MS/MS for high dynamic range characterization of the human plasma proteome. *Anal Chem* 2004;76:1134–1144.
13. Adamski M, Blackwell T, Menon R, et al. Data management and preliminary data analysis in the pilot phase of the HUPO Plasma Proteome Project. *Proteomics* 2005;5:3246–3261.
14. Peri S, Navarro JD, Amanchy R, et al. Development of human protein reference database as an initial platform for approaching systems biology in humans. *Genome Res* 2003;13:2363–2371.
15. Hanash S. Building a foundation for the human proteome: the role of the Human Proteome Organization. *J Proteome Res* 2004;3:197–199.
16. Schonberger J, Wang L, Shin JT, et al. Mutation in the transcriptional coactivator EYA4 causes dilated cardiomyopathy and sensorineural hearing loss. *Nat Genet* 2005;37:418–422.
17. Allen PG. Functional consequences of disulfide bond formation in gelsolin. *FEBS Lett* 1997;401:89–94.

7

2D PAGE Databases for Proteins in Human Body Fluids

Christine Hoogland, Khaled Mostaguir, Jean-Charles Sanchez, Denis F. Hochstrasser, and Ron D. Appel

Summary

With the development of the Internet, a growing number of proteomics databases have become available. The web is a powerful tool for data integration because it links the components constituting these databases, in general gel images and protein information, while offering rapid means to navigate from one database to another. Unfortunately, with only 15 maps available as electronic resources, the human body fluids do not really benefit from this development. This chapter summarizes the state of the art of proteomics databases, with an emphasis on human body fluids. Insights into one of these databases, SWISS-2DPAGE, available for more than 10 yr now, are given to show current functionalities and usage examples. Some general thoughts are also given on how to improve sharing and publication of proteomics data through electronic media.

Key Words: Proteomics database; web services; human body fluids.

1. Introduction

In the postgenomic era, there has been increasing interest in human body fluids as outstanding sources of circulating biological markers. In particular, the identification and characterization of proteins in human body fluids and the interpretation of their relationships with diseases have promoted proteomics to the rank of the technology of choice for biomedical applications. Many laboratories have identified proteins on 2D electrophoresis (2-DE) maps of human body fluids, either to build a reference gel or as a source of potential biological disease markers. The present book is certainly a good proof (*see* the 16 chapters of Part II, each dedicated to a specific body fluid). Another example is given by a simple search in the PubMed resource *(1)*. With appropriate key words, around 50 pertinent articles published during the last 2 yr can be retrieved in which

From: *Proteomics of Human Body Fluids: Principles, Methods, and Applications*
Edited by: V. Thongboonkerd © Humana Press Inc., Totowa, NJ

authors characterize proteomes of various human body fluids, not only to catalog proteins but also to compare their expression levels in different diseases. Unfortunately, very few of these data are available as electronic resources. In 2005, up to 50 2-DE databases were freely accessible on the Internet, for a total of nearly 300 annotated image maps (*see* http://www.expasy.org/ch2d/2d-index.html for an updated list *[2]*). However, less than 15 annotated image maps concern human body fluid samples (**Table 1**), with the most analyzed being plasma, urine and cerebrospinal fluid.

2. 2-DE Databases as Electronic Resources

The elements constituting a 2-DE database, i.e., gel images and protein information, can be conveniently integrated into a web database. Thanks to the active hypertext links, the user has a powerful tool for data integration, in addition to the opportunity to navigate from one database to another *(3)*. Images are scanned representations of the gels provided with annotations. These represent general information about the identified protein (such as protein name, gene name, taxonomy, references) and experimental data, including the physical localization on the gel (i.e., isoelectric point and molecular weight) and the protein identification method. In addition, cross-references to other biological databases might be offered in some cases. These databases can be accessed in several ways, including both textual and graphical searches. SWISS-2DPAGE was the first federated 2-DE database available as an electronic resource *(4)*.

2.1. A Detailed Example: The SWISS-2DPAGE Database

The SWISS-2DPAGE database started in 1993 and is still maintained as a collaborative work between the Swiss Institute of Bioinformatics and the Biomedical Proteomics Research Group of Geneva University. The first releases contained 10 human samples, i.e., plasma and cerebrospinal fluid (the others being human cells, tissues, or culture cell lines) *(4)*. At that time, the plasma map totalized 20% of the total accesses to the SWISS-2DPAGE database through the ExPASy server *(5)*. Ten years later, SWISS-2DPAGE contains 36 reference maps from various origins (seven species) *(6)*. Nonetheless, the plasma map is still one of the most accessed, with approximately 30% of the total hits. The major feature of the SWISS-2DPAGE database, still with no equivalent today, is the high level of annotations accessible either in the given protein information (mapping methods, experimental data, and so on) or as supplementary materials (protocols, release notes with validation criteria and statistical resume, and so on).

2.1.1. Annotation

SWISS-2DPAGE provides many annotations at various levels. The first level consists of general information about the protein, its name, function, taxonomy

Table 1
Electronic 2-DE Databases Containing Data from Human Body Fluids in 2005

Database name	Body fluid type[a]	Website URL	Ref. no.
BALF2D	Bronchoalveolar lavage fluid (–)	http://www.umh.ac.be/~biochim/BALF2D.html	*15*
HUPO Plasma Proteome Project	Plasma (+)	http://www.hupo.org/ http://www.bioinformatics.med.umich.edu/hupo/ppp/	*16*
Inner Ear Protein Database	Inner ear fluid (cochlear perilymph) (+)	http://oto.wustl.edu/thc/	*17*
ISPA gel gallery	Milk fat globule (+)	http://www.csaapz.to.cnr.it/proteoma/2DE/	*18*
LeeLab CSF	Cerebrospinal fluid (+/–)	http://www.leelab.org/csfmap/	*19*
LECB gel gallery	Serum (–), urine[b]	http://www.lecb.ncifcrf.gov/2DgelDataSets/	*20*
Proteomics Danish center	Urine (–)[b]	http://proteomics.cancer.dk/	*21*
SIENA-2DPAGE	Amniotic fluid (+)	http://www.bio-mol.unisi.it/2d/2d.html	*22*
SWISS-2DPAGE	Plasma (+), cerebrospinal fluid (+)	http://www.expasy.org/ch2d/	*6*
UAB database	Urine (–)[b]	http://www.uab.edu/proteinmenu	*23*
UCHSC gel gallery	Seminal fluid (+)[b]	http://proteomics.uchsc.edu/2Dexample/	*24*

[a](+), control; (–), disease.
[b]Not available at the time of writing.

origin, quaternary structure, posttranslational modifications, and so on. The second level relates to gel electrophoresis and spot identification. For each protein that has been experimentally identified in one or many gels, the information gathered includes its physical location on the gel (isoelectric point [p*I*]/molecular weight [Mw]), its unique spot serial number, and the identification method (i.e., gel matching, amino acid composition, immunoblotting, microsequencing, peptide mass fingerprinting, tandem mass spectrometry, or a combination of these) together with the experimental data itself (currently the amino acid composition in percent, the peptide masses that allowed the protein identification, and the peptide sequences identified from tandem mass spectrometry spectra). Annotations regarding protein expression levels, either normal, pathological, or after a treatment, as well as descriptions of variant polymorphisms, either physiological or related to disease, are also documented in some cases. A certain degree of data integration, the third level of annotations available in SWISS-2DPAGE, comes from database cross-references. Instead of gathering general or specific information existing in other resources, a choice has been made to provide pointers to this information directly in its original database, leading to a better reliability. Currently, the SWISS-2DPAGE links to about 10 different databases, either sequence related (i.e., the UniProt Knowledgebase *[7]*) or other 2-DE collections (COMPLUYEAST-2DPAGE, HSC-2DPAGE, LENS-2DPAGE, OGP-WWW, PHCI-2DPAGE, PMMA-2DPAGE, Siena-2DPAGE, and others). Last but not least, technical information related to proteomics experiments (protocols, chemicals, apparatus, identification software, and validation criteria, and so on), to database structure schema, and to the current content (number of gels, number of identified spots and proteins, species distribution, and so on) are supported by separate documents. These can be viewed either in the contextual help accessible when browsing the data or as full material for a better comprehension of the database content. Finally, in all cases, relevant literature references are provided, with full author lists, titles, and citations, accompanied by the PubMed identifier for further access to the article abstract.

2.1.2. Availability

Since the beginning, SWISS-2DPAGE was hosted by the ExPASy website (http://www.expasy.org/ch2d/), and users could explore the data with various search engines, either by clicking on an image map or by textual queries. Typical browsing may start in different ways:

1. The simplest approach, well suited for gel exploration, is to navigate on the reference map. After choosing the gel of interest (from http://www.expasy.org/cgi-bin/map1), the user can see an image with identified spots marked by crosses, which facilitates their localizations. (Alternatively, a raw image is also available.)

Moving the mouse over the spots of interest displays a summary of the experimental information in a floating box. To access the full data content related to a specific spot, one could just click on it.

2. The second approach toward interaction with the SWISS-2DPAGE database is through textual queries. Many search engines are provided from the top page (http://www.expasy.org/ch2d/). If the user is aware of identifiers (either protein accession number or spot number) for a spot or protein of interest, he/she can go directly to the corresponding database entry. However, very often, the user has only very partial knowledge about the protein(s) for which he/she wants to have details. In this case, the web interface offers searches either by protein description, gene name, UniProtKB/Swiss-Prot keyword, author name, identification methods, p*I*/Mw range, or a combination of all these. In any case, the result consists of a protein list matching the querying criteria with access to the detailed information for each protein entry.
3. The third way to browse the SWISS-2DPAGE data is to search out the protein list for the reference gel of interest (http://www.expasy.org/cgi-bin/get-ch2d-table.pl) and then to click on the specific protein entry to display the full information.

At the end of all possible ways of querying the database, the user will have the detailed information for a specific protein **(Fig. 1)**, in a friendly view, with all the annotations as described in the section above (*see* **Subheading 2.1.1.**). In addition, this view links to all the maps available in the database, on which the protein has been identified or not. When one clicks on the icons representing these maps, the entire gel images are displayed in which all the spots identified for that protein are highlighted. It displays a rectangle showing an estimated region where the protein is expected to migrate, according to its amino acid composition. Cross-references to external resources (PubMed, UniProt Knowledgebase, other 2-DE databases) are provided as hypertext links.

For people interested in having the SWISS-2DPAGE database locally, annotations and image files can be downloaded from the ExPASy FTP server (ftp://www.expasy.org/ftp/databases/swiss-2dpage/). Either through the website or as a download, the SWISS-2DPAGE database is free for nonprofit institutions. It has an annual subscription fee for commercial entities (for details, *see* http://www.genebio.com/).

2.1.3. Usage

With more than 1800 hits per day, the database is commonly used for preliminary investigations. Usually, users rely on the SWISS-2DPAGE reference maps annotations to start proteomics studies, ranging from biomarker discovery projects, to comparative studies, to theoretical analysis. A biomarker discovery project or a comparative study typically starts with a 2D gel electrophoresis, which is then compared with a reference map, established for a similar or related

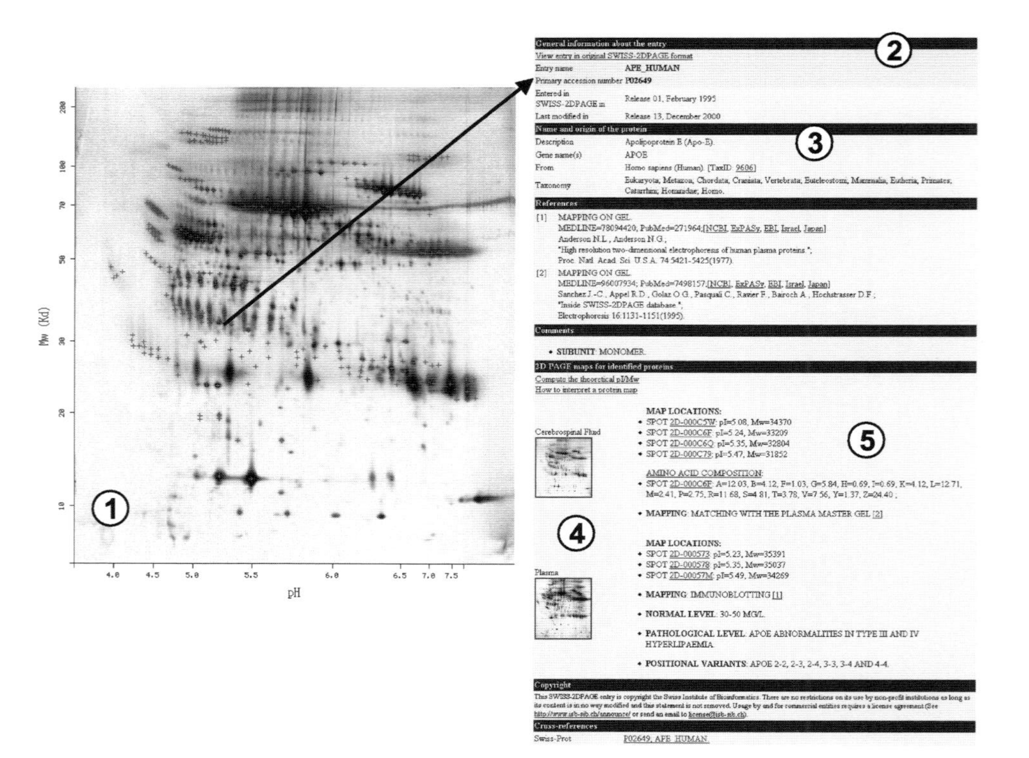

Fig. 1. Following a graphical search, the NiceView of a SWISS-2DPAGE entry shows its content in a user-friendly view. 1, Graphical search on the human cerebrospinal fluid reference map. Crosses show identified proteins, here nearly 300 spots. 2, NiceView of a SWISS-2DPAGE entry. 3, General information describing the protein entry (name, origin, references), provided with links to specific online databases such as PubMed or NCBI's taxonomic classification. 4, Icons of reference maps available for the protein shown. 5, Annotations including here experimental p*I*/Mw, amino acid composition, mapping methods, pathological and polymorphism information, and so on.

sample. When the reference map is well covered, with many identified spots, the comparative analysis with adequate software (e.g., ImageMaster™ 2D Platinum, developed by the Swiss Institute of Bioinformatics and commercialized by GeneBio and GE Healthcare) may reduce the number of spots that need further analysis. Such image analysis software is also able to highlight spots of interest that are over- or underexpressed, thanks to appropriate statistical differential analysis. With nearly 4000 identified spots on the 36 reference maps, SWISS-2DPAGE is widely used for such purposes.

As an example, Wattiez et al. ***(8)*** used the SWISS-2DPAGE human plasma reference map in a preliminary matching analysis to identify a number of proteins in their bronchoalveolar lavage fluid (BALF) 2-DE gel. They were able to identify 19 of 69 proteins of interest by matching the plasma and the BALF samples. The authors restricted the postseparation analysis (in this case microsequencing) to the only 50 remaining proteins. It was a first step to a better understanding of the normal BALF sample, which then allowed these researchers to analyze differential protein expression further between different lung pathological samples. In another example, Pietrogrande et al. ***(9)*** tested a mathematical method to extract information from 2-DE gels of complex protein mixture. The authors compared their simulated gels with three human reference maps from the SWISS-2DPAGE database (PLASMA_HUMAN, HEPG2_ HUMAN, DLD1_HUMAN) and thus confirmed their model hypothesis. As a third example (not from human body fluid samples but easily transferrable), Martens and co-workers ***(10)*** compared their human platelet protein list obtained with nongel technology with protein lists published earlier using gel technology, including the SWISS-2DPAGE platelet proteins. The low overlap (40% so far) observed between both technologies highlighted the advantage of nongel technology to identify proteins in complex mixtures even with low concentrations.

2.2. Other Useful Resources

Since 1993, a growing number of 2-DE images have been made publicly available, thanks to the web. Unfortunately, most of them are limited to one image and a protein list, thus not making the most of the Internet properties (data integration and navigation). With less than 20 image maps electronically available, temporarily or not, in the last 12 years, proteomics data from human body fluids are definitive examples of data regrettably lost or not fully exploited. Although bioinformatics tools have been developed to help scientists build their own database servers, it is clear that it is still not easy to do for institutes with limited access to the necessary infrastructure. These tools have similar functionalities. Starting from at least one image map with identified spots, they provide facilities to build a web server that allow easy browsing and searching sets of 2-DE data. Make2D-DB (http://www.expasy.org/ch2d/make2ddb.html), developed by the Swiss Institute of Bioinformatics, is an integrated tool that allows anyone to get a SWISS-2DPAGE-like database ***(11)***. PROTICdb has been designed especially for plant samples (http://moulon.inra.fr/~bioinfo/PROTICdb). The French National Institute for Agricultural Research laboratories use it ***(12)***. Both tools are freely available for download and offer useful features to import new data and to browse the data by keyword or graphical searches. Public repositories, in which users can upload their own 2-DE data through a web form, are not yet currently used. So far, two such repositories are available through the Internet, namely, GELBANK (http://gelbank.anl.gov/) from the Argonne National Lab ***(13)*** and GelBank (http://www.gelscape.ualberta.ca:8080/htm/gdbIndex.html) from the University of Alberta ***(14)***. Both systems are proposed as public archives for 2-DE image maps and thus are open to host any data. However, the current data available are still limited to their own local or test samples.

Further measures need to be taken if scientists really want to build a large proteomics data network, which would be one of the objectives of proteomics. In this perspective, in late 2005 the Proteome Informatics group at the Swiss Institute of Bioinformatics started to provide the scientific community with some bioinformatics services related to proteomics. In particular, the group proposes various solutions to build proteomics electronic databases, ranging from assistance in the setup to data preparation for further remote installation, up to full data preparation and website hosting (http://www.expasy.org/ch2d/service/). We hope that with this contribution, more and more institutes will be able to give access to their proteomics data and thus promote this global virtual database that we all have dreamed of.

3. Conclusions

Even though bioinformatics have been developed to help scientists publish their own proteomics data on the web, giving access to a large proteomics data network, one must admit that we are still far from this reality. Assistance from bioinformatics groups should be reinforced to provide scientists with new tools. Some of the available tools already contain functionalities for better data reliability and consistency, and then integration of different "omics" will be the next step to a further understanding of systems biology.

Acknowledgments

Our thanks go to Patricia M. Palagi, who read the manuscript and provided valuable comments.

References

1. Wheeler DL, Barrett, T, Benson DA, et al. Database resources of the National Center for Biotechnology Information. *Nucleic Acids Res* 2005;33:D39–D45.
2. Hoogland C, Sanchez JC, Walther D, et al. Two-dimensional electrophoresis resources available from ExPASy. *Electrophoresis* 1999;20:3568–3571.
3. Appel RD, Bairoch A, Sanchez JC, et al. Federated two-dimensional electrophoresis database: a simple means of publishing two-dimensional electrophoresis data. *Electrophoresis* 1996;17:540–546.
4. Appel RD, Sanchez JC, Bairoch A, et al. SWISS-2DPAGE: a database of two-dimensional gel electrophoresis images. *Electrophoresis* 1993;14:1232–1238.
5. Gasteiger E, Gattiker A, Hoogland C, Ivanyi I, Appel RD, Bairoch A. ExPASy: The proteomics server for in-depth protein knowledge and analysis. *Nucleic Acids Res* 2003;31:3784–3788.
6. Hoogland C, Mostaguir K, Sanchez JC, Hochstrasser DF, Appel RD. SWISS-2DPAGE, ten years later. *Proteomics* 2004;4:2352–2356.
7. Bairoch A, Apweiler R, Wu CH, et al. The Universal Protein Resource (UniProt). *Nucleic Acids Res* 2005;33:D154–D159.
8. Wattiez R, Hermans C, Cruyt C, Bernard A, Falmagne P. Human bronchoalveolar lavage fluid protein two-dimensional database: study of interstitial lung diseases. *Electrophoresis* 2000;21:2703–2712.
9. Pietrogrande MC, Marchetti N, Tosi A, Dondi F, Righetti PG. Decoding two-dimensional polyacrylamide gel electrophoresis complex maps by autocovariance function: a simplified approach useful for proteomics. *Electrophoresis* 2005; 26:2739–2748.
10. Martens L, Van Damme P, Van Damme J, et al. The human platelet proteome mapped by peptide-centric proteomics: a functional protein profile. *Proteomics* 2005;5:3193–3204.
11. Mostaguir K, Hoogland C, Binz PA, Appel RD. The Make 2D-DB II package: conversion of federated two-dimensional gel electrophoresis databases into a relational format and interconnection of distributed databases. *Proteomics* 2003;3:1441–1444.
12. Ferry-Dumazet H, Houel G, Montalent P, et al. PROTICdb: a web-based application to store, track, query, and compare plant proteome data. *Proteomics* 2005;5: 2069–2081.
13. Babnigg G, Giometti CS. GELBANK: a database of annotated two-dimensional gel electrophoresis patterns of biological systems with completed genomes. *Nucleic Acids Res* 2004;32:D582–D585.
14. Young N, Chang Z, Wishart DS. GelScape: a web-based server for interactively annotating, manipulating, comparing and archiving 1D and 2D gel images. *Bioinformatics* 2004;20:976–978.
15. Wattiez R, Hermans C, Bernard A, Lesur O, Falmagne P. Human bronchoalveoalr lavage fluid: two-dimensional gel electrophoresis, amino-acid microsequencing and identification of major proteins. *Electrophoresis* 1999;20:1634–1645.

16. Omenn GS. Advancement of biomarker discovery and validation through the HUPO plasma proteome project. *Dis Markers* 2004;20:131–134.
17. Thalmann I. Proteomics and the inner ear. *Dis Markers* 2001;17:259–270.
18. Fortunato D, Giuffrida MG, Cavaletto M, et al. Structural proteome of human colostral fat globule membrane proteins. *Proteomics* 2003;3:897–905.
19. Finehout EJ, Franck Z, Lee KH. Towards two-dimensional electrophoresis mapping of the cerebrospinal fluid proteome from a single individual. *Electrophoresis* 2004;25:2564–2575.
20. Robinson MK, Myrick JE, Henderson LO, et al. Two-dimensional protein electrophoresis and multiple hypothesis testing to detect potential serum protein biomarkers in children with fetal alcohol syndrome. *Electrophoresis* 1995;16: 1176–1183.
21. Rasmussen HH, Orntoft TF, Wolf H, Celis JE. Towards a comprehensive database of proteins from the urine of patients with bladder cancer. *J Urol* 1996;155: 2113–2119.
22. Liberatori S, Bini L, De Felice C, et al. A two-dimensional protein map of human amniotic fluid at 17 weeks' gestation. *Electrophoresis* 1997;18:2816–2822.
23. Hill A, Kim H. The UAB Proteomics Database. *Bioinformatics* 2003;19: 2149–2151.
24. Fung KY, Glode LM, Green S, Duncan MW. A comprehensive characterization of the peptide and protein constituents of human seminal fluid. *Prostate* 2004;61: 171–181.

8

Bioinformatics and Experimental Design for Biomarker Discovery

Marc R. Wilkins and Sybille M. N. Hunt

Summary

Proteomics is supremely well suited to the discovery of biomarkers in human body fluids. However, the diversity of the human population and the multistep nature of most proteomic techniques require great care to be taken in experimental design and data analysis. This chapter outlines strategies for biomarker discovery, using 2D polyacrylamide gel electrophoresis or other proteomic technologies, and explores issues of analytical and biological variation, data quality, and statistical analysis. Particular attention is given to the issue of experimental design and how this can influence the statistical outcomes of biomarker discovery efforts.

Key Words: Proteomics; biomarker; experimental design; bioinformatics; variance; statistics.

1. Introduction

Proteomics is supremely well suited to the analysis of human body fluids. Proteins in these fluids are typically soluble in nature, making them highly amenable to analysis using a variety of proteomic techniques. Whereas body fluids are well suited to analysis using proteomics, the application of mRNA-based techniques such as polymerase chain reaction or hybridization arrays is largely irrelevant for the study of such fluids. This is because the cells that produce proteins in body fluids are either not found in the fluids (e.g., urine) or are some distance away from where the fluids are usually sampled (e.g., cerebrospinal fluid). Additionally, proteins in some body fluids (e.g., serum or plasma) originate from cells that are present in multiple tissues. For these reasons, there is enormous interest in the use of proteomics for the study of human body fluids, in an attempt to understand better the normal physiology and pathophysiology of diseases and to discover biomarkers.

From: *Proteomics of Human Body Fluids: Principles, Methods, and Applications*
Edited by: V. Thongboonkerd © Humana Press Inc., Totowa, NJ

A number of proteomic technologies are commonly used for analysis of human body fluids. Typically, these technologies combine one or more means of high-resolution protein separation with high-sensitivity detection techniques. They are often linked with one or more types of mass spectrometer to allow high-precision protein identification and characterization. The mass spectrometer can also serve as a means of generating semiquantitative data for the measurement of protein abundance. Technologies in widespread use are detailed below. Some of their key attributes are compared in **Table 1**.

1.1. 2D Polyacrylamide Gel Electrophoresis (2D-PAGE)

Simultaneously described in 1975 by O'Farrell ***(1)*** and Klose ***(2)***, this technique combines a high-resolution separation of proteins on the basis of isoelectric point or p*I* (isoelectric focusing) with a 2D separation on the basis of apparent molecular mass (SDS-PAGE). Typically, 2D-PAGE can separate 1000 to 2000 proteins from a sample ***(3)***. However, the large-format 2D gels can separate up to 10,000 proteins from a single sample ***(4)***. Biomarker discovery via 2D-PAGE typically relies on the comparison of 2D gel images from normal vs diseased samples with the assistance of image analysis software. These biomarkers can then be identified using mass spectrometry (MS) of protein spots from the 2D gels.

1.2. Multidimensional Liquid Chromatography Coupled With Mass Spectrometry

Known as "shotgun proteomics" or multidimensional protein identification technology (MudPIT) ***(5)***, this approach first digests all proteins in a sample to peptides, and then uses multidimensional liquid chromatography (LC) to separate these peptides that are then subjected to fragmentation in an online mass spectrometer. The fragmentation data are matched against sequence databases, typically using the SEQUEST software ***(6)*** to map the peptides to the proteins present in the original sample. The approach has been applied increasingly to the analysis of human body fluids, and over 800 different gene products have been identified with confidence from human plasma ***(7,8)***. In the analysis of tissue samples, a maximum of approx 5300 proteins from a single sample has been described ***(9)***. Shotgun proteomics, or MudPIT analysis, does not provide a ready means for biomarker identification, as 7 to 10 runs are required of each sample to identify all proteins present ***(10,11)***. Qualitative data, over quantitative data, are typically the results. However, use of techniques such as isotope-coded affinity tag (ICAT) ***(12,13)*** or ^{18}O labelling ***(14)***, which produce heavy and light versions of peptides from two samples, allows for the quantitative

Table 1
Proteomic Technologies and Their Applications, Resolving Power, Dynamic Range, Compatibility With Type of Experimental Design, and Any Major Restrictions

Technology	Best application	Resolving power	Experimental design	Restrictions
2D-PAGE	Biomarker discovery and discovery of isoforms and splice variants High reproducibility is amenable to large experiments with many patient samples	Typically 1000 to 2000 proteins, but up to 10,000 have been reported	Hypothesis independent	Will not separate very basic, very acidic, very large, very small, or very hydrophobic proteins Low abundance proteins may require enrichment
LC-MS	High resolving power provides good capacity for identifying large numbers of proteins in a proteome Less useful for analysis of many samples	Up to 5000 proteins can be resolved from a sample Shown to be effective for up to 8 orders of magnitude	Hypothesis independent	Requires genomic data to be available in databases Up to 10 LC-MS runs may be required per sample to give maximal coverage of the proteome
Pattern-based discovery	Screening of many samples to yield a fingerprint for each sample, to allow sample classification	Spectra are converted to data points, from 8000 to 1,000,000 data points per spectra	Hypothesis independent	Analysis will allow patient classification but may not necessarily identify protein biomarker(s)
Protein "chips" or arrays	Screening samples when the protein(s) under study are known	Depends on density of array	Hypothesis dependent[a]	Array size is typically restricted by availability of antibodies In some cases, two antibodies are required per protein

[a]Most protein chips are low density and use antibodies in an array to measure expression levels. Because the investigator has to specify which antibodies are to be used, the experiments are typically testing hypotheses on the expression of certain proteins.

pairwise comparison of complex samples. Biomarkers are beginning to be discovered using these approaches ***(15)***.

1.3. Pattern-Based Discovery Techniques

There has been recent interest in the use of techniques that can produce a fingerprint of a complex protein sample, can screen large numbers of clinical samples, and can be coupled with statistical or artificial intelligence approaches to classify samples via the resulting fingerprint data. One particular approach is known as surface-enhanced laser desorption/ionization (SELDI), whereby samples are reacted with a chromatographic surface on a metal array. The bound peptides and proteins are then overlaid with an organic matrix and finally analyzed using a matrix-assisted laser desorption/ionization time-of-flight (MALDI-TOF) mass spectrometer (for reviews, *see* **refs. *16*** and ***17***). High-mass and low-mass ranges can be analyzed, and low- and high-resolution mass spectrometers have been used in this approach. The resulting spectra, although not a representation of the complete proteome, provide a complex signal of 15,000 to 1,000,000 *x,y* data points that determine patterns or profiles of proteins in the samples that have been studied. Because the spectra are treated as a complex signal and identities of the signal peaks are generally not known, the technique does not precisely yield information on which proteins are biomarkers. However, certain patterns of peaks, or combinations of spectral peaks, are usually found to be critical for the classification of samples. Although controversy has surrounded applications of the technique for the diagnosis of ovarian cancer and prostate cancer ***(18,19)***, pattern-based technology remains a popular approach for clinical proteomics.

1.4. Protein-Affinity Arrays

Protein-affinity arrays are different in concept to the approach just mentioned. Typically, they do not involve a high-resolution separation of proteins in the samples or use of a mass spectrometer for protein identification. Instead, the arrays or microarrays are first generated by the ordered immobilization of different affinity probes (e.g., antibody, DNA aptamer, or lectin) onto a surface (e.g., coated glass slides). The protein targets of the affinity probes are usually, but not always, known. The arrays are then incubated with protein samples under conditions that allow affinity reactions to occur. Frequently, the protein samples are prelabeled with fluorescent dyes such as Cy3 or Cy5 ***(20,21)*** that allow the presence of certain proteins to be measured with a high-resolution scanner. Alternatively, labeled secondary antibodies can provide high-sensitivity detection in the same manner as a sandwich ELISA, although the requirement of a pair of antibodies per antigen makes this undesirable for high-density arrays.

There are further iterations of protein-affinity arrays, as reviewed by Liotta et al. *(22)* and Haab *(23)*. High-density arrays are beginning to be produced, either through the use of large number of commercially available antibodies of known specificity *(20)*, or via the use of libraries of artificially generated antibody fragments *(21)*. Protein microarrays of body fluids and tissue homogenates, combined with data analysis using appropriate normalization and statistical techniques, are being used increasingly for the discovery of biomarkers *(24,25)*.

2. Types of Protein Biomarkers

Protein biomarkers are proteins that are associated with a phenotype. They can assist in disease diagnosis, prognosis, and stratification, in the choice of therapy and in monitoring the efficacy of treatment. Typically, biomarker proteins are differentially expressed in association with a phenotype. However, there are other aspects of proteins that can be considered and ultimately developed to be used as the biomarkers. Consideration of the type of biomarkers that are being sought can impact on the strategies used in proteomic analysis and data mining. Three major groups of biomarkers are outlined below, and technologies that are most suitable for their discovery are given in **Table 2**. A final consideration is whether a protein biomarker is primary or secondary in nature.

2.1. Differentially Expressed Proteins

These include proteins that are expressed uniquely in association with a phenotype, as well as proteins that show statistically significant over- or underexpression in patients' samples compared with normal controls or other groups of diseases. The well-known clinical biomarkers are prostate-specific antigen (PSA; discussed in **ref.** *26*) and the ovarian cancer marker CA125 (discussed in **ref.** *27*). Interestingly, neither of them has a well-understood function, highlighting the fact that elucidation of function is not essential for a molecule to be used as a biomarker. To date, differentially expressed proteins are the main focus of most biomarker discovery projects.

2.2. Differentially Posttranslationally Modified Proteins

Proteins that are differentially posttranslationally modified in association with a phenotype can also serve as biomarkers. The modifications include differential glycosylation, phosphorylation, methylation or other modifications (*see* **ref.** *28* for other common posttranslational modifications). For example, expression of the sialyl Lewis a, sialyl Lewis x, and sialyl Tn antigens, which are particular types of protein glycosylation found in association with adenocarcinoma (for review, *see* **ref.** *29*), have been the subject of intensive investigation as diagnostic and prognostic markers *(30)*. The diagnostic utility of abnormal protein phosphorylation, for example, the phosphorylation of serine 73 of

Table 2
Biomarker Types and Their Discovery Using Different Techniques

	Technology			
Biomarker type	2D-PAGE	LC-MS/MS	Pattern-based discovery	Affinity arrays
Differentially expressed proteins	Yes	Yes, via differential isotopic labeling	Yes, although the protein may not be easily identified	Yes
Posttranslationally modified proteins	Yes, if the modification produces a notable mass or charge shift	Yes; however, the search for particular types of modification is required	No; a change in a pattern may appear, but this is difficult to confirm as a modification	No
Novel protein fragments	Yes, particularly when fragments are of medium to large mass	Yes, particularly when fragments are of low mass and are being analyzed via peptidomics	No; a change in a pattern may appear, but this is difficult to confirm as a novel fragment	Fragments will not be detected unless a specific antibody to that fragment is available
Sequence isoforms and splice variants	Yes, if isoforms or variants produce a notable shift in charge or mass	Difficult to discover unless these are present in reference databases	No; a change in a pattern may appear, but this is difficult to confirm as an isoform or splice variant	Isoforms or splice variants will not be detected unless a specific antibody to that isoform is available

cytokeratin 8 in head and neck adenocarcinoma ***(31)*** or troponin I phosphorylation in myocardial injury ***(32)*** has also been discussed. The unusual methylation of histone H3 in myeloid leukaemia has been proposed as a putative marker of granulocyte abnormalities ***(33)***.

2.3. Isoforms, Splice Variants, and Protein Fragments

Isoforms, splice variants, and fragments of proteins can serve as biomarkers. Such biomarkers reflect changes in the genes themselves (inherited or spontaneous mutations), the mRNA splicing machinery, or the capacity of an individual's proteases to process or degrade a protein. For example, apolipoprotein E ε-4 isoform, which can be separated from other isoforms by isoelectric focusing, is predictive for the risk of Alzheimer's disease in homozygotes ***(34)***. However, the ε-4 isoform allele in heterozygotes is not predictive of Alzheimer's ***(35)***, highlighting the need to understand the dominance of an allele when investigating its application as a biomarker.

Splice variants, although mostly studied via the analysis of mRNA in expressed sequence tag (EST) libraries, have been investigated as biomarkers for cancer (for review, *see* **ref. *36***). The differential splicing of the Wilms' tumor 1 gene is one example ***(37)***. Although some technical challenges remain in the discovery of splice variants via proteomic techniques, it should be kept in mind that alternatively spliced forms of proteins have the capacity to serve as protein biomarkers. Protein spots that migrate anomalously on 2D gels should always be investigated in detail as potential novel splice variants.

Protein fragments also have the capacity to serve as biomarkers. These are being investigated either through MS-based approaches termed *peptidomics* (reviewed in **ref. *38***) or through the application of SELDI to the analysis of low-mass-range proteins and peptides. One interesting variation of this is to analyze the peptides that are found to interact with human serum albumin ***(39)***, which serves as a means of enriching for these low-abundance fragments. It is notable that protein fragments seen in the mouse proteome have been shown to be heritable ***(40)***, emphasizing that there is a genetic basis to these fragments.

2.4. Primary and Secondary Biomarkers

Biomarkers can be considered as either primary or secondary. Primary biomarkers are proteins whose expression levels, modification status, splice variants, or other isoforms are thought to be responsible for the phenotype under investigation. Reduced levels of the cystic fibrosis transmembrane conductance regulator (CFTR) protein in cystic fibrosis patients, although detected only in tissues, not in body fluids, is an example of the primary one ***(41)***. Secondary biomarkers are those that arise as a consequence of a disease but are not thought to be the primary trigger for the disease pathogenesis. Examples of the secondary biomarkers include CA125 in ovarian cancer and PSA in prostate cancer ***(26,27)***. Note that the utility of a biomarker is not necessarily determined by its primary or secondary status. Other metrics, including the specificity and sensitivity of a biomarker when used in clinical diagnostics, are more critical.

3. Bioinformatics for Biomarker Discovery

So far, this review has explored the technologies that are used for the discovery of biomarkers and the types of biomarkers that can be discovered using proteomics. It is apparent that many different technical approaches and strategies can be employed. The remainder of this chapter will focus on bioinformatics for biomarker discovery. In particular, areas such as discovery strategies, data acquisition and analysis, understanding variation, experimental design, and statistical analysis for biomarkers will be covered. It should be noted that biomarker discovery is distinct from biomarker validation, which usually involves different laboratory methods, data types, and data analysis issues. Biomarker validation is outside of the scope of this chapter; however, many aspects of validation are explored in reviews elsewhere ***(42,43)***.

3.1. Strategy

In the simplest case, a biomarker discovery strategy will use a proteomic technology to compare two clinical samples to find proteins that have different expression levels. These include proteins that are uniquely expressed in one sample but not in another. Typically, the experiment being undertaken is hypothesis independent, meaning that the question being asked is general (e.g., what proteins are changing?) instead of specific (e.g., is the level of pyruvate kinase changing?). However, it may appear that many of the differentially expressed proteins are ultimately of no interest. There are a number of reasons for this, including:

- The differences may be owing to genetically based variations in protein expression levels in the population that are not associated with the disease.
- The differences may be owing to environmental factors that affect protein expression but are not associated with the disease.
- The differences will not be statistically significant when they are evaluated with appropriate statistical tests.
- The changes in protein expression levels observed are owing to chance alone.
- The changes are owing to analytical variation alone.

To discover biomarkers that will ultimately be useful as clinical tools, a robust strategy for data generation, manipulation, and analysis is required. This needs to be executed in the context of an appropriate experimental design. The discussion below will concentrate on issues that are relevant to the generation and analysis of data obtained from gel-based proteomics. The relevance of this, however, will also be explored for other proteomic data types.

3.2. Experimental Design: Fundamental Principles

The purpose of biomarkers, as discussed earlier, is to assist in patient diagnosis, prognosis, and stratification, in the choice of therapy and, in monitoring the efficacy of treatment. For most of these applications, biomarkers will be used to classify the patient as a member of a certain population (e.g., disease vs normal). In a statistical sense, biomarker discovery experiments must analyze a sufficient number of samples from each population to ascertain that there is a true difference in the expression of certain proteins. Summary statistics for each population, typically the mean, standard deviation, coefficient of variation, or other measures of variance for the proteins being analyzed will be required. These statistics will estimate the *biological variation* (owing to genetic and environmental effects) in a population and will become more accurate when larger numbers of samples are analyzed. However, the accuracy of these estimates is also influenced by the quality of proteomic analyses that are used. An understanding of the *analytical variation* inherent in any proteomic technology is also essential to ultimately produce a good experimental design. It will affect, for example, the number of replicate analyses required per sample.

3.3. Managing Analytical Variation: Issues to Consider

A number of issues can be considered in the management of analytical variation. These are worked through below.

3.3.1. Randomization to Decrease Bias

Bias can have a profound effect on the validity of biomarker research. Many types of bias (explored in detail by Ransohoff *[44]*) may have an impact on proteomic research. These include bias introduced during specimen collection, sample handling, and storage, the technical analysis itself, and use of data analysis tools. The complete elimination of bias from an experimental system is almost impossible. However, an overriding principle is to use uniform collection, handling, and analysis techniques with randomizing the order in which samples are collected and examined on instrumentation and with randomizing the order in which data are analyzed (**Table 3**) *(45)*. Use of blinding will further help in removing bias in these steps, although this may be impractical during the statistical testing of certain hypotheses. These principles can be applied to biomarker research using any of the analytical techniques described in **Subheading 1**.

3.3.2. Dynamic Range, Linearity, and Detection Limits

Dynamic range, linearity, and detection limits will affect the quality of raw data; thus the degree of differences can ultimately be found between populations. The impact of these issues is explored in **Table 4.**

Table 3
How Bias Can Be Addressed in Experimental and Observational Studies

	Involving people	Involving samples
Design	Randomize allocation to compared groups	Arrange for uniform and, if possible, blinded collection, handling, and analysis of samples
Conduct	Measure and report baseline characteristics of groups	Check that uniform handling occurs and whether blinding is successful
Interpretation	If groups are unequal, discuss direction, magnitude, and potential impact of bias	If groups are unequal, discuss direction, magnitude, and potential impact of bias

From **ref. *44***.

Table 4
Issues To Be Faced in the Generation of High-Quality Proteomic Data[a]

Issue	Impact	Remedy
Low dynamic range	Small differences between samples may not be detected	Use dye, labeling, and detection systems that maximize dynamic range
Nonlinearity of detection	Unequal loading of samples for analysis will generate different results	Appropriate normalization of data can be applied if the structure of the nonlinearity is known
Poor limit of detection	Low-abundance proteins or peptides may not be detected in all analyses	Omit these proteins from data analysis, use greater numbers of replicates, or increase protein abundance by enrichment

[a]These issues can be applied to techniques that are gel-based as well as those that generate and analyze mass spectra.

In 2D-PAGE, the dynamic range is affected both by the stain used and by the scanning technology employed (i.e., CCD camera or laser-based scanner). Dynamic range can be visualized via a histogram of number of spots against the optical density of each spot **(Fig. 1)**. Note, however, that this is a different graph than one that plots a histogram of the numbers of pixels from an image against the optical density, which better reveals the level of background staining or signal present. **Figure 1** shows **(A)** an ideal distribution of data; **(B)** distribution of data obtained from an under-stained gel; and **(C)** distribution of data obtained from a saturated image. These types of graphs can also be generated for quantitative data obtained from MS or from protein chips. In a biomarker discovery

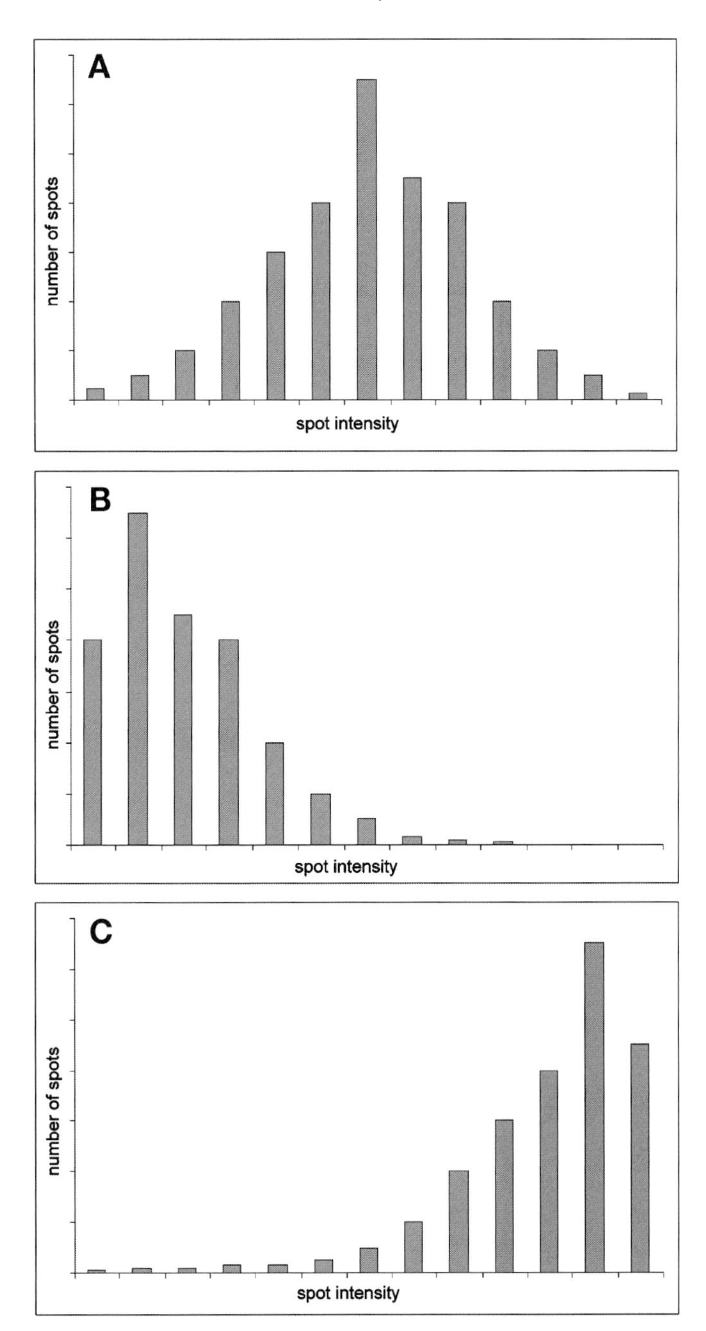

Fig. 1. Histograms of number of proteins plotted against a range of spot intensity. **(A)** An ideal distribution. **(B)** Data obtained from a gel that may be understained. **(C)** Data obtained from a saturated image.

study, it is of importance that all analyses of all samples show a similar dynamic range. However, normalization can be employed in some instances, for example, in the case of unequal protein loading (*see* next paragraph).

Linearity of detection is important to allow the mapping of an analytically detected difference from a true molar or copies-per-cell difference. In 2D-PAGE, linearity of spot density is influenced by both the dynamic range of the stains employed and the effectiveness of the 2D image analysis software for spot boundary detection on a digital 2D image. Fluorescent dyes such as SYPRO Ruby and Deep Purple, as well as the visible dye colloidal Coomassie Blue, all show good linearity of their dynamic range with $R^2 = 0.96$ to 0.99 over 3 orders of magnitude ***(46)***. Note that silver staining, however, is not linear in response over its entire dynamic range ***(46,47)***. An equally critical issue is the capacity of 2D image analysis software to detect protein spot boundaries accurately and therefore quantitate the protein spots present accurately. By the generation of artificial 2D gel images, Raman et al. ***(48)*** showed that some commercially available image analysis packages do not accurately detect protein spot boundaries to yield linear quantitative data. Rosengren et al. ***(49)*** and Hunt et al. ***(45)*** also illustrate similar results using other software packages. Users of image analysis packages are advised to test the linearity of their software with these artificial images, which are downloadable from the Internet.

In 2D gels, the detection and quantitation of low-abundance proteins raises certain challenges. When protein spots are faint, image analysis software will have difficulty in detecting them over background noise. If protein spots are also small and are defined by a small number of pixels, the inclusion or exclusion of neighboring pixels will have a profound impact on their quantitation. For example, the difference between having a spot of 4 pixels and 6 pixels is likely to be dramatic. The impact of this in large experiments is that the variance associated with the quantitative measurement of low-abundance proteins is typically high and increases the difficulty of finding a significant difference in protein expression between two populations.

3.3.3. Completeness of Analysis

A recently explored issue is that of completeness of analysis. Discussed to date in the context of MudPIT LC-MS/MS ***(10,11)***, it refers to a phenomenon whereby a technique used for the analysis of a complex protein mixture may only yield information for a fraction of all proteins in any single analytical run. For example, it is known that two replicate MudPIT analyses will yield two sets of protein identifications with approx 65% overlap. Thirty-five percent of the proteins in the second analysis will be novel compared with the first. A third MudPIT analysis will yield a set of identifications that has 80% overlap with those from the first two analyses, but with 20% new identifications **(Fig. 2)**.

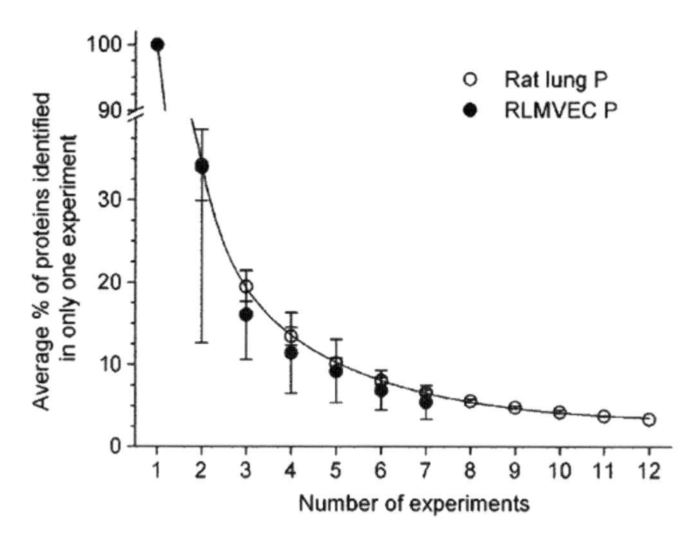

Fig. 2. Completeness of analysis in MudPIT. The analytical completeness increases when the greater numbers of LC-MS/MS runs are performed. Up to 10 runs may be required per sample to gain a complete understanding of its protein composition, in this case from rat lung. (From Durr et al. ***(10)***; reproduced with permission from Nature Biotechnology.)

Because of this, between 10 and 12 MudPIT analyses may be necessary before a near-complete list of protein identities is generated from a single sample ***(10,11)***. This phenomenon has a substantial impact on the use of MudPIT for biomarker discovery work, as the capacity to compare two samples comprehensively requires very high numbers of replicate analyses per sample.

Slightly different phenomena are faced in 2D-PAGE results including the very large number of spots visualized on 2D gels, changes in the number of spots from gel to gel ***(50)***, and the difficulty that software has in matching these spots. It means that the number of spots seen consistently across an experiment is much smaller than the number seen in any particular gel. Challapalli et al. ***(51)*** explain that the pairwise comparison of any two gels may show about 3500 spots in common. However, when three gels are matched to create an "average gel," only 1700 proteins are matched by the software across all gels. Statistically, this produces a series of missing values in data sets that can hinder analysis, although values can be imputed to address this issue ***(52)***. Fortunately, gel images can be examined manually for a particular spot of interest if it is required.

3.3.4. Normalization and Transformation of Data

The minimization of variation of protein expression between replicate 2D gels relies on effective normalization. Typically, the expression level or optical density of any particular spot on a 2D gel is calculated by summing the value of all pixels within the spot boundary. In any gel, this can be normalized by expressing the value of any spot as a percent of the total optical density of all spots on that gel (OD%). When comparing the normalized values from one 2D gel to another, a number of issues arise. These include:

- Differences in protein loading per gel.
- Different numbers of spots on each gel.
- Large changes in protein expression between gels.
- In larger sets of gels, only a subset of all spots will match across all gels.

Although the comparison of spot OD% from one gel to another is useful, better normalization can be achieved through (1) establishing which proteins are found to match across all gels in a gel set; (2) checking that these proteins are not dramatically over- or underexpressed between gels; and (3) normalizing the protein levels in each gel by comparison with the sum of the optical density of these matching proteins from the same gel. The efficiency of normalization can easily be viewed with box and whisker plots (**Fig. 3**). Alternatively, nonparametric models can also be used for normalization ***(53)***.

When using Cy dye techniques for differential display ***(54)***, more advanced normalization techniques may be needed. This is particularly true if a pooled standard, labeled with one of the Cy dyes, is run on all gels ***(55)***. Recent work has illustrated that normalization techniques using rescaling ***(56)*** or DNA microarray normalization methods ***(57)*** should be considered in addition to the methods employed in the DeCyder software (Amersham Biosciences).

When protein expression data are to be analyzed with standard statistical tests (e.g., Student's *t*-test), these tests will assume that the data come from a normal distribution. Data from 2D gels are typically not from a normal distribution, instead being highly skewed (**Fig. 4**). Thus, transformation is required before data can be used for statistical analysis. Logarithmic transformation is effective for this ***(45)***. However, other transformations, such as an inverse hyperbolic sine transformation, can be considered ***(58)***. Frequency histograms and quantile-quantile plots are utilized to demonstrate the success of transformation on protein expression data in **Fig. 4**. However, be aware that expression data may show kurtosis (a sharpening or flattening of a normal distribution) ***(56)***.

3.3.5. Analytical Reproducibility and Variance

Prior to the statistical analysis of 2D gel expression data for biomarker discovery, it is critical to understand the reproducibility of 2D-PAGE in a laboratory. This will have an impact on the appropriate experimental design (*see* **Subheading 3.4.** below) and also allow the evaluation of data in the context of best practice reported in the literature. Reproducibility of spot position can also be investigated ***(59,60)*** but will not be discussed in detail here.

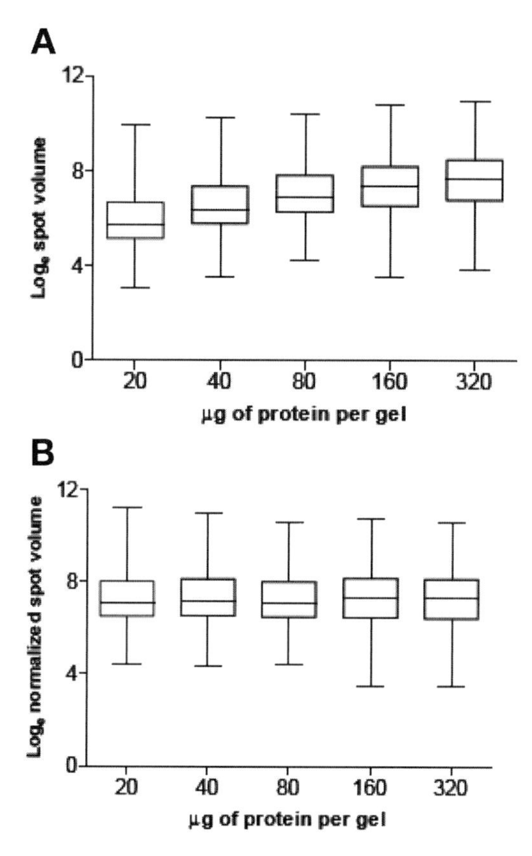

Fig. 3. Normalization of 2D-PAGE protein expression data, graphed with box and whisker plots. **(A)** Data from five gels prior to normalization. **(B)** The same data post normalization. (From Hunt et al. ***(45)***; reproduced with permission from the *Journal of Proteome Research.*)

Analytical reproducibility, otherwise termed analytical variation, is usually expressed in terms of Pearson's correlation coefficient (r and R^2) or percent coefficient of variation (CV%; calculated by dividing the standard deviation by the mean and multiplying by 100). **Table 5** summarizes some R^2 and CV% data from the literature. It can be seen that best practice for reproducibility of 2D-PAGE should yield CV% of 18 to 25 and R^2 values of greater than 0.90. The type of sample being analyzed, particularly the protein solubility, may affect these values. It is possible to assess variance with other statistical approaches, which are more sophisticated than R^2 and CV% and will better account for the multivariate nature of the data. To date, these have included random linear mixed effects models ***(45)*** as well as a quadratic variance model ***(61)***.

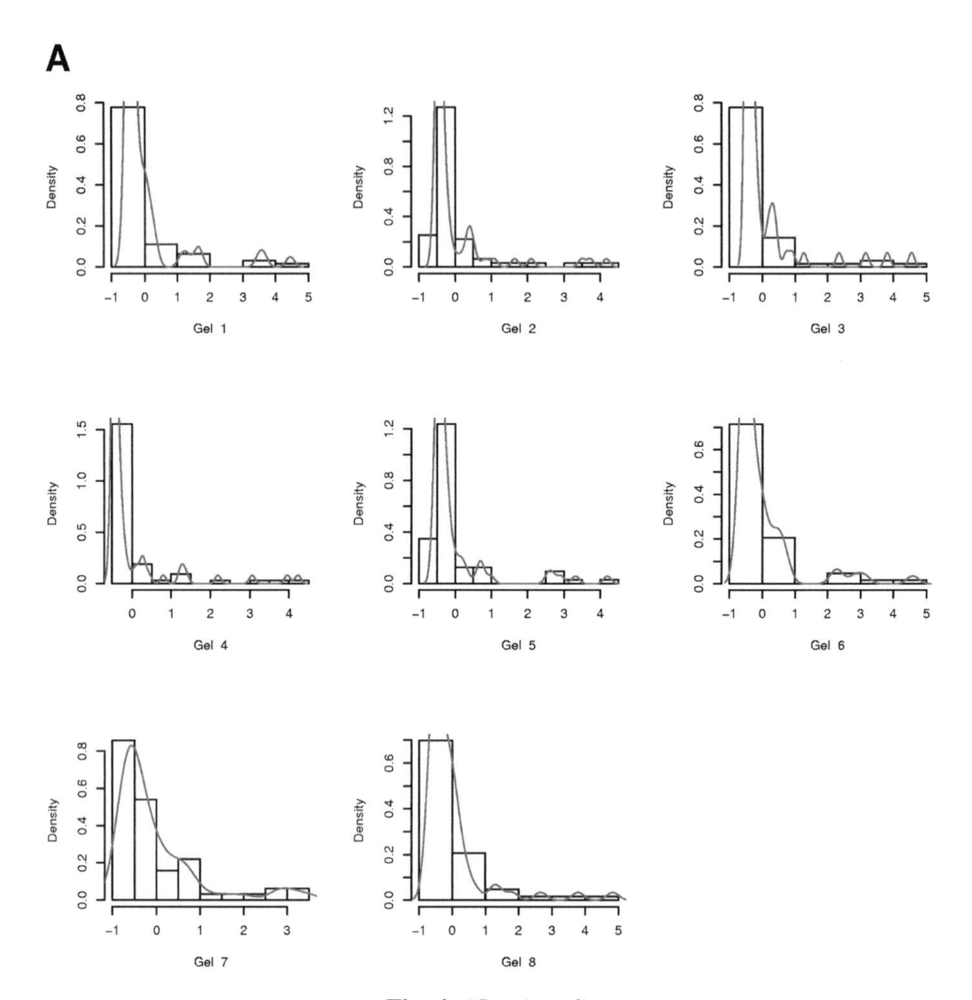

Fig. 4. *(Continued)*

3.4. Biological Variation, Power Analysis, and Advanced Experimental Design

The biological variation inherent in any experiment is a sum of variations from a number of sources. In human samples, these include genetic variation, environmental influences, effects of any medication, and disease load, as well as other factors that may arise in the acquisition, storage, and preparation of samples. In the sampling of human tissues, heterogeneity of cell type is a further source of variation. Because how a sample has been acquired and treated can have a profound influence on the results, particular care is needed to ensure that these processes are controlled as closely as possible via consistency of methods and through randomized experimental designs. Interestingly, if the pedigree or

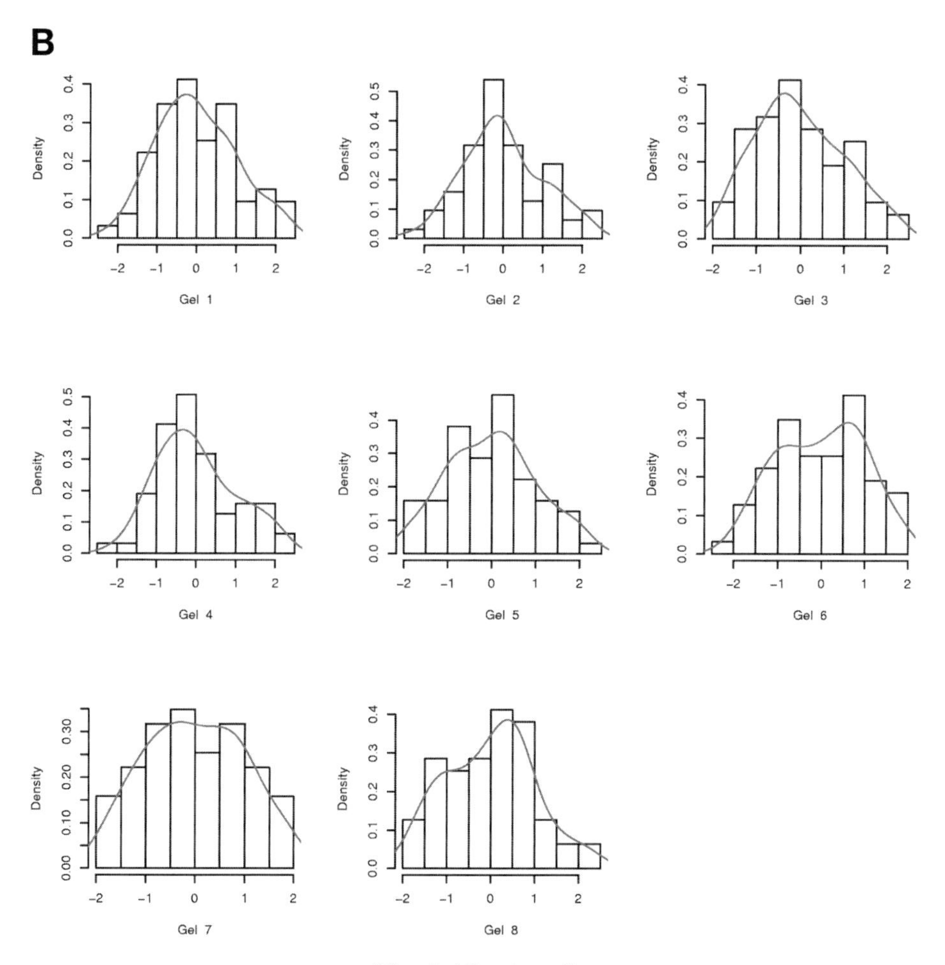

Fig. 4. *(Continued)*

genetic background of a cohort of experimental subjects is known, this may assist us in understanding and controlling biological variation ***(40)***. However, there are relatively few examples, to date, in which genetic data have been closely combined with proteomic results.

Biological variation can be measured in the same manner as for analytical variation, most simplistically by using correlation coefficient and percent coefficient of variation. **Table 6** summarizes some R^2 and CV% data for biological variation and shows the greater degree of variation in this data set compared with those corresponding values in the set of analytical variation (**Table 5**). Indeed, this is required for the effective discovery of biomarkers (*see* **Fig. 5**).

The analytical and biological variation in a data set can be used to design proteomic experiments through the use of a statistical power analysis ***(45,50)***.

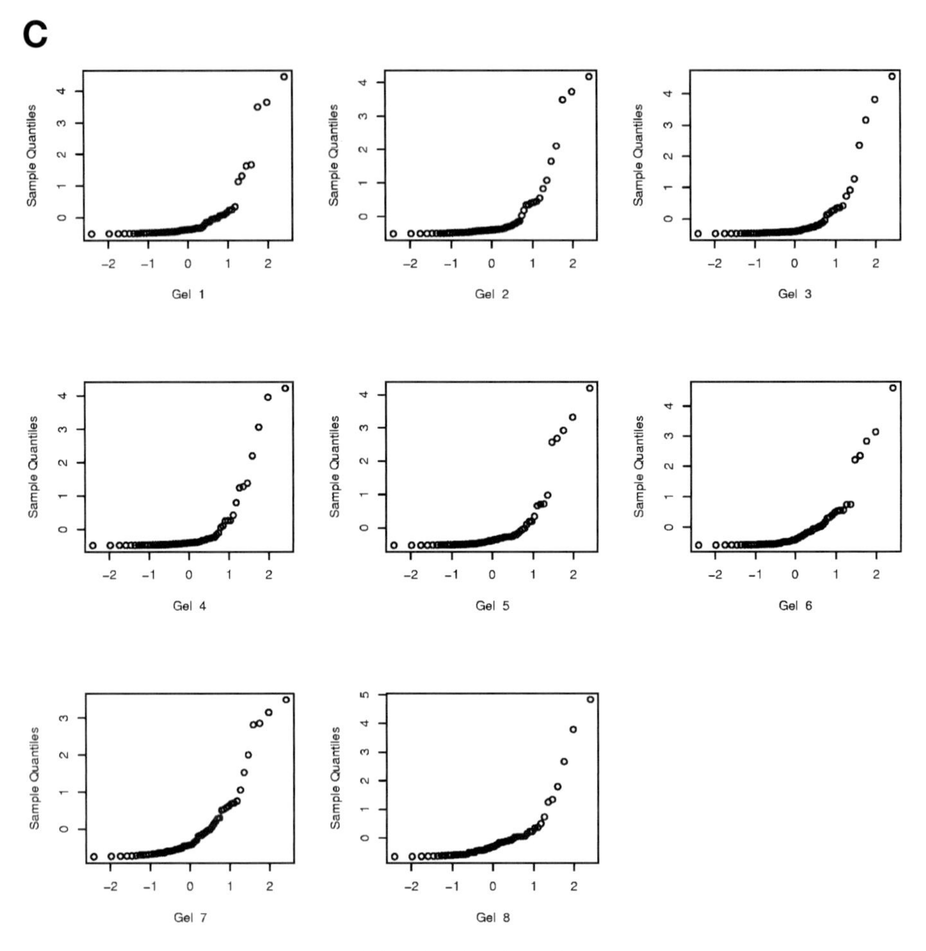

Fig. 4. *(Continued)*

This can predict the precise number of samples that should be analyzed in a pairwise differential display experiment, as well as the numbers of replicate analyses that are desirable, all in the context of the degree of differential expression one would like to see. Tools for this purpose have been made freely available on the Internet (www.emprhon.com) ***(45)***. **Figure 6** shows a power analysis curve from the 2D-PAGE analysis of a pilot set of cerebrospinal fluid samples. For this example, the graph shows that approximately seven clinical samples from control and experimental groups will be required to find statistically significant differences of twofold (effect size of 100%). More samples would need to be analyzed to detect smaller differences significantly. The number of replicates (one to four gels) will slightly, not dramatically, decrease the number of clinical samples required, and there will be little benefit from triplicate

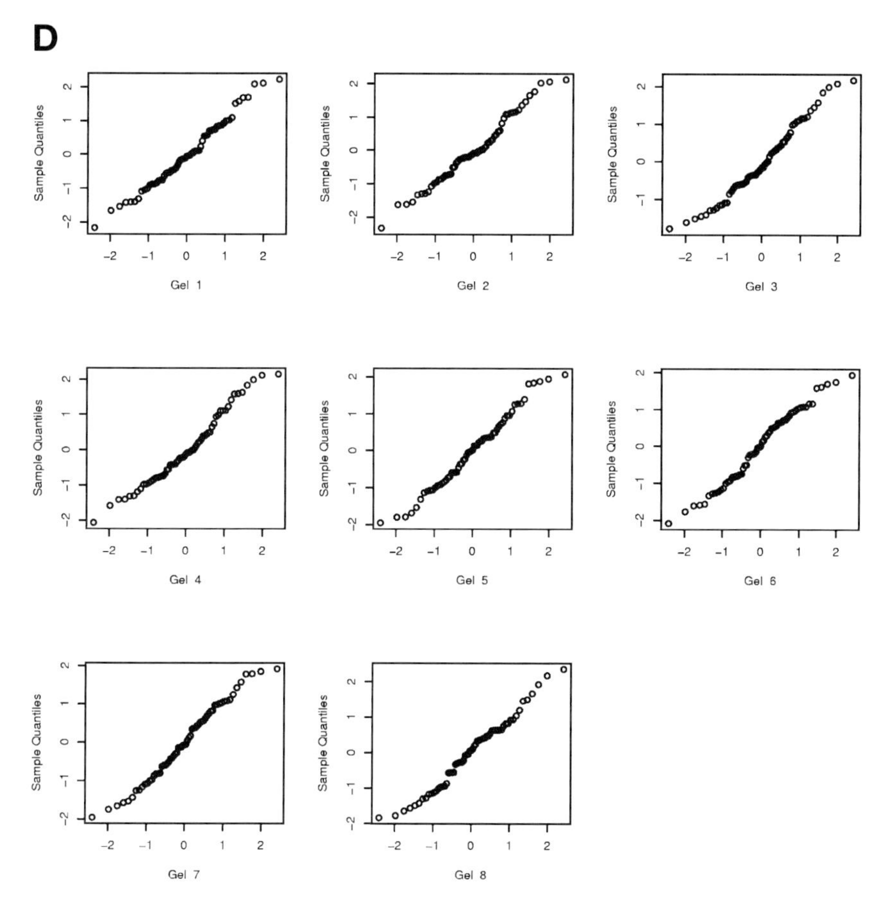

Fig. 4. Eight samples of human plasma analyzed with 2D-PAGE. **(A)** Histograms of protein expression data for each gel, prior to transformation. Note the highly skewed distribution. **(B)** The same data after log transformation, which has the data appearing normally distributed. (**C** and **D**) Quantile-quantile plots of the same data. **(D)** Clearly shows approximately normal distribution of the log transformed data.

over duplicate sample analysis. It can also be seen that there is an asymptote of the graph toward approx 40% effect size, indicating that even with access to an infinite number of samples, it will be impossible to detect differences of below 40% effect size.

3.5. Statistical Analysis for Biomarker Discovery

After the completion of a biomarker discovery experiment, there are a number of statistical tests that can be used for data analysis. In the broadest sense, these can be classified as tests that will examine a hypothesis (supervised tests), those that can explore trends or highlight associations in data without testing

Table 5
Some Recently Published Data on Replicate Analyses of 2D Gels[a]

Experimental replicates	CV%	R^2	Reference
Human plasma, two independent experiments	18, 19	0.97, 0.96	Hunt et al. ***(45)***
Human cerebrospinal fluid	4–23	0.82	Terry and Desiderio ***(62)***
Human synovial fluid	N.R.	0.89–0.95	Yamagiwa et al. ***(63)***
Colon tumor cell line	22.4	0.86	Molloy et al. ***(50)***
Mouse brain cytosol	N.R.	0.926	Challapalli et al. ***(51)***
Plant cell wall	16.2	N.R.	Asirvatham et al. ***(64)***

N.R., not reported.
[a]Note that these data consider only analytical variation, not biological variation.

Table 6
Some Recently Published Information on the Biological Variation Measured via 2D-PAGE Analysis

Biological sample to sample variation[a]	CV%	R^2	Reference
Human plasma	N.R.	0.43–0.90	Hunt et al. ***(45)***
Breast tumor cell line	28.4	0.82	Molloy et al. ***(50)***
Colon tumor cell line	23.6	0.74	Molloy et al. ***(50)***
Human synovial fluid	N.R.	0.85–0.93	Yamagiwa et al. ***(63)***
Mouse brain cytosol	N.R.	0.87	Challapalli et al. ***(51)***

N.R., not reported.
[a]From pairwise comparisons.

a hypothesis (unsupervised methods), tests that are designed for the analysis of univariate data, and those that are designed for multivariate data. Both unsupervised methods and supervised tests are useful for biomarker discovery in body fluids. **Table 7** summarizes the common tests in each category. It is outside the scope of this chapter to review comprehensively the variety of techniques that can be used for the statistical analysis of data sets; however, a few key points can be highlighted.

Quantitative proteomic data are inherently multivariate. This is because every measurement for any particular protein or peptide that is taken from a sample by 2D-PAGE, protein microarray, LC-MS/MS, or SELDI is made in the context of measurements of other proteins and peptides in the same analysis. The analytical and normalization approaches used (as discussed in **Subheading 3.3.**) mean that the level of a protein measured in one sample can have an effect on the quantitation of other proteins in that same sample. For this reason, it is desirable to use multivariate techniques for the analysis of proteomic data. This has the advantage

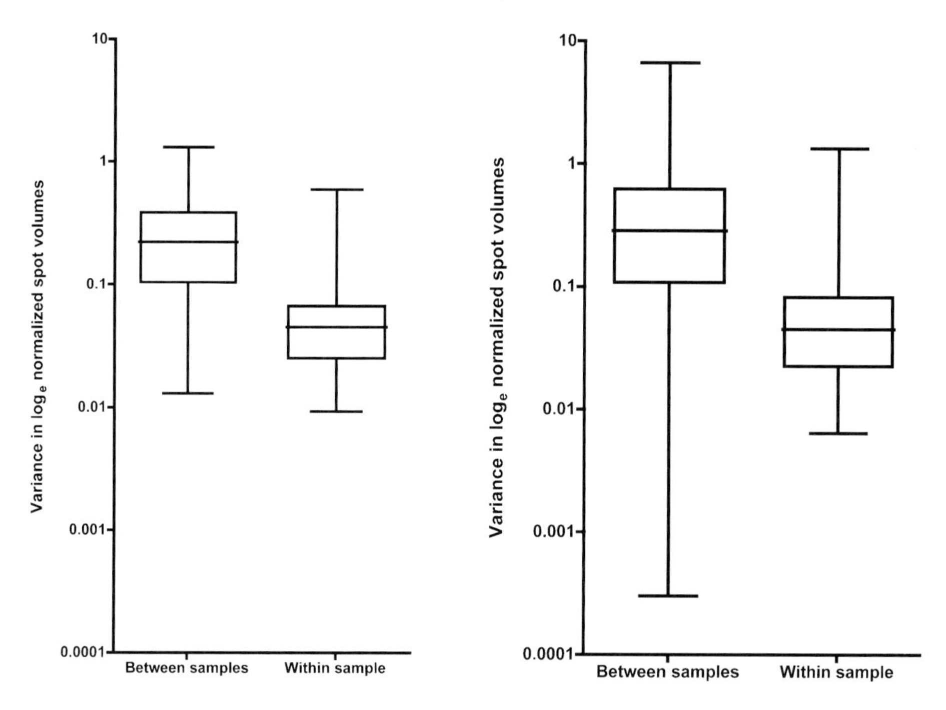

Fig. 5. Biological (between-samples) and analytical (within-sample) variation can be visualized with box and whisker plots. In this case, data were from two independent experiments in which replicate analyses of plasma samples from patients with ovarian cancer and control subjects were analyzed (*see* Hunt et al. *[45]* for further details). Note that in both cases, the degree of biological variation is greater than that of the analytical variation.

of being more accurate in the estimation of noise and variance in an experiment and being able to identify small coordinated changes in the expression of groups of proteins that by themselves may not appear to be significant. Furthermore, in the analysis of human body fluids, most clinical samples are provided with a comprehensive clinical research form. This form contains data for variables of many types (e.g., age, sex, height, weight, smoking status, alcohol intake, disease type, disease stage, current medication regime, and so on). These variables can themselves be used to explore proteomic data further and to ensure that trends seen in patients reflect the disease status rather than the effect of another variable (e.g., smoking status).

There is currently no single "best practice" for the analysis of proteomic expression data for biomarker discovery. In part, this reflects the diversity of techniques that are available to the statistician and some of the different statistical schools of thought (e.g., Beyesian vs frequentist). The use of Student's *t*-test, although widespread, ignores the inherent multivariate nature of proteomic

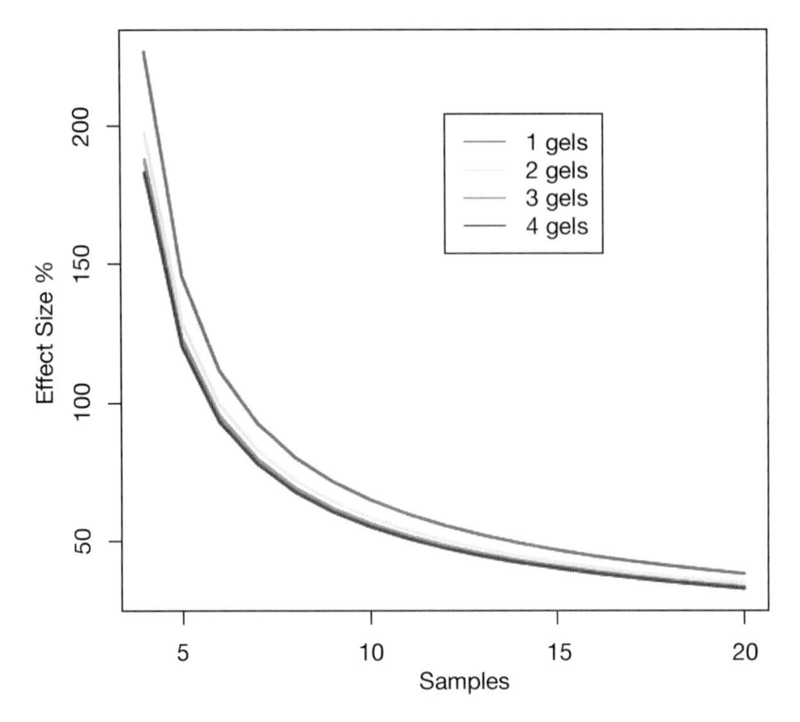

Fig. 6. Power analysis curve generated from duplicate 2D analysis of three controls and three diseased samples of human cerebrospinal fluid. The curve was generated by using power of 0.8 and a p value of 0.05. Tools at *www.emphron.com* were used for this analysis.

Table 7
Statistical Test Types and Their Suitability for the Analysis of Univariate or Multivariate Data

	Unsupervised methods	Supervised tests
Univariate	Data modality analysis[a]	*t*-test
	Kernel density estimates	Z-test
		ANOVA test
Multivariate	Hierarchical clustering	Discriminant analysis
	K-means	Support vector machines
	K-median	Survival analysis
	Principal component analysis	Gene shaving
	Gene shaving	

[a]Note that most unsupervised methods for exploring univariate data are to assist in the evaluation of the variance and modality of the data set. This is different in purpose from the unsupervised methods used in multivariate techniques that can classify groups and reveal trends.

Table 8
Some Examples of the Applications of Common Statistical Techniques to the Analysis of Proteomic Data

Statistical method	Application area	Reference
ANOVA	Testing capacity of biomarkers to categorize stroke patients	Allard et al. *(65)*
Partial least squares discriminant analysis	Discovery of differentially expressed proteins in schizophrenia patients	Karp et al. *(66)*
	Classification of ovarian cancer tumors	Alaiya et al. *(67)*
Principal components analysis	Identification of the regulatory proteins in human pancreatic cancers	Marengo et al. *(68)*
	Classification of ovarian cancer tumors	Alaiya et al. *(67)*
Hierarchical clustering analysis	Discrimination of rheumatoid arthritis patients from healthy individuals	Dotzlaw et al. *(69)*
Student's *t*-test	Testing of significance of cardiac transplant rejection marker in ELISA data[a]	Borozdenkova et al. *(70)*

[a]This marker was discovered via 2D-PAGE. Quantitative measurements using an ELISA assay were performed to determine clinical significance of the obtained data in the same patient cohort. This was an appropriate use of the *t*-test, as the data were univariate.

data, and it has been shown to be highly sensitive to different techniques used for data normalization and missing value substitution ***(52)***. In the future, multivariate techniques are anticipated to be used increasingly. **Table 8** provides examples of a number of statistical techniques used in proteomics from the recent literature.

4. Conclusions

This chapter has discussed some of the technologies that are available for biomarker discovery, has explored how these technologies can be applied to the discovery of different types of biomarkers, and, finally, has worked through many of the basic, but critical, issues associated with the generation and analysis of the proteomic expression data. The chapter has not addressed many of the more commonly discussed bioinformatics issues concerning protein identification or explored the body of work that is seeking to provide high confidence in those results. These issues have been explored in detail elsewhere ***(71,72)***. The bioinformatics and biostatistics discussed herein are required to ensure that the data being produced are of high quality, that the experiments are appropriately designed, and that the results are analyzed appropriately to identify one or more biomarkers in a biological system. We hope that this will be useful in assisting the biomarker discovery efforts of current and future proteomics researchers.

Acknowledgments

We thank Drs. Mervyn Thomas and David Nott for vigorous discussion on statistics and also the staff at Proteome Systems Limited, including Drs. Rebecca Harcourt, Sanne Pedersen, Lucille Sebastian, Andrew Sloan, Robyn Linder, and Sindhu Prasad for permission to use their data in Figures 4, 5, and 6.

References

1. O'Farrell PH. High resolution two-dimensional electrophoresis of proteins. *J Biol Chem* 1975;250:4007–4021.
2. Klose J. Protein mapping by combined isoelectric focusing and electrophoresis of mouse tissues. A novel approach to testing for induced point mutations in mammals. *Humangenetik* 1975;26:231–243.
3. Gorg A, Weiss W, Dunn MJ. Current two-dimensional electrophoresis technology for proteomics. *Proteomics* 2004;4:3665–3685.
4. Klose J, Kobalz U. Two-dimensional electrophoresis of proteins: an updated protocol and implications for a functional analysis of the genome. *Electrophoresis* 1995;16:1034–1059.
5. Liu H, Lin D, Yates JR, III. Multidimensional separations for protein/peptide analysis in the post-genomic era. *Biotechniques* 2002;32:898, 900, 902.
6. Eng JK, McCormack AL, Yates JRI. An approach to correlate tandem mass spectral data of peptides with amino acid sequences in a protein database. *J Am Soc Mass Spectrom* 1994;5:976–989.
7. Qian WJ, Jacobs JM, Camp DG, et al. Comparative proteome analyses of human plasma following in vivo lipopolysaccharide administration using multidimensional separations coupled with tandem mass spectrometry. *Proteomics* 2005;5:572–584.
8. Shen Y, Jacobs JM, Camp DG, et al. Ultra-high-efficiency strong cation exchange LC/RPLC/MS/MS for high dynamic range characterization of the human plasma proteome. *Anal Chem* 2004;76:1134–1144.
9. Liu T, Qian WJ, Chen WN, et al. Improved proteome coverage by using high efficiency cysteinyl peptide enrichment: the human mammary epithelial cell proteome. *Proteomics* 2005;5:1263–1273.
10. Durr E, Yu J, Krasinska KM, et al. Direct proteomic mapping of the lung microvascular endothelial cell surface in vivo and in cell culture. *Nat Biotechnol* 2004;22:985–992.
11. Liu H, Sadygov RG, Yates JR III. A model for random sampling and estimation of relative protein abundance in shotgun proteomics. *Anal Chem* 2004;76:4193–4201.

12. Gygi SP, Rist B, Gerber SA, Turecek F, Gelb MH, Aebersold R. Quantitative analysis of complex protein mixtures using isotope-coded affinity tags. *Nat Biotechnol* 1999;17:994–999.
13. Gygi SP, Rist B, Griffin TJ, Eng J, Aebersold R. Proteome analysis of low-abundance proteins using multidimensional chromatography and isotope-coded affinity tags. *J Proteome Res* 2002;1:47–54.
14. Heller M, Mattou H, Menzel C, Yao X. Trypsin catalyzed ^{16}O-to-^{18}O exchange for comparative proteomics: tandem mass spectrometry comparison using MALDI-TOF, ESI-QTOF, and ESI-ion trap mass spectrometers. *J Am Soc Mass Spectrom* 2003;14:704–718.
15. Zhang J, Goodlett DR, Peskind ER, et al. Quantitative proteomic analysis of age-related changes in human cerebrospinal fluid. *Neurobiol Aging* 2005;26:207–227.
16. Petricoin EF, Liotta LA. SELDI-TOF-based serum proteomic pattern diagnostics for early detection of cancer. *Curr Opin Biotechnol* 2004;15:24–30.
17. White CN, Chan DW, Zhang Z. Bioinformatics strategies for proteomic profiling. *Clin Biochem* 2004;37:636–641.
18. Ransohoff DF. Lessons from controversy: ovarian cancer screening and serum proteomics. *J Natl Cancer Inst* 2005;97:315–319.
19. Diamandis EP. Point: proteomic patterns in biological fluids: do they represent the future of cancer diagnostics? *Clin Chem* 2003;49:1272–1275.
20. Wingren C, Steinhauer C, Ingvarsson J, Persson E, Larsson K, Borrebaeck CA. Microarrays based on affinity-tagged single-chain Fv antibodies: sensitive detection of analyte in complex proteomes. *Proteomics* 2005;5:1281–1291.
21. Miller JC, Zhou H, Kwekel J, et al. Antibody microarray profiling of human prostate cancer sera: antibody screening and identification of potential biomarkers. *Proteomics* 2003;3:56–63.
22. Liotta LA, Espina V, Mehta AI, et al. Protein microarrays: meeting analytical challenges for clinical applications. *Cancer Cell* 2003;3:317–325.
23. Haab BB. Methods and applications of antibody microarrays in cancer research. *Proteomics* 2003;3:2116–2122.
24. Urbanowska T, Mangialaio S, Hartmann C, Legay F. Development of protein microarray technology to monitor biomarkers of rheumatoid arthritis disease. *Cell Biol Toxicol* 2003;19:189–202.
25. Hudelist G, Pacher-Zavisin M, Singer CF, et al. Use of high-throughput protein array for profiling of differentially expressed proteins in normal and malignant breast tissue. *Breast Cancer Res Treat* 2004;86:281–291.
26. Troyer DA, Mubiru J, Leach RJ, Naylor SL. Promise and challenge: markers of prostate cancer detection, diagnosis and prognosis. *Dis Markers* 2004;20:117–128.
27. Lewis S, Menon U. Screening for ovarian cancer. *Expert Rev Anticancer Ther* 2003;3:55–62.
28. Wilkins MR, Gasteiger E, Gooley AA, et al. High-throughput mass spectrometric discovery of protein post-translational modifications. *J Mol Biol* 1999;289:645–657.
29. Dall'Olio F, Chiricolo M. Sialyltransferases in cancer. *Glycoconj J* 2001;18: 841–850.

30. Akamine S, Nakagoe T, Sawai T, et al. Differences in prognosis of colorectal cancer patients based on the expression of sialyl Lewisa, sialyl Lewisx and sialyl Tn antigens in serum and tumor tissue. *Anticancer Res* 2004;24:2541–2546.
31. Gires O, Andratschke M, Schmitt B, Mack B, Schaffrik M. Cytokeratin 8 associates with the external leaflet of plasma membranes in tumour cells. *Biochem Biophys Res Commun* 2005;328:1154–1162.
32. McDonough JL, Van Eyk JE. Developing the next generation of cardiac markers: disease-induced modifications of troponin I. *Prog Cardiovasc Dis* 2004;47: 207–216.
33. Lukasova E, Koristek Z, Falk M, et al. Methylation of histones in myeloid leukemias as a potential marker of granulocyte abnormalities. *J Leukoc Biol* 2005;77:100–111.
34. Irizarry MC. Biomarkers of Alzheimer disease in plasma. *NeuroRx* 2004;1: 226–234.
35. Fukumoto H, Ingelsson M, Garevik N, et al. APOE epsilon 3/epsilon 4 heterozygotes have an elevated proportion of apolipoprotein E4 in cerebrospinal fluid relative to plasma, independent of Alzheimer's disease diagnosis. *Exp Neurol* 2003;183:249–253.
36. Brinkman BM. Splice variants as cancer biomarkers. Clin Biochem 2004;37: 584–594.
37. Baudry D, Hamelin M, Cabanis MO, et al. WT1 splicing alterations in Wilms' tumors. *Clin Cancer Res* 2000;6:3957–3965.
38. Schulte I, Tammen H, Selle H, Schulz-Knappe P. Peptides in body fluids and tissues as markers of disease. *Expert Rev Mol Diagn* 2005;5:145–157.
39. Zhou M, Lucas DA, Chan KC, et al. An investigation into the human serum "interactome". *Electrophoresis* 2004;25:1289–1298.
40. Klose J, Nock C, Herrmann M, et al. Genetic analysis of the mouse brain proteome. *Nat Genet* 2002;30:385–393.
41. Farinha CM, Penque D, Roxo-Rosa M, et al. Biochemical methods to assess CFTR expression and membrane localization. *J Cyst Fibros* 2004;3 Suppl 2:73–77.
42. Sargent DJ, Conley BA, Allegra C, Collette L. Clinical trial designs for predictive marker validation in cancer treatment trials. *J Clin Oncol* 2005;23:2020–2027.
43. Grizzle WE, Semmes OJ, Basler J, et al. The early detection research network surface-enhanced laser desorption and ionization prostate cancer detection study: a study in biomarker validation in genitourinary oncology. *Urol Oncol* 2004;22: 337–343.
44. Ransohoff DF. Bias as a threat to the validity of cancer molecular-marker research. *Nat Rev Cancer* 2005;5:142–149.
45. Hunt SM, Thomas MR, Sebastian LT, et al. Optimal replication and the importance of experimental design for gel-based quantitative proteomics. *J Proteome Res* 2005;4:809–819.
46. Chevalier F, Rofidal V, Vanova P, Bergoin A, Rossignol M. Proteomic capacity of recent fluorescent dyes for protein staining. *Phytochemistry* 2004;65: 1499–1506.

47. Patton WF. Detection technologies in proteome analysis. *J Chromatogr B Analyt Technol Biomed Life Sci* 2002;771:3–31.
48. Raman B, Cheung A, Marten MR. Quantitative comparison and evaluation of two commercially available, two-dimensional electrophoresis image analysis software packages, Z3 and Melanie. *Electrophoresis* 2002;23:2194–2202.
49. Rosengren AT, Salmi JM, Aittokallio T, et al. Comparison of PDQuest and Progenesis software packages in the analysis of two-dimensional electrophoresis gels. *Proteomics* 2003;3:1936–1946.
50. Molloy MP, Brzezinski EE, Hang J, McDowell MT, VanBogelen RA. Overcoming technical variation and biological variation in quantitative proteomics. *Proteomics* 2003;3:1912–1919.
51. Challapalli KK, Zabel C, Schuchhardt J, Kaindl AM, Klose J, Herzel H. High reproducibility of large-gel two-dimensional electrophoresis. *Electrophoresis* 2004;25: 3040–3047.
52. Meleth S, Deshane J, Kim H. The case for well-conducted experiments to validate statistical protocols for 2D gels: different pre-processing = different lists of significant proteins. *BMC Biotechnol* 2005;5:7.
53. Almeida JS, Stanislaus R, Krug E, Arthur JM. Normalization and analysis of residual variation in two-dimensional gel electrophoresis for quantitative differential proteomics. *Proteomics* 2005;5:1242–1249.
54. Unlu M, Morgan ME, Minden JS. Difference gel electrophoresis: a single gel method for detecting changes in protein extracts. *Electrophoresis* 1997;18: 2071–2077.
55. Alban A, David SO, Bjorkesten L, et al. A novel experimental design for comparative two-dimensional gel analysis: two-dimensional difference gel electrophoresis incorporating a pooled internal standard. *Proteomics* 2003;3: 36–44.
56. Karp NA, Kreil DP, Lilley KS. Determining a significant change in protein expression with DeCyder during a pair-wise comparison using two-dimensional difference gel electrophoresis. *Proteomics* 2004;4:1421–1432.
57. Kreil DP, Karp NA, Lilley KS. DNA microarray normalization methods can remove bias from differential protein expression analysis of 2D difference gel electrophoresis results. *Bioinformatics* 2004;20:2026–2034.
58. Gustafsson JS, Ceasar R, Glasbey CA, Blomberg A, Rudemo M. Statistical exploration of variation in quantitative two-dimensional gel electrophoresis data. *Proteomics* 2004;4:3791–3799.
59. Zhan X, Desiderio DM. Differences in the spatial and quantitative reproducibility between two second-dimensional gel electrophoresis systems. *Electrophoresis* 2003;24:1834–1846.
60. Corbett JM, Dunn MJ, Posch A, Gorg A. Positional reproducibility of protein spots in two-dimensional polyacrylamide gel electrophoresis using immobilised pH gradient isoelectric focusing in the first dimension: an interlaboratory comparison. *Electrophoresis* 1994;15:1205–1211.

61. Anderle M, Roy S, Lin H, Becker C, Joho K. Quantifying reproducibility for differential proteomics: noise analysis for protein liquid chromatography-mass spectrometry of human serum. *Bioinformatics* 2004;20:3575–3582.
62. Terry DE, Desiderio DM. Between-gel reproducibility of the human cerebrospinal fluid proteome. *Proteomics* 2003;3:1962–1979.
63. Yamagiwa H, Sarkar G, Charlesworth MC, McCormick DJ, Bolander ME. Two-dimensional gel electrophoresis of synovial fluid: method for detecting candidate protein markers for osteoarthritis. *J Orthop Sci* 2003;8:482–490.
64. Asirvatham VS, Watson BS, Sumner LW. Analytical and biological variances associated with proteomic studies of *Medicago truncatula* by two-dimensional polyacrylamide gel electrophoresis. *Proteomics* 2002;2:960–968.
65. Allard L, Lescuyer P, Burgess J, et al. ApoC-I and ApoC-III as potential plasmatic markers to distinguish between ischemic and hemorrhagic stroke. *Proteomics* 2004;4:2242–2251.
66. Karp NA, Griffin JL, Lilley KS. Application of partial least squares discriminant analysis to two-dimensional difference gel studies in expression proteomics. *Proteomics* 2005;5:81–90.
67. Alaiya AA, Franzen B, Hagman A, et al. Classification of human ovarian tumors using multivariate data analysis of polypeptide expression patterns. *Int J Cancer* 2000;86:731–736.
68. Marengo E, Robotti E, Cecconi D, Hamdan M, Scarpa A, Righetti PG. Identification of the regulatory proteins in human pancreatic cancers treated with trichostatin A by 2D-PAGE maps and multivariate statistical analysis. *Anal Bioanal Chem* 2004;379:992–1003.
69. Dotzlaw H, Schulz M, Eggert M, Neeck G. A pattern of protein expression in peripheral blood mononuclear cells distinguishes rheumatoid arthritis patients from healthy individuals. *Biochim Biophys Acta* 2004;1696:121–129.
70. Borozdenkova S, Westbrook JA, Patel V, et al. Use of proteomics to discover novel markers of cardiac allograft rejection. *J Proteome Res* 2004;3:282–288.
71. Johnson RS, Davis MT, Taylor JA, Patterson SD. Informatics for protein identification by mass spectrometry. *Methods* 2005;35:223–236.
72. Chamrad DC, Korting G, Stuhler K, Meyer HE, Klose J, Bluggel M. Evaluation of algorithms for protein identification from sequence databases using mass spectrometry data. *Proteomics* 2004;4:619–628.

9

Integrative Omics, Pharmacoproteomics, and Human Body Fluids

K. K. Jain

Summary

Proteomics of human body fluids for pharmaceutical applications covers a broad spectrum of technologies. A major challenge is the integration of data from various omics technologies: proteomics, genomics, and metabolomics. Applications relevant to the pharmaceutical industry include diagnostics, drug discovery, and drug development. Special features of these applications are that they are done on a large commercial scale. Currently, sample volumes require large-scale separation prior to analysis by mass spectrometry. Several new technologies, particularly those on the nanoscale, are refining this process and reducing the volume of samples required as well as expense. Refinements of technologies for fluid proteomics, including microfluidics and nanobiotechnology, for detection of biomarkers have applications in clinical diagnostics as well as drug discovery. Analysis of body fluids by metabolomic technologies may uncover biomarkers of drug toxicity that may help in avoiding clinical trials for such drugs, which are then discontinued after considerable expense. Finally, by facilitating the integration of diagnostics with therapeutics, these technologies will allow the development of personalized medicine.

Key Words: Pharmacoproteomics; omics; proteomics; metabolomics; metabonomics; drug discovery; biomarkers; bioinformatics; microfluidics; nanobiotechnology.

1. Introduction

Following the introduction of the terms *genomics* and *transcriptomics* (and later on *proteomics*), numerous other "omics" have been coined. They fit in with a systems biology concept. These terms will be used in the discussion of body fluids in relation to pharmacoproteomics, which refers to the applications of proteomics for drug discovery and development. Various omic technologies used for drug discovery have been described elsewhere ***(1,2)***. The applications of these technologies in the pharmaceutical sector will be considered in this chapter, which in a broad sense includes integration of diagnostics and therapeutics as well as development

From: *Proteomics of Human Body Fluids: Principles, Methods, and Applications*
Edited by: V. Thongboonkerd © Humana Press Inc., Totowa, NJ

of personalized medicine. Detection of biomarkers is a link for various applications. Chemoproteomics, glycoproteomics, metabonomics, and metabolomics are described briefly in this section.

1.1. Chemoproteomics

Chemoproteomics (or chemical proteomics) is complementary to chemical genomics and involves the use of proteomic approaches to study how small molecules interact with cells. It also deals with the "chemome" which is nonenzymatic, chemical modifications of biomolecules in the body. Chemical proteomics has been applied to target identification and drug discovery ***(3)***. Activity-based proteomics is an important approach for chemical proteomics and was established in an attempt to focus proteomic efforts on subsets of physiologically important protein targets. This new approach to proteomics is centered around the use of small molecules termed activity-based probes (ABPs) as a means to tag, enrich, and isolate distinct sets of proteins based on their enzymatic activity ***(4)***. Chemical probes can be "tuned" to react with defined enzymatic targets through the use of chemically reactive warhead groups and fused to selective binding elements that control their overall specificity. As a result, ABPs function as highly specific, mechanism-based reagents that provide a direct readout of enzymatic activity within complex proteomes. Modification of protein targets by an ABP facilitates their purification and isolation, thereby eliminating many of the confounding issues of dynamic range in protein abundance. This technology can be applied to advance the fields of biomarker discovery, in vivo imaging, and small molecule screening and drug target discovery.

1.2. Glycoproteomics

Glycoproteomics is the study of glycoproteins, which have a predominant role in cell-cell and cell-substratum recognition events in multicellular organisms. There is increasing recognition of the importance of posttranslational modifications such as glycosylation as diversifiers of proteins and as potential modulators of their function in health as well as in disease. The term *glycome* is defined, in analogy to the genome and proteome, as a whole set of glycans produced in a single organism. Modifications of proteomic technologies are applied for the analysis of glycoproteins. Isolation of glycopeptides followed by mass spectrometry (MS) analysis was shown to characterize efficiently the structures of β_2-glycoprotein I with four *N*-glycosylation sites and was applied to an analysis of total serum glycoproteins ***(5)***.

1.3. Metabonomics and Metabolomics

Metabonomics is a systems approach to investigate the metabolic consequences of drug exposure, disease processes, and genetic modification, whereas metabolomics is the measurement of metabolite concentrations in cell systems. The technical advantage of metabolomics is its versatility in analyzing all bodily fluids such as whole blood, plasma, cerebrospinal fluid, and urine. Various metabolite-oriented approaches have been described such as metabolite target analysis, metabolite profiling, and metabolic fingerprinting ***(6)***. The capability of analyzing large arrays of metabolites means that biochemical information can be extracted reflecting true functional end points of overt biological events; other functional genomics technologies such as transcriptomics and proteomics, although highly valuable, merely indicate the potential cause for phenotypic response. Therefore, they cannot necessarily predict drug effects, toxicological response, or disease states at the phenotype level unless functional validation is added.

Metabolomics bridges this information gap by depicting in particular such functional information since metabolite differences in biological fluids and tissues provide the closest link to the various phenotypic responses. Such changes in the biochemical phenotype are of direct interest to the pharmaceutical, biotechnology, and health-care industries once the appropriate technology allows the cost-efficient mining and integration of this information. Phenotype is not necessarily predicted by genotype. In this chain of biomolecules, from the genes to phenotype, metabolites are the quantifiable molecules with the closest link to the phenotype. Many phenotypic and genotypic states, such as a toxic response to a drug or disease prevalence, are predicted by differences in the concentrations of functionally relevant metabolites within biological fluids and tissues.

2. Omic Technologies for Analysis of Human Body Fluids for Drug Discovery

Selected omic technologies for analysis of human body fluids for drug discovery are listed in **Table 1**. Proteomic technologies for analysis of body fluids are described in other chapters of this book. The emphasis in this section will be on other omics, particularly metabonomics.

For the measurement of proteins within the range of 10 to 200 kDa, protocols using 2D gel electrophoresis are well established. Peptides between 0.5 and 20 kDa in human body fluids can be analyzed by Differential Peptide Display, from BioVisioN (Hannover, Germany). Proteomics has the potential to identify and compare complex protein profiles, which can be used to generate sensitive molecular fingerprints of proteins present in a body fluid at any given time.

A high-throughput MS immunoassay system has been described for the analysis of proteins directly from plasma *(7)*. A 96-well format robotic workstation was used to prepare antibody-derivatized affinity pipet tips for subsequent use in the extraction of specific proteins from plasma and deposition onto 96-well format matrix-assisted laser desorption ionization-time of flight

Table 1
Selected Omic Technologies for Analysis of Human Body Fluids for Drug Discovery

Proteomic technologies
Two-dimensional polyacrylamide gel electrophoresis (2D-PAGE)
Mass spectrometry (MS)
Matrix-assisted laser desorption/ionization (MALDI) MS
Liquid chromatography-tanden MS (LC-MS/MS)
Surface-enhanced laser desorption/ionization (SELDI) MS
Electrospray ionization (ESI)
Peptide mass fingerprinting
Isotope-coded affinity tag peptide labeling (ICAT)
Chemoproteomics technologies
Activity-based proteomics
Peptide probes
Glycoproteomic technologies
HPLC/MS
Metabonomics/metabolomics technologies
Magnetic resonance imaging (MRI)
Nuclear magnetic resonance (NMR) spectroscopy
Mass spectrometry (MS)
Miniaturized technologies
Microfluidics
Nanofluidics
Nanobiotechnology
Bioinformatics
Protein pattern analysis
Biomarker Amplification Filter (BAMF™) Technology

(MALDI-TOF)-MS targets. This is used to screen samples from multiple individuals with regard to the plasma protein transthyretin, followed by analysis of the same plasma samples for the transthyretin-associated transport protein, retinol-binding protein. The approach represents a rapid and accurate means of characterizing specific proteins present in large numbers of individuals.

2.1. Technologies for Metabolomics

Metabolomics technologies provide a comprehensive quantitative measurement of key metabolites, which represent the whole range of pathways of intermediary metabolism. In a systems biology approach, it provides a functional readout of changes determined by genetic blueprint, regulation, protein abundance and modification, and environmental influence. The importance of metabolomic studies is indicated by the finding that a large proportion of the 6000 genes present in the genome of *Saccharomyces cerevisiae*, and of those sequenced in other organisms, encode proteins of unknown function. Many of these genes are "silent," i.e. they show no overt phenotype, in terms of growth rate or other fluxes, when they are deleted from the genome. Quantification of the change in several metabolite concentrations relative to the concentration change of one selected metabolite can reveal the site of action, in the metabolic network, of a silent gene. In the same way, comprehensive analyses of metabolite concentrations in mutants, providing "metabolic snapshots," can reveal functions when snapshots from strains deleted for unstudied genes are compared with those deleted for known genes.

Apart from being an essential component of the drug discovery and development process, metabolomics can be used to diagnose or predict disease, to stratify patient populations by their specific metabolism, or to determine the safety or efficacy of a therapeutic intervention *(8)*. Specific metabolites are already used to identify drugs with liver or kidney toxicity. The technical advantages of metabolomics are the versatility to analyze:

- All bodily fluids such as whole blood, plasma, cerebrospinal fluid, and urine as well as cultured or isolated cells and biopsy materials.
- Biological samples with high-throughput capability, allowing simultaneous monitoring of multiple experimental alterations.
- Multiple pathways and arrays of metabolites simultaneously from microliter sample quantities.
- Large metabolite data sets by an observation-driven approach with unprecedented speed and markedly enhanced cost effectiveness.

Commercial technologies are being introduced for metabolomic studies. An approach combining proteomics with metabolomics is being pursued by a collaboration of the Thermo Electron Corporation and Paradigm Genetics Inc (Research Triangle Park, NC). The companies will jointly design and develop the next generation of chromatography/MS systems to create a new platform for identifying and validating metabolite biomarkers important to the development of safe and effective drugs. The unique combination of gene expression profiling (determining the level of activity of genes in an organism at a specific time), metabolic profiling (determining the identities and quantities of chemicals in an organism at a specific time), and phenotypic profiling (measuring the physical and chemical characteristics of an organism at a specific time), with data from all systems being managed and analyzed in Paradigm Genetics FunctionFinder™ bioinformatics system, will create a new paradigm for industrializing functional genomics.

The MS-based technology of Biocrates Life Sciences (Innsbruck, Austria) quantifies thousands of metabolites simultaneously from microliter quantities of biological material with high speed, precision, and sensitivity using proprietary preanalytical steps. Quality-assured data are generated from individual samples in a matter of minutes and interpreted employing cutting-edge statistical software tools. Integration of all components into a new technology platform provides the key for widespread utilization and commercialization of metabolomic information.

Paradigm Genetics is building a reference library of all stable metabolites in the human body, under its Human Metabolome Project. This reference library will include biochemical profiling data of human cell lines, tissues, and fluids, which Paradigm believes will be a valuable tool to complement and enhance traditional genomic technologies used for drug discovery and development.

2.2. Technologies for Metabonomics

Although a number of spectroscopic methods have been used for metabonomic studies, nuclear magnetic resonance (NMR) spectroscopy is considered to be one of the most powerful methods for generating multivariate metabolic data ***(9)***. An NMR-based systems approach is used for drug toxicity screening to aid lead compound selection. Metabolic phenotyping (metabotyping) is also used for investigating the metabolic effects of genetic modification and modeling of human disease processes. One deliberate gene knockout can produce several metabolic disturbances. Metabonomics can thus be used as a functional genomics tool with applications in various stages of drug discovery and development.

Metabometrix (London, UK) has a proprietary platform of metabonomics technology for generating, classifying, and interpreting metabolic information obtained from biological fluids and tissues using NMR spectroscopy and advanced chemometric methods. The applications of this technology include drug toxicity screening, drug efficacy screening, and clinical diagnostics.

2.3. Miniaturized Technologies

2.3.1. Microfluidic and Nanofluidic Approaches

Microfluidics involves the handling of small quantities (e.g., microliters, nanoliters, or even picoliters) of fluids flowing in channels the size of a human hair (approx 50 μm thick) or even narrower. Fluids in this environment show very different properties than in the macro world. This new field of technology was allowed by advances in microfabrication—the etching of silicon to create very small features. Microfluidics is one of the most important innovations of biochip technology. Typical dimensions of microfluidic with chips are 1 to 50 cm^2, with channels of 5 to 100 μm. Usual volumes are 0.01 to 10 μL but they can be less. Microfluidics is the link between microarrays and nanoarrays as we reduce the dimensions and volumes.

Reduction in size with microfluidics allows a corresponding increase in the throughput rate of handling, processing, and analyzing the sample. Other advantages of microfluidics include increased reaction rates, enhanced detection sensitivity, and control of adverse events. Lab-on-a-chip (Caliper's LabChip), a miniaturized and integrated liquid handling and biochemical-processing device, is used for computer-aided analytical laboratory procedures that can be performed automatically in seconds. It is used for proteomics as well.

Microfluidic chips can be combined with MS analysis. Electrospray interface to a mass spectrometer can be integrated with a capillary in microfluidic devices, providing a convenient platform for automated sample processing in proteomics applications. The Microfluidic eTag™ Assay System of ACLARA Biosystems (Mountain View, CA) contains eTag reporters that are fluorescent labels with unique and well-defined electrophoretic mobilities; each label is coupled to biological or chemical probes via cleavable linkages. When an eTag reporter-labeled probe binds to its target, the coupling linkage is cleaved and the eTag is released. The distinct mobility address of each eTag reporter allows mixtures of these tags to be rapidly deconvoluted and quantitated by capillary electrophoresis. ACLARA has synthesized eTag reporter libraries in spectrally distinct colors. It is possible to use both mobility and color to increase the degree of multiplexing dramatically. This allows concurrent gene expression, protein expression, and protein function analyses from the same sample. Multiplexed assays can be configured to monitor various types of molecular recognition events such as protein-protein interactions and protein-small molecule binding. This technology is suited for automated, high-throughput application in drug discovery.

2.3.2. Nanofluidics

Nanofluidics implies extreme reduction in the quantity of the fluid analyte in a microchip to nanoliter levels, and the chips used for this purpose are referred to as nanochips. Two well-known commercially available techniques are lab-on-chip, based on nanoliter microfluidics, from Nanolytics (Raleigh, NC) and NanoChip, based on microelectronics, from Nanogen (San Diego, CA).

In one technique, chemical compounds within individual nanoliter droplets of glycerol are microarrayed onto glass slides at 400 spots/cm^2 ***(10)***. Using aerosol deposition, subsequent reagents and water are metered into each reaction center to assemble diverse multicomponent reactions rapidly without cross-contamination or the need for surface linkage. This proteomics technique allows the kinetic profiling of protease mixtures, protease-substrate interactions, and high-throughput screening reactions. The rapid assembly of thousands of nanoliter reactions per slide using a small biological sample (2 mL) represents a new functional proteomics format implemented with standard microarraying and spot-analysis tools.

The use of liquid chromatography (LC) in analytical chemistry is well established, but the relatively low sensitivity associated with conventional LC makes it unsuitable for the analysis of certain biological samples. Furthermore, the flow rates at which it is operated are not compatible with the use of specific detectors, such as electrospray ionization mass spectrometers. Therefore, owing to the analytical demands of biological samples, miniaturized LC techniques were developed to allow for the analysis of samples with greater sensitivity than that afforded by conventional LC. In nanoflow LC (nanoLC) chromatographic separations are performed using flow rates in the range of low nanoliters per minute, which result in high analytical sensitivity owing to the large concentration efficiency afforded by this type of chromatography. NanoLC, in combination with tandem mass spectrometry (MS/MS), was first used to analyze peptides and as an alternative to other MS methods to identify gel-separated proteins. Gel-free analytical approaches based on LC and nanoLC separations have been developed and are allowing proteomics to be performed in a faster and more comprehensive manner than by using strategies based on the classical 2D gel electrophoresis approaches ***(11)***.

Protein identification using nanoflow LC coupled with MS/MS provides reliable sequencing information for the low femtomole level of protein digests. However, this task is more challenging for subfemtomole peptide levels.

2.3.3. Nanobiotechnology

Nanotechnology is the creation and utilization of materials, devices, and systems through the control of matter on the nanometer-length scale, i.e., at the level of atoms, molecules, and supramolecular structures. It is the popular term for the construction and utilization of functional structures with at least one characteristic dimension measured in nanometers. (A nanometer is one billionth of a meter—(10^{-9} m.) Proteins are 1 to 20 nm in size. The application of nanotechnology in life sciences is referred to as nanobiotechnology and is described in detailed in a special report on this topic ***(12)***.

Several nanotechnologies have been used for the study of proteomics. Some of the technologies based on nanoparticles have refined the protein diagnostics of body fluids. Other nanotechnologies have an impact on the application of proteomics in drug discovery. One example is fluorescence planar wave guide technology (PWG) technology, which has demonstrated exceptional performance in terms of sensitivity, making it a viable method for detection in the ZeptoMARK protein profiling system—a chip-based microarray from Bayer Technology Services (Leverkusen, Germany).

Thin-film PWGs consist of a 150- to 300-nm-thin film of a material with a high refractive index, which is deposited on a transparent support with a lower refractive index (e.g., glass or polymer). A parallel laser light beam is coupled to the wave-guiding film by a diffractive grating, which is etched or embossed into the substrate. The light propagates within this film and creates a strong evanescent field perpendicular to the direction of propagation into the adjacent medium. It has been shown that the intensity of this evanescent field can be enhanced dramatically by increasing the refractive index of the wave-guiding layer and equally by decreasing the layer thickness. Compared with confocal excitation, the field intensity close to the surface can be increased by a factor of up to 100. The field strength decays exponentially with the distance from the waveguide surface, and its penetration depth is limited to about 400 nm. This effect can be utilized to excite selectively only fluorophores located at or near the surface of the wave guide. By taking advantage of the high field intensity and the confinement of this field to the close proximity of the wave guide, PWG technology combines highly selective fluorescence detection with the highest sensitivity. For bioanalytical applications, specific capture probes or recognition elements for the analyte of interest are immobilized on the wave-guide surface. The presence of the analyte in a sample applied to a PWG chip is detected using fluorescent reporter molecules attached to the analyte or one of its binding partners in the assay. Upon fluorescence excitation by the evanescent field, excitation and detection of fluorophores are restricted to the sensing surface, whereas signals from unbound molecules in the bulk solution are not detected. The result is a significant increase in the signal/noise ratio compared with conventional optical detection methods.

A variety of proteins can be immobilized on PWG microarrays as selective recognition elements for the investigation of specific ligand-protein interactions such as antigen-antibody, protein-protein, and protein-DNA interactions. Protein microarrays based on PWG allow the simultaneous, qualitative, and quantitative analysis of protein interactions with high sensitivity in a massively parallel manner. This method allows cost-effective determination of efficacy of drug candidates in a vast number of preclinical study samples.

2.4. Role of Bioinformatics for Integration and Analysis of Various Omic Technologies

The role of bioinformatics in the analysis of proteomics data is well recognized, and several tools are available for this purpose. The application of bioinformatics is particularly important in large-scale analysis of fluids in the pharmaceutical industrial setting. Bioinformatics is also important for integration of data from various omic technologies and applications for various purposes including clinical diagnostics and drug discovery. A detailed discussion of this topic is beyond the scope of this chapter, and only one example will be given here.

Biomarker AMplification Filter (BAMF™), a proprietary technology of Predictive Diagnostics (Vacaville, CA), is a comprehensive suite of *in silico* machine learning technologies and advanced informatics tools that examines multiple protein biomarkers found in the blood utilizing high-resolution MS data. The biomarkers, which have been derived from profiling thousands of potential markers, allow a comprehensive reading of the "fingerprint" left by diseased cells. Most of the proteomics-based diagnostic tests have focused on single-biomarker analysis, but the potential for improved test accuracy (sensitivity and specificity) is greatly increased when the emphasis is on multiple disease biomarkers. BAMF Technology not only improves diagnostics but also facilitates treatment options during the earliest stages of disease. BAMF technology has been used to diagnose accurately cancers such as breast, lung, pancreatic, and prostate, as well as other disorders such as those of the central nervous system (CNS).

3. Search for Protein Biomarkers in Body Fluids

The first decision to be made in the search for a biomarker is whether to look in a body fluid or a tissue. Body fluids have the advantage of being more easily accessible and are more likely to be of clinical use because serum or urine can be obtained by noninvasive methods routinely.

3.1. Plasma vs Serum for Industrial Scale Proteomic Studies

Plasma, obtained by centrifuging blood to remove red blood cells (RBCs) and filtering to remove white blood cells (WBCs), contains fibrinogen. Plasma represents the noncellular components of the blood and is used in preference to serum by the Plasma Proteome Institute and by the Human Proteome Organization (HUPO). The HUPO Plasma Proteome Project has started a pilot phase study focused on key problems essential for standardization of specimen collection, specimen handling, choice of fractionation and analysis technologies, and search engines and databases for protein identifications ***(13)***.

Serum is obtained by coagulation of the blood, which traps the RBCs in fibrin—a degradation product of fibrinogen. One disadvantage of serum is that clotting can release proteins and peptides back into the serum. The advantages of using serum are that it is more stable and can be stored frozen for many years. Serum and plasma are preferred for large-scale proteomic analysis for the following reasons:

- These are already the most frequent fluids examined in clinical laboratories.
- Both potential therapeutic proteins/peptides and biomarkers of disease are present in plasma/serum.
- They are available in large amounts from patients as well as controls.

- They can be pooled, whereas pooling of tissues is difficult.
- These fluids are devoid of cells, the protein contents of which complicate a cell-based approach.

Blood serum is the favored source for investigators interested in large-scale proteomics, because it has the most proteins. However, so far only about 500 of the 30,000 proteins in serum have been identified. Removal of albumin and the other five major proteins allows further study of the proteome.

3.2. Challenges Facing the Use of Biofluid Proteomics for Drug Discovery

Many factors make this research challenging, beginning with the lack of standardization of sample collection and continuing through the entire analytical process. Identification of rare proteins in blood is often hindered by highly abundant proteins, such as albumin and immunoglobulin, which obscure less plentiful molecules. A solution to this problem is an immunoaffinity column, the Multiple Affinity Removal System from Agilent Technologies (Palo Alto, CA), which can handle antibodies to the six most abundant proteins found in human blood. By merely running a sample over the matrix, one can specifically remove all six proteins at once, unveiling lower abundance species that may represent new biomarkers for disease diagnosis and therapy. The process removes about 85% of the total protein mass. The multiple affinity removal system works with blood, cerebrospinal fluid, and urine, all of which contain the same major proteins. High-abundant protein removal, combined with 2D differential gel electrophoresis, is a practical approach for enriching and characterizing lower abundant proteins in human serum *(14)*. This method offers advances in proteomic characterization and therefore in the identification of biomarkers from human serum.

Plasma proteins are unstable in postcollection samples. Once the plasma proteins are removed from their protected in vivo environment, ex vivo losses start immediately and it may be difficult to measure proteins after a few hours or even a few minutes. BD Diagnostics (Franklin Lakes, NJ) is trying to address the problem of protein degradation. MS was used to find proteins that are damaged, and quantitative analysis revealed the intensity of peptide changes with time, suggesting that some plasma proteins are digested over the course of time. A protein inhibitor cocktail included in blood collection tubes at the time of blood withdrawal stabilizes the peptide mass spectra over time.

4. Biomarker Discovery by Omic Technologies

Proteomic profiling of serum is an emerging technique to identify new biomarkers indicative of disease severity and progression. This has clinical diagnostic applications, and the biomarkers can also be used for drug discovery.

4.1. Cancer Biomarkers in Blood

MS analysis of serum proteins has revealed biomarker patterns in a variety of cancers that provide useful information for drug discovery. Preliminary studies have shown this to be a promising direction.

4.1.1. Biomarkers of Ovarian Cancer

Serum samples from patients with ovarian cancer, patients with benign tumors, and healthy donors have been analyzed on strong anion-exchange surfaces of protein biochips using surface-enhanced laser desorption/ionization (SELDI)-TOF MS technology ***(15)***. Univariate and multivariate statistical analysis was applied to the protein profiling data obtained. The investigators discovered three ovarian cancer biomarker protein panels that, when used together, effectively distinguished serum samples from healthy controls and patients with either benign or malignant ovarian neoplasia.

A study from the National Cancer Institute (Bethesda, MD) showed that a new method of assessing protein patterns in blood samples, using MS, has 100% sensitivity and 95% specificity ***(16)***. Although proteomic profiling of serum initially appeared to be dramatically effective for diagnosis of early-stage ovarian cancer, these results have proved difficult to reproduce. A later analysis revealed that the pattern allowing successful classification was biologically implausible and that the method does not classify the data accurately ***(17)***. The reproducibility of the proteomic profiling approach has yet to be established.

4.1.2. Biomarkers of Breast Cancer

Proteomic approaches such as SELDI MS, in conjunction with bioinformatics tools, could greatly facilitate the discovery of new and better biomarkers. The high sensitivity and specificity achieved by the combined use of the selected biomarkers show great potential for the early detection of breast cancer ***(18)***.

Mammary ductal cells are the origin of 70 to 80% of breast cancers. Nipple aspirate fluid (NAF) contains proteins secreted directly by the ductal and lobular epithelia in nonlactating women. NAF has been used for many years as a potentially noninvasive method to identify markers for breast cancer risk or early detection. SELDI-TOF-MS can identify patterns of proteins that might define a proteomic signature for breast cancer. SELDI analysis of NAF is rapid, reproducible, and capable of identifying protein signatures that appear to differentiate NAF samples from breast cancer patients and healthy controls.

Proteomic approaches offer a largely unbiased way to evaluate NAF as a source of biomarkers and are sufficiently sensitive for analysis of small NAF volumes (10 to 50 μL). In a study at the Pacific Northwest National Laboratory, this process resulted in a volume of NAF that was suitable for analysis in approximately 90% of subjects. Proteomic characterization of NAF identified

64 proteins *(19)*. Although this list primarily includes abundant and moderately abundant NAF proteins, very few of these proteins have previously been reported in NAF. At least 15 of the NAF proteins identified have previously been reported to be altered in serum or tumor tissue from women with breast cancer, including cathepsin D and osteopontin. This study provides the first characterization of the NAF proteome and identifies several candidate proteins for future studies on breast cancer markers in NAF.

Another study has examined proteomic changes in response to paclitaxel chemotherapy or 5-fluorouracil, doxorubicin, and cyclophosphamide (FAC) chemotherapy in plasma from patients with stage I to III breast carcinoma *(20)*. A single chemotherapy-inducible SELDI-MS peak and five other peaks that distinguished plasma obtained from patients with breast carcinoma from plasma obtained from normal, healthy women were identified. The proteins represented by these peaks are candidate markers for micrometastatic disease after surgery.

4.1.3. Biomarkers of Head and Neck Cancer

MS-based techniques have been applied to the study of serum as well as tissue proteomics for head and neck squamous cell carcinoma (HNSCC). In addition, the evaluation of salivary fluids provides new sources for HNSCC biomarkers. A MALDI profiling study of HNSCC and lung cancer serum has also been reported using SELDI-TOF-MS screening for differentially expressed proteins in serum from patients with HNSCC and normal controls *(21)*. The serum samples are processed for SELDI analysis using an automated robotic sample prior to incubation with copper-coated IMAC3 ProteinChip arrays from Ciphergen Biosystems (Fremont, CA), which are analyzed using a SELDI PBS-II instrument *(22)*. The scheme of this procedure is shown in **Fig. 1**.

MALDI and SELDI protein chip profiling can be used to screen patient serum to identify protein expression profiles consistent with HNSCC. The serum profiling approaches that were described for HNSCC serum are based on a dilution strategy, the premise being that minimizing sample processing would facilitate assay reproducibility and higher throughput, thus making an eventual diagnostic test more feasible. A strategy of fractionating serum into multiple enriched fractions prior to SELDI or MALDI, or removal of major serum proteins such as albumin prior to analysis, remains a viable alternative to this approach. Another possibility is selectively enriching for major serum carrier proteins such as albumin, as these could be the primary carriers of the low-mass peptides evaluated by SELDI/MALDI profiling strategies.

4.1.4. Biomarkers of Lung Cancer

Proteomic technologies have been used to search for lung disease markers in bronchoalveolar lavage fluid. Epithelial lining fluid is sampled by bronchoalveolar lavage during bronchoscopy. The protein contents of lavage fluid are very

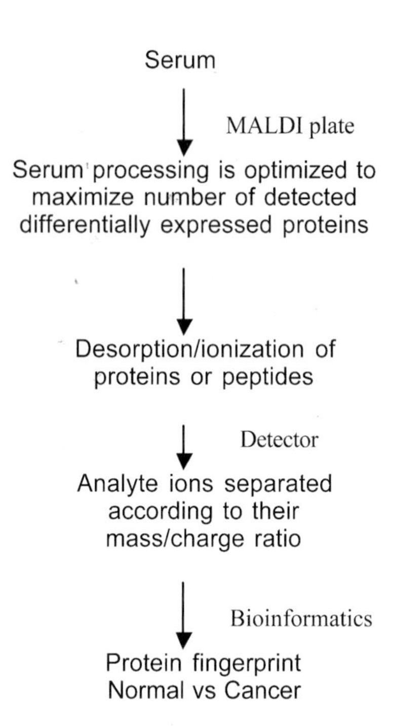

Fig. 1. Sequence of steps in the use of ProteinChip SELDI for detection of serum biomarkers.

complex, not only because of the wide variety of the proteins but also because of the great diversity of their cellular origins. Proteomic approaches are helpful in sorting out the analysis of these complex proteins *(23)*. Results of a study using the SELDI technique suggest that lavage fluid is a promising resource for detection of specific lung cancer biomarkers when combined with an artificial intelligence classification algorithm *(24)*.

4.1.5. Biomarkers of Prostate Cancer

A panel of serum proteins has been identified retrospectively that can discriminate between men with prostate cancer (clinically confined to the organ) and men with benign prostate disease *(25)*. Serum proteomics mass spectra of these patients were generated using ProteinChip arrays and a ProteinChip Biomarker System II SELDI-TOF mass spectrometer from Ciphergen Biosystems. A combination of bioinformatics tools including ProPeak from 3Z Informatics (Charleston, SC) was used to reveal the optimal panel of biomarkers for maximum separation of the prostate cancer and the benign prostate disease cohorts. The results showed that serum proteomics patterns may potentially aid in the early detection of prostate cancer.

4.1.6. Biomarkers of Pancreatic Cancer

Proteins associated with pancreatic cancer identified through proteomic profiling technologies could be useful as biomarkers for early diagnosis, for therapeutic targets, and for disease response markers. For the discovery phase, protein profiling can be used to analyze pancreatic tissue, pancreatic juice, and serum to identify candidate biomarkers. Candidate biomarkers can be further validated by immunohistochemistry in large cohorts of patient samples, using as tissue arrays, and subsequently developed into diagnostic biomarkers in serum or pancreatic juice, using enzyme-linked immunosorbent assay (ELISA), protein array, and/or high-throughput proteome-screening technology ***(26)***.

4.2. Biomarkers in Hepatitis

Analysis of the polypeptide profile in tissues, cells, and sera by high-resolution 2D-polyacrylamide gel electrophoresis (2D-PAGE) offers promise in the identification of biomarkers that correlate with disease. However, sera contain many polypeptides bearing *N*-linked glycosylation that can complicate interpretation. To simplify serum proteome profiles, polypeptides present in human serum were left untreated or subjected to de-*N*-glycosylation by incubation with PNGase F and resolved by high-resolution 2D-PAGE ***(27)***. de-*N*-glycosylation reduced the number of glycoform variants, enhanced the resolution of many polypeptides, and allowed other polypeptides to become visible. As an initial test of concept, clinically relevant serum samples from individuals with or without hepatocellular carcinoma were compared. Several polypeptides, apparent only after de-*N*-glycosylation, were shown to correlate with disease. The data suggest that de-*N*-glycosylation offers a method to enhance the resolution of serum polypeptide profiles and has value in comparative proteomic studies.

4.3. Biomarkers in Rheumatoid Arthritis

An objective of current research for developing new management approaches to rheumatoid arthritis (RA) is the discovery of protein biomarkers of that can predict which patients will develop erosive, disabling disease. A two-step proteomic approach was used for biomarker discovery and verification: (1) 2D LC-MS/MS was used to generate protein profiles of synovial fluid from patients with RA; and (2) verification of the selected candidate markers using quantitative multiple reaction monitoring MS in sera of patients and of healthy controls ***(28)***. Several protein marker candidates have been identified for prognosis of the erosive form of RA. This study demonstrates the usefulness of MS for global discovery and verification of clinically relevant sets of disease biomarkers in synovial fluid and serum.

4.4. Biomarkers in CNS Disorders

Use of BAMF technology in combination with the BioXPRESSION Biomarker platform of PerkinElmer (Boston, MA) has led to discovery of biomarkers based on patterns of proteins and peptides that distinguish Alzheimer's disease patients from those without clinical signs. MALDI-TOF-MS, in combination with serum proteomic pattern analysis, could be useful in the diagnosis of multiple sclerosis, and a larger, masked trial to identify proteomic spectral patterns characteristic of relapsing-remitting, primary progressive, and secondary progressive variants of this disease is justified ***(29)***.

4.5. Applications of Biomarkers in Drug Discovery and Development

The advantage of applying biomarkers to early drug development is that they might aid in preclinical and early clinical decisions such as dose ranging, definition of treatment regimen, or even a preview of efficacy ***(30)***. Later in the clinic, biomarkers could be used to facilitate patient stratification, selection, and the description of surrogate end points. Information derived from biomarkers should result in a better understanding of preclinical and clinical data, which will ultimately benefit patients and drug developers. If the promise of biomarkers is realized, they will become a routine component of drug development and companions to newly discovered therapies.

5. Conclusions

The proteomics of human body fluids for pharmaceutical applications covers a broad spectrum of technologies. A major challenge is the integration of data from various omics technologies: proteomics, genomics, and metabolomics. Applications relevant to the pharmaceutical industry include diagnostics, drug discovery, and drug development. A special features of these applications is that they are done on a large commercial scale. Currently, sample volumes require large-scale separation prior to analysis by MS. Several new technologies, particularly those on the nanoscale, are refining this process and reducing the volume of samples required as well as the expense. Refinements of fluid proteomics, technologies, including microfluidics and nanobiotechnology, for detection of biomarkers have applications in clinical diagnostics as well as drug discovery. Analysis of body fluids by metabolomic technologies may reveal biomarkers of drug toxicity that may help to prevent the discontinuance of clinical trials after considerable expense has been incurred. Finally, by facilitating the integration of diagnostics with therapeutics, these technologies will allow the development of personalized medicine.

References

1. Jain KK. Proteomics and drug discovery. *Contrib Nephrol* 2004;141:308–327.
2. Jain KK. *Proteomics: Technologies, Markets and Companies*. Basel: Jain PharmaBiotech Publications, 2007:1–536.
3. Verhelst SH, Bogyo M. Chemical proteomics applied to target identification and drug discovery. *Biotechniques* 2005;38:175–177.
4. Berger AB, Vitorino PM, Bogyo M. Activity-based protein profiling: applications to biomarker discovery, in vivo imaging and drug discovery. *Am J Pharmacogenomics* 2004;4:371–381.
5. Wada Y, Tajiri M, Yoshida S. Hydrophilic affinity isolation and MALDI multiple-stage tandem mass spectrometry of glycopeptides for glycoproteomics. *Anal Chem* 2004;76:6560–6565.
6. Fiehn O. Combining genomics, metabolome analysis, and biochemical modelling to understand metabolic networks. *Comp Funct Genom* 2001;2:155–168.
7. Kiernan UA, Tubbs KA, Gruber K, et al. High-throughput protein characterization using mass spectrometric immunoassay. *Anal Biochem* 2002;301: 49–56.
8. Watkins SM, German JB. Metabolomics and biochemical profiling in drug discovery and development. *Curr Opin Mol Ther* 2002;4:224–228.
9. Nicholson JK, Connelly J, Lindon JC, Holmes E. Metabonomics: a platform for studying drug toxicity and gene function. *Nat Drug Discov* 2002;1:153–161.
10. Gosalia DN, Diamond SL. Printing chemical libraries on microarrays for fluid phase nanoliter reactions. *Proc Natl Acad Sci U S A* 2003;100:8721–8726.
11. Cutillas PR. Principles of nanoflow liquid chromatography and applications to proteomics. *Curr Nanosci* 2005;1:65–71.
12. Jain KK. *Nanobiotechnology: Applications, Markets and Companies*. Basel: Jain PharmaBiotech Publications, 2007:1–605.
13. Omenn GS. Advancement of biomarker discovery and validation through the HUPO plasma proteome project. *Dis Markers* 2004;20:131–134.
14. Chromy BA, Gonzales AD, Perkins J, et al. Proteomic analysis of human serum by two-dimensional differential gel electrophoresis after depletion of high-abundant proteins. *J Proteome Res* 2004;3:1120–1127.
15. Kozak KR, Amneus MW, Pusey SM, et al. Identification of biomarkers for ovarian cancer using strong anion-exchange ProteinChips: potential use in diagnosis and prognosis. *Proc Natl Acad Sci U S A* 2003;100:12,343–12,348.
16. Petricoin EF, Ardekani AM, Hitt BA, et al. Use of proteomic patterns in serum to identify ovarian cancer. *Lancet* 2002;359:572–577.
17. Baggerly KA, Morris JS, Edmonson SR, Coombes KR. Signal in noise: evaluating reported reproducibility of serum proteomic tests for ovarian cancer. *J Natl Cancer Inst* 2005;97:307–319.
18. Li J, Zhang Z, Rosenzweig J, et al. Proteomics and bioinformatics approaches for identification of serum biomarkers to detect breast cancer. *Clin Chem* 2002;48: 1296–1304.

19. Varnum SM, Covington CC, Woodbury RL, et al. Proteomic characterization of nipple aspirate fluid: identification of potential biomarkers of breast cancer. *Breast Cancer Res Treat* 2003;80:87–97.
20. Pusztai L, Gregory BW, Baggerly KA, et al. Pharmacoproteomic analysis of prechemotherapy and postchemotherapy plasma samples from patients receiving neoadjuvant or adjuvant chemotherapy for breast carcinoma. *Cancer* 2004;100: 1814–1822.
21. Sidransky D, Irizarry R, Califano JA, et al. Serum protein MALDI profiling to distinguish upper aerodigestive tract cancer patients from control subjects. *J Natl Cancer Inst* 2003;95:1711–1717.
22. Drake RR, Cazare LH, Semmes OJ, Wadsworth JT. Serum, salivary and tissue proteomics for discovery of biomarkers for head and neck cancers. *Expert Rev Mol Diagn* 2005;5:93–100.
23. Noel-Georis I, Bernard A, Falmagne P, Wattiez R. Proteomics as the tool to search for lung disease markers in bronchoalveolar lavage. *Dis Markers* 2001;17: 271–284.
24. Xiao X, Liu D, Tang Y, et al. Development of proteomic patterns for detecting lung cancer. *Dis Markers* 2003–2004;19:33–39.
25. Li J, White N, Zhang Z, et al. Detection of prostate cancer using serum proteomics pattern in a histologically confirmed population. *J Urol* 2004;171:1782–1787.
26. Chen R, Pan S, Brentnall TA, Aebersold R. Proteomic profiling of pancreatic cancer for biomarker discovery. *Mol Cell Proteomics* 2005;4:523–533.
27. Comunale MA, Mattu TS, Lowman MA, et al. Comparative proteomic analysis of de-*N*-glycosylated serum from hepatitis B carriers reveals polypeptides that correlate with disease status. *Proteomics* 2004;4:826–838.
28. Liao H, Wu J, Kuhn E, et al. Use of mass spectrometry to identify protein biomarkers of disease severity in the synovial fluid and serum of patients with rheumatoid arthritis. *Arthritis Rheum* 2004;50:3792–3803.
29. Avasarala JR, Wall MR, Wolfe GM. A distinctive molecular signature of multiple sclerosis derived from MALDI-TOF/MS and serum proteomic pattern analysis. *J Mol Neurosci* 2005;25:119–126.
30. Lewin DA, Weiner MP. Molecular biomarkers in drug development. *Drug Discov Today* 2004;9:976–983.

II

Proteomic Analysis of Specific Types of Human Body Fluids

Methods, Findings, Applications, Perspectives, and Future Directions

10

The Human Plasma and Serum Proteome

Gilbert S. Omenn, Rajasree Menon, Marcin Adamski, Thomas Blackwell, Brian B. Haab, Weimin Gao, and David J. States

Summary

Human plasma and serum are the preferred specimens for noninvasive studies of normal and disease-associated proteins in the circulation and arising from cells throughout the body. The attributes of extreme complexity, very wide dynamic range, genetic and physiological variation, endogenous and ex vivo modifications, and incompleteness of sampling by mass spectrometry all represent major challenges to reproducible, high-resolution, high-throughput analyses of the plasma proteome. This chapter summarizes the major reports to date identifying proteins in normal individuals and identifies paths to increased use of proteomics methods with human specimens for biomarker discovery and application in various diseases.

Key Words: Plasma proteome; complexity; dynamic range; abundance; variation; posttranslational modifications; splice isoforms; ex vivo modification; high-throughput; error estimation.

1. Introduction

Plasma or serum specimens are the most available samples from patients and from participants in clinical research (**Table 1**). These specimens offer the best prospects for minimally invasive sampling of proteins that are discovered and validated to be biomarkers for physiological and pathological changes and responses to pharmacologic agents. The proteins of the noncellular fraction of blood represent a combination of those essential to circulatory functions (osmotic pressure, coagulation, complement, immunity) and those secreted or released into the circulation from cells in organs throughout the body both during normal cell turnover *(1,2)* and in diseases with specific injury, such as myocardial infarction *(3)*. They may also include proteins derived from common saprophytic microorganisms in our bodies. Both the plasma and various types of cells contain numerous proteases, so there is an expectation that many proteins

From: *Proteomics of Human Body Fluids: Principles, Methods, and Applications*
Edited by: V. Thongboonkerd © Humana Press Inc., Totowa, NJ

Table 1
The Plasma Proteome

Advantages
The most available human specimen
The most comprehensive sample of tissue-derived proteins
Specific disadvantages
Extreme complexity/enormous dynamic range
High risk of ex vivo modifications
Lack of highly standardized protocols
General challenges
Inadequate appreciation of incomplete sampling by MS/MS evolving
Annotations and unstable, evolving databases

will be found at lower molecular weight than that of the full-length protein. Although the gene and protein databases commonly give protein lengths and molecular weights for the primary gene product, often a precursor protein, that form of the protein may not be found in the plasma at all, nor is genetic variation in primary sequence represented in the most widely used protein sequence databases. The action of proteases and other enzymes that create posttranslational modifications of proteins may continue ex vivo while specimens are collected, processed, stored, thawed, and processed again before analysis.

1.1. Dynamic Range of Concentration or Abundance

The concentrations of proteins in plasma range from the most abundant by far, albumin at 40 mg/mL, to cytokines and kallikreins (including prostate-specific antigen [PSA]) at about 1 ng/mL, to proteins of tissue origin down to much lower concentrations. From 40 mg/mL to 4 pg/mL is 10 orders of magnitude (*see* **Fig. 1**). The most common technology for fractionating and identifying proteins, 2D gel electrophoresis (2DE), has a range of detection of not more than 3 orders of magnitude. Other methods have similar or slightly larger ranges. Plasma has a far larger dynamic ranges than tissue or other body-fluid specimens.

Albumin accounts for approx 50% of the total mass of proteins in the plasma. Another several proteins account for about 40% of total protein mass *(2)*. The most abundant 22 proteins are estimated to account for 99% of total protein mass *(4)*. **Table 2** lists particularly abundant and readily detected proteins, with their usual concentrations. Abundance is an extremely important variable, not only because more abundant proteins are more readily detected, but also because the peptides and peptide ions from these proteins compete against peptides from other proteins in identification by mass spectrometry (MS). Thus, several methods have been developed for depletion of highly abundant proteins,

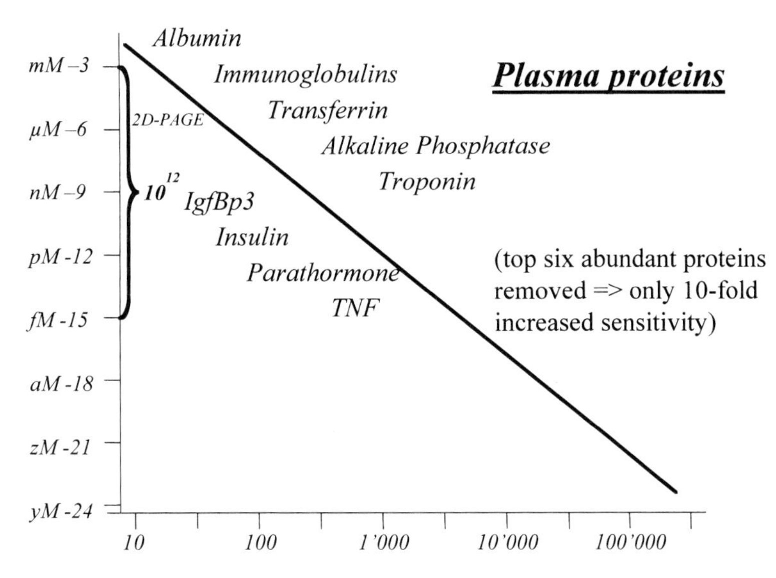

Fig. 1. Dynamic range of proteins in plasma. (Adapted from figure kindly provided by Dr. Denis Hochstrasser.)

Table 2
Most Abundant Plasma Proteins

Protein	Plasma level (mg/mL)
Albumin	35–45
IgG, IgA, IgM	12–18
α-2 and β-lipoproteins (LDL)	4–7
Fibrinogen	2–6
α_1-Antitrypsin	2–5
α_2-Macroglobulin	2–4
Transferrin	2–3
α-1-Acid glycoprotein	1
Hemopexin	1
α-lipoproteins (HDL)	0.6–1.5
Haptoglobin	0.3–2
Prealbumin	0.3–0.4
Ceruloplasmin	0.3

Adapted from *Chemical Rubber Handbook of Biochemistry*, 1970:C-36–39, and current data from GenWay, at the September 2002 PPP Workshop.

including binding to dye molecules or proteins with affinity for albumin and for immunoglobulins and binding on immunoaffinity columns to polyclonal antibodies raised against the abundant proteins. One column from Agilent removes six sets of proteins (albumin, IgG, IgA, α_1 antitrypsin, haptoglobin, and transferrin; a new version also removes fibrinogen). Other columns from GenWay Biotech remove 6 (albumin, IgG, IgM, IgA, transferrin, and fibrinogen) or 12 proteins (also α_1 antitrypsin, α_2 macroglobulin, haptoglobin, apolipoproteins A-I and A-II, and orosomucoid/α_1 acid glycoprotein) ***(5)***. A Sigma-Aldrich product removes 20 plasma proteins.

Concentration is dependent on stability, binding, and clearance of proteins once in the circulation. Proteins small enough to pass through the glomerular filter of the kidney will be lost into the urine, unless they are bound to other proteins; such binding occurs commonly (*see* **Subheading 3.4.**). Protein-protein interactions are important to functions of proteins and are often modified under physiological, pathological, and pharmacological influences ***(6,7)***.

1.2. Complexity and Variation

The current estimate of the number of protein-coding genes in the human genome is about 22,000. The number of corresponding proteins is, however, much higher. Multiple proteins are generated owing to alternative transcription of genes (splice-isoforms), single-nucleotide polymorphisms (SNPs) and other mutation-derived and chromosomal genetic variations, and numerous posttranslational modifications, starting with processing of precursors. In pathological conditions such as cancers, derangements in splicing and posttranslational processing mechanisms may further increase the number of protein products produced. The many isoforms, especially of abundant proteins, generate families of protein spots on gels and in any other method. The number of proteins and their isoforms in normal plasma is unknown but is presumed to be several hundred thousands, possibly including some amount of nearly every protein in the body. As we shall document below, intracellular and intranuclear proteins are readily detected in plasma with sensitive MS or antibody capture methods. MS methods that have extraordinary sensitivity to detect a single purified protein or a mixture of small numbers of proteins run up against their limits at much higher concentrations when the specimen has such complexity. Plasma has far greater complexity than tissue proteomes. Immunoglobulins represent a particularly diverse set of related proteins.

1.3. The Inherent Incompleteness of Peptide and Protein Identifications and the Risk of False-Positive Identifications

Because of competition and suppression by peptide ions, as well as vast differences in protein concentrations, MS is an inherently incomplete sampling process ***(8,9)***. One may compare this process with placing a fishing rod in a fast-moving stream with lots of fish. Furthermore, the automated search engine algorithms that match spectra and mass/charge ratio peaklists to peptide sequences have a substantial risk of false-positive identifications ***(10–13)***. Even repeating the same analysis may give a different result owing to nearly identical mass values for certain expected or modified amino acids and stochastic features of the algorithm. When peptides are later matched to protein sequences in gene or protein databases, multiple high-confidence assignments may occur, reflecting the high homology of protein sequences, especially for protein families and protein isoforms owing to SNPs, or splicing, or other modifications. The combination of dynamic range and complexity of plasma makes the sampling problem much greater with plasma than with organ/tissue proteomes.

1.4. Ex Vivo Sources of Variation

When blood is collected, many more changes in proteins may occur due to proteolytic enzymes (proteases) and other enzymes that are active in the blood sample during handling and processing. Proteins from blood's cellular components may be released ex vivo, as well. It may be difficult to differentiate in vivo from ex vivo hemolysis (breakdown of red blood cells with release of hemoglobin and other proteins), platelet activation (enhanced at 4°C, with release of platelet basic protein, thymosin-β-4, platelet factor 4, and other platelet markers), or white blood cell degranulation or breakdown with release of other proteins.

Plasma is converted to serum by permitting or activating clot formation, usually at room temperature, which involves the very active protease action of thrombin on fibrinogen and related protein targets and other proteases on other proteins of the coagulation cascade ***(14)*** (*see* also **Subheading 7.**). The forming clot itself provides a physical scaffold for attachment of proteins. Plasma is protected from clotting by use of one of three different anticoagulants: sodium citrate, K2-EDTA, or lithium heparin. Each has its own characteristics; citrate and EDTA have desirable features of antiprotease activity. Some investigators add a protease inhibitor cocktail, often proprietary in composition; generally these cocktails include both peptide and small molecule inhibitors ***(15)***. The peptide(s) may compete directly with and interfere with the detection of peptides in the mass spectrometer, whereas such small molecules as ABESF, a sulfonyl fluoride, have been shown to form covalent bonds with proteins and thereby shift the isoelectric point (p*I*) of the protein ***(16)***.

2. Methods for Detection and Identification of Proteins

2.1. Gel Electrophoresis

The workhorse method remains 2-DE of protein mixtures, followed by marking of protein spots of interest for physical picking of those spots, which are then digested with trypsin and subjected to liquid chromatography-tandem MS (LC-MS/MS) analysis. The online fractionation of the tryptic peptide digest greatly reduces the complexity presented to the MS/MS ion trap and enhances detection of distinctive peptide features in mass spectra. Gel methods are poor at separating and visualizing hydrophobic proteins, high-molecular-weight (MW) proteins, very low-MW proteins/peptides, and proteins at the extremes of p*I*. According to Anderson et al. ***(17)***, 2-DE was able to resolve 40 distinct plasma proteins in 1976; because of the dynamic range limitations and the many isoforms, that number was still only 60 in 1992 and remains about the same today. Multidimensional fractionation has greatly expanded this resolution, including combinations of depletion of abundant proteins, fractionation of intact proteins, chromatographic resolution of peptides, and separation of peptide fragments in the mass spectrometer. Variations on this approach include liquid-phase 2D (p*I*, MW) and 3D (p*I*, hydrophobicity, MW) fractionation of the proteins, often preceded by immunoaffinity depletion of the most abundant proteins. An example is the intact protein analytical system of Wang et al. ***(18)***.

2.2. Quantitation of Proteins

Quantitation is a complex challenge. Differential labeling of two specimens to be compared (like before and after treatment), using fluorescent Cy dyes, can yield relative concentration measurements when the two labeled specimens are pooled and processed together. Mass isotope-coded affinity tag (ICAT) labeling methods can generate precise ratios of concentrations after analogous labeling and pooling experiments ***(19)***. A recently introduced isobaric tags for relative and absolute quantitation (iTRAQ) method, using a set of four tags that fragment in the mass spectrometer to release fragments and retain labels of 114, 115, 116, and 117 mass, permits comparing concentrations of four different specimens simultaneously ***(20)***.

A particularly promising approach using "proteotypic peptides" has been introduced recently to score proteins for identification and for relative or absolute quantification ***(21,22)***. The principle is that peptides are identified from databases or from large empirical studies, including pooled studies (www.peptideatlas.com), that have sequences uniquely matching a single protein and physiochemical properties that favor detectability in MS/MS experiments ***(21)***. Synthesis and isotopic labeling of these peptides then permits their use through large-scale spiking to generate paired peptide ions and peptide ion fragments from which the relative concentration can be calculated ***(23)***.

2.3. Protein Capture Methods

Enzyme-linked immunosorbent assay (ELISA) is a protein capture method that has been proved to be of enormous utility. Dependent on the availability of high-quality antibodies, ELISA assays can be sensitive, robust, and reliable. A range of automated, parallel, and multiplexed formats have been developed to enhance the throughput of ELISA assays. Protein microarrays are well suited to detection of antigens with antibodies, or of antibodies with antigens *(24,25)*. The specificity of the antibodies is critical. The choice of concentration at which to spot them robotically, potentially hundreds or even thousands of antibodies, on glass slides is important to the successful capture of protein antigens in a specimen with a very wide range of concentrations, like plasma. Antigen-antibody reactions are highly sensitive to affinity of the antibody and concentrations of both proteins. Alternatively, the method can be used to spot antigens fractionated from cell lysates, for example, and then to test plasma or serum specimens for autoantibodies against those proteins. If the spotted proteins are known, then the immunoreactivity can be related to specific proteins. If the proteins are not known, the initial result of the experiment is a "pattern" of immunoreactivity, which can be compared for sera from different patients. Aptamers and other protein capture agents can be used similarly.

2.4. Patterns of Proteins Without Protein Identification

Pattern recognition methods have been used extensively for diagnostic discovery research, including numerous papers describing surface-enhanced laser desorption/ionization (SELDI) MS *(26)*. Up to 15,000 low molecular *m/z* peak "squiggles" may be noted, of which as few as 5 will be sufficient for artificial intelligence algorithms to use for discriminant analysis between a set of sera from patients with a particular kind of cancer and a set of sera from normal volunteers. Generally, no proteins are actually identified, just pattern differences from certain *m/z* peaks. In a few cases, an abundant protein like haptoglobin has been found to account for at least one of the *m/z* peak differences *(27)*. Precise matching of all specimen collection, handling, and analytical variables for the controls and patients is essential.

3. How Many Proteins Can Be Detected in Human Plasma or Serum?

There is a rapidly growing literature of results from extensive analyses of human plasma and serum specimens. Here we summarize certain features of each published study and then compare each study with the 3020 protein list recently compiled in the Human Proteome Organization (HUPO) Plasma Proteome Project (PPP), as described below and shown in **Table 3**. The numbers of proteins depend on many factors, including the extent of fractionation and number of MS/MS runs with the sample, the number of peptide ions sequenced in the tandem MS, the stringency of criteria for identification of peptides and minimization of false positives, the restriction to tryptic or semitryptic peptides, the exclusion or inclusion of immunoglobulins and keratins, and the

Table 3
Websites With HUPO Plasma Proteome Project Data Sets

Group	Website
University of Michigan	www.bioinformatics.med.umich.edu/hupo/ppp
European Bioinformatics Institute	www.ebi.ac.uk/pride
Institute for Systems Biology	www.peptideatlas.org

tolerance for multiple ambiguous assignments of the peptides to proteins in gene or protein databases.

Anderson et al. ***(17)*** published a compilation of 1175 nonredundant proteins reported in at least one of four sources (a literature review plus three recent experimental data sets ***[28–30]***. Of the 1175, only 195 were reported in any two of the four input data sets; only 46 proteins were reported in all four sources; 284 reported in the literature were not found in any of the three experimental data sets; only 3 of the 46 were not already known in the literature. Patterson and colleagues ***(14)*** have suggested that such discordance reflects high false-positive rates from reliance on single-peptide hits. The three experimental papers used immunodepletion/ion exchange/size exclusion multidimensional protein fractionation followed by 2-DE and then MS ***(28)***; immunoglobulin depletion followed by tryptic digestion and then ion-exchange/reversed-phase 2D-LC-MS/MS analysis of the tryptic digest ***(29)***; or molecular mass fractionation followed by multidimensional protein identification technology (MudPIT) 2D-LC-MS/MS of a tryptic digest of low-MW plasma fractions ***(30)***.

The first used serum from two healthy male donors and depleted albumin, haptoglobin, transferrin, transthyretin, α_1 antitrypsin, α-1-acid glycoprotein, hemopexin, and α_2 macroglobulin; LC Q-MS/MS results were searched against the NCBI database using Sequest. The second used serum from a healthy female donor, depletion of immunoglobulins with protein A/protein G, analysis with LC-Q-Deca XP, and Sequest search with the NCBI (May 2002) database. The third used a standard human serum purchased from the National Institute of Standards and Technology (NIST), centrifugal filters to capture proteins with a molecular mass cutoff at 30 kDa, analysis with Ciphergen PBS-II time of flight (TOF)/MS, and search with Sequest against a human FASTA database that contained 100,000 viral sequences useful for detection of false-positive matches. Any sequences (fragments, splice variants) that shared a region greater than 15 amino acids with more than 95% identity were assigned to the same cluster and reported as a single entry in the nonredundant set of 1175 proteins, from the total of 1680 initial human accessions. The largest cluster had 109 immunoglobulins.

Various methods were used to annotate signal sequences, transmembrane domains, and Gene Ontology-based descriptors of cell localization and biological and molecular functions. The nonproteomic list of proteins historically accrued in the literature ($N = 468$) had a predominance of signal sequence-containing extracellular proteins, whereas the data sets from the three proteomics publications had a much higher representation of cellular proteins, including many with nuclear, as well as cytoplasmic, localization. At successively deeper layers of detection, the distribution of proteins shifted from mostly extracellular to a distribution more like the primarily cellular total proteome. At least 10 transcription factors were reported, each by only a single method, and none in the literature set; in contrast, only 4 of 39 cytokines and growth factors were found in any experimental set, whereas 37 occurred in the literature. The authors recommend that multiple analytical methods should be utilized to enhance the depth of identification of plasma proteins.

Shen et al. ***(31)*** used high-efficiency nanoscale reversed-phase liquid chromatography (RP-LC) and strong cation exchange LC in conjunction with ion-trap MS/MS and then applied conservative Sequest peptide identification criteria (with or without considering chymotryptic or elastic peptides) and peptide LC normalized elution time constraints. Between 800 and 1682 human proteins were identified, depending on the criteria used for identification (δC_n cutoff of >0.05 or >0.10), from a total of 365 μg (5 μL) of a single human plasma sample. They deliberately did not deplete albumin or immunoglobulins owing to variable and selective losses of other proteins in the process. The sample was diluted for denaturation and reduction, desalting, and then tryptic digestion, followed by RPLC-MS/MS; 110 proteins were identified from 428 different peptides in the first process, lower than expected, which was attributed to the interference from peptides from highly abundant proteins.

Shen and Smith ***(32)*** have reviewed recent developments in combined separations with MS for sensitive and high-throughput proteomic analyses. As illustrated above, these developments primarily involve high-efficiency separation with peak capacities of approximately 10^3, nanoscale LC with flow rates of approx 20 nL/min at optimal liquid mobile-phase separation linear velocities through narrow packed capillaries, and high-sensitivity, high-resolution Fourier transform-ion cyclotron resonance (FT-ICR)-MS. Such approaches allow analysis of low-nanogram-level proteomic samples (i.e., nanoscale proteomics) with individual protein identification sensitivity at the low zeptomole level. The resultant protein measurement dynamic range can approach 10^6 for nanogram-sized proteomic samples, whereas more abundant proteins can be detected from subpicogram-sized (total) proteome samples.

Chan et al. ***(33)*** resolved trypsin-digested protein peptides from a Sigma pooled standard serum into 20 fractions by ampholyte-free liquid-phase isoelectric focusing. These 20 peptide fractions were submitted to μRP-LC-MS/MS to identify 957 unique peptides assigned to 473 proteins. Aliquots of these

20 fractions were subjected to strong cation-exchange chromatography, generating 7 × 20 fractions, which were then analyzed by microcapillary RP-LC-MS/MS with an LC-Q-DecaXP, yielding 2071 peptides and 1143 protein matches. Dynamic exclusion was utilized to prevent redundant acquisition of the sets of three peptides previously selected for MS/MS.

In total, they identified 1444 unique proteins from 2646 unique peptides after searching with Sequest against the Expert Protein Analysis System (www.expasy.org) database. The filters applied were Xcorr ≥ 1.9, 2.2, 3.5 for fully tryptic peptides with 1+, 2+, and 3+ charge state, $\delta C_n \geq 0.08$, plus slightly higher Xcorr cutpoints for nontryptic peptides. The number based on single-peptide hits was not stated; a high percentage in the supplementary table was, indeed, based on just a single peptide (http://bpp.nci.nih.gov). These authors did not utilize a depletion step, to avoid the risk of removing nontarget low-abundance proteins. Proteins from all functional classes, cellular localizations, and abundance levels were identified, the majority attributed to secretion or shedding of proteins by cells during signaling, necrosis, apoptosis, and hemolysis. They estimated the confidence of peptide identifications to be more than 90% based on matching of spectra to peptides from an *Archea* database with 12,038 nonhuman protein sequences. All of the 22 most abundant proteins were identified, along with the expected coagulation and complement factors, transport and binding proteins, growth factors, and hormones. Intracellular and membrane-associated proteins made up 40 and 32% of the identified proteins, respectively. The website (above) appears still to be limited to this report (as of July 27, 2005).

Zhou et al. ***(34)*** identified an aggregate of 210 low-MW proteins or peptides after multiple immunoprecipitation steps with antibodies against albumin, IgA, IgG, IgM, transferrin, and apolipoprotein, followed by RP-LC-MS/MS. This aggregate result comprises nine different experimental methods, based on the notion that these abundant proteins may act as "protein scaffolds" or "molecular sponges" for low-MW proteins and peptides, including hormones. Of these proteins, 73 and 67% were not found by the same lab in previous studies of the low-MW or whole serum proteome ***(30,33)***. Unfortunately, there was no duplicate analysis in any of these three studies to ascertain the percent concordance with the same method, same sample, and same lab; other evidence suggests that concordance might be 50% or much less.

The four different albumin removal methods (protein G-coupled anti-HSA antibody and Millipore Montage Albumin Deplete Kit affinity columns with three different buffer conditions) led to identification of only 33, 63, 24, and 56 proteins, respectively. There were 38, 19, 38, and 13 proteins identified as bound with protein G-coupled specific antibody against IgA, IgM, apolipoprotein A-I, and transferrin, respectively, and 53 bound to IgG removed with protein G itself. Presumably, the abundant proteins/peptides were discarded in the analysis, although no evidence was presented to show that these proteins were fully eliminated with use of the YM-30 Centricon ultrafilters with an MW cutoff at 30 kDa followed by centrifugation at 1000*g*. In fact, proteins (or fragments) with nominal MW as high as 565 kDa (ryanodine receptor 2) and 181 kDa (pregnancy plasma protein A) were reported.

From the nine experiments, a total of only 378 unique peptides (not limited to tryptic peptides) were identified, which matched to 210 proteins. Clearly, many or most of the protein IDs were based on just one peptide (data not given). The Sequest filter criteria for peptide identifications were based only on Xcorr values, with no use of δC_n or Rsp filters. Of the 210 proteins, 57 (26%) had been identified by Tirumalai et al. ***(30)*** in a low-MW serum proteome study; 39 of the 210 were hypothetical proteins in the EMBL-nr database utilized. By contrast, only 1 was identified by Adkins et al. ***(29)*** in serum depleted of IgG and then analyzed by LC-MS/MS; only 4 were identified in plasma by 2-DE followed by MS ***(25)***; and only about 70 of the 1500 claimed by Chan et al. ***(33)*** were identified.

Rose et al. ***(35)*** reported fractionation with an industrial-scale approach, starting with 6 L of blood and 2.5 L of plasma from 53 healthy males that were depleted of albumin and IgG with affinity resin and protein G, respectively, to yield 53 g total protein. Then smaller proteins (MW < approx 40 kDa) and polypeptides, after gel filtration (1.5 g), were separated into 12,960 fractions by chromatographic techniques. Fragments of larger proteins could not be excluded. Electrospray ionization (ESI)-MS and matrix-assisted laser desorption/ionization (MALDI)-TOF-MS were performed on the small proteins on MALDI plates, and then aliquots following tryptic digestion were subjected to LC-ESI-MS/MS (Bruker Esquire 3000 ion trap instrument). The protocol was driven by the estimate that 100 fmol of a protein is required for successful separation, digestion, and MS identification of peptides; the corresponding concentration must be at least 1 n*M* for a sample size of 100 μL, or 100 f*M* if 1L of sample is the starting material. Some 8533 of the final fractions had no identifiable protein; 994 had a single protein ID; and the most complex fraction had 21 identifiable proteins. About 1.5 million MS/MS spectra were analyzed with commercial algorithms and then a new Olav engine, with six different databases. From thousands of peptide identifications, 502 different proteins and polypeptides were matched, 405 of which were included in the publication, of which 115 were based on just a single peptide. When their criteria were applied to the list of 490 proteins created by Adkins et al. ***(29)***, only 164 of the more common proteins were retained. One hundred peptides were identified from areas of the genome where no proteins had previously been predicted.

3.1. The HUPO Plasma Proteome Project (PPP)

HUPO has initiated several proteome initiatives, including liver, brain, and plasma proteomes and antibody and protein bioinformatics/standards initiatives ***(36)***. Here we summarize the PPP pilot phase **(Fig. 2)**. Eighteen of the

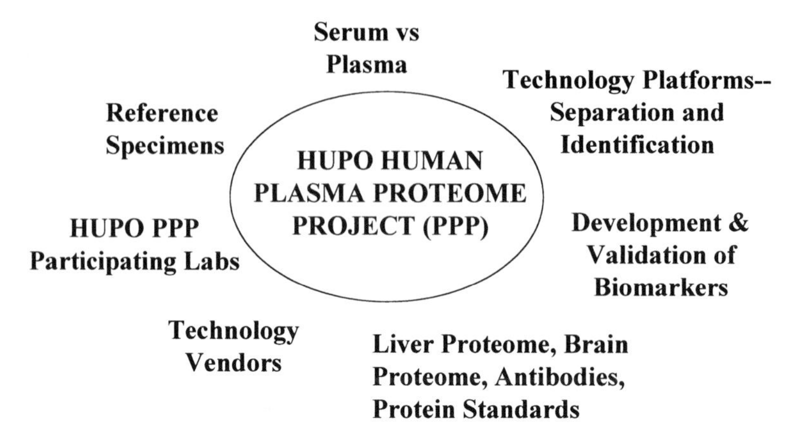

Fig. 2. Scheme showing aims and linkages of the HUPO Plasma Proteome Project. (Adapted from **ref.** ***37*** with permission.)

laboratories participating in the HUPO PPP conducted a wide range of depletion and fractionation protocols combined with tandem MS/MS or FT-ICR-MS with HUPO PPP reference specimens ***(1,37)***. They submitted a total of 42,306 protein identifications using various search engines and databases to handle spectra and generate peptide sequence lists from the specimens analyzed. Peptides with six amino acids or more matched to 15,519 nonredundant proteins in the International Protein Index (IPI) of the European Bioinformatics Institute in Hinxton, UK ***(38)***. IPI version 2.21 (July 2003) was chosen as the standard reference database for this project. We designed an integration algorithm, which selected one representative protein among multiple proteins (homologs and isoforms) to which the identified peptides gave 100% sequence matches ***(39,40)***. This integration process resulted in 9504 proteins in the IPI v2.21 database. The PPP database is conservative from this point of view, counting homologous proteins and all isoforms of particular proteins (and their corresponding genes) just once, unless the sequences actually could differentiate additional matches.

We included at this stage proteins identified by matches to one or more peptide sequences of "high" or "lower" confidence according to cutpoints utilized with the various search engines used by different MS/MS instruments. We have reported ***(1)*** details of the cutpoints or filters used by each investigator. All laboratories utilizing Sequest were asked to reanalyze their results using the PPP-specified filters of Xcorr values ≥ 1.9, 2.2, and 3.75 for singly, doubly, and triply charged ions, with δC_n value ≥ 0.1 and Rsp ≤ 4 for fully tryptic peptides for "high-confidence" identifications; most did so. No equivalency rules were applied across all the search algorithms for all the cutpoints. However,

Kapp et al. ***(13)*** provided such a cross-algorithm analysis for three specified false-positive rates using one laboratory data set and five search algorithms.

Data management for this project comprised guidance and protocols for data collection, centralized integration and analysis, and dissemination of findings worldwide. Key challenges were integration of heterogeneous data sets, reduction of redundant information, and data set annotation. Multiple factors had to be balanced, including when to "freeze" on a particular release of the ever-changing database selected for the PPP. Freezing of the database was essential to conduct extensive comparisons of complex data sets and annotations of the data set as a whole. However, freezing complicates the work of linking findings of the current study to evolving knowledge of the human genome and its annotation. Many of the entries in the IPI protein sequence database(s) available at the initiation of the project or even the analytical phase were revised, replaced, or withdrawn over the course of the project. This fact complicates all cross-study comparisons. Our policies and practices anticipated the guidelines issued recently by Carr et al. ***(41)***, as documented by Adamski et al. ***(40)***.

Since the approaches and analytical instruments used by the various laboratories were far too diverse to utilize a standardized set of mass spec/search engine criteria, we created a relatively stringent, defined subset of protein IDs from the 9504 above by requiring that the same protein be identified with at least a second peptide. In a peptide chromatography run for MS, not all peaks are selected for MS/MS analysis, and the identification of peptide fragment ions is a low-percentage sampling process. Thus, additional analyses in the same lab and in other labs would be expected to enhance the yield of peptide IDs. Consequently, MS data from the individual laboratories were combined to increase the probability of peptide identification and protein assignment. Of the total of 9504 protein IDs, 6484 were based on one peptide, whereas 3020 were based on two or more peptides. The list of 3020 proteins (5102 before integration) has been utilized as the Core Protein Dataset for the HUPO PPP knowledge base. Full details with unique IPI accession numbers for each protein are accessible for examination and reanalysis at the websites at the University of Michigan, the European Bioinformatics Institute, and the Institute for Systems Biology, as listed in **Table 3**. As a sample of use, during the period 5 to 31 January, 2005, there were 5000 hits and 1000 downloads of PPP data from PRIDE. The 3020 proteins represent a very broad sampling of the IPI proteins in terms of characterization by p*I* and by MW of the transcription product (often a "precursor" protein).

The publicly available PPP database permits future users to choose their own cutpoints for subanalyses. For examples, 2857 proteins were identified at least once with "high-confidence" criteria; 1555 proteins were based on two or more peptides, at least one of which was reported as high confidence (from the intersection of the 3020 and the 2857); and 1274 protein IDs were based on matching to three or more peptides. We have subjected the PPP results to further very logical, stringent analyses, adjusting for protein length and for multiple comparisons testing (a Bonferroni correction for potential random matching to the entire 43,730 proteins in IPI v2.21), yielding a data set of 889 proteins ***(42)***. These adjustments may be overly stringent, since they assume equivalent random matching to all proteins, whereas proteins occur in various families and have considerable homologies. Two laboratories reported glycoprotein enrichment, one with hydrazide chemistry and the other with binding to three lectins ***(43,44)***. Together they had 254 protein IDs, of which 164 were reported also by other laboratories, whereas 90 were identified only after glycoprotein enrichment ***(1)***.

4. Cross-Study Comparisons With HUPO PPP Core Database

Across studies, as well as across the PPP-participating laboratories, incomplete sampling of proteins is a dominant feature. A substantial depth of analysis is achieved with depletion of highly abundant proteins, and fractionation of intact proteins followed by digestion and two or more MS/MS runs for each fraction ***(45)***. Haab et al. ***(46)*** showed that the number of peptides identified for a protein in this collaborative data set correlates highly with the measured concentration of the presumed same protein by immunoassays (correlation = 0.90 for 76 proteins in the 9504 data set and 0.86 for 49 proteins in the 3020 data set) (*see* **Fig. 3**).

Table 4 presents the degree of congruence for the five published studies described in detail above in **Subheading 3.** in relation to the HUPO PPP findings. Standardized, statistically sound criteria for peptide identification, protein matching, and estimation of error rates are necessary features for comprehensive profiling studies and especially for cross-study comparisons.

Of the 990 proteins that have IPI v2.21 identifiers in the four studies compiled by Anderson et al. ***(17)***, 316 are found in the PPP 3020 protein Core Dataset. When we relaxed the integration requirement (5102 IPI IDs), this figure rose only to 356 matches. Using the full 9504 data set, the corresponding matches were 471 with integration and 539 without integration. With the cooperation of Shen et al. ***(31)***, we reran their raw spectra using HUPO PPP Sequest parameters (high confidence: Xcorr $\geq$ 1.9/2.2/3.75 [for charges +1/+2/+3], $\delta C_n \geq 0.1$, and RSp $\leq$ 4; and lower confidence: XCorr $\geq$ 1.5/2.0/2.5 [for charges +1/+2/+3], $\delta C_n \geq 0.1$) and obtained 1842 IPI protein matches. Of these, 526 and 213 were found in the PPP 9504 and 3020 data sets, respectively. When we mapped the 1444 proteins reported by Chan et al. ***(33)*** against the IPI v2.21 database, there were 1019 distinct proteins. From this set, 402 and 257 proteins matched with the 9504 and 3020 data sets, respectively. With the Zhou et al. ***(34)*** protein-bound proteins, 148 proteins were mapped with IPI identifiers, of which 88 and 62 were found in the 9504 and 3020 PPP protein lists, respectively. Finally, of the 287 low-MW proteins (<40 kDa) from Rose et al. ***(35)*** that

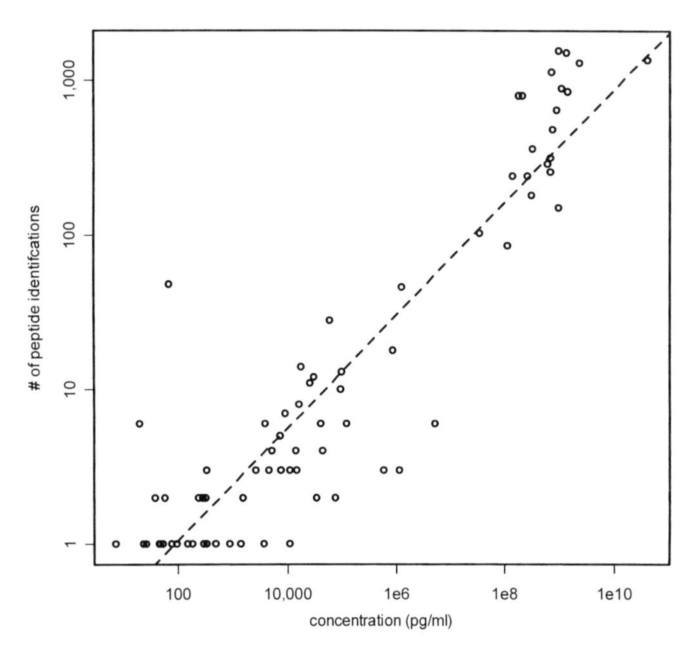

Fig. 3. Correlation of total number of peptide identifications across the collaboration for each protein with the immunoassayed concentration of the presumed same protein in plasma or serum in the HUPO Plasma Proteome Project. The log-log correlation fit the linear relationship log 10 (N) = 0.365 log 10 (C)–0.711, with r = 0.90 for 76 proteins in the 9504 dataset. Note that the lowest set of matches is for proteins with one peptide match. (Adapted form ***42*** with permission and from ***46***).

Table 4
Comparison of HUPO Plasma Proteome Project (PPP) Protein Identification Lists With Published Data Sets for Plasma or Serum

Published data	Total IDs	No. of IPI proteins	PPP_9504 Data set	PPP_3020 Data set
Anderson et al. ***(17)***	1175	990	471	316
Shen et al. ***(31)***	1682	1842	526	213
Chan et al. ***(33)***	1444	1019	402	257
Zhou et al. ***(34)***	210	148	88	62
Rose et al. ***(35)***	405	287	159	142

IPI, International Protein Index.

mapped to IPI identifiers, 159 and 142 are included in our 9504 and 3020 protein data sets, respectively.

These data sets vary remarkably in the protocols for depletion and/or fractionation, the criteria for protein IDs, and the inclusion or depletion of immunoglobulins. All claim some relatively low abundant proteins. For example, the PPP reported 10 proteins, identified from two or more peptides, in the concentration range of 200 pg/mL to 16 ng/mL, among 49 identified and measured with quantitative immunoassays ***(46)***: α-fetoprotein, tumor necrosis factor receptor-8 (TNF-R-8), TNF-ligand-6, platelet-derived growth factor receptor-α (PDGF-Rα), leukemia inhibitory factor receptor, matrix metalloproteinase-2 (MMP-2)/gelatinase, epidermal growth factor receptor (EGFR), tissue inhibitor of metalloproteinase-1 (TIMP-1), insulin-like growth factor binding protein-2 (IGFBP-2), and activated leukocyte adhesion molecule. Selectin L, at 17 ng/mL, was identified with 10 peptides across the collaborating laboratories ***(1)***. Nevertheless, abundance remains the single strongest determinant of protein detectability by MS, and essentially all the proteins detected in common across multiple studies are present at high concentrations in blood.

Error rate estimation is a nascent aspect of the literature. Methods include use of statistical criteria, as in PeptideProphet/ProteinProphet ***(10,11)***; matching to nonhuman protein sequence databases (Archea); matching to reversed sequence ***(33)*** or shuffled sequence human databases; Poisson distribution methods ***(39, 40,42)***; and modeling of random matches to length of protein sequences ***(42)***.

5. Diversity of the Proteins Detected in Individual Plasma/Serum Specimens

The Core data set of 3020 proteins was annotated with use of Gene Ontology (GO) for subcellular localization, molecular processes, and biological functions **(Fig. 4)** ***(1)***. There is very broad representation of cellular proteins. Subcellular component classification of the 1276 IPI-3020 proteins included in GO showed 26% of proteins from membrane compartments, 19% from nuclei, 11% from cytoskeleton, and 23% from other sites, compared with just 14% for the expected predominance of traditional plasma proteins. GO analyses of molecular processes showed 39% binding, 28% catalytic, 7% signal transducer, 6% transporter, 4% transcription regulator, and 3% enzyme regulator. GO biological functions included 36% metabolism, 25% cell growth and maintenance, 5% immune response, 1% blood coagulation, and 1% complement activation.

Examination of specific GO terms against a random sample of 3020 proteins from the Human Genome showed some proteins more than 3 SD from the expected line based on the distribution of such classes of proteins in the entire 56,530 human protein IPI v2.21 data set **(Fig. 5A)**. Overrepresented categories include extracellular, immune response, blood coagulation, complement, lipid transport, and blood pressure regulation, as might be expected; surprisingly large numbers of cytoskeletal proteins, receptors, and transporters were also identified.

InterPro analyses similarly compared the 3020 proteins with the fine-grained protein families and domains described for the full IPI v2.21 database **(Fig. 5B)**.

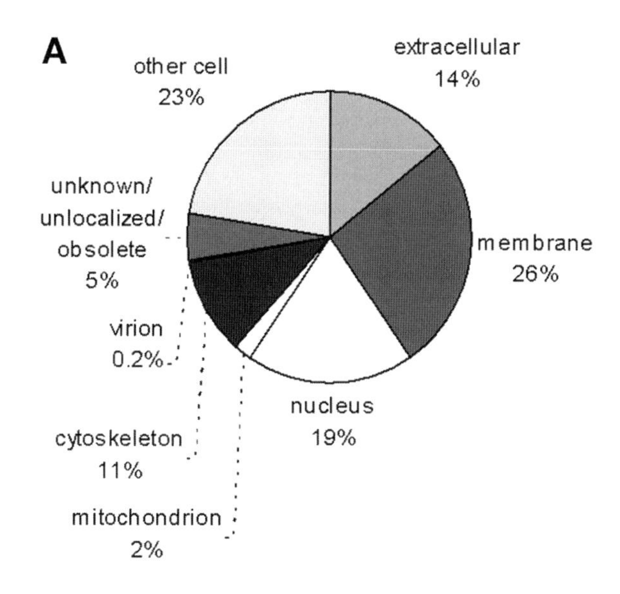

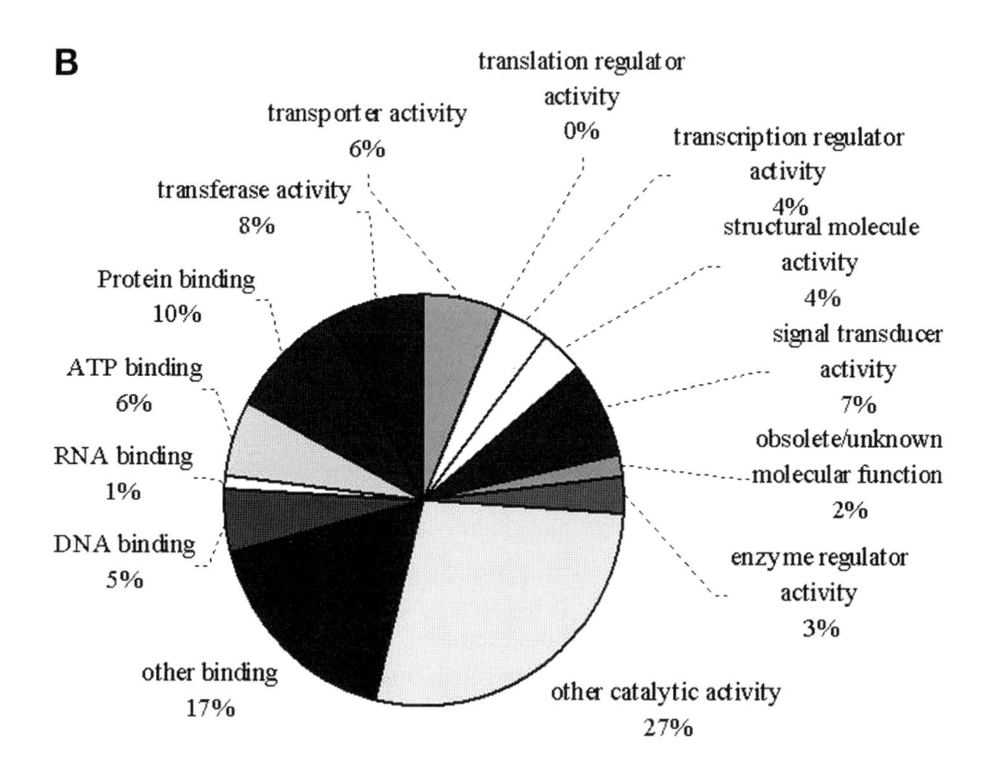

Fig. 4. *(Continued)*

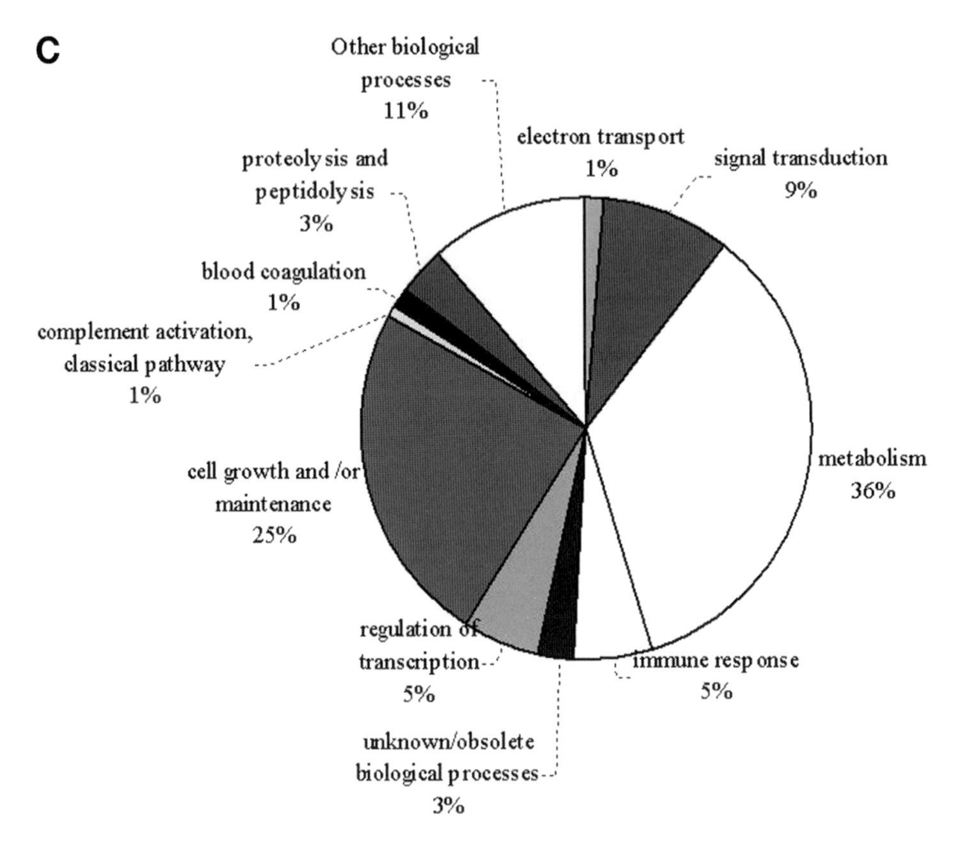

Fig. 4. Gene Ontology annotations for the HUPO PPP reference specimens of plasma and serum. **(A)** Analysis with Gene Ontology for cellular components/subcellular localization. Of 3020 proteins identified with at least two peptides, 1276 are annotated with 207 distinct GO cellular component terms, giving 1886 annotations. **(B)** Analysis with Gene Ontology for molecular functions. Of 3020 proteins identified with at least two peptides, 1475 are annotated with 678 distinct GO molecular function terms, giving 3470 annotations. **(C)** Analysis with Gene Ontology for biological processes. Of the 3020 proteins, 1383 are annotated with 668 distinct GO biological process terms, giving 3121 annotations.

Domains associated with EGF, intermediate filament protein, sushi, thrombospondin, complement C1q, and cysteine protease inhibitor were overrepresented (>3 SD) compared with random occurrence, whereas zinc finger RING protein, tyrosine protein phosphatase, tyrosine and serine/threonine kinases, helix-turn-helix motif, and IQ calmodulin-binding region were underrepresented.

Muthusamy et al. *(47)* subjected protein IDs and Ensembl gene matches to BLAST queries to identify splice isoforms; they report that 51% of the genes encoded more than one isoform (a total of 4932 products for the 2446 genes).

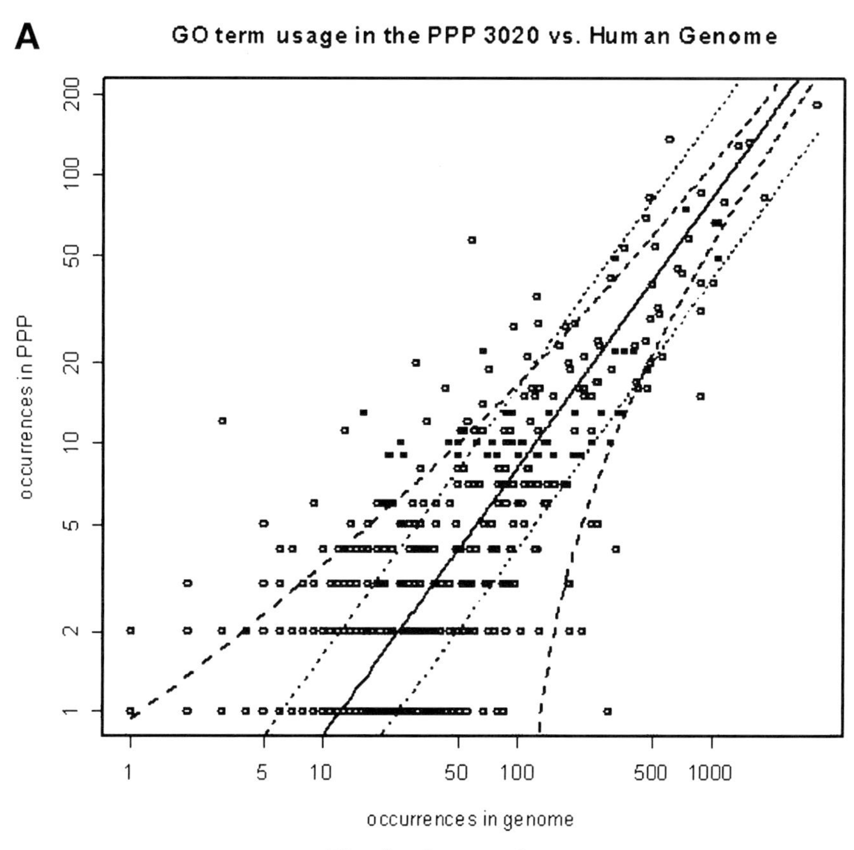

Fig. 5. *(Continued)*

They mapped 11,381 SNPs involving protein-coding regions onto the protein sequences. Berhane et al. ***(48)*** found 345 proteins in plasma/serum of particular interest for cardiovascular research among the 3020 proteins. The collaborating investigators ***(1)*** noted that 338 of the 3020 IPI proteins matched Ensembl genes in the Online Mendelian Inheritance in Man database, including such interesting disease-associated genes as RAG 2 for severe combined immunodeficiency (SCID/Omenn syndrome), polycystin 1 for polycystic kidney disease, and breast cancer BRCA1 and BRCA2, multicancer p53, and colon cancer APC for inherited cancer syndromes.

6. Plasma vs Serum as the Sample of Choice

Preanalytical variables are often ignored or reported casually in the proteomic analysis and comparison of samples. These variables will be critically important in disease marker research. The HUPO PPP Specimens Committee ***(16)*** and the collaborating investigators ***(1)*** concluded that plasma is preferable

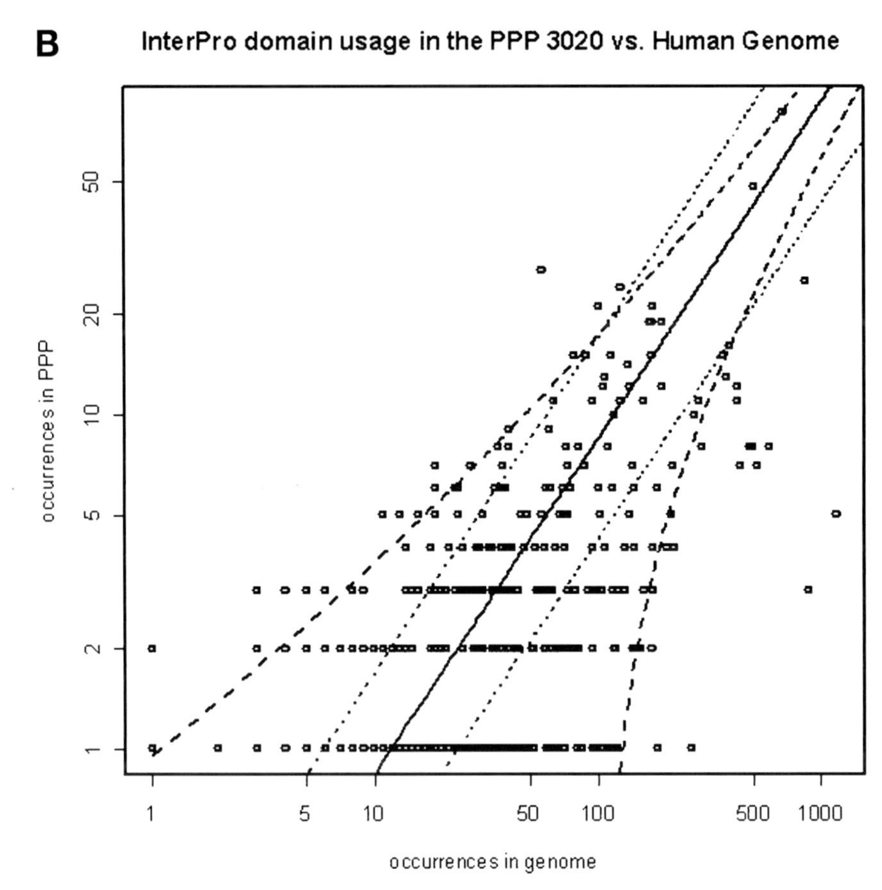

Fig. 5. Comparison of PPP 3020 proteins *vs* full IPI v2.21 protein set. **(A)** GO term usage in the PPP 3020 *vs* human genome. Shown in the figure are the rates of occurrence of Gene Ontology terms in the HUPO PPP 3020 set relative to the frequency of occurrence of the same terms in the human genome. The solid line shows a linear regression estimate for the frequency that would be expected if the 3020 uniformly sampled the genome as reflected in the IPI database. The parallel dotted lines show twofold over- and underrepresentation relative to uniform sampling. The curved dashed lines show over- and underrepresentation by 3 standard deviations. **(B)** InterPro domain usage in the PPP 3020 *vs* human genome. Shown in the figure are the rates of occurrence of InterPro domains in the HUPO PPP 3020 set relative to the frequency of occurrence of the same terms in the human genome. The solid line shows a linear regression estimate for the frequency that would be expected if the 3020 uniformly sampled the genome as reflected in the IPI database. The parallel dotted and dashed lines show twofold, and threefold, respectively, over- and underrepresentation relative to uniform sampling.

to serum. The reasons are less degradation ex vivo and much less variability than arises in the protease-rich process of clotting. Tammen et al. ***(49)*** reported that at least 40% of the peptides detected in serum were serum specific. Clotting is unpredictable, owing to influences of temperature, time, and medications, which are hard to standardize. They also noted the risk that platelet activation, especially at 4°C, may release large amounts of platelet proteins not normally present in vivo in the plasma. We did not find a striking abundance of platelet proteins (thymosin-β-4, zyxin, platelet factor 4, and platelet basic protein). We recommend EDTA or citrate anticoagulated plasma over heparinized plasma. Heparin acts through activation of antithrombin III, whereas citrate and EDTA inhibit coagulation and other enzymatic processes by chelate formation with ion-dependent enzymes. Since citrate introduces a 10 to 15% dilution effect, we recommend EDTA-plasma as the preferred specimen for future plasma or serum biomarker research.

We did not, in the end, recommend inclusion of protease inhibitors in the collection tubes or buffers; the peptide inhibitor aprotinin requires µg/mL concentrations, which interfere with the analyses, whereas the small molecule inhibitor ABESF forms covalent bonds with proteins that alter the mobility of the protein ***(16)***, as noted in **Subheading 1.4.** above. A surprising finding that disease-related patterns of ex vivo proteolysis in plasma undergoing clotting to form serum may be clinically useful has recently been reported ***(50,51)***. Protease inhibitors are not used in such experiments.

7. Use of Plasma or Serum for Protein Biomarkers

Biomarkers detected in the circulation that provide indications of disease processes in particular organs and cell types may arise from any quantitative or qualitative changes in protein expression in any of at least three sources: (1) the primary cells for the disease, such as neoplastic cells or endothelial cells; (2) the microenvironment of the primary cells; and (3) systemic responses to the altered protein expression, namely, acute-phase reactant proteins (i.e., C-reactive protein, amyloid A, and haptoglobin) or immunoglobulins (i.e., antibodies), reflecting immune responses to altered proteins.

The first category may be studied most efficiently by first determining changes in protein expression in the primary tissue, simultaneously obtaining blood in experimental and clinical studies, and then examining plasma for corresponding detectable changes. A valuable resource, developed by Uhlen in Stockholm ***(52)***, is the Human Protein Atlas. The Atlas has immunohistochemistry results for antibodies against 718 proteins for 48 normal human tissues (three individuals each) and 20 different cancer types by organ of origin (12 patients each; sometimes 4 patients; www.proteinatlas.org). These antibodies can also be utilized as specific affinity reagents for functional protein assays and as capture (pull-down) reagents for purification of specific proteins and their associated complexes for structural and biochemical analyses. Sometimes, the protein changes correlate with mRNA changes in gene expression; however, more often the two are uncorrelated or discordant, reflecting the differential lifetimes of these molecules and the posttranslational modifications of proteins through cleavage, aggregation, side-chain reactions, and ubiquitination and other processes of elimination and excretion of proteins. The third category has the special advantage that the immune response provides a kind of "biological amplification." We have no in vitro manipulation for proteins to match the spectacularly successful polymerase chain reaction (PCR) amplification for tiny amounts of nucleic acids. However, autoantibodies against tumor proteins seem to have particular potential ***(53,54)***. Autoantibody signatures for prostate cancers have been reported using a 22-phage-peptide detector panel; at least four of the phage-protein microarrays were derived from in-frame, identifiable protein coding sequences, with initial results appearing far superior to the widely used PSA test **(Fig. 6)** *(55)*.

Heterogeneity of disease limits the sensitivity and specificity of tests. Thus, there is a huge literature about such single protein biomarkers for cancers as PSA for prostate cancers, CA 125 for ovarian cancers, carcinoembryonic antigen (CEA) for pancreatic cancers, and CA 19-9 for breast cancers. Unfortunately, these tests are notoriously poor in sensitivity (hence giving false negatives) and specificity (hence giving false positives) in the screening situation; they have more value in monitoring patients for recurrence. Other single protein tests include C-reactive protein for cardiovascular disease, presumably reflecting an inflammatory component in atherogenesis; this very old test has been rejuvenated and become popular in cardiovascular epidemiology and clinical studies, although it lacks specificity. The triumvirate of serum amyloid A, haptoglobin, and trans-thyretin has been called "the stubbed toe syndrome" by Naylor (CHI PepTalk Conference, January 2005), since these acute-phase proteins seem to increase, often strikingly, in a host of clinical situations without any apparent specificity. We have recently shown striking increases in C-reactive protein and statistically significant increases in amyloid A, MUC1, and α_1 antitrypsin in sera of patients with lung cancer, compared with patients with chronic obstructive lung disease (COPD) or normals *(56)*.

Sometimes, a particular protein is modified so markedly that it can be a potential biomarker. In addition, its alteration may be recognized immunologically, leading to detectable autoantibodies. Such a situation applies to a truncated form of calreticulin in patients with liver cancers (L. Beretta, personal communication). Autoantibodies against isoforms of calreticulin were reported in sera of patients with pancreatic cancers *(57)*.

The approach of identifying and validating individual proteins with incomplete sensitivity for a particular condition can be expanded combinatorially to

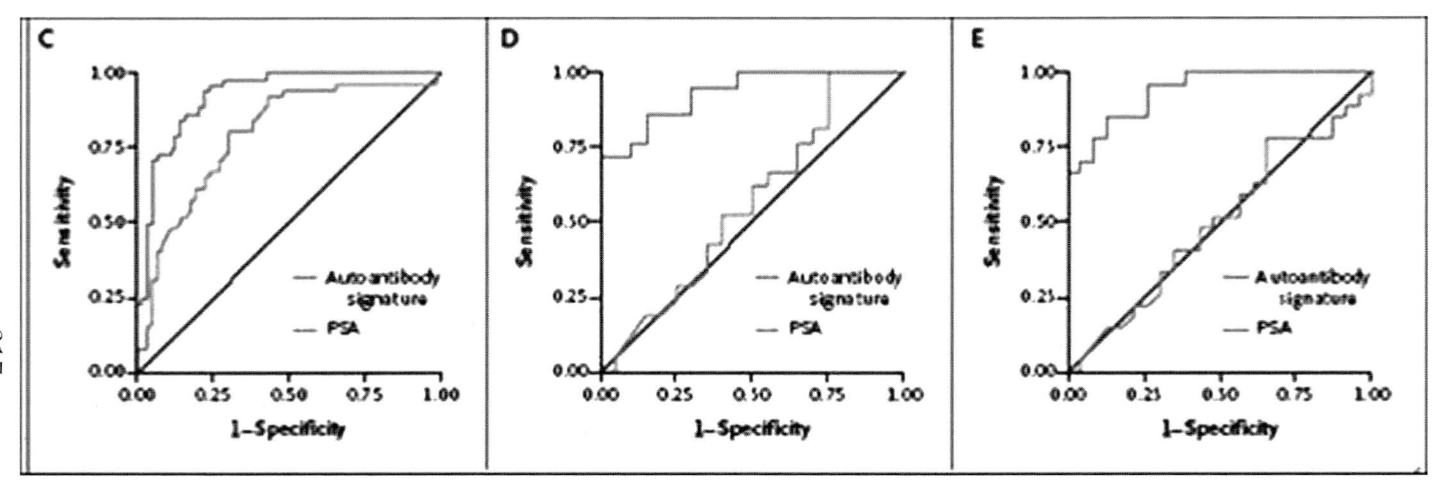

Fig. 6. Autoantibody signatures in prostate cancer. The figure shows receiver-operating characteristic (ROC) curves based on multiplex analysis of 22-phage peptide biomarkers and prostate-specific antigen (PSA) in sera of 60 patients with prostate cancer and 68 controls. Identified proteins in this set are BRD2, eIF4G1, RPL22, and RPL13a, all involved in transcription/translation and all overexpressed in prostate tumors by immunohistochemistry. Left panel, full range of PSA values; center panel, PSA levels between 4 and 10 ng/mL, with 22 patient and 20 control sera; right panel, PSA levels between 2.5 and 10 ng/mL, with 28 patients and 232 controls. The PSA line is close to the diagonal in each panel. (From **ref. 55** with permission.)

generate a more sensitive and still specific panel of protein biomarkers. This approach underlies the work of the NCI Early Detection Research Network and many individual laboratories and offers the practical avenue of combining laborious discovery and validation research with high-throughput assays once the combinations of specific protein markers have been identified, selected, and validated. Those assays are likely most often to be multiplex ELISA, antibody microarrays, or analogous protein capture methods.

An entirely complementary approach involves "blind" recognition of "patterns of differences" between tissue lysate or plasma proteins from patients vs normals. This approach has rapidly generated a very large literature with direct MS/SELDI methods, using Ciphergen protein chips with distinctive surfaces for quick fractionation of complex protein mixtures. Given the marked heterogeneity of all tumor types, many observers have been skeptical that such "pattern recognition" of perhaps five small *m/z* peaks chosen by an artificial-intelligence algorithm and not identified with any particular protein could possibly yield the claimed nearly perfect sensitivity and specificity ***(58)***. Most troubling has been the lack of evidence for specificity across tumor types and the reliance on training-and-testing routines on a batch basis with limited reliability across laboratories. Efforts to improve cross-lab concordance have now been reported ***(27,59)***, along with advances in the instruments that will facilitate identification of at least some of the protein *m/z* peaks. So far, these seem to be fragments of highly abundant acute-phase reactants, like the well-known haptoglobin ***(24,60)***. It would also enhance credibility if findings in tissue lysates could be related to findings in serum (or plasma).

Finally, rapid advances in microarray technologies have been reported. Hundreds or thousands of fractions of cell lysate antigens or of selected antibodies can be spotted robotically and then reacted with corresponding autoantibodies or circulating antigens. Patterns of immunoreactive spots have been analyzed for prostate cancer ***(61)***, lung cancer ***(62)***, and other conditions; in principle, these reacting fractions can be analyzed by sensitive MS/MS methods to identify the responsible proteins. Reverse-phase microarrays have been used to study phosphorylation events in EGFR-mediated signaling pathways ***(63)***. In practice, there are many challenges, including insufficient protein to be detected and identified; multiple proteins identified, making assignment of the protein responsible for the immunoreactivity a continuing validation process; and risk of false-positive identifications, especially at the limit of detectability with few or single peptides identified.

In general, much progress is needed with highly standardized specimens and with standard proteins, protein mixtures, and biological specimens as reference materials ***(64)***. Examples of standard mixtures include the CRM 470 of 15 human plasma proteins, widely used in Europe; the 18-protein nonhuman commercial proteins combined into a standard mixture by Keller et al. ***(65)*** at the Institute for Systems Biology; and most recently, over 300 commercially available proteins, for each of which at least 25 spectra have been generated with MALDI-TOF-TOF for use as standards in calibrating instruments and spiking samples ***(66)***. Such challenges were the focus of a NIST/NCI Workshop in August 2005 in Gaithersburg, MD ***(64)***. Much more stringent criteria for documentation in publications will add value and consistency, as well ***(42,67,68)***.

8. Goals for Progress on the Plasma Proteome

The Holy Grail for plasma proteomics is high-resolution, high-sensitivity, and high-throughput analysis. It is likely that most of the protein biomarkers of greatest interest originating from disease processes in specific tissues will be at quite low abundance after dilution into 4 L of blood and 17 L of extracellular fluid. To perform proteomics analyses on dozens, hundreds, or thousands of specimens from participants in clinical trials or epidemiological studies and to do so with replicates to assess intraindividual variation, new strategies with high throughput must complement the presently laborious methods for discovery and even validation of potential protein biomarkers.

The following options have been suggested, as summarized recently ***(1,69)***:

1. LC-MS with highly accurate mass and elution time parameters for identification of peptides already characterized in an extensive database with adequate sequence coverage of proteins to differentiate variants owing to splicing ***(70)***. Specific depletion of abundant proteins, tryptic digestion, nanoflow LC for elution time standardization, and highly accurate MS may make it feasible to base identifications on such enhanced mass fingerprints.
2. High-accuracy LC-MS/MS/MS has recently been applied to intracellular localization and discovery-phase identification of posttranslational modifications ***(71)***. Such findings might then be converted into a protein capture microarray methodology or multiplex ELISA reactions once the most useful biomarkers for differentiation of disease from normal or advanced disease from localized disease were validated.
3. Protein affinity microarrays with ligands designed to recognize conserved regions in each open reading frame, with sensitive readout technologies, could improve protein identifications over a wide dynamic range ***(72)***.
4. Since it is likely that differences between specimens from patients with disease vs specimens from normals may be quantitative, methods of quantitative proteomics, such as ICAT may be applicable. Aebersold ***(73)*** has proposed developing standard mixtures of chemically synthesized peptides tagged with heavy isotope for each gene and even each protein isoform needed for high discrimination as biomarkers. These "proteotypic peptides" would permit spiking of specimens to facilitate identification of mass pairs with the same peptides in the biological specimen and quantitation of the peptide and its protein ***(21–23)*** (*see* **Subheading 2.2.**).

5. A multiplexed multiple reaction monitoring (MRM) approach with 1D-LC-MS/MS can be exploited to identify and quantitate high- and medium-abundant proteins, which may have value as biomarkers. This approach has many potential variations, including use of sensitive antipeptide antibody-enhanced assays for lower abundant proteins ***(74)***.

In summary, work on all aspects of proteomics has generated renewed confidence that such approaches will reveal important features of normal biology and physiology and also assist in the discovery, validation, and application of protein biomarker panels for early diagnosis of disease and monitoring responses to therapies.

Acknowledgments

This work was supported by grant MTTC GR687 for the Michigan Proteomics Alliance for Cancer Research, NCI/SAIC contract UOM 23XS110A on Mouse Models of Human Cancers, and an NIH cooperative agreement U54DA021519 for the National Center for Integrative Biomedical Informatics.

References

1. Omenn GS, States DJ, Adamski M, et al. Overview of the HUPO Plasma Proteome Project: results from the pilot phase with 35 collaborating laboratories and multiple analytical groups, generating a core dataset of 3020 proteins and a publicly-available database. *Proteomics* 2005;5:3226–3245. Also, Omenn GS. Exploring the Human Plasma Proteome. Wiley-Liss, 2006.
2. Ping P, Vondriska TM, Creighton CJ, et al. A functional annotation of subproteomes in human plasma. *Proteomics* 2005;5:3506–3519.
3. Putnam FW. *The Plasma Proteins: Structure, Function, and Genetic Control.* New York: Academic, 1975.
4. Anderson NL, Anderson NG. The human plasma proteome: history, character, and diagnostic prospects. *Mol Cell Proteomics* 2002;1.11:845–867.
5. Echan LA, Tang H-Y, Ali-Khan N, Lee K, Speicher DW. Depletion of multiple high abundance proteins improves protein profiling capacities of human serum and plasma *Proteomics* 2005;5:3292–3303.
6. Rhodes DR, Tomlins SA, Varambally S, et al. Comprehensive model of the human protein-protein interaction network. *Comput Biol* 2005;8:1–9.
7. Peri S, Navarro JD, Amanchy R, et al. Development of human protein reference database as an initial platform for approaching systems biology in humans. *Genome Res* 2003;13:2363–2371.
8. Mann M, Aebersold R. Mass spectrometry-based proteomics. *Nature* 2003; 422;198–207.
9. Sadygov R, Yates JR. A hypergeometric probability model for protein identification and validation using tandem mass spectral data and protein sequence databases. *Anal Chem* 2003;75:3792–3798.

10. Keller A, Nesvizhskii AI, Kolker E, Aebersold R. Empirical statistical model to estimate the accuracy of peptide identifications made by MS/MS and database search. *Anal Chem* 2002;74:5383–5392.
11. Nesvizhskii AI, Keller A, Kolker E, et al. A statistical model for identifying proteins by tandem mass spectrometry. *Anal Chem* 2003;75:4646–4658.
12. Cargile BJ, Bundy JL, Stephenson JL, Jr. Potential for false positive identifications from large databases through tandem mass spectrometry. *J Proteome Res* 2004;3:1082–1085.
13. Kapp EA, Schutz F, Connolly LM, et al. An evaluation, comparison and accurate benchmarking of several publicly-available MS/MS search algorithms: sensitivity and specificity analysis. *Proteomics* 2005;5:3475–3490.
14. Johnson RS, Davis MT, Taylor JA, Patterson SD. Informatics for protein identification by mass spectrometry. *Methods* 2005;35:223–236.
15. Marshall J, Kupchak P, Zhu W, et al. Processing of serum proteins underlies the mass spectral fingerprinting of myocardial infarction. *J Proteome Res* 2003;2:361–372.
16. Rai AJ, Gelfand CA, Haywood BC, et al. Human Proteome Organization—Plasma Proteome Project specimen collection and handling: towards the standardization of parameters for plasma proteome samples. *Proteomics* 2005;5:3262–3277.
17. Anderson NL, Polanski M, Pieper R, et al. The human plasma proteome: a nonredundant list developed by combination of four separate sources. *Mol Cell Proteomics* 2004;3:311–316.
18. Wang H, Clouthier SG, Galchev V, et al. Intact-protein based high-resolution three-dimensional quantitative analysis system for proteome profiling of biological fluids. *Mol Cell Proteomics* 2005;4:618–625.
19. Gygi SP, Rist B, Gerber SA, Turecek F, Gelb MH, Aebersold R. Quantitative analysis of complex protein mixtures using isotope-coded affinity tags. *Nat Biotechnol* 1999;17:994–999.
20. Ross PL, Huang YN, Marchese JN, et al. Multiplexed protein quantitation in *Saccharomyces cerevisiae* using amine-reactive isobaric tagging reagents. *Mol Cell Proteomics* 2004;3:1154–1169.
21. Kuster B, Schirle M, Mallick P, Aebersold R. Scoring proteomes with proteotypic peptide probes. *Nat Rev Mol Cell Biol* 2005;6:577–583.
22. Craig R, Cortens UJP, Beavis RC. The use of proteotypic peptide libraries for protein identification. *Rapid Commun Mass Spectrom* 2005;19:1844–1850.
23. Ong SE, Mann M. Mass spectrometry-based proteomics turns quantitative. *Nat Chem Biol* 2005;1:252–262.
24. Haab B. Methods and applications of antibody microarrays in cancer research. *Proteomics* 2003;3:2116–2122.
25. Zhu H, Bilgin M, Bangham R, et al. Global analysis of protein activities using proteome chips. *Science* 2001;293:2101–2105.
26. Petricoin EF, Ardekani AM, Hitt BA, et al. Use of proteomic patterns in serum to identify ovarian cancer. *Lancet* 2002;359:572–577.
27. Rai AJ, Stemmer PM, Zhang Z, et al. Analysis of HUPO PPP reference specimens using SELDI-TOF mass spectrometry: multi-institution correlation of spectra and identification of biomarkers. *Proteomics* 2005;5:3467–3474.

28. Pieper R, Gatlin CL, Makusky AJ, et al. The human serum proteome: display of nearly 3700 chromatographically separated protein spots on two-dimensional electrophoresis gels and identification of 325 distinct proteins. *Proteomics* 2003; 3:1345–1364.
29. Adkins JN, Varnum SM, Auberry KJ, et al. Toward a human blood serum proteome: analysis by multidimensional separation coupled with mass spectrometry. *Mol Cell Proteomics* 2002;1:947–952.
30. Tirumalai RS, Chan KC, Prieto DA, Issaq HJ, Conrads TP, Veenstra TD. Characterization of the low molecular weight human serum proteome. *Mol Cell Proteomics* 2003;2:1096–1103.
31. Shen Y, Jacobs JM, Camp DG II, et al. Ultra-high-efficiency strong cation exchange LC/RPLC/MS/MS for high dynamic range characterization of the human plasma proteome. *Anal Chem* 2004;76:1134–1144.
32. Shen Y, Smith RD. Advanced nanoscale separations and mass spectrometry for sensitive high-throughput proteomics. *Expert Rev Proteomics* 2005;2:431–447.
33. Chan KC, Lucas DA, Hise D, et al. Analysis of the human serum proteome. *Clin Proteomics* 2004;1:101–226.
34. Zhou M, Lucas DA, Chan KC, et al. An investigation in the human serum interactome. *Electrophoresis* 2004;25:1289–1298
35. Rose K, Bougueleret L, Baussant T, et al., Industrial-scale proteomics: from liters of plasma to chemically synthesized proteins. *Proteomics* 2004;4:2125–2150.
36. Hanash SM, Celis JE. The Human Proteome Organization: a mission to advance proteome knowledge. *Mol Cell Proteomics* 2002;1:413–414.
37. Omenn GS. The Human Proteome Organization Plasma Proteome Project pilot phase: reference specimens, technology platform comparisons, and standardized data submissions and analyses. *Proteomics* 2004;4:1235–1240.
38. Kersey PJ, Duarte J, Williams A, Karavidopoulou Y, Birney E, Apweiler, R. The International Protein Index: an integrated database for proteomics experiments. *Proteomics* 2004;4:1985–1988. http://www.ebi.ac.uk/IPI/IPIhelp.html.
39. Adamski M, States DJ, Omenn GS. Data standardization and integration in collaborative proteomics studies. In: Srivastava S, ed. *Informatics in Proteomics*. New York: Marcel Dekker, 2004:169–194.
40. Adamski M, Blackwell T, Menon R, et al. Data management and preliminary data analysis in the pilot phase of the HUPO Plasma Proteome Project. *Proteomics* 2005;5:3246–3261.
41. Carr S, Aebersold R, Baldwin M. The need for guidelines in publication of peptide and protein identification data. *Mol Cell Proteomics* 2004;3:351–353.
42. States DJ, Omenn GS, Blackwell TW, et al. Deriving high confidence protein identifications from a HUPO collaborative study of human serum and plasma. *Nat Biotechnol* 2006;24:333–338.
43. Zhang H, Li XJ, Martin DB, Aebersold R. Identification and quantification of N-linked glycoproteins using hydrazide chemistry, stable isotope labeling and mass spectrometry. *Nat Biotechnol* 2003;21:660–666.

44. Yang Z, Hancock WS, Richmond-Chew T, Bonilla L. A Study of glycoproteins in human serum and plasma reference standards (HUPO) using multi-lectin affinity chromatography coupled with RPLC-MS/MS. *Proteomics* 2005;5: 3353–3366.
45. Tang HY, Ali-Khan N, Echan LA, et al. A novel 4-dimensional strategy combining protein and peptide separation methods enables detection of low abundance proteins in human plasma and serum proteomes. *Proteomics* 2005;5:3329–3342.
46. Haab BB, Geierstanger BH, Michailidis G, et al. Immunoassay and antibody microarray analysis of the HUPO PPP reference specimens: systematic variation between sample types and calibration of mass spectrometry data. *Proteomics* 2005;5:3278–3291.
47. Muthasamy B, Hanumanthu G, Reshmi R, et al. Plasma proteome database as a resource for proteomics research. *Proteomics* 2005;5:3531–3536.
48. Berhane BT, Zong C, Liem, et al. Cardiovascular-related proteins identified in human plasma by the HUPO Plasma Proteome Project pilot phase. *Proteomics* 2005;5:3520–3530.
49. Tammen H, Schulte I, Hess R, et al. Peptidomic analysis of human blood specimens: comparison between plasma specimens and serum by differential peptide display. *Proteomics* 2005;5:3414–3422.
50. Villanueva J, Shaffer DR, Philip J, et al. Differential exoprotease activities confer tumor-specific serum peptidome patterns. *J Clin Invest* 2006;116:271–284.
51. Liotta LA, Petricoin EF. Serum peptidome for cancer detection: spinning biologic trash into diagnostic gold. *J Clin Invest* 2006;116:226–230.
52. Uhlen M, Ponten F. Antibody-based proteomics for human tissue profiling. *Mol Cell Proteomics* 2005;4.4:384–393.
53. Hanash S. Harnessing immunity for disease proteomics. *Nat Biotechnol* 2003;21: 37–38.
54. Imafuku Y, Omenn GS, Hanash S. Proteomics approaches to identify auto-immune antibodies as cancer biomarkers. *Dis Markers* 2004;20:149–153.
55. Wang X, Yu J, Sreekumar A, et al. Autoantibody signatures in prostate cancer. *N Engl J Med* 2005;353:1224–1235.
56. Gao WM, Kuick R, Orchekowski RP, et al. Distinctive serum protein profiles involving abundant proteins in lung cancer patients based upon antibody microarray analysis. *BMC Cancer* 2005;5:110.
57. Hong SH, Misek DE, Wang H, et al. Auto-antibody-mediated immune response to calreticulin isoforms in pancreatic cancer. *Cancer Res* 2004;64:5504–5510.
58. Ransohoff DF. Lessons from controversy: ovarian cancer screening and serum proteomics. *J Natl Cancer Inst* 2005;97:315–319.
59. Zhang Z, Bast RC Jr, Yu Y, et al. Three biomarkers identified from serum proteomic analysis for the detection of early stage ovarian cancer. *Cancer Res* 2004; 64:5882–5890.
60. Semmes OJ, Feng Z, Adam BL, et al. Evaluation of serum protein profiling by surface-enhanced laser desorption/ionization time-of-flight mass spectrometry

for the detection of prostate cancer: I. Assessment of platform reproducibility. *Clin Chem* 2005;51:102–112.
61. Bouwman K, Qiu J, Zhou H, et al. Microarrays of tumor cell derived proteins uncover a distinct pattern of prostate cancer serum immunoreactivity. *Proteomics* 2003;3:2200–2207.
62. Qiu J, Madoz-Gurpide J, Misek DE, et al. Development of natural protein microarrays for diagnosing cancer based on an antibody response to tumor antigens. *J Proteome Res* 2004;3:261–267.
63. Espina V, Woodhouse EC, Wulfkuhle J, et al. Protein microarray detection strategies: focus on direct detection technologies. *J Immunol Methods* 2004;290: 121–133.
64. Barker PE, Wagner PD, Stein SE, Bunk DM, Srivastava S, Omenn GS. Standards for plasma and serum proteomics in early cancer detection: a needs assessment from the NIST-NCI SMART workshop. *Clin Chem* 2006;52:1669–1674.
65. Keller A, Purvine S, Nesvizhskii AI, Stolyar S, Goodlett DR, Kolker E. Experimental protein mixture for validating tandem mass spectral analysis OMICS. *J Integr Biol* 2002;6:207–212.
66. Strahler JR, Veine D, Walker A, et al. A publicly available dataset of MALDI-TOF/TOF and LTQ mass spectra of known proteins ASMS Session 2005: Bioinformatics Code: TP22 Time Slot/Poster Number: 398.
67. Nesvizhskii AI, Aebersold R. Interpretation of shotgun proteomic data-the protein inference problem. *Mol Cell Proteomics* 2005;4.10:1419–1440.
68. Rauch A, Bellew M, Eng J, et al. Computational Proteomics Analysis System (CPAS): an extensible, open-source analytic system for evaluating and publishing proteomic data and high throughput biological experiments. *J Proteome Res* 2006;5:112–121.
69. Omenn GS. Strategies for proteomics profiling of human tumors. *Proteomics* 2006;6:5662–5673.
70. Adkins JN, Monroe ME, Auberry KJ, et al. A proteomic study of HUPO's Plasma Proteome Project pilot samples using an accurate mass and time tag strategy. *Proteomics* 2005;5:3454–3466.
71. Olsen JV, Mann M. Improved peptide identification in proteomics by two consecutive stages of mass spectrometric fragmentation. *Proc Natl Acad Sci U S A* 2004; 101:13,417–13,422.
72. Humphery-Smith I. A human proteome project with a beginning and an end. *Proteomics* 2004;4:2519–2521.
73. Aebersold R. Constellations in a cellular universe. *Nature* 2003;422:115–116.
74. Anderson L, Hunter CL. Quantitative mass spectrometric multiple reaction monitoring assays for major plasma proteins. *Mol Cell Proteomics* 2006;5(4):573–588.

11

Proteomics of Human Urine

Visith Thongboonkerd, Pedro R. Cutillas, Robert J. Unwin, Stefan Schaub, Peter Nickerson, Marion Haubitz, Harald Mischak, Dobrin Nedelkov, Urban A. Kiernan, and Randall W. Nelson

Summary

Urine is one of the most interesting and useful human body fluids for clinical proteomics studies because of its availability in almost all patients and ease of collection without any invasive or painful procedure. Proteomic analysis of human urine offers invaluable information not only for a better understanding of the normal renal physiology and pathophysiology of kidney/kidney-related diseases, but also for biomarker discovery. This chapter provides principles, brief methods, and applications of various proteomic technologies (i.e., gel-based method, liquid chromatography-based approach, surface-enhanced laser desorption/ionization time-of-flight mass spectrometry, capillary electrophoresis coupled to mass spectrometry, and mass spectrometric immunoassay) to examine the human urine. At the end, perspectives and future directions in the field of urinary proteomics are discussed.

Key Words: Urine; kidney; proteome; pathophysiology; biomarker discovery; 2D-PAGE; liquid chromatography; SELDI; capillary electrophoresis; mass spectrometric immunoassay.

1. Introduction

The urine is derived from the plasma that flows into the kidney (350–400 mL per 100 g of tissue per min). The glomerulus is a specialized intrarenal microstructure that can selectively permeabilize some components of the plasma to pass through as ultrafiltrate (150–180 L per day). Proximal renal tubules then selectively reabsorb components in the ultrafiltrate until approximately 1% of ultrafiltrate volume is excreted with waste products as the urine *(1)*. In the normal state, urinary proteins can originate from any of these mechanisms: plasma ultrafiltration, tubular secretion, shedding of apical surface of renal tubular epithelial cells, shedding of whole cells along urinary passage, and exosome secretion.

From: *Proteomics of Human Body Fluids: Principles, Methods, and Applications*
Edited by: V. Thongboonkerd © Humana Press Inc., Totowa, NJ

In addition, proteins originating from the seminal gland (in males) and those contaminated by leukorrhea (in females) may be present in the urine and should be of concern during data interpretation. Recent urinary proteome analyses have indicated that approximately 30% of proteins in the normal human urine are plasma proteins, whereas the other 70% are proteins excreted from mechanisms other than plasma ultrafiltration *(2–4)*. In disease states, the normal physiology of protein excretion is disrupted, resulting in an altered urinary proteome profile and/or increased urinary protein excretion. Therefore, proteomic analysis of human urine should provide important information for a better understanding of the normal renal physiology, unraveling the pathophysiology of renal diseases, and identification of biomarker candidates that can be used as diagnostic and/or prognostic markers. Several approaches (i.e., gel based, liquid chromatography [LC] based approach, surface-enhanced laser desorption/ionization time-of-flight mass spectrometry [SELDI-TOF-MS], capillary electrophoresis coupled to mass spectrometry [CE-MS], and mass spectrometric immunoassay [MSIA]) can be employed to examine the urinary proteome. In this chapter, principles, brief methods, and applications of these technologies in human urinary proteomics are summarized, and future directions of proteomic analysis of human urine are discussed.

2. Analytical Approaches for Urinary Proteomics

Proteomic analyses of human urine and other body fluids can be performed using different approaches based on the study purposes **(Fig. 1)**. The *classical approach* is to examine proteins extensively and systematically for their expression and function to understand better the normal physiology and pathophysiology of diseases *(5)*. The analytical strategies in this approach involve expression proteomics (using 2D polyacrylamide gel electrophoresis [2D-PAGE] or LC for protein separation), quantitative analysis, bioinformatics, and functional proteomics. The *alternative approach* is to examine proteome profiles/patterns to differentiate types or groups of biofluids (e.g., normal vs diseases; a specific disease vs others), rather than focusing on detailed characterizations (identities, posttranslational modifications [PTMs], interactions, and so on] of particular proteins *(5)*. Common analytical methods used in the latter approach are SELDI-TOF-MS *(6)*, microarrays *(7)*, and CE-MS *(8)*. The alternative approach is suitable for clinical diagnostics and biomarker discovery, especially in cases of multifactorial diseases for which one marker may not be sufficient for effective detection or diagnosis. In addition, information about dynamic changes of the proteome profile during or after treatment may be useful to predict therapeutic outcome and/or prognosis of the disease. The alternative approach may be used as a complementary diagnostic tool for some medical diseases in which clinical diagnosis relies only on invasive procedures that may be limited on some occasions. For

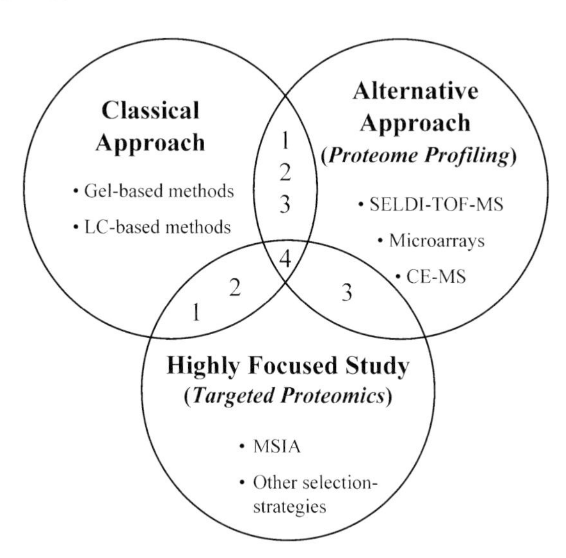

Fig. 1. Analytical approaches for urinary proteomics. The *classical approach* is to examine extensively and systematically urinary proteins for their expression and function. Analytical strategies in this approach include expression proteomics (using gel-based or LC-based methods), quantitative analysis, bioinformatics, and functional proteomics. The *alternative approach* or *proteome profiling*, which bypasses complicated analytical procedures in the classical approach, is to examine proteome profiles/patterns to differentiate types or groups of samples, rather than focusing on detailed characterizations of particular proteins. Common analytical methods used in the latter approach are SELDI-TOF-MS, microarrays, and CE-MS. *Targeted proteomics* is the highly focused study targeting an interesting set of proteins, which can be selected using MSIA or other selection strategies. All these approaches are complementary and, as a result, lead to achievement of ultimate goals that are designated as numbers: (1) better understanding of the normal physiology; (2) unraveling the pathophysiology of diseases; (3) biomarker discovery; and (4) better therapeutic outcome.

example, renal biopsy, which remains the gold standard for the diagnosis of glomerular diseases, may not be possible in patients whose indications for this invasive procedure are not fulfilled and those with a bleeding tendency or flank skin infection; thus diagnosis is delayed. In these cases, identification of urinary biomarkers via the alternative approach may result in an earlier diagnosis that ultimately leads to a better therapeutic outcome.

There are two levels of analytical strategies in urinary proteomics. Analysis of the entire urinary proteome or *global analysis*, using either the classical or alternative approach, is suitable for screening for *global changes* of urinary proteins affected by an experimental condition or disease state. The major limitation of the global analysis is that identification of low-abundance proteins may be obscured by major-abundance components, making the analysis difficult. Analysis of selected compartment(s) of urinary proteome or *targeted proteomics* is suitable for the highly focused study ***(5)***. This term may be confused with subproteome analysis that is sometimes used interchangeably. Although the strategy of subproteome analysis is to examine selectively a particular ultrastructure or microstructure of cell or organ ***(9,10)***, selection strategy of targeted proteomics has a wider spectrum, e.g., selection of a particular set of renal salt and water transport proteins ***(11–14)***; low-abundance proteins ***(15)***; hydrophilic or hydrophobic proteins ***(16–18)***; peptides with specific amino acid (Cys, His, Met, and so on.) ***(19–21)***; N- or C-terminal peptides ***(22,23)***; posttranslationally modified proteins (by glycosylation, phosphorylation, oxidation, nitration, and so on.) ***(24)***; and affinity-purified proteins (lectin affinity, co-immunoprecipitation, tandem affinity purification, mass spectrometric immunoassay, and so on) ***(25)***. In this chapter, mass spectrometric immunoassay is used as a model for the discussion of targeted proteomics study of human urine.

3. Gel-Based Urinary Proteomics

3.1. Principles and Methods

To date, 2D-PAGE has been the most commonly used method in proteomic studies because of its simplicity and availability in most of proteomic laboratories. In urinary proteomics, this approach is suitable for constructing the reference urinary proteome map for further use, defining differentially expressed proteins in different sets of samples, and screening for PTMs. Principles and basic methodologies of 2D-PAGE can be found elsewhere ***(26)***. Separated proteins in a 2D gel can be visualized by various types of staining (Coomassie Brilliant Blue, silver, fluorescence, and others). For quantitative intensity analysis in differential proteomics, the stain to be used should have these properties: (1) wide linear dynamic range of intensity-concentration correlation; (2) high sensitivity to visualize low-abundance protein spots; and (3) compatibility with subsequent mass spectrometric analysis. Silver staining is one of the most sensitive staining methods employed for visualization ***(27,28)***. Although modifications of the conventional protocol make it more compatible with mass spectrometric analysis ***(29)***, silver stain has a narrow dynamic range of intensity-concentration correlation and should not be used in quantitative intensity analysis and differential proteomics study ***(30)***. Coomassie Brilliant Blue is less sensitive but more widely employed because of its simplicity. Fluorescent stains such as SYPRO Ruby (a ruthenium-based compound) and Deep Purple (a compound isolated from the fungus *Epicoccum nigrum*) are superior in their ability to quantify proteins across a broad concentration range and should be used as the standard stains for quantitative intensity analysis and differential proteomics study ***(30,31)***. The recent development of multiplexed proteomic technologies such as differential gel electrophoresis (DIGE) using more than one dye to stain differential sets of protein samples in the same gel has allowed more accurate quantitative intensity analysis *(32)*.

After spot matching and quantitative intensity analysis, the spots of interest are excised, in-gel digested with proteolytic enzymes (trypsin, chymotrypsin, Arg-C, Asp-N, Lys-C, PepsinA, V8-E, V8-DE, and so on) and identified, mostly by matrix-assisted laser desorption ionization (MALDI)-TOF-MS, which provides a high-throughput manner of protein identification; hundreds of proteins can be identified within a day *(33)*. In MALDI analysis, proteins or tryptic peptide fragments are mixed with a chemical matrix that has the property of facilitating ion activation by laser beam. The most commonly used matrix in MALDI is α-cyano-4-hydroxycinnamic acid (for peptide analysis) and trans-3,5-dimethoxy-4-hydroxycinnamic acid or sinapinic acid (for protein analysis) *(34)*. The mixture of analytes and matrix is spotted onto the target plate and allowed to air-dry. Thereafter, crystals of matrix are formed together with the analytes. The target plate is then placed into the MS instrumentation, which is equipped with a laser-beam generator. After laser firing, the matrix adsorbs the energy from laser and then undergoes rapid solid-to-gas phase transition or the exciting state. The analytes embedded in the matrix crystals are then ionized by the excited matrix with a poorly understood process. Most of ions formed in the activated analytes are single charged *(35)*. These ions are then ejected from the target plate to the mass analyzer. The most commonly employed mass analyzer in MALDI analysis is TOF. With a fixed distance of ion passage, the time to the target of the activated ions is different because of variations in mass and charge of the ionized peptides. Various peptide fragments can then be distinguished and identified by peptide mass fingerprinting, based on differential mass per charge (*m/z*) values.

3.2. Applications to Human Urine

2D-PAGE has been employed to examine human urine since 1979 *(36)*. Its applications to urinary proteomics are summarized in **Table 1**. Commonly, 2D-PAGE is applied to global, classical analysis of urinary proteome. However, it can be applied to targeted urinary proteomics, as demonstrated in highly focused studies using 2D Western blotting.

4. LC-MS-Based Methods for Urinary Proteomics

2D-PAGE and LC-based methods for proteomic studies are complementary in nature, and although the abundance of most proteins can in principle be determined using either method, there are instances in which the choice can determine which protein classes will be detected (*see* **refs. *73–76*** for reviews). 2D-PAGE should be the method of choice for detecting protein isoforms, because this technique can separate proteins according to molecular mass and

Table 1
Applications of 2D-PAGE to Human Urinary Proteomics

Global analysis
- 2D proteomic analysis (2D-PAGE with mass spectrometric analysis)
  - 2D proteome maps of normal human urine ***(2–4,37–39)***
  - Glomerular diseases ***(40)***
  - Gammopathy ***(41)***
  - Bladder cancers ***(42,43)***
  - Prostate cancer ***(44)***
  - Lung cancer ***(45)***
  - Reconstruction of urinary bladder with intestinal transposition ***(46)***
  - Pilonidal abscess ***(47)***
- 2D spot profiling (2D-PAGE without mass spectrometric analysis)
  - Normal human urine ***(36,48–52)***
  - Asymptomatic proteinuria ***(53)***
  - Prostate cancer and benign prostatic hyperplasia ***(54,55)***
  - Leukemia ***(56)***
  - Multiple myeloma ***(57,58)***
  - Nephrolithiasis ***(59,60)***
  - Chronic renal failure ***(61,62)***
  - Renal transplantation ***(63,64)***
  - Cadmium toxicity ***(65)***

Highly focused study
- 2D Western blotting
  - Malaria—for malarial antigens ***(66)***
  - Chagas' disease—for *Trypanosoma cruzi* antigens ***(66)***
  - Trophoblastic tumors—for human chorionic gonadotropin ***(67)***
  - IgD myeloma—for monoclonal δ-chains of immunoglobulin ***(68)***
  - γ-1-heavy chain disease—for γ-1-heavy chains of immunoglobulin ***(69)***
  - Dialysis-related amyloidosis—for β_2-microglobulin ***(70,71)***
  - Bladder squamous cell carcinoma—for psoriasin ***(72)***

isoelectric point, two physicochemical properties of proteins that are altered as a result of PTMs, such as proteolysis or covalent modifications, respectively. In contrast, LC- and LC-MS-based methods are more effective for high-throughput analyses and detection of small proteins and peptides, which are a relatively large proportion of the urinary proteome ***(77)***. LC-MS is also more effective than 2D-PAGE in detecting hydrophobic and membrane proteins ***(78)***. Indeed, the difficulties in resolving most hydrophobic and membrane proteins by 2D-PAGE have been reviewed ***(79)***. Although it can be argued that this limitation of 2D-PAGE may not matter when one is analyzing water-soluble proteins like those found in biological fluids, it has been found that urine also contains a *membrane fraction* consisting of small vesicles or *exosomes*, which may prove of value as potential diagnostic markers ***(80,81)***. Because of its hydrophobic nature, the proteome of such urinary vesicles will probably be more comprehensively analyzed by LC-MS than by 2D-PAGE ***(81)***.

The use of 2D-LC-MS/MS as a proteomic tool for the large-scale analysis of proteins was pioneered by the group of Yates and colleagues ***(82)***, who developed an approach called *multidimensional protein identification technology* (MudPIT) to detect more than 1000 proteins in yeast. This approach involves proteolytic digestion of the total protein mixture to obtain a set of protein-derived peptides that are then separated by strong cation exchange chromatography (SCX). Peptides present in fractions from this SCX step are separated further by reversed-phase (RP) LC and then sequenced by MS/MS. Several thousand peptides can be sequenced in this way in a relatively short time. For analysis of the urinary proteome, Pang and co-workers ***(47)*** described the use of 2D-LC-MS/MS to detect changes in urinary proteins in the setting of inflammation. Spahr et al. ***(83)*** used LC-MS/MS to detect more than 100 proteins in commercial lyophilized urine that was free of proteins below 10 kDa. This number of identifications is quite remarkable, since their study involved only a single RP separation (i.e., without a prior SCX step) before MS/MS.

An alternative (although related) approach to MudPIT (in which the protein mixture is digested before two orthogonal modes of HPLC separation) for LC-MS-based proteomics is to separate the undigested protein mixture by high-performance liquid chromatography (HPLC) and then analyze the proteins present in each chromatographic fraction by MS and LC-MS ***(84,85)***. Proteins and peptides in these fractions can then be analyzed using modern biological MS. Proteins can be digested after HPLC separation and prior to LC-MS/MS analysis, a requirement for peptide sequence determination by MS/MS. Moreover, peptides and proteins can also be analyzed by MALDI-TOF MS for measuring intact molecular mass. This approach has been used to identify peptides and proteins in the urine of renal Fanconi patients ***(86,87)***. In this study, proteins and peptides were separated into fractions by SCX and then sequenced by LC-MS/MS. In addition to the common and known urinary proteins, renal Fanconi urine contained several cytokines, chemokines, and other bioactive peptides with molecular mass of less than 15 kDa. (The smallest was bradykinin, with a molecular mass of 1 kDa.)

In an independent investigation ***(77)***, urinary proteins from normal donors and renal Fanconi patients were separated by RP-HPLC, and the peptides and proteins present in these fractions were analyzed by MALDI-TOF MS before and by LC-MS/MS after digestion with proteolytic enzymes. This study was aimed to address a puzzling issue raised by the Comper's group ***(88)***, which had proposed that albumin and other plasma proteins are processed mainly by renal proximal tubules in such a way that they are degraded into peptides, which are then reexcreted into the urine. Norden et al. *(77)* found that there are indeed many different peptides below 10 kDa in the urine. Additional findings observed by Cutillas and Unwin (unpublished data) showed that more than 100 different molecular species could be detected in the normal urine within the 1- to 10-kDa molecular mass range. However, the largest UV absorbance of an undigested RP-HPLC separation corresponded to the fraction in which full-length albumin eluted. Intact molecular mass determination by MALDI-TOF MS and sequencing of proteolytic peptides by LC-MS/MS confirmed that full-length albumin was the predominant protein in this fraction. Thus, this study led to the conclusion that although normal urine contains a large number of peptides and protein fragments below 10 kDa, full-length proteins of plasma and renal origins are quantitatively the major components of the normal human urinary proteome *(77)*. Nevertheless, a question that has not yet been fully addressed is the role, if any, these small peptides may have in regulating renal function. It is well known that proteolytic processing of protein precursors can lead to the formation of active peptides with potential effects on renal tubular cell function. Some of these peptides, such as epidermal growth factor and bradykinin, are found in the normal urine *(89,90)* and could have a role in controlling nephron function by acting via luminal membrane receptors of tubular epithelial cells. The issues that remain to be investigated are: (1) which of the many small urinary peptides are bioactive; (2) what their actions on the renal tubules might be; and (3) whether they are merely inactive byproducts of renal protease activity.

The complementary nature of 2D-PAGE- and LC-MS-based methods for proteomic analysis of the human urine was also illustrated in a study on the urinary proteome of Dent's disease *(91)*, a genetic form of the renal Fanconi syndrome. This condition results from mutations of the chloride ion channel ClC-5, which is involved in megalin-mediated endocytosis along renal proximal tubules *(92)*. Mutations in *CLCN5*, the gene that encodes this channel protein, cause impaired endocytosis of filtered proteins by human and murine renal proximal tubular cells in vivo and in vitro *(93,94)*. Therefore, it is likely that the protein composition of the Dent's urine is close to that of the glomerular ultrafiltrate *(95,96)*. Using 2D-PAGE and LC-MS approaches, more than 100 proteins could be detected and quantified in control and Dent's urine samples *(91)*. The results obtained using these two methods were found to be highly consistent; however, certain protein classes such as small cytokines and chemokines were detected only by the LC-MS method. In general, it was found that some proteins of plasma origin like cytokines and vitamin-binding proteins were present in Dent's urine in relatively higher amounts than other plasma proteins like albumin or immunoglobulins. These findings indicate that the endocytic apparatus responsible for the reabsorption of filtered proteins, which is present at the

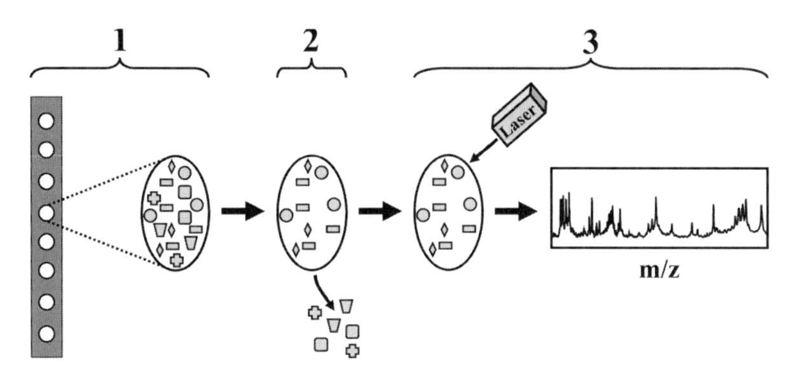

Fig. 2. Description of SELDI-TOF-MS technology. Sample analysis with SELDI-TOF-MS involves three steps: (1) the sample is applied to the chip covered with a chromatographic surface (i.e., ProteinChip); (2) a subset of proteins binds to the surface while all others are removed by a wash step; and (3) bound proteins are analyzed by TOF-MS.

apical membrane of renal tubular cells and is defective in Dent's patients, has a greater affinity for cytokines and vitamin-binding proteins than for other protein classes (e.g., albumin and immunoglobulins). This study is an example of how urinary proteomics of well-defined genetic models of human renal pathophysiology can provide valuable information on normal renal physiology. From the methodological standpoint, this study demonstrated that several complementary proteomic approaches are necessary to make a comprehensive comparative analysis of proteomes, even when these contain a relatively small number of components, as in the urinary proteome.

5. SELDI-TOF-MS for Urinary Proteomics

5.1. Description of the Method: Advantages and Limitations

SELDI-TOF-MS combines MALDI-TOF-MS with surface retentate chromatography. Specifically, a sample is applied to a chip surface carrying a functional group (e.g., normal phase, hydrophobic, cation- or anion-exchange). After incubation, proteins that do not bind to the surface are removed by a simple wash step, and bound peptides/proteins are analyzed by TOF-MS (**Fig. 2**). This approach reduces the complexity of the sample being analyzed by selecting only a subset of the total proteins. However, a significant fraction of the more abundant proteins can be analyzed by varying the chip surfaces and the binding conditions. SELDI-TOF-MS offers some advantages for protein analysis in the urine. First, only 5 to 10 μL of sample is needed for one analysis. Second, this method can be readily automated, making it particularly useful for the high-throughput analysis. Third, the reproducibility is high regarding the detection of proteins when standardized protocols are used *(97)*.

For comparative analysis of urine samples, it is important to know the factors that can influence the composition and relative abundance of urinary proteins but may not be related to the process/disease investigated. The urine can have changing physicochemical properties (e.g., pH, dilution, and activity of proteases) and a variable amount of cellular components (e.g., erythrocytes, leukocytes, and epithelial cells), which can affect the protein content and its detection by SELDI-TOF-MS. In an analysis of various intrinsic factors (e.g., blood in urine, urine dilution, and first-void vs midstream urine) and extrinsic factors (e.g., urine storage and freeze-thaw cycles) concerning their impact on urinary proteome profiles acquired by SELDI-TOF-MS, it was found that (1) up to five freeze-thaw cycles had minimal impact on the proteome profiles, (2) midstream urine samples did not undergo changes when stored for 3 d at 4°C, (3) significant hematuria confounded the detection of normal urinary proteins either because of hemolysis or addition of plasma proteins, (4) urine output above 2 to 3 L per day resulted in protein concentrations that led to a loss of detectable urinary proteins, and (5) female first-void urine had a different population of proteins compared with the midstream urine ***(97)***. Furthermore, recent studies highlight that urinary proteins are present not only as intact proteins but also as smaller cleaved fragments. These are derived from proteolytic degradation in tubular epithelial cell lysosomes with subsequent regurgitation into the urine ***(98,99)*** or from intraluminal cleavage of urinary proteins by secreted/released proteases (e.g., from tubular epithelial cells or inflammatory cells) ***(100)***. This highlights the fact that standardization of sample collection and handling is essential, and both intrinsic and extrinsic factors must be taken into account for accurate data interpretation.

An important issue for any proteomic platform is to determine the detection threshold, which allows one to estimate the *visible* part of the analyzed proteome ***(101)***. The detection of a protein by SELDI-TOF-MS is critically determined by its concentration in the sample, its binding to the chromatographic surface, and its ionization process within the mass spectrometer. The sensitivity of a mass spectrometer can be determined by analyzing single proteins. The detection threshold ranges from 100 ng/L to 1 mg/L and is higher with increasing molecular mass. The higher detection threshold for large molecular mass proteins is well known and is thought to be related to their inferior ionization. When a known quantity of a protein is added to the complex protein mixture of urine, the detection threshold increases by roughly 10- to 100-fold compared with the detection threshold for the pure form of the protein (**Fig. 3**). This decrease in the sensitivity is mainly caused by binding site competition on the chip and competition for ionization (i.e., ion suppression). Whereas the former is unique to the SELDI-TOF-MS platform, the latter is a common problem for all mass spectrometers. By changing the conditions for protein binding to

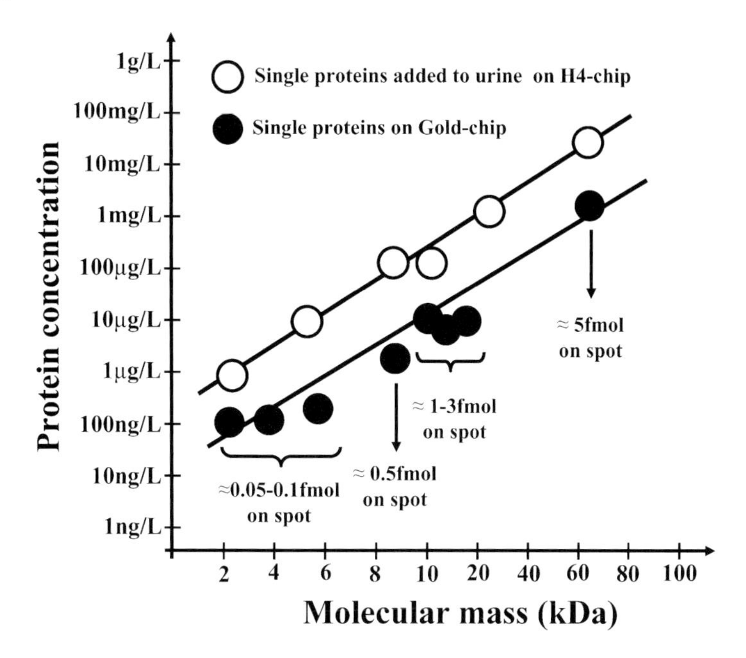

Fig. 3. Detection threshold of SELDI-TOF-MS (ProteinChip Reader II). The detection threshold for single proteins was determined by applying proteins with molecular masses from 2 to 67 kDa in increasing dilutions to Gold-chips (i.e., MALDI application). The detection threshold for single proteins in the urine was determined by adding proteins with molecular masses from 2 to 67 kDa in increasing dilutions to the urine from a healthy individual with a protein concentration of 110 mg/L and subsequent analysis on chips covered with a hydrophobic surface (i.e., H4-ProteinChips). For single proteins, the detection threshold increases from 100 ng/L (equals 0.05–0.1 fmol on the ProteinChip spot) for proteins with molecular masses from 2–6 kDa to 1 mg/L (equals 5 fmol on the ProteinChip spot) for a protein with a molecular mass of 67 kDa. The same correlation exists for single proteins added to the urine; however, the detection threshold is 10 to 100 times higher than for single proteins alone.

different chromatographic surfaces, some proteins may be selected and enriched, whereas others may be excluded, allowing the detection threshold to approximate that of the pure form of the single protein.

5.2. Applications to Human Urine

To date, SELDI-TOF-MS has been used mainly as a platform for differential urinary proteome profiling. Rogers et al. ***(102)*** and Zhang et al. ***(103)*** found urinary proteome patterns that were more than 80% sensitive and specific for renal cell carcinoma or bladder transitional cell carcinoma, respectively. Both groups used bioinformatic algorithms to define a *classifier* comprised of a group of protein peaks in a training set and subsequently validated their classifier in a smaller but *blinded* group. Interestingly, Rogers et al. ***(102)*** performed another validation of their classifier on a different set of blinded samples 10 mo later and found a marked decline in sensitivity and specificity, ranging from 41 to 76%. This illustrates that the robustness of this diagnostic approach needs to be substantially improved before it can be applied to a clinical setting and likely requires large data sets to ensure that the range of biological variation, which exists for a given condition, is appropriately represented in the training set ***(101)***.

Recently, three research groups have published their experiences using SELDI-TOF-MS to define urinary proteins/peptides associated with acute renal allograft rejection ***(104–106)***. Interestingly, each group found different urinary proteome patterns associated with the rejection response. To understand these apparent discrepancies, one must first consider that disease definition, sample collection, sample handling, protocol for protein separation/visualization, and data analysis were not identical in individual studies. Indeed, the three studies had major differences in (1) the definition of rejection and control groups, (2) the selection of chromatographic chip surfaces, and (3) the analysis of the generated spectra. Because of the power of the SELDI-TOF-MS technique to detect different populations of proteins depending on the chip surface used, it is not surprising on the basis of this fact alone that the patterns do not match. In an attempt to ensure reproducibility and accuracy, the report by Schaub et al. ***(105)*** defined all groups based on allograft function, clinical course and histology, included sequential urine analysis to enhance specificity, showed all individual spectra, and separated the groups based on the visual presence or absence of protein peaks.

Although some propose to use sophisticated bioinformatic analysis of the proteome profiles generated by SELDI-TOF-MS as a diagnostic tool, we see the application of SELDI-TOF-MS primarily as a high-throughput biomarker discovery platform, which implies subsequent protein identification as the logical next step. To date, protein identification strategy is based on complete in-solution purification (ion-exchange beads and RP-HPLC) and digestion with subsequent analysis of the peptides by RP-HPLC-MALDI-MS(/MS). This method allows one to follow the target protein(s) by SELDI-TOF-MS during all steps and to acquire very detailed information regarding the protein sequence ***(100)***. Subsequently, this information can be used to develop quantitative, high-throughput assays (i.e., enzyme-linked immunosorbent assay [ELISA]) for the potential protein biomarker. Nevertheless, it is clear at this point that irrespective of the approach (protein pattern diagnostics or protein identification/ELISA assay development), the results of all studies, to date, require validation in an independent sample set. Ultimately, the success of any unbiased proteomic approach will depend on the quality of the patient cohorts used for the comparative analysis. A rather rare but useful application of SELDI-TOF-MS is the detection of microheterogeneities of urinary proteins ***(100,107)***. Up to a molecular mass of 15 kDa, changes in the sequencc of the whole protein down to the amino acid level can be detected and studied.

6. CE-MS for Urinary Proteomics

6.1. Technical Aspects of CE-MS Coupling

High resolution in polypeptide separation can be accomplished by either LC-based methods (HPLC or 2D-LC) or CE. However, CE appears to offer several advantages over LC: (1) it is fairly robust and uses inexpensive capillaries and (2) it is compatible with essentially all buffers and analytes. Given these advantages, it appears surprising that CE has not been used more widely. A possible explanation might be the relatively small sample volume that can be loaded onto the CE, accentuating a problem of sensitive detection ***(108)***. In contrast, LC allows the analysis of rather large sample volume and hence is less demanding on sensitivity. With improving methods of ionization, the development of micro- and nanoflow ionsprays, and improvement of the detection limits of mass spectrometers, sensitivity has become less of an issue. Owing to the advantages of CE compared with LC, efforts have been made to achieve stable CE-MS coupling ***(109–112)***. The different methods of coupling have recently been reviewed in detail by Kolch et al. ***(113)***. In contrast to LC, CE generally has no flow rate but requires a closed electric circuit. Sheath-flow coupling is the most widely used approach, whereby the sheath liquid closes the electrical circuit and provides a constant flow rate. This form of ionization is essentially comparable to the micro- or nanoflow ionspray (depending on the liquid flow rate). The sample is diluted by the use of sheath liquid but, to a lesser extent than expected, probably owing to incomplete mixing in the Taylor cone ***(112)***.

The sensitivity is generally superior with sheathless interfaces, but these appear to be less stable, mostly owing to deposits on the metal or graphite coating used. The enhanced stability of sheath-flow coupling is an enormous benefit and outweighs its lower sensitivity. Even with sheath-flow coupling, detection limits in the high attomole range can be achieved, particularly when flow rates of the sheath flow are reduced ***(114,115)***. Another important consideration is coating of capillaries, which has also been reviewed by Kolch et al. ***(113)***. Generally, the analytes in the sample interact with the inner wall of the capillaries. This results in peak broadening or even deposits of analytes. As a solution, several types of coating and different coating protocols for the capillaries have been described; many of them are quite tedious and time consuming ***(116,117)***. Experience with different types of coating is yet not fully satisfactory ***(115)***; consequently the ideal approach appears to be the use of uncoated capillaries. Currently, the best results can be obtained at low pH in the presence of acetonitrile. However, a new coating protocol that has recently been described by Ullsten et al. ***(118)*** appears to be quite stable and applicable to even crude samples.

In contrast to direct CE-MS coupling using electrospray, coupling of CE to MALDI-MS by deposition of the analytes onto different targets has also been reported. Musyimi et al. ***(119)*** recently described a rotating ball interface, which allows an online coupling of CE to MALDI-MS. The main advantages of MALDI appear to be the enhanced stability and an easier handling in comparison with electrospray ionization (ESI). In addition, once the analytes are deposited on the target, they can be reanalyzed several times without the need of a new CE run. Moreover, the deposited analytes can be subsequently manipulated. This has recently been described by Zuberovic et al. ***(120)***: the authors spotted a CE run onto a MALDI target and were able to digest the deposited sample. Such an approach holds promise that even higher molecular mass polypeptides and proteins in clinical samples can be identified. The disadvantages of MALDI are certainly the decreased dynamic range in comparison with ESI and the higher sensitivity toward signal suppression. This leads to the detection of far fewer polypeptides in a complex sample. Whereas approximately 1500 polypeptides can generally be detected in a CE-ESI-MS run, the number of detectable different polypeptides drops to 300 when the same sample is spotted onto a MALDI target (Wittke et al., manuscript in preparation).

Taking all these technical problems into account, CE using uncoated fused silica capillaries and coupling via a sheath-liquid interface to the mass spectrometer appears to be a suitable tool for the analysis of a large number of complex biological samples. Although the CE-MS analysis itself is not demanding, several challenges become obvious when hundreds of samples need to be analyzed: (1) sample preparation must be highly reproducible (irrespective of the protein content); (2) overloading of any system used must be omitted; (3) larger polypeptides and proteins are frequently denatured and precipitated at low pH condition in the CE capillaries; and (4) the software solutions provided by manufacturers of mass spectrometers are inadequate to analyze the pattern of numerous complex samples for clinical use. A thorough evaluation of one single CE-MS run, which results in over 1000 consecutive individual MS spectra, would require several days. For the detection of the narrow CE-separated analyte zones, a fast and sensitive mass spectrometer is required. Thus, both ion trap and TOF systems appear to be adequate. Although ion trap MS acquires data over a suitable mass range at the rate of several spectra per second, the resolution is generally too low to resolve the single isotope peaks of more than threefold charged molecules. Consequently, assignment of charge to these peaks is hampered. Modern ESI-TOF instruments record up to 20 spectra per second and provide resolution (M/dM) of more than 10,000 and a mass accuracy better than 5 ppm. Thus, the most suitable instrument for this type of analysis is currently ESI-TOF-MS.

6.2. Sample Preparation

When one is analyzing complex biological samples, major concerns are losses of polypeptides and information as well as reproducibility. Consequently, a crude and unprocessed sample should be the ideal specimen. However, as all body fluids contain interfering components like different ions, lipids, carbohydrates, and so on, these samples cannot be analyzed in the native *as is* form in a mass spectrometer. Furthermore, separation is a prerequisite in order to cope with the complexity and dynamic range. CE is quite insensitive to interfering substances. This allows the injection even of crude urine samples. However, high salt content of the crude urine sample interferes with separation and seems to stress the glass capillary, causing it to break after a few runs. Consequently, it is advantageous to remove salts and other low-molecular-mass compounds. Nevertheless, the results on crude urine samples allow one to judge all sample manipulation with respect to recovery based on polypeptides detectable in these samples. An efficient method to remove unwanted substances in the samples initially appears to be the use of anion-exchange ***(121)*** or RP materials ***(122)***. Since larger polypeptides do not elute efficiently for RP-C-18 columns, it is advantageous to use RP-C-2 columns. Both RP-C-2 and anion-exchange give satisfactory results, but the reproducibility of the analyses using RP columns is higher ***(118)***.

6.3. Software and Data Analysis

One of the main obstacles using high-resolution separation in combination with mass spectrometry is the lack of suitable software. Each individual CE-MS run consists of about 1200 spectra. A typical contour plot (graphical depiction of the raw data file) of a urine sample from a healthy volunteer is shown in **Fig. 4**. A magnification of a small area reveals the enormous amounts of information contained in these spectra. The essential information that needs to be extracted from an entire CE-MS run is the identity and quantity of detected polypeptides. Consequently, MosaiquesVisu has been developed as one approach toward extracting information from these highly complex spectra ***(123)***.

This software package performs a stepwise examination of CE (or LC) MS spectra. In the first step, signals are collected by fitting the criteria of a certain signal-to-noise level and presented as several consecutive spectra. Next, the collected MS peaks are charge-deconvoluted based on isotopically resolved peaks. In the last step, the MS peaks are examined for conjugated peaks up to a charge of 50. This is especially important for nonresolved MS peaks. The conjugated

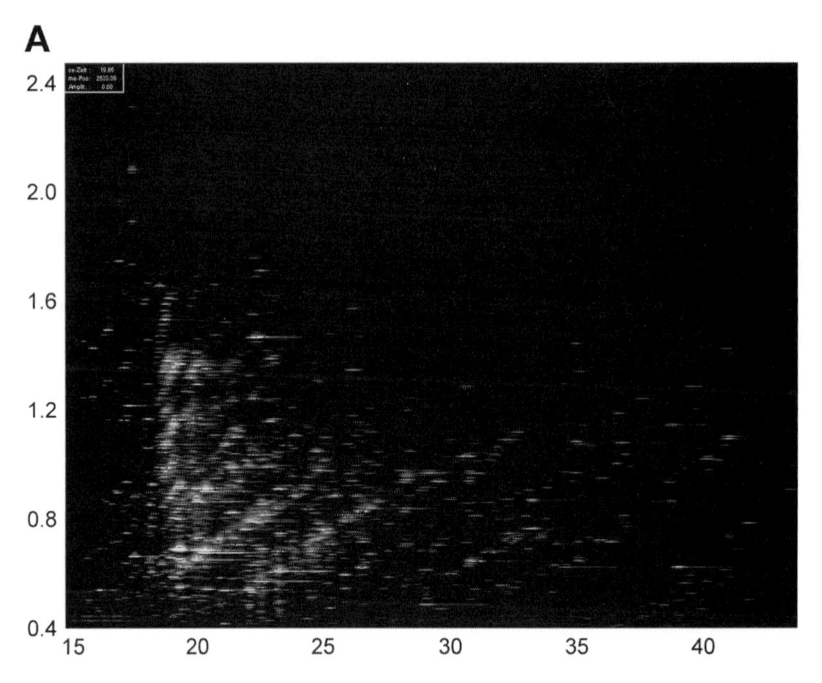

Fig. 4. CE-MS analysis of the urine from a healthy volunteer. **(A)** A graphical illustration of the raw data is shown: mass per charge on the *y*-axis (in Daltons/*z*) against the CE migration time (in min) on the *x*-axis. **(B)** Eight consecutive individual spectra (from 18.98 to 19.34 min) from **(A)** are shown. As evident, each single spectrum generally holds information on more than 50 different analytes. The high resolution is further illustrated by the magnification (X5) of one of the lower abundant polypeptide signals shown in the left upper trace.

peaks (and their amplitudes) are combined into a single peak, which is characterized by mass and migration time, as well as the combined amplitude of the conjugated peaks. Based on the mass deviation of the conjugated peaks and the match of isotopic distribution, the software also gives the probability that the peptide, which is evaluated based on calculated charge, really exists. The evaluation of an entire CE-MS run, comprising 1000 to 1500 individual spectra, is accomplished within 5 min with an error rate below 2% as judged by the 200 most abundant polypeptides. Hence, this software allows the timely evaluation of highly complex spectra. Migration time and mass are used to assess the identity of a peptide. Since migration time varies depending on the ion load in the sample and signal intensity varies depending on the efficiency of ionization as well as the detector gain, the polypeptide list is normalized with respect to migration time and signal intensity utilizing a set of 200 internal standard polypeptides, which are found with high frequency in the urine *(8)*.

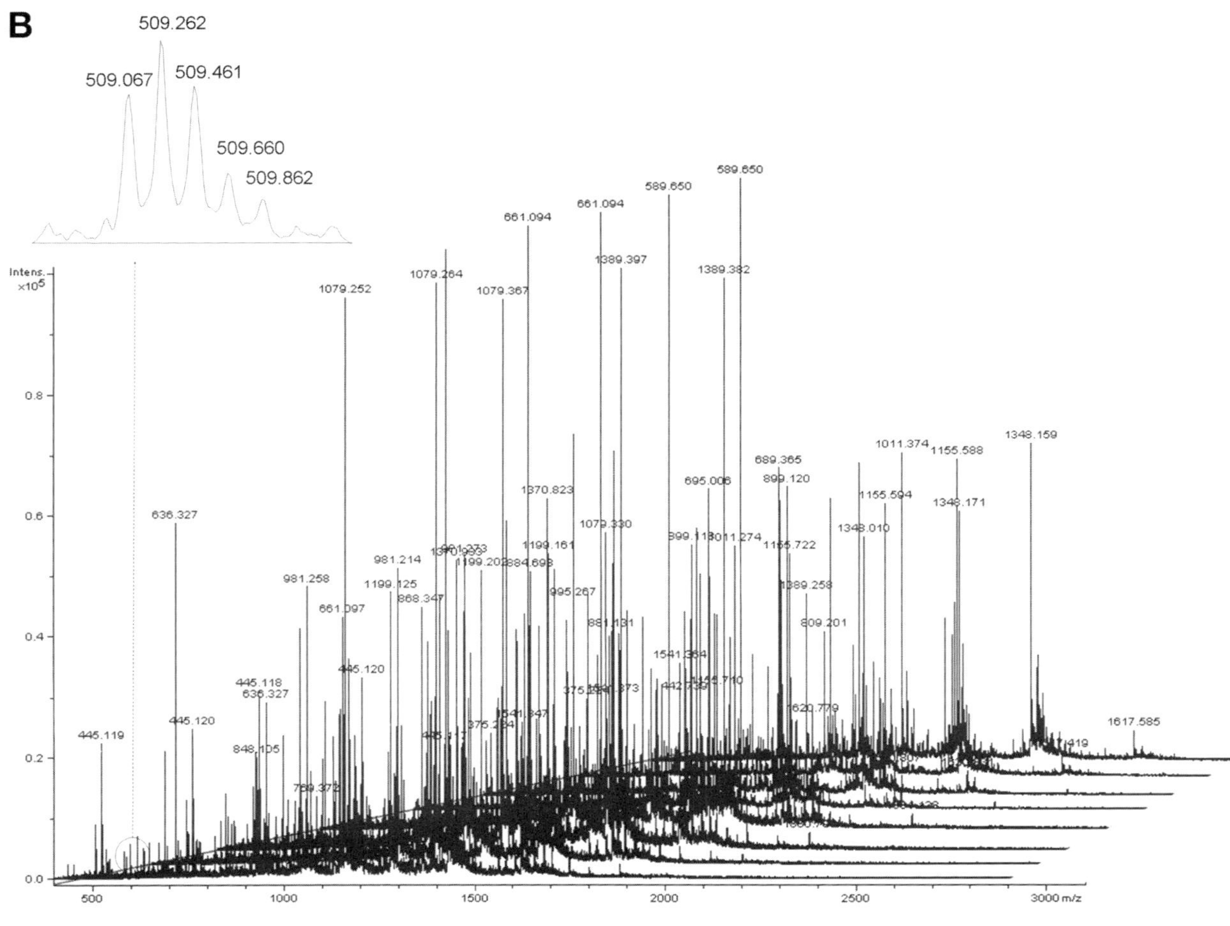

Fig. 4. *(Opposite Page)*

6.4. Applications to Human Urine

The development of CE-MS in combination with efficient sample preparation and software establishment has suggested that this proteomic technology can be applied to examine human urine for clinical diagnostics. Using CE-MS, Weissinger et al. ***(8)*** have initially shown that typical *housekeeping* urinary polypeptides, which can be found in almost all human urine samples, do in fact exist. Not surprisingly, polypeptides in the urine show significant inter- and intraindividual variabilities, as well as changes during the day. This can be attributed, at least in part, to physical activity, diet, smoking, and so on ***(8)***. The data on comparability also underscore the fact that single biomarkers are always subjected to variability. The results further indicate that a *common biomarker* that should display unique characteristics (presence/absence of signal intensity) in each sample from patients with a defined disease does not exist. Hence, the clinical utility of a single biomarker measurement may be limited in some instances, even if test accuracy and reproducibility are optimal. In these cases, proteome profiling will make the diagnosis more robust and specific, since changes in individual analytes do not result in gross change of the proteome pattern.

The highest reproducibility of individual polypeptides can be found when urine samples from infants are evaluated. Generally, the reproducibility decreases when the age increases even in apparently healthy individuals. Recently, Decramer et al. have shown that CE-MS examination of urinary polypeptides can predict the clinical outcome of uteropelvic junction obstruction with high accuracy (ms. submitted). The data also revealed that each of more than 500 individual polypeptides was present in more than 90% of all urine samples examined. This is the highest reproducibility ever observed in a proteomics study using patient samples. Additionally, Mischak and colleagues ***(8,123–125)*** have established characteristic urinary polypeptide patterns for various renal diseases, e.g., minimal change disease, focal-segmental glomerulosclerosis, membranous glomerulopathy, IgA nephropathy, and diabetic nephropathy. These patterns can be used for differential diagnosis, since each renal disease results in a specific and typical urinary polypeptide pattern (**Fig. 5**). The discrimination between different primary renal diseases can be done with high sensitivity and specificity (generally above 90%). Prospective studies to underline these results are on the way.

The large number of biomarkers characteristic for a specific disease also suggests that this approach may allow evaluation of disease progression and therapy. The latter has recently been shown in patients with diabetic nephropathy, whose treatment with an angiotensin II receptor antagonist significantly changed the urinary polypeptide excretion ***(126)***. The recently published data on patients with type 2 diabetes with and without nephropathy indicate that more than one

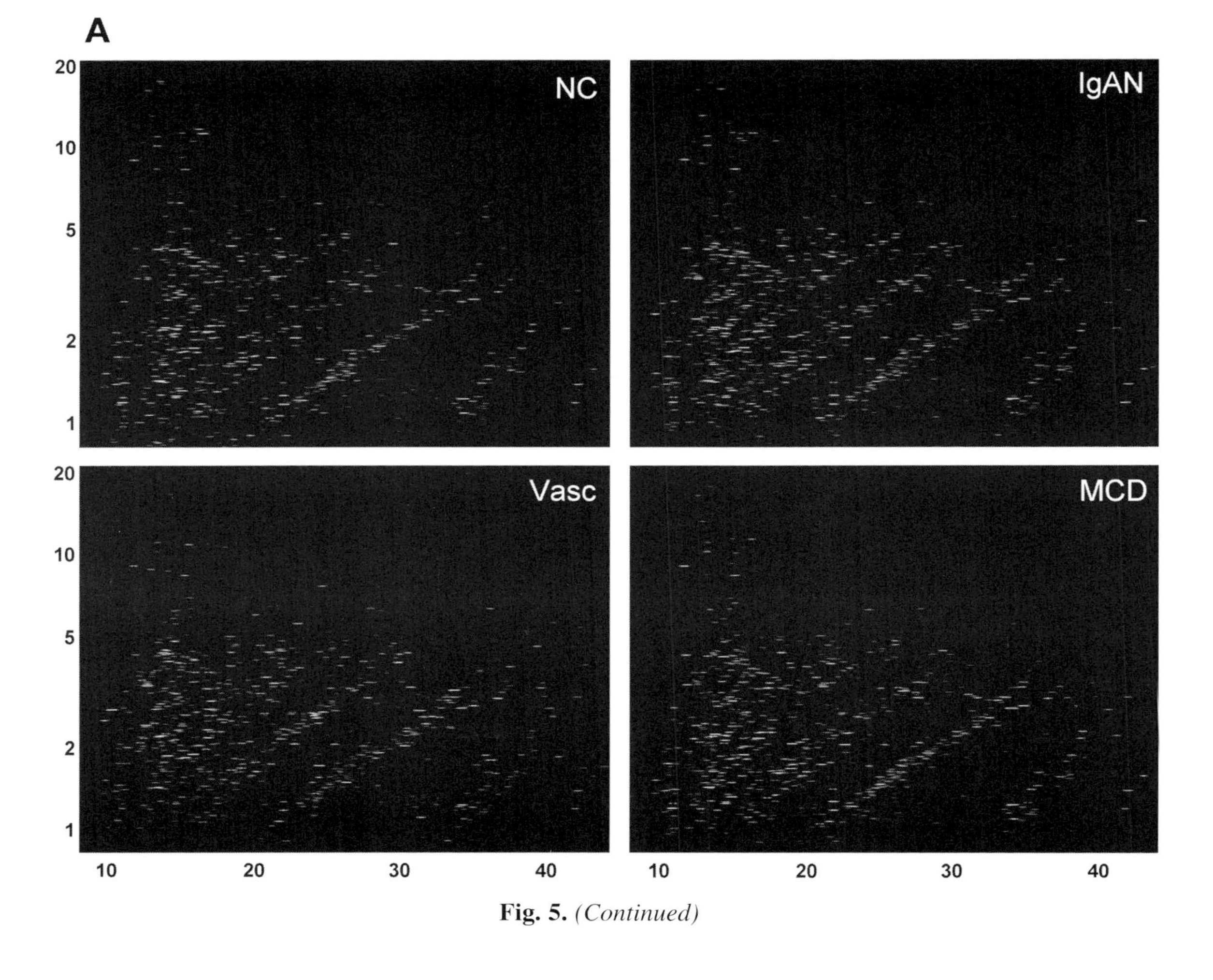

Fig. 5. *(Continued)*

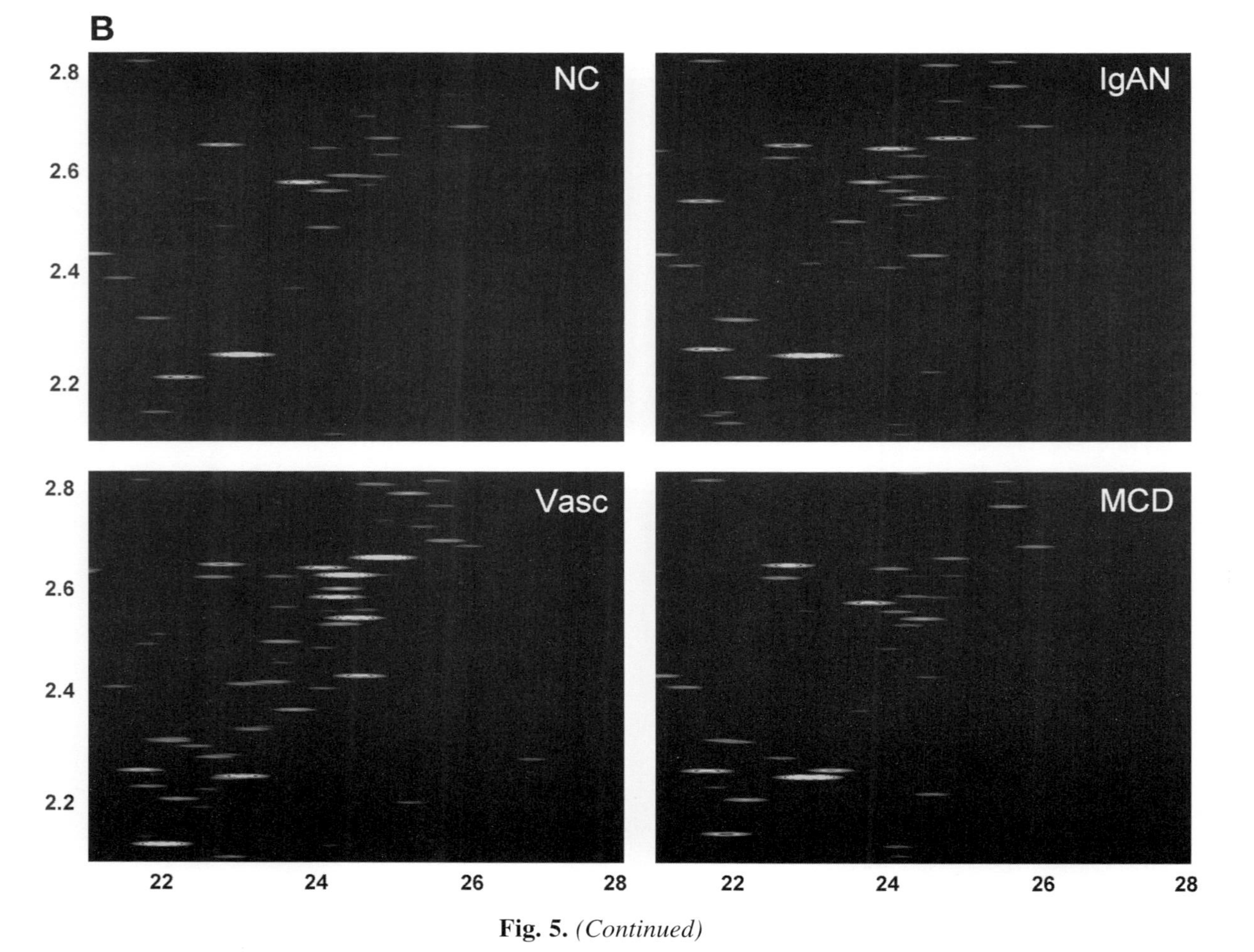

Fig. 5. *(Continued)*

clinical condition can be examined simultaneously and independently using CE-MS analysis ***(124)***. This is further underlined in another report by Gwinner et al. (ms. submitted) in which numerous biomarkers have been defined to recognize the transplanted kidney as an allograft and to distinguish among stable transplants, urinary tract infection, and allograft rejection. As also illustrated in **Fig. 5**, the data from one CE-MS analysis can display multiple diseases. These promising results suggest that CE-MS analysis of urinary polypeptides can be widely applied to detect pathological changes in the kidney.

Certainly, the clinical use of biomarkers requires knowledge of their identity. Because of the complexity of biological samples, it may be difficult to purify a single polypeptide of interest for further examination using conventional techniques. Consequently, polypeptides of interest must initially be uniquely defined by highly accurate mass and, if possible, additional identifying parameters (e.g., migration time) to allow sequencing by MS/MS. CE can be directly interfaced with an MS/MS instrument for sequencing, and the entire CE run can also be spotted onto a MALDI target plate; subsequent polypeptides of interest can then be analyzed ***(119,124)***. With the latter approach, once the entire CE-MS run is immobilized, single fractions (spots) can be analyzed repeatedly without the need for a new CE run. In addition, MALDI-TOF/TOF spectra can easily be generated using state-of-the-art MS systems. The major limitation of this or similar approaches is that polypeptides above 3 kDa generally provide unsatisfactory data. This problem can be solved by either digesting the sample on the target as described ***(120)*** or using Fourier transform ion cyclotron resonance (FT-ICR) instruments, which allow the identification of urinary polypeptides below 10 kDa (Chalmers et al., ms. submitted).

A large number of polypeptides that were identified in the urine are derived from highly abundant proteins like albumin, β_2-microglobulin, and others. These peptides, which are most likely the results of proteolytic cleavage, appear to carry more information than would held in a mere reflection of the abundance of the parental proteins. At first sight, an array of different fragments of these proteins (which would cover most of the sequence, favoring peptides that would be

Fig. 5. *(Opposite page)* CE-MS analysis of the urine from patients with various renal diseases. **(A)** The compiled contour plots of urinary polypeptides from normal controls (NC), IgA nephropathy (IgAN), active vasculitis (Vasc), and minimal change disease (MCD) that allow the tentative identification of discriminatory polypeptides. Although the general appearance of these patterns is quite similar owing to the high reproducibility, an array of disease-specific polypeptides can be pinpointed upon closer examination of the data. **(B)** Magnification of the contour plots in the area of 2.1 to 2.85 kDa and 21 to 28 min shows differences in urinary polypeptide patterns among groups. Mass (in kDa on a logarithmic scale) is indicated on the *y*-axis, whereas normalized migration time (in minutes) is labeled on the *x*-axis.

subjected to proteolysis) is expected. It appears that this is not the case, and distinct peptides present at high abundance in certain renal diseases can be found. The data available to date suggest that each disease results in different proteolytic fragments. Although most chronic renal diseases result in increased albumin secretion into the urine, apparently disease-specific albumin peptides can also be found. It is tempting to speculate that this might represent variable activity of distinct proteases in different disease processes. Consequently, it is important to know exact sequences of these peptides. Furthermore, the nature of the proteolytic cleavage sites might enable one to pinpoint the underlying proteases, thereby allowing a better understanding of the disease pathophysiology.

7. Mass Spectrometric Immunoassay for Urinary Proteomics

7.1. Background and Technical Aspects

In the last decade, the scientific community has realized the potential and usefulness of MALDI-TOF MS in protein structural investigations ***(127,128)***. Consequently, MALDI-TOF MS has become an integral part of today's *modus operandi* in proteome analysis, often used in combination with *in situ* enzymatic digestion of proteins separated by 2D-PAGE for putative gene product identification via database searches ***(129)***. However, there is substantially more to proteome analysis than the initial identification step. The complexity of the proteome (stemming from the 3D protein structure, protein–protein interactions, PTMs, point mutations, splice variants, and so on) necessitates the assaying of individual proteins from their native environment with as much information content as possible—preferably in their full-length intactness subsequently followed by increasingly more detailed fragment analyses. Thus, such *top-down* MALDI-TOF MS characterization (as opposed to *bottom-up* analysis of *a priori* fragmented gene products) of proteins present in complex biological mixtures becomes a fairly difficult proposition. These difficulties arise in part because the targeted analytes are present at less than optimal concentrations while in the presence of a huge abundance of other proteins and biomolecules. Additionally, biofluids (especially urine) are of general high-salt content, which gives rise to suppression effects and matrix neutralization during the MALDI process. It is therefore clearly necessary to fractionate target proteins from biofluids prior to top-down MALDI-TOF MS characterization; in the event that the target proteins are known, affinity isolation is the preferred means of fractionation.

Over the last 10 yr, Nelson and colleagues ***(130–143)*** have demonstrated that immunoassays, in which the captured antigen is detected qualitatively and quantitatively using MALDI-TOF MS, have the potential to greatly extend the range, utility, and speed of biological research and clinical assays. The approach is termed *mass spectrometric immunoassay* (MSIA) **(Fig. 6)** and is, in essence, the rational combination of microscale affinity capture techniques with quantitative

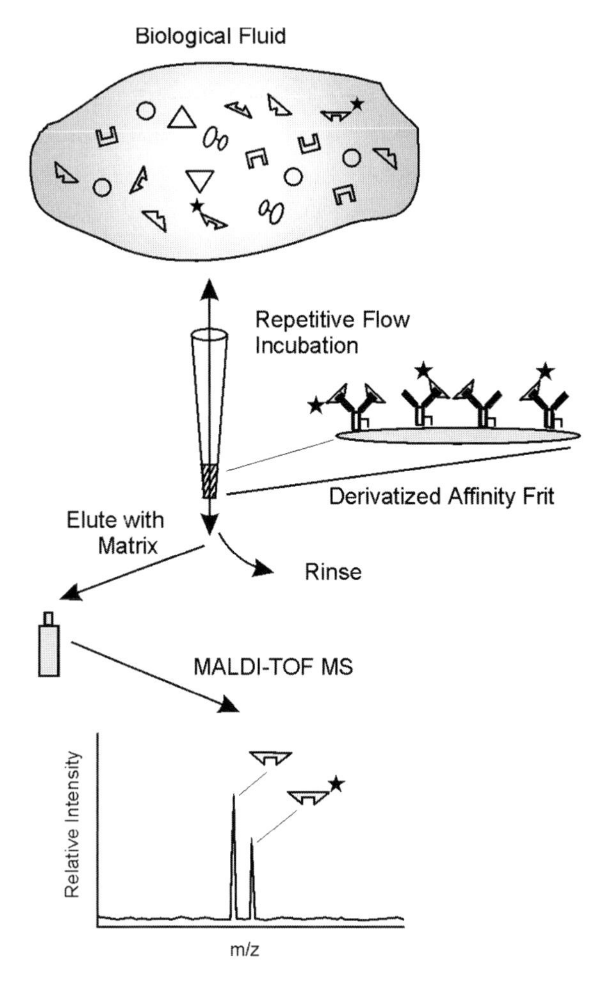

Fig. 6. Illustration of the MSIA approach. Analytes are selectively retrieved from solution by repetitive flow through a receptor-derivatized porous frit at the entrance of a wide-bore P-200 pipetor tip (MSIA-Tip). Once washed off nonspecifically bound compounds, the retained species are eluted onto a mass spectrometer target using a MALDI matrix. MALDI-TOF MS then follows, with analytes detected at precise *m/z* values. The MSIA analyses are qualitative by nature but can be made quantitative by incorporating mass-shifted variants of the analyte into the procedure for use as internal standards.

MALDI-TOF MS ***(144)***. In the initial stage of development, agarose beads (derivatized with an affinity ligand) were used to create a microliter-volume column inside a micropipetor tip (thus creating an affinity pipet) ***(130,132)***. More recently, tailored, high-flow-rate, high-binding-capacity affinity micropipetes have been manufactured and used in combination with robotic platforms for the preparation of up to 96 samples in parallel ***(134,138,142)***. Using this approach, target proteins are selectively retained and concentrated by repeatedly causing a biofluid to flowthrough the affinity pipet. After washing to remove unspecified compounds, retained species are eluted onto a mass spectrometer target, generally using a solution of the MALDI matrix. The eluted species are then mass analyzed, and the analytes and their variants can be detected at characteristic *m/z* values. For quantification, a known variant of the analyte (chemically modified to shift the molecular mass without significantly altering the affinity for the immobilized ligand) is introduced into the biofluid at a constant concentration, ultimately to be retained, eluted, and analyzed simultaneously with the target protein. The target protein is quantified by normalizing the signal of the analyte to that of the internal reference and (after proper calibration) equating the normalized signal intensity with analyte concentration.

Overall, the specificity of MSIA is two-dimensional: the first is implied during the affinity capture step (through selective interaction) and the second during the mass spectrometric detection (determination of species at defined *m/z* values). Thus analyte detection comes with a high degree of precision and accuracy—all but eliminating false positives during screening—and can also rapidly evaluate proteins for PTMs, point mutations, splice variants, protein degradation, and interacting partners. Also, because analytes are detected at unique *m/z* values, MSIA can be used to detect and identify multiple analytes simultaneously, i.e., multiplexing for unrelated species using multiple affinity ligands. By incorporating quantitative methods into the analysis, the assay is expanded to allow MSIA to be used relatively in screening biological fluids for biomarker discovery and in more directed applications (highly focused studies) in which well-characterized proteins indicative of disease state are rigorously quantified. Given these qualities and the fact that MSIA analyses require as little as 1 min per analysis when applied in the high-throughput manner, it stands to reason that the approach will find use in the proteomics and clinical analysis of urinary proteins. Nelson and colleagues have previously reported on the development of MSIA for the targeted analysis of a number of urinary proteins ***(133,137,139,140)***. Selected examples are given here illustrating the use of MSIA in such targeted urinary proteomics studies.

7.2. Applications to Human Urine

7.2.1. Example 1: Quantification of Urinary β_2-Microglobulin

β_2-microglobulin (β_2m) is a low-molecular-mass protein identified as the light chain of the class I major histocompatibility complex synthesized in all nucleated cells. Upon activation of the immune system, both B- and T-lymphocytes actively release β_2m into the circulation from where it is later eliminated via glomerular filtration and tubular reabsorption. Serum levels of β_2m have been

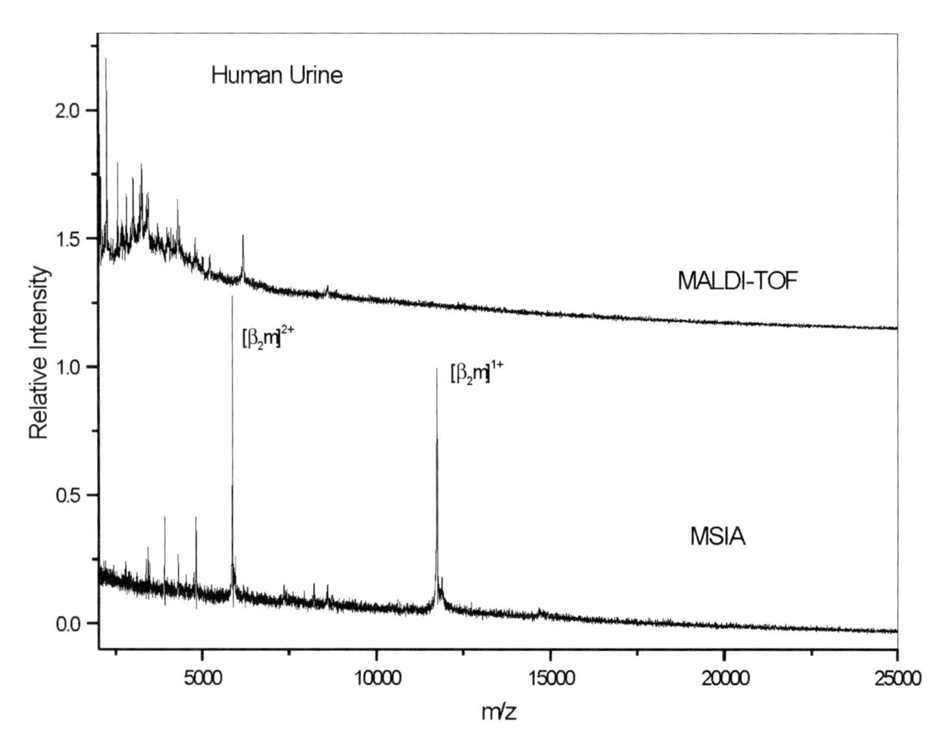

Fig. 7. β_2m MSIA screening of a human urine sample. Upper trace: direct MALDI-TOF mass spectrum of diluted, unfractionated human urine sample. Bottom trace: mass spectrum of the human urine sample following the MSIA screen.

measured and nominally correlated to a number of ailments. On the other hand, β_2m levels in urine are indicators of glomerular filtration rate and tubular reabsorption ***(145)***. Conventionally, β_2m levels are monitored using a variety of immunological assays, including ELISA, radioimmunoassay, and particle-enhanced turbidimetry assay. The quantitative dynamic range (spanning β_2m concentrations of approx 0.2–20 mg/L) and the accuracy (1–10%) of these assays are sufficient to cover the normal and elevated levels of β_2m in a variety of biological fluids.

Previously, Tubbs et al. ***(133)*** have reported on the development of a urine-based MSIA assay targeting β_2m. **Figure 7** shows spectra resulting from the MSIA analysis of a human urine sample. For comparison, a MALDI-TOF mass spectrum of whole unfractionated human urine is shown. The MALDI-TOF spectrum of the whole urine shows a number of signals in the peptide region and an absence of signals for β_2m. On the other hand, the results obtained during MSIA are dominated by signals from the β_2m, with few additional signals from nonspecified compounds and of suitable quality to develop a fully quantitative assay. For the quantification of β_2m in human urine, internal standards need to be incorporated into the MSIA protocol. Equine β_2m (eβ_2m) was chosen as an internal reference for quantification because of its high degree of similarity to human β_2m (hβ_2m) (approx 75% sequence homology), resolvable mass difference from hβ_2m ($M_{r\,(e\beta_2m)} = 11{,}396.6$; $M_{r\,(h\beta_2m)} = 11{,}729.7$) and its availability. Briefly:

1. Fresh horse urine was collected (at a local stable) and treated immediately with a protease inhibitor cocktail.
2. Low solubility compounds were removed from the urine by overnight refrigeration (at 4°C) followed by centrifugation at 5000*g* for 5 min.
3. The urine was then concentrated 20-fold over a 10-kDa molecular mass cutoff filter with repetitive rinses with HEPES-buffered saline (HBS) and water and several filter exchanges (4 filters/200 mL urine).
4. Treatment of 200 mL fresh urine resulted in 10 mL of β_2m-enriched horse urine, which served as the stock internal reference solution for approx 100 analyses.
5. Standards were prepared by stepwise dilution (i.e., X 0.8, 0.6, 0.4, 0.2, and 0.1, in HBS) of a 1.0 mg/L hβ_2m stock solution to a concentration of 0.1 mg/L; the 0.1 mg/L solution served as stock for an identical stepwise dilution covering the second decade in concentration (0.01–0.1 mg/L).
6. A blank solution containing no hβ_2m was also prepared.
7. The samples for MSIA were prepared by mixing 100 µL of each of the hβ_2m standards with 100 µL of stock horse urine and 200 µL of HBS buffer.
8. MSIA was performed on each sample as described above, resulting in the simultaneous extraction of both eβ_2m and hβ_2m.
9. Ten 65-laser-shot MALDI-TOF spectra were taken from each sample, with each spectrum taken from a different location on the target.
10. Care was taken during data acquisition to maintain the ion signals in the upper 50 to 80% of the *y*-axis range and to avoid driving individual laser shots into saturation.
11. Spectra were normalized to the eβ_2m signal through baseline integration, and the integral hβ_2m peak was determined.
12. Integrals from the 10 spectra taken for each calibration standard were averaged, and the standard deviation was calculated.
13. A calibration curve was constructed by plotting the average of the normalized integrals for each standard vs the hβ_2m concentration.
14. **Figure 8A** shows spectra representing MSIA analyses of hβ_2m standards in a concentration range of 0.01 to 1.0 mg/L.
15. Each spectrum, normalized to the eβ_2m signal, is one of 10 65-laser-shot spectra taken for each calibration point.
16. The calibration curve is shown in **Fig. 8B**. Linear regression fitting of the data yields $\mathrm{Int}_{h\beta_2m}/\mathrm{Int}_{e\beta_2m} = 4.09$ (hβ_2m in mg/L) + 0.021 ($R^2 = 0.983$), with a working LOD (limit of detection) of 0.0025 mg/L (210 p*M*) and a LOQ (limit of quantitation) of 0.01 mg/L (850 p*M*).

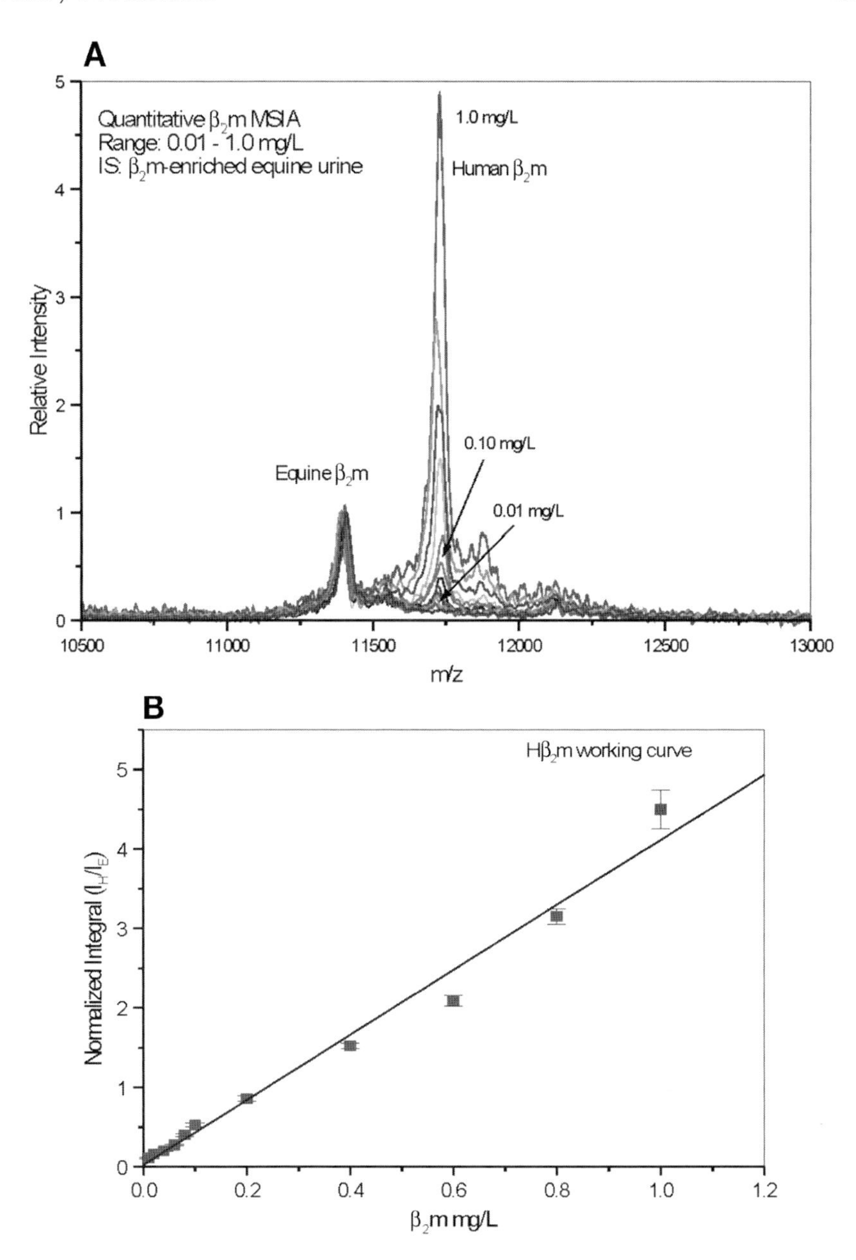

Fig. 8. Quantitative β_2m-MSIA: working curve. **(A)** Representative spectra of data used to generate the working curve. Human β_2m concentrations of 0.01 to 1.0 mg/L were investigated. Equine β_2m (M_r = 11,396.6) was used as an internal standard. **(B)** The working curve was generated using the data represented in **(A)**. The two-decade range was spanned with good linearity ($R^2 = 0.983$) and low standard error (approx 5%). Error bars reflect the standard deviation of 10 repetitive 65-laser-shot spectra taken from each sample.

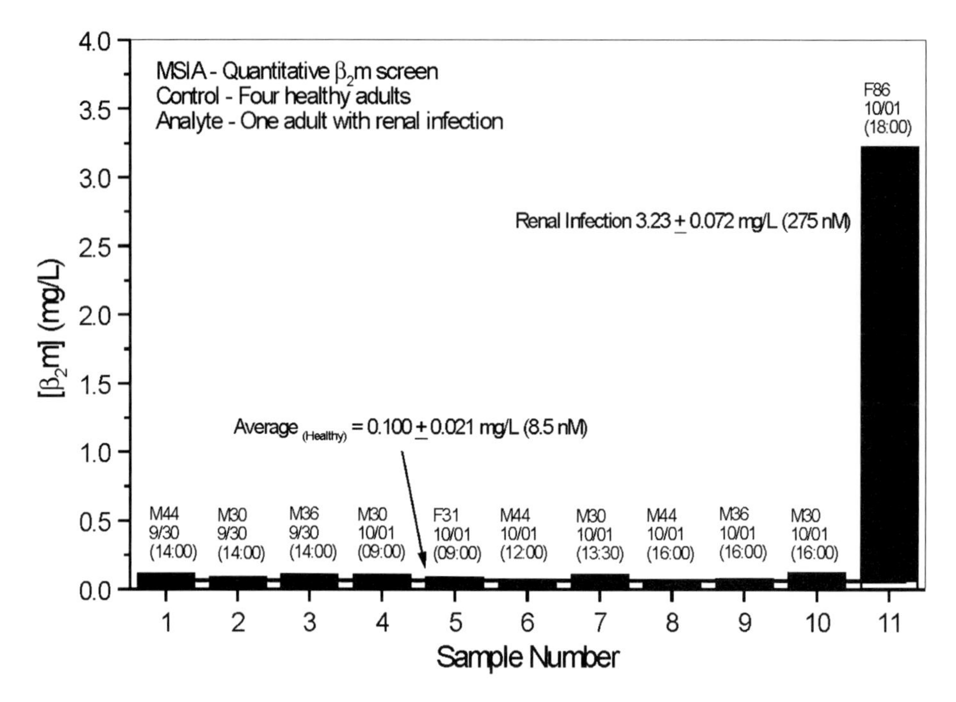

Fig. 9. Quantitative β_2m-MSIA: screening. Human urine samples from five individuals were screened over a period of 2 days. The average value determined for healthy individuals (10 samples from four individuals [three males and one female]; ages 30–44 yr) was 0.100 ± 0.021 mg/L. The level determined for an 86-yr-old woman with a recent urinary tract infection indicated a significant increase in β_2m concentration (3.23 ± 0.072 mg/L).

For the quantitative MSIA screens:

1. Ten samples were collected from four individuals:
   a. Female, 31 yr old, pregnant; 1 sample (F31).
   b. Male, 30 yr old; four samples over 2 d (M30).
   c. Male, 36 yr old; two samples over 2 d (M36).
   d. Male, 44 yr old; three samples over 2 d (M44).

   All the individuals were healthy when the samples were collected.
2. In preparation for MSIA, 100 µL of each urine sample was mixed with 100 µL of stock horse urine and 200 µL of HBS.
3. Results from the MSIA of the 10 urine samples are shown in **Fig. 9**. The bars depict the β_2m concentration determined for each sample.
4. The data for the 10 samples show remarkable consistency, with an average β_2m concentration of 0.100 ± 0.021 mg/L (high = 0.127 mg/L; low = 0.058 mg/L).
5. An additional analysis was performed on a urine sample obtained from an 86-yr-old woman (F86) who had recently suffered a renal infection. Because of the significantly higher level of β_2m found in this sample, it was necessary to dilute

the urine quantitatively by a factor of 10 to keep the β_2m signal inside the dynamic range of the working curve and accurately establish the β_2m concentration in F86 (at 3.23 ± 0.02 mg/L).

7.2.2. Example 2: Qualitative Comparison of Urinary Proteins among Individuals

This second example is focused on applying urinary MSIA assays to urine samples of two different individuals collected throughout the course of a single day. Of particular note is the ability to view differences between individuals, as well as the ability to screen for intraindividual variability during the day. Because urine expulsion from humans can be viewed as in a constant state of flux, such *real-time* analyses have the potential to define the range of *normal* protein predispositions in individuals more accurately—basal information of great value when compared against disease states. Briefly:

1. Urine samples from two individuals were collected from the first four voids of the day (over approx 8 h).
2. Eight MSIA assays (β_2m, cystatin C [cysC], transthyretin [TTR], urinary protein 1 [UP1], retinol binding protein [RBP], IgG light-chain λ [IgGLCL], albumin [Alb], and transferrin [TRFE]) were performed by addressing the samples in parallel using a repeating octapette.
3. The samples were assayed between voids (i.e., void1, assay1; void2, assay2; void3, assay3; and void4, assay4), requiring a total analysis time (void-to-data) of approximately 1 h for the eight assays.
4. **Figure 10** shows the results of the exercise. Overall, the assays show good reproducibility as applied to the individuals throughout the day.
5. It is worth noting that, collectively, 64 assays were performed by hand, in (near) real time over the course of approximately 8 h. This rate is not limited by any foreseeable obstacles (ironically, in this example, it was the nature of the experiment—the time needed between voids—that was the rate-limiting step) and can certainly be increased given the addition of even more samples, assays, and the incorporation of parallel robotic preparation.
6. Importantly, an outcome of the screening was the observance of differences in the target proteins at both the intra- and inter-individual levels.
7. **Figure 11** shows differences in the RBP assay of Individual1 observed between void1 and subsequent voids (void3 is shown). In the first void of the day (i.e., approx 6 h after the last void of the previous day), the relative contribution of the lowest molecular mass breakdown product of RBP (RBP-[C-terminal]—RNLL) is observed to increase. Similar results were not observed for the second individual however.
8. **Figure 12** shows the results of the cysC MSIA of Individual1, void3. Overall, high-quality spectra were obtained with good signal-to-noise ratio and generally free of nonspecified proteins. Inspection of the singly charged region revealed the presence of wild-type cysC (m/z = 13,343 Daltons) and an oxidized version (dm = +16 Daltons) that has been noted previously in the literature (oxidation of proline at residue 3 ***[146]***). Two other species were also observed, one at m/z = 13,256

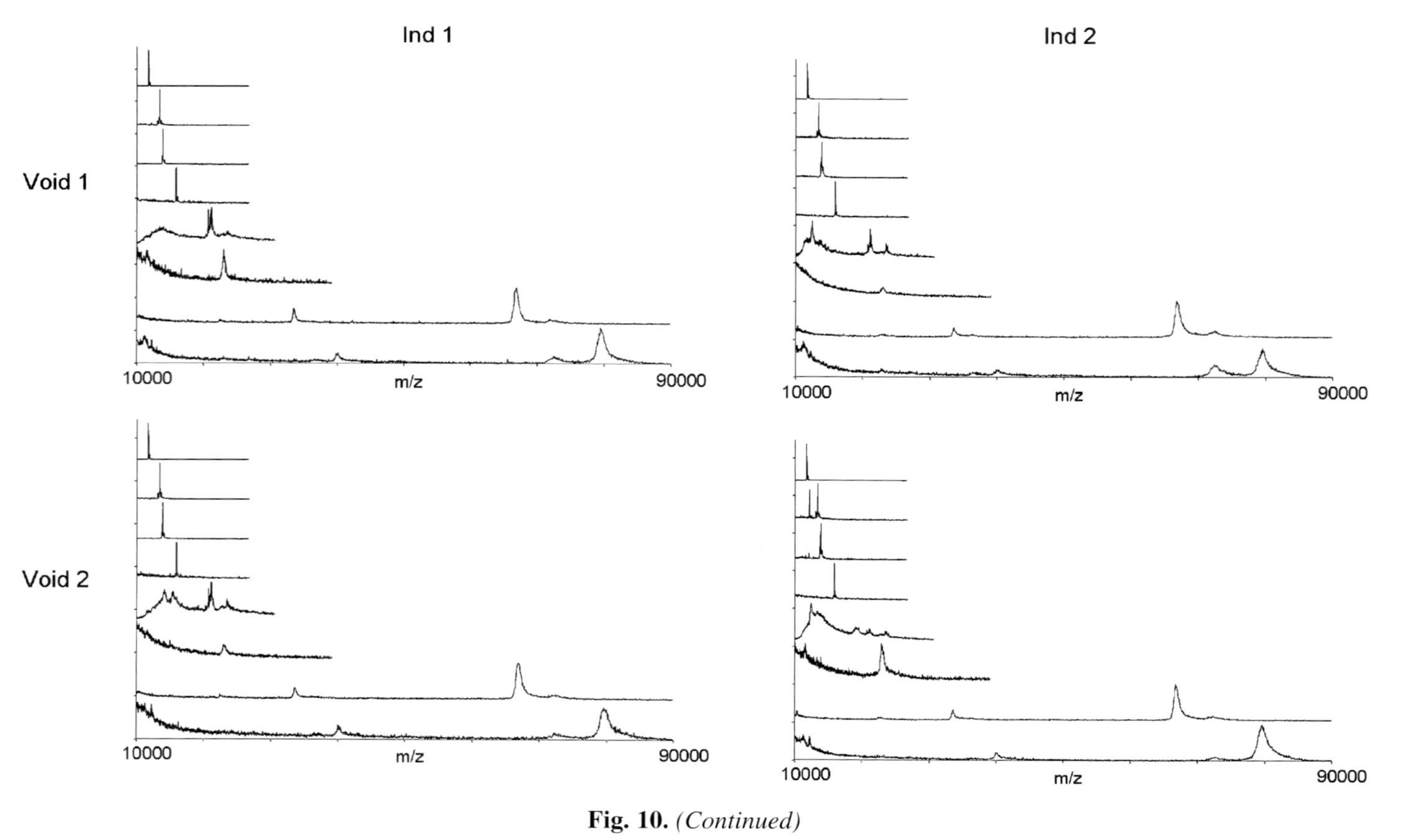

Fig. 10. *(Continued)*

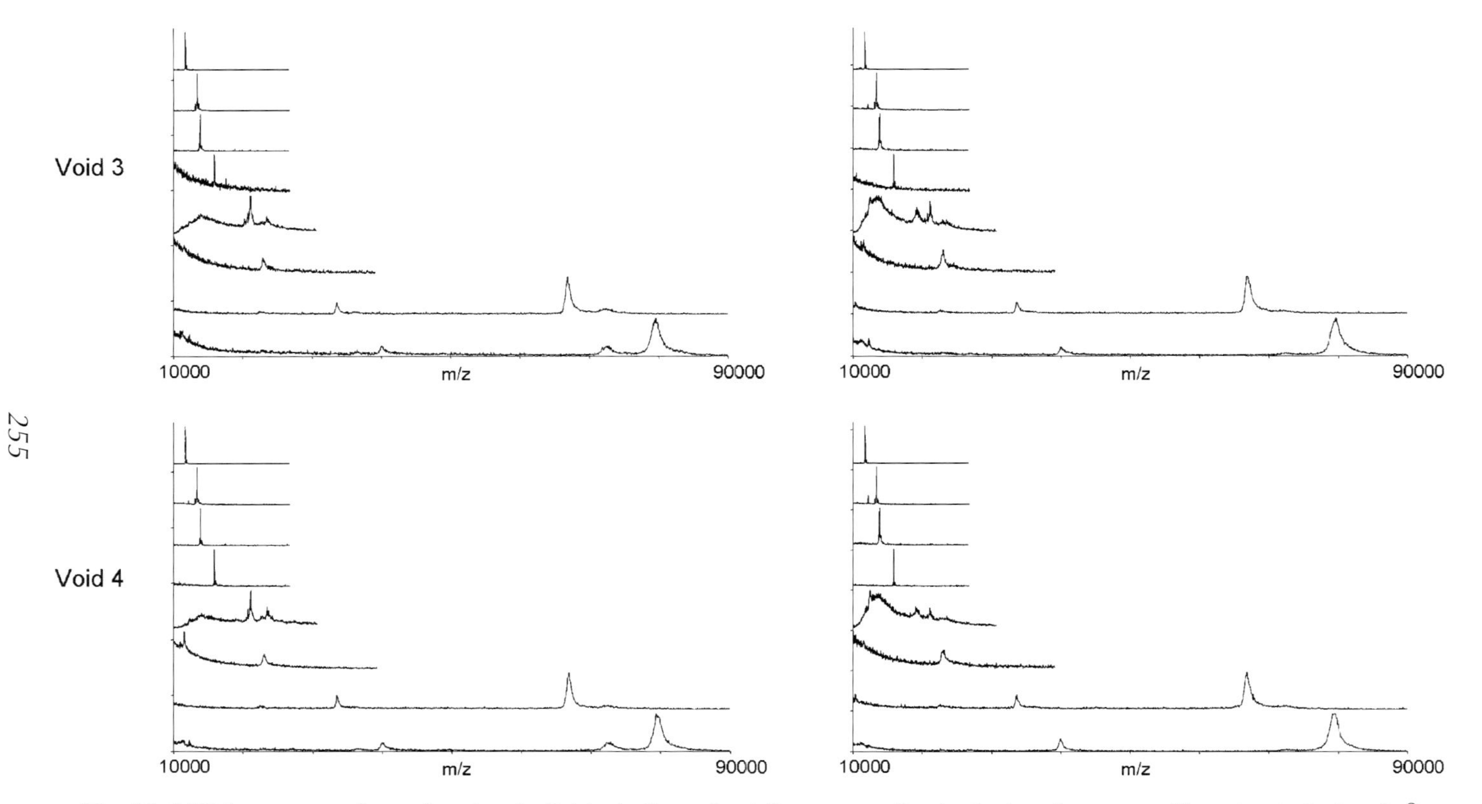

Fig. 10. MSIA assays performed on two individuals throughout the course of a single day. Assays are (from top to bottom): β_2m, CysC, TTR, UP1, RBP, LCL, Alb, and TRFE.

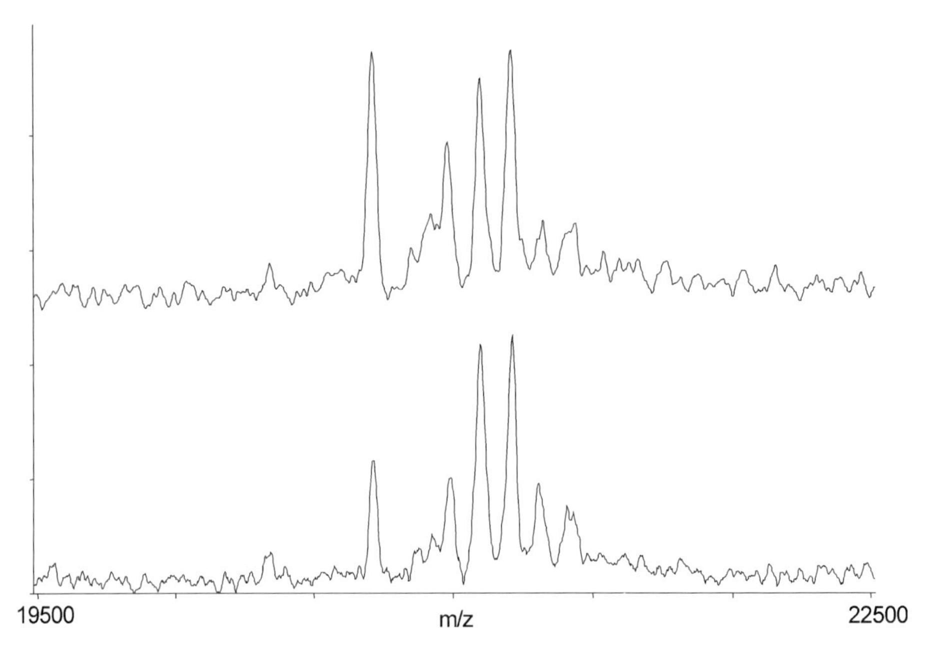

Fig. 11. MSIA RBP spectra obtained from the urine of Individual1, void1 (upper trace) and Individual1, void3 (lower trace). A slight change in the relative contribution of the truncated variants—from left (C-term), des-RNLL, des-LL, des-L, and intact—is observed.

Daltons (still sharing the oxidation) and the second at *m/z* = 13,072 Daltons. These signals were most likely owing to truncation of the N-terminal Ser and Ser-Ser-Pro, respectively (measured molecular masses within 1.5 Daltons of calculated) and were consistently observed in both individuals for all voids. However, the two other variants, owing to loss of four or eight residues from the N-terminus, were observed in only the third void of Individual1.

9. **Figure 13** shows three TTR MSIA spectra, two from Individual2 (void1 and void3) and one from Individual1 (void3). The dominant signal in all three is owing to the cysteinylated version of TTR (119 Daltons higher in molecular mass than the wild type). Individual2 shows an approx +30-Dalton split, indicative of point mutation (upper and middle spectra) relative to homozygous wild type (lower spectrum). Also observed in Individual2 on the first void of the day is a great deal of heterogeneity in the signal surrounding the wild-type molecular mass, making an accurate molecular mass determination difficult. Analysis of later voids improves the quality of the spectrum, making accurate molecular mass determination of the intact species possible. Using mass mapping and reflectron DE-MALDI-TOF MS, this particular approx +30 Da variant was subsequently measured as a mass shift of Δm = 29.988 Daltons in tryptic fragment 104–127, which identified the point mutation as Thr119Met (Δm = 29.992 Daltons).

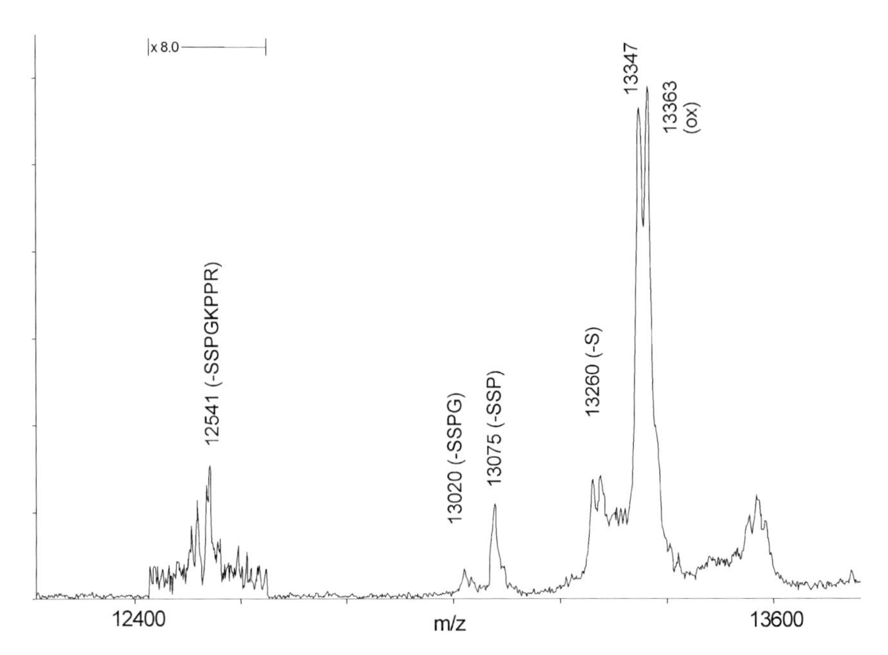

Fig. 12. Cystatin C MSIA obtained from the urine of Individual1, void3. Two variants (N-term), des-S and des-SSP, are observed in all voids from both individuals, two variants, des-SSPG and des-SSPGKPPR, are observed unique to this void in this individual.

These examples demonstrate the use of MSIA in two urinary-based applications: the rigorous quantification of a urinary protein and the qualitative relative profiling of multiple urine proteins/samples. Regarding protein quantification, it is accurate from a general analytical biochemistry point of view to state that there has always been a pressing need for new analytical techniques able to determine protein concentrations in biofluids accurately. This statement is essentially meant to transcend the boundaries of proteomics and to point out bluntly that there are altogether too few quantitative approaches in use today that are able to live up to the complications and challenges derived from the concerted analysis of biological systems. Even so, and learning from the examples given above, it is often an error to presume the exact nature of a protein when one is using quantitative approaches that are not able to discern microheterogeneity present in a given protein. At the very least, MSIA can be used to define multiple, related species qualitatively such that this knowledge may be used in concert with assays that yield only a single quantitative value. However, given the appropriate experimentally design, which includes mass selective detection, MSIA can be used to determine proteins levels in a manner that is largely free of such complicating artifacts. Viewed differently, these artifacts (in the form of

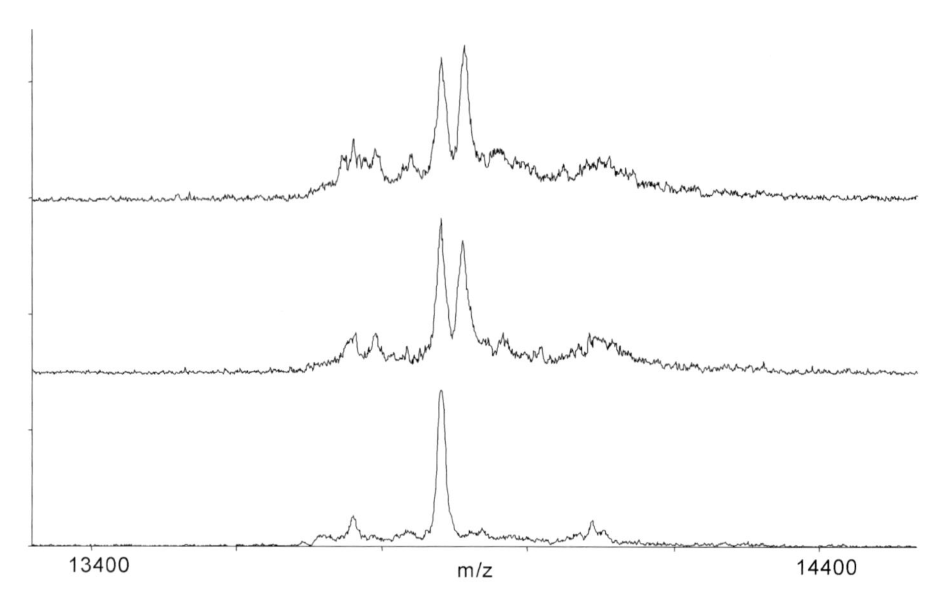

Fig. 13. TTR variants observed during the MSIA analysis of Individual2, void1 (upper trace) and void3 (middle trace) and Individual1, void3 (lower trace).

point mutations, PTMs, protein truncations, and so on) yield a wealth of information on the true nature of a protein in its natural environment, as well as on the individual in whom the proteins reside. Notably, when all data given above are considered, 19 variants (collectively for the eight targeted proteins) were observed in just the small longitudinal study between the two individuals.

Clearly, more information is contained in the detailed analysis of these species than would be produced using approaches that are limited by an *a priori* assumption of a protein qualitative makeup. This ability to view changes in protein structure, both in a single individual with time and between individuals (and with time) with great speed, sensitivity, and accuracy results in a promising approach not only to identify putative biomarkers but also to form the basis of studies geared to a more fundamental understanding of protein behavior in human biofluids. Obviously, additional studies involving the assaying of a greater number of proteins in larger populations are needed to define more precisely such qualitative and quantitative modulations expected in both healthy and diseased individuals. Given such pursuits, MSIA has great potential for characterizing urinary proteins and upon further development will find significant use in biomarker discovery/identification/validation and subsequent quantitative clinical and diagnostic applications.

8. Perspectives and Future Directions

Most urinary proteomics studies to date have utilized a single approach in each individual study. Each approach has different advantages, disadvantages, and limitations, as pointed out in individual sections above. Therefore, it is unlikely that complete information on proteins in the urine or other body fluids will be obtained using only one analytical approach. The complexity of protein expression and function in these biofluids requires integration of several approaches to fulfill the dynamic image of the physiology and pathophysiology of such body fluids, as well as their related organs. Because alterations in protein expression levels and microheterogeneities can cause diseases, an ideal approach should have the capability of both quantification and qualitative analysis.

Because the urine is one of the most interesting and useful human body fluids for clinical proteomics studies applied to biomarker discovery, experimental design has become a crucial issue. To define any urinary protein(s) as biomarker candidate(s) of a medical disease, one must be aware of the pathophysiology of how excretion of such protein(s) is altered during the disease state; otherwise the data can be *misleading*. For a renal disease, in addition to normal healthy individuals, the controls must include patients who have renal disorders that result in a clinical syndrome similar to the disease of interest. For example, the control groups for diabetic nephropathy must include other glomerular disorders; the controls for acute renal allograft rejection must include acute tubular necrosis and urinary tract infection. For a disease in which the primary affected organ is not the kidney, a remote kidney effects that can cause changes in urinary composition that are not specific to such a disease/disorder, must be of serious concern. An excellent example is *paraneoplastic syndrome* in various types of cancers, which has systemic effects, leading to alterations in general metabolic and biochemical profiles. Urinary protein composition in this condition can be changed (either up- or downregulated or involving changes that are present or absent) but may not be specific for a particular type of cancer. When the investigators wish to define urinary biomarker(s) for a particular type of cancer, the control groups to be evaluated must include other types of cancers that can cause the paraneoplastic syndrome in addition to the healthy controls. The proteomics investigators must work closely together with health-care professionals to define groups of patients to be examined. After the biomarker candidates are identified, they need to be examined in a larger number of patients to evaluate sensitivity and specificity before they can be used in clinical practice. To date, no urinary biomarker has been identified by proteomic analysis that has been validated and can be used in clinical diagnostics. Hopefully, with the integration of different approaches and improved quantitative and qualitative techniques, identification of validated urinary biomarkers for medical diseases will be more feasible.

References

1. Moe OW, Baum M, Berry CA, Rector FC. Renal transport of glucose, amino acids, sodium, chloride and water. In: Brenner BM, ed. *Brenner & Rector's The Kidney*, 7th ed. Philadelphia: WB Saunders, 2004:413–452.
2. Thongboonkerd V, McLeish KR, Arthur JM, Klein JB. Proteomic analysis of normal human urinary proteins isolated by acetone precipitation or ultracentrifugation. *Kidney Int* 2002;62:1461–1469.
3. Pieper R, Gatlin CL, McGrath AM, et al. Characterization of the human urinary proteome: a method for high-resolution display of urinary proteins on two-dimensional electrophoresis gels with a yield of nearly 1400 distinct protein spots. *Proteomics* 2004;4:1159–1174.
4. Oh J, Pyo JH, Jo EH, et al. Establishment of a near-standard two-dimensional human urine proteomic map. *Proteomics* 2004;4:3485–3497.
5. Thongboonkerd V. Proteomic analysis of renal diseases: Unraveling the pathophysiology and biomarker discovery. *Expert Rev Proteomics* 2005;2:349–366.
6. Fung E, Diamond D, Simonsesn AH, Weinberger SR. The use of SELDI ProteinChip array technology in renal disease research. *Methods Mol Med* 2003;86:295–312.
7. Templin MF, Stoll D, Schwenk JM, Potz O, Kramer S, Joos TO. Protein microarrays: promising tools for proteomic research. *Proteomics* 2003;3:2155–2166.
8. Weissinger EM, Wittke S, Kaiser T, et al. Proteomic patterns established with capillary electrophoresis and mass spectrometry for diagnostic purposes. *Kidney Int* 2004;65:2426–2434.
9. Nouwens AS, Cordwell SJ, Larsen MR, et al. Complementing genomics with proteomics: the membrane subproteome of Pseudomonas aeruginosa PAO1. *Electrophoresis* 2000;21:3797–3809.
10. Neverova I, Van Eyk JE. Application of reversed phase high performance liquid chromatography for subproteomic analysis of cardiac muscle. *Proteomics* 2002;2:22–31.
11. Knepper MA, Masilamani S. Targeted proteomics in the kidney using ensembles of antibodies. *Acta Physiol Scand* 2001;173:11–21.
12. Knepper MA. Proteomics and the kidney. *J Am Soc Nephrol* 2002;13:1398–1408.
13. Ecelbarger CA. Targeted proteomics using immunoblotting technique for studying dysregulation of ion transporters in renal disorders. *Expert Rev Proteomics* 2004;1:219–227.
14. Ecelbarger CA. Proteomics and sodium transport. In: Thongboonkerd V, Klein JB, eds. *Proteomics in Nephrology.* Basel: Karger, 2004:124–141.
15. Govorukhina NI, Keizer-Gunnink A, van der Zee AG, de Jong S, de Bruijn HW, Bischoff R. Sample preparation of human serum for the analysis of tumor markers. Comparison of different approaches for albumin and gamma-globulin depletion. *J Chromatogr A* 2003;1009:171–178.
16. Rabilloud T. Solubilization of proteins in 2-D electrophoresis. An outline. *Methods Mol Biol* 1999;112:9–19.
17. Santoni V, Rabilloud T, Doumas P, et al. Towards the recovery of hydrophobic proteins on two-dimensional electrophoresis gels. *Electrophoresis* 1999;20:705–711.

18. Seigneurin-Berny D, Rolland N, Garin J, Joyard J. Technical advance: differential extraction of hydrophobic proteins from chloroplast envelope membranes: a subcellular-specific proteomic approach to identify rare intrinsic membrane proteins. *Plant J* 1999;19:217–228.
19. Gevaert K, Ghesquiere B, Staes A, et al. Reversible labeling of cysteine-containing peptides allows their specific chromatographic isolation for non-gel proteome studies. *Proteomics* 2004;4:897–908.
20. Ren D, Penner NA, Slentz BE, Mirzaei H, Regnier F. Evaluating immobilized metal affinity chromatography for the selection of histidine-containing peptides in comparative proteomics. *J Proteome Res* 2003;2:321–329.
21. Gevaert K, Van Damme J, Goethals M, et al. Chromatographic isolation of methionine-containing peptides for gel-free proteome analysis: identification of more than 800 Escherichia coli proteins. *Mol Cell Proteomics* 2002;1:896–903.
22. Gevaert K, Goethals M, Martens L, et al. Exploring proteomes and analyzing protein processing by mass spectrometric identification of sorted N-terminal peptides. *Nat Biotechnol* 2003;21:566–569.
23. Kumazaki T, Terasawa K, Ishii S. Affinity chromatography on immobilized anhydrotrypsin: general utility for selective isolation of C-terminal peptides from protease digests of proteins. *J Biochem (Tokyo)* 1987;102:1539–1546.
24. Mann M, Jensen ON. Proteomic analysis of post-translational modifications. *Nat Biotechnol* 2003;21:255–261.
25. Mirzaei H, Regnier F. Structure specific chromatographic selection in targeted proteomics. *J Chromatogr B Analyt Technol Biomed Life Sci* 2005;817:23–34.
26. Gorg A, Weiss W, Dunn MJ. Current two-dimensional electrophoresis technology for proteomics. *Proteomics* 2005;5:826–827.
27. Swain M, Ross NW. A silver stain protocol for proteins yielding high resolution and transparent background in sodium dodecyl sulfate-polyacrylamide gels. *Electrophoresis* 1995;16:948–951.
28. Rabilloud T, Carpentier G, Tarroux P. Improvement and simplification of low-background silver staining of proteins by using sodium dithionite. *Electrophoresis* 1988;9:288–291.
29. Shevchenko A, Wilm M, Vorm O, Mann M. Mass spectrometric sequencing of proteins silver-stained polyacrylamide gels. *Anal Chem* 1996;68:850–858.
30. Lopez MF, Berggren K, Chernokalskaya E, Lazarev A, Robinson M, Patton WF. A comparison of silver stain and SYPRO Ruby Protein Gel Stain with respect to protein detection in two-dimensional gels and identification by peptide mass profiling. *Electrophoresis* 2000;21:3673–3683.
31. Smejkal GB, Robinson MH, Lazarev A. Comparison of fluorescent stains: relative photostability and differential staining of proteins in two-dimensional gels. *Electrophoresis* 2004;25:2511–2519.
32. Unlu M, Morgan ME, Minden JS. Difference gel electrophoresis: a single gel method for detecting changes in protein extracts. *Electrophoresis* 1997;18:2071–2077.
33. Henzel WJ, Watanabe C, Stults JT. Protein identification: the origins of peptide mass fingerprinting. *J Am Soc Mass Spectrom* 2003;14:931–942.

34. Pierce WM, Cai J. Applications of mass spectrometry in proteomics. In: Thongboonkerd V, Klein JB, eds. *Proteomics in Nephrology.* Basel: Karger, 2004:40–58.
35. Karas M, Gluckmann M, Schafer J. Ionization in matrix-assisted laser desorption/ionization: singly charged molecular ions are the lucky survivors. *J Mass Spectrom* 2000;35:1–12.
36. Anderson NG, Anderson NL, Tollaksen SL. Proteins of human urine. I. Concentration and analysis by two- dimensional electrophoresis. *Clin Chem* 1979;25:1199–1979.
37. Tracy RP, Young DS, Hill HD, Cutsforth GW, Wilson DM. Two-dimensional electrophoresis of urine specimens from patients with renal disease. *Appl Theor Electrophor* 1992;3:55–65.
38. Grover PK, Resnick MI. Two-dimensional analysis of proteins in unprocessed human urine using double stain. *J Urol* 1993;150:1069–1072.
39. Lafitte D, Dussol B, Andersen S, et al. Optimized preparation of urine samples for two-dimensional electrophoresis and initial application to patient samples. *Clin Biochem* 2002;35:581–589.
40. Thongboonkerd V, Klein JB, Jevans AW, McLeish KR. Urinary proteomics and biomarker discovery for glomerular diseases. In: Thongboonkerd V, Klein JB, eds. *Proteomics in Nephrology.* Basel: Karger, 2004:292–307.
41. Miller I, Teinfalt M, Leschnik M, Wait R, Gemeiner M. Nonreducing two-dimensional gel electrophoresis for the detection of Bence Jones proteins in serum and urine. *Proteomics* 2004;4:257–260.
42. Rasmussen HH, Orntoft TF, Wolf H, Celis JE. Towards a comprehensive database of proteins from the urine of patients with bladder cancer. *J Urol* 1996;155:2113–2119.
43. Iwaki H, Kageyama S, Isono T, et al. Diagnostic potential in bladder cancer of a panel of tumor markers (calreticulin, gamma-synuclein, and catechol-O-methyltransferase) identified by proteomic analysis. *Cancer Sci* 2004;95:955–961.
44. Rehman I, Azzouzi AR, Catto JW, et al. Proteomic analysis of voided urine after prostatic massage from patients with prostate cancer: a pilot study. *Urology* 2004; 64:1238–1243.
45. Tantipaiboonwong P, Sinchaikul S, Sriyam S, Phutrakul S, Chen ST. Different techniques for urinary protein analysis of normal and lung cancer patients. *Proteomics* 2005;5:1140–1149.
46. Nabi G, N'dow J, Hasan TS, Booth IR, Cash P. Proteomic analysis of urine in patients with intestinal segments transposed into the urinary tract. *Proteomics* 2005;5:1729–1733.
47. Pang JX, Ginanni N, Dongre AR, Hefta SA, Opitek GJ. Biomarker discovery in urine by proteomics. *J Proteome Res* 2002;1:161–169.
48. Lapin A. A practicable two-dimensional electrophoretic method for routine analysis of urinary proteins. *J Clin Chem Clin Biochem* 1989;27:81–86.
49. Lapin A, Gabl F, Kopsa H. Diagnostic use of an analysis of urinary proteins by a practicable sodium dodecyl sulfate-electrophoresis method and rapid two-dimensional electrophoresis. *Electrophoresis* 1989;10:589–595.
50. Bueler MR, Wiederkehr F, Vonderschmitt DJ. Electrophoretic, chromatographic and immunological studies of human urinary proteins. *Electrophoresis* 1995;16:124–134.

51. Marshall T, Williams K. Two-dimensional electrophoresis of human urinary proteins following concentration by dye precipitation. *Electrophoresis* 1996;17: 1265–1272.
52. Harrison HH. The "ladder light chain" or "pseudo-oligoclonal" pattern in urinary immunofixation electrophoresis (IFE) studies: a distinctive IFE pattern and an explanatory hypothesis relating it to free polyclonal light chains. *Clin Chem* 1991;37:1559–1564.
53. Watanabe T, Takahashi S. Asymptomatic low molecular weight proteinuria: qualitative urinary protein analysis. *Acta Paediatr Jpn* 1992;34:28–35.
54. Edwards JJ, Anderson NG, Tollaksen SL, von Eschenbach AC, Guevara J Jr. Proteins of human urine. II. Identification by two-dimensional electrophoresis of a new candidate marker for prostatic cancer. *Clin Chem* 1982;28:160–163.
55. Grover PK, Resnick MI. High resolution two-dimensional electrophoretic analysis of urinary proteins of patients with prostatic cancer. *Electrophoresis* 1997;18:814–818.
56. Gebauer W, Lindl T. Detection of urinary glycoprotein GP 41 in leukemic patients and in healthy donors. *Blut* 1990;60:301–303.
57. Tichy M, Stulik J, Kovarova H, Mateja F, Urban P. Analysis of monoclonal immunoglobulin light chains in urine using two-dimensional electrophoresis. *Neoplasma* 1995;42:31–34.
58. Williams KM, Williams J, Marshall T. Analysis of Bence Jones proteinuria by high resolution two-dimensional electrophoresis. *Electrophoresis* 1998;19: 1828–1835.
59. Grover PK, Resnick MI. Evidence for the presence of abnormal proteins in the urine of recurrent stone formers. *J Urol* 1995;153:1716–1721.
60. Pillay SN, Asplin JR, Coe FL. Evidence that calgranulin is produced by kidney cells and is an inhibitor of calcium oxalate crystallization. *Am J Physiol* 1998;275:F255–F261.
61. Miyata T, Oda O, Inagi R, et al. Molecular and functional identification and purification of complement component factor D from urine of patients with chronic renal failure. *Mol Immunol* 1990;27:637–644.
62. Oda O, Shinzato T, Ohbayashi K, et al. Purification and characterization of perlecan fragment in urine of end-stage renal failure patients. *Clin Chim Acta* 1996;255:119–132.
63. Lapin A, Kopsa H, Smetana R, Ulrich W, Perger P, Gabl F. Modified two-dimensional electrophoresis of urinary proteins for monitoring early stages of kidney transplantation. *Transplant Proc* 1989;21:1880–1881.
64. Lapin A, Feigl W. A practicable two-dimensional electrophoresis of urinary proteins as a useful tool in medical diagnosis. *Electrophoresis* 1991;12:472–478.
65. Myrick JE, Caudill SP, Robinson MK, Hubert IL. Quantitative two-dimensional electrophoretic detection of possible urinary protein biomarkers of occupational exposure to cadmium. *Appl Theor Electrophor* 1993;3:137–146.
66. Katzin AM, Kimura ES, Alexandre CO, Ramos AM. Detection of antigens in urine of patients with acute *Falciparum* and *Vivax malaria* infections. *Am J Trop Med Hyg* 1991;45:453–462.
67. Hoermann R, Spoettl G, Grossmann M, Saller B, Mann K. Molecular heterogeneity of human chorionic gonadotropin in serum and urine from patients with trophoblastic tumors. *Clin Invest* 1993;71:953–960.

68. Stulik J, Kovarova H, Tichy M, Urban P. Electrophoretic analysis of microheterogeneity of paraproteins in a patient with IgD myeloma. *Neoplasma* 1995;42: 105–108.
69. Tichy M, Stulik J, Osanec J, Skopek P. Electrophoretic characterization of a gamma-1-heavy chain disease. *Neoplasma* 2000;47:118–121.
70. Argiles A, Garcia-Garcia M, Derancourt J, Mourad G, Demaille JG. Beta 2 microglobulin isoforms in healthy individuals and in amyloid deposits. *Kidney Int* 1995;48:1397–1405.
71. Miyata T, Oda O, Inagi R, et al. beta 2-Microglobulin modified with advanced glycation end products is a major component of hemodialysis-associated amyloidosis. *J Clin Invest* 1993;92:1243–1252.
72. Celis JE, Rasmussen HH, Vorum H, et al. Bladder squamous cell carcinomas express psoriasin and externalize it to the urine. *J Urol* 1996;155:2105–2112.
73. Herbert BR, Harry JL, Packer NH, Gooley AA, Pedersen SK, Williams KL. What place for polyacrylamide in proteomics? *Trends Biotechnol* 2001;19:S3–S9.
74. Mann M, Hendrickson RC, Pandey A. Analysis of proteins and proteomes by mass spectrometry. *Annu Rev Biochem* 2001;70:437–473.
75. Cutillas P, Burlingame A, Unwin R. Proteomic strategies and their application in studies of renal function. *News Physiol Sci* 2004;19:114–119.
76. Cutillas PR. Principles of nanoflow liquid chromatography and applications to proteomics. *Curr Nanosci* 2005;1:65–71.
77. Norden AG, Sharratt P, Cutillas PR, Cramer R, Gardner SC, Unwin RJ. Quantitative amino acid and proteomic analysis: very low excretion of polypeptides >750 Da in normal urine. *Kidney Int* 2004;66:1994–2003.
78. Wu CC, MacCoss MJ, Howell KE, Yates JR III. A method for the comprehensive proteomic analysis of membrane proteins. *Nat Biotechnol* 2003;21:532–538.
79. Santoni V, Molloy M, Rabilloud T. Membrane proteins and proteomics: un amour impossible? *Electrophoresis* 2000;21:1054–1070.
80. McKee JA, Kumar S, Ecelbarger CA, Fernandez-Llama P, Terris J, Knepper MA. Detection of Na(+) transporter proteins in urine. *J Am Soc Nephrol* 2000;11: 2128–2132.
81. Pisitkun T, Shen RF, Knepper MA. Identification and proteomic profiling of exosomes in human urine. *Proc Natl Acad Sci U S A* 2004;101:13368–13373.
82. Washburn MP, Wolters D, Yates JR III. Large-scale analysis of the yeast proteome by multidimensional protein identification technology. *Nat Biotechnol* 2001;19: 242–247.
83. Spahr CS, Davis MT, McGinley MD, et al. Towards defining the urinary proteome using liquid chromatography-tandem mass spectrometry. I. Profiling an unfractionated tryptic digest. *Proteomics* 2001;1:93–107.
84. Mabuchi H, Nakahashi H. Profiling of urinary medium-sized peptides in normal and uremic urine by high-performance liquid chromatography. *J Chromatogr* 1982;233:107–113.
85. Heine G, Raida M, Forssmann WG. Mapping of peptides and protein fragments in human urine using liquid chromatography-mass spectrometry. *J Chromatogr A* 1997;776:117–124.

86. Cutillas PR, Norden AG, Cramer R, Burlingame AL, Unwin RJ. Detection and analysis of urinary peptides by on-line liquid chromatography and mass spectrometry: application to patients with renal Fanconi syndrome. *Clin Sci (Lond)* 2003;104:483–490.
87. Cutillas PR, Norden AGW, Cramer R, Burlingame AL, Unwin RJ. Urinary proteomics of renal Fanconi syndrome. In: Thongboonkerd V, Klein JB, eds. *Proteomics in Nephrology.* Basel: Karger, 2004:142–154.
88. Greive KA, Balazs ND, Comper WD. Protein fragments in urine have been considerably underestimated by various protein assays. *Clin Chem* 2001;47:1717–1719.
89. Kato H, Matsumura Y, Maeda H. Isolation and identification of hydroxyproline analogues of bradykinin in human urine. *FEBS Lett* 1988;232:252–254.
90. Mount CD, Lukas TJ, Orth DN. Purification and characterization of epidermal growth factor (beta-urogastrone) and epidermal growth factor fragments from large volumes of human urine. *Arch Biochem Biophys* 1985;240:33–42.
91. Cutillas PR, Chalkley RJ, Hansen KC, et al. The urinary proteome in Fanconi syndrome implies specificity in the reabsorption of proteins by renal proximal tubule cells. *Am J Physiol Renal Physiol* 2004;287:F353–F364.
92. Lloyd SE, Pearce SH, Fisher SE, et al. A common molecular basis for three inherited kidney stone diseases. *Nature* 1996;379:445–449.
93. Christensen EI, Devuyst O, Dom G, et al. Loss of chloride channel ClC-5 impairs endocytosis by defective trafficking of megalin and cubilin in kidney proximal tubules. *Proc Natl Acad Sci U S A* 2003;100:8472–8477.
94. Gunther W, Piwon N, Jentsch TJ. The ClC-5 chloride channel knock-out mouse—an animal model for Dent's disease. *Pflugers Arch* 2003;445:456–462.
95. Norden AG, Lapsley M, Lee PJ, et al. Glomerular protein sieving and implications for renal failure in Fanconi syndrome. *Kidney Int* 2001;60:1885–1892.
96. Norden AG, Scheinman SJ, Deschodt-Lanckman MM, et al. Tubular proteinuria defined by a study of Dent's (CLCN5 mutation) and other tubular diseases. *Kidney Int* 2000;57:240–249.
97. Schaub S, Wilkins J, Weiler T, Sangster K, Rush D, Nickerson P. Urine protein profiling with surface-enhanced laser-desorption/ionization time-of-flight mass spectrometry. *Kidney Int* 2004;65:323–332.
98. Russo LM, Bakris GL, Comper WD. Renal handling of albumin: a critical review of basic concepts and perspective. *Am J Kidney Dis* 2002;39:899–919.
99. Gudehithlu KP, Pegoraro AA, Dunea G, Arruda JA, Singh AK. Degradation of albumin by the renal proximal tubule cells and the subsequent fate of its fragments. *Kidney Int* 2004;65:2113–2122.
100. Schaub S, Wilkins JA, Antonovici M, et al. Proteomic-based identification of cleaved urinary beta2-microglobulin as a potential marker for acute tubular injury in renal allografts. *Am J Transplant* 2005;5:729–738.
101. Diamandis EP. Mass spectrometry as a diagnostic and a cancer biomarker discovery tool: opportunities and potential limitations. *Mol Cell Proteomics* 2004;3:367–378.
102. Rogers MA, Clarke P, Noble J, et al. Proteomic profiling of urinary proteins in renal cancer by surface enhanced laser desorption ionization and neural-network analysis: identification of key issues affecting potential clinical utility. *Cancer Res* 2003;63:6971–6983.

103. Zhang YF, Wu DL, Guan M, et al. Tree analysis of mass spectral urine profiles discriminates transitional cell carcinoma of the bladder from noncancer patient. *Clin Biochem* 2004;37:772–779.
104. Clarke W, Silverman BC, Zhang Z, Chan DW, Klein AS, Molmenti EP. Characterization of renal allograft rejection by urinary proteomic analysis. *Ann Surg* 2003;237:660–664.
105. Schaub S, Rush D, Wilkins J, et al. Proteomic-based detection of urine proteins associated with acute renal allograft rejection. *J Am Soc Nephrol* 2004;15: 219–227.
106. O'Riordan E, Orlova TN, Mei JJ, et al. Bioinformatic analysis of the urine proteome of acute allograft rejection. *J Am Soc Nephrol* 2004;15:3240–3248.
107. Schweigert FJ, Wirth K, Raila J. Characterization of the microheterogeneity of transthyretin in plasma and urine using SELDI-TOF-MS immunoassay. *Proteome Sci* 2004;2:5.
108. Moini M, Huang H. Application of capillary electrophoresis/electrospray ionization-mass spectrometry to subcellular proteomics of *Escherichia coli* ribosomal proteins. *Electrophoresis* 2004;25:1981–1987.
109. Jensen PK, Pasa-Tolic L, Peden KK, et al. Mass spectrometric detection for capillary isoelectric focusing separations of complex protein mixtures. *Electrophoresis* 2000;21:1372–1380.
110. Gelpi E. Interfaces for coupled liquid-phase separation/mass spectrometry techniques. An update on recent developments. *J Mass Spectrom* 2002;37:241–253.
111. Schmitt-Kopplin P, Frommberger M. Capillary electrophoresis-mass spectrometry: 15 years of developments and applications. *Electrophoresis* 2003;24: 3837–3867.
112. Hernandez-Borges J, Neususs C, Cifuentes A, Pelzing M. On-line capillary electrophoresis-mass spectrometry for the analysis of biomolecules. *Electrophoresis* 2004;25:2257–2281.
113. Kolch W, Neususs C, Pelzing M, Mischak H. Capillary electrophoresis-mass spectrometry as a powerful tool in clinical diagnosis and biomarker discovery. *Mass Spectrom Rev* 2005.
114. Neususs C, Pelzing M, Macht M. A robust approach for the analysis of peptides in the low femtomole range by capillary electrophoresis-tandem mass spectrometry. *Electrophoresis* 2002;23:3149–3159.
115. Kaiser T, Wittke S, Just I, et al. Capillary electrophoresis coupled to mass spectrometer for automated and robust polypeptide determination in body fluids for clinical use. *Electrophoresis* 2004;25:2044–2055.
116. Belder D, Deege A, Husmann H, Kohler F, Ludwig M. Cross-linked poly(vinyl alcohol) as permanent hydrophilic column coating for capillary electrophoresis. *Electrophoresis* 2001;22:3813–3818.
117. Gelfi C, Vigano A, Ripamonti M, Righetti PG, Sebastiano R, Citterio A. Protein analysis by capillary zone electrophoresis utilizing a trifunctional diamine for silica coating. *Anal Chem* 2001;73:3862–3868.
118. Ullsten S, Zuberovic A, Wetterhall M, Hardenborg E, Markides KE, Bergquist J. A polyamine coating for enhanced capillary electrophoresis-electrospray ionization-mass spectrometry of proteins and peptides. *Electrophoresis* 2004;25: 2090–2099.

119. Musyimi HK, Narcisse DA, Zhang X, Stryjewski W, Soper SA, Murray KK. Online CE-MALDI-TOF MS using a rotating ball interface. *Anal Chem* 2004;76: 5968–5973.
120. Zuberovic A, Ullsten S, Hellman U, Markides KE, Bergquist J. Capillary electrophoresis off-line matrix-assisted laser desorption/ionisation mass spectrometry of intact and digested proteins using cationic-coated capillaries. *Rapid Commun Mass Spectrom* 2004;18:2946–2952.
121. Kaiser T, Hermann A, Kielstein JT, et al. Capillary electrophoresis coupled to mass spectrometry to establish polypeptide patterns in dialysis fluids. *J Chromatogr A* 2003;1013:157–171.
122. Wittke S, Fliser D, Haubitz M, et al. Determination of peptides and proteins in human urine with capillary electrophoresis-mass spectrometry, a suitable tool for the establishment of new diagnostic markers. *J Chromatogr A* 2003;1013: 173–181.
123. Neuhoff N, Kaiser T, Wittke S, et al. Mass spectrometry for the detection of differentially expressed proteins: a comparison of surface-enhanced laser desorption/ionization and capillary electrophoresis/mass spectrometry. *Rapid Commun Mass Spectrom* 2004;18:149–156.
124. Mischak H, Kaiser T, Walden M, et al. Proteomic analysis for the assessment of diabetic renal damage in humans. *Clin Sci (Lond)* 2004;107:485–495.
125. Haubitz M, Wittke S, Weissinger EM, et al. Urine protein patterns can serve as diagnostic tools in patients with IgA nephropathy. *Kidney Int* 2005;66.
126. Rossing K, Mischak H, Walden M, et al. The impact of diabetic nephropathy and angiotensin II receptor blocker treatment on urinary polypeptide patterns in type 2 diabetic patients. *Kidney Int* 2005;66.
127. Cotter RJ. *Time-of-Flight Mass Spectrometry. Instrumentation and Applications in Biological Research*. Washington, DC: American Chemical Society, 1997.
128. Fenselau C. MALDI MS and strategies for protein analysis. *Anal Chem* 1997;69:661A–665A.
129. Bakhtiar R, Nelson RW. Electrospray ionization and matrix-assisted laser desorption ionization mass spectrometry. Emerging technologies in biomedical sciences. *Biochem Pharmacol* 2000;59:891–905.
130. Nelson RW, Krone JR, Bieber AL, Williams P. Mass spectrometric immunoassay. *Anal Chem* 1995;67:1153–1158.
131. Nelson RW, Krone JR, Dogruel D, Tubbs KA, Granzow R, Jansson O. Interfacing biomolecular interaction analysis with mass spectrometry and the use of bioreactive mass spectrometer probe tips in protein characterization. In: Marshak DR, ed. *Techniques in Protein Chemistry VIII*. San Diego, CA: Academic Press, 1997:493–504.
132. Krone JR, Nelson RW, Williams P. Mass spectrometric immunoassay. *SPIE* 1996;2680:415–421.
133. Tubbs KA, Nedelkov D, Nelson RW. Detection and quantification of beta-2-microglobulin using mass spectrometric immunoassay. *Anal Biochem* 2001;289: 26–35.

134. Niederkofler EE, Tubbs KA, Gruber K, et al. Determination of beta-2 microglobulin levels in plasma using a high-throughput mass spectrometric immunoassay system. *Anal Chem* 2001;73:3294–3299.
135. Niederkofler EE, Tubbs KA, Kiernan UA, Nedelkov D, Nelson RW. Novel mass spectrometric immunoassays for the rapid structural characterization of plasma apolipoproteins. *J Lipid Res* 2003;44:630–639.
136. Kiernan UA, Nedelkov D, Tubbs KA, Niederkofler EE, Nelson RW. Selected expression profiling of full-length proteins and their variants in human plasma. *Clin Proteomics* 2004;1:7–16.
137. Kiernan UA, Nedelkov D, Tubbs KA, Niederkofler EE, Nelson RW. Proteomic characterization of novel serum amyloid P component variants from human plasma and urine. *Proteomics* 2004;4:1825–1829.
138. Kiernan UA, Tubbs KA, Gruber K, et al. High-throughput protein characterization using mass spectrometric immunoassay. *Anal Biochem* 2002;301:49–56.
139. Kiernan UA, Tubbs KA, Nedelkov D, Niederkofler EE, McConnell E, Nelson RW. Comparative urine protein phenotyping using mass spectrometric immunoassay. *J Proteome Res* 2003;2:191–197.
140. Kiernan UA, Tubbs KA, Nedelkov D, Niederkofler EE, Nelson RW. Comparative phenotypic analyses of human plasma and urinary retinol binding protein using mass spectrometric immunoassay. *Biochem Biophys Res Commun* 2002;297:401–405.
141. Kiernan UA, Tubbs KA, Nedelkov D, Niederkofler EE, Nelson RW. Detection of novel truncated forms of human serum amyloid A protein in human plasma. *FEBS Lett* 2003;537:166–170.
142. Nedelkov D, Tubbs KA, Niederkofler EE, Kiernan UA, Nelson RW. High-throughput comprehensive analysis of human plasma proteins: a step toward population proteomics. *Anal Chem* 2004;76:1733–1737.
143. Nelson RW, Nedelkov D, Tubbs KA, Kiernan UA. Quantitative mass spectrometric immunoassay of insulin like growth factor 1. *J Proteome Res* 2004;3:851–855.
144. Nelson RW, McLean MA, Hutchens TW. Quantitative determination of proteins by matrix-assisted laser-desorption ionization time-of-flight mass spectrometry. *Anal Chem* 1994;66:1408–1415.
145. Schardijn GH, Statius van Eps LW. Beta 2-microglobulin: its significance in the evaluation of renal function. *Kidney Int* 1987;32:635–641.
146. Grubb AO. Cystatin C—properties and use as diagnostic marker. *Adv Clin Chem* 2000;35:63–99.

12

Proteomics of Human Cerebrospinal Fluid

Margareta Ramström and Jonas Bergquist

Summary

Examination of human cerebrospinal fluid (CSF) has been performed for over 100 years. Since CSF is in direct contact with the extracellular surface of the brain, the biochemical composition of this fluid is altered in disorders related to the central nervous system. Hence, it is of great interest to investigate thoroughly the human CSF proteome, to identify proteins, and to examine possible biomarkers of, for example, neurodegenerative disorders. This chapter will provide a short introduction to proteins found in CSF and describe designs and applications of proteomics methods used for the analysis of CSF.

Key Words: Cerebrospinal fluid; 2D-PAGE; LC-MS; neurodegenerative disorders; central nervous system.

1. Introduction

Cerebrospinal fluid (CSF) is a colorless body fluid surrounding the brain and the spinal chord in vertebrates. Its main functions are mechanical protection of the brain and transportation of metabolically active substances and waste products. Owing to its close contact with the brain, CSF is often investigated when one is examining disorders related to the central nervous system (CNS). Analysis of CSF has been performed for diagnostic purposes for over 100 yr *(1)*. The body fluid is secreted at a rate of 0.35 mL/min, mainly from the choroid plexus in the brain. The total volume of CSF is around 100 to 150 mL in adults, which means that the fluid is replaced three to four times every day *(2)*.

Most of the chemical compounds found in CSF originate from the blood. The levels of electrolytes, e.g., Na^+ and Cl^-, are approximately the same in both body fluids. Water, gases, and lipid-soluble compounds move freely from the blood into CSF; glucose, amino acids, and cations are transported by carrier-mediated processes. The protein content in CSF is typically 350 mg/L, which is

From: *Proteomics of Human Body Fluids: Principles, Methods, and Applications*
Edited by: V. Thongboonkerd © Humana Press Inc., Totowa, NJ

Table 1
The Most Abundant Proteins in CSF and Plasma[a]

Protein	Concentration (mg/L)
CSF	
Albumin	200
Prostaglandin D-synthase	26
IgG	22
Transthyretin	17
Transferrin	14
α_1-antitrypsin	8
Apo-lipoprotein A	6
Cystatin C	6
Plasma	
Albumin	45,000
IgG	9900
α-Lipoprotein	3000
Fibrinogen	3000
Transferrin	2300
β-Lipoprotein	2000
α_2-macroglobulin	2000
α_1-antitrypsin	1400

[a]Prostaglandin D-synthase, transthyretin, and cystatin C are examples of brain-derived proteins that are present in high relative concentrations in CSF. Molecules of larger sizes, such as α- and β-lipoprotein and α_2-macroglobulin, do not enter the blood-CSF barrier readily and are hence present in lower proportions in CSF than in plasma.

approximately 200 times lower than in plasma. Around 20% of the proteins in CSF are brain derived, and the remaining part is derived from blood via the blood-CSF barrier. Hence, the protein repertoire resembles that of plasma, but there are significant divergences, and it is not accurate to describe CSF solely as being an ultrafiltrate of plasma. This is evident, for example, when one compares the rank order of the most abundant proteins in the two body fluids *(3)* (**Table 1**). Three of the eight most abundant proteins in CSF (prostaglandin D-synthase, transthyretin, and cystatin C) are brain derived and therefore not found at high proportions in plasma. Proteins referred to as brain derived are formed at different sites; prostaglandin D-synthase and cystatin C are released from leptomeninges, whereas other proteins, e.g., tau protein and S-100, originate from neurons and glial cells. Transthyretin in CSF is mainly formed in the brain, but a nonnegligible part also originates from the blood *(4)*. Therefore, the expression of transthyretin in CSF is more complex to study, and mathematical corrections are necessary, e.g., when comparing the levels of brain-derived transthyretin in different compartments of the CNS *(4)*. The relative amounts of blood-derived proteins in CSF do not truly correlate to the relative concentrations in plasma. For example, the molecular weight of high-abundant proteins in CSF is typically lower. It has been demonstrated that smaller proteins generally enter the CSF more readily from blood than larger ones and that protein charge also influences the ability of proteins to cross the blood-CSF barrier *(3)*.

The CSF proteome is a unique biochemical environment and a rich source of potential biomarkers for various conditions. This chapter will describe and discuss proteomic approaches applied in the analysis of CSF. Although proteomics can be divided into several subdisciplines *(5)*, this chapter will focus on expression proteomics, including qualitative, quantitative, and comparative studies that have been presented in the literature.

2. Methods

2.1. Sample Collection

2.1.1. Lumbar Puncture

The most common way to obtain CSF is via lumbar puncture, which can be performed at the lateral decubitus position at the L3-L4, L4-L5, or L5-S1 vertebral interspaces *(6)*. Routinely, 12 mL CSF is collected for analysis from adults. It is important to mix the samples gently after collection to avoid gradient effects. The choice of tubes and storage conditions also influences the results and should be optimized prior to analysis. It is vital to avoid contamination from blood during collection. Owing to the large difference in protein concentrations of the two body fluids, a minor contamination of blood will alter the protein content of CSF dramatically. To check for possible contaminations, cell counts and total protein analysis should be performed.

2.1.2. Ventricular CSF

Analysis of ventricular CSF is less frequently performed. It should be noted that the protein concentration in ventricular and lumbar CSF differ. The protein concentrations given in **Table 1** refer to lumbar CSF. The total protein concentration is lower, but proteins released from brain cells are more abundant in ventricular CSF *(4)*.

2.2. CSF Proteomics Using 2D-PAGE

The dominating approach for expression proteomics studies of CSF is 2D polyacrylamide gel electrophoresis (2D-PAGE). An overview of a standard experimental procedure is given in **Fig. 1**. As mentioned earlier, the protein concentration in CSF is rather low, whereas the salt concentration is high (>150 m*M*) *(7)*. Sample preparation prior to 2D-PAGE involving a desalting

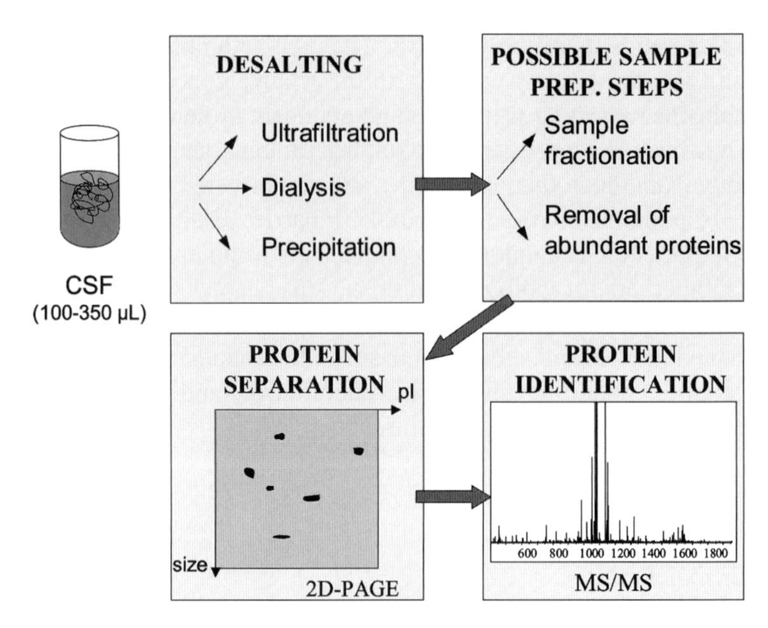

Fig. 1. Experimental procedure for CSF analysis by 2D-PAGE. Typically, 100 to 350 µL of CSF are desalted prior to sample loading on-gel. Different approaches for sample fractionation and removal of high-abundant proteins can also be applied to reduce sample complexity.

step is necessary in order to avoid negative effects on the electrophoretic separation. Several different methods have been used for this purpose, including ultrafiltration ***(8)***, dialysis ***(9)***, and protein precipitation using cold acetone ***(10)*** or trichloroacetic acid/acetone ***(11,12)***. When one is choosing a method for desalting, it is important to estimate the protein recovery and the reproducibility of the method. Yuan and Desiderio ***(7,11)*** compared the recoveries and performances of different methods for desalting. Their studies report on the highest recovery for acetone precipitation and an alternative method for salt removal, based on spin columns.

After removal of salts, the protein pellet is resolved in an appropriate buffer and, typically, sample volumes corresponding to 100 to 350 µL of native CSF are loaded on gel prior to separation. An example of a 2D gel of CSF is shown in **Fig. 2**. The most intense spots correspond to the dominating proteins given in **Table 1**. Human serum albumin (HSA) is by far the dominating protein, which is also seen on gel. From the spots, 20 to 30 different proteins can typically be identified ***(8,10,12)***, and the standard approach for identification involves tryptic digestion of each protein spot prior to matrix-assisted laser desorption/ionization (MALDI) mass spectrometry (MS) ***(10,11)***.

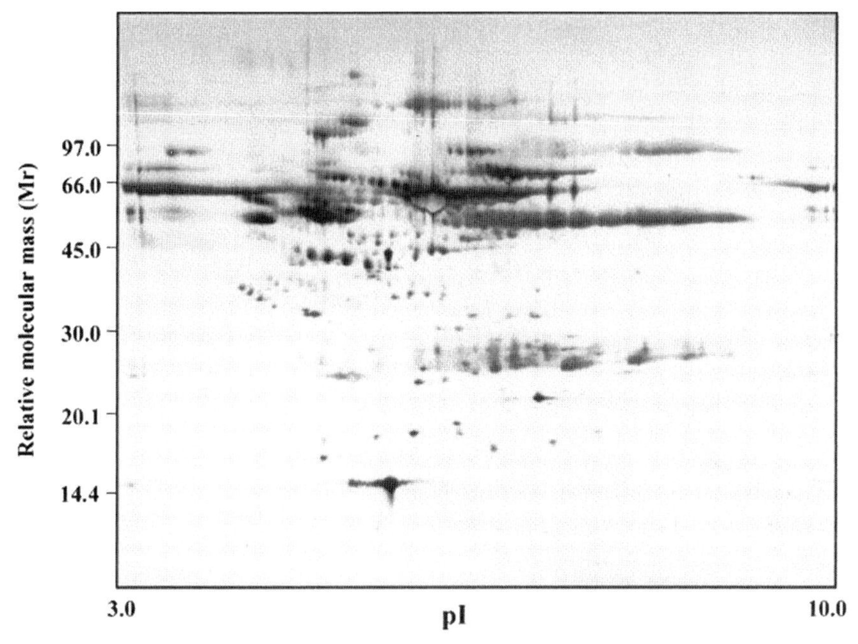

Fig. 2. A representative 2D gel of human CSF as presented by Yuan *et al.* ***(11)***. (*Reproduced with permission from WILEY-VCH Verlag GmbH.*)

The concentration range of proteins in CSF is very wide; a few highly abundant proteins limit the amount of total proteins that can be loaded on gel, and thus the possibility of detecting lower abundant components is limited. Therefore, different approaches have been developed to fractionate CSF or to remove high-abundant proteins prior to analysis using 2D-PAGE. Davidsson and colleagues ***(13,14)*** reported on a prefractionation method relying on liquid-phase isoelectric focusing. After fractionation with respect to isoelectric points, selected fractions were pooled, and each pool was then applied individually on separate 2D gels. It was reported that, generally, more protein spots of larger sizes could be found on 2D gels of fractionated samples, and additional proteins could be identified in these analyses. The sample consumption for one experiment was 3 mL native CSF, which is a clinically available volume, but approximately 10 times higher than in ordinary 2D-PAGE studies. An alternative approach for sample prefractionation was presented by Yuan and Desiderio ***(15)***. They applied reversed-phase (RP) solid-phase extraction, whereby the proteins are separated with respect to hydrophobicity. One milliliter of CSF was loaded on the RP column, and three fractions were collected from the sample. The fractions were loaded on separate 2D gels; in total, 42 different proteins could be identified using MALDI-time-of-flight (TOF) analysis of tryptic digested protein spots.

More specific removal of selected proteins can be achieved by applying various approaches. The most commonly used strategies are based on Cibachon Blue-Sepharose media ***(16)*** or monoclonal antibodies ***(17–19)***. In expression proteomics studies of CSF, variants of both these approaches have been applied prior to 2D gel electrophoresis ***(12,20)***. Signals from lower abundant proteins were enhanced, but some spots corresponding to nonabundant proteins were also removed in the initial studies ***(12)***.

In conclusion, a removal or fractionation step is often advantageous when one is studying lower abundant proteins in CSF. However, the risk of, at the same time, removing low-abundant proteins via binding to the target proteins or the affinity column should be estimated and investigated. Also, the sample consumption and the analysis time need to be taken into account when one is introducing an extra step for sample preparation.

2.3. CSF Proteomics Using Online Liquid Separation and High-Resolution Mass Spectrometry

As alternatives to gel-based proteomics, methods relying on liquid separations and mass spectrometry have been applied in CSF analysis (**Fig. 3**). These approaches are often faster to perform, easier to automate, and require smaller sample volumes. In standard strategies, the proteins in CSF are enzymatically digested, all at the same time, and the peptides are then separated in one or several dimensions. After this, the fragments are electrosprayed into a mass spectrometer using online electrospray ionization.

Our group has reported on strategies for CSF proteomics based on Fourier transform ion cyclotron resonance mass spectrometry (FT-ICR MS) ***(21–25)***. The FT-ICR mass spectrometer provides mass accuracy on the ppm level, ultrahigh resolution, and high sensitivity. Therefore, it is a powerful instrument to be applied for the identification of components of various concentrations in complex samples. In the approaches described, the proteins were first globally digested using trypsin. After desalting, the tryptic peptides were either analyzed by direct infusion electrospray ionization (ESI)-FTICR MS ***(21)*** or separated using packed capillary liquid chromatography (LC) ***(22–24)*** or capillary electrophoresis (CE) ***(25)*** prior to MS detection. The subsequent protein identification procedure relied on the high mass measurement accuracy of the mass spectrometer. Packed capillary LC was performed on a standard high-performance liquid chromatography (HPLC) system; the separation took place on in-house packed RP columns of 200-μm inner diameter. CE experiments were conducted on an in-house-designed CE instrument. Pressurized injection of tryptic digested CSF sample was applied, and the separation took place in 100-cm-long fused silica capillaries (inner diameter 25 μm). The electrospray emitters used in all three experimental categories were Black Dust (polyimide-graphite) sheathless emitters ***(26)***. In direct

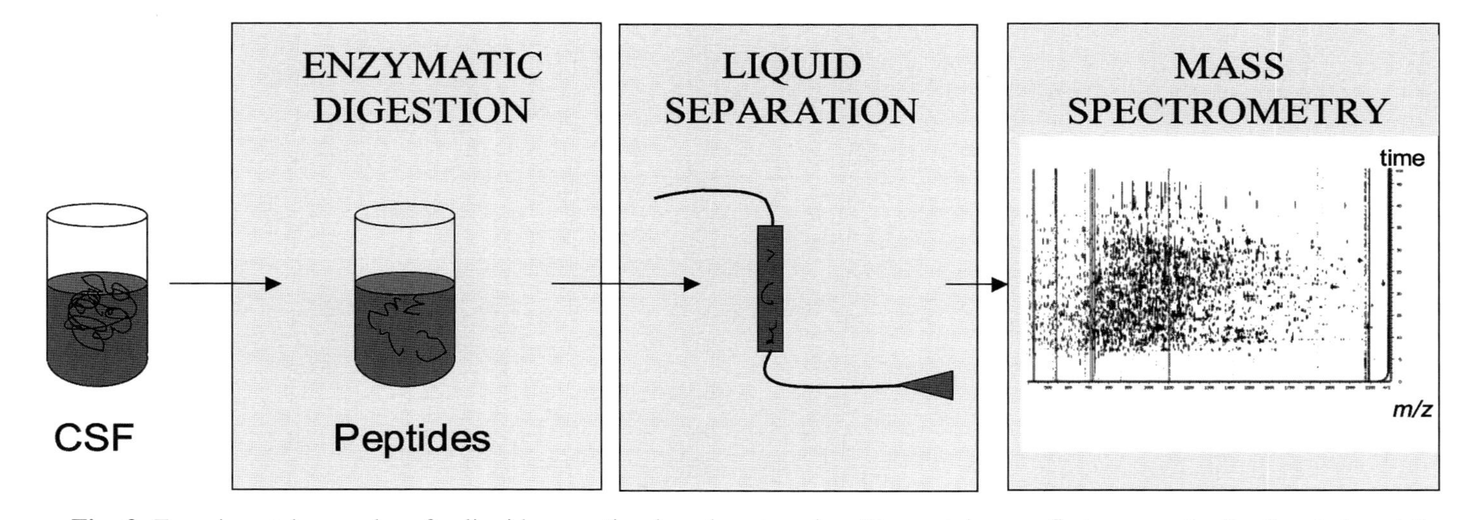

Fig. 3. Experimental procedure for liquid separation-based proteomics. The proteins are first enzymatically digested, and the resulting peptides are then separated in one or several dimensions using liquid chromatography or capillary electrophoresis. Peptides are electrosprayed online into a mass spectrometer. The peptides are identified based on sequence-tag information by MS/MS, or accurate mass information using FT-ICR MS.

Table 2
A Comparison of Three Methods for Sample Introduction Into an FT-ICR Mass Spectrometer[a]

	Direct infusion	Capillary electrophoresis (CE)	Liquid chromatography (LC)
Sample consumption (µL)	7.8	0.016	32
Experimental time (min)	40	15	60
No. of detected peptides	<1000	1500	6600
No. of identified proteins	13	30	39

[a]The LC approach is the most powerful one in terms of the number of detected fragments and the number of identified proteins. However, the CE method is attractive, since it is a faster approach and consumes less amount of sample.

infusion experiments, flow rates of 60 nL/min were chosen; the flow rates in the LC and CE experiments were adjusted to conditions appropriate for the separation.

A comparison of CSF analysis by the three approaches is shown in **Table 2**. In all cases, the sample consumption was minor compared with 2D-PAGE. Using CE, only 16 nL were consumed in one experiment. However, usually a larger volume is required for sample preparation and sample infusion. As expected, more peptides were detected and more proteins could be identified applying a separation step prior to FT-ICR MS. A typical mass chromatogram from an LC-FT-ICR experiment of CSF is shown in **Fig. 4**. Each spot in this figure corresponds to a detected tryptic peptide. The number of identified proteins in the LC and CE experiments was quite comparable to the results from 2D-PAGE analysis. The advantage of the direct infusion experiment is its simplicity and ease of use. Informative results can also be generated very fast. Frequently, data collection during 1 to 5 min is sufficient to get an overview of a complex sample, whereas longer analysis times permit increased signal-to-noise ratios of components in the mass spectra.

Also, when one is applying LC-MS-based proteomics, it is advantageous to include a removal step for HSA and other high-abundant proteins. In a recent study, we investigated the effects of two commercially available depletion kits combined with the LC-FT-ICR MS approach *(24)*. It was shown that the number of identified proteins in CSF increased dramatically, when an antibody-based method, developed for the removal of HSA and IgG, was used **(Fig. 5)**. The study also demonstrated that methods initially created for depletion of proteins from plasma or serum were compatible with other body fluids, such as CSF. A trend in this field is to develop antibody-based methods that specifically remove many of the high-abundant proteins. However, for this purpose, it would be advantageous to construct CSF-specific columns, since the high-abundant brain-derived proteins would otherwise limit the sensitivity of the method.

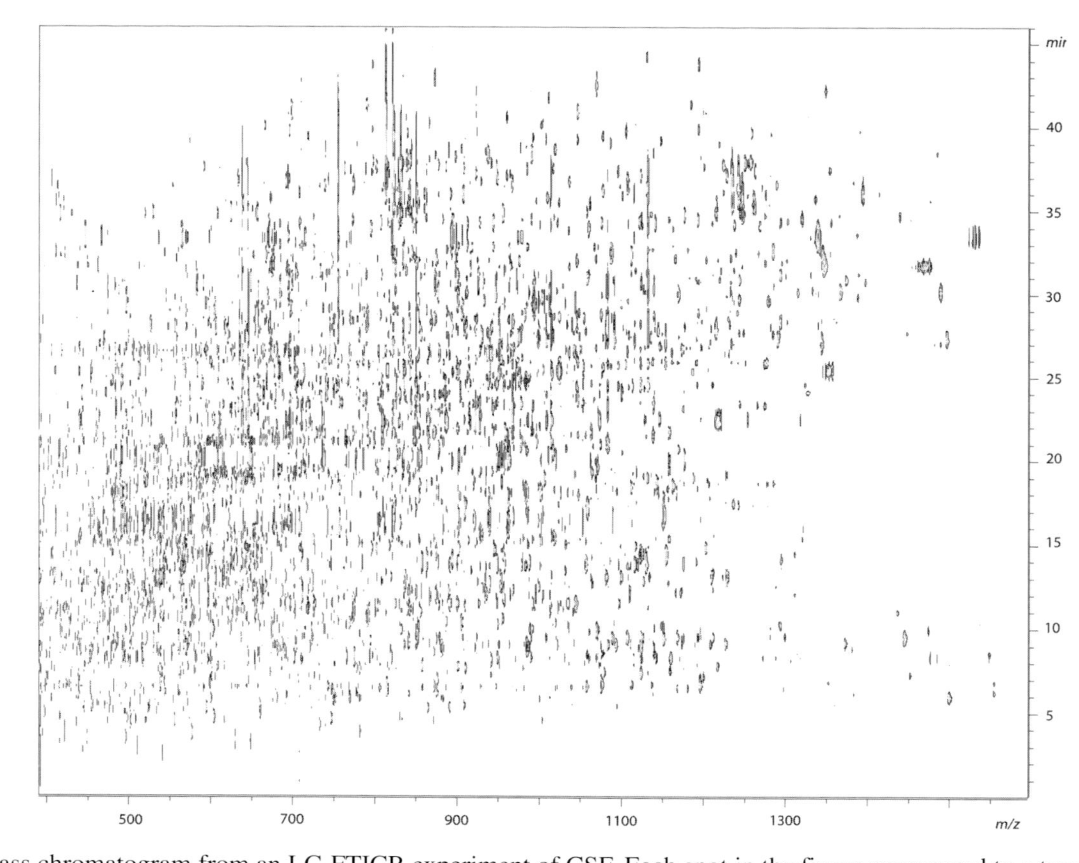

Fig. 4. A mass chromatogram from an LC-FTICR experiment of CSF. Each spot in the figure correspond to a tryptic peptide. The *m/z* is given on the *x*-axis, and elution time is given on the *y*-axis. Approximately 6600 peptides were detected in one experiment.

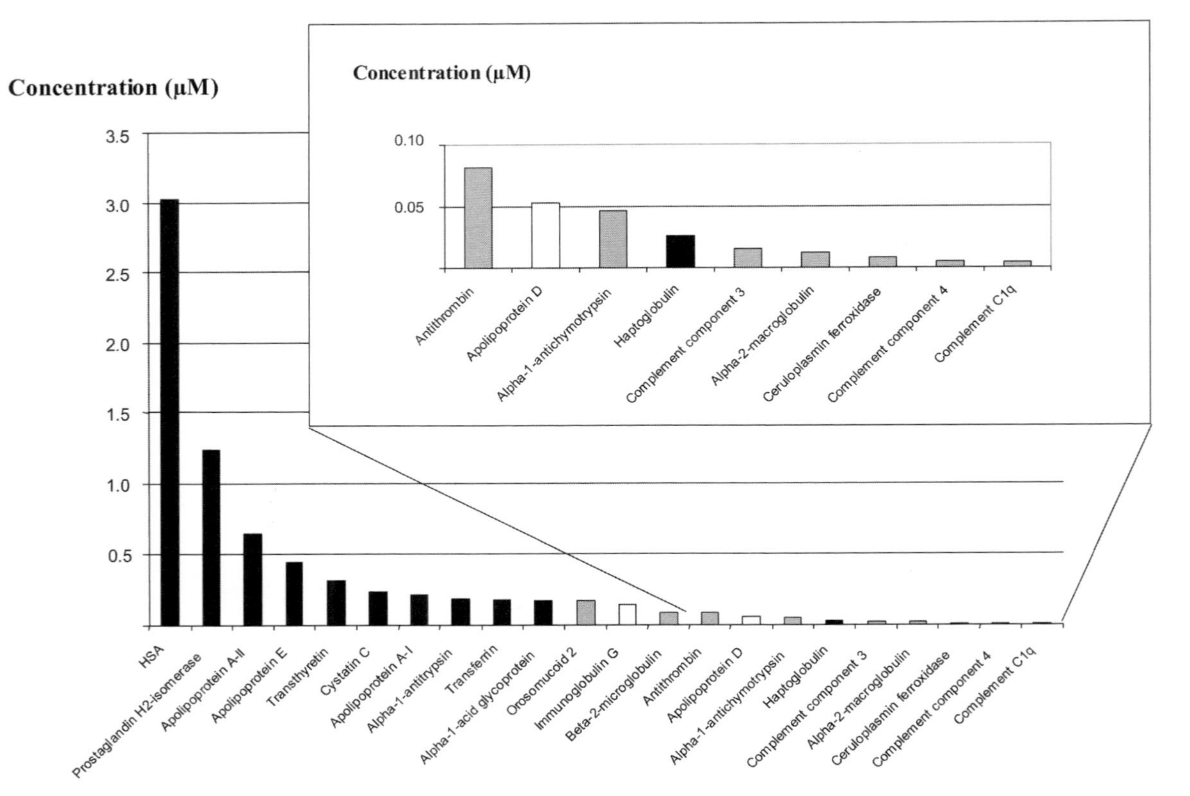

Fig. 5. Proteins identified in CSF using a proteomic approach based on LC-FT-ICR MS, prior to and after removal of albumin and immunoglobulin. Black denotes proteins identified using both methods; gray proteins and white correspond to the proteins only detected after depletion of abundant proteins, and in native CSF, respectively. The identification of lower abundant components was greatly enhanced using a sample preparation step.

In so-called shotgun proteomics, tandem mass spectrometry is performed to fragment the peptides, and protein identification is then based on sequence-tag information of the tryptic peptides. Shotgun proteomics has been applied in analysis of the lumbar ***(27)*** and ventricular CSF proteome ***(28)***. In the study of lumbar CSF, gradual precipitation with acetonitrile was performed to separate albumin from lower abundant proteins. The tryptic peptides were then separated in a 2D microcapillary column LC system, consisting of a strong cation-exchange column and two alternating RP capillary columns. Tandem mass spectrometry was performed using an ion trap. The authors report on the identification of 165 proteins from the CSF proteome ***(27)***. A similar experimental setup was applied for separation and identification of tryptic peptides from the ventricular CSF proteome ***(28)***. In this study, HSA and IgG were removed using an in-house constructed antibody-based resin. Identification of 249 unique proteins was reported. However, only 6% of these proteins were common to all 10 subjects under the study, and the identification of 67% of all proteins was based on the sequential information from one single peptide.

3. Findings and Applications

The ultimate goal of many proteomic studies is to identify possible biomarkers for certain conditions. Traditionally, up- and downregulation of candidate biomarkers has been investigated using methods for quantification of preselected proteins one at a time. Immunoassays, such as enzyme-linked immunosorbent assay (ELISA) and radioimmunoassay (RIA), are some of the most commonly applied strategies. Analyses by standard immunoassay techniques have revealed alterations in CSF proteins correlated with injuries and disorders of the CNS ***(29–35)***. A few examples are the increase in tau protein and decrease in $A\beta_{1-42}$ in Alzheimer's disease (AD) ***(35)*** and the elevation of neurofilament protein (NFL) levels in amyotrophic lateral sclerosis, AD, and miscellaneous neurodegenerative disorders ***(34)***.

Many of the alterations found in CSF are not specific to a certain disorder but rather general for similar conditions. An optimal biomarker should be disorder specific and selective. Such findings would allow for the development of diagnostic tools based on biochemical analysis and would provide a deeper understanding of the etiology of the specific disorder rather than the process of, e.g., neurodegeneration in general. As the biochemical environment in a body fluid should be expected to change in many correlated ways during a certain condition, it is advantageous to achieve a more global view of the proteins in the samples. Comparative proteomics is indeed a more discovery-oriented approach to the problem and, therefore, has been applied to study various conditions.

Most commonly, 2D-PAGE-based proteomics has been applied to compare CSF proteomes. A list of potential biomarkers for several disorders related to the

CNS identified using this approach is given in a recent review *(7)*. For example, protein alterations correlated with AD have been investigated in several proteomic studies *(36–38)*. In the reports, a few proteins including apolipoproteins have been suggested to be affected in AD. The level of apolipoprotein E (ApoE) has been demonstrated to be significantly altered in several studies *(37,38)*, consistent with the data obtained from ELISA-based studies *(39)*. The concentration of ApoE was also reported to be reduced, but to a lesser degree, in frontotemporal dementia *(40)*. Choe et al. *(36)* demonstrated that the proteins significantly altered in CSF of AD patients compared with controls varied for the same data set depending on the choice of statistical method for comparison of 2D gels. This illustrates one of the greatest challenges associated with proteomics and the comparison of complex samples. The comparison of 2D gels is not straightforward and powerful, and accurate tools are required for statistical analysis.

Gel-free methods have also been applied to solve similar problems. Recently, a study was conducted in order to search for possible biomarkers of amyotrophic lateral sclerosis *(23)* using the previously described LC-FT-ICR MS approach. Mass chromatograms of CSF from 12 patients and 10 controls were collected. Significant differences in protein expression could be observed when the patterns were compared, and the differentially expressed peptide masses were extracted. Unfortunately, the identities of the characteristic peptides could not be revealed by the technique applied. When one is trying to classify samples as unknown, 80% of these samples were sorted into the correct group. Hence, the study is very promising, but more samples need to be analyzed before any conclusion can be drawn. In order to assign the characteristic peptides, experiments applying complementary MS/MS techniques have also been initiated.

An LC-MS/MS study was performed to compare protein levels in CSF from elderly and younger patients. A strategy for relative quantification called isotope-coded affinity tags (ICAT) was used for this purpose. It was shown that the relative concentration of 30 proteins were changed by more than 20% between the group of younger persons (aged 22–36 yr) and older subjects (aged 66–85 yr) *(27)*. This study demonstrates the importance of choosing age-matched control samples when screening CSF for possible biomarkers.

4. Perspectives and Future Directions

The biochemistry of CSF reflects the status of the brain and the CNS. The protein content in CSF is low and contains both brain- and plasma-derived components. Recently, proteomics has evolved to become a useful tool for identification of proteins in complex samples and also for the comparison of protein expression patterns. Interest in biomarkers for various pathologies is presently enormous, and the need for new technologies in screening and discovery is obvious. Today's approaches to CSF proteomics will most probably not end up as routine analytical tools in the clinical arena or as point-of-care screening devices. However, we see great potential to our approach in the discovery phase of biomarker research: new markers can be identified in a nonbiased way.

Acknowledgments

The authors thank all colleagues at Analytical Chemistry and the Division of Ion Physics, Uppsala University who have contributed to this work. Financial support from Knut and Alice Wallenberg, the Swedish Foundation for Strategic Research, the Swedish Society for Medical Research, and the Swedish Research Council (grants 621-2002-5261 and 629-2002-6821) is gratefully acknowledged. Jonas Bergquist has a senior research position at the Swedish Research Council (VR).

References

1. Green AJE. Cerebrospinal fluid brain-derived proteins in the diagnosis of Alzheimer's disease and Creutzfeldt-Jakob disease. *Neuropathol Appl Neurobiol* 2002;28:427–440.
2. Betz AL, Goldstein GW, Katzman R. Blood-brain-cerebrospinal fluid barriers. In: Siegel GJ, ed. *Basic Neurochemistry: Molecular, Cellular and Medical Aspects,* 5th ed. New York: Raven Press, 1994:681–699.
3. Thompson EJ. *The CSF Proteins: A Biochemical Approach.* Amsterdam: Elsevier, 1988.
4. Reiber H. Dynamics of brain-derived proteins in cerebrospinal fluid. *Clin Chim Acta* 2001;310:173–186.
5. Marko-Varga G, Fehniger TE. Proteomics and disease—the challenge for technology and discovery. *J Proteome Res* 2004;3:167–178.
6. Wikkelsö C. Likvorundersökningar. In: Aquilonius S-M, Fagius J, eds. *Neurologi.* Stockholm: Almqvist & Wiksell Medicin, 1994:92–96.
7. Yuan X, Desiderio D. Proteomics analysis of human cerebrospinal fluid. *J Chromatogr B* 2005;815:179–189.
8. Sickmann A, Dormeyer W, Wortelkamp S, Woitalla D, Kuhn W, Meyer HE. Towards a high resolution separation of human cerebrospinal fluid. *J Chromatogr B* 2002; 771:167–196.
9. Hesse C, Nilsson CL, Blennow K, Davidsson P. Identification of the apolipoprotein E4 isoform in cerebrospinal fluid with preparative two-dimensional electrophoresis and matrix-assisted laser desorption/ionization-time of flight mass spectrometry. *Electrophoresis* 2001;22:1834–1837.
10. Sickmann A, Dormeyer W, Wortelkamp S, Woitalla D, Kuhn W, Meyer HE. Identification of proteins from human cerebrospinal fluid, separated by two-dimensional polyacrylamide gel electrophoresis. *Electrophoresis* 2000;21: 2721–2728.
11. Yuan X, Russell T, Wood G, Desiderio D. Analysis of the human lumbar cerebrospinal fluid proteome. *Electrophoresis* 2002;23:1185–1196.

12. Raymacker J, Daniels A, De Brabandere V, Missiaen C. Identification of two-dimentionally separated human cerebrospinal fluid proteins by N-terminal sequencing, matrix-assisted laser desorption/ionization-mass spectrometry, nano-liquid chromatography-electrospray ionization-time of flight mass spectrometry, and tandem mass spectrometry. *Electrophoresis* 2000;21:2266–2283.
13. Davidsson P, Folkesson S, Christiansson M, et al. Identification of proteins in human cerebrospinal fluid using liquid-phase isoelectric focusing as a prefractionation step followed by two-dimensional gel electrophoresis and matrix-assisted laser desorption/ionisation mass spectrometry. *Rapid Commun Mass Spectrom* 2002;16:2083–2088.
14. Folkesson Hansson S, Puchades M, Blennow K, Sjögren M, Davidsson P. Validation of a prefractionation method followed by two-dimensional electrophoresis—applied to cerebrospinal fluid proteins from frontotemporal dementia patients. *Proteome Sci* 2004;2:1–11.
15. Yuan X, Desiderio DM. Proteomics analysis of prefractionated human lumbar cerebrospinal fluid. *Proteomics* 2005;5:541–550.
16. Travis J, Bowen J, Tewksbury D, Johnson D, Pannell R. Isolation of albumin from whole human plasma and fractionation of albumin-depleted plasma. *Biochem J* 1976;157:301–306.
17. Wang YY, Cheng P, Chan DW. A simple affinity spin tube filter method for removing high-abundant common proteins or enriching low-abundant biomarkers for serum proteomic analysis. *Proteomics* 2003;3:243–248.
18. Greenough C, Jenkins RE, Kitteringham NR, Pirmohamed M, Park BK, Pennington SR. A method for the rapid depletion of albumin and immunoglobulin from human plasma. *Proteomics* 2004;4:3107–3111.
19. Steel LF, Trotter MG, Nakajima PB, Mattu TS, Gonye G, Block T. Efficient and specific removal of albumin from human serum samples. *Mol Cell Protein* 2003;2:262–270.
20. Rueggeberg S, Bathke A, Li X, Franz T. Removal of albumin and immunoglobulin fron human cerebrospinal fluid (CSF) prior to 2-D gel electrophoresis using the Aurum Serum Protein Mini Kit. Bio-Rad, Tech Note 2004;3061.
21. Bergquist J, Palmblad M, Wetterhall M, Håkansson P, Markides KE. Peptide mapping of proteins in human body fluids using electrospray ionization Fourier transform ion cyclotron resonance mass spectrometry. *J Mass Spectrom Rev* 2002;21:2–15.
22. Ramström M, Palmblad M, Markides KE, Håkansson P, Bergquist J. Protein identification in cerebrospinal fluid using packed capillary liquid chromatography Fourier transform ion cyclotron resonance mass spectrometry. *Proteomics* 2003;3:184–190.
23. Ramström M, Ivonin I, Johansson A, et al. Cerebrospinal fluid protein patterns in neurodegenerative disease revealed by liquid chromatography fourier transform ion cyclotron resonance mass spectrometry. *Proteomics* 2004;4: 4010–4018.

24. Ramström M, Hagman C, Mitchell JK, Derrick PD, Håkansson P, Bergquist J. Depletion of high-abundant proteins in body fluids prior to liquid chromatography fourier transform ion cyclotron resonance mass spectrometry. *J Proteome Res* 2005;4:410–416.
25. Wetterhall M, Palmblad M, Håkansson P, Markides KE, Bergquist J. Rapid analysis of tryptically digested cerebrospinal fluid using capillary electrophoresis-electrospray ionization-Fourier transform ion cyclotron resonance-mass spectrometry. *J Proteome Res* 2002;1:361–366.
26. Nilsson S, Wetterhall M, Bergquist J, Nyholm L, Markides KE. A simple and robust conductive graphite coating for sheathless electrospray emitters used in capillary electrophoresis/mass spectrometry. *Rapid Commun Mass Spectrom* 2001; 15:1997–2000.
27. Zhang J, Goodlett DR, Peskind ER, et al. Quantitative proteomic analysis of age-related changes in human cerebrospinal fluid. *Neurobiol Aging* 2005;26:207–227.
28. Wenner BR, Lovell MA, Lynn BC. Proteomics analysis of human ventricular cerebrospinal fluid from neurologically normal, elderly subjects using two-dimensional LC-MS/MS. *J Proteome Res* 2004;3:97–103.
29. Kay AD, Petzold A, Kerr M, Keir G, Thompson EJ, Nicoll JA. Cerebrospinal fluid apolipoprotein E concentration decreases after traumatic brain injury. *J Neurotrauma* 2003:243–250.
30. Terrisse L, Poirier J, Bertrand P, et al. Increased levels of apolipoprotein D in cerebrospinal fluid and hippocampus of Alzheimer's patients. *J Neurochem* 1998;71:1643–1650.
31. Smyth MD, Cribbs DH, Tenner AJ, et al. Decreased levels of C1q in cerebrospinal fluid of living Alzheimer patients correlate with disease state. *Neurobiol Aging* 1994;15:609–614.
32. Tenhunen R, Iivanainen M, Kovanen J. Cerebrospinal fluid beta 2-microglobulin in neurological disorders. *Acta Neurol Scand* 1978;58:366–373.
33. Brisby H, Olmarker K, Rosengren L, Cederlund C-G. Markers of nerve tissue injury in the cerebrospinal fluid in patients with lumbar disc hernation and sciatica. *Spine* 1999;24:742–746.
34. Rosengren LE, Karlsson J-E, Karlsson J-O, Persson LI, Wikkelsö C. Patients with amytrophic lateral sclerosis and other neurodegenerative diseases have increased levels of neurofilament protein in CSF. *J Neurochem* 1996;67:2013–2018.
35. Blennow K, Hampel H. CSF markers for incipient Alzheimer's disease. *Lancet Neurol* 2003;2:605–613.
36. Choe LH, Dutt MJ, Relkin N, Lee KH. Studies of potential cerebrospinal fluid molecular markers for Alzheimer's disease. *Electrophoresis* 2002;23:2247–2251.
37. Puchades M, Folkesson Hansson S, Nilsson C, et al. Proteomic studies of potential cerebrospinal fluid protein markers for Alzheimer's disease. *Brain Res Mol Brain Res* 2003;118:140–146.
38. Davidsson P, Westman-Brinkmalm A, Nilsson CL, et al. Proteome analysis of cerebrospinal fluid proteins in Alzheimer patients. *Neuroreport* 2002:5.

39. Merched A, Serot JM, Visvikis S, Aguillon D, Faure G, Siest G. Apolipoprotein E, transthyretin and actin in the CSF of Alzheimer's patients: relation with the senile plaques and cytoskeleton biochemistry. *FEBS Lett* 1998;425:225–228.
40. Davidsson P, Sjögren M, Andreasen N, et al. Studies of the pathophysiological mechanisms in frontotemporal dementia by proteome analysis of CSF proteins. *Mol Brain Res* 2002;109:128–133.

13

Proteomics of Pleural Effusion

Joost Hegmans, Annabrita Hemmes, and Bart Lambrecht

Summary

A pleural effusion is the abnormal accumulation of fluid between the two layers of pleura that line the chest cavity and surround the lung. Pleural effusion can be the result of several causes. Proteomic analysis may be useful in indicating the pathogenic mechanisms involved in pleural fluid accumulation and might pinpoint specific diagnosis. Differential gel electrophoresis (DIGE) has been employed to compare directly the proteome profile of serum and pleural effusion of mesothelioma patients to identify unique proteins by observing concentration changes and modifications on the single protein level. This technique involves the preelectrophoretic labeling of complex protein samples using different cyanine-based fluorescent tags prior to carrying out separation by 2D polyacrylamide gel electrophoresis. Several proteins have been found to be differentially or uniquely expressed in the serum or pleural effusion of patients.

Key Words: 2D-DIGE; apolipoprotein A1; CyDyes; DeCyder; mesothelioma; pleural effusion; quantitative proteomics.

1. Introduction

1.1. Pleural Effusion

The human lung is surrounded by an outer parietal layer and inner visceral layer of pleura **(Fig. 1A)**. In the pleural cavity, which is the space between these two layers, a small amount of fluid is present under normal physiological conditions consisting of a few milliliters (approx 0.3 mL per kg body mass). The forces operating on the pleura with respect to movement of liquid are: (1) the oncotic pressure exerted by the blood in the pleural capillaries and by the liquid in the pleural space; and (2) the hydrostatic pressure within pleural capillaries and in the pleural space. The fluid is produced continuously and reabsorbed mainly through the lymphatic system ***(1,2)***. Its function is to reduce friction of the lungs during respiratory movements. In abnormal conditions, the pleural space can be filled with air, blood, plasma, serum, lymph, or pus. This expansion

From: *Proteomics of Human Body Fluids: Principles, Methods, and Applications*
Edited by: V. Thongboonkerd © Humana Press Inc., Totowa, NJ

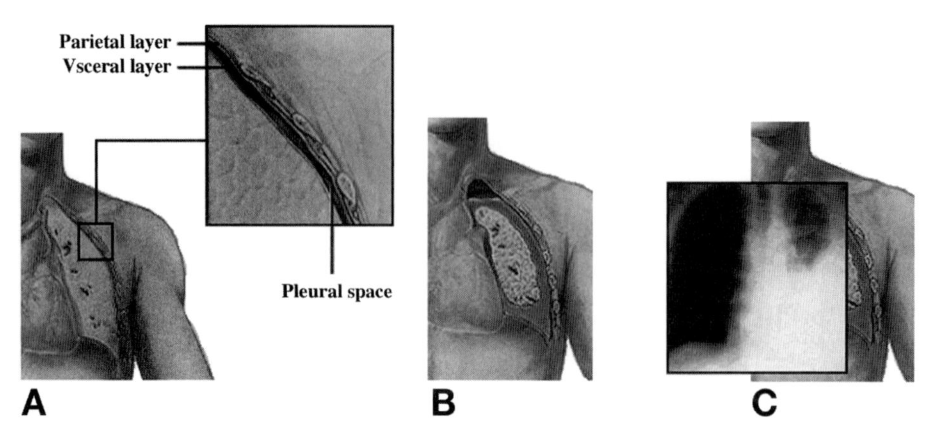

Fig. 1. (A) Human lung under normal conditions. **(B)** With the accumulation of pleural fluid between the two layers that line the lungs and chest cavity. **(C)** Chest radiography shows the presence of pleural effusion.

of the pleural space can compress the underlying tissue and causes partial collapse of the lung. There are two main types of pleural effusions: transudates and exudates. Pleural transudates pass membranes or squeeze through tissue into the extracellular space when imbalances in hydrostatic or oncotic pressures occur. In contrast, pleural exudates are slowly discharged from blood vessels as a result of an alteration in vascular permeability. Pleural effusions can be from several causes, such as congestive cardiac failure and low protein content in the blood, as in liver diseases, severe malnutrition, and certain kidney disorders. Physical trauma, infection, blockage of blood supply to the lung, and cancer can also result in accumulation of fluid in the pleural space **(Fig. 1B)**.

1.2. Diagnosing Pleural Effusions

Pleural effusions can usually be seen in an X-ray image of the chest **(Fig. 1C)**. For diagnostic purposes, the fluid is removed with fine needle aspiration inserted in the pleural cavity. Numerous biochemical criteria with different cutoff values have been used to separate pleural transudates from exudates, but most clinicians use the three criteria proposed by Light et al. *(3)* to determine whether a pleural fluid is an exudate. These criteria include: (1) a pleural fluid-to-serum protein ratio more than 0.5; (2) a pleural fluid-to-serum lactate dehydrogenase (LDH) ratio more than 0.6; and (3) pleural fluid LDH concentration more than 200 IU/I. Most transudates result from congestive heart failure, with the next most common cause being hepatic hydrothorax. Defining an effusion as a transudate narrows the differential diagnosis to a small number of disorders. It also ends the need for further diagnostic evaluation of the pleural effusion itself *(4)*. Exudates have a much larger differential diagnosis of over 50 causes, predominantly caused by infectious conditions, lymphatic abnormalities, inflammatory processes, and malignant conditions.

The most frequent etiology of malignant pleural effusion is bronchogenic carcinoma, which causes over one-third of all such cases. Other frequent causes of malignant pleural effusion include metastatic breast cancer, lymphoma, mesothelioma, gastric or esophageal cancer, and ovarian carcinoma. The diagnosis of a malignant pleural effusion is established by demonstrating malignant cells in the pleural fluid or in the pleura itself. Numerous papers have recommended various diagnostic tests, such as cytological and chromosomal analysis of pleural cells, measurement of pH, glucose, amylase, or measurement of proteins as carcinoembryonic antigen (CEA) or LDH in the effusions to discriminate malignant from nonmalignant pleural exudates. However, the diagnosis of disease based on a pleural effusion is often difficult, and to confirm the cause, or to rule out other possible causes, proteomics may be useful in indicating the pathogenic mechanism involved in the production of the effusion.

1.3. Complexity of Proteomic Studies

Proteins that are overexpressed and shed into pleural effusions have been studied for many decades for diagnosing the specific cause of their formation. We utilized proteomics to analyze pleural effusions in order to discover changes in expression of pleural proteins and to elucidate the basic molecular mechanisms that either cause, or result from, malignant mesothelioma. Malignant mesothelioma is a tumor of mesodermally derived tissue lining the coelomic cavities accompanied by diagnostic problems and for which no satisfactory curative treatment is available ***(5)***. Mesothelioma cells release proteins into the pleural effusion that may have diagnostic value as biomarkers on their own and will provide further insights into pathological processes. Ultimately, these proteins could be valuable in cancer research, e.g., as targets for the design of drug treatments. However, tumor-associated proteins are of low abundance and therefore difficult to detect. A low total-protein concentration, a high amount of albumin and immunoglobulins (IgG), and a wide dynamic range (several orders of magnitude) of protein concentration cause several difficulties in the identification of tumor-associated proteins. It is also apparent that, in most diseases, proteins are subjected to numerous changes including posttranslational modifications (PTMs) and/or proteolytic cleavage.

Proteins present in effusions are produced and secreted by many different types of human cells (e.g., inflammatory cells, tumor cells), each of which contains at least 2000 to 6000 different primary proteins ***(6,7)***. PTMs such as glycosylation, phosphorylation, acetylation, nitration, and ubiquitation multiply this number ***(8–11)***. Therefore, it is not surprising that no approaches currently come close in detecting all proteins present in complex biological systems.

Table 1
Characteristics of Differential Gel Electrophoresis (DIGE)

Advantages
Application of up to three samples on one 2D gel
Four orders of magnitude dynamic range and good correlation between spot density and protein content
Internal standard allows for quantitative comparison of multiple gels
Compatible with mass spectrometry
Disadvantages
Mass shift of approx 500 Daltons, impractical for subsequent MS (poststaining required especially for low-molecular-weight proteins)
Labeling dependent on lysine content
Gel spots only visible under fluorescent light, equipment required for visualization and spot excision

Often the highly expressed proteins ("housekeeping proteins") are detected, but the expression and modification changes of less abundant proteins may be most interesting. Therefore, the development of quantitative proteomics such as differential gel electrophoresis has widened the applicability for detecting proteins that undergo modifications in order to produce a phenotypic change.

1.4. Differential Gel Electrophoresis

A powerful quantitative technique currently available is differential gel electrophoresis (DIGE) ***(12–16)***. This technology is commercially available (GE Healthcare, formerly Amersham Biosciences). It has the potential to overcome many of the limitations of proteomic studies by allowing the direct comparison in a proteome profile of different samples at a particular time, under a particular set of conditions **(Table 1)** ***(17)***. DIGE encompasses a simple strategy involving three molecular weight- and charge-matched cyanine dyes (Cy2, Cy3, and Cy5) possessing unique absorption and emission spectra. The dyes are used to label fluorescently up to three different protein samples prior to mixing them together and running them simultaneously on the same 2D gel ***(18,19)***. The fluorescent dyes bind to the terminal amino group of lysine side chains in proteins with no change in protein charge and add only 0.5 kDa to the mass of the protein, thereby minimizing dye-induced shifting during electrophoresis. Because of a minimal labeling (only 2–5% of the total number of lysine residues are labeled), binding of the dye to the protein appears to have no effect on mass spectrometry analyses.

Two different samples are labeled with Cy3 and Cy5, and a third sample, labeled with Cy2, is introduced as an internal control for each gel. The internal control is often a pooled sample comprising equal amounts of each of the samples within the study. This allows normalization and both inter- and intragel matching of proteins and is imperative for accurate protein quantification. Once labeled, samples are mixed and isoelectrically focused on an immobilized pH gradient (IPG) strip and coelectrophoresed on a 2D polyacrylamide gel under denaturing conditions. Each dye is then scanned using different emission filters, and images are analyzed with DeCyder Differential In-gel Analysis software. This software allows accurate protein alignment and quantification between scanned images. Spots may be directly picked through an automated system. When DIGE is combined with mass spectrometry, proteins undergoing relevant changes in the context of development, pathology, and experimental manipulation can be detected and identified.

2. Methods

2.1. Sample Collection and Preparation

1. Serum and pleural effusion samples are collected from patients who present with large pleural effusions. Removal of effusion is performed to treat a patient's shortness of breath. Prior to the pleural fluid removal procedure, patients are given a local anesthetic (lidocaine 1%).
2. After a metallic needle is introduced into the pleural cavity, fluid is gently aspirated and collected in sterile tubes without anticoagulant or other additives.
3. Because many components of biological samples may interfere with analysis, they are removed before storage. Insoluble substances are removed by centrifugation at 400*g* and 4°C for 10 min.
4. The supernatant is then subjected to a second centrifugation at 3000*g* and 4°C for 20 min, and the resulting supernatant is stored in aliquots at −80°C until further analysis.
5. Abundant proteins, such as albumin and IgG, are removed by using the ProteoPrep Blue Albumin Depletion Kit (Sigma-Aldrich, St. Louis, MO), according to the manufacturer's instructions. In this procedure, the samples are applied on a highly specific albumin and IgG binding medium in small spin columns.
6. After incubation, nonbinding proteins are washed off in a small volume of washing buffer.
7. Samples are further purified by removal of salts, lipids, and other interfering substances and concentrated by using the 2D clean-up kit (GE Healthcare, Fairfield, CT).
8. Before applying samples to 2D electrophoresis, the protein concentration is determined using the 2D Quant kit (GE Healthcare).
   a. First, proteins are precipitated and then resuspended in a copper-containing solution.
   b. Unbound copper is visualized with a colorimetric agent, and the color density is measured on a spectrophotometer at wavelength 480 nm. The color density is thus inversely related to the protein concentration.
   c. This assay has the advantage that it is compatible with samples containing reagents that are often used in protein sample preparation, like detergents, reductants, chaotropes, and carrier ampholytes.

2.2. 2D Polyacrylamide Gel Electrophoresis (2D-PAGE)

2D-PAGE can be used to separate proteins in the samples in two ways, according to their isoelectric point or p*I* (first dimension) and to their molecular mass (second dimension).

1. For the first dimension, the samples are dissolved, after precipitation, in a rehydration buffer containing 8 *M* urea, 2% (w/v) 3-[(3-cholamidopropyl)dimethylamonio]-1-propanesulfonate (CHAPS), 0.5% (v/v) IPG buffer (pH 4–7) containing carrier ampholytes (GE Healthcare), and 2.8 mg/mL dithiothreitol (DTT).
2. Samples are applied onto 18-cm Immobiline Drystrips pH 4–7 (GE Healthcare) and focused using an IPGphor isoelectric focusing unit (GE Healthcare). This flatbed system uses thermally conductive ceramic stripholders with built-in platinum electrodes as focusing and rehydration chambers.
3. After rehydration of the strips for 12 h at 30 V, a program is run with these parameters: 500 V for 1 h; 1000 V for 1 h; slowly ramped up to 8000 V for 3 h; and then maintained at 8000 V for an additional 2 h to achieve more than 35,000 V/h.
4. Following focusing, the strips are incubated for 15 min in an equilibration buffer containing 6 *M* urea, 50 m*M* Tris-HCl, pH 8.8, 20% (v/v) glycerol, and 2% (w/v) sodium dodecyl sulfate (SDS), and 10 mg/mL DTT.
5. A second incubation of 15 min is performed in the same solution, but replacing DTT by 25 mg/mL iodoacetamide. This step alkylates the proteins, preventing reoxidation during electrophoresis.
6. After equilibration, the strips are transferred onto 12% uniform SDS polyacrylamide gels for resolution in the second dimension.
7. Gels are poured between low-fluorescent glass plates; the inner side of one plate is treated with Bind-Silane (GE Healthcare) so that the gel remains attached to this plate after electrophoresis and disassembly of the gel sandwich. This minimizes gel distortion and simplifies handling during subsequent fixing, staining, imaging, and automatic spot picking steps.
8. Gels are run on the Ettan Dalt system (GE Healthcare) at a constant power of 180 W at room temperature (RT) until the dye front reaches the bottom of the gel.

2.3. Gel Staining and Visualization

1. For visualization of the separated proteins, gels are stained using SYPRO Ruby protein gel stain (Molecular Probes, Leiden, The Netherlands) according to the manufacturer's instructions.
2. Gels are scanned on the Typhoon 9410 laser-scanner (GE Healthcare) using 457 nm as the excitation wavelength and 610 nm, band pass (BP) 30 nm, as the emission filter.

2.4. 2D-DIGE

2D-DIGE allows detection and quantitation of differences in protein abundances between different samples using cyanine dye labeling with spectrally resolvable CyDye DIGE Fluor minimal dyes (GE Healthcare). This technique has been used to detect differences in protein profiles of serum and pleural fluid of mesothelioma patients. For this procedure, samples are prepared and protein concentration is determined as described in **Subheading 2.1**.

1. Typically, 50 µg of sample is minimally labeled with 400 pmol of either Cy3 or Cy5 freshly dissolved in anhydrous dimethyl formamide.
2. Labeling reactions are performed on ice in the dark for 30 min and then quenched with an excess of free lysine (1 µL of 10 m*M* L-lysine solution [Sigma-Aldrich]) for 10 min on ice.
3. A pool of all samples is also prepared and labeled with Cy2 to be used as a standard on all gels to aid image matching and cross-gel statistical analysis.
4. Differentially labeled samples, which will be run on one 2D gel, are mixed, and 50 µg of Cy2-labeled pooled samples is added to 50 µg of Cy3-labeled sample and 50 µg of Cy5-labeled sample.
5. In preparative gels, 350 µg of pooled unlabeled proteins are additionally loaded.
6. The final volume is adjusted to 350 µL rehydration buffer, the volume needed for loading of an 18-cm IPG strip.
7. Isoelectric focusing and second-dimension gel electrophoresis are performed as described in **Subheading 2.2**.
8. Gels are scanned on a Typhoon 9410 laser scanner for visualization. The scanning settings used are: 100 µm resolution, PMT values between 500 and 520. The laser settings for the CyDyes (laser [nm], emission filter [nm]) are: Cy2 (488, 520BP40), Cy3 (532, 580BP30), and Cy5 (633, 670BP30).
9. After scanning, one glass plate is removed, and the gel is fixed in 10% (v/v) methanol, 7.5% (v/v) acetic acid overnight.
10. A poststaining is performed with SYPRO Ruby dye for 4 h at RT, and the gel is scanned at 457 nm with emission filter 610 nm with a band pass of 30 nm.
11. Gel analysis is performed using DeCyder DIA V5.02 for intragel comparison; for intergel matching, the DeCyder BVA V5.02 software (GE Healthcare) is used.

2.5. Automatic Spot Picking

Fluorescently stained protein spots of interest are excised from 2D gels using an automated Ettan spot picker (GE Healthcare) **(Fig. 2)** following the manufacturer's instructions. This robotic system automatically picks selected protein spots from gels, using a pick list from the image analysis, and transfers them onto 96-well low-protein-binding microplates (Nunc, Roskilde, Denmark).

2.6. In-Gel Tryptic Digestion

1. The excised plugs are washed with milli-Q water for 5 min.
2. Then gel plugs are alternately treated twice for 5 min each with Milli-Q water and acetonitrile.
3. Gel plugs are dried in a rotary evaporator (Savant, Farmingdale, NY) for 30 min.
4. Proteins are digested overnight at RT in 4 µL of 100 µg/mL sequencing grade modified trypsin (Promega, Madison, WI).

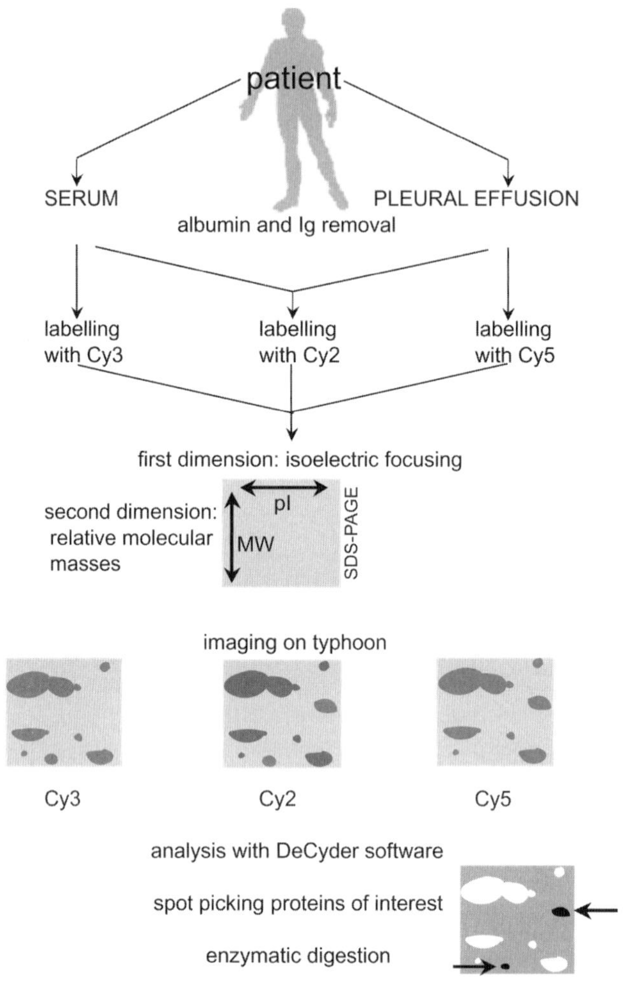

Fig. 2. Flowchart of DIGE analysis of serum and malignant pleural fluid derived from a patient suffering from malignant mesothelioma. Samples to be compared are labeled with either Cy3 or Cy5, whereas Cy2 is employed to label a pooled sample comprising equal amounts of serum and effusion within the study. The labeled samples are combined and then run on a single 2D gel. Proteins are detected using a dual laser-scanning device equipped with different excitation/emission filters in order to generate three separate images. The images are matched by a computer-assisted overlay method, signals are normalized using the corresponding Cy2 spot intensities, and spots of interest are excised and analyzed by mass spectrometry. Differentially expressed proteins in pleural effusions can be useful in diagnosis or for the detection of biomarkers in cancer.

2.7. Matrix-Assisted Laser Desorption/Ionization Time-of-Flight Mass Spectrometry (MALDI-TOF MS)

1. After digestion, 7 μL of 0.7% trifluoroacetic acid in 30% acetonitrile is added to the gel plugs.
2. Then 1 μL of this mixture is then added to 2 μL of 2 mg/mL ionization enhancing material, α-cyano-4-hydroxy-trans-cinnamic acid (Bruker Daltonics, Billerica, MA), in acetonitrile.
3. Of the sample-matrix mixture, 1 μL is applied onto a 600 μm 384-spot metal anchor chip plate and crystallized in air.
4. Peptide mass spectra are acquired on a MALDI-TOF mass spectrometer equipped with a 337-nm nitrogen laser (ULTRAFLEX, Bruker Daltonics).
5. The instrument is calibrated with a peptide calibration standard (Bruker Daltonics).

2.8. Database Search

1. Peptide mass fingerprinting is based on mass measurement of peptide fragments derived from a single protein, digested with trypsin.
2. The anticipated mass values of peptides in virtual digests of all proteins are calculated and listed in the MSDB database of the National Center for Biotechnology Information (NCBI).
3. A mass list of peptides, from each sample analyzed on the MALDI-TOF MS, is generated in the Flexanalysis software (Bruker Daltonics). Peaks obtained from autolytic fragments of trypsin are omitted from the spectra.
4. The mass lists are submitted to the Mascot software (Matrix Science, UK) to identify the proteins in the MSDB database.
5. The criteria used for the search are as follows: (1) maximum allowed peptide mass error of 200 ppm; (2) at least five matching peptide masses; (3) molecular weight of identified protein should match estimated values by comparing with marker proteins; and (4) top-scores given by software should be higher than 61 ($p < 0.05$).

2.9. Western Blotting

1. For detection of apolipoprotein spots in the 2D-PAGE, proteins are electroblotted onto Immobilon-P membranes (Millipore, Billerica, MA) with the Criterion Blotter (Bio-Rad, Hemel Hempstead, UK).
2. To saturate nonspecific protein binding sites, the membranes are incubated for 1 h with blocking buffer (TTBS containing 5% [w/v] low fat milk powder) and then incubated overnight with 1:10,000 diluted rabbit antihuman apolipoprotein A1 (Calbiochem, San Diego, CA) in blocking buffer.
3. Blots are washed and incubated with 1:1000 diluted horseradish peroxidase-conjugated swine antirabbit antibody (DAKO, Glostrup, Denmark) and visualized by the SuperSignal West Pico chemiluminescent substrate (Pierce Perbio, Rockford, IL) according to the manufacturer's instructions.

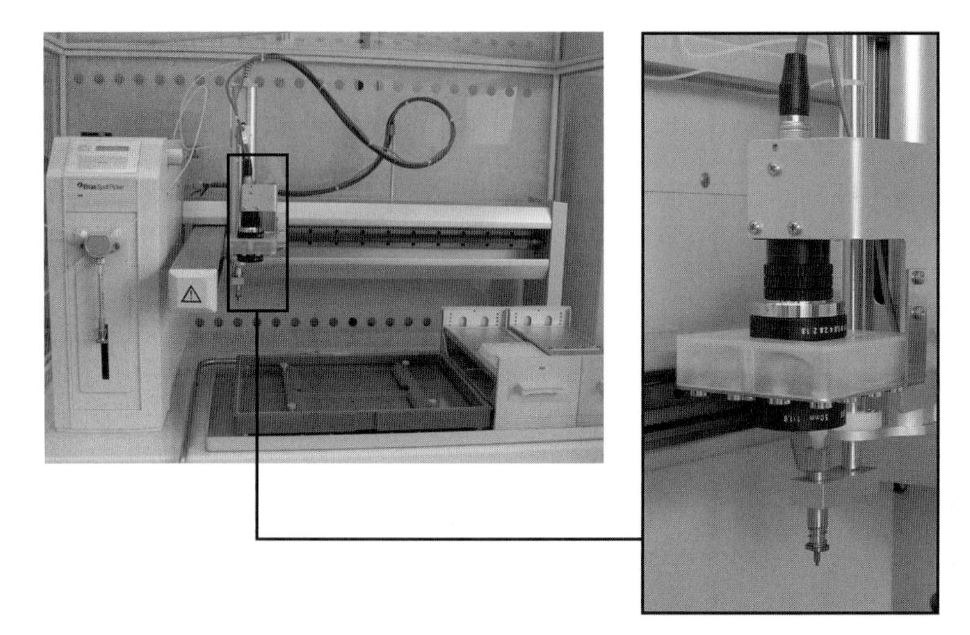

Fig. 3. The automated Ettan spot picker instrument (GE Healthcare) with the picker head enlarged (right).

3. Findings and Applications

Evaluation of pleural effusion is useful for improving the diagnosis and research of several inflammatory diseases and malignancies. This section describes a strategy for the comparative analysis of serum proteome and pleural effusion proteome to elucidate the basic molecular mechanisms that either cause, or result from, cancerous disorders. Pleural effusion and serum obtained from the same patients with malignant mesothelioma have been analyzed using a strategy that combined analytical techniques in electrophoresis, mass spectrometry, and Western blotting **(Fig. 3)**. The resulting data provide fundamental information on the composition and difference of protein contents in these body fluids.

Serum and pleural effusions were collected according to special procedures for a reliable and consistent sample collection. Note that many factors (such as time of collection, containers used, preservatives and other additives, transport to the laboratory, and storage) affect the quality of the samples, and the stability of the proteins of interest must be considered at the initial collection stage. For comparative studies between patients, it is crucial that body fluids be handled and stored consistently throughout the study. Pleural effusions were separated by 2D-PAGE using isoelectric focusing (IEF) as the first-dimensional separation followed by SDS-PAGE for the second dimension. Because most

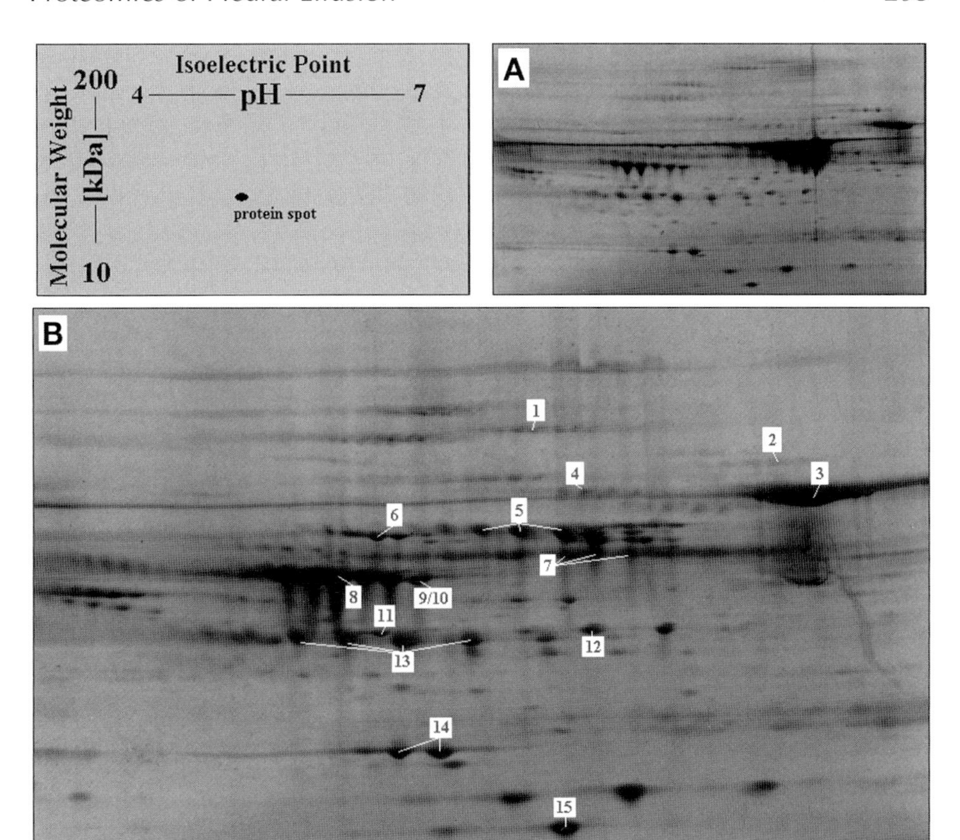

Fig. 4. Human pleural effusion was processed with the ProteoPrep albumin/IgG removal kit, and proteins from the unbound fractions were collected. Pleural proteins (50 μg) from a cancer patient were resolved by 2DE using a pH 4 to 7 gradient IPG strip in the first dimension and a 4 to 12% gradient polyacrylamide SDS gel in the second dimension. Gels were stained with SYPRO Ruby and imaged using a Typhoon 9410. 2D map using equal protein amounts for crude human pleural effusion **(A)** and after depletion of albumin and IgG **(B)** were obtained. Numbers correspond to the excised protein spot and were analyzed by mass spectrometry (*see* **Table 2**).

proteins focus within the pH 4 to 7 range, narrow-range IPG strips were used. **Figure 4A** demonstrates a typical 2D proteome profile of pleural effusion visualized by SYPRO Ruby staining.

More than 300 individual protein spots have been detected in the molecular mass range of 20 to 200 kDa. The albumin smear at around 67 kDa and IgG fragments are the major protein components of pleural effusions, representing 50 to 70% and 10 to 20% of the total protein in pleural effusions, respectively (mg/mL range). Together with transferrin, fibrinogen, complement components, and a few other proteins, the top 20 proteins are responsible for approximately 99% of the total protein mass. Pleural effusions contain comparatively small quantities of cytokines detected by cytokine arrays and enzyme-linked immunosorbent assay (ELISA). These include transforming growth factor-β (TGF-β), interleukins (IL-1, IL-6, IL-8, and IL-10), and vascular endothelial growth factor (VEGF), which are in the ng/mL to pg/mL range, a difference of 9 orders of magnitude or more. Therefore, removal of abundantly expressed proteins is a key element of proteome research to allow the visualization of comigrating proteins on 1D and 2D gels and to allow a higher sample load for improved visualization of lower copy number proteins. A convenient approach to remove high-abundance proteins from body fluids is affinity chromatography with resins carrying highly efficient and specific ligands for these proteins. Removal of albumin and IgG using the commercially available ProteoPrep kit (Sigma-Aldrich) clearly improves resolution and increases spot count in depleted samples **(Fig. 4B)**.

Mass spectrometry can be used to identify proteins in a sample by providing the molecular mass to electric charge (m/z ratio) of peptides in the femtomole to attomole range with an accuracy of less than 10 ppm. With the completion of the Human Genome Project *(20)*, it is now possible to identify proteins by using search algorithms that interrogate protein sequence databases in an automated fashion. A panel of 15 spots (labeled in **Fig. 4B**) has been selected as an example and subjected to protein identification by trypsin digestion, MS analysis, and database search **(Table 2)**. All these proteins have been described to be present in pleural fluid and are likely to originate from serum *(21–27)*. The large number of spots in a 2D gel is partly owing to posttranslational and proteolytic modifications of proteins; one protein may, therefore, be present in several locations in the gel **(Fig. 4B)**.

This approach has recently been published for a composite pleural effusion sample from seven lung adenocarcinoma patients. This study has revealed at least 472 silver-stained protein spots to be present in a 2D map, half of which could be identified by liquid chromatography-tandem MS *(21)*. Although the results of these studies provide information for a basic understanding of the protein composition of pleural effusions, the value for clinical medicine is limited. The approach is time consuming, and the interpretation of results is hampered by additional factors introduced by variables in experimental parameters. For example, difficulties in the detection of low-abundance proteins due to limitations in dynamic range, diversity within a complex biological sample, and the typical changes associated with different causes of effusion as well as timing are variables that impede the interpretation.

Many of the proteins present in pleural fluid are likely to originate from serum. Of interest are the proteins that have not previously been reported in the literature to be present in serum. These proteins can originate from infiltrating

Table 2
Identified Proteins From the Human Pleural Effusion Proteome

No.	Protein	Accession no.	Nominal mass (kDa)	Coverage (%)	Score[a]
1	Complement factor H	NBHUH	139.034	13	95
2	Complement factor B	BBHU	86.847	27	152
3	Transferrin	TFHUP	77.000	36	182
4	Fibrinogen precursor	FGHUA	69.756	40	111
5	Hemopexin	CAA26382	51.643	18	93
6	α-1B glycoprotein	OMHU1B	52.479	38	100
7	Fc-α receptor I	MGC27165	53.358	17	104
8	α_1-antitrypsin	1THU	46.707	56	209
9	Vitamin D-binding protein	AAA61704	54.513	60	228
10	Antithrombin	1AZXI	47.656	35	85
11	Apolipoprotein A4	Q13784	28.141	38	68
12	Chain of fibrinogen	1FZAB	36.331	81	267
13	Haptoglobin	HPHU1	38.941	36	133
14	Apolipoprotein A1	CAA00975	28.061	69	192
15	Transthyretin	2ROXA	12.996	86	124

[a]Scores higher than 61 are significant ($p < 0.05$).

cells or from the parenchymal interstitial linings of the lung. They also represent potential candidates for useful biomarkers that are concentrated or only measurable in pleural effusions. To discover the proteins of interest, we have employed a strategy of comparative analysis of serum proteome and pleural effusion proteome from the same mesothelioma patient using the DIGE technology (**Fig. 3**). Comparing body fluids of the same patient gives less individual variation, as the genetic component is no longer a variable, and samples are completely matched for age, sex, ethnic origin, and other parameters (smoking, alcohol consumption, medication, and many others). The spectrally distinct dyes allow coseparation of different CyDye DIGE fluorescence-labeled samples in the same gel and ensure that all samples will be subjected to exactly the same electrophoretic running conditions. This limits the experimental variation and thus ensures accuracy within gel matching.

Samples were differentially labeled with spectrally different fluorescent dyes: serum was labeled with the cyanine dye Cy3 and pleural fluid with cyanine dye Cy5. A Cy2-labeled pool of samples was used as the standard with all Cy3- and Cy5-labeled sample pairs to facilitate cross-gel quantitative analysis. Once labeled, all samples (Cy2, Cy3, and Cy5) were mixed. For mass spectrometry, a preparative gel was run. Therefore, 350 μg unlabeled pooled sample was added to the CyDye-labeled mixture. The protein mixture was isoelectrically

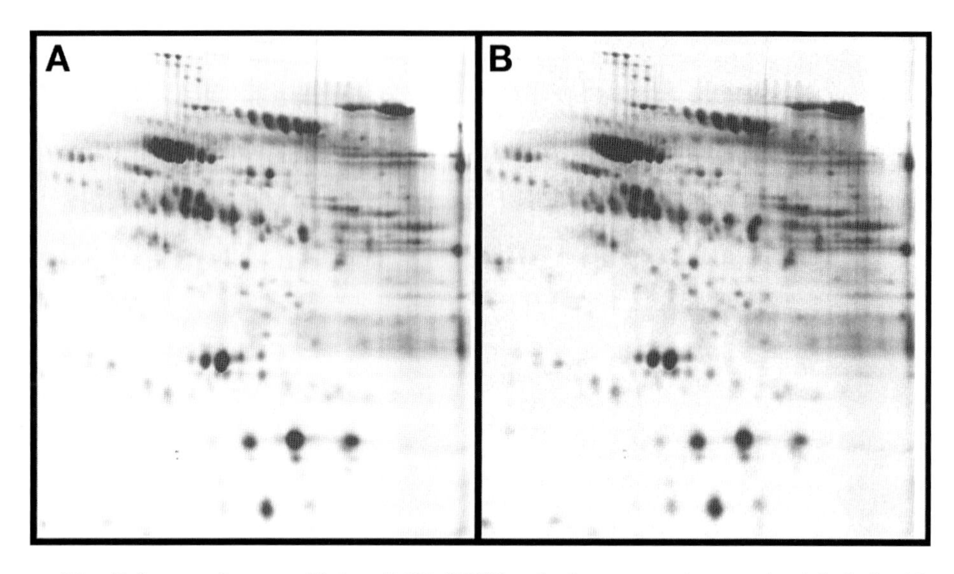

Fig. 5. Images from a pH 4 to 7 2D-DIGE gel of two protein samples labeled with minimal CyDye DIGE fluors. **(A)** The Cy3 image corresponds to serum. **(B)** The Cy5 image corresponds to pleural effusion.

focused on an IPG strip and coelectrophoresed on a 2D polyacrylamide gel. Each dye was then scanned using a Typhoon gel scanner equipped with different emission filters, and quantitation of differential protein expression was analyzed with DeCyder DIGE Analysis software (**Fig. 5**).

In DeCyder analysis, the Cy2, Cy3, and Cy5 images were merged for each gel, and spot boundaries were detected for the calculation of normalized spot volumes or protein abundances. At this stage, dust particles, scratches, and other features resulting from nonprotein sources were filtered out. The analysis was performed to calculate abundance differences rapidly between serum and effusion run in the same gel. The 2D-DIGE map is shown in **Fig. 5**, and 1436 spots of various intensities have been detected by the software in this map. Statistical analysis was performed on each protein spot with a twofold-change criterion of significant difference. Most of the spots (1304 spots, 90.8% of total) showed small changes between effusion and serum. The normalized volumes of 76 spots (5.3%) were decreased, whereas 56 spots (3.9%) had volumes increased by more than twofold, i.e., the change in protein level in pleural effusion compared with serum proteins (**Fig. 6**). To ensure consistency in the observed DIGE profile, the whole experiment was repeated a second time to eliminate confounding factors that may arise during the practical procedure. Although the total amount of spots detected differed, there was no substantial variation in the profile of the protein spots when the whole experiment was

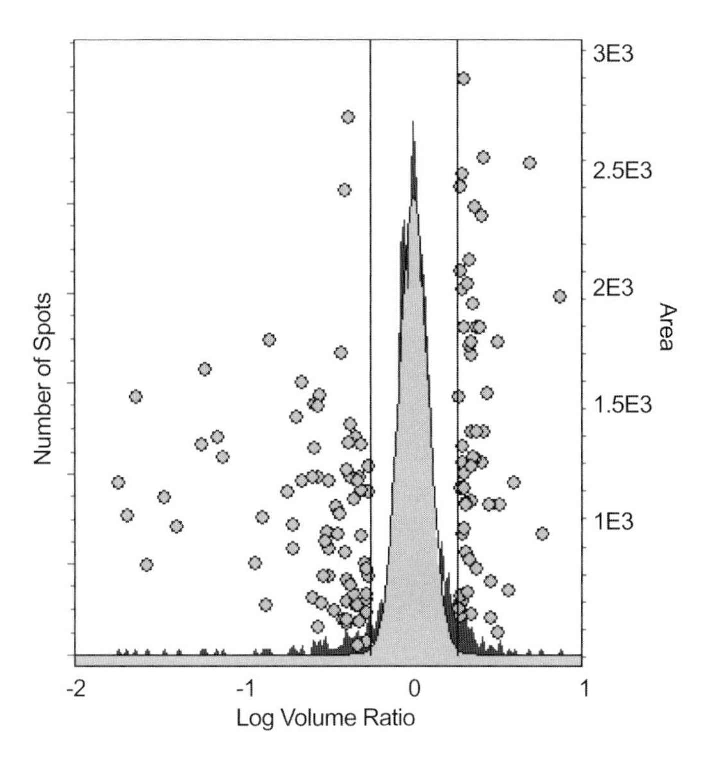

Fig. 6. DeCyder software statistical output of twofold changed protein spots. Dots on the left were decreased in pleural effusion, whereas dots on the right were increased in effusion compared with serum of the same patient. Spots could be highlighted individually for detailed information display, e.g., 3D intensity view and table view of quantitative data.

repeated using the same samples on a different day. Expression differences identified by 2D DIGE can therefore be confidently assigned to biological differences and are not caused by system variability. Serum and pleural effusion from other mesothelioma patients were analyzed to confirm consistency and relevancy of the differentially expressed proteins in serum and pleural effusion using Cy2-labeled pooled sample as a cross-gel standard. Every difference has been confirmed by a confidence value. The DeCyder BVA software allows protein alignment and quantification between scanned images.

Overexpression of proteins by mesothelioma cells can result from their shedding into the pleural effusion and will lead to enhanced-intensity spots compared with serum of the same patient. Absence of proteins in the effusion may be caused by specific proteolysis or by specific absorption from the circulation by tumor cells. A protein spot with the approximate molecular mass of 30 kDa and a p*I* of 5.5, significantly expressed in the serum but not in the effusion, was

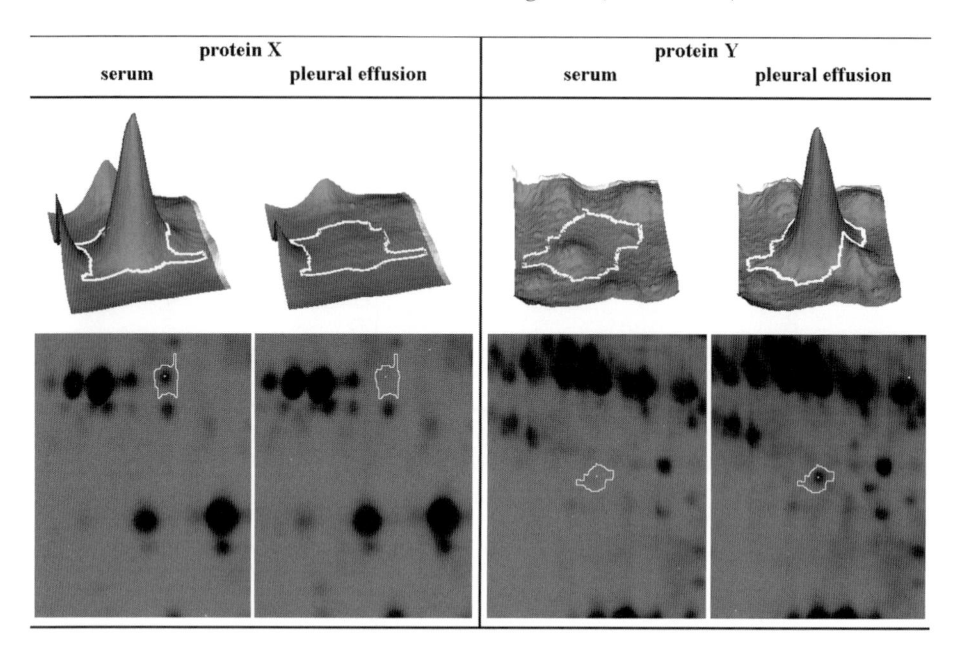

Fig. 7. DeCyder software output for a 3D fluorescence intensity profile of the circled spots in the magnified gel image with Cy3-labeled serum and Cy5-labeled pleural effusion. 3D simulation of the protein spots allows an objective view for the comparison of spot intensity between the two images. Protein X was decreased, whereas protein Y was increased in pleural effusion compared with serum from the same patient.

selected for identification and further analyses (**Fig. 7**, protein X). Selection of proteins for excision requires poststaining of gels with SYPRO Ruby because most of the protein will not exactly comigrate with the CyDye-labeled protein. (Dyes add approx 0.5 kDa to the total molecular mass.) Thus, the SYPRO Ruby-stained proteins of interest, rather than the CyDye-labeled protein spots, were excised from the gel through an automated system using a pick list (**Fig. 8**). When DIGE is combined with mass spectrometry, proteins undergoing relevant changes in the context of development, pathology, and experimental manipulation can be detected and identified.

The resultant mass fingerprinting spectra of the tryptic digest were used for protein search against the MSDB database of the NCBI. Nineteen matched peptides with a total coverage of 75% were the bases of the identification of the decreased spot in effusions as apolipoprotein A1 (accession number CAA00975; with a molecular mass of 28.061 Da and a p*I* of 5.27). This spot (isoform I) migrated differently from the major apolipoprotein A1 spots (indicated by arrows in **Fig. 9**) and represented a small fraction of the total serum

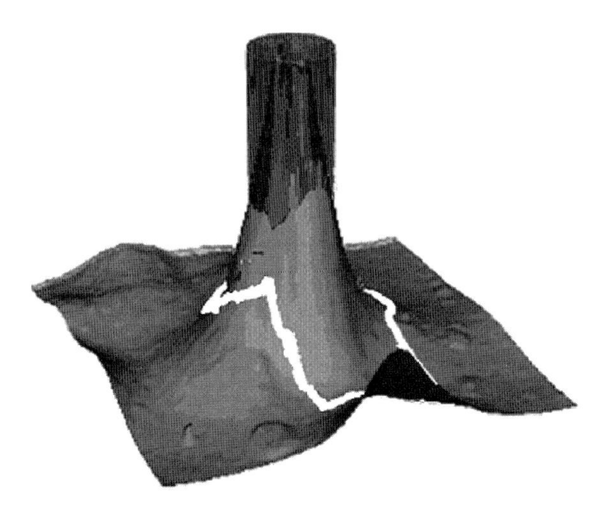

Fig. 8. 3D intensity plot of a protein selected for MS identification showing the area used in the analysis (white line) and the selected area for automated spot picking (circular volume).

apolipoprotein A1 but was absent in pleural effusion. This directly illustrates the advantage of 2D gel-based approaches in visualizing changes in the molecular weight and p*I* of a protein. The different p*I* and slightly different molecular mass reflect biological significant processing and charge-altered PTMs, possibly owing to phosphorylation, sulfation, or (de)acetylation. Thus, comparison of the protein spots from serum and effusion by 2D-DIGE provides a striking quantitative picture of proteins absorbed or shed into body fluids.

The identity of apolipoprotein A1 has been further confirmed by 2D Western blot analysis using a rabbit antihuman apolipoprotein A1 **(Fig. 9)**. The smaller isoform (isoform II) was not detected by Western blot analysis. However, this isoform was identified by MS with a total of 13 matched peptides that covered 57% of full-length, unprocessed apolipoprotein A1, as shown in **Fig. 10**. The matched peptides were clustered at the C-terminal region of the protein. A cleavage site before amino acid 27 could produce a protein with a smaller molecular weight, consistent with the different migration in the 2D gel. The absence of the smaller isoform II in Western blotting was probably caused by the N-terminal binding of the antibody to apolipoprotein A1. This truncation product has not been reported previously, and it is not known whether the fragmentation was owing to in vivo biological processing or protease activity. It may be the product of cleavage by one or more proteases, including kallikrein or matrix metalloproteases. These results highlight one of the advantages of 2D-DIGE, as the proteins are separated according to their p*I* and molecular masses.

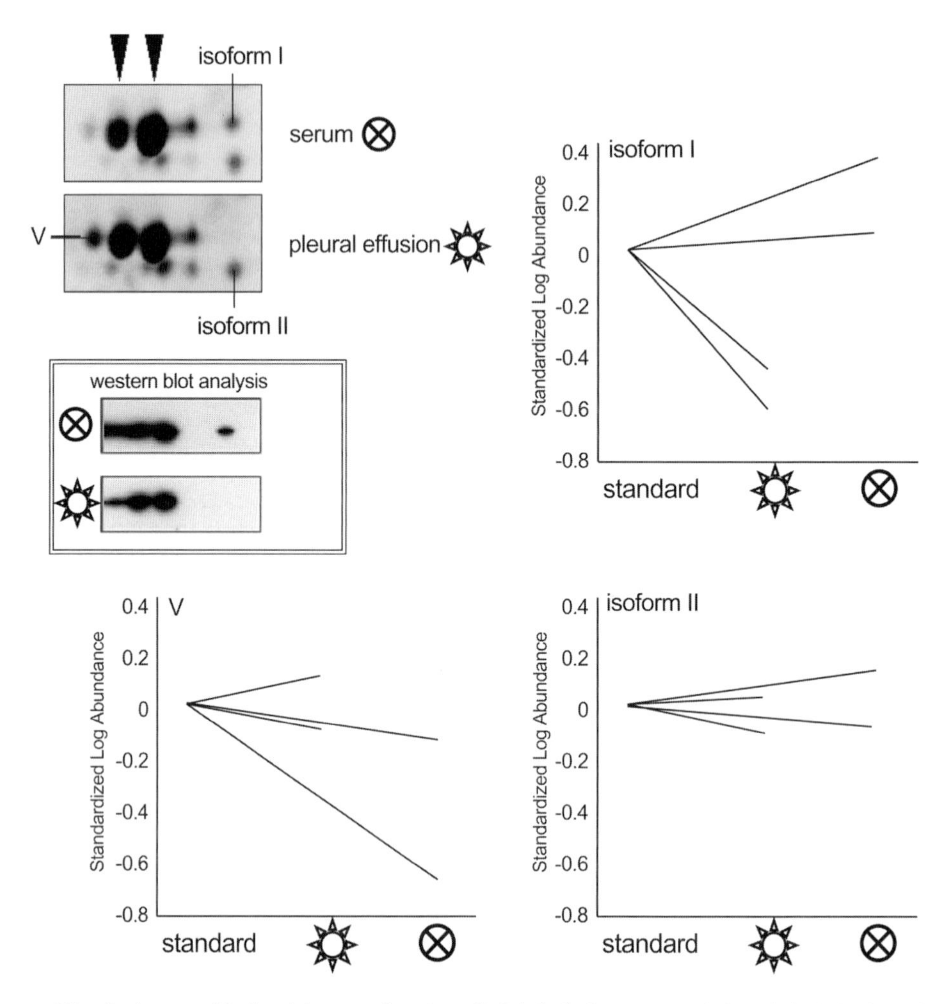

Fig. 9. A magnified gel image showing Cy3-labeled serum and Cy5-labeled pleural effusion. Isoform I represented only a small fraction of the total serum apolipoprotein A1 (arrowheads, upper left) but was not detectable in pleural effusion. This was confirmed by 2D Western blot analysis. DyCyder BVA output illustrates graphs of standardized log abundances (*y*-axis) for the isoforms (I and II) and protein V in sera (group 2; squares) and pleural effusions (group 1; dots) after intra- and intergel matching.

With regard to the apolipoprotein A1 isoform I identified in this study, it is difficult to speculate on the physiological meaning of the change in their relative abundance between the serum and pleural effusion of the same patient. Reduction in the serum levels of apolipoprotein A1 has been correlated with hepatitis B virus-induced diseases ***(28–30)***. An isoform of apolipoprotein A1 has been detected by 2D-PAGE of serum obtained from individuals with high risk for the

Apolipoprotein A1 isoform I

1	**DEPPQSPWDR**	VK**DLATVYVD**	**VLK**DSGR**DYV**	**SQFEGSALGK**	QLNLK**LLDNW**
51	**DSVTSTFSK**L	R**EQLGPVTQE**	**FWDNLEKETE**	**GLR**QEMSKDL	EEVKAK**VQPY**
101	**LDDFQK**K**WQE**	**EMELYR**QKVE	PLR**AELQEGA**	**RQKLHELQEK**	**LSPLGEEMR**D
151	RAR**AHVDALR**	**THLAPYSDEL**	**R**QLAARLEA	LKENGGAR**LA**	**EYHAKATEHL**
201	**STLSEKAKPA**	**LEDLRQGLLP**	**VLESFKVSFL**	**SALEEYTK**KL	NTQ

Apolipoprotein A1 isoform II

1	*DEPPQSPWDR*	*VKDLATVYVD*	*VLK*DSGR**DYV**	**SQFEGSALGK**	QLNLK**LLDNW**
51	**DSVTSTFSK**L	R**EQLGPVTQE**	**FWDNLEKETE**	**GLR**QEMSKDL	EEVKAK**VQPY**
101	**LDDFQK**K**WQE**	**EMELYR**QKVE	PLR**AELQEGA**	**RQKLHELQEK**	**LSPLGEEMR**D
151	RAR**AHVDALR**	**THLAPYSDEL**	**R**QLAARLEA	LKENGGAR**LA**	**EYHAKATEHL**
201	**STLSEKAKPA**	**LEDLRQGLLP**	**VLESFKVSFL**	**SALEEYTK**KL	NTQ

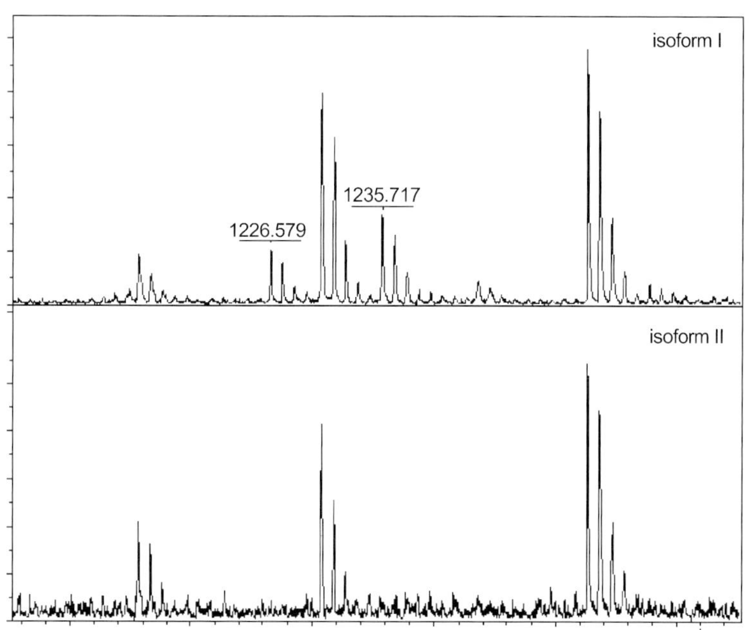

Fig. 10. Identification of apolipoprotein A1 by MS. The amino acid sequences of apolipoprotein A1 isoforms are shown. Peptides derived from isoforms I and II were matched by peptide mass fingerprinting, printed in bold and underlined. The peptides DEPPQSPWDR (with *m/z* 1226.579) and DLATVYVDVLK (with *m/z* 1235.717), which were identified in isoform I, were not detected in isoform II (bold in the spectral illustration and sequence alignment for isoform I, but italicized in the sequence alignment for isoform II).

development of or who were diagnosed with hepatocellular carcinoma *(31)*. Apolipoprotein A1 is a potential marker of the aggression in colonic adenocarcinoma *(32)* and is upregulated in primary carcinoma tissue of the vagina *(33)*. However, a downregulation of apolipoprotein A1 in serum is described in early-stage ovarian cancer *(34,35)*. The isoform described in this study could be induced by PTMs, but experiments using both biochemical and biological approaches are necessary to assess further the role of apolipoprotein A1. PTMs (e.g., phosphorylation, glycosylation) play a crucial role in cell signaling and protein function ***(8)***, and more than 200 different protein modifications have already been described ***(10,11)***. We are currently trying to identify the additional proteins whose expression is significantly altered, as revealed by DeCyder analysis.

4. Perspectives and Future Directions

Novel proteomic methods have the advantage that the results are less biased by the theories or beliefs of the investigators. The proteomic technologies are only limited by the sensitivity of the methods; they can give rise to new discoveries and can generate new hypotheses. These are contrary to the traditional reductionist, one-stimulus, one-protein investigations, e.g., an ELISA or Western blotting, in which the researchers have to decide beforehand on which antibodies to be used. 2D-DIGE technology is gaining acceptance in the field of proteomics and has enabled the detection of more subtle changes in protein expression than the conventional 2D-PAGE ***(36)***. Quantitative comparison by 2D-DIGE and proteome profiling techniques allow the rapid comparison of different complex samples, making a study of specific diseases or biological processes under clinically relevant conditions possible. Recent DIGE studies comparing the difference between normal and cancerous tissue have been successful for demonstrating changes in protein expression levels ***(37–39)***.

We have provided a brief example of comparing serum and pleural effusion of the same mesothelioma patient to identify unique proteins by observing concentration changes and modifications on a single protein level. Several proteins have been found to be differentially or uniquely expressed in the serum or pleural effusion. In our study, the number of patients is too small to permit conclusive associations between disease biomarkers and the unique mode of biosynthesis as well as processing of, e.g., isoform I of apolipoprotein A1. Nonetheless, the results are provocative, and further work into the function and expression of this protein and other differentially expressed proteins (either up- or downregulated) is thus suggested. New possibilities with the DIGE technology can range from single variable comparisons to complex multivariable comparisons and/or time-course studies (e.g., transudates vs exudates, mesothelioma effusions vs effusions of other origins).

In conclusion, we show that 2D-DIGE with DeCyder analysis is a sensitive, MS-compatible technique for identifying statistically significant differences in protein expression profiles of multiple samples. The ongoing rapid development in separation techniques, quantitative analysis, mass spectrometry, and bioinformatics will continue to stimulate the investigation of pleural effusions and will lead to new insights into the mechanisms of disease in the near future.

The area of clinical proteomics and the study of the disease mechanism will have a major impact on the way diseases are diagnosed and treated.

References

1. Noppen M, De Waele M, Li R, et al. Volume and cellular content of normal pleural fluid in humans examined by pleural lavage. *Am J Respir Crit Care Med* 2000; 162:1023–1026.
2. Miserocchi G. Physiology and pathophysiology of pleural fluid turnover. *Eur Respir J* 1997;10:219–225.
3. Light RW, Macgregor MI, Luchsinger PC, Ball WC Jr. Pleural effusions: the diagnostic separation of transudates and exudates. *Ann Intern Med* 1972;77: 507–513.
4. Bartter T, Santarelli R, Akers SM, Pratter MR. The evaluation of pleural effusion. *Chest* 1994;106:1209–1214.
5. Hoogsteden HC, Langerak AW, van der Kwast TH, Versnel MA, van Gelder T. Malignant pleural mesothelioma. *Crit Rev Oncol Hematol* 1997;25:97–126.
6. Celis JE, Ostergaard M, Jensen NA, Gromova I, Rasmussen HH, Gromov P. Human and mouse proteomic databases: novel resources in the protein universe. *FEBS Lett* 1998;430:64–72.
7. Duncan R, McConkey EH. How many proteins are there in a typical mammalian cell? *Clin Chem* 1982;28:749–755.
8. Mann M, Jensen ON. Proteomic analysis of post-translational modifications. *Nat Biotechnol* 2003;21:255–261.
9. Miklos GL, Maleszka R. Protein functions and biological contexts. *Proteomics* 2001;1:169–178.
10. Wold F. In vivo chemical modification of proteins (post-translational modification). *Annu Rev Biochem* 1981;50:783–814.
11. Wold F, Moldave K. A short stroll through the posttranslational zoo. *Methods Enzymol* 1984;107:xiii–xvi.
12. Tonge R, Shaw J, Middleton B, et al. Validation and development of fluorescence two-dimensional differential gel electrophoresis proteomics technology. *Proteomics* 2001;1:377–396.
13. Von Eggeling F, Gawriljuk A, Fiedler W, et al. Fluorescent dual colour 2D-protein gel electrophoresis for rapid detection of differences in protein pattern with standard image analysis software. *Int J Mol Med* 2001;8:373–377.
14. Yan JX, Devenish AT, Wait R, Stone T, Lewis S, Fowler S. Fluorescence two-dimensional difference gel electrophoresis and mass spectrometry based proteomic analysis of *Escherichia coli*. *Proteomics* 2002;2:1682–1698.
15. Nordvarg H, Flensburg J, Ronn O, et al. A proteomics approach to the study of absorption, distribution, metabolism, excretion, and toxicity. *J Biomol Tech* 2004; 15:265–275.
16. Alban A, David SO, Bjorkesten L, et al. A novel experimental design for comparative two-dimensional gel analysis: two-dimensional difference gel electrophoresis incorporating a pooled internal standard. *Proteomics* 2003;3:36–44.

17. Unlu M, Morgan ME, Minden JS. Difference gel electrophoresis: a single gel method for detecting changes in protein extracts. *Electrophoresis* 1997;18: 2071–2077.
18. Patton WF. Detection technologies in proteome analysis. *J Chromatogr B Analyt Technol Biomed Life Sci* 2002;771:3–31.
19. Lilley KS, Friedman DB. All about DIGE: quantification technology for differential-display 2D-gel proteomics. *Expert Rev Proteomics* 2004;1:401–409.
20. Venter JC, Adams MD, Myers EW, et al. The sequence of the human genome. *Science* 2001;291:1304–1351.
21. Tyan YC, Wu HY, Su WC, Chen PW, Liao PC. Proteomic analysis of human pleural effusion. *Proteomics* 2005;5:1062–1074.
22. Pieper R, Gatlin CL, Makusky AJ, et al. The human serum proteome: display of nearly 3700 chromatographically separated protein spots on two-dimensional electrophoresis gels and identification of 325 distinct proteins. *Proteomics* 2003; 3:1345–1364.
23. Sloane AJ, Duff JL, Wilson NL, et al. High throughput peptide mass fingerprinting and protein macroarray analysis using chemical printing strategies. *Mol Cell Proteomics* 2002;1:490–499.
24. Wu SL, Amato H, Biringer R, Choudhary G, Shieh P, Hancock WS. Targeted proteomics of low-level proteins in human plasma by LC/MSn: using human growth hormone as a model system. *J Proteome Res* 2002;1:459–465.
25. Choudhary G, Wu SL, Shieh P, Hancock WS. Multiple enzymatic digestion for enhanced sequence coverage of proteins in complex proteomic mixtures using capillary LC with ion trap MS/MS. *J Proteome Res* 2003;2:59–67.
26. Sanchez JC, Appel RD, Golaz O, et al. Inside SWISS-2DPAGE database. *Electrophoresis* 1995;16:1131–1151.
27. Adkins JN, Varnum SM, Auberry KJ, et al. Toward a human blood serum proteome: analysis by multidimensional separation coupled with mass spectrometry. *Mol Cell Proteomics* 2002;1:947–955.
28. Nayak SS, Kamath SS, Kundaje GN, Aroor AR. Diagnostic significance of estimation of serum apolipoprotein A along with alpha-fetoprotein in alcoholic cirrhosis and hepatocellular carcinoma patients. *Clin Chim Acta* 1988;173: 157–164.
29. Matsuura T, Koga S, Ibayashi H. Increased proportion of proapolipoprotein A-I in HDL from patients with liver cirrhosis and hepatitis. *Gastroenterol Jpn* 1988;23:394–400.
30. Fujii S, Koga S, Shono T, Yamamoto K, Ibayashi H. Serum apoprotein A-I and A-II levels in liver diseases and cholestasis. *Clin Chim Acta* 1981;115:321–331.
31. Steel LF, Shumpert D, Trotter M, et al. A strategy for the comparative analysis of serum proteomes for the discovery of biomarkers for hepatocellular carcinoma. *Proteomics* 2003;3:601–609.
32. Tachibana M, Ohkura Y, Kobayashi Y, et al. Expression of apolipoprotein A1 in colonic adenocarcinoma. *Anticancer Res* 2003;23:4161–4167.

33. Hellman K, Alaiya AA, Schedvins K, Steinberg W, Hellstrom AC, Auer G. Protein expression patterns in primary carcinoma of the vagina. *Br J Cancer* 2004;91:319–326.
34. Zhang Z, Bast RC Jr, Yu Y, et al. Three biomarkers identified from serum proteomic analysis for the detection of early stage ovarian cancer. *Cancer Res* 2004; 64:5882–5890.
35. Kuesel AC, Kroft T, Prefontaine M, Smith IC. Lipoprotein(a) and CA125 levels in the plasma of patients with benign and malignant ovarian disease. *Int J Cancer* 1992;52:341–346.
36. Gade D, Thiermann J, Markowsky D, Rabus R. Evaluation of two-dimensional difference gel electrophoresis for protein profiling. Soluble proteins of the marine bacterium *Pirellula* sp. strain 1. *J Mol Microbiol Biotechnol* 2003;5:240–251.
37. Zhou G, Li H, DeCamp D, et al. 2D differential in-gel electrophoresis for the identification of esophageal scans cell cancer-specific protein markers. *Mol Cell Proteomics* 2002;1:117–124.
38. Gharbi S, Gaffney P, Yang A, et al. Evaluation of two-dimensional differential gel electrophoresis for proteomic expression analysis of a model breast cancer cell system. *Mol Cell Proteomics* 2002;1:91–98.
39. Friedman DB, Hill S, Keller JW, et al. Proteome analysis of human colon cancer by two-dimensional difference gel electrophoresis and mass spectrometry. *Proteomics* 2004;4:793–811.

14

Proteomics of Bronchoalveolar Lavage Fluid and Sputum

Ruddy Wattiez, Olivier Michel, and Paul Falmagne

Summary

Bronchoalveolar lavage fluid (BALF) and sputum, obtained through more or less noninvasive techniques, contain cells and many soluble compounds from respiratory tract secretions and are, therefore, important sources for the study of the lung and its pathologies. The protein composition of these physiological fluids faithfully reflects cellular and molecular changes induced by lung disorders and are, to date, best investigated by proteome analysis. This powerful experimental approach using differential-display proteomics allows not only the molecular characterization of the BALF and sputum proteomes, but also definition of changes in the proteomes correlated specifically to different pathophysiological states. They have been successfully applied to the study of interstitial lung diseases and allergic asthma. The increasing success of BALF and sputum proteome analysis is essentially owing to recent progress in sample preparation and the fast development of leading-edge and high-throughput biotechnologies like 2D gel electrophoresis, multidimensional liquid chromatography, and biochips coupled to mass spectrometry. They offer the unique opportunity to explore the molecular mechanisms of lung diseases and to define new specific biomarkers for early diagnosis, prevention, and optimal therapy of lung injury.

Key Words: Bronchoalveolar lavage fluid; sputum; lung; proteome; interstitial lung diseases; biomarker.

1. Introduction

The cellular interface between the lung and the environment is composed of heterologous epithelia including pseudostratified epithelium in the proximal airways, cuboidal epithelium in the distal airways, and very thin epithelium in the alveoli; the latter represents more than 95% of the lung surface. It produces secretions such as mucus, host-defence proteins and surfactants ***(1–4)***. The airways, particularly the alveoli, are covered with a thin layer of epithelial lining fluid, which is a rich source of many different soluble components of the lung

From: *Proteomics of Human Body Fluids: Principles, Methods, and Applications*
Edited by: V. Thongboonkerd © Humana Press Inc., Totowa, NJ

(proteins, lipids, and so on) that play important roles in airway integrity and pulmonary defense. The protein composition of this fluid most faithfully reflects the state of the lung and effects of external factors that influence the lung, and is of primary importance in early diagnosis, assessment, and characterization of lung disorders as well as in the search for disease markers ***(5)***.

Diverse techniques are currently used for sampling the distal airways. Induction of sputum with hypertonic saline and bronchoalveolar lavage with isotonic saline during fiberoptic bronchoscopy is a more or less noninvasive technique that opens the way to a small part of the lung proteome ***(6–10)***. If the main constituents of the bronchoalveolar lavage fluid (BALF) originate from the alveoli, sputum is thus a source of secretions closer to the tracheobronchial tree. Expectorated sputum is a sign of disease and indicates the excessive production and retention that occur in patients with respiratory infection, bronchitis, asthma, and cystic fibrosis. BALF and sputum contain cells (lymphocytes, neutrophils, eosinophils, and so on) and a wide variety of soluble compounds (proteins, lipids, and so on) originating from respiratory tract secretions ***(11–14)***. Therefore, BALF and sputum are important diagnostic sources to investigate cellular and molecular changes during the course of lung disorders. During the last 10 yr, proteomic research of BALF and sputum has been boosted by recent advances in leading-edge technologies, especially mass spectrometry, and numerous studies have shown the outstanding interest for more accurate diagnosis, follow-up, prognosis, and treatment of all lung disorders. Moreover, the pattern of proteins in BALF and sputum from patients at different stages of respiratory disorders may also provide deeper insights into the molecular mechanisms of the diseases.

2. Methods

2.1. BALF and Sputum Sampling

Bronchoalveolar lavage is considered a safe procedure (with no lethal complication reported) compared with more invasive techniques such as transbronchial lung biopsy. The most recent publications describe a standardized washing procedure of the right middle or lower lobe of the lung with 5X 20 mL of sterile 0.9% (w/v) NaCl (normal saline solution) during fiberoptic bronchoscopy to obtain BALF ***(6)***. Induction of sputum offers the potential of a rapid, direct, noninvasive, and inexpensive alternative way of collecting multiple samples of airway secretions. It involves inhalation of sterile hypertonic saline aerosol to induce expectoration ***(15)***. Two different techniques have been used for processing the sputum sample once obtained. One technique culls out all viscid or dense portions of the sample to separate airway fluid from saliva ***(16)***. In the other technique, saliva can be separated from sputum at the time of collection, and the resulting sample may be processed in its entirety ***(17)***. Both sputum processing techniques are reproducible and valid in their ability to provide differential cell counts and measurements of soluble compounds. For proteomic research, sputum may be processed with dithiothreitol (DTT) to enhance cell dispersal from the surrounding mucus. Briefly, using the protocol described by Fahy et al. ***(18)***.

1. The subjects are nebulized with sterile hypertonic saline (3%) solution for 12 min using an Ultra-Neb 99 ultrasonic nebulizer (DeVilbiss, Somerset, PA). This nebulizer generates particles with a mean mass diameter of 4.5 μm and has an output of 2.4 mL/min.
2. After expelling saliva, the subjects are encouraged to cough at 2-min intervals.
3. The sputum is collected, and viscid plugs are selected to minimize saliva contamination.
4. The volume of sputum collected is measured and an equal volume of 0.1% DTT is added.
5. The homogenized sputum is then centrifuged at 16,000*g* rpm for 5 min.
6. The supernatant is collected and frozen at –70°C for further analysis.

2.2. Methods for Proteomic Analysis of BALF and Sputum

The 2D gel electrophoresis (2-DGE) method is presently the most widely used protein separation technique. It has the power to separate thousands of proteins simultaneously and to visualize them with a level of sensitivity that makes computer analysis feasible ***(19)***. A number of methodological improvements have been made since the introduction of 2-DGE technology by O'Farrell in 1975 ***(20)***. The development of immobilized pH gradients of different ranges has made the technique more reproducible ***(21)***. 2-DGE has been used in several studies to analyze the protein content of BALF ***(22–28)***. It has been shown that profiling of low-abundance BALF proteins is limited because of the higher abundance albumin, immunoglobulins, and mucopolysaccharides. However, analysis of these samples by gel systems coupled to desalting and concentration steps could significantly improve the detection of less abundant proteins.

Mass spectrometry (MS)-based techniques play important roles in the proteomic era and are most commonly used to identify proteins separated and visualized on 2-DGE gels. The MS techniques include matrix-assisted laser desorption/ionization time-of-flight MS (MALDI-TOF-MS), electrospray ionization-MS (ESI-MS), ESI tandem-MS (ESI-MS/MS), ESI-quadrupole-TOF-MS (ESI-Q-TOF-MS), reversed-phase high-performance liquid chromatography-ESI-MS (RP-HPLC-ESI-MS), and the recently developed Fourier-transform MS (FT-MS) ***(29)***. The recent development of MS coupled to multidimensional liquid chromatography (multi-LC-MS or multi-LC-MS/MS) or to Biochips (i.e., surface-enhanced laser desorption/ionization, SELDI-MS) allows the analysis of biofluids like BALF. The use of Biochips or LC systems coupled to MS provides the advantage of reducing the complexity of the BALF protein mixture, with the consequence that greater numbers of low-abundance proteins and peptides can be detected ***(30)***. Recent progress in SELDI allows the collection of protein profiles of different biological samples, especially physiological fluids. Protein profiling is the rapid screening of samples by MS with limited or no sample preparation. The resulting profiles of *m/z* ratio peaks of different samples can then be compared, and differences in the relative abundance of proteins can be evaluated. Another progress in SELDI techniques is the integration of on-chip capture, purification, and quantitative detection of targeted proteins on a single platform ***(31)***. These techniques provide a complementary method to 2-DGE for protein visualization. Recently, Nelsestuen and colleagues ***(32)*** used MALDI-TOF profiling approach to analyze BALF samples. Clearly, this technology has the advantage of improving the differential display of peptides and smaller proteins (molecular mass <20 kDa), which are underdetected or undetectable in 2-DGE.

2.2.1. Sample Preparation for Gel-Based Analysis

Two major problems associated with 2-DGE of BALF proteins are the low protein concentration of BALF and its high salt content, which comes from the phosphate-buffered saline used for the lavage procedure. A high level of salts interferes with the isoelectric focusing (IEF) of proteins. Therefore, it is necessary to remove salts from a BALF sample and preferentially to enrich proteins in the sample prior to 2-DGE ***(33)***. Four different desalting and concentration methods have been reported: protein precipitation, ultrafiltration, microdialysis, and the Bio-Spin® column.

2.2.1.1. Protein Precipitation

Desalting by precipitation of proteins using trichloroacetic acid (TCA), cold acetone, TCA combined with acetone, or a two-step combination of precipitant and coprecipitant such as the PlusOne 2-D Clean-Up kit (Amersham Biosciences, Uppsala, Sweden) are the most common sample preparation methodologies for protein analyses ***(26,33)***. Briefly:

1. BALF or sputum samples are incubated with ice-cold acetone or TCA.
2. Precipitated proteins are pelleted by centrifugation and then further washed to remove nonprotein contaminants.
3. Finally, pellets are solubilized in the appropriate sample buffer for the IEF.

Protein precipitation leads to a relatively low protein recovery (24–85%) owing to protein losses and difficulties in solubilization of precipitated proteins. Moreover, horizontal/vertical streaks are observed on the 2-DGE gel, causing a high level of background. Actually, protein precipitation is not the most appropriate methodology for comparative BALF and sputum proteomics.

2.2.1.2. Ultrafiltration

Ultrafiltration can be performed using different commercial systems, i.e., the Ultrafree-4 Centrifugal Filter Unit. Desalting by ultramembrane centrifugation is a highly effective technique for salt removal, resulting in minimal sample loss compared with protein precipitation. Protein adsorption onto the filtering membrane can be avoided by repeated washing steps with sample solution after centrifugation, as shown by Plymoth and colleagues ***(33)***.

2.2.1.3. Microdialysis

Overnight dialysis is performed using a membrane with a molecular mass cutoff at 3.5 kDa in a volatile buffer such as 12 m*M* NH_4HCO_3-HCl buffer (pH 7.0). BALF and sputum samples are then volume-reduced by SpeedVac centrifugation ***(26,34)***. Desalting by dialysis is a method equivalent to ultramembrane centrifugation with regard to protein yield, but it requires a further concentration step.

2.2.1.4. Bio-Spin Column

BALF and sputum samples are loaded onto a polyacrylamide microcolumn (molecular mass cutoff at 6 kDa). After centrifugation at 1000*g* for 4 min, salts and other impurities are bound to the column. Peptides and proteins can then be purified. Clearly, salt removal with Bio-Spin columns is preferred because of the high levels of protein recovery and improved gel resolution ***(35)***.

2.2.2. Sample Preparation for Gel-Free Analysis

MALDI-TOF protein profiling, without prior 2-DGE separation, has been used by Nelsestuen and colleagues ***(32)*** to analyze BALF samples. In this approach, BALF samples for MS analysis are diluted if needed, acidified with 0.5 µL of 10% TFA per 20 µL of sample, and subjected to ZipTip (C18) processing according to standard procedures. After elution, the peptides/proteins are spotted onto the MALDI target plate. In SELDI, proteins are differentially adsorbed on a ProteinChip array composed of various chromatographic surfaces that allow target proteins to be fractionated or enriched in combination with a selected wash conditions and subsequently detected by TOF-MS. The preparation of the BALF samples depends on the nature of the chromatographic surfaces. For the hydrophobic ProteinChip, the BALF samples are spotted directly onto the surface, dried, and washed with water. For the anionic or cationic surfaces, the BALF samples are processed in Bio-Spin columns before transfer into the appropriate buffers.

2.2.3. Strategies to Enrich Low-Abundance Proteins in BALF and Sputum Samples

An important drawback is the impaired detection and identification of low-abundance proteins in BALF and sputum hindered by major-abundance proteins like albumin (50%), immunoglobulins (30%), transferrin (5–6%), and α_1-antitrypsin (3–5%) ***(13)***. Strategies for separation and detection of low-abundance proteins in BALF and sputum samples are identical to those applied to other physiological fluids such as plasma. In this context, different methods have been developed based on sample prefractionation.

2.2.3.1. 2-DGE in Narrow pH Ranges

To detect proteins of low abundance, it is crucial to simplify the profiling of such proteins so that individual spots can be visualized. Additional improvements of the 2-DGE protein maps of human BALF samples can be obtained by using narrow-range IPG strips for the first dimensional separation. With the use of narrow-range immobilized pH gradient (IPG) strips covering the pH interval 4.5 to 6.7 in combination with the "paper bridge" method, greater numbers of low-abundance proteins in BALF can be detected ***(27)***. Whereas 409 spots have been detected with a pH 4.5 to 5.5 strip, and 425 have been visualized using a pH 5.5 to 6.7 strip, only 678 spots can be detected using a 3 to 7 pH gradient. Splitting a p*I* range into multiple narrower p*I* ranges using equal strip lengths produces a proportionally broader reconstituted 2-DGE map. Accordingly, the use of these narrow-range IPG strips for IEF improves the overall quality of BALF 2-DGE proteome maps by increasing the resolution, since proteins with very closed p*I*, which cannot be separated with a wide-range gradient, can then be resolved. Therefore, using narrow-range gradient gels is similar to prefractionation. Since more protein extract can be loaded on the narrow-range IPG strips, as a consequence, the probability of detecting scarce proteins will be higher. Using this approach, Sabounchi-Schutt and colleagues ***(27)*** have identified 49 proteins in the narrow pH range of 4.5 to 5.2 from an individual healthy BALF sample. Among these proteins, 17 have not been detected in plasma, and 12 have not previously been described in the BALF 2D proteome map.

2.2.3.2. Affinity Chromatography

The depletion of albumin and IgG from biological fluids can be obtained by adsorption on affinity resins such as a Cibacron Blue-based matrix. However, removing the albumin fraction by binding to Affi-Gel Blue gives quite unsatisfactory results because other proteins are also lost. In an effort to remove albumin from BALF, Plymoth and colleagues ***(33)*** have examined an antihuman serum albumin column. Clearly, removing the albumin using this approach improves the pattern of BALF proteins resolved on 2D gels and makes possible the mapping of some minor protein spots. Nevertheless, this study has shown that a few protein spots, especially proteins interacting with human serum albumin, are also lost in the preparation process. Hence, reactive dyes or immunoglobulins in affinity-based depletion methods for the removal of high-abundance proteins should be used with caution. However, under optimized conditions, these approaches can be useful to deplete high-abundant proteins from BALF and sputum samples, as shown by numerous studies on plasma samples.

3. Findings and Applications

3.1. Proteome Description of BALF and Sputum

BALF and sputum proteomics, defined as the study of all the proteins present in these biological fluids, involves the comprehensive description of all protein species as well as the set of protein isoforms and posttranslational modifications. The development of leading-edge technologies like MS and recent progress in sample preparation methods for biological fluids have boosted BALF and sputum proteome research, which is now in an exponential growth phase *(29)*. Proteome analysis of BALF was performed for the first time in 1979 by Bell et al. *(36)*. The emergence of high-throughput technologies associated with protein identification from gels has led to a tremendous increase in the number of identified proteins from BALF. Classically, master gels of BALF proteins comprise more than 1200 protein spots visualized by silver staining. Among these protein spots, 150 unique proteins have been identified corresponding to 800 protein spots (including immunoglobulin isoforms). The major investigators of the BALF proteome are the groups of Lindahl *(23–25)*, Wattiez *(26,35,37)*, Magi *(28,34)*, and Sabounchi-Schutt *(27,38)*. The best characterizations of the BALF proteome show the presence of numerous different classes of proteins that reflect the great diversity of their cellular origins and functions. Proteins in BALF may originate from a broad range of sources, such as diffusion from serum across the air-blood barrier (i.e., albumin), secretion by different lung cell types, or release from cellular debris. A comparison between BALF and plasma proteomes reveals that a certain number of proteins are characterized by higher levels in BALF than in plasma, suggesting that they are specifically produced in the airways (**Fig. 1**). These proteins are good candidates for becoming lung-specific biomarkers. Recently, the same observations have been shown for the sputum proteome.

3.2. BALF and Sputum Differential-Display Proteomics

One of the major interests of the proteome approach is the association of changes in the proteome with different pathophysiological states using differential-display proteomics, i.e., computer-assisted comparisons between

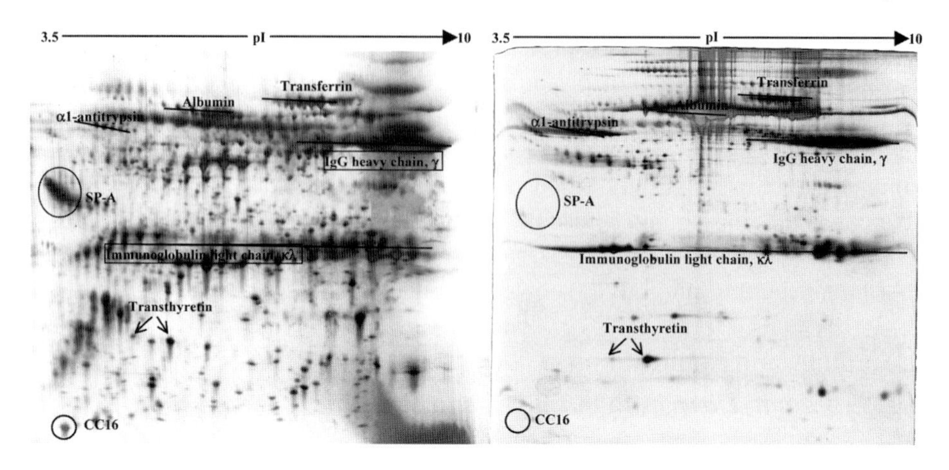

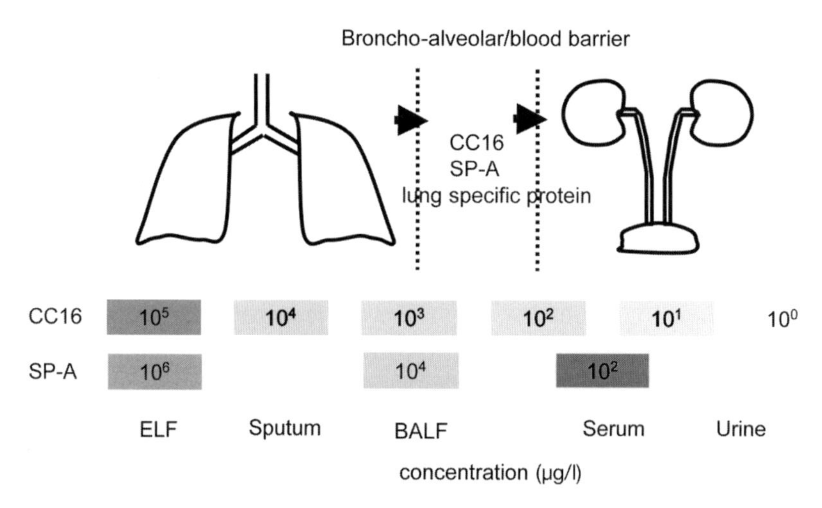

Fig. 1. Comparison between bronchoalveolar lavage fluid (BALF) and plasma proteomes. Several proteins are present in BALF but not in plasma, suggesting that they are specifically produced in the airways. These proteins are, therefore, good candidates for lung-specific biomarkers. Lung-specific proteins, such as surfactant protein A (SP-A) and Clara cell protein 16 (CC16), are elevated in BALF proteome. Concentrations of both proteins are indicated in different physiological fluids. ELF, epithelial lining fluid. *(From Wattiez et al.* ***[35]****, reproduced with permission from Elsevier.)*

2D proteome maps. The aim of this investigation is to define proteins or group of proteins significantly associated with a specific disease. Thus, differential-display proteomics offers the opportunity to understand disease mechanisms and to develop new biomarkers for early diagnosis/disease prediction.

3.2.1. Mechanisms of Lung Pathologies

With the goal of gaining a better understanding of lung disease mechanisms, during these last 10 yr many BALF proteome analyses have been dedicated to different lung pathologies such as cystic fibrosis ***(39)***, alveolar proteinosis ***(40)***, chronic eosinophilic pneumonia ***(37)***, hypersensitivity pneumonitis ***(37)***, Wegener's granulomatosis ***(37)***, lupus erythematosis ***(37)***, bacterial pneumonia ***(26)***, sarcoidosis ***(28,34,37,38)***, chronic lung allograft rejection ***(32)***, and, recently, pulmonary fibrosis associated with systemic sclerosis (SSc) as well as idiopathic pulmonary fibrosis (IPF) ***(34)***.

3.2.1.1. Interstitial Lung Diseases

Among all lung pathologies, fibrosing interstitial lung diseases like sarcoidosis, IPF, and pulmonary fibrosis associated with SSc have been most studied by a differential-display proteomics approach ***(28,34,37,38)***. The studies of diffuse interstitial lung diseases are a challenge owing to the complexity and heterogeneity of various underlying pathogenic mechanisms. Although fibrosis is a common feature, the pathogenesis, prognosis, and degree of fibrosis vary between the different types. Fibrosis is generally considered the final event of complex immunoinflammatory reactions.

To obtain a more comprehensive picture of these complex pathogenic mechanisms at the alveolar level, BALF protein compositions have been analyzed by 2-DGE coupled to MS. These studies have shown that there are statistically quantitative and qualitative differences between the three diseases ***(28,34,37,38)***; the major differences have been observed between sarcoidosis and IPF (**Fig. 2**). Researches have observed that many plasma proteins (such as albumin, immunoglobulins, α_1-antitrypsin, α_1-antichymotrypsin, haptoglobin β, α_1-antiplasmin, ceruloplasmin, α_1-B-glycoprotein, and others were more abundant in BALF from sarcoidosis than that from IPF patients, possibly owing to altered alveolar membrane integrity during alveolitis in the course of sarcoidosis. However, some plasma proteins (such as α_2-macroglobulin, a major antiprotease) were more abundant in IPF compared with sarcoidosis. In contrast, many of the low-molecular-weight proteins produced locally were more abundant in IPF compared with sarcoidosis. Clearly, if some of them are products of cell damage, which is severe in IPF, others are produced by active secretion and have various functions, e.g., proteins involved in inflammatory processes (like calgranulin A, calgranulin B, and macrophage migration inhibitory factor [MIF]) and antioxidant proteins (like thioredoxin, peroxisomal antioxidant enzyme [AOPP], and thioredoxin peroxidase 2).

Recently, Rottoli and colleagues ***(34)*** have performed a complete proteome analysis combined with a cytokine profile study of BALFs from patients with

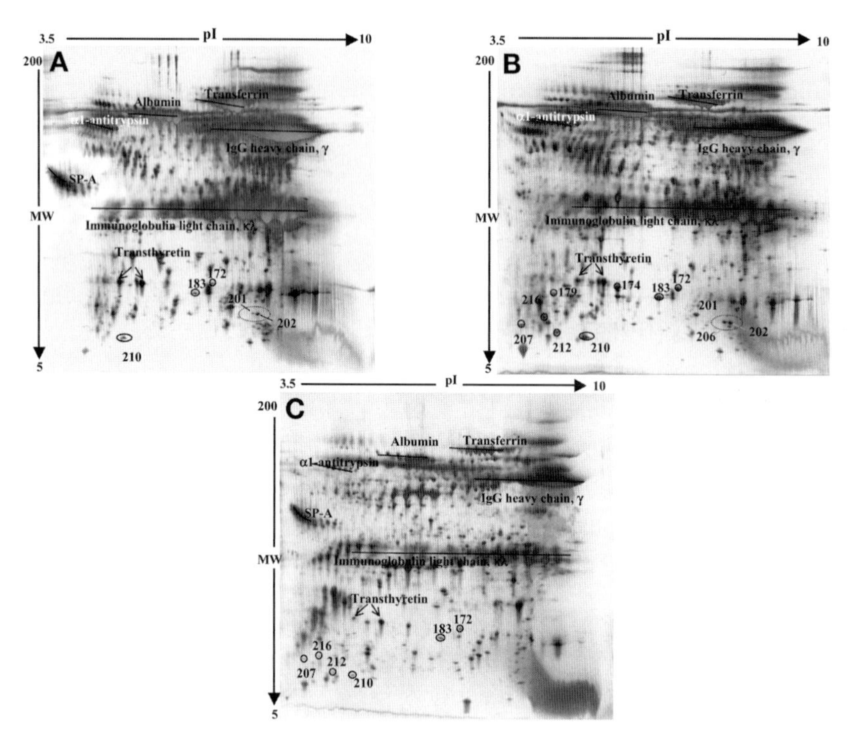

Fig. 2. Silver-stained 2D gels of human BALF from patients with sarcoidosis (**A**), patients with idiopathic pulmonary fibrosis (**B**), and healthy subjects (**C**). Twenty-five micrograms of sample was dissolved in 9 *M* urea, 0.5% (v/v) Triton X-100, 2% (v/v) ampholytes 3 to 10, 65 m*M* DTT, and 8 m*M* phenylmethylsulfonyl fluoride (PMSF), and loaded on pH 3 to 10, nonlinear IPG strips for isoelectric focusing. In the second dimension, proteins were separated on ExcelGel XL 12 to 14% and visualized by silver staining. #172, fatty acid binding protein, epidermal; #174, cathepsin D light chain; #179, intestinal trefoil factor; #183, fatty acid binding protein, adipocyte; #201, 202, and 206, calgranulin A; #207, saposin D light chain; #210, ubiquitin like; #212, calcyclin; #216, calvasculin. *(From Wattiez et al.* ***[26]****, reproduced with permission from WILEY-VCH Verlag GmbH.)*

sarcoidosis, IPF, and pulmonary fibrosis associated with SSc. The results showed that pulmonary fibrosis associated with SSc had an intermediate protein profile between IPF and sarcoidosis, with the following peculiarity, however. Among the low-molecular-weight proteins, which are more abundant in IPF than in sarcoidosis, the investigators showed that six of them (cyclophilin A, calgranulin B, translationally controlled tumor proteins [TCTPs], MIF, galectin 1, and ubiquitin) were also significantly increased with respect to SSc, whereas the other five (thioredoxin, AOPP, calgranulin A, L-fatty acid binding protein, and thioredoxin peroxidase 2) were increased in SSc compared with sarcoidosis. These findings are in agreement with the different pathogenesis of these diseases: IPF is considered a prevalently fibrotic disorder limited to the lung with intense local production of functionally different proteins, whereas sarcoidosis and SSc are systemic immunoinflammatory diseases. These different studies based on comparison of the proteome profiles have led to a better understanding of the mechanisms of interstitial lung diseases. Moreover, the proteomic differences found between these pathologies strongly suggest that, in the near future, a group of proteins can be used as the indicators of progression toward lung fibrosis.

3.2.1.2. Allergic Asthma

Allergic asthma is a chronic inflammatory disease involving a multitude of cell types. In asthma, reversible airway obstruction is characterized by airway inflammation and hyperresponsiveness, bronchoconstriction, increased mucus secretion, and increased permeability of lung vessels ***(41,42)***. Despite the great research efforts made, the underlying mechanisms for the development of severe symptoms remain poorly defined. The first report on the asthmatic state of human BALF and nasal lavage fluid proteomes has been given by Lindhal et al. ***(25)***. These investigators have shown that levels of lipocalin-1, cystatin S, IgBF, and transthyretin were changed in patients who suffer from asthma.

Recent studies have investigated the modifications of proteome profile that occur in lung tissues and airspaces following challenges using animal models. In this context, Signor et al. ***(43)*** have evaluated the BALF proteome of rats treated with an allergen (ovalbumin [OVA]) or endotoxin (lipopolysaccharide [LPS]). Their study showed that some proteins produced locally were significantly increased (calgranulin A) or decreased (Clara cell 16-kDa secretory protein, pulmonary surfactant protein B) in both treatments compared with controls. In contrast, pulmonary surfactant-associated protein A was decreased in the OVA treatment and was not significantly affected in the LPS challenge. Moreover, some proteins derived from plasma were also different after exposure. Clearly, these results obtained from animal models are essential for the identification of potential biomarkers for inflammation. In this context, our group has analyzed the sputum proteome to evaluate the inflammatory response to inhaled LPS in healthy volunteers. As observed in the animal studies, our preliminary results have shown an increase in some proteins, such as calgranulin A, that is produced locally and is involved in the inflammatory process, 7 and 24 h following LPS inhalation (**Fig. 3**). These preliminary results have demonstrated the potential of the proteome analysis of sputum induced with hypertonic saline to study inflammatory lung diseases.

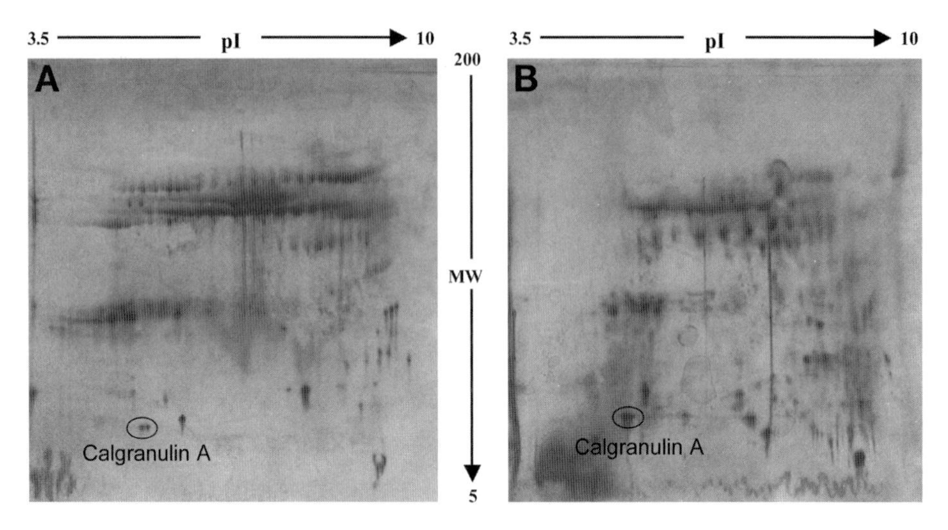

Fig. 3. Silver-stained 2D gels of human sputum from healthy subjects before **(A)** and after **(B)** 24 h of LPS inhalation. Twenty-five micrograms of sample was dissolved in 9 *M* urea, 0.5% (v/v) Triton X-100, 2% (v/v) ampholytes 3 to 10, 65 m*M* DTT, and 8 m*M* phenylmethylsulfonyl fluoride (PMSF) and loaded on pH 3 to 10 nonlinear IPG strips for isoelectric focusing. In the second dimension, proteins were separated on ExcelGel XL 12 to 14% and detected by silver staining.

3.2.2. Biomarker Research

The analysis of BALF proteome profiles may be helpful to quantify the overall molecular changes associated with lung injuries and also offers the possibility to search for biomarkers of a specific lung pathology. A good example of this approach has been demonstrated by Nelsestuen et al. ***(32)*** with the application to chronic lung allograft rejection.

3.2.2.1. Chronic Lung Allograft Rejection

Chronic allograft rejection remains a leading cause of morbidity and mortality in lung transplant recipients. Currently, diagnosis is based on lung biopsies or the presence of bronchiolitis obliterans syndrome (BOS). To search for biomarkers of chronic rejection, Nelsestuen et al. ***(32)*** performed a global study of the proteins in BALFs from lung transplant recipients using gel-free technology based on MALDI-TOF protein profiling **(Fig. 4)**. A total of 126 BALF samples from 57 individuals was tested. MS analysis revealed abundant changes in many components in most patients who experienced BOS, but three peaks at m/z = 3373, 3444, and 3488 were unusually intense. These were identified as human neutrophil peptides (HNPs) 1 to 3. HNPs belong to a family of antimicrobial peptides referred to as defensins. Quantification of these peptides by enzyme-linked

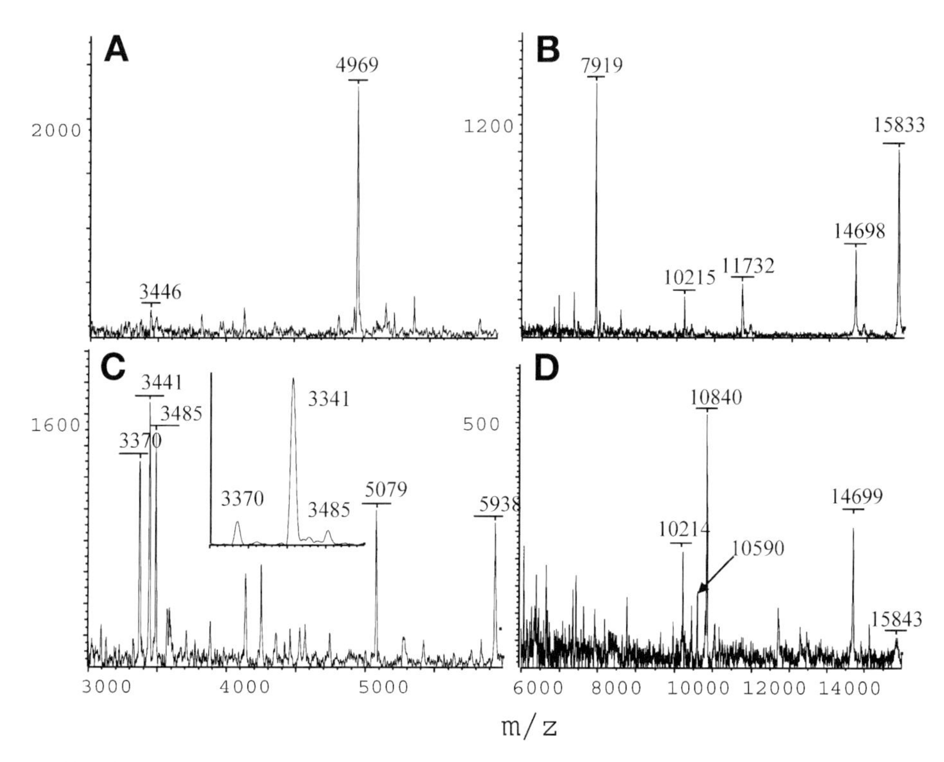

Fig. 4. MALDI-TOF profiles. MALDI-TOF spectra are shown for the *m/z* 3000 to 6000 (**A** and **C**) and *m/z* 6000 to 16,000 (**B** and **D**) regions of BALF from a control subject who did not develop BOS (**A** and **B**; 2.48 µg total protein) and another one at 5 mo before the diagnosis of BOS2 (**C** and **D**; 2.76 µg total protein). The MALDI-TOF spectra were split into two parts to illustrate the different intensities better (e.g., range 2000 for **A**, 1200 for **B**, 1600 for **C**, and 500 for **D**). (**C**, **inset**) A portion of the spectra after addition of 5 ng of HNP1 to the sample before extraction and analysis. *(From Nelsestuen et al.* ***(32)****, reproduced with permission from WILEY-VCH Verlag GmbH.)*

immunosorbent assay corroborated the extraordinary levels of HNPs detected by MS. To quantify HNPs via MS, the investigators introduced known amounts of commercial HNP1 and -2 into BALF samples. Clearly, they found an elevated HNP level (>0.3 ng/µg protein) in 89% of patients who developed BOS2-3 within 15 mo. In control patients, 35% demonstrated a slightly elevated HNP level that declined in all who had subsequent BALF available for testing. HNP levels did not correlate with episodes of acute rejection, cytomegalovirus, or fungal infection. The investigators concluded that elevated HNP levels are associated with the onset of BOS and can predict the clinical onset of disease up to 15 mo. This work showed the high potential of the MS approach applied to BALF analysis to search for new biomarkers of lung injuries.

4. Perspectives and Future Directions

Numerous studies during the last 5 yr have demonstrated the fundamental and clinical potentials of the proteome analysis of BALF and sputum samples, despite their complexity and wide dynamic ranges. These advances have reflected the considerable progress made in sample preparation and MS technologies. The rapid evolution in MS, especially that associated with LC, and the development of novel methods for comparison and quantification, such as protein profiling techniques and the isotope-coded affinity tag (ICAT) method, have boosted the proteome analysis of biological fluids such as BALF and sputum. Clearly, these advances offer a more complete image of the proteomes of BALF and sputum and of the function of their different constituents. Recent studies of BALF samples have already shown that proteome analysis is relevant in the detection of biomarkers for lung alterations and in a better understanding of the molecular mechanisms involved in the pathogenesis of lung diseases. Nevertheless, bronchoscopy is an invasive procedure that cannot be performed in all patients to obtain BALF. The advantages of inducing sputum with hypertonic saline are that the samples are readily accessible and that this technique can be applied to children or adult patients with asthma. To date, there is no proteomics study on induced sputum samples, partly because of the presence of high concentrations of salts and contaminants. However, recent developments in some important sample preparation steps associated with BALF might soon lead to further breakthroughs in the proteome analysis of induced sputum, as shown in our preliminary studies.

One difficulty in the interpretation of the data obtained from differential-display proteome studies of BALF and induced sputum is the great diversity of cellular origins of different proteins identified in these fluids. Moreover, it is clear that the level of one protein represents the integration of a multitude of different mechanisms involved in its synthesis, release, and/or clearance. Therefore, analysis of the proteomes of different cell types such as alveolar macrophages and neutrophils present in these fluids is required to understand better the roles of altered proteins in the BALF or induced sputum that are associated with lung pathologies. In this context, recent work has already focused on proteome analysis of the alveolar macrophages in patients' BALF samples ***(44,45)***. Clearly, the integration of these proteome analyses of different cell types in the same fluid will give a better picture of the complex mechanisms of lung pathologies.

Finally, to be most efficient, proteome research on BALF and induced sputum definitely needs better rationales for the sample preparation (especially for the induced sputum samples) and the patient selection criteria (age, sex, smoking or not smoking, and so on.). Moreover, the application of leading-edge novel technologies, like ICAT and proteome profiling methods, opens new perspectives for research on BALF and induced sputum.

Acknowledgments

R. Wattiez is a Research Associate at the National Funds for Scientific Research.

References

1. Robbins R, Rennard S. Biology of airway epithelial cells. In: Crystal R, West J, Weibel E, Barnes P, eds. *The Lung: Scientific Foundations* Philadelphia: Lippincott-Raven 1996:445–457.
2. Shak S. Mucins and lung secretions. In: Crystal R, West J, Weibel E, Barnes P, eds. *The Lung: Scientific Foundations*. Philadelphia: Lippincott-Raven, 1996:479–486.
3. Lubman R, Kim K, Crandall E. Alveolar epithelial barrier properties. In: Crystal R, West J, Weibel E, Barnes P, eds. *The Lung: Scientific Foundations*. Philadelphia: Lippincott-Raven, 1996:585–602.
4. Hawgood S. Surfactant: composition, structure, and metabolism. In: Crystal R, West J, Weibel E, Barnes P, eds. *The Lung: Scientific Foundations*. Philadelphia: Lippincott-Raven, 1996:557–571.
5. Griese M. Pulmonary surfactant in health and human lung diseases: state of the art. *Eur Respir J* 1999;13:1455–1476.
6. Reynolds HY. Use of bronchoalveolar lavage in humans—past necessity and future imperative. *Lung* 2000;178:271–293.
7. Capron F. Lavage Bronchoalvéolaire. *Arch Anat Cytol Pathol* 1997;45:255–260.
8. Hargreave FE, Leigh R. Induced sputum, eosinophilic bronchitis, and chronic obstructive pulmonary disease. *Am J Respir Crit Care Med* 1999;160:S53–S57.
9. Henig NR, Tonelli MR, Pier MV, Burns JL, Aitken ML. Sputum induction as a research tool for sampling the airways of subjects with cystic fibrosis. *Thorax* 2001;56:306–311.
10. Olivieri D, D'Ippolito R, Chetta A. Induced sputum: diagnostic value in interstitial lung disease. *Curr Opin Pulm Med* 2000;6:411–414.
11. Chlap Z, Jedynak U, Sladek K. Mast cell: its significance in bronchoalveolar lavage fluid cytologic diagnosis of bronchial asthma and interstitial lung disease. *Pneumonol Alergol Pol* 1998;66:321–329.
12. Jacobs JA, De Brauwer E. BAL fluid cytology in the assessment of infectious lung disease. *Hosp Med* 1999;60:550–555.
13. Noel-Georis I, Bernard A, Falmagne P, Wattiez R. Database of bronchoalveolar lavage fluid proteins. *J Chromatogr B Analyt Technol Biomed Life Sci* 2002; 771:221–236.
14. Thomas RA, Green RH, Brightling CE, et al. The influence of age on induced sputum differential cell counts in normal subjects. *Chest* 2004;126:1811–1814.
15. Beier J, Beeh K, Kornmann O, Buhl R. Induced sputum methodology: validity and reproducibility of total glutathione measurement in supernatant of healthy and asthmatic individuals. *J Lab Clin Med* 2004;144:38.
16. Gibson PG, Henry RL, Thomas P. Noninvasive assessment of airway inflammation in children: induced sputum, exhaled nitric oxide, and breath condensate. *Eur Respir J* 2000;16:1008–1015.

17. Gibson PG, Girgis-Gabardo A, Morris MM, et al. Cellular characteristics of sputum from patients with asthma and chronic bronchitis. *Thorax* 1989;44: 693–699.
18. Fahy JV, Liu J, Wong H, Boushey HA. Cellular and biochemical analysis of induced sputum from asthmatic and from healthy subjects. *Am Rev Respir Dis* 1993;147:1126–1131.
19. Klose J. Protein mapping by combined isoeletric focusing and electrophoresis of mouse tissues. A novel approach to testing for induced ponit mutations in mammals. *Humandenetik* 1975;26:231–243.
20. O'Farrell PH. High resolution two-dimensional electrophoresis of proteins. *J Biol Chem* 1975;250:4007–4021.
21. Westermeier R, Postel W, Weser J, Gorg A. High-resolution two-dimensional electrophoresis with isoelectric focusing in immobilized pH gradients. *J Biochem Biophys Methods* 1983;8:321–330.
22. Lenz AG, Meyer B, Costabel U, Maier K. Bronchoalveolar lavage fluid proteins in human lung disease: analysis by two-dimensional electrophoresis. *Electrophoresis* 1993;14:242–244.
23. Lindahl M, Stahlbom B, Tagesson C. Two-dimensional gel electrophoresis of nasal and bronchoalveolar lavage fluids after occupational exposure. *Electrophoresis* 1995;16:1199–1204.
24. Lindahl M, Ekstrom T, Sorensen J, Tagesson C. Two dimensional protein patterns of bronchoalveolar lavage fluid from non-smokers, smokers, and subjects exposed to asbestos. *Thorax* 1996;51:1028–1035.
25. Lindahl M, Stahlbom B, Tagesson C. Newly identified proteins in human nasal and bronchoalveolar lavage fluids: potential biomedical and clinical applications. *Electrophoresis* 1999;20:3670–3676.
26. Wattiez R, Hermans C, Bernard A, Lesur O, Falmagne P. Human bronchoalveolar lavage fluid: two-dimensional gel electrophoresis, amino acid microsequencing and identification of major proteins. *Electrophoresis* 1999;20:1634–1645.
27. Sabounchi-Schutt F, Astrom J, Eklund A, Grunewald J, Bjellqvist B. Detection and identification of human bronchoalveolar lavage proteins using narrow-range immobilized pH gradient Drystrip and the paper bridge sample application method. *Electrophoresis* 2001;22:1851–1860.
28. Magi B, Bini L, Perari MG, et al. Bronchoalveolar lavage fluid protein composition in patients with sarcoidosis and idiopathic pulmonary fibrosis: a two-dimensional electrophoretic study. *Electrophoresis* 2002;23:434–444.
29. Hirsch J, Hansen KC, Burlingame AL, Matthay MA. Proteomics: current techniques and potential applications to lung disease. *Am J Physiol Lung Cell Mol Physiol* 2004;287:L1–L23.
30. Link AJ, Eng J, Schieltz DM, et al. Direct analysis of protein complexes using mass spectrometry. *Nat Biotechnol* 1999;17:676–682.
31. Merchant M, Weinberger SR. Recent advancements in surface-enhanced laser desorption-ionization-time of flight-mass spectrometry. *Electrophoresis* 2000;21: 1164–1177.

32. Nelsestuen GL, Martinez MB., Hertz ML, Savik K, Wendt CH. Proteomic identification of human neutrophil alpha defensins in chronic lung allograft rejection. *Proteomics* 2005;5:1705–1713.
33. Plymoth A, Lofdahl CG, Ekberg-Jansson A, et al. Biofluid analysis with special emphasis on sample preparation. *Proteomics* 2003;3:962–968.
34. Rottoli P, Magi B, Perari MG, et al. Cytokine profile and proteome analysis in bronchoaveolar lavage of patients with sarcoidosis, pulmonary fibrosis associated with systemic sclerosis and idiopathic pulmonary fibrosis. *Proteomics* 2005; 5:1423–1430.
35. Wattiez R, Falmagne P. Proteomics of bronchoalveolar lavage fluid. *J Chromatogr B Anal Technol Biomed Life Sci* 2005;815:169–178.
36. Bell DY, Haseman JA, Spock A, McLennan G, Hook GE. Plasma proteins of the bronchoalveolar surface of the lungs of smokers and nonsmokers. *Am Rev Respir Dis* 1981;124:72–79.
37. Wattiez R, Hermans C, Cruyt C, Bernard A, Falmagne P. Human bronchoalveolar lavage fluid protein two-dimensional database: study of interstitial lung diseases. *Electrophoresis* 2000;21:2703–2712.
38. Sabounchi-Schutt F, Astrom J, Hellman U, Eklund A, Grunewald J. Changes in bronchoalveolar lavage fluid proteins in sarcoidosis: a proteomic approach. *Eur Respir J* 2003;21:414.
39. Griese M, von Bredow C, Birrer P. Reduced proteolysis of surfactant protein A and changes of the bronchoalveolar lavage fluid proteome by inhaled alpha 1-protease inhibitor in cystic fibrosis. *Electrophoresis* 2001;22:165–171.
40. He C. Proteomic analysis of human bronchoalveolar lavage fluid: expression profiling of surfactant-associated protein A isomers derived from human pulmonary alveolar proteinosis using immunoaffinity detection. *Proteomics* 2003;3:87–94.
41. Kroegel C, Virchow JC, Luttmann W, Walker C, Warner JA. Pulmonary immune cells in health and disease: the eosinophil leucocyte (Part I). *Eur Respir J* 1994;7:519–543.
42. Jackson AD. Airway goblet-cell mucus secretion. *Trends Pharmacol Sci* 2001; 22:39–45.
43. Signor L, Tigani B, Beckmann N, Falchetto R, Stoeckli M. Two-dimensional electrophoresis protein profiling and identification in rat bronchoalveolar lavage fluid following allergen and endotoxin challenge. *Proteomics* 2004;4:2101–2110.
44. Wu HM, Jin M, Marsh CB.Toward functional proteomics of alveolar macrophages. *Am J Physiol Lung Cell Mol Physiol* 2005;288:L585–L595.
45. Fessler MB, Malcolm KC, Duncan MW, Worthen GS.A genomic and proteomic analysis of activation of the human neutrophil by lipopolysaccharide and its mediation by p38 mitogen-activated protein kinase. *J Biol Chem* 2002 30; 277:31,291–31,302.

15

Proteomics of Sinusitis Nasal Lavage Fluid

Begona Casado, Simona Viglio, and James N. Baraniuk

Summary

This chapter provides a brief survey of the two principal techniques applied in the study of the nasal lavage fluid (NLF) proteome. Lindahl et al. *(11)* used 2D gel electrophoresis (2-DE) to analyze the proteome of NLFs from subjects exposed to methyltetrahydrophthalic anhydride (MHHPA) or dimethylbenzylamine (DMBA) and from healthy nonsmokers and smokers. Casado et al. *(2)* used liquid chromatography coupled to electrospray ionization-tandem mass spectrometry (LC-ESI-MS/MS) to study the NLF proteome in normal subjects and individuals affected by sinusitis before and after pharmacological treatment. New proteins involved in the acquired and innate immune response in the nose against microbial infections were identified with both techniques. A comparison between the normal and sinusitis NLF proteome facilitates understanding of changes in the protective mechanisms of the nasal mucosa against pathogens and pollutants. Acute sinusitis was associated with a large increase in plasma, glandular, and cellular components. Treatment successfully reduced the complexity of the nasal proteome. Finally, the presence of carbohydrate sulfotransferase is an indication of the acidic mucin synthesis.

Key Words: Sinus; sinusitis; rhinitis; nasal lavage fluid; glandular; mucus; proteome.

1. Introduction

Nasal secretions humidify, heat, cool, and clean inhaled air. Proteins of the innate and acquired immune systems are secreted from glandular mucous and serous cells and leukocytes, as well as by plasma extravasation. Distinct patterns of secretion are seen during allergen exposure (allergic rhinitis), in rhinovirus, adenovirus, influenza, and bacterial rhinosinusitis, in cystic fibrosis, and in occupational exposures. The last is the most common problem at work. In acute sinusitis, an inflammation of the paranasal cavities within the facial bones generally follows an acute or chronic nasal inflammatory process such as a common cold viral infection or allergic rhinitis. About 0.5% of viral rhinitis is complicated by acute bacterial rhinosinusitis with mucopurulent discharges.

From: *Proteomics of Human Body Fluids: Principles, Methods, and Applications*
Edited by: V. Thongboonkerd © Humana Press Inc., Totowa, NJ

Chronic sinusitis subjects may develop either nasal and sinus polyps or glandular hyperplasia. These distinct pathophysiological processes lead to differences in the properties of mucus and its protein constituents (the proteome). The identification of specific profiles of mucous proteins may thus give important insights into the innate and acquired immune defense mechanisms involved in each disease.

Nasal lavage is a simple method for collecting samples from the upper airways. Determination of the "normal" profile of mucous proteins of human nasal lavage fluid (NLF) is vital. The functions of normal nasal mucus can be determined and disease-specific changes can be identified. Proteomic analysis of human NLF profiles the mucosal secretions and alterations in innate and acquired immune defense mechanisms. Different approaches have been used for analyzing the NLF proteome, including (1) capillary liquid chromatography (CapLC) with electrospray ionization (ESI) tandem mass spectrometry (MS/MS) to assess the NLF of normal and acute sinusitis patients before and after pharmacological treatment; and (2) 2D electrophoresis (2-DE) with matrix-assisted laser desorption/ionization (MALDI) MS to analyze the effects of methyltetrahydrophthalic anhydride (MHHPA) and dimethylbenzylamine (DMBA) on the NLF proteomes of healthy nonsmokers and smokers. The two approaches gave complementary information about the proteins involved in host protection and defence against microorganisms and occupational exposure. Many proteins of clinical interest have been identified, including the PLUNC (palate-lung-nasal epithelium clone) family of proteins, Ig binding factor, and cystatin S. Future studies of the mucosal secretory responses to microorganisms and chemicals are anticipated to define specific patterns of biomarkers (biosignatures) that can be used for advanced diagnosis and as the targets for novel rhinitis therapies.

2. Methods

2.1. LC-ESI-MS/MS

Casado et al. analyzed the NLF proteome in normal subjects ***(1)*** and in individuals affected by sinusitis before and after pharmacological treatment ***(2)***. Four "series" of NLFs were collected from subjects affected by sinusitis. The procedure employed was as follows.

2.1.1. Nasal Washing, Lavage and Hypertonic Saline Provocation

1. On d1, normal saline solution (1X NSS) was sprayed into the nostrils of normal subjects as an initial washing, and the lavage fluid was discarded.
2. Then preexisting secretions were removed from subjects' nostrils by spraying 0.9% NaCl (1X NSS) into each nostril using a Beconase AQ pump aspirator spray device (first "series" collection) ***(3,4)***. This provided the proteins indicative of the basal state of the nasal mucosa.

3. After a 5-min interval, the same subjects received provocations with 100 μL of hypertonic NaCl (21.6%; 24 times the tonicity of normal saline; 24X NSS). Previous studies have established that this provocation stimulates pain-carrying nonmyelinated nerves and local mucosal neural axon responses that cause glandular secretion without vascular permeability ***(5,6)***.
4. A second "series" collection of NLF was then performed.

2.1.2. Acute Rhinosinusitis

Acute rhinosinusitis (sinusitis) is an infection of the paranasal cavities within the facial bone ***(6,7)***. The diagnosis was made clinically using contemporary criteria ***(6)***.

1. Subjects underwent baseline (1X NSS) and hypertonic saline (24X NSS) provocations in the same fashion as the normal subjects.
2. Sinusitis subjects were then treated for 6 d with amoxicillin-clavulanic acid (antibiotic), fluticasone proprionate nasal spray (steroid), oxymetazoline nasal spray (α2-adrenergic receptor agonist, vasoconstrictor), and saline nasal sprays (0.9% NaCl).
3. The third and fourth "series" of NLFs were finally collected before and after the subjects had received the nasal provocations described above.
4. All samples collected were stored at –20°C until analysis ***(2)***.

2.1.3. NLF Preparation Prior to MS Analysis

1. The total protein concentration was measured by the Lowry method ***(8)***.
2. Proteins were precipitated with an equal volume of 50% ethanol, 50% acetic acid containing 0.02% sodium bisulfite.
3. Peptides, lipids, salts, and other low-molecular-weight materials were extracted into the supernatant; proteins larger than 10 to 15 kDa were precipitated. This step was essential to prevent lipid blockage of the CapLC process.
4. Precipitation was performed overnight at –20°C.
5. The protein pellet was resuspended in 10 μL of 0.1 *M* ammonium bicarbonate buffer, pH 7.8, and digested with trypsin overnight at 37°C (20:1 protein/trypsin ratio) ***(1,2)***.

2.1.4. CapLC-Q-TOF Analysis

1. Tryptic peptides were desalted and concentrated in a Bio-Basic C18 precolumn (35 × 0.32 mm) and then separated using a reversed-phase ZorbaxC18 column (100 mm × 150 μm ID).
2. Elution was perfomed using a gradient of 95% solvent A (H_2O with 0.2% formic acid) to 95% solvent B (acetonitrile with 0.2% formic acid) at a flow rate of 1 μL/min by CapLC operating with Masslynx, Version 3.5 software.
3. Proteins were identified using the Mascot search engine.
4. Manual data interpretation and identification of proteins were performed using the Peptide Match program in the NREF database of PIR. Proteins identified were compared on the basis of their origin and functions ***(1,2)***.

2.2. Two-Dimensional Electrophoresis

Lindahl et al. analyzed NLF after chemical exposure to MHHPA ***(11–13)*** and DMBA ***(9,14)***, as well as in healthy nonsmokers vs smokers ***(10,15)***.

2.2.1. Nasal Provocation and NLF Preparation Prior to 2-DE

1. NLF was obtained using a modified "nasal pool" device ***(16)***.
2. Samples were desalted in a PD-10 gel filtration column and lyophilized ***(9–11,13–15)***.
3. Protein concentrations were measured by the Bradford method ***(17)***.
4. Proteins were resuspended in a denaturing buffer containing 9 *M* urea, 65 m*M* dithiothreitol (DTT), 2% v/v ampholytes (3–10), 0.5% v/v Triton X-100, and 0.004% w/v bromophenol blue and then centrifuged ***(9–11,13–15,18)***.

2.2.2. 2-DE Procedures

1. Horizontal 2-DE was performed as described by Görg et al. ***(18)***.
2. Immobilized pH gradient (IPG) strips (0.50 × 3 × 180 mm) containing Immobilines NL 3-10 were rehydrated with 8 *M* urea, 0.5% v/v Triton X-100, 0.5% ampholytes (3–10), 13 m*M* DTT, and 0.2 mg/mL Orange G ***(9,11,13–15)***.
3. In analytical gels, samples were applied on the anodic side, and proteins were allowed to enter into IPG strips applying a low-voltage gradient, and then run overnight at approximately 2500 V to get a final focusing of 45,000 V/h.
4. In preparative gels used for mass spectrometric analysis, samples were applied by in-gel rehydration for 6 h using low voltage (30 V) in pH 4 to 7 or 3 to 10 NL IPGs.
5. The proteins were then focused for 55,000 to 60,000 V/h at a maximum voltage of 8000 ***(9–11,14)***.
6. The second dimension analysis (sodium dodecyl sulfate-polyacrylamide gel electrophoresis; SDS-PAGE) was performed in various gel formats running at 20 to 40 mA for 4 h ***(9–11,13–15,18)***.
7. Protein spots were detected by silver staining ***(9–11,13–15,19)***.

2.2.3. Protein Spot Identification

1. Specific proteins were identified by combinations of Western blotting, comparison of 2-DE images, and MS in different experiments.
2. The proteome profiles of silver-stained analytical gels were analyzed as digitized images using a CCD camera (1024 × 1024 pixels) combined with a computerized imaging system (Visage 4.6).
3. Proteins were identified by comparison with 2-DE plasma gels in SWISS-2DPAGE (ExPASy server) or with their own 2-DE NLF reference gels.
4. Protein spot concentrations were determined as background-corrected optical density.
5. Wilcoxon's rank sum test was used to calculate the significant differences ***(9,11, 13–15)***.

6. Western blotting was performed using different methods *(9,13–15,20)*.
7. The protein spots were cut for N-terminal sequence analysis with the Edman technique *(10,14,21)*.
8. For MS analysis, gels were stained with 0.1% Coomassie Brilliant Blue R-250 for 1 h or SYPRO Ruby fluorescent staining and visualized/analyzed using a CCD camera (1340 × 1340 pixels).
9. In-gel digestion was performed using trypsin.
10. Peptides from gel pieces were recovered and dried.
11. Samples were resuspended in 10 μL of 10% acetonitrile and purified by Zip-Tip.
12. The matrix (saturated solution of α-cyano-4-hydroxycinnamic acid) and extracted peptides were mixed on the target plate using the thin-layer method for MALDI-TOF analysis *(9,10,22)*.
13. The peptide mass list generated from the major peaks of the spectra was submitted to a database search (NCBI or Swiss-Prot) using PeptIdent, MS-Fit, and Mascot search engines.
14. The mass of PSD composite spectrum was submitted to the MS-Tag search engine.
15. The N-terminal amino acid sequences were analyzed with BLAST *(9,10)*.

3. Findings and Applications

In the study reported here and described by Casado et al. *(1,2)*, the proteome profiles of the NLF collected from healthy controls and subjects with sinusitis (before and after stimulating glandular secretion as well as after 6 d of pharmacological treatment) have been compared. The proteome profile obtained is shown in **Table 1**. A scheme summarizing the sources of secretions from nasal mucosa is reported in **Fig. 1.**

3.1. Anatomy of Human Nasal Mucosa and Origins of Nasal and Sinus Secretions

The nose is a complex structural barrier designed to protect the upper and lower airways against dehydration, extremes of inhaled air temperature, viruses, bacteria, fungi, irritants, chemicals, pollutants, and fine particulate materials *(23–25)*. The mucosa is specifically designed to regulate nasal airflow and permit exudation of interstitial fluid to provide water for humidification and the sol phase for ciliary action. The glands contain serous and mucous cells. Between the glands and the periostium are the erectile venous sinusoids that regulate thickness of the nasal mucosa and nasal airflow resistance. The nasal mucosa is the virtual prototype for bronchial, gastrointestinal, and other mucosal surfaces, with the exception that erectile vessels replace smooth muscle to regulate lumen diameter. The major protein classes detected in published proteomic studies are cytoskeletal proteins (33%), innate (27%) and acquired immunity (21%) system proteins, and other cellular proteins (20%) **(Table 1)**. The cytoskeletal proteins may have been derived from the turnover of epithelial

Table 1
Qualitative Detection of Proteins in Nasal and Sinusitis Mucus Proteomes

					CapLC-Q-TOF					
					Normal		Sinusitis			
					Day 1		Day 1		Day 6	
	2-DE				1X	24X	1X	24X	1X	24X
Protein name	HC[a–c]	SM[b]	Exp1[c]	Exp2[d]	NSS	NSS	NSS	NSS	NSS	NSS
ACTB protein (actin)							•			
Actin mutant β-actin					•	•				
Actin prepeptide					•	•	•			
Actin α2					•	•				
Actin β							•			
Actin γ1 propeptide							•	•		
Actin γ2 propeptide							•	•		
α-1B-adrenergic receptor							•			
Albumin	•	•	•	•	•	•	•	•	•	•
Aldehyde dehydrogenase					•	•				
Amyloid P	•		•							
Annexin A2					•					
α_1-Antichymotrypsin	•		•	•				•		
Antithrombin III	•		•	•						
α_1-Antitrypsin	•	•	•	•	•	•	•	•		
Apolipoprotein D				•	•	•				
Apolipoprotein J (clusterin)					•	•		•		
Apolipoprotein-AI	•	•	•		•	•	•	•	•	
Pro-apolipoprotein-AI	•		•		•	•				
Apoptosis-inducing factor						•				
Calgranulin B	•	•	•				•			
Calgranulin C							•			
Carbohydrate sulfotransferase									•	
Carbonic anhydrase						•				
Ceruloplasmin	•		•							

(Continued)

Table 1 *(Continued)*

					CapLC-Q-TOF					
					Normal		Sinusitis			
					Day 1		Day 1		Day 6	
	2-DE				1X	24X	1X	24X	1X	24X
Protein name	HC[a–c]	SM[b]	Exp1[c]	Exp2[d]	NSS	NSS	NSS	NSS	NSS	NSS
Chymotrypsin inhibitor							•			
Clara cell protein 16	•	•	•	•						
C3	•		•	•	•	•	•	•		
C4	•		•	•	•	•				
Cystatin S	•	•	•	•						
Cystatin SN	•	•	•							
Cystatin C	•	•								
DMBT1					•	•				
DNA methyltransferase 2f								•		•
Fibrinogen β-chain	•		•	•	•	•	•			
Fibrinogen γ-chain	•		•	•			•			
Gelsolin								•		
Gene OXA1 protein							•			
Glial fibrillary acidic protein					•	•				
α2-HS-glycoprotein	•		•	•						
α2-glycoprotein 1, zinc							•			
α1-B-glycoprotein	•		•	•						
Zinc-α-2-glycoprotein	•	•	•		•	•	•			
Haptoglobin					•	•	•			
Haptoglobin Hp2					•	•				
Haptoglobin α-chain	•	•	•		•	•				
Haptoglobin α1-chain	•									
Haptoglobin α2-chain				•						
Haptoglobin β-chain	•		•	•						
Hemoglobin α2									•	•

(Continued)

Table 1 *(Continued)*

					CapLC-Q-TOF					
					Normal		Sinusitis			
					Day 1		Day 1		Day 6	
	2-DE				1X	24X	1X	24X	1X	24X
Protein name	HC[a–c]	SM[b]	Exp1[c]	Exp2[d]	NSS	NSS	NSS	NSS	NSS	NSS
Hemoglobin α					•	•				
Hemoglobin β	•			•	•	•			•	•
Hemopexin	•		•	•						
Histon H2A					•	•	•			
Histon H2B					•	•	•			
Histon H4					•	•	•		•	
Ig-α	•		•	•	•	•				
Ig-α1	•	•	•	•	•	•	•	•	•	•
Ig-α2	•	•	•	•	•	•	•			
Ig binding factor	•	•	•	•			•	•	•	•
Ig-γ	•		•	•	•	•	•			
Ig-γ1					•	•				
Ig-γ2					•	•				
Ig-κ	•		•		•	•	•	•		
Ig light chain				•	•	•	•			
Ig-λ	•		•		•	•	•			
Ig-J	•	•			•	•				
Ig-μ	•		•	•	•	•				
IL-1 receptor antagonist							•			
IL-16							•			
IL-17E							•		•	
Keratin 1					•	•	•	•	•	
Keratin 2a					•	•				
Keratin 3					•					
Keratin 4					•	•				
Keratin 5					•		•	•		
Keratin 6a					•		•		•	•
Keratin 6b					•	•				
Keratin 6F					•	•	•			
Keratin 6L					•	•				
Keratin 7					•	•				
Keratin 8					•	•				
Keratin 9					•	•				
Keratin 10					•		•			

(Continued)

Table 1 *(Continued)*

					CapLC-Q-TOF					
					Normal		Sinusitis			
					Day 1		Day 1		Day 6	
	2-DE				1X	24X	1X	24X	1X	24X
Protein name	HC[a–c]	SM[b]	Exp1[c]	Exp2[d]	NSS	NSS	NSS	NSS	NSS	NSS
Keratin 12					•	•				
Keratin 13						•	•			
Keratin 14					•	•				
Keratin 16					•	•				
Keratin 19					•	•				
Keratin 25D					•	•				
Lacrimal proline-rich					•	•		•		
Lacrimal proline-rich4 (NCAPR4)					•	•				
Lacritin precursor								•		
Lactoferrin	•	•	•	•	•	•	•	•	•	•
Lipocalin-1	•	•	•	•	•	•	•	•		
Lipocortin I (annexin-1)	•	•	•		•			•		
5-Lipoxygenase							•			
Lsm1 protein					•	•		•		•
LUNX protein					•	•		•		
Lysozyme	•	•	•		•	•	•	•	•	•
α_2-macroglobulin	•		•		•	•				
Mammaglobin	•	•			•					
Matrix metalloprotease 27							•			
α_1-microglobulin	•		•							
β_2-microglobulin	•	•			•	•				
Mucin 5					•	•				
Mucin 5AC					•	•				
Mucin 5B					•	•				
Myeloperoxidase								•		
Nesprin-1					•	•				
NGAL				•	•	•				
Orosomucoid	•		•	•		•				
Plasminogen	•		•		•	•				
Short-PLUNC1 (PLUNC)	•	•	•		•	•				

(Continued)

Table 1 *(Continued)*

					CapLC-Q-TOF					
					Normal		Sinusitis			
					Day 1		Day 1		Day 6	
	2-DE				1X	24X	1X	24X	1X	24X
Protein name	HC[a–c]	SM[b]	Exp1[c]	Exp2[d]	NSS	NSS	NSS	NSS	NSS	NSS
Long-PLUNC1 (VEMSGP)						•	•	•		
Long-PLUNC2 (BPI)					•	•				
Poly-Ig receptor					•	•	•	•		
Prolactin-induced protein	•	•	•	•	•					
Prothrombin	•		•							
RBBP8					•	•				
Serine proteinase inhibitor							•			
Similar to cytoplasmic β-actin					•	•				
Similar to CSP					•	•	•			
SCCA					•	•				
Statherin			•							
τ-tubulin kinase						•				
TCN1						•				
TGF-β receptor II							•			
TSAP				•		•				
Transferrin	•		•	•	•	•	•			
Transthyretin	•	•	•			•				

Abbreviations: HC, healthy nonsmoker subjects; SM, Smoker subjects; Exp1, dimethylbenzylamine (DMBA); Exp2, methyltetrahydrophthalic anhydride (MHHPA); NGAL, neutrophil gelatinase-associated protein; SCCA, squamous cell carcinoma antigen; CSP, common salivary protein; TSAP, thiol-specific antioxidant protein; NCAPR4, nasopharyngeal carcinoma-associated proline rich 4; DMBT1, deleted in malignant brain tumors 1; PLUNC, palate-lung-nasal epithelium clone; BPI, bactericidal/permeability-increasing protein-like 1; VEMSGP, Von Ebner minor salivary gland protein; IL, interlukin; TCN 1, Transcobalamin I; TEF-β, transforming growth factor-β; RBBP8, Retinoblastoma binding protein 8; 2-DE, 2D electrophoresis; CapLC-Q-TOF, capillary liquid chromatography quantitative time of flight; NSS, normal saline solution.

[a]**Ref.** ***15***.
[b]**Ref.** ***10***.
[c]**Ref.** ***9***.
[d]**Ref.** ***11***.

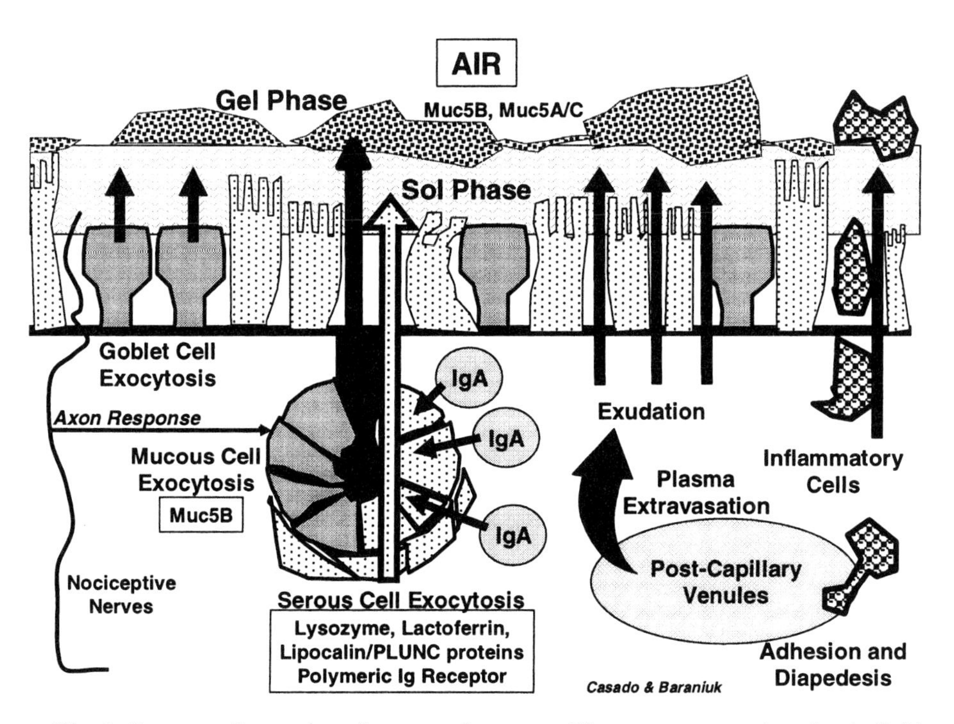

Fig. 1. Sources of secretions from nasal mucosa. The mucosa contains glands (left) and superficial postcapillary venules (right). Plasma exudation from the venules and fenestrated capillaries (not shown) form the interstitial fluid. This fluid is drawn between the ciliated (crenulated) and goblet (gray) epithelial cells into the liquid sol phase. Serous cells of the mucous glands (dotted) also contribute antimicrobial proteins to the sol phase. IgA produced by periglandular plasma cells (center) binds to serous cell polymeric immunoglobulin receptors and transports across the serous cells into the glandular mucus. Epithelial goblet and glandular mucous cells secrete mucin proteins (Muc5A/C and Muc5B, respectively) into the gel phase that floats above the sol phase. Keratins are supplied by the various types of epithelial cells. Neutrophils are present in normal and bacterial sinusitis nasal lavage fluids (far right). Stimulation of nociceptive nerves can cause substance P release and exocytosis from glands (far left). Sinus mucosa is thinner and lacks glands. Deep, erectile venous sinusoids are not shown.

goblet and ciliated cells, granule exocytosis from glands, and neutrophils that normally patrol the nasal cavity *(1)*. Quantitative assays divide the proteins into three roughly equal sources: exocytosis from glandular serous and mucous cells and plasma exudation.

3.1.1. Serous Cells

Serous cells are factories for the synthesis of antimicrobial proteins and produce about one-third of the nasal protein mass. Lysozyme cleaves bacterial peptidoglycans. Sir Stanford Fleming was actually studying lacrimal lysozyme when he serendipitously noted the effects of *Penicillium*. Lysozyme accounts for approx 15% of the total NLF protein. Lactoferrin binds free iron so this essential growth factor is unavailable for bacterial growth. Lactoferrin accounts for about 2 to 4% of total protein but was detected in 11/15 of our normal nasal samples.

3.1.2. IgA

Serous cells also participate in adaptive immunity by transporting locally synthesized immunoglobulin A (IgA). IgA-producing plasma cells are clustered around serous cells. Mucosal plasma cells secrete IgA as a dimer linked by the joining chain (IgJ). The dimer binds to the polymeric immunoglobulin receptor (PIgR; secretory component) on the exterior membrane of serous cells. The PIgR$(IgA)_2$IgJ complex (secretory IgA, sIgA) is endocytosed and transported through the serous cells to be exocytosed into the gland duct mucus with the other serous cell granule proteins. sIgA accounts for about 15% of the total nasal protein. PIgR was detected in 7/15 normal nasal lavage samples.

3.1.3. Lipocalin Proteins

The lipocalin superfamily of proteins is an important product of serous and other cells. Lipocalin-1 is the prototypic protein. It forms an eight-sheet β-barrel that curves to contain a hydrophobic inner space. Different bacterial lipopolysaccharides, lipotechoic acids, β-glucans, and other lipids are selectively bound within these β-barrel proteins. Several members of the PLUNC family were detected. Short-PLUNC-1 (SPLUNC1) has been studied as the prototypic PLUNC. It forms a single β-barrel. SPLUNC1 is expressed in submucosal glands of the upper airway, nose, and sinuses ***(26)***. In contrast, mRNA for SPLUNC1, -2, and -3 has been previously described for human nasal mucosa ***(27)***. However, we have not detected SPLUNC2 and SPLUNC3, possibly because of low concentrations, ineffective trypsin digestion, or other problems in MS/MS detection.

The long-PLUNC proteins have a "lid" that may be a ligand for poorly characterized cellular receptors that phagocytose these proteins and their microbial lipids. LPLUNC1 (von Ebner minor salivary gland protein) and LPLUNC2 (bacterial permeability inducer [BPI]) have been detected ***(28)***. Apolipoprotein D belongs to this family. Neutrophil granules contain lipocalin-2, also known as neutrophil gelatinase-associated lipocalin (NGAL) ***(1)***. Lipocalin-2 helps in packaging metalloproteases in neutrophil granules. When secreted, it acts as an antimicrobial protein. Interestingly, its production is induced by the same growth factors (insulin-like growth factor I and transforming growth factor-α) that also stimulate mucosal tissue regeneration and repair of the physical barrier against microorganisms ***(29)***. Lipocalin-2 may be a marker for dysregulated keratinocyte differentiation in human skin *(30)* and potentially may be released from epithelia as well as neutrophils and macrophages.

3.1.4. Exocytosis

Evidence of exocytosis was provided by the presence of common salivary protein 1 (CSP1). Also known as zymogen granule membrane protein and GP-2, this protein binds to the exterior of intracellular secretory granules so they can fuse to the cell membrane, become hydrated, and extrude their contents like toothpaste being squeezed out of a tube.

3.1.5. Mucous Cells

Mucous cells and epithelial goblet cells secrete acidic, sulfated, sialic acid-rich mucin proteins (muc5A/C from goblet cells and muc5B from mucous cells). These proteins have low densities because they are densely covered by serine or threonine *O*-linked carbohydrate side chains. Their cysteine-rich N- and C-terminal regions form intricately disulfide-bonded, cross-linking mucin rafts that float above the aqueous sol phase. Mucins are estimated to account for approx 28% of total protein.

3.1.6. Plasma Exudation

Plasma extravasation from the superficial fenestrated capillaries and postcapillary venous flow provides the interstitial fluid that exudes across the epithelium. This process contributes the albumin (approx 15% of all protein), IgG subclasses (approx 2% of all protein), IgM, α_1-antitrypsin, α_2-macroglobulin, Zn^{2+}-α_2-glycoprotein, apolipoprotein-A1, and other plasma proteins present in NLF. IgG is commonly detected by proteomic methods. This is likely because of its ubiquitous presence in NLF, its large size, and its many tryptic digestion sites. As a result, IgG peptides, especially those from Igγ1, have been readily detectable in our proteomic analysis. The other immunoglobulin heavy chains, Igγ2, Igγ3, Igγ4, and Igμ, and light chains (Igκ > Igλ) were detected in frequencies roughly the same as their plasma concentrations. Epithelial ion and water transport systems regulate the composition of the sol phase in health and disease.

3.1.7. Cytokeratins and Nasal Mucosa Epithelium

Cytokeratins (CK) polypeptides form intermediate filaments (IF) of the epithelial cell cytoskeleton. The patterns of cytoskeletal proteins can help us to understand and predict the turnover and status of the ciliated, transitional, and squamous respiratory epithelium in rhinosinusitis. At least 49 human keratin genes can be divided into five categories. The predominant categories in the human nasal mucosa are type I (9, 10, 10c1, 14, 16, and 19c1) and type II (1, 1c10, 2, 2A, 2c8, 2c6e, 2c1, 4, 6B, and 7) keratins *(31)*. Keratin filaments are composed of a type I protein and a type II protein. Specific keratin pairs are expressed in distinct epithelial cell types ***(32)***. K1, K10, and K9 are present in the epidermis of anterior nares. K4, K6, and K16 are present in other squamous cells. Nonkeratinized squamous cells of the transitional cutaneous-mucosal boundary express K5, K14, and K19. Transitional cuboidal and pseudostratified respiratory epithelial cells express K7, K8, K18, and K19, whereas basal cells are characterized by K5 and K14. K19 is also a marker of epithelial inflammation ***(33)***.

Actin filaments are composed of two strands of polymerized actin protein. These cytoskeletal elements are localized to the cellular periphery under the plasma membranes, where they maintain and support the cellular surface structures. The presence of actin, intermediate filament, and microtubule proteins demonstrates that these intracellular filamentous structures can be readily released through exocytosis or epithelial cell turnover or may be derived from nasal neutrophils and other inflammatory cells.

3.2. Nasal Provocations and Toxic Exposure

Lindahl et al. ***(14,15)*** have identified different patterns of expression of IgA, SPLUNC1, calgranulin B, statherin, and others. It is possible that exposures to toxic agents cause alterations in innate immune system, mucosal defense, and protein patterns or concentrations that place workers at risk for subsequent development of virus-induced upper or lower respiratory tract infections, sinusitis, or asthma.

3.2.1. Hypertonic Saline

Hypertonic saline stimulates type C neurons, which mediate centrally appreciated burning pain and local mucosal-neural axon responses. The neurotransmitter substance P (as well as others) is released locally and may act via substance P receptors on the glands to generate glandular exocytosis. Although quantitative studies have demonstrated an increase in glandular secretion with no change in vascular permeability ***(5,6)***, our proteomic studies have not demonstrated any qualitative changes in the detectable proteins.

3.2.2. Dimethylbenzylamine (DMBA)

Two hours of exposure to DMBA caused several significant changes in NLF protein spot intensities when assessed by 2-DE. In healthy controls not previously exposed to DMBA, IgA was increased, whereas SPLUNC1 and Clara cell-16 (uteroglobin) were decreased. In contrast, epoxy workers previously exposed to occupational DMBA demonstrated increased spot intensities for SPLUNC1, ceruloplasmin, α_2-macroglobulin, and Zn^{2+}-α_2-glycoprotein. Spot intensity of the cytoskeletal protein statherin was also significantly decreased ***(9,12)***. Lindahl et al.

(12) suggested that SPLUNC1 may serve as a potential irritant biomarker under these circumstances.

3.2.3. Methyltetrahydrophthalic Anhydride (MHHPA)

Subjects exposed to MHHPA showed increased IgA and cystatin S spot intensity with a decrease in lactoferrin. The mechanisms are unclear since these are all serous cell products.

3.2.4. Cigarette Smoking

NLFs from smokers showed decreases in Clara cell protein (uteroglobin), a truncated variant of lipocortin-1- (annexin-1), and three acidic forms of α_1-antitrypsin. One phosphorylated isoform of cystatin S was increased ***(10)***.

3.3. Sinusitis

The nasal sinuses are aerated cavities within the bones of the face and skull. They are connected to the nasal cavity through 1- to 3-mm-diameter ostia. The combination of obstruction of these openings and viral or bacterial infection can lead to a transient sterile sinusitis (postrhinovirus) or purulent rhinosinusitis ***(34)***.

3.3.1. Acute Sinusitis

The mucus collected from acute sinusitis subjects, including those with acute exacerbations of chronic sinusitis, had some similarities with normal nasal secretions but also had a substantial number of lower abundance proteins. Albumin and Igγ1, Igλ, and Igκ were prominent. The presence of acute-phase antiprotease reactants including α_1-antitrypsin, α_1-antichymotrypsin, α_2-Zn^{2+}-glycoprotein, and α_2-sulfated-glycoprotein suggested that inhibition of bacterial proteases is an important host defense mechanism. Again, iron sequestration was evident from the detection of both haptoglobin and transferrin. Of interest was the presence of β- and γ-fibrinogens. These coagulants combined with cross-linked mucins, albumin, and neutrophilic DNA would form the tenacious mucus that is characteristic of acute sinusitis. The presence of DNA was suggested by detection of DNA methyltransferase 2f and histones H2A, H2B, and H4 ***(2)***.

3.3.2. Acute Sinusitis Inflammation

The presence of neutrophils is also supported by identification of myeloperoxidase, the enzyme that gives the greenish tinge to mucopurulent secretions. 5-Lipoxygenase, which is present only on myeloid series cells, has also been detected, indicating the probable production of leukotriene B4 and possibly cysteinyl leukotrienes. Chemoattractants and inflammatory cell activators including calgranulins B and C, interleukin-16 (IL-16) and IL-17E were also found. Complement factor C3 was identified. Its proteolytic fragment, C3b, is a major neutrophil chemoattractant. The detection of actin and the actin-binding proteins ACTB-β, -γ1, and -γ2 and gelsolin suggests chemotactically active cells. Transforming growth factor-β receptor II may play a role in epithelial or fibroblast repair processes or antiinflammatory effects. IL-1 receptor antagonist (IL-1-RA) may play a similar counterregulatory role *(2)*.

3.3.3. Acute Sinusitis Glandular Secretion

During infection, the anticipated serous cell proteins lactoferrin, lysozyme, LPLUNC1, lipocalin-1, and common salivary protein 1 are augmented by lacrimal proline-rich protein, lacritin, LUNX (lung, upper airways, nasal clone), and a serine/cysteine protease inhibitor (likely cystatin C). LUNX is probably an isoform of SPLUNC1 since they differ by only one amino acid. The putative serine/cysteine protease inhibitor may play a role in the inhibition of apoptosis (capsases) or bacterial cysteine proteases *(2)*.

3.3.4. Acute Sinusitis Keratins and Cell Types

A less extensive set of keratins (1, 5, 6A, 6F, 10, and 13) is expressed more predominantly in the sinusitis mucus than in normal NLF. These are generally related to squamous epithelia. This is compatible with the progressive changes in epithelial phenotype as chronic sinusitis worsens *(33)*. Normally, ciliated cells predominate. As the inflammation of rhinosinusitis begins, ciliated cells are replaced by goblet cells. This leads to reduced mucociliary transport and mucostasis. As the disease worsens, the microvillous cells become prominent. In pansinusitis and nasal-sinus polyposis, squamous metaplasia occurs. Scanning electron microscopy indicates that the epithelium is breached, with patches of naked basement membranes and erythrocytes on the mucosal surface that indicate bleeding *(2)*.

3.4. Sinusitis Mucus After Six Days of Treatment

The total protein collected and number of proteins detected were significantly reduced by the antibiotic, nasal steroid, decongestant, and saline lavage therapy. Expected proteins were albumin, IgG, IgA, lysozyme, lactoferrin, common salivary protein 1, and keratins 1 and 6A. Novel proteins included hemoglobins α2 and β. These indicated that bleeding was present either because of the rhinosinusitis or as a side effect of the topical nasal steroid sprays *(35)*. An important novel protein induced by therapy was carbohydrate sulfotransferase. This is the enzyme responsible for adding sulfate (sulfuric acid) to the sialated carbohydrate side chains of muc5A/C and muc5B. The acid mucins have an isoelectric point in the range of 2 to 2.5. This may increase their antimicrobial activity. However, detrimental effects would increase the tenacity that may lead to the formation of sequestered mucoclots that could become the sanctuaries for biofilm-forming bacteria. These acid mucins are also very irritating and induce nausea when swallowed. Hypertonic saline had little effect on glandular secretion or vascular permeability in acute sinusitis *(6)* or after therapy.

4. Perspectives and Future Directions

The glandular and epithelial secretion of innate immune system antimicrobial proteins and highly adherent mucins form a strong barrier to protect the airways. These defenses are augmented by neutrophilic and acquired immune mechanisms (immunoglobulin production and cellular immune mechanisms). Toxic inhalants and acute sinusitis cause marked changes in the nasal proteome that provide insights into the host defense mechanisms. Serial assessments after provocations may identify the chronology of changes during viral, allergic, or toxic exposures. The results can be readily adapted to the bronchial, salivary, and gastrointestinal systems since the epithelial, glandular, and vascular architecture are quite similar to the nasal mucosa *(26)*.

The nasal mucus proteome is a challenge to analyze. It contains lipids that can interfere with LC methods. The mucous globules are very tightly bound, especially given the additional adherent proteins found in sinusitis. The expression of protein families requires conservative bioinformatics analysis so that quite similar isoforms such as LUNX and SPLUNC1 can be properly identified. The diversity of immunoglobulin variable regions creates a large number of unique peptides that may interfere with separation or MS/MS detection methods after tryptic digestion. The high concentrations of albumin and IgG in NLF, and the large number of high-concentration tryptic peptides may limit the number of lower abundance peptides to be sequenced and identified. Future studies will require removal of these high-abundance proteins for optimal CapLC-MS/MS and 2-DE identification of significant, disease-specific low-abundance proteins.

The search for disease biomarkers seems to be more efficient with 2-DE at the present time. However, its limitations are that replicate runs and pool samples to detect low-abundance proteins as well as multiple proteins that may be hidden behind single spots or albumin and immunoglobulin smears are required. Continued analysis and comparison of results obtained by complementary methods will lead to advances in our understanding of the innate immune responses to toxic and other exposures, as well as the development of sets of diagnostic biomarkers for rhinosinusitis diseases *(27)*. These studies will surely be extended to examination of the olfactory mucosa *(28)*.

Acknowledgments

This work was supported by PHS Award AI 42403 and by U.S. Public Health Service Award 2 P50 DC000214-18 to the Monell/Jefferson Chemosensory Clinical Research Center (CCRC).

References

1. Casado B, Pannell LK, Iadarola P, Baraniuk JN. Identification of human nasal mucous proteins using proteomics. *Proteomics* 2005;5:2949–2959.
2. Casado B, Pannell LK, Viglio S, Iadarola P, Baraniuk JN. Analysis of the sinusitis nasal lavage fluid proteome using capillary liquid chromatography interfaced to electrospray inonization quandrupole time of flight tandem mass spectrometry. *Electrophoresis* 2004;25:1386–1393.
3. Ali M, Maniscalco J, Baraniuk JN. Spontaneous release of submucosal gland serous and mucous cell macromolecules from human nasal explants in vitro. *Am J Physiol* 1996;270:L595–L600.
4. Baraniuk JN, Silver PB, Kaliner MA, Barnes PJ. Perennial rhinitis subjects have altered vascular, glandular, and neural responses to bradykinin nasal provocation. *Int Arch Allergy Immunol* 1994;103:202–208.
5. Baraniuk JN, Ali M, Yuta A, Fang Sheen-Ien, Naranch C. Hypertonic saline nasal provocation stimulates nociceptive nerves, substance P release and glandular mucous exocytosis in nomal humans. *Am J Respir Crit Care Med* 1999;160: 655–662.
6. Baraniuk JN, Petrie KN, Le U, et al. Neuropathology in rhinosinusitis. *Am J Respir Crit Care Med* 2005;71:5–11.
7. Rhinosinusitis: establishing definitions for clinical research and patient care. Meltzer EO, Hamilos DL, Hadley JA, Lanza DC, Marple BF, Nucklas RA, eds. *J Allergy Clin Immunol* 2004;114:1A–S212.
8. Lowry OH, Rosebrough NJ, Farr AL, Randall RJ. Protein measurement with the Folin phenol reagent. *J Biol Chem* 1951;193:265–275.
9. Lindahl M, Ståhlbom B, Tagesson C. Identification of new potential airway irritation marker, palate lung nasal epithelial clone protein, in human nasal lavage fluid with two- dimensional electrophoresis and matrix-assisted laser desorption/ionization-time of flight. *Electrophoresis* 2001;22:1795–1800.
10. Ghafouri B, Ståhlbom B, Tagesson C, Lindahl M. Newly identified proteins in human nasal lavage fluids from non-smokers and smokers using two-dimensional gel electrophoresis and peptide mass fingerprinting. *Proteomics* 2002;2:112–120.
11. Lindahl M, Ståhlbom B. Tagesson C. Two-dimensional gel electrophoresis of nasal and bronchoalveolar lavage fluids after occupational exposure. *Electrophoresis* 1995;16:1199–1204.
12. Lindahl M, Irander K, Tagesson C, Ståhlbom B. Nasal lavage fluid and proteomics as means to identify the effects of the irritating epoxy chemical dimethylbenzylamine. *Biomarkers* 2004;9:56–70.
13. Lindahl M, Ståhlbom B, Tagesson C. Demonstration of different forms of the anti-inflammatory proteins lipocalin-1 and Clara cell protein-16 in human nasal and bronchoalveolar lavage fluids. *Electrophoresis* 1999;20:881–890.
14. Lindahl M, Ståhlbom B, Tagesson C. Newly identified proteins in human nasal and bronchoalveolar lavage fluids: potential biomedical and clinical applications. *Electrophoresis* 1999;20:3670–3676.

15. Lindahl M, Ståhlbom B, Svartz J, Tagesson C. Protein pattern of human nasal and bronchoalveolar lavage fluids analyzed with two-dimensional gel electrophoresis. *Electrophoresis* 1998;19:3222–3229.
16. Greiff L, Pipcorn U, Alkner U, Persson CG. The 'nasal pool' device applies controlled concentrations of solutes on human nasal airway mucosa and samples its surface exudations/secretions. *Clin Exp Allergy* 1990;20:253–259.
17. Bradford MM. A rapid and sensitive method for the quantitation of microgram quantities of protein utilizing the principle of protein-dye binding. *Anal Biochem* 1976;72:248–254.
18. Gorg A, Postel W, Gunther S. The current state of two-dimensional electrophoresis with immobilized pH gradients. *Electrophoresis* 1988;9:531–546.
19. Bjellqvist B, Basse B, Olsen E, Celis JE. Reference points for comparisons of two dimensional maps of proteins from different human cell types defined in a pH scale where isoelectric points correlate with polypeptide compositions. *Electrophoresis* 1994;15:529–539.
20. Towbin H, Staehelin T, Gordon J. Electrophoretic transfer of proteins from polyacrylamide gels to nitrocellulose sheets: procedure and some applications. *Proc Natl Acad Sci U S A* 1979;76:4350–4354.
21. Matsudaira P. Sequence from picomole quantities of proteins electroblotted onto polyvinylidene difluoride membranes. *J Biol Chem* 1987;262:10,035–10,038.
22. Roepstorff P, Larsen MR, Rahlbek-Nielsen H, Nordhorff E. In: Celis, JE, ed. *Cell Biology: A Laboratory Handbook.* New York: Academic, 1998:556–565.
23. Jackson AD. Airway goblet-cell mucus secretion. *Trends Pharmacol Sci* 2001; 221:39–45.
24. van Eeden SF, Yeung A, Quinlam K, Hogg JC. Systemic response to ambient particulate matter. *Proc Am Thorac Soc* 2005;2:61–67.
25. Hunter DD, Satterfield BE, Huang J, Fedan JS, Dey RD. Toluene diisocyanate enhances substance P in sensory neurons innerveting the nasal mucosa. *Am J Respir Crit Care Med* 2000;161:543–549.
26. Bingle L, Cross SS, High AS, et al. SPLUNC1 (PLUNC) is expressed in glandular tissues of the respiratory tract and in lung tumours with a glandular phenotype. *J Pathol* 2005;205:491–497.
27. Bingle CD, Craven CJ. PLUNC: a novel family of candidate host defense proteins expressed in the upper airways and nasopharynx. *Hum Mol Genet* 2002;11:937–943.
28. Flower DR. The lipocalin protein family: structure and function. *Biochem J* 1996; 318:1–14.
29. Sorensen OE, Cowland JB, Theilgaard-Monch K, Liu L, Ganz T, Borregaard N. Wound healing and expression of antimicrobial peptides/polypeptides in human keratinocytes, a consequence of common growth factors. *J Immunol* 2003;170: 5583–5589.
30. Mallbris L, O'Brien KP, Hulthen A, et al. Neutrophil gelatinase-associated lipocalin is a marker for dysregulated keratinocyte differentiation in human skin. *Exp Dermatol* 2002;11:584–591.

31. Hesse M, Magin TM, Weber K. Genes for intermediate filament proteins and the draft sequence of the human genome: novel keratin genes and a surprisingly high number of pseudogenes related to keratin genes 8 and 18. *J Cell Sci* 2001;114: 2569–2575.
32. Rugg EL, Leigh IM. The keratins and their disorders. *Am J Med Genet C Semin Med Genet* 2004;131C:4–11.
33. Chu PG, Weiss LM. Keratin expression in human tissues and neoplasms. *Histopathology* 2002;40:403–439.
34. Low DE, Desrosiers M, McSherry J, et al. A practical guide for the diagnosis and treatment of acute sinusitis. *CMAJ* 1997;156:1–14.
35. Trangsrud AJ, Whitaker AL, Small RE. Intranasal corticosteroids for allergic rhinitis. *Pharmacotherapy* 2002;22:1458–1467.
36. Passalacqua G, Ciprandi G, Canonica GW. The nose-lung interaction in allergic rhinitis and asthma: united airways disease. *Curr Opin Allergy Clin Immunol* 2001;1:7–13.
37. Passalacqua G, Canonica GW. Impact of rhinitis on airway inflammation: biological and therapeutic implications. *Respir Res* 2001;2:320–323.
38. Schwob JE, Saha S, Youngentob SL, Jubelt B. Intranasal inoculation with the olfactory bulb line variant of mouse hepatitis virus causes extensive destruction of the olfactory bulb and accelerated turnover of neurons in the olfactory epithelium of mice. *Chem Senses* 2001;26:937–952.

16

Proteomics of Human Saliva

Francisco M. L. Amado, Rui M. P. Vitorino, Maria J. C. Lobo, and Pedro M. D. N. Domingues

Summary

The salivary proteome is a complex protein mixture resulting from the activity of salivary glands with the contribution of other components that form the oral environment such as oral tissues and microorganisms. Knowledge of the salivary proteome will bring not only improvement in the comprehension and diagnosis of oral pathologies but also the evaluation of systemic status. In fact, for diagnosis purposes, saliva collection has the great advantage of being an easy and noninvasive technique. This chapter gives an overview of the most frequently used proteomics methodologies for the isolation and identification of salivary proteins and peptides, such as the separation techniques (2D polyacrylamide gel electrophoresis, high-performance liquid chromatography, and 2D liquid chromatography) in conjunction with mass spectrometry. Currently, more than 300 salivary proteins have been identified. It is expected that this number will quickly increase in the near future, particularly when the actual major limitations of protein identification in saliva, such as the high content of mucins, debris, and bacteria, as well as its high proteolytic activity, are overcome.

Key Words: Saliva; salivary glands; oral cavity; proteins; peptides; mass spectrometry; proteome; proteomics.

1. Introduction

Compared with other human body fluids, saliva has been less studied and used in analytical and clinical applications because of its high mucin content, which confers particular physicochemical properties, and an apparent complex circadian rhythm. However, the interest in saliva is increasing. Efforts have been made during the past few years to determine salivary proteins and their physiological roles. Although progress has been made, this adventure is in an initial phase compared with the study of other human body fluids. In human health, development of new analytical methods for the analysis of saliva will bring not

From: *Proteomics of Human Body Fluids: Principles, Methods, and Applications*
Edited by: V. Thongboonkerd © Humana Press Inc., Totowa, NJ

only improvement in the comprehension and diagnosis of oral pathologies but also in the evaluation of systemic status. Additionally, saliva collection has the great advantage of being an easy and noninvasive technique.

Saliva is produced and secreted into the oral cavity by salivary glands, both major and minor, which together produce an impressive amount of 1000 to 2000 mL/d ***(1–3)***. The major salivary glands include parotid, submandibular, and sublingual glands. Minor salivary glands are situated on the tongue, palate, and buccal and labial mucosa. They are small mucosal glands with primarily mucous secretion ***(4)***. The working part of salivary glands consists of the secretory end pieces (acini) and the branched ductal system. The fluid first passes through intercalated ducts, which have a low cuboidal epithelium and narrow lumen. The secretion then enters the striated ducts, which are lined with more columnar cells rich in mitochondria. Finally, the saliva passes through the excretory ducts, where the cell type is cuboidal with stratified squamous epithelium ***(5)***. The acinar cells first secrete isotonic primary saliva and then the striated duct cells actively extract ions to render the saliva progressively more hypotonic as it passes down the ducts toward the mouth ***(2,6,7)***.

Numerous physiological studies have concluded that salivary acinar cells release the primary components of saliva ***(2,8,9)***. Acinar cells comprise almost 90% of salivary glands and synthesize and secrete nearly all the salivary proteins, using a complex array of ion pumps and channels at the cell surface to drive the transepithelial transport of interstitial fluid in order to provide water and electrolyte components of the secretion ***(10,11)***. Parotid acinar cells produce mainly a group of enzymes (amylase, peroxidase, and others) and elongate polypeptides (proline-rich proteins [PRPs], histatins, and others), being in this way considered amicrobial ***(12)***. Sublingual acinar cells produce mainly mucins, whereas submandibular cells produce mucins, enzymes, and elongate polypeptides similar to those produced by both sublingual and parotid glands ***(2)***.

The release of salivary components by acinar cells is regulated by neuronal stimuli. Acinar cells are richly innervated by both sympathetic and parasympathetic nerve fibers, which have the classic neurotransmitters and selected bioactive peptides (such as substance P) as the main stimuli of secretion ***(2)***. These two types of stimuli have significantly different effects on the composition of salivary output. Parasympathetic stimulation evokes output of saliva that has a large volume and low protein concentration, whereas sympathetic stimulation has the opposite effect, causing release of saliva that has a relatively small volume and high protein concentration ***(2,13)***.

The principal pathway by which salivary proteins are released from acinar cells is exocytosis of secretory granules. Within the cytoplasm of acinar cells, most of the membrane-bound organelles are part of the intracellular transport or secretory pathway. This pathway has the classic orientation toward the apical secretory surface, at which the extensive endoplasmic reticulum and nucleus fill the basal cytoplasm beneath a central Golgi complex and an overlying accumulation of secretory granules. In both resting and stimulated cells, almost all newly synthesized proteins (85%) follow this secretory pathway, whereas the remaining 15% of secretory proteins release without storage. These proteins enter the endoplasmic reticulum during translation; most are transported to the Golgi complex, where posttranslational modifications (PTMs) are largely completed, and then they are routed into forming granules, where they are condensed for storage at concentrations that exceed 300 mg/mL ***(2,8,13)***.

Salivary secretion is mainly controlled by the autonomic nervous system. The normal daily imbalance of the cholinergic and adrenergic systems or exogenous stimulatory conditions, such as medication, radiation, and food ingestion, can produce changes in salivary flow and/or composition ***(14)***. All salivary functions are based on a wide spectrum of multifarious components. These components not only contribute to oral cavity homeostasis but also provide clues for systemic diseases. Human saliva is an attractive diagnostic fluid because its collection is noninvasive and simple, as opposed to blood collection for serum/plasma analyses. Blood concentrations of many components are reflected in saliva. In addition, the cost for sample handling is low. Moreover, since it is a noninvasive collection technique, it can dramatically reduce anxiety and discomfort in patients, simplifying collection of serial samples for monitoring general health and disease states over time (*see* reviews by Kaufman and Lamster ***[13]***, Lawrence ***[15]***, and Dodds et al. ***[16]***). Efforts have been made recently to find a correlation between salivary proteins and different systemic diseases; *salivary biomarkers* may potentially be used for monitoring general health and for early diagnosis of disease ***(15,17,18)***.

2. Brief Overview of Salivary Proteomics

2.1. Methodology

Proteomics has emerged in the past decade as a multidisciplinary and technology-driven science focusing on analysis of the proteome, including the complex of proteins expressed in a biological system at a given moment, as well as their structures, interactions, and PTMs ***(19)***. In saliva, most proteins originate from salivary glands. However, blood ***(20)***, oral tissues ***(21,22)***, and microorganisms, particularly bacteria ***(23)***, can also be other important sources of proteins. Thus, all extrinsic and intrinsic salivary proteins should be included as parts of the salivary proteome. Recent advances in mass spectrometry (MS) with different separation techniques, such as 2D polyacrylamide gel electrophoresis (2D-PAGE), high-performance liquid chromatography (HPLC), and 2D liquid chromatography (2D-LC), have allowed the development and exponential growth of proteomics, which has attained the status of one of the most commonly used techniques for protein analysis *(24)*. All these techniques have already been applied to the study of salivary proteins. The result is the identification of more than 300 proteins in saliva. This number will quickly increase in the near future, particularly when the actual major limitations of protein identification in saliva (such as the high content of mucins, debris, and bacteria, as well as its high proteolytic activity) are overcome *(25–30)*. **Figure 1** gives an overview of the state of the art of proteomic methodologies applied to salivary protein characterization. The different experimental steps will be discussed below.

2.2. Sample Collection

The first step in analyzing human saliva is its collection and preservation. Factors that may affect the compositions of saliva include circadian rhythm *(31,32)*, different gland contributions *(33–36)*, stimulations *(35,36)*, health status, physical exercise, medications, food intake, gender, and age *(37–39)*. The volume that is required for a proteomics study varies from few *(40)* to several milliliters *(41)*. Methods for human saliva collection at resting state or under stimulating conditions include intraoral duct cannulation *(42)* and Lashley cups (and their modifications) *(43)* as well as Schneyer's device *(44)*. The subjects are usually required to cease eating, drinking, or using oral hygiene products for at least 1 h prior to saliva collection. Collection of nonstimulated whole saliva is based on draining (into a reservoir), aspiration, or using an absorbent material that is chewed or placed somewhere in the mouth *(45,46)*. Stimulated saliva (whole saliva, saliva from parotid or submandibular glands) can be obtained under conditions of gustatory stimulation, such as sour candies *(31)*, lemon juice or citric acid (approximately 2% w/v) *(18,47,48)*, and chewing paraffin wax *(49)*. Parotid secretion is relatively easy to collect with a Carlson-Crittenden device or a Lashley cup *(50)*. For submandibular/sublingual secretion, several devices can be employed including custom-made field-collecting devices (placed at the opening aperture of the Wharton's/Bartholin's ducts) *(9,44)*, Lashley cups *(45)*, gentle suction *(51)*, or polyethylmethacrylate devices *(52)*. In all collection methods, the reservoirs for saliva samples are kept on ice during the collection and immediately frozen in liquid nitrogen thereafter.

3. Analyses of Salivary Proteins

Saliva contains a wide spectrum of proteins, namely, amylase, carbonic anhydrase VI, lysozyme, lactoferrin, lactoperoxidase, immunoglobulins, agglutinin, mucins, and others, which have biological functions of particular importance to oral health *(12,13,16)*. The major salivary proteins, which account for approximately 50% of total salivary protein, are amylase, immunoglobulins, mucins, and PRPs. The salivary immunoglobulins are mainly IgA (85%) and, in a much smaller proportion, IgG. Together, they make up about 5 to 15% of

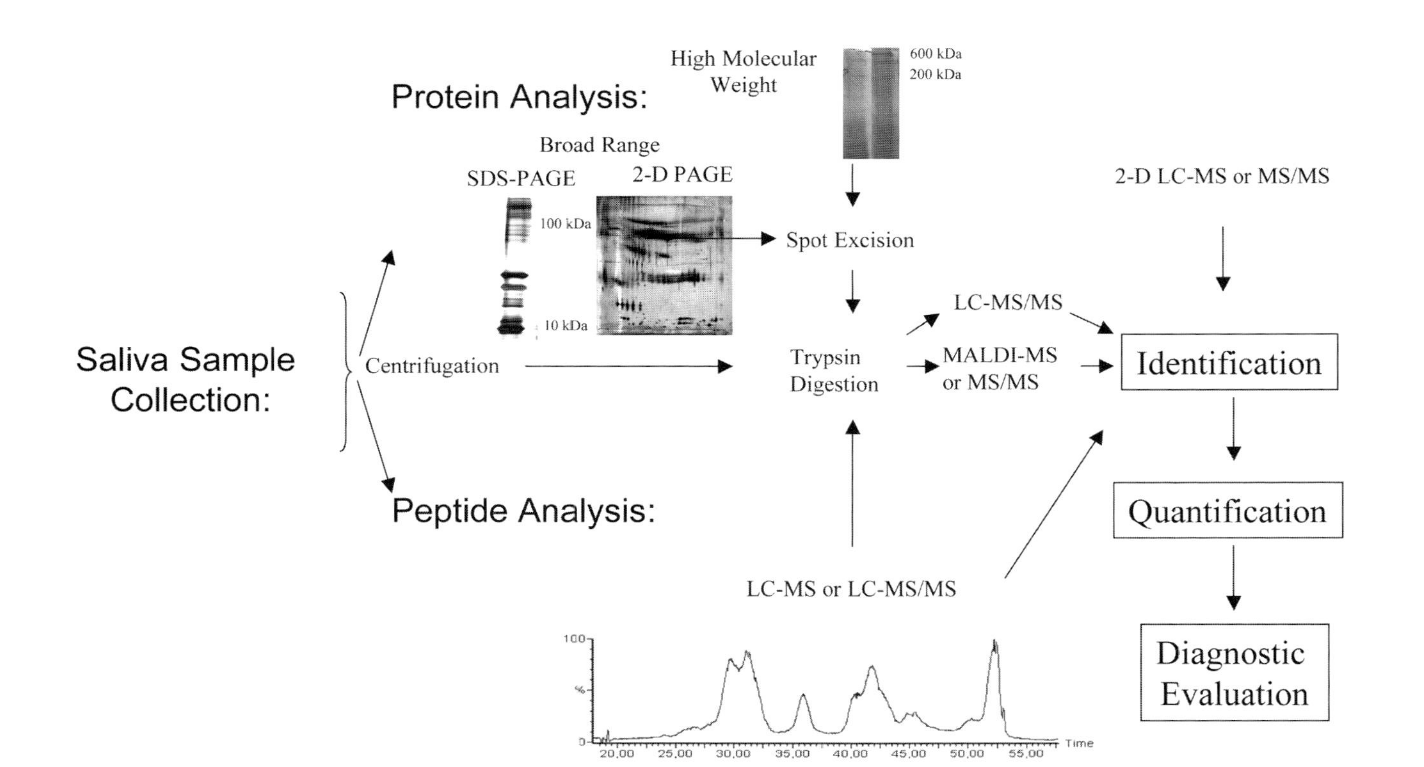

Fig. 1. Overview of proteomic methodologies applied to salivary protein characterization.

total salivary protein. Mucins constitute another important class of salivary glycoproteins. In unstimulated whole saliva, they are the major components, making up 20 to 30% of the total protein. The low-abundant salivary proteins are metabolic enzymes with antimicrobial activity, such as lactoperoxidase and lysozyme ***(12)***. In addition, saliva contains several low-molecular-weight (MW) components with important bactericidal activity and remineralizing property. These include histatins, defensins, PRPs, and cystatins.

To analyze a complex organic fluid like saliva and to evaluate its protein composition, several separation techniques have been used. Both proteomic methodologies and traditional analytical techniques, such as sodium dodecyl sulfate (SDS) PAGE and isoelectric focusing (IEF), have been widely employed in separation, characterization, and identification of salivary proteins. Moreover, other methods involving various chromatographic techniques (size exclusion, ion exchange, affinity chromatography, and others) and methodologies incorporating chemically modified solid surfaces for protein analysis have also been applied by different researchers ***(34,53,54)***.

Prior to protein separation, sample treatment depends on the technique to be used. The first step is to centrifuge the sample to remove insoluble materials (10,000–14,000*g*, 10–25 min) ***(25–27,55–57)***; the supernatant is then saved at –20°C to –80°C until analysis. Although centrifugation is considered a critical step in sample preparation for removing bacteria, cellular debris, and glycoprotein aggregates, concomitantly several proteins such as cystatins, amylase, PRPs, and statherin are coprecipitated ***(50)***. This step may be avoided in the case of clean samples, such as those obtained by cannulation. Long storage time as well as freeze-thaw cycles can induce protein precipitation, in particular of low-MW components ***(58,59)***. Some investigators thus use liquid nitrogen for snap-freezing the sample for its homogeneity.

In cases of electrophoretic techniques, such as IEF and 2D-PAGE, special care should be taken in sample treatment because these methods are salt sensitive. To remove salts, dialysis of the samples is usually performed overnight against 12 m*M* ammonium bicarbonate or deionized water ***(60)*** using a 10,000-kDa molecular mass cutoff dialysis membrane ***(28)***. If necessary, proteins can be concentrated by acid precipitation, e.g., with 10% trichloroacetic acid (TCA)/ 0.1% dithiothreitol (DTT) ***(26)*** or 10% TCA/90% acetone/20 m*M* DTT ***(29)***. After the centrifugation step, the protein pellet can be resuspended in a rehydration buffer and quantified by conventional methodologies, e.g., Lowry ***(61)***, bicinchoninic acid (BCA) ***(62)***, or Biuret ***(63)***.

3.1. 1D Polyacrylamide Gel Electrophoresis (1D PAGE)

SDS-PAGE or 1D PAGE is an easy-to-use separation procedure and has the advantage of having a low sensitivity to salivary salt concentration. It has been employed in the study of the protein composition of whole saliva and specific-gland saliva (parotid, submandibular, or sublingual), with or without stimulation. More than 30 protein bands can be resolved from whole saliva ***(36,64)***. SDS-PAGE analysis has also been used in the characterization of salivary proteins that interact with the tooth surface, forming the acquired enamel pellicle ***(65,66)***. To expand the range of observed salivary proteins, both anionic and cationic gels may be used to obtain well-resolved bands. Using this technique, several investigators have achieved a fine separation of histatins, cystatins, and PRPs ***(67,68)***. To maximize the MW range separation (from 5 to 300 kDa), Francis et al. ***(50)*** and Bardow et al. ***(69)*** have used gradient gels, ranging from 5 to 20%.

An interesting approach to the study of salivary complexes has been used by Oho et al. ***(70)*** and by Iontcheva et al. ***(71)*** using SDS-PAGE of salivary proteins under native and denaturating conditions. These electrophoretic fractionation studies have allowed the verification of the existence of complexes of mucin MUC5B (MG1) with other salivary proteins like histatins, PRPs, amylase, and statherins and have also demonstrated that agglutinin is comprised of a high-molecular-mass glycoprotein (later confirmed as the glycoprotein GP340) ***(72)*** attached to secretory immunoglobulin A (sIgA) components ***(70)***. The recent work of Soares and co-workers is another reference to the study of salivary complexes ***(73,74)***. These authors have used SDS-PAGE and blotting techniques to show the existence, in vivo and in vitro, of complexes of the mucin MG2 with several other proteins like lactoferrin, amylase, PRPs, and lysozyme.

IEF using carrier ampholytes (without 2D separation) has also been applied to salivary protein analysis during the last decade. With this approach, based on the isoelectric point (p*I*) of each protein, more than 20 bands can be visualized with Coomassie Blue stain, including basic PRPs that are difficult to separate by SDS-PAGE because of their basic p*I* (>8) ***(56)***. A weakness of IEF using carrier ampholytes is related to the need for a prior desalting step, which is time consuming and can lead to protein loss.

3.2. 2D Polyacrylamide Gel Electrophoresis (2D-PAGE)

Using 2D-PAGE, more than 600 protein spots per gel can be visualized in a saliva sample ***(75)***. Nowadays, owing to its extraordinary resolution, availability, and abundantly accumulated knowledge, 2D-PAGE is widely used for separation of protein mixture ***(76)***. Moreover, 2D-PAGE allows the separation of different protein isoforms that are likely generated by PTMs (glycosylation, phosphorylation, or proteolytic processing). Typically, the two dimensions correspond to the use of IEF in the first dimension to separate proteins according to their p*I*, followed by SDS-PAGE in the second dimension to resolve proteins according to their molecular size (M_r). Since the immobilized pH gradient (IPG) technique was introduced and better reproducibility was combined with the higher protein load capacity ***(77)***, 2D-PAGE has been used more widely, and increasing information has been obtained.

2D-PAGE was first applied to saliva samples by Beeley et al. ***(56)***. Initially, only a few major components of saliva such as amylase and immunoglobulins were identified. With the introduction of the IPG technique ***(56)***, substantial resolution can be achieved allowing the detection of more than 50 spots visualized with Coomassie Blue and more than 100 spots using silver stain ***(56)***. Some of these spots (from parotid secretion) have been identified as lactoferrin, secretory component, amylase, PRPs, Ig heavy chain, and albumin. Improvement of protein solubilization using a high concentration of urea results in better resolution of a 2D-PAGE map in the pH 4 to 7 range, allowing the detection of more than 200 spots visualized with a fluorescent dye (SYPRO Ruby) ***(25,26)***. The combination of a high concentration of urea and thiourea ***(78)*** further results in better resolution, which allows the separation of more than 200 spots within the pH range 3 to 10.

Although it is one of the most powerful protein separation techniques, 2D-PAGE has some drawbacks that must be highlighted. In fact, high-MW salivary proteins, glycoproteins (e.g., mucins), and low-MW proteins/peptides are out of range of the separation capability of common 2D-PAGE techniques (typically resolved proteins with molecular masses between 10 and 120 kDa). Moreover, in spite of the use of robots and precast gels, this technique is still laborious and time consuming. Even though reproducible spot patterns are usually achieved in the same laboratory, data comparison between different laboratories is often difficult because gel reproducibility is highly dependent on experimental conditions and procedures. On the other hand, although not identical, similar results can be observed when one compares the results obtained by different groups ***(20,27,29)***.

3.2.1. 2D-PAGE Procedures

1. Whole saliva is centrifuged at 12,000 rpm and 4°C for 20 min to remove cell debris and particulate matter. To minimize degradation of proteins, protease inhibitor cocktail and phenylmethylsulfonyl fluoride (PMFS) may be added to the supernatant.
2. Protein content can be concentrated by adding 10% TCA/0.1% DTT ***(20)*** or 10% TCA/90% acetone/20 m*M* DTT ***(29)***.
3. The supernatant is discarded and the pellet is then washed with cold 20 m*M* DTT/acetone (or 10 m*M* phosphate buffer) and resuspended with a rehydration buffer containing 8 *M* urea, 2 *M* thiourea, 2% 3-[(3-cholamidopropyl)dimethylammonio]-1-propanesulfonate (CHAPS), 0.1% ampholytes (pH 3–10 or 4–7), 2 m*M* DTT, and a trace of bromophenol blue.
4. Samples can be analyzed immediately or stored at –20°C to –80°C until analysis.
5. The addition of thiourea increases protein solubility and improves resolution on the 2D gel map ***(78)***. Moreover, the increment of protein solubility by thiourea

diminishes the interference of mucins and other high-MW glycoproteins present in saliva that can block the current passage ***(20,27,29)***.

6. In case of high salt content, improvement of the IEF can be made by introducing a small voltage step (e.g., 100 V, 2 h for 7-cm-long IPG strips and 3.5 h for 13-cm-long IPG strips).
7. Details of 2D-PAGE procedures can be found elsewhere or in the manufacturers' manuals. (Note that the initial centrifugation step can be skipped in the case of parotid or submandibular saliva).

3.3. Detection or Visualization of Protein Bands/Spots

Experimental protocols used for staining salivary proteins are different from those applied to other samples. The most commonly used stains are Coomassie Blue, silver, and fluorescent dyes. Coomassie dyes are moderately sensitive and have a linear dynamic range, making them suitable for quantitative analysis, which is essential for proteomic analysis. Silver staining is more sensitive than Coomassie Blue but has some limitations including the narrow dynamic range and the tendency of the dye to stain differently, based on amino acid composition and PTMs ***(79–81)***. Fluorescent dyes have recently been developed with the advantage of linear dynamic range. Both silver and fluorescent stains have led to substantial improvement in the sensitivity of protein detection/visualization ***(82)***. Nevertheless, PRPs, especially the acidic ones, are difficult to stain with silver, most likely because of their insufficiency in basic sulfur-containing amino acids thought to be responsible for binding to silver ions ***(80)***.

Once proteins are separated by SDS-PAGE or 2D-PAGE, they are visualized and the protein spot profiles are analyzed using image analysis software in order to determine qualitative and quantitative changes in protein expression. The type of staining procedure (organic dyes, metal ion reducing, fluorescence, or radioactive isotopes) provides different sensitivity on protein spot detection. Consequently, the amounts of proteins to be loaded differ (**Table 1**). For example, detection by Coomassie Blue R-250 needs greater amounts of protein compared with silver staining, SYPRO Ruby ***(25,26)***, or radioactive methods.

In addition to sensitivity, linear dynamic range, and reproducibility, visualization methods should be fully capable of interfacing with modern proteomic tools. For protein identification, proteins of interest are selected, excised from the gel, and subjected to enzymatic digestion. For the acquisition of good mass spectra from low-abundance proteins, it is imperative that the dye does not interfere with MS analysis. Interfering dyes can cause ion suppression, resulting in fewer peptides obtained and a reduction in signal intensity ***(79,81)***.

As noted before, high mucin content may block the current passage in the first-dimensional separation, interfering with the IEF of salivary proteins. To avoid this problem, one can load smaller amount of proteins in combination with

Table 1
Comparisons Among Different Staining Methods

Method	Sensitivity	MS	Cost	View
CBB R-250	50–100 ng	Yes	Low	Visible
Silver	1 ng	±	Low	Visible
SYPRO Ruby	1 ng	Yes	Medium	Fluorescence
DIGE	1 ng	Yes	High	Fluorescence
Radioactivity	1–10 pg	±	Low	Film/phosphoimager

more sensitive stains (e.g., silver, fluorescence, or Coomassie Blue G-250, which is 10 times more sensitive than R-250) ***(82)***. Silver staining, in particular reversible silver stain according to Yan et al. ***(83)***, may be used as a routine detection method. This method has some advantages over conventional silver staining because it is less prone to spot saturation and thus has a higher dynamic range and is compatible with MS analysis.

For acidic PRPs, important salivary proteins that are hardly detected with silver stain ***(80)***, visualization may be achieved with Coomassie staining, which provides pink-violet colorization after destaining without organic solvents. This effect is owing to a metachromatic effect and allows the precise identification of these proteins ***(84,85)***.

3.3.1. Methods for PRPs Detection

1. Prepare Coomassie Blue solution (0.25% w/v Coomassie Blue R-250 in 10% acetic acid) and remove insoluble materials with filtration.
2. Stain the gel for 30 min.
3. Destain the gel using several changes of 10% glacial acetic acid.

For mucins and other glycoproteins, Alcian Blue and periodic acid Schiff (PAS) are typically used ***(86)***. Because both have several limitations in terms of concentration and sensitivity, Jay et al. ***(87)*** have demonstrated that the visualization of mucins and other salivary glycoproteins can be enhanced by subsequent silver staining.

3.4. Identification of Salivary Proteins

Saliva is a complex mixture of proteins mainly composed of amylase, PRPs, immunoglobulins, albumin, and cystatins. Nevertheless, this fluid comprises many other types of proteins recently identified by proteome analysis including PLUNC (palate-lung-nasal-epithelial clone), cystatin A and B ***(27)***, interleukin ***(26)***, anti-TNF-α antibody light-chain Fab, anti-HBs antibody light-chain Fab, and lipocortin ***(29)***. It must be considered that immunoblotting and amino acid sequencing were the unquestionable methods used to identify salivary proteins prior to MS ***(65,66,69)***.

3.4.1. Immunoblotting

Immunoblotting provides information on the presence, MW, and/or quantity of an antigen by combining protein separation via gel electrophoresis with specific recognition of antigens by antibodies. This system, in combination with the resolving power of 2D-PAGE, provides a quick and sensitive way to monitor a single protein with known characteristics and to compare major differences in paired samples. Moreover, several protein isoforms can be identified. As referred to above, Beeley et al. ***(56)*** initially identified salivary proteins on a 2D gel map using this approach. Immunoblotting is still used as a routine procedure on saliva protein characterization ***(18,88)***. Only a relatively small amount of protein is required to obtain a positive identification.

3.4.2. Mass Spectrometry

During the past decade, a rapid revolution has occurred in the field of protein identification. The knowledge obtained from the genome projects of several species has made the construction of protein databases possible. With advanced MS techniques (which require small amount of materials, in the femtomole to attomole range) and bioinformatic tools, these databanks have given us a quick and easy way to identify proteins. The combination of MS with different separation techniques such as 2D-PAGE, HPLC, or 2D-LC has already been introduced to saliva analysis, making the study (separation, identification, and/or quantitation) of a wide variety of salivary proteins feasible ***(24,89)***.

3.4.2.1. 2D-PAGE Followed by MS

2D-PAGE followed by mass spectrometric protein identification is the typical approach for salivary proteome analysis. After protein separation by 2D-PAGE, each spot of interest is excised and digested with a protease (usually trypsin), which cleaves proteins at specific amino acid residues (C-terminal lysine and arginine for trypsin). The peptide fragments obtained from digestion can be analyzed using MS and proteins are identified by peptide mass fingerprinting, by which the molecular mass of each peptide is compared with those obtained from theoretical digestion on available genome and protein databases. A confirmation of identification can be obtained using the sequence information obtained by tandem mass spectrometry (MS/MS), in which the peptide of interest is selected and fragmented within the mass spectrometer and, again, protein databases are used to search for an identical tryptic peptide sequence. In case of a negative identification, *de novo* sequencing can be performed, and the amino acid sequence can be interpreted on nucleotide probes using BLAST analysis ***(90)***. Salivary protein identification based on peptide mass fingerprinting and MS/MS analyses of SDS-PAGE bands or 2D protein spots have been performed on saliva samples by several groups ***(20,25–27,29)***. Moreover, the identification of several spots corresponding to the same protein shows the ability of 2D-PAGE to detect PTMs such as glycosylation, truncation, and phosphorylation.

3.4.2.2. Methods for 2D-PAGE Followed by MS Analysis ***(27)***

1. After 2D-PAGE, resolved protein spots are visualized with various stains and analyzed, typically with 2D image analysis software.
2. Protein spots of interest are then excised from the gel manually using a small knife or a pipet tip or automatically using a spot-excision robot.
3. The gel pieces are washed twice with 25 m*M* ammonium bicarbonate/50% acetonitrile (ACN) and dried.
4. These 25 µL of 10 µg/mL trypsin in 50 m*M* ammonium bicarbonate is added to the dried residues, and the samples are incubated overnight at 37°C.
5. Tryptic peptides are then lyophilized and resuspended in 10 µL of 50% ACN/0.1% formic acid.
6. The samples are mixed (1:1) with a matrix solution containing α-cyano-4-hydroxycinnamic acid prepared in 50% ACN/0.1% formic acid. Aliquots of samples (0.35 µL) are then spotted onto the MALDI (matrix-assisted laser desorption/ionization) sample target plate.
7. Peptide mass spectra are obtained from a MALDI-TOF (time-of-flight) or MALDI-TOF/TOF mass spectrometer in the positive ion reflector mode. Trypsin autolytic peaks are used for internal calibration of the mass spectra, allowing a mass accuracy of better than 25 ppm.
8. Further analysis in respect to peptide sequence is achieved by performing MS/MS experiments. In the case of MALDI-TOF/TOF, both MS spectra and MS/MS spectra can be obtained. MS/MS data can also be obtained by nano-LC (PepMap C18 column, 75 µm × 150 mm; particle size 5 µm) electrospray ionization (ESI)-MS. Peptides are separated using a linear gradient from 5% of the solution containing 95% ACN/5% H_2O/0.1% formic acid/0.01% trifluoroacetic acid (TFA) to 60% in 70 min and to 95% in 15 min.
9. The MS and MS/MS data obtained are searched against the human NCBI nonredundant protein database using the Mascot search tool (www.matrixscience; Matrix Science, UK). Other software such as ProteinProspector (http://prospector.ucsf.edu; UCSF Mass Spectrometry Facility, San Francisco, CA) and PROWL (www.prowl.rockfeller.edu; ProteoMetrics, New York) can also be used for protein identification.

3.4.2.3. LC Coupled to MS (LC-MS)

Although 2D-PAGE is currently the most common method used for protein separation, there are problems associated with this technique, including poor recovery of hydrophobic proteins and limited resolving power for proteins with extremely low or high M_r and those with basic p*I* ***(91)***. A new approach made possible by recent advances in LC coupled to ESI-MS instrumentation allows greater protein identification from the complex mixture of proteins ***(92–94)***.

This advance in the 2D-LC-MS technique is often referred to as MudPIT (multidimensional protein identification technology). The application of cation exchange and a reversed-phase column in sequence allows one to separate complex peptide mixture such as protein digest. Recently, Wilmarth et al. ***(28)*** identified 102 proteins using this approach toward whole saliva. Hu et al. ***(29)*** have identified 309 proteins using a similar method.

Salivary protein identification by 2D-LC covers a wider range of p*I* and M_r than 2D-PAGE-MS, making possible the identification of high-MW proteins (e.g., mucin 5B, with a molecular mass of 590 kDa) and extremely basic proteins (e.g., salivary proline-rich glycoprotein PRB2, with a p*I* of 12.03). Proteins with low molecular mass (<10 kDa), such as PRPs, statherin, histatins, and defensins, have also been identified by 2D-LC proteomics ***(28,29)***. According to Hu et al. ***(29)***, from the 64 proteins identified by 2D-LC-MS, only 21 have been identified by both methods. 2D-PAGE-MS and 2D-LC-MS should be considered complementary techniques at the moment.

3.4.2.4. Methods for LC-MS ***(28,29)***

1. A portion of whole saliva supernatant is dissolved in a solution containing 8 *M* urea, 0.4 *M* ammonium bicarbonate, and 80 m*M* methylamine.
2. Sample reduction is performed by adding 10 m*M* DTT and subsequent alkylation by adding iodoacetamide.
3. Ammonium bicarbonate and sequencing-grade modified trypsin (ProMega, cat. no. V5113) with a ratio of 1:20 of protein are added, and the mixture is incubated at 37°C overnight.
4. Two approaches can be used for tryptic peptide analysis:

   a. Peptides can be purified by application of each fraction to a Sep-Pak Light solid-phase extraction cartridge (Waters, Milford, MA, cat. no. WAT051910) as described by Wilmarth et al. ***(28)***.
      i. The combined digest is injected into a 100 × 2.1-mm polysulfoethyl A cation exchange column at 200 μL/min using a linear gradient of 0 to 50% of 10 m*M* sodium phosphate (pH 3.0)/25% ACN/350 m*M* KCl) over 40 min, followed by a linear gradient of 50 to 100% of the same solution over 20 min.
      ii. Collected fractions are dried, redissolved in 5% formic acid, and analyzed with an LC-MS system (3.0 cm × 180-μm trap cartridge containing 5-μm particle size C18 switched online to a 10 cm × 180 μm) using a standard electrospray source.
      iii. Peptides are then eluted using a gradient 0 to 10% of the solution containing 0.2% acetic acid/75% ACN/0.2% acetic acid over 5 min, 10 to 40% over 60 min, and then 40 to 100% over 25 min.
   b. In the case of the procedure used by Hu et al. ***(29)***:
      i. After digestion, samples are dried and first loaded into a C18 precolumn (300 × 1 mm; particle size 5 μm) and washed for 5 min with the loading solvent (3% ACN/97% H_2O/0.1% formic acid/0.01% TFA).

   ii. The samples are then injected into an LC PepMap C18 column (75 μm × 150 mm; particle size 5 μm) for nano-LC separation at a flow rate of 180 nL/min.
   iii. Peptides are separated using a linear gradient from 5% of the solution containing 95% ACN/5% H_2O/0.1% formic acid/0.01% TFA to 60% in 55 min and ramped to 95% in 0.1 min.

5. Peptides are identified by comparing the observed MS/MS spectra with theoretical fragmented spectra of peptides generated utilizing the Mascot database search engine (Matrix Science).

4. Analyses of Salivary Peptides

One of the disadvantages in the use of polyacrylamide gels is its inability to resolve and detect the low-MW proteome (<15 kDa). The Tris-Tricine gels are useful in the separation of small proteins and peptides, but this technique is not adequate for displaying the whole low-MW proteome ***(95)***. HPLC is particularly suitable for separating peptides and small-MW components with good resolution and with a relatively short separation time. Multidimensional LC techniques employing ion exchange combined with reversed-phase LC have considerably improved the resolving power of LC, as described in several papers ***(96–98)***. The absence of technically induced modifications of protein residues (such as those observed on cysteine in polyacrylamide gel) is another important attribute of HPLC.

Several authors have used C8 and C18 reversed-phase HPLC columns for the separation of major salivary peptides with similar linear gradients, from 0 to 55% of ACN/water in TFA ***(40,99)*** or acetic acid ***(100)***, with comparable results. The use of HPLC has allowed the separation of intact or fragmented peptides. For instance, Perinpanayagam et al. ***(101)*** have used laborious solvent extraction procedures, HPLC, and Edman sequencing to identify several histatins, statherin, and basic PRPs fragments. Moreno et al. ***(102)*** have applied hydrophobic-interaction HPLC columns to the analysis of different classes of PRPs. Hydrophobic chromatography has been also useful for the separation of amylase isoenzymes ***(103)***. Hay et al. ***(104)*** have employed anion-exchange HPLC for characterization of PRPs, showing the presence of polymorphisms with different PRPs phenotypes. A similar approach, using cation-exchange HPLC, has allowed Ayad et al. ***(105)*** to detect 19 different peptide species, which are considered proteolytic products from basic PRPs.

Salivary small protein/peptide characterization has been achieved using conventional purification approaches that include separation of peptides by exclusion chromatography on Sephadex columns followed by ion exchange or HPLC. After collecting the sample peaks, peptides are subjected to Edman sequencing ***(106–110)***. During the past two decades, several authors have performed peptide characterization using this methodology ***(106–110)***. HPLC-MS has been used by Vitorino et al. ***(40)*** for the identification of different salivary peptides and by Castagnola and co-workers ***(32,111–113)*** for characterization of the acidic soluble peptide fraction. HPLC-MS/MS has been used to sequence histatin fragments and to identify low-MW salivary peptides.

4.1. Characterization of Human Salivary Peptides

Based on the results of earlier work, the salivary peptides are usually classified into five different main classes according to their intrinsic properties: PRPs, histatins, cystatins, statherin, and defensins ***(12)***. PRPs are named according to their high content of proline residues and are characterized in three different forms: acidic, basic, and glycosylated ***(75)***. Basic PRPs have been characterized by Kauffman et al. ***(110)*** and classified as IB-1, IB-4, IB-5, IB-6, IB-7, IB-8a, IB-8b, IB-8c, IB-9, and II-2. IB-1 and II-2 contain a phosphorylation site at the C-terminus that has the ability to inhibit hydroxyapatite formation in vitro ***(110)***. So far, it is not possible to determine a specific role for basic PRPs in the oral cavity. However, their ability to bind tannins can contribute as a protective factor in the oral cavity ***(114)***. Recently, it has been shown that an unidentified basic PRP seems to inhibit HIV-1 infection ***(115)***. Our group has found an increase in the relative abundances of II-2 and IB-1 in caries-free subjects ***(116)***. Acidic PRPs take part, together with statherin ***(104,107)***, IB-1, and II-2, in the chemical balance of calcium phosphate on teeth surfaces and are components of the enamel pellicle. There are six principal acidic PRPs isoforms, including PRP-1, PRP-2, PRP-3, PRP-4, PIF-s, and PIF-f ***(106)***. PRP-2 and PRP-4 differ from PRP-1 and PRP-3 by the substitution of an Asp residue by an Asn at position 50. PRP-3 and PRP-4 are derived from PRP-1 and from PRP-2, respectively, by cleavage at position 150. PIF-s differs from PRP-1 by the substitution of an Asn residue by an Asp at the position 4 ***(117)***.

Another group of low-MW salivary peptides are histatins, characterized by high histidine content ***(118)***. Histatins show strong antifungal activities, and histatin 1 seems to participate actively in teeth remineralization processes ***(119)***. Troxler et al. ***(120)*** have reviewed the current peptide chemical information and nomenclature for histatins. Briefly, histatin 2 contains the last 26 C-terminal residues of histatin 1 (composed of 38 amino acids, phosphorylated at Ser-2), and nine peptides, all related to the sequence of histatin 3, are named histatins 4 to 12. With the exception of histatin 2, the other minor histatins probably originated from proteolytic cleavage of histatin 3. Compared with the other minor histatins, histatin 5 has the greater amount in saliva ***(121)***.

The cystatin class contains five major isoforms (S, C, D, SA, and SN). They are powerful inhibitors of cysteine peptidases such as cathepsins B, C, H, and L and have strong bactericidal and virucidal properties ***(122,123)***. There is an association between increased levels of cystatin C and periodontal inflammatory diseases ***(124)***.

By combining HPLC with MS, it is possible to separate and characterize several salivary peptides in the same run. For basic PRPs, IB-8b, IB-5, IB-9, IB-8c, IB-4, IB-6, II-2, IB-1, glycosylated IB-8a, IB-7, and D1A have been identified by this method ***(40,111,112)***. For acidic PRPs, PRP-1 and PRP-3 have been detected as the predominant isoforms ***(116)***. Other isoforms have been also found with lower amounts including triphosphorylated derivatives of PRP-1/PRP-2/PIF-s and DB-s with an additional phosphate group at Ser-17; monophosphorylated forms of PRP-1/PRP-2/PIF-s, PRP-3/PRP-4/PIF-f, and DB-s/f; and a nonphosphorylated form of PRP-3/PRP-4/PIF-f ***(113,125)***. In the case of histatins, isoforms corresponding to histatins 5 to 12 have been identified. In addition, 13 new fragments corresponding to 1-11, 1-12, 1-13, 5-13, 6-11, 6-13, 7-11, 7-12, 7-13, 14-24, 14-25, 15-25, and 28-32 residues of histatin-3 have been detected ***(121)***. All these fragments show a complex proteolytic pathway involving histatin 3. The absence of proteolytic activity on histatin 1 can be attributed to the lack of an amino acid equivalent to the Arg-25 residue of histatin 3, which seems to represent the first crucial cleavage site for this peptide ***(121)***.

Statherin, a multifunctional molecule that possesses a high affinity for calcium phosphate minerals such as hydroxyapatite, contributes to maintenance of the appropriate mineral solution dynamics of enamel. Jensen et al. ***(126)*** have characterized three isoforms of statherin: SV1 (lacks of one phenylalanine residue at the C-terminus), SV2 (lacks residues 6–15), and SV3 (SV2 that lacks a phenylalanine residue at the C-terminus). The existence of an additional statherin isoform corresponding to the loss of one aspartic acid residue on the N-terminal has been recently suggested ***(67)***. The defensin family, detected in plasma, wound fluid, intestine, and skin of humans, is generally recognized by its antibiotic, antifungal, and antiviral properties ***(100)***. Salivary defensins have been identified by HPLC-MS with previous treatment on SPE cartridges ***(100)*** or after passing through membranes with a molecular mass cutoff 30 kDa ***(40)***. Morevover, HPLC-MS allows the detection of histatins, statherin, and basic PRP fragments derived from proteolytic activity.

4.2. Methods for Analyzing Salivary Peptides Using HPLC-MS

1. Sample treatment is different depending on the type of analysis. To analyze salivary peptides, which are the free form in saliva, the sample treatment should be (a) or (b), as described below. To analyze salivary peptides involving in salivary complexes, the sample treatment should be (a) and (c).
    a. Centrifuge whole saliva as described previously. PMSF or antiprotease cocktail can be added to avoid proteolysis. Pass the supernatant through a defined membrane filter (10, 30, or 50 kDa cutoff) to prefractionate salivary peptides.

b. Add 0.2% TFA to saliva in the proportion 1:1 (v/v), and centrifuge at 9100*g* (to induce the precipitation of high-MW salivary complexes, mucins, amylase, and IgAs and to decrease sample viscosity with the inhibition of intrinsic protease activity). Pass it through a defined membrane filter (10, 30, or 50 kDa cutoff) to prefractionate salivary peptides.
c. Add 300 µL of 6 *M* guanidine (a denaturating agent) to 1 mL of saliva for a final concentration of more than 1.5 *M* (which promotes salivary complex disruption, permitting analysis of the majority of low-MW salivary components). Incubate at 37°C with continuous stirring for 2 h. PMSF or antiprotease cocktail can be added to avoid proteolytic activity. Pass it through a defined membrane filter (10, 30, or 50 kDa cutoff) to prefractionate salivary peptides.

2. Peptide separation is performed using a C18 column (20 × 1.5 mm, 5-µm particle size) at a flow rate of 15 µL/min, compatible with the used column and the ESI ion source or a C8 column (150 × 2.1 mm, 5-µm particle size). Peptides are eluted with a gradient from eluent A (water/0.05% TFA) to eluent B (ACN/0.05% TFA).
3. Usually, the identification of salivary peptides is based on theoretical MWs available from Swiss-Prot and NCBI databases. Monoisotopic experimental values are obtained after deconvolution using appropriate software. Detection of PTMs is based on comparison between experimental molecular mass and information available in the database referred to above in addition to the values for these modifications found in the ExPASy-FindMod tool (http://us.expasy.org/tools/findmod/findmod_masses.html). Further confirmation of the salivary peptides evolves peak collection, direct spotting on MALDI-TOF plates, or tryptic digestion of each fractionated analysis by MS.
   a. Lyophilized powders are dissolved with 140 µL of 0.1 *M* ammonium bicarbonate (pH 8.0) and incubated with 20 µL trypsin at 37°C for 5 h.
   b. Analysis of tryptic digest is performed using MALDI-TOF, or the mixture is loaded again on an LC-ESI-MS system as described above.
4. *Note: In the case of parotid or submandibular saliva, skip the centrifugation step.*

5. Analyses of High-MW Components

Mucins constitute an important class of salivary glycoproteins. There are two types of mucins, MUC7 (also called MG2) and MUC5B (also called MG1), present in glandular secretions ***(12)***. They are characterized as multimerizing macromolecules with an extensive *O*-linked glycosylation and about 60% (for MG2) and 80% (for MG1) carbohydrate content ***(127)***. Salivary mucins are major constituents of the viscous layer coating hard and soft tissues in the oral cavity and serve as a lubricant and permselective barrier, which protects underlying oral surfaces from a potentially harmful external environment ***(128,129)***. Several publications ***(130–132)*** report interactions among mucins, carbohydrate side chains, and salivary proteins, as well as oral bacteria ***(133)***. Such interactions can be involved in the formation of stable covalently heterotypic complexes, which contribute to homeostasis and to maintenance of the integrity of the oral cavity ***(131)***. Another high-MW glycoprotein is agglutinin, which forms a complex with secretory immunoglobulin A (sIgA) and plays an important role in the interaction between the agglutinin and PAc of *Streptococcus mutans*, as suggested by Oho et al. ***(70)***. The binding between bacteria and sIgA appears to be mediated by a relatively short peptide stretch in the scavenger receptor domains ***(134)***. Analysis of these molecules and their related heterotypic complexes may be performed under nondenaturing and denaturing conditions.

5.1. Separation of High-MW Components Under Nondenaturing Conditions

Various methodologies have been described for mucin analysis. However, a consensus is clear in respect to application of cesium density-gradient centrifugation for isolation of MG1.

5.1.1. Method Described by Raynal et al. ***(135)***

1. Fresh saliva is gently stirred overnight at 4°C with an equal volume of 0.2 *M* NaCl and centrifuged at 4400*g* at 4°C for 30 min.
2. The supernatant is fractionated by equilibrium density-gradient centrifugation (starting density of 1.45 mg/mL) in CsCl/0.1 *M* NaCl (pH 6.5) with a Beckman Ti45 rotor at 65,000*g* at 15°C for 65 h.
3. The high-density fractions containing MG1 are pooled, dialyzed against 0.1 *M* NaCl (pH 6.5), and stored at 4°C after the addition of 0.05% (w/v) sodium azide.
4. A fraction of this preparation is chromatographed on a size exclusion column (1000 × 52 mm, Sepharose CL-2B) eluted with 0.1 *M* NaCl/10 m*M* EGTA (pH 8.0) at a flow rate of 24 mL/h. MG1 fractions, in the void volume of the column, are pooled and dialyzed against 0.1 *M* NaCl (pH 6.5).

5.1.2. Method Described by Iontecheva et al. ***(130)***

1. Saliva supernatant (after centrifugation at 10,000*g* for 10 min) is loaded onto a Sepharose CL-2B column (5 cm × 42 cm), and pools are collected, dialyzed against water, and concentrated by ultrafiltration on an XM 300 membrane.
2. This sample is then adjusted to a density of 1.4 g/cm^3 by the addition of solid CsCl and centrifuged with a Ti70.1 rotor at 70,000*g* and 15°C for 72 h.
3. Separation of MG2 can be achieved using an approach similar to that applied to MG1 with the exception of the application of ultrafiltration through a YM30 membrane (Amicon) ***(136)***. The retentate is centrifuged at 10,000*g* at 4°C for 15 min, chromatographed with 15-mL aliquots in a Sephadex G-200 column (2.5 × 50 cm), equilibrated, and developed with a buffer containing 50 m*M* Tris-HCl (pH 7.5), 5 m*M* EDTA, and 0.02% sodium azide (buffer A).

5.1.3. Method for Analysis of Agglutinin According to Ligtenberg et al. ***(137)*** *and Yamaguchi et al.* ***(138)***

1. Incubate clarified saliva at 4°C (which results in the formation of precipitates).
2. Centrifuge at 5000*g* at 4°C for 15 min, and resuspend the pellet in 0.1 vol of PBS supplemented with 10 m*M* EDTA.

3. Load onto a Sephacryl S-400 (Amersham Biosciences) gel filtration column (15 × 490 mm) and run with PBS.
4. The eluate is subsequently filtered (0.2-μm pore size), dialyzed against an aggregation buffer containing 0.02% NaN_3, and subjected to gel filtration chromatography on Superdex 200 HR (Pharmacia, Uppsala, Sweden) equilibrated with aggregation buffer. The eluate at the void volume is collected and used as salivary agglutinin.

Native-PAGE is performed using 3 to 15% gradient polyacrylamide slab gels according to the method described by Davis ***(139)*** or 7% polyacrylamide slab gels. These allow the visualization of high-MW components.

5.2. Separation of High-MW Components Under Denaturing Conditions

Denaturing conditions are characterized by the addition of denaturing agents to the samples. The most commonly used denaturing agent for the analysis of high-MW components of saliva is guanidine. To promote denaturation of the salivary complex, a solution of 4 to 8 *M* guanidinc ***(130,131,134–136)*** is added to saliva supernatant (at 4400*g* at 4°C for 30 min), and samples are incubated from 2 h to overnight under continuous stirring. After this step, samples are loaded onto described molecular exclusion columns and/or CsCl gradient centrifugation (as mentioned in **Subheading 5.1.2.**) ***(140)***.

Isolated components from nondenaturing conditions can be analyzed under denaturing conditions by addition of mercaptoethanol and 2% SDS and boiled for 5 min. SDS-PAGE is performed with 7.5 and 12.5% separating polyacrylamide gels ***(135)***. Since these components contain a high carbohydrate content, specific staining methods such as PAS can be used for protein characterization.

In all these cases, proteins in eluted fractions can be identified by enzyme-linked immunosorbant assay (ELISA) using antibodies against MG1, MG2, sIgA, and agglutinin or analysis of tryptic digest by MS ***(135)***.

6. Clinical Applications

Salivary peptides participate in several functions contributing to homeostasis of the oral cavity. For some, like the basic PRPs, a specific role has not yet been defined. However, the ability of these proteins to bind tannins ***(141)*** can contribute as a protective factor in the oral cavity. Recently, it has been shown that an unidentified basic PRP seems to inhibit HIV-1 infectivity ***(142)***. Moreover, it has also been reported that IB-1 and II-2 contain a phosphorylation site at the C-terminus, which has the ability to inhibit hydroxyapatite formation in vitro ***(110)***. Acidic PRPs take part, together with statherin ***(143,144)***, IB-1, and II-2, in the chemical balance of calcium phosphate on teeth surfaces and are components of the enamel pellicle. An increase in the relative abundances of acidic PRPs, statherin, II-2, and IB-1 have been found in caries-free subjects in a study related to dental caries ***(145)***. An extensive proteolytic fragmentation has also been observed in caries-susceptible subjects ***(145)***. Since dental caries is perhaps the most widespread disease in humans, the huge number of papers describing a possible relationship between saliva and dental caries is not surprising. For instance, Tabak and co-workers have suggested that there is an inverse relationship between the levels of cystatins (potent inhibitors of cysteine peptidases with strong bactericidal and virucidal properties ***[41,146]***) in resting whole saliva of children and their past and present caries experience. Recently, Banderas-Tarabay et al. ***(64)*** found a higher content of MG1 and PRPs in caries-free subjects.

Other oral pathologies like periondotitis, Sjögren's syndrome, diabetes mellitus, xerostemia, and oral cancer have been largely studied. Baron et al. ***(147)*** and Henskens et al. ***(148)*** have found increased levels of cystatin C and cystatin S in association with periondotitis. Patients with colon cancer have high levels of cystatin B and cystatin C in serum and show a significantly higher risk of death than those with low levels of inhibitors, suggesting a potential role of cystatin C in the progression of cancer ***(149)***.

Recently, Contucci et al. ***(150)*** have found a sensible reduction in statherin levels in the saliva of patients with precancerous and cancerous lesions of the oral cavity compared with healthy subjects. Other proteins, detectable in 2D-PAGE maps, can be correlated with cancer and other pathologies. For example, an increase in blood calgranulin concentration has been reported in patients with rheumatoid arthritis, colorectal carcinoma, cystic fibrosis, and HIV infection ***(151)***. Salivary calgranulin levels are also found to be elevated in subjects with candidiasis and Sjögren's syndrome, meaning that calgranulin production, or release, may be increased in patients with these pathologies ***(152,153)***. Short PLUNC (SPLUNC1) protein has been found in the entire respiratory tract, and its overexpression is related to primary pulmonary neoplasm, mostly adenocarcinoma and metastasis ***(154)***. Since PLUNC protein has also been detected in whole saliva ***(27–29)***, it may be possible to explore a clinical application.

7. Perspectives

Although much progress has been made, we are just beginning to discern protein salivary composition at the molecular level and its importance in the oral environment. The information obtained from the methodologies described for analysis of salivary proteins/peptides and the promise of rapid improvements of new methods will make salivary proteomics more relevant to clinical applications in the near future.

References

1. Turner RJ, Sugiya H. Understanding salivary fluid and protein secretion. *Oral Dis* 2002;8:3–11.

2. Castle D, Castle A. Intracellular transport and secretion of salivary proteins. *Crit Rev Oral Biol Med* 1998;9:4–22.
3. Bardow A, Nyvad B, Nauntofte B. Relationships between medication intake, complaints of dry mouth, salivary flow rate and composition, and the rate of tooth demineralization in situ. *Arch Oral Biol* 2001;46:413–423.
4. Ferguson DB. The flow rate and composition of human labial gland saliva. *Arch Oral Biol* 1999;44:S11–S14.
5. Whelton H. The anatomy and physiology of salivary glands. In: Edgar WM, O´Mullane DM, eds. *Saliva and Oral Health*, 2nd ed, London: British Dental Association, 1996:1–8.
6. Smith PM. Mechanisms of secretion by salivary glands. In: Edgar WM, O´Mullane DM, eds. *Saliva and Oral Health*, 2nd ed, London: British Dental Association, 1996:9–25.
7. Martinez JR. Cellular mechanisms underlying the production of primary secretory fluid in salivary glands. *Crit Rev Oral Biol Med* 1990;1:67–78.
8. Huang AY, Castle AM, Hinton BT, Castle JD. Resting (basal) secretion of proteins is provided by the minor regulated and constitutive-like pathways and not granule exocytosis in parotid acinar cells. *J Biol Chem* 2001;276: 22,296–22,306.
9. Castle JD, Guo Z, Liu L. Function of the t-SNARE SNAP-23 and secretory carrier membrane proteins (SCAMPs) in exocytosis in mast cells. *Mol Immunol* 2002;38:1337–1340.
10. Baum BJ. Neurotransmitter control of secretion. *J Dent Res* 1987;66:628–632.
11. Melvin JE. Chloride channels and salivary gland function. *Crit Rev Oral Biol Med* 1999;10:199–209.
12. Van Nieuw Amerongen A, Bolscher JG, Veerman EC. Salivary proteins: protective and diagnostic value in cariology? *Caries Res* 2004;38:247–253.
13. Kaufman E, Lamster IB. The diagnostic applications of saliva—a review. *Crit Rev Oral Biol Med* 2002;13:197–212.
14. Jensen JL, Barkvoll P. Clinical implications of the dry mouth. Oral mucosal diseases. *Ann N Y Acad Sci* 1998;842:156–162.
15. Lawrence HP. Salivary markers of systemic disease: noninvasive diagnosis of disease and monitoring of general health. *J Can Dent Assoc* 2002;68:170–174.
16. Dodds MW, Johnson DA, Yeh CK. Health benefits of saliva: a review. *J Dent* 2005; 33:223–233.
17. Kaufman E, Lamster IB. Analysis of saliva for periodontal diagnosis—a review. *J Clin Periodontol* 2000;27:453–465.
18. Hu S, Denny P, Denny P, et al. Differentially expressed protein markers in human submandibular and sublingual secretions. *Int J Oncol* 2004;25:1423–1430.
19. de Hoog CL, Mann M. Proteomics. *Annu Rev Genomics Hum Genet* 2004; 5:267–293.
20. Huang CM. Comparative proteomic analysis of human whole saliva. *Arch Oral Biol* 2004;49:951–962.
21. Madden RD, Sauer JR, Dillwith JW. A proteomics approach to characterizing tick salivary secretions. *Exp Appl Acarol* 2002;28:77–87.

22. Kojima T, Andersen E, Sanchez JC, et al. Human gingival crevicular fluid contains MRP8 (S100A8) and MRP14 (S100A9), two calcium-binding proteins of the S100 family. *J Dent Res* 2000;79:740–747.
23. Macarthur DJ, Jacques NA. Proteome analysis of oral pathogens. *J Dent Res* 2003;82:870–876.
24. Ferguson PL, Smith RD. Proteome analysis by mass spectrometry. *Annu Rev Biophys Biomol Struct* 2003;32:399–424.
25. Yao Y, Berg EA, Costello CE, Troxler RF, Oppenheim FG. Identification of protein components in human acquired enamel pellicle and whole saliva using novel proteomics approaches. *J Biol Chem* 2003;278:5300–5308.
26. Ghafouri B, Tagesson C, Lindahl M. Mapping of proteins in human saliva using two-dimensional gel electrophoresis and peptide mass fingerprinting. *Proteomics* 2003;3:1003–1015.
27. Vitorino R, Lobo MJ, Ferrer-Correira AJ, et al. Identification of human whole saliva protein components using proteomics. *Proteomics* 2004;4:1109–1115.
28. Wilmarth PA, Riviere MA, Rustvold DL, Lauten JD, Madden TE, David LL. Two-dimensional liquid chromatography study of the human whole saliva proteome. *J Proteome Res* 2004;3:1017–1023.
29. Hu S, Xie Y, Ramachandran P, et al. Large-scale identification of proteins in human salivary proteome by liquid chromatography/mass spectrometry and two-dimensional gel electrophoresis-mass spectrometry. *Proteomics* 2005;5:1714–1728.
30. Hardt M, Thomas LR, Dixon SE, et al. Toward defining the human parotid gland salivary proteome and peptidome: identification and characterization using 2D SDS-PAGE, ultrafiltration, HPLC, and mass spectrometry. *Biochemistry* 2005; 44:2885–2899.
31. Gusman H, Leone C, Helmerhorst EJ, et al. Human salivary gland-specific daily variations in histatin concentrations determined by a novel quantitation technique. *Arch Oral Biol* 2004;49:11–22.
32. Castagnola M, Cabras T, Denotti G, et al. Circadian rhythms of histatin 1, histatin 3, histatin 5, statherin and uric acid in whole human saliva secretion. *Biol Rhythm Res* 2002;33:213–222.
33. Veerman EC, van den Keybus PA, Vissink A, Nieuw Amerongen AV. Human glandular salivas: their separate collection and analysis. *Eur J Oral Sci* 1996; 104:346–352.
34. Jensen JL, Lamkin MS, Oppenheim FG. Adsorption of human salivary proteins to hydroxyapatite: a comparison between whole saliva and glandular salivary secretions. *J Dent Res* 1992;71:1569–1576.
35. Proctor GB, Carpenter GH. Chewing stimulates secretion of human salivary secretory immunoglobulin A. *J Dent Res* 2001;80:909–913.
36. Schwartz SS, Zhu WX, Sreebny LM. Sodium dodecyl sulphate-olyacrylamide gel electrophoresis of human whole saliva. *Arch Oral Biol* 1995;40:949–958.
37. Filaire E, Bonis J, Lac G. Relationships between physiological and psychological stress and salivary immunoglobulin A among young female gymnasts. *Percept Mot Skills* 2004;99:605–617.

38. Walsh NP, Montague JC, Callow N, Rowlands AV. Saliva flow rate, total protein concentration and osmolality as potential markers of whole body hydration status during progressive acute dehydration in humans. *Arch Oral Biol* 2004;49:149–154.
39. Akimoto T, Kumai Y, Akama T, et al. Effects of 12 months of exercise training on salivary secretory IgA levels in elderly subjects. *Br J Sports Med* 2003;37:76–79.
40. Vitorino R, Lobo MJ, Duarte JA, Ferrer-Correia AJ, Domingues PM, Amado FM. Analysis of salivary peptides using HPLC-electrospray mass spectrometry. *Biomed Chromatogr* 2004;18:570–575.
41. Baron A, Barrett-Vespone N, Featherstone J. Purification of large quantities of human salivary cystatins S, SA and SN: their interactions with the model cysteine protease papain in a non-inhibitory mode. *Oral Dis* 1999;5:344–353.
42. Mogi M, Hiraoka BY, Fukasawa K, Harada M, Kage T, Chino T. Two-dimensional electrophoresis in the analysis of a mixture of human sublingual and submandibular salivary proteins. *Arch Oral Biol* 1986;31:119–125.
43. Mogi M, Hiraoka BY, Harada M, Kage T, Chino T. Analysis and identification of human parotid salivary proteins by micro two-dimensional electrophoresis and Western-blot techniques. *Arch Oral Biol* 1986;31:337–339.
44. Wolfe A, Begleiter A, Moskona DA. Novel system of human submandibular/sublingual saliva collection. *J Dent Res* 1997;76:1782–1786.
45. Stephen KW, Speirs CF. Methods for collecting individual components of mixed saliva: The relevance to clinical pharmacology. *Br J Clin Pharmacol* 1976;3:315–319.
46. Saunte C. Quantification of salivation, nasal secretion and tearing in man. *Cephalalgia* 1983;3:159–173.
47. Sugiyama K, Ogata K. High-performance liquid chromatographic determination of histatins in human saliva. *J Chromatogr* 1993;619:306–309.
48. Khoo KS, Beeley JA. Isoelectric focusing of human parotid salivary proteins in hybrid carrier ampholyte-immobilized pH gradient polyacrylamide gels. *Electrophoresis* 1990;11:489–494.
49. Rayment SA, Liu B, Offner GD, Oppenheim FG, Troxler RF. Immunoquantification of human salivary mucins MG1 and MG2 in stimulated whole saliva: factors influencing mucin levels. *J Dent Res* 2000;79:1765–1772.
50. Francis CA, Hector MP, Proctor GB. Precipitation of specific proteins by freeze-thawing of human saliva. *Arch Oral Biol* 2000;45:601–606.
51. Johnson DA, Yeh CK, Dodds MW. Effect of donor age on the concentrations of histatins in human parotid and submandibular/sublingual saliva. *Arch Oral Biol* 2000;45:731–740.
52. Truelove EL, Bixler D, Merritt AD. Simplified method for collection of pure submandibular saliva in large volumes. *J Dent Res* 1967;46:1400–1403.
53. Liu B, Rayment S, Oppenheim FG, Troxler RF. Isolation of human salivary mucin MG2 by a novel method and characterization of its interactions with oral bacteria. *Arch Biochem Biophys* 1999;364:286–293.
54. Spielman AI, Bennick A. Isolation and characterization of six proteins from rabbit parotid saliva belonging to a unique family of proline-rich proteins. *Arch Oral Biol* 1989;34:117–130.

55. Nishita T, Sakomoto M, Ikeda T, Amasaki H, Shino M. Purification of carbonic anhydrase isozyme VI (CA-VI) from swine saliva. *J Vet Med Sci* 2001;63: 1147–1149.
56. Beeley JA, Sweeney D, Lindsay JC, Buchanan ML, Sarna L, Khoo KS. Sodium dodecyl sulphate-polyacrylamide gel electrophoresis of human parotid salivary proteins. *Electrophoresis* 1991;12:1032–1041.
57. Beeley JA, Khoo KS. Salivary proteins in rheumatoid arthritis and Sjogren's syndrome: one-dimensional and two-dimensional electrophoretic studies. *Electrophoresis* 1999;20:1652–1660.
58. Nurkka A, Obiero J, Kayhty H, Scott JA. Effects of sample collection and storage methods on antipneumococcal immunoglobulin A in saliva. *Clin Diagn Lab Immunol* 2003;10:357–361.
59. Ng V, Koh D, Fu Q, Chia SE. Effects of storage time on stability of salivary immunoglobulin A and lysozyme. *Clin Chim Acta* 2003;338:131–134.
60. Helmerhorst EJ, Flora B, Troxler RF, Oppenheim FG. Dialysis unmasks the fungicidal properties of glandular salivary secretions. *Infect Immun* 2004;72:2703–2709.
61. Lowry OH, Rosebrough N, Farr A, Randall R. Protein measurement with Folinphenol reagent. *J Biol Chem* 1951;193:265–275.
62. Smith PK, Krohn RI, Hermanson GT, et al. Measurement of protein using bicinchoninic acid. *Anal Biochem* 1985;150:76–85.
63. Gornall AG, Bardawill CJ, David MM. Determination of serum protein by means of biuret reaction. *J Biol Chem* 1949;177:751–766.
64. Banderas-Tarabay JA, Zacarias-D'Oleire IG, Garduno-Estrada R, Aceves-Luna E, Gonzalez-Begne M. Eletrophoretic analysis of whole saliva and prevalence of dental caries. A study in Mexican dental students. *Arch Med Res* 2002;33: 499–505.
65. Carlen A, Borjesson AC, Nikdel K, Olsson J. Composition of pellicles formed in vivo on tooth surfaces in different parts of the dentition, and in vitro on hydroxyapatite. *Caries Res* 1998;32:447–455.
66. Vacca Smith AM, Bowen WH. In situ studies of pellicle formation on hydroxyapatite discs. *Arch Oral Biol* 2000;45:277–291.
67. Jensen JL, Lamkin MS, Oppenheim FG. Adsorption of human salivary proteins to hydroxyapatite: a comparison between whole saliva and glandular salivary secretions. *J Dent Res* 1992;71:1569–1576.
68. Troxler RF, Offner GD, Xu T, Vanderspek JC, Oppenheim FG. Structural relationship between human salivary histatins. *J Dent Res* 1990;69:2–6.
69. Bardow A, Hofer E, Nyvad B, et al. Effect of saliva composition on experimental root caries. *Caries Res* 2005;39:71–77.
70. Oho T, Yu H, Yamashita Y, Koga T. Binding of salivary glycoprotein-secretory immunoglobulin A complex to the surface protein antigen of *Streptococcus mutans*. *Infect Immun* 1998;66:115–121.
71. Iontcheva I, Oppenheim FG, Troxler RF. Human salivary mucin MG1 selectively forms heterotypic complexes with amylase, proline-rich proteins, statherin, and histatins. *J Dent Res* 1997;76:734–743.

72. Frangsmyr L, Holmskov U, Leffler H, et al. Salivary agglutinin, which binds Streptococcus mutans and Helicobacter pylori, is the lung scavenger receptor cysteine-rich protein gp-340. *J Biol Chem* 2000;275:39,860–39,866.
73. Soares RV, Siqueira CC, Bruno LS, Oppenheim FG, Offner GD, Troxler RF. MG2 and lactoferrin form a heterotypic complex in salivary secretions. *J Dent Res* 2003;82:471–475.
74. Soares RV, Lin T, Siqueira CC, et al. Salivary micelles: identification of complexes containing MG2, sIgA, lactoferrin, amylase, glycosylated proline-rich protein and lysozyme. *Arch Oral Biol* 2004;49:337–343.
75. Rabilloud T. Two-dimensional gel electrophoresis in proteomics: old, old fashioned, but it still climbs up the mountains. *Proteomics* 2002;2:3–10.
76. Gorg A, Weiss W, Dunn MJ. Current two-dimensional electrophoresis technology for proteomics. *Proteomics* 2004;4:3665–3685.
77. Gorg A, Postel W, Gunther S, et al. Approach to stationary two-dimensional pattern: influence of focusing time and immobiline/carrier ampholytes concentrations. *Electrophoresis* 1988;9:37–46.
78. Rabilloud T. Use of thiourea to increase the solubility of membrane proteins in two-dimensional electrophoresis. *Electrophoresis* 1998;9:758–760.
79. Patton WF. Detection technologies in proteome analysis. *J Chromatogr B Analyt Technol Biomed Life Sci* 2002;771:3–31.
80. Beeley JA, Newman F, Wilson PH, Shimmin IC. Sodium dodecyl sulphate-polyacrylamide gel electrophoresis of human parotid salivary proteins: comparison of dansylation, coomassie blue R-250 and silver detection methods. *Electrophoresis* 1996;17:505–506.
81. Lauber WM, Carroll JA, Dufield DR, Kiesel JR, Radabaugh MR, Malone JP. Mass spectrometry compatibility of two-dimensional gel protein stains. *Electrophoresis* 2001;22:906–918.
82. Candiano G, Bruschi M, Musante L, et al. Blue silver: a very sensitive colloidal Coomassie G-250 staining for proteome analysis. *Electrophoresis* 2004;25:1327–1333.
83. Yan JX, Wait R, Berkelman T, et al. A modified silver staining protocol for visualization of proteins compatible with matrix-assisted laser desorption/ionization and electrospray ionization-mass spectrometry. *Electrophoresis* 2000;21:3666–3672.
84. Bennick A. Chemical and physical characteristics of a phosphoprotein from human parotid saliva. *Biochem J* 1975;145:557–567.
85. Humphreys-Beher MG, Wells DJ. Metachromatic staining patterns of basic proline-rich proteins from rat and human saliva in sodium dodecyl sulfate-polyacrylamide gels. *J Appl Biochem* 1984;6:353–360.
86. Becerra L, Soares RV, Bruno LS, et al. Patterns of secretion of mucins and non-mucin glycoproteins in human submandibular/sublingual secretion. *Arch Oral Biol* 2003;48:147–154.
87. Jay GD, Culp DJ, Jahnke MR. Silver staining of extensively glycosylated proteins on sodium dodecyl sulfate-polyacrylamide gels: enhancement by carbohydrate-binding dyes. *Anal Biochem* 1990;185:324–330.

88. Ruhl S, Rayment SA, Schmalz G, Hiller KA, Troxler RF. Proteins in whole saliva during the first year of infancy. *J Dent Res* 2005;84:29–34.
89. de Hoog CL, Mann M. Proteomics. *Annu Rev Genomics Hum Genet* 2004; 5:267–293.
90. Banks E, Dunn MJ, Hochstrasser DF, et al. Proteomics: new perspective, new biomedical opportunities. *Lancet* 2000;18:1749–1756.
91. Washburn MP, Wolters D, Yates JR 3rd. Large-scale analysis of the yeast proteome by multidimensional protein identification technology. *Nat Biotechnol* 2001;19:242–247.
92. McCormack AL, Schieltz DM, Goode B, et al. Direct analysis and identification of proteins in mixtures by LC/MS/MS and database searching at the low-femtomole level. *Anal Chem* 1997;69:767–776.
93. Peng J, Elias JE, Thoreen CC, Licklider LJ, Gygi SP. Evaluation of multidimensional chromatography coupled with tandem mass spectrometry (LC/LC-MS/MS) for large-scale protein analysis: the yeast proteome. *J Proteome Res* 2003;2:43–50.
94. Stone KL, DeAngelis R, LoPresti M, Jones J, Papov VV, Williams KR. Use of liquid chromatography-electrospray ionization-tandem mass spectrometry (LC-ESI-MS/MS) for routine identification of enzymatically digested proteins separated by sodium dodecyl sulfate-polyacrylamide gel electrophoresis. *Electrophoresis* 1998; 19:1046–1052.
95. Yao Y, Grogan J, Zehnder M, et al. Compositional analysis of human acquired enamel pellicle by mass spectrometry. *Arch Oral Biol* 2001;46:293–303.
96. Nagele E, Vollmer M, Horth P. Improved 2D nano-LC/MS for proteomics applications: a comparative analysis using yeast proteome. *J Biomol Tech* 2004; 15:134–143.
97. Wolters DA, Washburn MP, Yates JR 3rd. An automated multidimensional protein identification technology for shotgun proteomics. *Anal Chem* 2001;73: 5683–5690.
98. Le Bihan T, Duewel HS, Figeys D. On-line strong cation exchange micro-HPLC-ESI-MS/MS for protein identification and process optimization. *J Am Soc Mass Spectrom* 2003;14:719–727.
99. Dodds MW, Johnson DA, Mobley CC, Hattaway KM. Parotid saliva protein profiles in caries-free and caries-active adults. *Oral Surg Oral Med Oral Pathol Oral Radiol Endod* 1997;83:244–251.
100. Goebel C, Mackay LG, Vickers ER, Mather LE. Determination of defensin HNP-1, HNP-2, and HNP-3 in human saliva by using LC/MS. *Peptides* 2000; 21:757–765.
101. Perinpanayagam HE, Van Wuyckhuyse BC, Ji ZS, Tabak LA. Characterization of low-molecular-weight peptides in human parotid saliva. *J Dent Res* 1995;74:345–350.
102. Moreno EC, Kresak M, Hay DI. Adsorption thermodynamics of acidic proline-rich human salivary proteins onto calcium apatites. *J Biol Chem* 1982;257: 2981–2989.

103. Liang H, Wang Y, Wang Q, Ruan MS. Hydrophobic interaction chromatography and capillary zone electrophoresis to explore the correlation between the isoenzymes of salivary alpha-amylase and dental caries. *J Chromatogr B Biomed Sci Appl* 1999;724:381–388.
104. Hay DI, Oppenheim FG. The isolation from human parotid saliva of a further group of proline-rich proteins. *Arch Oral Biol* 1974;19:627–632.
105. Ayad M, Van Wuyckhuyse BC, Minaguchi K, et al. The association of basic proline-rich peptides from human parotid gland secretions with caries experience. *J Dent Res* 2000;79:976–782.
106. Schlesinger DH, Hay DI. Complete covalent structure of a proline-rich phosphoprotein, PRP-2, an inhibitor of calcium phosphate crystal growth from human parotid saliva. *Int J Pept Protein Res* 1986;27:373–379.
107. Oppenheim FG, Offner GD, Troxler RF. Phosphoproteins in the parotid saliva from the subhuman primate *Macaca fascicularis*. Isolation and characterization of a proline-rich phosphoglycoprotein and the complete covalent structure of a proline-rich phosphopeptide. *J Biol Chem* 1982;257:9271–9282.
108. Schlesinger DH, Hay DI. Complete covalent structure of statherin, a tyrosine-rich acidic peptide which inhibits calcium phosphate precipitation from human parotid saliva. *J Biol Chem* 1977;252:1689–1695.
109. Baron A, Barrett-Vespone N, Featherstone J. Purification of large quantities of human salivary cystatins S, SA and SN: their interactions with the model cysteine protease papain in a non-inhibitory mode. *Oral Dis* 1999;5:344–353.
110. Kauffman DL, Bennick A, Blum M, Keller PJ. Basic proline-rich proteins from human parotid saliva: relationships of the covalent structures of ten proteins from a single individual. *Biochemistry* 1991;30:3351–3356.
111. Messana I, Cabras T, Inzitari R, et al. Characterization of the human salivary basic proline-rich protein complex by a proteomic approach. *J Proteome Res* 2004;3:792–800.
112. Messana I, Loffredo F, Inzitari R, et al. The coupling of RP-HPLC and ESI-MS in the study of small peptides and proteins secreted in vitro by human salivary glands that are soluble in acidic solution. *Eur J Morphol* 2003;41:103–106.
113. Castagnola M, Cabras T, Inzitari R, et al. Determination of the post-translational modifications of salivary acidic proline-rich proteins. *Eur J Morphol* 2003;41:93–98.
114. Lu Y, Bennick A. Interaction of tannin with human salivary proline-rich proteins. *Arch Oral Biol* 1998;43:717–728.
115. Robinovitch MR, Ashley RL, Iversen JM, Vigoren EM, Oppenheim FG, Lamkin M. Parotid salivary basic proline-rich proteins inhibit HIV-I infectivity. *Oral Dis* 2001;7:86–93.
116. Vitorino R, Lobo MJ, Duarte JR, Ferrer-Correia AJ, Domingues PM, Amado FM. The role of salivary peptides in dental caries. *Biomed Chromatogr* 2005;19:214–222.
117. Schlesinger DH, Hay DI, Schluckebier SK, Ahern JM. Primary structure of a novel human salivary acidic proline-rich protein. *Pept Res* 1994;7:242–247.

118. Oppenheim FG, Xu T, McMillian FM, et al. Histatins, a novel family of histidine-rich proteins in human parotid secretion. Isolation, characterization, primary structure, and fungistatic effects on *Candida albicans*. *J Biol Chem* 1988;263: 7472–7477.
119. Edgerton M, Koshlukova SE. Salivary histatin 5 and its similarities to the other antimicrobial proteins in human saliva. *Adv Dent Res* 2000;14:16–21.
120. Troxler RF, Offner GD, Xu T, Vanderspek JC, Oppenheim FG. Structural relationship between human salivary histatins. *J Dent Res* 1990;69:2–6.
121. Castagnola M, Inzitari R, Rossetti DV, et al. A cascade of 24 histatins (histatin 3 fragments) in human saliva. Suggestions for a pre-secretory sequential cleavage pathway. *J Biol Chem* 2004;279:41,436–41,443.
122. Baron A, Barrett-Vespone N, Featherstone J. Purification of large quantities of human salivary cystatins S, SA and SN: their interactions with the model cysteine protease papain in a non-inhibitory mode. *Oral Dis* 1999;5:344–353.
123. Dickinson DP. Cysteine peptidases of mammals: their biological roles and potential effects in the oral cavity and other tissues in health and disease. *Crit Rev Oral Biol Med* 2002;13:238–275.
124. Baron AC, Gansky SA, Ryder MI, Featherstone JD. Cysteine protease inhibitory activity and levels of salivary cystatins in whole saliva of periodontally diseased patients. *J Periodontal Res* 1999;34:437–344.
125. Inzitari R, Cabras T, Onnis G, et al. Different isoforms and post-translational modifications of human salivary acidic proline-rich proteins. *Proteomics* 2005; 5:805–815.
126. Jensen JL, Lamkin MS, Troxler RF, Oppenheim FG. Multiple forms of statherin in human salivary secretions. *Arch Oral Biol* 1991;36:529–534.
127. Thomsson KA, Schulz BL, Packer NH, Karlsson NG. MUC5B glycosylation in human saliva reflects blood group and secretor status. *Glycobiology* 2005; 15:791–804.
128. Tabak LA. In defense of the oral cavity: structure, biosynthesis, and function of salivary mucins. *Annu Rev Physiol* 1995;57:547–564.
129. Thomsson KA, Prakobphol A, Leffler H, et al. The salivary mucin MG1 (MUC5B) carries a repertoire of unique oligosaccharides that is large and diverse. *Glycobiology* 2002;2:1–14.
130. Iontcheva I, Oppenheim FG, Troxler RF. Human salivary mucin MG1 selectively forms heterotypic complexes with amylase, proline-rich proteins, statherin, and histatins. *J Dent Res* 1997;76:734–743.
131. Soares RV, Siqueira CC, Bruno LS, Oppenheim FG, Offner GD, Troxler RF. MG2 and lactoferrin form a heterotypic complex in salivary secretions. *J Dent Res* 2003;82:471–475.
132. Soares RV, Lin T, Siqueira CC, et al. Salivary micelles: identification of complexes containing MG2, sIgA, lactoferrin, amylase, glycosylated proline-rich protein and lysozyme. *Arch Oral Biol* 2004;49:337–343.
133. Scannapieco FA, Torres G, Levine MJ. Salivary alpha-amylase: role in dental plaque and caries formation. *Crit Rev Oral Biol Med* 1993;4:301–307.

134. Bikker FJ, Ligtenberg AJ, van der Wal JE, et al. Immunohistochemical detection of salivary agglutinin/gp-340 in human parotid, submandibular, and labial salivary glands. *J Dent Res* 2002;81:134–139.
135. Raynal BD, Hardingham TE, Sheehan JK, Thornton DJ. Calcium-dependent protein interactions in MUC5B provide reversible cross-links in salivary mucus. *J Biol Chem* 2003;278:28703–28710.
136. Liu B, Rayment S, Oppenheim FG, Troxler RF. Isolation of human salivary mucin MG2 by a novel method and characterization of its interactions with oral bacteria. *Arch Biochem Biophys* 1999;364:286–293.
137. Ligtenberg AJ, Bikker FJ, De Blieck-Hogervorst JM, Veerman EC, Nieuw Amerongen AV. Binding of salivary agglutinin to IgA. *Biochem J* 2004;383: 159–164.
138. Yamaguchi T. Purification of saliva agglutinin of *Streptococcus intermedius* and its association with bacterial aggregation and adherence. *Arch Microbiol* 2004; 181:106–111.
139. Davis BJ. Disc electrophoresis—II. Method and application to human serum proteins. *Ann N Y Acad Sci* 1964;121:404–427.
140. Sheehan JK, Howard M, Richardson PS, Longwill T, Thornton DJ. Physical characterization of a low-charge glycoform of the MUC5B mucin comprising the gel-phase of an asthmatic respiratory mucous plug. *Biochem J* 1999;338:507–513.
141. Lu Y, Bennick A. Interaction of tannin with human salivary proline-rich proteins. *Arch Oral Biol* 1998;43:717–728.
142. Robinovitch MR, Ashley RL, Iversen JM, Vigoren EM, Oppenheim FG, Lamkin M. Parotid salivary basic proline-rich proteins inhibit HIV-I infectivity. *Oral Dis* 2001;7:86–93.
143. Oppenheim FG, Offner GD, Troxler RF. Phosphoproteins in the parotid saliva from the subhuman primate *Macaca fascicularis*. Isolation and characterization of a proline-rich phosphoglycoprotein and the complete covalent structure of a proline-rich phosphopeptide. *J Biol Chem* 1982;257:9271–9282.
144. Hay DI, Oppenheim FG. The isolation from human parotid saliva of a further group of proline-rich proteins. *Arch Oral Biol* 1974;19:627–632.
145. Vitorino R, Lobo MJ, Duarte JR, Ferrer-Correia AJ, Domingues PM, Amado FM. The role of salivary peptides in dental caries. *Biomed Chromatogr* 2005;19:214–222.
146. Dickinson DP. Cysteine peptidases of mammals: their biological roles and potential effects in the oral cavity and other tissues in health and disease. *Crit Rev Oral Biol Med* 2002;13:238–275.
147. Baron AC, Gansky SA, Ryder MI, Featherstone JD. Cysteine protease inhibitory activity and levels of salivary cystatins in whole saliva of periodontally diseased patients. *J Periodontal Res* 1999;34:437–444.
148. Henskens YM, Veerman EC, Nieuw Amerongen AV. Cystatins in health and disease. *Biol Chem Hoppe Seyler* 1996;377:71–86.
149. Kos J, Krasovec M, Cimerman N, Nielsen HJ, Christensen IJ, Brunner N. Cysteine proteinase inhibitors stefin A, stefin B and cystatin C in sera from patients with colorectal cancer: relation to prognosis. *Clin Cancer Res* 2000; 6:505–511.

150. Contucci AM, Inzitari R, Agostino S, et al. Statherin levels in saliva of patients with precancerous and cancerous lesions of the oral cavity: a preliminary report. *Oral Dis* 2005;11:95–99.
151. Yui S, Nakatani Y, Mikami M. Calprotectin (S100A8/S100A9), an inflammatory protein complex from neutrophils with a broad apoptosis-inducing activity. *Biol Pharm Bull* 2003;26:753–760.
152. Kleinegger CL, Stoeckel DC, Kurago ZB. A comparison of salivary calprotectin levels in subjects with and without oral candidiasis. *Oral Surg Oral Med Oral Pathol Oral Radiol Endod* 2001;92:62–67.
153. Sweet SP, Denbury AN, Challacombe SJ. Salivary calprotectin levels are raised in patients with oral candidiasis or Sjogren's syndrome but decreased by HIV infection. *Oral Microbiol Immunol* 2001;16:119–123.
154. Bingle L, Cross SS, High AS, et al. SPLUNC1 (PLUNC) is expressed in glandular tissues of the respiratory tract and in lung tumours with a glandular phenotype. *J Pathol* 2005;205:491–497.

17

Proteomics of Human Pancreatic Juice

Mads Grønborg, Anirban Maitra, and Akhilesh Pandey

Summary

Pancreatic juice has recently been characterized in detail using proteomic methods. The cataloging of proteins from healthy individuals and those diagnosed with pancreatic cancer has revealed the presence of a number of proteins in pancreatic juice that could serve as potential biomarkers for cancer. Because obtaining pancreatic juice is not trivial, it is possible that these biomarkers can be detected in serum using more sensitive methods like ELISA. Here, we discuss the protein constituents of pancreatic juice with special reference to cancer biomarkers.

Key Words: Pancreatic juice; LC-MS/MS; quantitative proteomics; biomarker discovery; mass spectrometry.

1. Introduction

Proteomic analyses of various body fluids have recently been reported. The pancreas is involved in secretion of different digestive enzymes, which facilitate the breakdown of carbohydrates, fats, and proteins in the duodenum. In addition, hormones are secreted by the pancreas (e.g., insulin and glucagon), which directly regulate the blood sugar level. This chapter describes the constituents of pancreatic juice and the regulation of their secretion. The hormones secreted by the endocrine tissue are also enumerated. In addition, a summary of key quantitative and nonquantitative proteomic studies of pancreatic juice that have led to identification of potential biomarkers for pancreatic cancer are presented.

2. The Pancreas and Pancreatic Juice

The pancreas is an elongated tapered organ, which is divided into five different parts (uncinate process, head, neck, body, and tail) and consists of two different types of tissue (exocrine and endocrine) **(Fig. 1)**. The exocrine part accounts

From: *Proteomics of Human Body Fluids: Principles, Methods, and Applications*
Edited by: V. Thongboonkerd © Humana Press Inc., Totowa, NJ

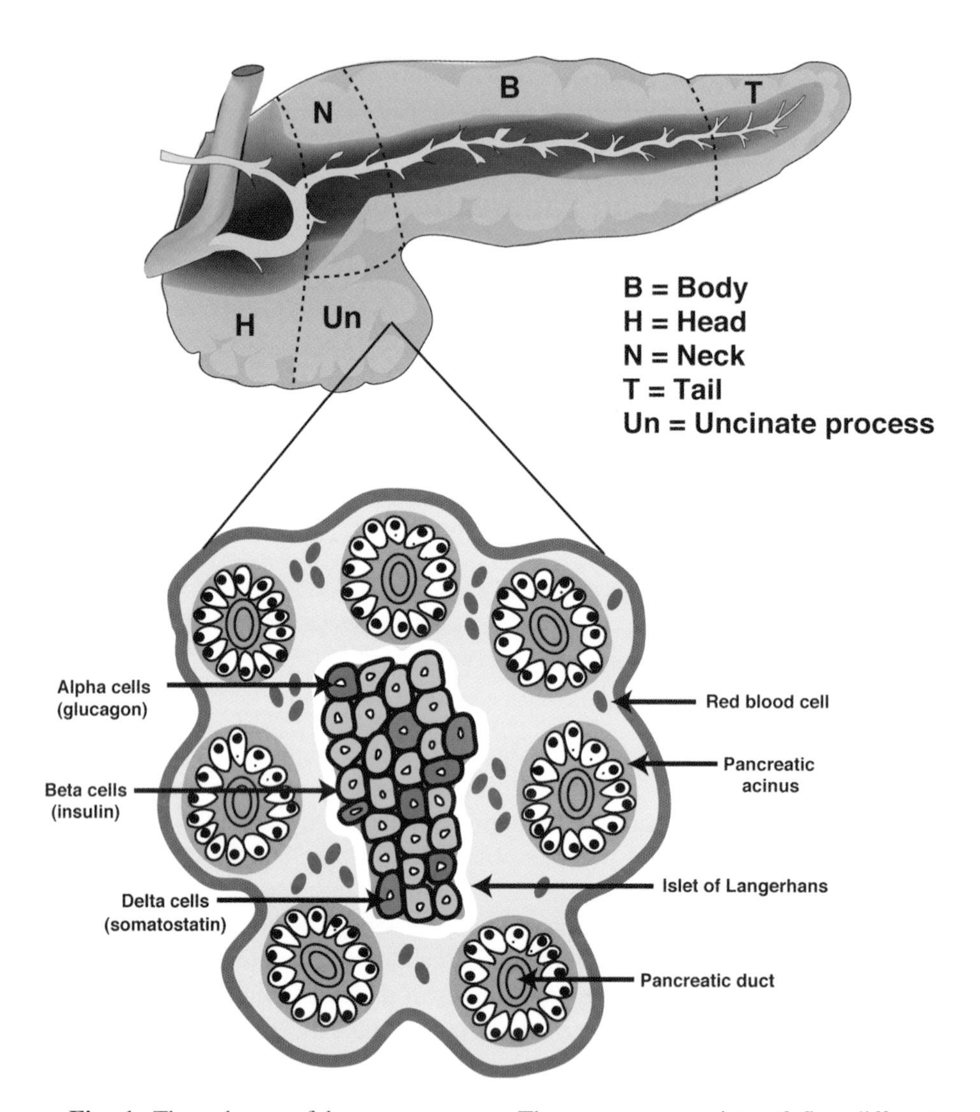

Fig. 1. The schema of human pancreas. The pancreas consists of five different regions including tail, body, neck, head, and the uncinate process. It contains both exocrine and endocrine tissues. The exocrine tissue contains two major cell types (acinar cells and ductal cells), which secrete digestive enzymes and water/bicarbonate, respectively. The endocrine tissue, which consists of approximately 1 million small clusters (named islets of Langerhans), secretes several hormones including insulin, glucagon, and somatostatin.

for approx 80% of glandular volume and contains at least two functional units: (1) acinar cells, which secrete digestive enzymes that facilitate the breakdown of carbohydrates, fats, proteins, and acids in the duodenum; and (2) ductal cells, which secrete fluids and electrolytes (e.g., water and bicarbonate). The endocrine cells, which consist of approximately 1 million small clusters (called islets of Langerhans), are embedded in the exocrine tissue and secrete hormones (e.g., insulin and glucagon) directly into the bloodstream (**Fig. 1**).

2.1. Exocrine Secretion From the Pancreas

The acinar cells have membrane-bound secretory granules, which contain digestive enzymes that are exocytosed into the lumen of the acinus. The secreted enzymes are ultimately collected by the main pancreatic duct, which drains directly into the duodenum. Three main classes of enzymes, which are critical for proper digestion (proteases, lipases, and amylases), are secreted by the acinar cells. The two major pancreatic proteases are trypsin and chymotrypsin. Both enzymes are synthesized as proenzymes (inactive) and released into the lumen of the small intestine, where trypsin is first activated (cleaved) by enterokinase, another protease embedded in the intestinal mucosa. Trypsin in its active form subsequently activates chymotrypsin and other trypsin precursor molecules, resulting in a dramatic increase in active proteases in the small intestine. Other proteolytic enzymes include carboxypeptidase A and B and elastase.

Triglycerides, the major component of dietary fat, cannot be absorbed directly across the intestinal mucosa. Pancreatic lipase, a major component of pancreatic juice, catalyzes or hydrolyzes the triglycerides into monoglycerides and free fatty acids. In addition, bile salts (produced by the liver) must be present in sufficient quantities in the intestinal lumen to ensure efficient digestion of triglycerides. The bile salts facilitate emulsification of fat droplets, which enhance lipase activity. In contrast to the proteolytic enzymes (e.g., trypsin and chymotrypsin), the lipolytic enzymes are secreted as active forms. Amylase is also secreted from the acinar cells to the lumen and facilitates the hydrolysis of starch to maltose (a glucose disaccharide), maltotriose, and dextrin (small polysaccharides). The major source of amylase is pancreatic secretion, although it is also a component of saliva.

As mentioned previously, the exocrine part contains two different functional units including the acinar cells, which secrete digestive enzymes, and the ductal cells, which are the primary source of bicarbonate and water. Bicarbonate (HCO_3^-) is produced from water (H_2O) and carbon dioxide (CO_2) by the enzyme carbonic anhydrase in the ductal cells and is subsequently secreted into the lumen of the pancreatic duct to maintain the pH level of pancreatic juice (normally 8.0–8.5). Approximately 1500 to 3000 mL of pancreatic juice is secreted daily by the pancreas.

2.2. Regulation of Exocrine Secretion

Exocrine secretion from the pancreas is subjected to neural and endocrine regulation. However, the most important stimulus for pancreatic secretion comes from three well-studied hormones, cholecystokinin, gastrin, and secretin, which are produced within the enteric endocrine system (endocrine cells within the gastrointestinal tract) and subsequently secreted into the bloodstream. Cholecystokinin is synthesized in the duodenum, and its secretion is stimulated by the presence of partially digested proteins and fat in the small intestine. Cholecystokinin is released into the bloodstream as chyme floods into the small intestine, where it binds to receptors on the pancreatic acinar cells, leading to secretion of digestive enzymes. Gastrin, a hormone similar to cholecystokinin, is secreted in large amounts by the stomach in response to gastric distension and irritation and stimulates the secretion of acid by gastric parietal cells and of digestive enzymes by pancreatic acinar calls. Secretin is secreted by the endocrinocytes located in the epithelium of the small intestine in response to acid in the duodenum. Secretin stimulates exocrine ductal cells to secrete water and bicarbonate, which flush the digestive enzymes from the pancreatic acinar cells into the lumen.

2.3. Endocrine Secretion From the Pancreas

As mentioned above, the human pancreas contains approximately 1 million small clusters called islets of Langerhans. These islets are the endocrine part of the pancreas and correspond to 1 to 2% of the total pancreatic tissue. Owing to a high degree of vascularization, the islets receive 10 to 15% of the pancreatic blood flow, which ensures the secreted hormones ready access to be circulated. In addition, the islets are innervated by different types of neurons that facilitate the initial stimulus for secretion of hormones. Three major cell types (α-, β- and δ-cells), each of which secretes specific hormones, are present in the islets **(Fig. 1)**. β-Cells are the most abundant and occupy the central portion of the islets. Insulin is synthesized and secreted from the β-cells and controls many biological processes, including glucose uptake from the blood. The β-cells are surrounded by a ring of α- and δ-cells, which secrete glucagon and somatostatin, respectively. Glucagon is secreted when the level of glucose in the blood is low. Insulin and glucagon exert opposite effects to maintain blood glucose at a constant level. Somatostatin is secreted by a broad range of tissues including the δ-cells in the pancreas. Two active forms of somatostatin are synthesized—a long and a short form, namely, SS-28 and SS-14, respectively. SS-28 is predominantly produced in the nervous system and pancreas, whereas SS-14 is secreted mostly by the intestine. Somatostatin's primary effect in the pancreas is to inhibit the secretion of insulin and glucagon from β- and α-cells, respectively. In addition, it also plays a role in suppressing pancreatic exocrine secretion by inhibiting cholecystokinin and secretin.

3. Proteomic Analysis of Pancreatic Juice

The availability and complexity of different body fluids for proteomic studies vary greatly. Serum is an example of a body fluid that is easy to obtain and contains thousands of different proteins *(1,2)*. One drawback of serum is that it contains a small number of proteins in very high abundance, which makes it difficult to identify the low-abundant proteins. The high-abundant proteins (e.g., albumin, transferrin, and immunoglobulins) represent as much as 80% of the total protein content in serum. For example, serum albumin is present at 35 to 55 mg/mL in serum, whereas cytokines are present at the low pg/mL level *(3)*. Several methods including ultracentrifugation *(4)*, immunodepletion, solvent precipitation, and size exclusion *(5)* have been used to reduce the complexity of serum prior to mass spectrometric analysis. One potential caveat of depletion strategies is that one might lose proteins that are bound to the depleted proteins. Even though the total number of proteins identified in serum is in the thousands *(2)*, it still represents only a fraction of the total estimated number of serum proteins.

Pancreatic juice is one example of a body fluid that is not easily obtained because it involves an invasive surgical procedure. It is usually collected from patients undergoing pancreatectomy or endoscopic retrograde cholangiopancreatography (ERCP) because of suspected pancreatic diseases such as cancer or pancreatitis. Pancreatic juice from healthy individuals (controls) is, therefore, even harder to obtain. Because of this, proteomic studies of pancreatic juice are limited, and knowledge about the actual constituents of human pancreatic juice is still inadequate. Early study to characterize pancreatic juice used a combination of 2D electrophoresis (2-DE) and silver staining to visualize the resolved proteins obtained from juice from normal and cancer patients *(6)*. Only three proteins (α_1-antitrypsin, transferrin, and albumin) were identified as overexpressed proteins in pancreatic juice in cancer patients. The proteins were quantified by densitometry and subsequently identified by Western blotting. Subsequent study used reversed-phase high-performance liquid chromatography (RP-HPLC) for initial fractionation and sodium dodecyl sulfate-polyacrylamide gel electrophoresis (SDS-PAGE) for obtaining molecular weight information about the resolved proteins *(7)*. Again, only a few enzymatic proteins were identified in this study.

We recently reported a comprehensive analysis of pancreatic juice from pancreatic cancer patients by first fractionating the sample using 1D electrophoresis (1-DE), followed by liquid chromatography coupled to tandem mass spectrometry (LC-MS/MS) *(8)*. A total of 170 unique proteins were identified, including proteins known to be constituents of normal pancreatic juice (e.g., pancreatic lipase, pancreatic carboxypeptidase, pancreatic amylase, and trypsinogens; *see* **Table 1** for a partial list of the proteins identified in this study) and proteins previously shown to be elevated in pancreatic cancer (e.g., annexin I, carcinoembryonic antigen-related cell adhesion molecule 5 [CEACAM5], clusterin, deleted in

Table 1
A Partial List of Proteins Identified From Pancreatic Juice

RefSeq no.	Protein name	Biological process	Molecular function	Cellular component
NP_000468	Albumin	Transport	Transporter activity	Extracellular
NP_001076	α_1-antichymotrypsin	Protein metabolism	Protease inhibitor activity	Extracellular
NP_000508	α_2-globin (hemoglobin, α2)	Transport	Transporter activity	Extracellular
NP_001125	α-Fetoprotein	Transport	Transporter activity	Extracellular
NP_443204	α_2-glycoprotein	Transport	Transporter activity	Extracellular
NP_001141	Aminopeptidase N	Protein metabolism	Enzyme (aminopeptidase)	Plasma membrane
NP_000691	Annexin I	Cell communication	Receptor ligand	Plasma membrane
NP_000479	Antithrombin III	Protein metabolism	Protease inhibitor activity	Extracellular
NP_001634	Apolipoprotein AII	Transport	Transporter activity	Extracellular
NP_036202	Attractin	Immune response	Complement	Plasma membrane
NP_000509	β-Globin	Transport	Transporter activity	Extracellular
NP_000033	β_2-glycoprotein I	Transport	Transporter activity	Extracellular
NP_001703	Carcinoembryonic antigen-related cell adhesion molecule 1 (CEACAM1)	Cell communication	Cell adhesion molecule activity	Plasma membrane
NP_004354	Carcinoembryonic antigen-related cell adhesion molecule 5 (CEACAM5)	Immune response	Cell adhesion molecule activity	Plasma membrane
NP_002474	Carcinoembryonic antigen-related cell adhesion molecule 6 (CEACAM6)	Immune response	Cell adhesion molecule activity	Plasma membrane
NP_001902	Cathepsin G	Protein metabolism	Enzyme (serine-type peptidase)	Extracellular
NP_000087	Ceruloplasmin	Transport	Transporter activity	Extracellular
NP_003456	Chitotriosidase	Energy pathways	Enzyme (hydrolase)	Extracellular

NP_009203	Chymotrypsin C	Protein metabolism	Enzyme (serine-type peptidase)	Extracellular
NP_001898	Chymotrypsin-like protease	Protein metabolism	Enzyme (serine-type peptidase)	Extracellular
NP_001897	Chymotrypsinogen B1	Protein metabolism	Enzyme (serine-type peptidase)	Endoplasmic reticulum
NP_001822	Clusterin	Immune response	Complement	Extracellular
NP_000053	Complement component 1 inhibitor	Protein metabolism	Protease inhibitor activity	Plasma membrane
NP_000055	Complement component 3	Immune response	Complement	Extracellular
NP_000558	C-reactive protein	Cell communication	Unclassified	Extracellular
NP_000090	Cystatin C	Protein metabolism	Protease inhibitor activity	Extracellular
NP_005208	Defensin, $\alpha 3$	Immune response	Complement	Extracellular
NP_015568	Deleted in malignant brain tumors 1 isoform b (DMBT-1b)	Immune response	Unclassified	Extracellular
NP_001963	Elastase neutrophil	Protein metabolism	Enzyme (serine-type peptidase)	Extracellular
NP_001954	Epidermal growth factor	Cell organization and biogenesis	Receptor ligand	Extracellular
NP_071317	Estrogen-regulated gene 1	Cell communication	Plasma membrane organization and biogenesis	Plasma membrane
NP_000500	Fibrinogen, γ-chain	Protein metabolism	Blood coagulation factor	Extracellular
NP_002017	Fibronectin 1	Cell organization and biogenesis	Structural molecule activity	Extracellular
NP_005558	Galectin 3 binding protein	Immune response	Receptor activity	Extracellular
NP_000168	Gelsolin	Cell communication	Calcium ion binding	Extracellular
NP_000604	Hemopexin	Transport	Transporter activity	Extracellular
NP_000403	Histidine-rich glycoprotein	Transport	Transporter activity	Extracellular
NP_653247	Immunoglobulin J chain	Immune response	Unclassified	Extracellular
NP_000588	Insulin-like growth factor binding protein 2 (IGFBP-2)	Cell communication	Cell adhesion molecule activity	Extracellular

(Continued)

Table 1 *(Continued)*

RefSeq no.	Protein name	Biological process	Molecular function	Cellular component
NP_000878	Integrin-αX	Cell communication	Receptor activity	Plasma membrane
NP_002248	Kallikrein 1	Protein metabolism	Enzyme (serine-type peptidase)	Extracellular
NP_002334	Lactotransferrin	Transport	Transporter activity	Extracellular
NP_005550	Laminin α-1	Cell organization and biogenesis	Extracellular matrix binding	Extracellular
NP_005555	Lipocalin 2 (oncogene 24p3)	Transport	Transporter activity	Extracellular
NP_002336	Lumican	Cell organization and biogenesis	Structural molecule activity	Extracellular
NP_000230	Lysozyme	Energy pathways	Enzyme (hydrolase)	Extracellular
NP_000005	Macroglobulin, α2	Protein metabolism	Protease inhibitor activity	Extracellular
NP_056146	Nicastrin	Cell communication	Plasma membrane organization and biogenesis	Plasma membrane
NP_000599	Orosomucoid	Immune response	Transporter activity	Extracellular
NP_000690	Pancreatic amylase, α2A	Energy pathways	Enzyme (amylase)	Extracellular
NP_001862	Pancreatic carboxypeptidase B1	Protein metabolism	Enzyme (carboxypeptidase)	Extracellular
NP_001823	Pancreatic colipase	Energy pathways	Enzyme (colipase)	Extracellular
NP_254275	Pancreatic elastase IIA	Protein metabolism	Enzyme (serine-type peptidase)	Extracellular
NP_000927	Pancreatic lipase	Energy pathways	Enzyme (lipase)	Extracellular
NP_000919	Pancreatic phospholipase A2	Energy pathways	Enzyme (phospholipase)	Extracellular
NP_001493	Pancreatic zymogen granule membrane protein 2 (GP2)	Unclassified	Plasma membrane organization and biogenesis	Plasma membrane
NP_002571	Pancreatitis-associated protein/hepatocarcinoma-intestine-pancreas (HIP/PAP)	Cell communication	Receptor ligand	Extracellular

NP_000437	Paraoxonase 1	Energy pathways	Enzyme (esterase)	Plasma membrane
NP_002621	Pepsinogen C	Protein metabolism	Enzyme (aspartic protease)	Extracellular
NP_005082	Peptidoglycan recognition protein	Immune response	Unclassified	Extracellular
NP_002606	Pigment epithelium-derived factor	Cell communication	Enzyme (serine-type peptidase)	Extracellular
NP_002635	Polymeric immunoglobulin receptor	Immune response	Receptor activity	Plasma membrane
NP_002764	Prostasin	Protein metabolism	Enzyme (serine-type peptidase)	Extracellular
NP_006498	Reg I α	Cell organization and biogenesis	Extracellular matrix binding	Extracellular
NP_065148	Resistin	Cell communication	Receptor ligand	Extracellular
NP_002955	S100 calcium-binding protein A8	Cell communication	Calcium ion binding	Extracellular
NP_004090	Stomatin	Cell communication	Plasma membrane organization and biogenesis	Plasma membrane
NP_001054	Transferrin	Transport	Transporter activity	Extracellular
NP_000362	Transthyretin	Transport	Transporter activity	Extracellular
NP_002761	Trypsinogen 2	Protein metabolism	Enzyme (serine-type peptidase)	Extracellular
NP_003290	Tumor rejection antigen (gp96)	Protein metabolism	Heat shock protein activity	Endoplasmic reticulum
NP_001176	Zinc α-2 glycoprotein	Regulation of nucleobase, nucleoside, nucleotide, and nucleic acid metabolism	Enzyme (ribonuclease)	Extracellular

Data from ref. ***8***.

malignant brain tumor 1 [DMBT1], lipocalin [oncogene 24p3], Mac-2 binding protein, and members of the S100 family of Ca^{+}-binding proteins; *see* **Table 2**). In addition, a functional annotation (including information on the biological process, molecular function, and cellular localization) of 170 identified proteins was carried out. As expected, enzymes (in particular, serine proteases) turned out to be the major subgroup (37%) of the identified proteins. Proteins involved in transport activity (13%), structure (10%), and complement components (6%) were also identified. Information about the cellular localization revealed that 62% of the identified proteins were extracellular or membrane bound.

Alternatively, pancreatic juice can be analyzed by the surface-enhanced laser desorption/ionization (SELDI) technique. SELDI generates a mass-to-charge (*m/z*) peak pattern or "signature," which can be used to compare individual samples. One drawback of SELDI is that the protein or peptide corresponding to the characteristic peak pattern is not identified in the experiment. Therefore, one has to carry out additional experiments to identify the protein/peptide giving rise to a specific *m/z* ratio. In a recent study, SELDI was used to evaluate pancreatic juice obtained from patients diagnosed with pancreatic cancer ***(9)***. One peak was identified to be present in 67% of the pancreatic cancer samples and was subsequently identified as hepatocarcinoma intestine pancreas/pancreatitis-associated protein I (HIP/PAP). In addition, HIP/PAP was subsequently quantified by enzyme-linked immunosorbent assay (ELISA) in 43 patients (28 with pancreatic cancer and 15 controls).

The isotope-coded affinity tagging (ICAT) method has been used to identify and quantify differentially expressed proteins in cancer, chronic pancreatitis, and normal pancreatic juice ***(10)***. A total of 78 and 61 proteins were identified in pancreatic juice from cancer and normal patients, respectively. When proteins in cancer were compared with normal samples, 30 proteins were identified to be differentially expressed (>twofold) in cancer (24 proteins upregulated and 6 proteins downregulated).

4. Identification of Biomarkers for Pancreatic Cancer

Pancreatic cancer is the fourth leading cause of cancer, and approx 30,000 new patients are diagnosed each year in the United States ***(11,12)***. Most patients diagnosed with pancreatic cancer die from their disease within months, and only 4% are alive 5 yr after diagnosis. At present, a cure for pancreatic cancer is only possible with surgical resection; however, more than 80% of the patients present with locally advanced disease or distant metastases rendering the tumor inoperable ***(13)***. Unfortunately, the clinical pancreatic tumor markers that are available today lack the sensitivity and specificity to diagnose potentially curable, small lesions ***(9)***. One biomarker widely used to diagnose pancreatic cancer is CA-19-9. However,

Table 2
A Partial List of Proteins Identified in Pancreatic Juice Whose Expression Has Previously Been Associated With Pancreatic and Other Cancers

Protein name	Cancer	Ref.
Annexin 5	Pancreas, pituitary	*43*
Annexin 1	Pancreas, esophagus, pituitary, lung	*41,43–46*
Carcinoembryonic antigen-related cell adhesion molecule 1 (CEACAM1)	Pancreas, lung, gastrointestinal, colorectal	*47–49*
Carcinoembryonic antigen-related cell adhesion molecule 5 (CEACAM 5)	Pancreas, lung, gastrointestinal, colorectal	*47,48*
Carcinoembryonic antigen-related cell adhesion molecule 6 (CEACAM6)	Pancreas, lung, gastrointestinal, colorectal	*47,48,50,51*
Cystatin C	Pancreas	*52*
Deleted in malignant brain tumors 1 (DMBT 1)	Pancreas, brain, lung, colon, gastric	*53–55*
Fibronectin 1	Pancreas	*41*
Galectin 3 binding protein (Mac-2 binding protein)	Pancreas, breast, colon	*56–59*
Hepatocarcinoma-intestine-pancreas/pancreatitis-associated protein precursor I (HIP/PAP I)	Pancreas, liver, cholangiocarcinoma cells	*60–62*
Insulin-like growth factor binding protein 2 (IGFBP-2)	Pancreas, CSF, serum, prostate gland, liver	*52,63–66*
Lipocalin 2 (oncogene 24p3)	Pancreas	*52*
Lysosomal-associated membrane protein 2 (CD107b)	Pancreas	*67*
Lumican	Pancreas	*68*
Melanoma inhibitory activity (MIA)	Pancreas	*40,52*
Pepsinogen C	Pancreas, breast, prostate	*52,69,70*
Profilin 1	Pancreas	*52*
Prostasin	Prostate, breast, ovarian	*71–73*
S100 calcium binding protein P	Pancreas	*40,52*

several problems are associated with the use of CA-19-9 including its elevation in benign conditions such as acute and chronic pancreatitis and inflammatory diseases of liver and biliary tree. Furthermore, the sensitivity and specificity of CA-19-9 is only approx 80%, which diminish its use for screening purposes ***(14)***. The poor prognosis and late presentation of pancreatic cancer emphasize the importance of an early detection strategy, especially for patients at high risk of developing pancreatic cancer. It is clear that changes in protein expression, aberrant localization, or differential protein modification can be observed during transformation of a healthy cell into a neoplastic cell ***(15–19)***. Proteomic profiling of cancer involves cataloging these changes, which can help one to understand the underlying mechanisms involved in cancer.

4.1. Pancreatic Juice as a Source of Biomarker Discovery for Pancreatic Cancer

Pancreatic juice is an excellent clinical specimen for identification of novel biomarkers for pancreatic cancer for several reasons. The relatively low complexity of pancreatic juice compared with other body fluids (e.g., serum) makes it easier to identify low-abundant proteins, which are potentially elevated in cancer patients compared with controls. The fact that most of the proteins secreted from the exocrine tissue of the pancreas are found in the pancreatic juice makes it more likely that one can identify changes in protein expression owing to the transformed cancer cells. It is likely that many of these proteins cannot be identified easily by proteomic profiling of bulk tissue.

In order to use pancreatic juice as a source for biomarker discovery, one has to compare the protein expression profile of juice obtained from control subjects against cancer patients. However, one problem is the high degree of heterogeneity among the different samples that is obvious even when juice is fractionated by 1-DE (**Fig. 2**). It is clear that the protein expression profiles are very different even when juice samples from multiple controls are compared. (The same is true when samples from different cancer patients are compared.) This problem is clearly pointed out in a recent proteomics study in which pancreatic juice from three individual patients diagnosed with pancreatic cancer was analyzed by a combination of 1-DE and LC-MS/MS. Out of 170 proteins identified, only 23 were found to be common to all three samples ***(8)***. The same observation was seen in a new quantitative ICAT analysis in which pancreatic juice from normal, chronic pancreatitis and cancer subjects were compared ***(10)***. When normal was compared with normal, 15 proteins (25%) showed greater than twofold difference in expression between the two specimens. The high degree of biologic variation in pancreatic juice should, therefore, be taken into consideration for biomarker assessment. Further validation of potential biomarkers identified from pancreatic juice by proteomic profiling is therefore

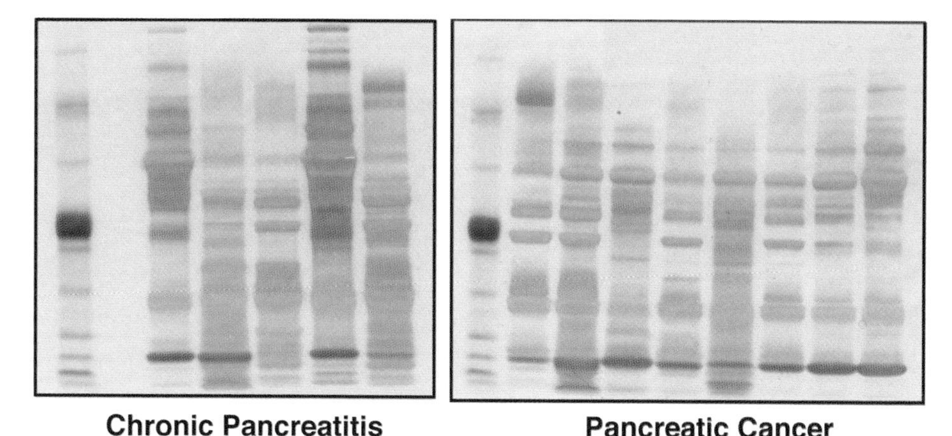

Fig. 2. Gel electrophoresis of pancreatic juice samples. One problem associated with proteomic analysis of pancreatic juice is the high degree of heterogeneity of samples. For illustration, approximately 10 μg of pancreatic juice obtained from controls (chronic pancreatitis, left panel) and pancreatic cancer patients (right panel) was loaded on the gel and stained with silver. The variation in expression profiles among different individuals is clearly seen in both controls and cancer patients.

needed. Validation can be conducted by immunohistochemistry (IHC) on tissue microarray (TMA) or by ELISA of serum.

4.2. Quantitative Proteomics for Identification of Pancreatic Cancer Biomarkers

During the last several years, the proteomic community has focused on developing quantitative strategies and platforms needed to visualize/detect differences in normal tissue vs cancer. 2-DE is a widely used method to study differentially expressed proteins owing to its robustness and ability to detect thousands of proteins in one gel *(20–22)*. 2-DE has several advantages including the capability to separate closely related isoforms of a particular protein species and the ability to differentiate posttranslationally modified proteins (e.g., different phosphorylation and glycosylation patterns) *(23,24)*. The drawbacks of 2-DE are that it does not work well with small proteins, very basic, and/or hydrophobic proteins. Several recent studies have used 2-DE to identify proteins differentially expressed in normal pancreas vs pancreatic cancer tissue *(25,26)*. Owing to the high degree of heterogeneity in the tumor tissue (i.e., mixture of normal and cancer cells), laser capture microdissection (LCM) can be performed prior to 2-DE *(27)*. One drawback of LCM is that only very small amounts of sample can be generated (microgram amounts), which makes enrichment (e.g., with antibodies) or fractionation prior to 2-DE difficult, as only the most abundant proteins will be identified.

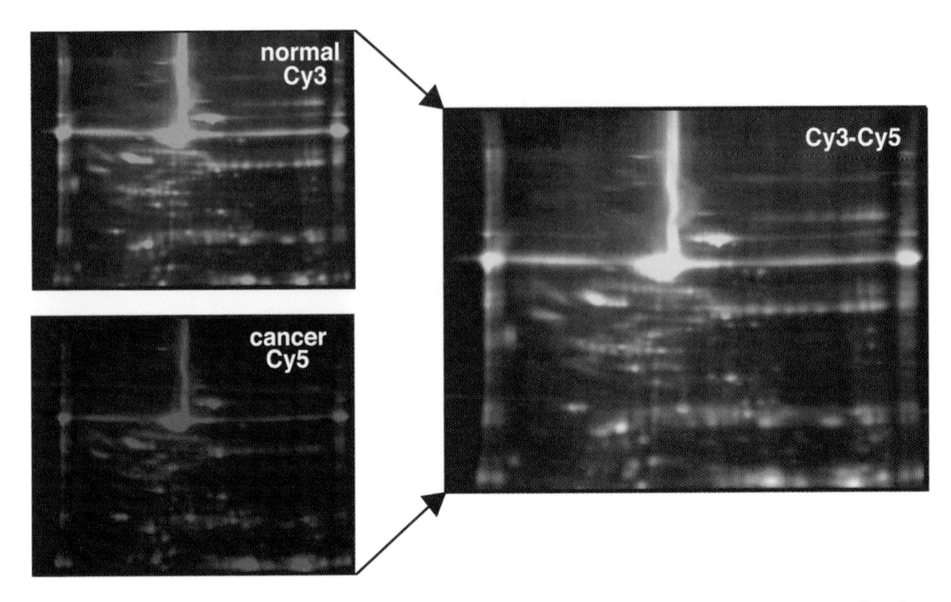

Fig. 3. Identification of differentially expressed proteins by DIGE. For identification of differentially expressed proteins in pancreatic juice from control patients (chronic pancreatitis) vs cancer patients, the pancreatic juice can be analyzed by difference gel electrophoresis (DIGE). Proteins from the control sample are labeled with Cy3 dye, whereas proteins from the cancer sample are labeled with a different dye, Cy5. The two images are then superimposed to determine proteins that are upregulated in cancer (red spots), unchanged (yellow spots), or downregulated in cancer (green).

Alternatively, differential gel electrophoresis (DIGE) can be used to identify differentially expressed proteins in pancreatic tissue or pancreatic juice *(28)*. Proteins from normal and cancer samples are labeled with different fluorescent dyes (e.g., Cy3 and Cy5) and subsequently separated by 2-DE. The two images are then superimposed, enabling one to pinpoint the proteins, which are up- or downregulated in cancer vs normal samples **(Fig. 3)**. In recent years, mass spectrometry-based approaches such as SELDI-time of flight (TOF)-MS have been applied for the identification of biomarkers in body fluids from diseased individuals compared with controls *(9,29,30)*. As mentioned previously, SELDI-TOF MS uses differences in peak pattern of the mass spectra to distinguish normal and disease samples. Although characteristic patterns of peptide peaks that differentiate normal from diseased individuals can be obtained easily by SELDI-TOF-MS, the identification of specific proteins corresponding to those specific protein peaks is still a difficult task. A combination of gel-based purification methods and LC-MS/MS analysis provides a way to generate comprehensive catalogs of the physiological proteome of the analyzed sample/fluid *(8,31–33)*. Such catalogs can serve as a starting point for further testing and validation of potential biomarker candidates.

The use of mass tagging (e.g., ICAT) has proved to be a powerful technique to obtain relative quantitation data on a large set of proteins ***(34,35)***. Protein expression profiles from cancer and normal tissues or fluids can then be compared, allowing one to identify specific proteins that are elevated in cancer compared with normal tissue ***(10,36,37)***. Alternatively, stable isotope labeling (e.g., stable isotope labeling with amino acids in cell culture [SILAC]) can be used to identify proteins that are differentially expressed in normal vs cancer tissue. SILAC has recently been used for relative quantitation of lysates from two prostate cancer cell lines with different metastatic potential ***(38)***. Over 400 proteins were identified and quantified in this study, out of which 60 were elevated in the highly metastatic cell line. We have recently used SILAC to compare the pancreatic secretome (secreted proteins only) of a normal human pancreatic ductal epithelium (HPDE) cell line with its neoplastic counterpart (Panc1) ***(39)***. Quantitative data were obtained from almost 200 proteins, in which 39 proteins were found to be elevated (>threefold) in cancer compared with normal cells. In addition, several novel proteins not previously known to be upregulated were identified and subsequently validated in pancreatic tissue by IHC on TMAs. The use of quantitative proteomics as an initial screening tool for identification of differentially expressed proteins has been shown to be a very important tool for the discovery of potential novel biomarkers. Promising candidates can then be tested in large-scale quantitative studies for further clinical use.

5. Methods

5.1. Sample Collection

1. Human pancreatic juice can be collected during surgery from the pancreatic duct of patients undergoing pancreatectomy or during an ERCP procedure. Approximately 20 to 500 µL is generally collected and should be stored immediately at –80°C with or without protease inhibitors.
2. The pancreatic juice should be kept on ice (4°C) at all times when not stored at –80°C.
3. Small debris can be present in the collected juice requiring centrifugation of the sample.
4. The protein concentration should be determined using a protein assay kit (e.g., Modified Lowry or Bradford). The protein concentration in pancreatic juice ranges from 2 to 15 mg/mL depending on the specific sample.

5.2. Fractionation and In-Gel Digestion

The complexity of human pancreatic juice is still not fully elucidated, but studies using gene expression array ***(40–42)*** and proteomics studies ***(8,10,39)*** indicate that hundreds of proteins should be present in pancreatic juice. It is, therefore, recommended to fractionate proteins by 1-DE or 2-DE prior to LC-MS/MS analysis to reduce the complexity of the sample. Alternatively, in-solution digest in combination with 2D LC-MS/MS can be performed. For 1-DE combined with tryptic digestion and LC-MS/MS, it is recommended that approximately 20 to 30 μg of protein be used as starting material. The proteins can be resolved by 10% homogeneous gel or 4 to 12% gradient gel. The proteins can be visualized by either silver staining or colloidal Coomassie staining and digested by trypsin as previously described *(8)*.

5.3. LC-MS/MS Analysis

Depending on the available hardware (liquid chromatography systems and mass spectrometers), different LC-MS/MS setups can be used. One setup utilizes a precolumn and an analytical column in tandem, whereas a different one uses a single analytical column. If the sample to be analyzed needs extensive desalting or if a large volume (10–40 μL) is required for loading, then a tandem column setup is preferred. The tandem column is very robust and can be used for most of the samples. A drawback is that one might lose sensitivity owing to broader chromatographic peaks. Several column materials can be used for precolumns (e.g., YMC ODS-A, 5–15-μm beads; Kanematsu USA, New York, NY) and analytical columns (e.g., Vydac MS218, 5-μm beads; Vydac, Columbia, MD). The performance (i.e., separation and resolution) of the LC system is dependent on the type of reversed-phase material, and it is strongly suggested that several trial runs including different reversed-phase materials should be tested for optimal results. In the following section, the conditions for a single column are outlined.

1. After tryptic digestion, the peptides are dried down in a vacuum centrifuge and redissolved in 10 μL of 5% formic acid.
2. The redissolved peptides are automatically loaded during 6 to 8 min onto the reversed-phase column by the autosampler in the HPLC system using a linear gradient of 5 to 10% mobile phase B (90% acetonitrile, 0.4% acetic acid, 0.005% heptafluorobutyric acid in water).
3. After loading, the peptides are subsequently separated and eluted from the column by 10 to 45% mobile phase B (35–40-min gradient), followed by 45 to 90% mobile phase B for 3 min.
4. The column is finally rinsed with 90% mobile phase B for 1 to 2 min and equilibrated in 5% mobile phase B for 3 min.
5. The flow rate delivered from the pumps to the column during loading is 1 to 1.5 μL/min, and during peptide separation/elution the flow rate is decreased to 250 to 300 nL/min.

6. Perspectives and Future Directions

Proteomic analysis of human pancreatic juice is still at an early stage, and only limited studies have been carried out thus far. Only recently, more comprehensive studied have been initiated and have provided valuable insights into the physiological proteome of human pancreatic juice ***(8,10)***. Quantitative proteomics can be used to identify proteins that are differentially expressed in juice obtained from normal and cancer patients. Proteins found to be elevated in cancer can be used as potential biomarkers for early detection. However, one obstacle in identifying differentially expressed proteins is the high degree of heterogeneity among the samples. This makes it difficult to pinpoint the exact changes caused by cancer as opposed to biological variation between samples. Control experiments using large sample sizes are, therefore, very important to confirm the results obtained (e.g., ELISA). The results from small-scale quantitative proteomics studies will help in initiating additional studies, which should lead to the discovery of novel biomarkers for the early detection of pancreatic cancer.

Acknowledgments

Anirban Maitra is supported by an AACR-PanCAN Career Development Award, a grant from the Lustgarten Foundation for Pancreatic Cancer Research, NCI 1R01CA113669, and the Michael Rolfe Foundation. Akhilesh Pandey is supported by a pilot project award from the Sol Goldman Trust and a grant (U54 RR020839) from the National Institutes of Health.

References

1. Anderson NL, Polanski M, Pieper R, et al. The human plasma proteome: a nonredundant list developed by combination of four separate sources. *Mol Cell Proteomics* 2004;3:311–326.
2. Muthusamy B, Hanumanthu G, Suresh S, et al. Plasma Proteome Database as a resource for proteomics research. *Proteomics* 2005;5:3531–3536.
3. Anderson NL, Anderson NG. The human plasma proteome: history, character, and diagnostic prospects. *Mol Cell Proteomics* 2002;1:845–867.
4. Tirumalai RS, Chan KC, Prieto DA, Issaq HJ, Conrads TP, Veenstra TD. Characterization of the low molecular weight human serum proteome. *Mol Cell Proteomics* 2003;2:1096–1103.
5. Pieper R, Gatlin CL, Makusky AJ, et al. The human serum proteome: display of nearly 3700 chromatographically separated protein spots on two-dimensional electrophoresis gels and identification of 325 distinct proteins. *Proteomics* 2003; 3:1345–1364.
6. Satoh J, Darley-Usmar VM, Kashimura H, Fukutomi H, Anan K, Ohsuga T. Analysis of pure pancreatic juice proteins by two-dimensional gel electrophoresis in cases of pancreatic cancer. *Gastroenterol Jpn* 1986;21:623–629.
7. Goke B, Keim V, Dagorn JC, Arnold R, Adler G. Resolution of human exocrine pancreatic juice proteins by reversed-phase high performance liquid chromatography (HPLC). *Pancreas* 1990;5:261–266.
8. Gronborg M, Bunkenborg J, Kristiansen TZ, et al. Comprehensive proteomic analysis of human pancreatic juice. *J Proteome Res* 2004;3:1042–1055.

9. Rosty C, Goggins M. Early detection of pancreatic carcinoma. *Hematol Oncol Clin North Am* 2002;16:37–52.
10. Chen R, Pan S, Brentnall TA, Aebersold R. Proteomic profiling of pancreatic cancer for biomarker discovery. *Mol Cell Proteomics* 2005;4:523–533.
11. Greenlee RT, Hill-Harmon MB, Murray T, Thun M. Cancer statistics, 2001. *CA Cancer J Clin* 2001;51:15–36.
12. Jemal A, Thomas A, Murray T, Thun M. Cancer statistics, 2002. *CA Cancer J Clin* 2002;52:23–47.
13. Yeo TP, Hruban RH, Leach SD, et al. Pancreatic cancer. *Curr Probl Cancer* 2002;26:176–275.
14. Steinberg W. The clinical utility of the CA 19-9 tumor-associated antigen. *Am J Gastroenterol* 1990;85:350–355.
15. Hanash S. Mining the cancer proteome. *Proteomics* 2001;1:1189–1190.
16. Petricoin EF, Ardekani AM, Hitt BA, et al. Use of proteomic patterns in serum to identify ovarian cancer. *Lancet* 2002;359:572–577.
17. Srinivas PR, Srivastava S, Hanash S, Wright GL Jr. Proteomics in early detection of cancer. *Clin Chem* 2001;47:1901–1911.
18. Srivastava S, Verma M, Henson DE. Biomarkers for early detection of colon cancer. *Clin Cancer Res* 2001;7:1118–1126.
19. Verma M, Srivastava S. Epigenetics in cancer: implications for early detection and prevention. *Lancet Oncol* 2002;3:755–763.
20. Shevchenko A, Jensen ON, Podtelejnikov AV, et al. Linking genome and proteome by mass spectrometry: large-scale identification of yeast proteins from two dimensional gels. *Proc Natl Acad Sci U S A* 1996;93:14,440–14,445.
21. Gygi SP, Corthals GL, Zhang Y, Rochon Y, Aebersold R. Evaluation of two-dimensional gel electrophoresis-based proteome analysis technology. *Proc Natl Acad Sci U S A* 2000;97:9390–9395.
22. Rabilloud T. Two-dimensional gel electrophoresis in proteomics: old, old fashioned, but it still climbs up the mountains. *Proteomics* 2002;2:3–10.
23. Veenstra TD. Proteome analysis of posttranslational modifications. *Adv Protein Chem* 2003;65:161–194.
24. Mann M, Jensen ON. Proteomic analysis of post-translational modifications. *Nat Biotechnol* 2003;21:255–261.
25. Shen J, Person MD, Zhu J, Abbruzzese JL, Li D. Protein expression profiles in pancreatic adenocarcinoma compared with normal pancreatic tissue and tissue affected by pancreatitis as detected by two-dimensional gel electrophoresis and mass spectrometry. *Cancer Res* 2004;64:9018–9026.
26. Lu Z, Hu L, Evers S, Chen J, Shen Y. Differential expression profiling of human pancreatic adenocarcinoma and healthy pancreatic tissue. *Proteomics* 2004;4: 3975–3988.
27. Shekouh AR, Thompson CC, Prime W, et al. Application of laser capture microdissection combined with two-dimensional electrophoresis for the discovery of differentially regulated proteins in pancreatic ductal adenocarcinoma. *Proteomics* 2003;3:1988–2001.

28. Seike M, Kondo T, Fujii K, et al. Proteomic signature of human cancer cells. *Proteomics* 2004;4:2776–2788.
29. Koopmann J, Fedarko NS, Jain A, et al. Evaluation of osteopontin as biomarker for pancreatic adenocarcinoma. *Cancer Epidemiol Biomarkers Prev* 2004;13: 487–491.
30. Verma M, Wright GL, Jr., Hanash SM, Gopal-Srivastava R, Srivastava S. Proteomic approaches within the NCI early detection research network for the discovery and identification of cancer biomarkers. *Ann N Y Acad Sci* 2001;945:103–115.
31. Mann M, Hendrickson RC, Pandey A. Analysis of proteins and proteomes by mass spectrometry. *Annu Rev Biochem* 2001;70:437–473.
32. Pandey A, Mann M. Proteomics to study genes and genomes. *Nature* 2000;405: 837–846.
33. Kristiansen TZ, Bunkenborg J, Gronborg M, et al. A proteomic analysis of human bile. *Mol Cell Proteomics* 2004;3:715–728.
34. Gygi SP, Rist B, Griffin TJ, Eng J, Aebersold R. Proteome analysis of low-abundance proteins using multidimensional chromatography and isotope-coded affinity tags. *J Proteome Res* 2002;1:47–54.
35. Zhou H, Ranish JA, Watts JD, Aebersold R. Quantitative proteome analysis by solid-phase isotope tagging and mass spectrometry. *Nat Biotechnol* 2002;20:512–515.
36. Li C, Hong Y, Tan YX, et al. Accurate qualitative and quantitative proteomic analysis of clinical hepatocellular carcinoma using laser capture microdissection coupled with isotope-coded affinity tag and two-dimensional liquid chromatography mass spectrometry. *Mol Cell Proteomics* 2004;3:399–409.
37. Meehan KL, Sadar MD. Quantitative profiling of LNCaP prostate cancer cells using isotope-coded affinity tags and mass spectrometry. *Proteomics* 2004;4: 1116–1134.
38. Everley PA, Krijgsveld J, Zetter BR, Gygi SP. Quantitative cancer proteomics: stable isotope labeling with amino acids in cell culture (SILAC) as a tool for prostate cancer research. *Mol Cell Proteomics* 2004;3:729–735.
39. Gronborg M, Kristiansen TZ, Iwahori A, et al. Biomarker discovery from pancreatic cancer secretome using a differential proteomics approach. *Mol Cell Proteomics* 2006;5:157–176.
40. Iacobuzio-Donahue CA, Ashfaq R, Maitra A, et al. Highly expressed genes in pancreatic ductal adenocarcinomas: a comprehensive characterization and comparison of the transcription profiles obtained from three major technologies. *Cancer Res* 2003;63:8614–8622.
41. Iacobuzio-Donahue CA, Maitra A, Olsen M, et al. Exploration of global gene expression patterns in pancreatic adenocarcinoma using cDNA microarrays. *Am J Pathol* 2003;162:1151–1162.
42. Iacobuzio-Donahue CA, Maitra A, Shen-Ong GL, et al. Discovery of novel tumor markers of pancreatic cancer using global gene expression technology. *Am J Pathol* 2002;160:1239–1249.
43. Mulla A, Christian HC, Solito E, Mendoza N, Morris JF, Buckingham JC. Expression, subcellular localization and phosphorylation status of annexins 1 and

5 in human pituitary adenomas and a growth hormone-secreting carcinoma. *Clin Endocrinol (Oxf)* 2004;60:107–119.

44. Wang Y, Serfass L, Roy MO, Wong J, Bonneau AM, Georges E. Annexin-I expression modulates drug resistance in tumor cells. *Biochem Biophys Res Commun* 2004;314:565–570.
45. Fang MZ, Liu C, Song Y, et al. Over-expression of gastrin-releasing peptide in human esophageal squamous cell carcinomas. *Carcinogenesis* 2004;25:865–876.
46. Pencil SD, Toth M. Elevated levels of annexin I protein in vitro and in vivo in rat and human mammary adenocarcinoma. *Clin Exp Metastasis* 1998;16:113–121.
47. Kodera Y, Isobe K, Yamauchi M, et al. Expression of carcinoembryonic antigen (CEA) and nonspecific crossreacting antigen (NCA) in gastrointestinal cancer; the correlation with degree of differentiation. *Br J Cancer* 1993;68:130–136.
48. Hasegawa T, Isobe K, Tsuchiya Y, et al. Nonspecific crossreacting antigen (NCA) is a major member of the carcinoembryonic antigen (CEA)-related gene family expressed in lung cancer. *Br J Cancer* 1993;67:58–65.
49. Sienel W, Dango S, Woelfle U, et al. Elevated expression of carcinoembryonic antigen-related cell adhesion molecule 1 promotes progression of non-small cell lung cancer. *Clin Cancer Res* 2003;9:2260–2266.
50. Scholzel S, Zimmermann W, Schwarzkopf G, Grunert F, Rogaczewski B, Thompson J. Carcinoembryonic antigen family members CEACAM6 and CEACAM7 are differentially expressed in normal tissues and oppositely deregulated in hyperplastic colorectal polyps and early adenomas. *Am J Pathol* 2000;156:595–605.
51. Duxbury MS, Ito H, Zinner MJ, Ashley SW, Whang EE. CEACAM6 gene silencing impairs anoikis resistance and in vivo metastatic ability of pancreatic adenocarcinoma cells. *Oncogene* 2004;23:465–473.
52. Sato N, Fukushima N, Maitra A, et al. Gene expression profiling identifies genes associated with invasive intraductal papillary mucinous neoplasms of the pancreas. *Am J Pathol* 2004;164:903–914.
53. Mori M, Shiraishi T, Tanaka S, et al. Lack of DMBT1 expression in oesophageal, gastric and colon cancers. *Br J Cancer* 1999;79:211–213.
54. Wu W, Kemp BL, Proctor ML, et al. Expression of DMBT1, a candidate tumor suppressor gene, is frequently lost in lung cancer. *Cancer Res* 1999;59:1846–1851.
55. Sasaki K, Sato K, Akiyama Y, Yanagihara K, Oka M, Yamaguchi K. Peptidomics-based approach reveals the secretion of the 29-residue COOH-terminal fragment of the putative tumor suppressor protein DMBT1 from pancreatic adenocarcinoma cell lines. *Cancer Res* 2002;62:4894–4898.
56. Iacobelli S, Arno E, D'Orazio A, Coletti G. Detection of antigens recognized by a novel monoclonal antibody in tissue and serum from patients with breast cancer. *Cancer Res* 1986;46:3005–3010.
57. Iacobelli S, Arno E, Sismondi P, et al. Measurement of a breast cancer associated antigen detected by monoclonal antibody SP-2 in sera of cancer patients. *Breast Cancer Res Treat* 1988;11:19–30.
58. Iacobelli S, Bucci I, D'Egidio M, et al. Purification and characterization of a 90 kDa protein released from human tumors and tumor cell lines. *FEBS Lett* 1993;319:59–65.

59. Iacobelli S, Sismondi P, Giai M, et al. Prognostic value of a novel circulating serum 90K antigen in breast cancer. *Br J Cancer* 1994;69:172–176.
60. Christa L, Simon MT, Brezault-Bonnet C, et al. Hepatocarcinoma-intestine-pancreas/pancreatic associated protein (HIP/PAP) is expressed and secreted by proliferating ductules as well as by hepatocarcinoma and cholangiocarcinoma cells. *Am J Pathol* 1999;155:1525–1533.
61. Christa L, Carnot F, Simon MT, et al. HIP/PAP is an adhesive protein expressed in hepatocarcinoma, normal Paneth, and pancreatic cells. *Am J Physiol* 1996;271:G993–G1002.
62. Rosty C, Christa L, Kuzdzal S, et al. Identification of hepatocarcinoma-intestine-pancreas/pancreatitis-associated protein I as a biomarker for pancreatic ductal adenocarcinoma by protein biochip technology. *Cancer Res* 2002;62:1868–1875.
63. Muller HL, Oh Y, Lehrnbecher T, Blum WF, Rosenfeld RG. Insulin-like growth factor-binding protein-2 concentrations in cerebrospinal fluid and serum of children with malignant solid tumors or acute leukemia. *J Clin Endocrinol Metab* 1994;79:428–434.
64. Cohen P. Serum insulin-like growth factor-I levels and prostate cancer risk—interpreting the evidence. *J Natl Cancer Inst* 1998;90:876–879.
65. Ho PJ, Baxter RC. Insulin-like growth factor-binding protein-2 in patients with prostate carcinoma and benign prostatic hyperplasia. *Clin Endocrinol (Oxf)* 1997;46:333–342.
66. Cariani E, Lasserre C, Kemeny F, Franco D, Brechot C. Expression of insulin-like growth factor II, alpha-fetoprotein and hepatitis B virus transcripts in human primary liver cancer. *Hepatology* 1991;13:644–649.
67. Kunzli BM, Berberat PO, Zhu ZW, et al. Influences of the lysosomal associated membrane proteins (Lamp-1, Lamp-2) and Mac-2 binding protein (Mac-2-BP) on the prognosis of pancreatic carcinoma. *Cancer* 2002;94:228–239.
68. Ryu B, Jones J, Hollingsworth MA, Hruban RH, Kern SE. Invasion-specific genes in malignancy: serial analysis of gene expression comparisons of primary and passaged cancers. *Cancer Res* 2001;61:1833–1838.
69. Sanchez LM, Freije JP, Merino AM, Vizoso F, Foltmann B, Lopez-Otin C. Isolation and characterization of a pepsin C zymogen produced by human breast tissues. *J Biol Chem* 1992;267:24,725–24,731.
70. Diaz M, Rodriguez JC, Sanchez J, et al. Clinical significance of pepsinogen C tumor expression in patients with stage D2 prostate carcinoma. *Int J Biol Markers* 2002;17:125–129.
71. Takahashi S, Suzuki S, Inaguma S, et al. Down-regulated expression of prostasin in high-grade or hormone-refractory human prostate cancers. *Prostate* 2003;54:187–193.
72. Chen LM, Chai KX. Prostasin serine protease inhibits breast cancer invasiveness and is transcriptionally regulated by promoter DNA methylation. *Int J Cancer* 2002;97:323–329.
73. Mok SC, Chao J, Skates S, et al. Prostasin, a potential serum marker for ovarian cancer: identification through microarray technology. *J Natl Cancer Inst* 2001;93:1458–1464.

18

Proteomics of Human Bile

Troels Zakarias Kristiansen, Anirban Maitra, and Akhilesh Pandey

Summary

Bile is an important body fluid with crucial functions ranging from fat absorption to excretion of metabolic breakdown products. Although the chemical composition of human bile is well understood, its protein constituents are beginning to be unravelled only recently. Here, we provide an overview of the proteomics of human bile, both in health and in diseases such as cholangiocarcinoma. A comprehensive catalog of proteins in the bile should facilitate better understanding of its role in physiology and the development and validation of biomarkers for hepatobiliary disorders.

Key Words: Bile; LC-MS/MS; proteomics; biomarker discovery; mass spectrometry; hepatobiliary cancer.

1. Introduction

Proteomic analysis of human body fluids, including bile, holds the promise of providing insights into homeostasis as well as systemic responses to disease. Thus far, the focus has been directed toward analyzing the proteome of serum and plasma, although recently the community has begun investigating many of the more specialized human body fluids.

The process of bile secretion starts in the liver, where hepatocytes continuously secrete bile into a specialized network of small ductules, or canaliculi. Bile is subsequently drained from the canaliculi into larger ductules called cholangioles, which again merge to form the interlobular bile ducts. The interlobular bile ducts unite to form septal bile ducts, which progressively merge until the bile eventually exits the liver through the left and right hepatic bile ducts *(1)*. Outside the liver, the left and right hepatic bile ducts join to form the common hepatic bile duct. The common hepatic duct joins the cystic duct from the gallbladder to form the common bile duct, which drains into the duodenum through the sphincter of Oddi **(Fig. 1)**. In approx 85% of individuals, the common bile duct and the

From: *Proteomics of Human Body Fluids: Principles, Methods, and Applications*
Edited by: V. Thongboonkerd © Humana Press Inc., Totowa, NJ

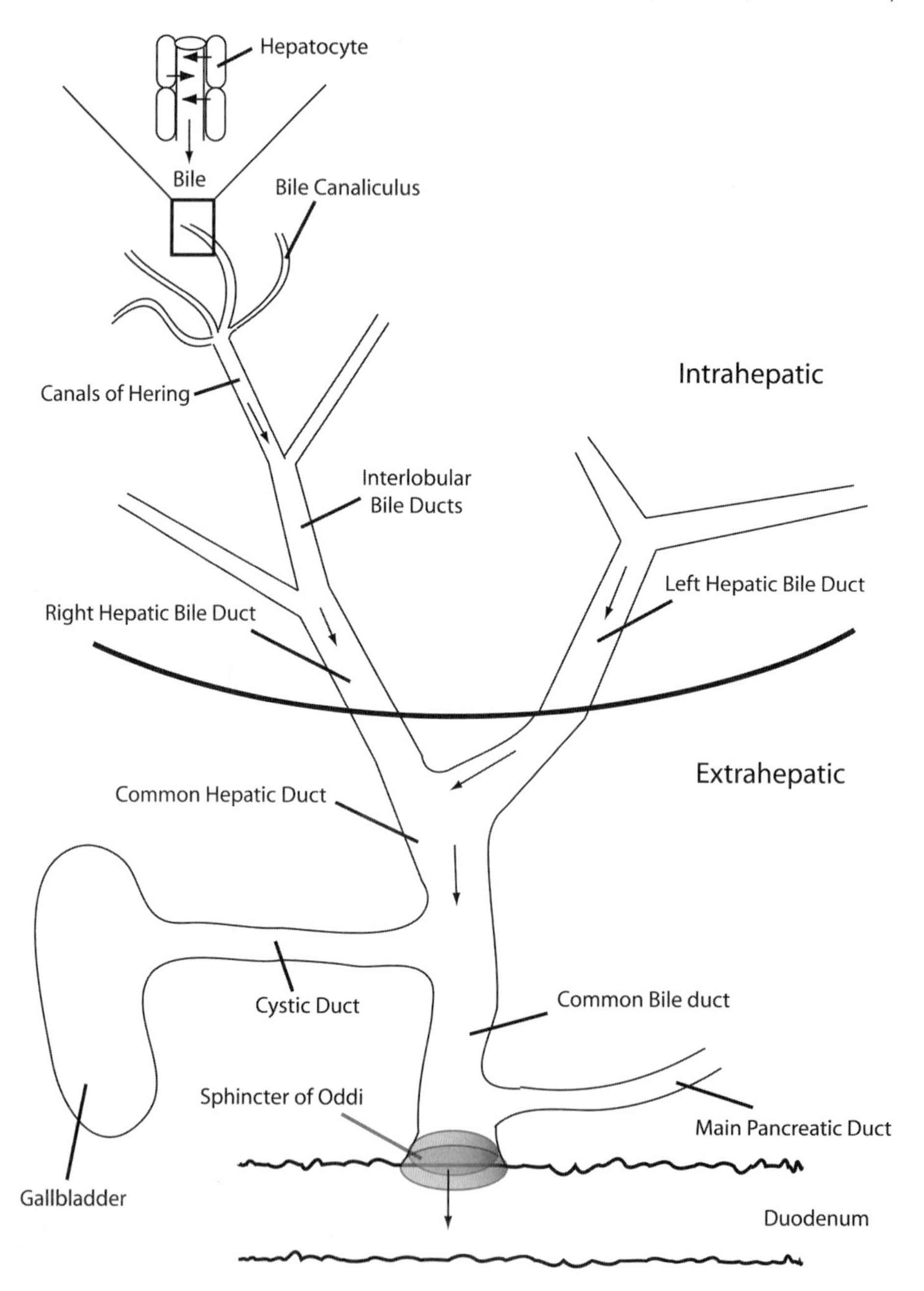

Fig. 1. Schema of the anatomy of the hepatic biliary tree. The arrows indicate the flow of bile from the liver to the intestine.

main pancreatic duct join to form a common channel. In the remaining 15%, the two ducts enter the duodenum as two separate channels *(2)*.

The bile serves several important functions in the body. First, it participates in the adsorption of fat from the intestine, which takes place in the duodenum. The emulsification of fat in the duodenum, owing to detergent-like properties of bile salts, aids in its breakdown by making it accessible to lipases secreted from pancreas *(3)*. Second, bile secretion serves to excrete bilirubin, cholesterol, phospholipids (mainly lecithin/phosphatidylcholine), and inorganic ions (such as iron and copper) from the body. Bilirubin is the breakdown product of heme and is conjugated to glucoronic acid in the liver, rendering it water soluble prior to secretion into bile *(4)*.

The flow of bile into the duodenum is regulated by contraction of the gallbladder and relaxation of the sphincter of Oddi. After food intake, the duodenum releases the main gallbladder-contracting hormone, cholecystokinin, which in combination with other hormones causes an increase in the bile flow, gallbladder contraction, and relaxation of the sphincter of Oddi. This results in the excretion of most of the stored bile in the gallbladder into the duodenum *(2)*. Although cholecystokinin is the major hormone responsible for the discharge of bile in the intestine, its effect is modified by several other peptide hormones; a partial list of these hormones is provided in **Table 1**.

2. Bile Salts

Bile salts play a crucial role in human physiology. They are the end product of cholesterol breakdown in the liver. Bile acids are amphipathic detergent-like molecules synthesized from cholesterol in hepatocytes *(4)*. About 50% of cholesterol degradation occurs through its catabolism to bile acids *(5)*. Cholic and chenodeoxycholic acids are the primary human bile acids produced directly by the liver. They are synthesized in hepatocytes from cholesterol through an elaborate system of enzymes located in the endoplasmic reticulum, mitochondria, cytosol, microsomes, and peroxisomes *(6)*. Deoxycholic and lithocholic acids are termed secondary bile acids and are formed by 7α-dehydroxylation by intestinal bacterial enzymes from primary bile acids (*see* **Tables 2** and **3** for the general composition of bile and bile acids, respectively). Bile salts are conserved in the body through enterohepatic circulation. Both primary and secondary bile acids are effectively reabsorbed through the portal vein in the ileum, and only 5% of the bile salt pool is lost through feces on a daily basis *(7)*. The liver, thus, only needs to synthesize bile acids lost through feces in order to maintain the bile salt pool. Bile salts are essential for the formation of mixed micelles, which solubilize both cholesterol and phospholipids for the transport

Table 1
A List of Proteins and Hormones Commonly Found in Bile

Protein/hormone	Concentration in gallbladder bile (mg/mL)	Concentration in hepatic bile (mg/mL)
Common proteins		
Albumin	1.9	0.2
Orosomucoid	0.07	0.02
Aminopeptidase N	1.6	0.2
α_1-antitrypsin	N.A.	N.A.
α_2-macroglobulin	N.A.	N.A.
Apolipoproteins (A-I, A-II, B, C-II, C-III)	N.A.	N.A.
Carcinoembryonic antigen-related cell adhesion molecules (CEACAMs)	N.A.	N.A.
Ceruloplasmin	N.A.	N.A.
Haptoglobin	0.04	0.01
Hemoglobin	N.A.	N.A.
Immunoglobulin IgG	0.3	0.04
Immunoglobulin IgM	0.09	0.04
Immunoglobulin IgA	0.2	0.09
Transferrin	N.A.	N.A.
Mucin	0.09	0.04
Alkaline phosphatase	N.A.	N.A.
Alkaline phosphodiesterase I	N.A.	N.A.
Leucine β-naphtylamidase	N.A.	N.A.
5′-Nucleotidase	N.A.	N.A.
β-Galactosidase	N.A.	N.A.
β-Glucuronidase	N.A.	N.A.
N-acetyl-β-glucosaminidase	N.A.	N.A.
Common peptide hormones		
CCK	N.A.	N.A.
Gastrin	N.A.	N.A.
Motilin	N.A.	N.A.
Secretin	N.A.	N.A.
Ocreotide	N.A.	N.A.
Enkephalin	N.A.	N.A.
Gastrin-releasing peptide	N.A.	N.A.
Vasoactive intestinal peptide	N.A.	N.A.

N.A., not available.
Adapted from **refs. *2*, *14*, and *17***.

Table 2
General Composition of Bile

Composition	Hepatic bile	Gallbladder bile
Water		
Bile salts	11 mg/mL	60 mg/mL
Bilirubin	0.4 mg/mL	3 mg/mL
Cholesterol	1 mg/mL	9 mg/mL
Fatty acids	1.2 mg/mL	12 mg/mL
Lecithin	0.4 mg/mL	3 mg/mL
Na^+	145 m*M*	130 m*M*
K^+	5 m*M*	12 m*M*
Ca^{2+}	2.5 m*M*	12 m*M*
Cl^-	100 m*M*	25 m*M*
HCO_3^-	28 m*M*	10 m*M*

Adapted from **ref.** ***3***.

Table 3
Composition of Bile Salts (in Percent of Total Moles)

Bile salts	% of total bile salt pool
Sulfoglycolithocholates	2
Taurodeoxycholates	4
Glycochenodeoxycholate	25
Glycodeoxycholates	13
Chenodeoxycholates	37
Deoxycholates	17
Sulfolithocholates	2

Adapted from **refs.** ***44*** and ***45***.

of otherwise insoluble cholesterol from the liver to the intestinal tract. In the intestinal tract, bile salts assist in solubilizing dietary compounds such as fatty acids and fat-soluble vitamins.

Bile formation is the result of active secretion by hepatocytes into the canalicular space followed by water through the tight junctions. Transport of biliary constituents from blood to bile is a process that includes uptake from the sinusoidal side, intracellular transport through the hepatocyte, and canalicular secretion via active membrane transportation **(Fig. 2)**. Bile salt uptake is mediated by Na^+-dependent and -independent mechanisms. The Na^+ taurocholate co-transporting polypeptide (NTCP) is the major transporter involved in transport across

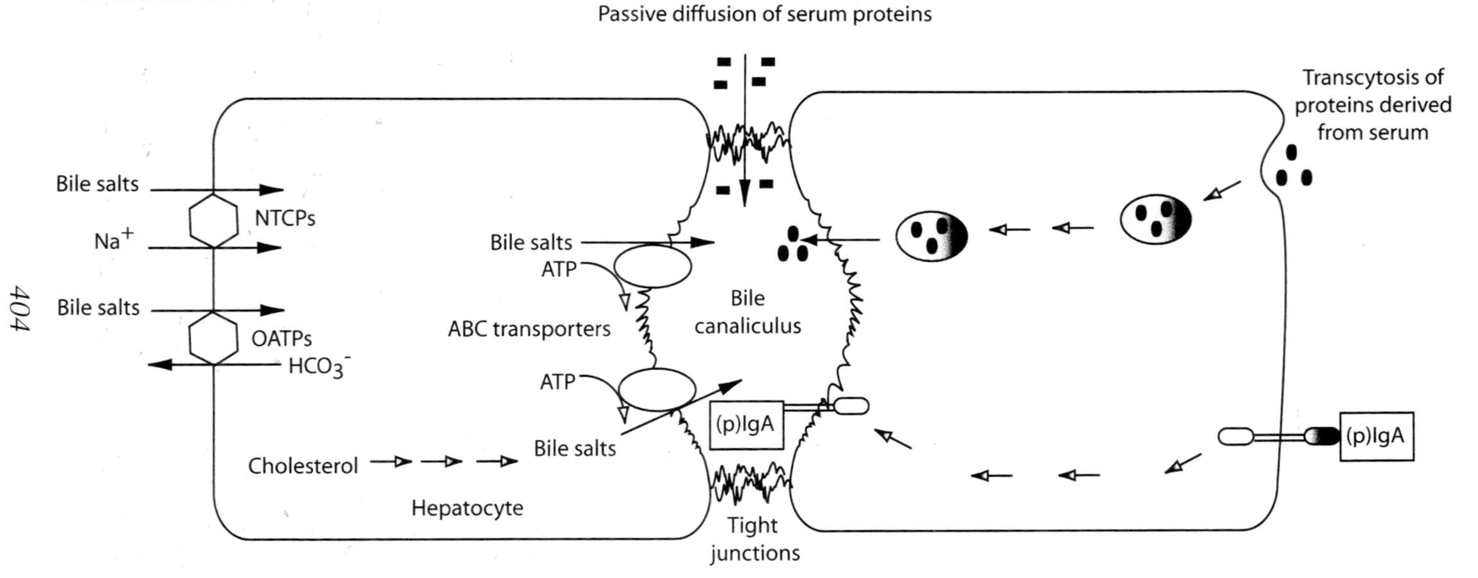

Fig. 2. Schema showing the basic mechanisms for transporting bile salts and bile proteins into the bile canaliculus. See text for discussion of the different mechanisms.

the sinusoidal membrane. The NTCP transports mainly conjugated bile salts using the Na^+ gradient over the membrane to facilitate the uptake ***(8–10)***. Na^+-independent uptake of bile salts is mediated by the organic anion transport polypeptides (OATPs). This family of transporters is responsible for the uptake of unconjugated bile salts. Also, the OATP family mediates the uptake of other compounds such as thyroid hormones, neutral steroids, and bilirubin ***(11)***. After transport across the sinusoidal membrane and intracellular transport (which might involve further modifications), the bile salts are secreted into the canaliculi. This process is the driving force for the bile flow. ATP-binding cassette transporters are responsible for that part of the process ***(12)***. Several ABC transporters with different specificities can be found in the apical membrane. Their primary role is to secrete bile salts into the lumen of the canaliculi, but they also play a pivotal role in clearing away harmful substances and drugs metabolized by the liver. After the initial bile flow has been established in the liver through secretion into the canaliculi, the bile can be further modified both by the cholangiocytes lining the bile ducts and by processes taking place in the gallbladder.

3. Sources of Proteins in Bile

Proteins in bile are derived either from plasma or from secretion by the hepatobiliary system. A major fraction of the serum proteins found in bile are believed to cross the tight junctions separating the sinusoids from the canaliculi. This is supported by the notion that there is an inverse relationship between the molecular weight of serum proteins and the abundance in the bile (the idea being that smaller proteins transverse the tight junctions easier) ***(13)***. Since the liver is responsible for producing a large majority of serum proteins, it is difficult to determine whether a given bile protein component is derived from serum crossing the tight junctions or whether it is the result of active secretion from the hepatocytes (or bile duct cells). A list of common proteins found in bile is given in **Table 1** (which is adapted from **refs.** ***13–17***).

Since bile is concentrated in the gallbladder, the total protein concentration varies according to the nutritional status and the site of collection. The total protein concentration of bile in gallbladder has been reported to range from 0.2 to 31 mg/mL, whereas that obtained from the common duct bile ranges from 0.34 to 13 mg/mL ***(18–20)***. Unfortunately, most common protein assays, such as BCA, Lowry's, and Bradford's methods, are somewhat incompatible with bile salts, lipids, and bilirubin, which makes it necessary to process the sample before such measurements can be done ***(21–23)***. This usually involves removal of lipids, dialysis, protein precipitation, or some types of chromatography.

Proteins secreted by hepatocytes can roughly be divided into two categories: those actively produced by the hepatocyte and those resulting from transcytosis from the sinusoidal side to the canaliculi. In general, very little is known about the identities and origins of proteins that are specific to bile ***(13,15–17)***. The best described group of proteins derived from hepatocytes is biliary enzymes. These can either be membrane-bound biliary enzymes, such as alkaline phosphatase and 5′-nucleotidase, or lysomal enzymes, such as β-galactosidase (*see* **Table 1** for a list of the most common enzymes found in bile). Transcytosis of proteins from the sinusoidal side to the apical side of the hepatocyte is the other route by which hepatocytes deliver proteins to bile. This process can be nonspecific or specific. Nonspecific transcytosis results from internalization and endosome formation of vesicles containing protein from the sinusoidal side. As opposed to passive diffusion over the tight junctions, this type of transport does not discriminate against larger size serum proteins. Specific or receptor-mediated transport is best illustrated by the unusual high concentration of IgA found in biliary juice. Bile plays a crucial role in controlling the delivery of immunoglobulin A to the intestine ***(24)***. IgA, in both its monomeric and polymeric form, is one of those proteins whose concentration is considerably higher than what would be expected from passive crossing of the tight junctions ***(15)***.

Both monomeric and polymeric/secretory IgA do not pass over the lining of hepatocytes whose sides as mentioned are closed by tight junctions. Instead, they belong to the class of biliary proteins that are actively transported across the hepatocyte from the sinusoidal side to the canaliculi. IgA antibody binds to the polymeric Ig receptor on the basolateral membranes of the epithelial cells, passes through the cell by transcytosis, and is displayed on the apical side; the polymeric Ig receptor is cleaved, resulting in release of secretory IgA into the biliary ductules ***(25)***. Polymeric Ig receptor is highly expressed on biliary epithelial cells, and quantitatively IgA is present in substantial amounts in bile. (Usually bile contains the highest concentration of IgA compared with any other body fluid.)

4. Proteomic Studies of Bile

Recently, proteomic studies of human body fluids have gained increased attention owing to the effects of the Human Plasma Protein Organization (HUPO) to analyze thoroughly the protein constituents of human plasma and serum. A large and comprehensive catalog of human serum proteins is now available, and it is evident that the complexity is far greater than expected. For an initial effort to characterize a body fluid, serum was the sample of choice, both because it is readily available and because most clinical diagnostic tests measure the values of markers found in blood. Nevertheless, compared with serum, little attention has been paid to the analysis of bile. Unlike serum, bile is difficult to obtain from healthy individuals (or individuals with diseases unrelated to the hepatobiliary tree). Proteomic analysis of biliary juice, however, poses some of the same problems as serum because of the presence of a few of high-abundance proteins ***(26,27)***.

Proteomic analysis by mass spectrometry is especially sensitive to having different proteins existing at vastly different concentrations. Typically, a mass spectrometry setup has a limited dynamic range and thus it becomes a difficult task to detect proteins present in concentrations differing by several orders of magnitude.

Fractionation or depletion of high-abundance protein species has been the most commonly used strategies to circumvent the problem of the limited dynamic range in mass spectrometry. Multidimensional fractionation can reduce the sample complexity by many orders of magnitude, and researchers have successfully coupled several different types of chromatography to achieve this goal. One- and two-dimensional separation techniques using ion-exchange, reversed-phase, or lectin affinity chromatography are some of the methods that have been successfully used to decrease sample complexity before reversed-phase separation and analysis by mass spectrometry. Although adding more dimensions to the fractionation scheme can help to increase the coverage of a particular proteome, a major drawback is the increase in workload. Each dimension typically increases the workload and data output by 1 to 2 orders of magnitude. Thus, coupling more than three fractionation-dimensions to a given experiment often requires enormous effort and generates huge amounts of data. Multidimensional approaches, therefore, frequently require both a dedicated mass spectrometric laboratory and good bioinformatics support, which might not always be accessible to the investigator. Another problem when one is dealing with biliary juice (or any other body fluid) is the definition of what constitutes "normal" bile juice. The composition of the bile proteome most likely changes with the feeding state of the individual, and sample variability can thus be a major problem. The problem of natural variability also affects the way samples can be normalized. Potentially, one could normalize according to protein concentration, bile salt concentration, or a specific marker protein. Alternatively, one could choose to compare samples on a "per volume" basis. All these methods vary independently of each other.

Hitherto, a limited number of proteomic studies have aimed at a global analysis of the protein constituents of bile. In a study designed to investigate proteins associated with mixed micelles found in human bile, He et al. ***(28)*** compared bile samples with reference 2D gel electrophoresis (2-DE) maps and identified eight common serum proteins in the mixed micelle fraction. The 2-DE comparison of total bile proteins against proteins associated with mixed micelle from human gallbladder bile also revealed that most proteins in bile are associated with mixed micelles. The process of cholesterol crystallization in the formation of human gallstones is thought to be influenced by many factors. One determinant has been postulated to be the balance of nucleating and nucleating-inhibitory polypeptides, in which the hydrophilic nature of the protein would be a determining trait. In order to identify hydrophobic peptides specifically in bile, Stark et al. ***(29)*** used chloroform/methanol extraction followed by specialized reversed-phase chromatography, gel filtration, and matrix-assisted laser desorption/ionization-time of flight (MALDI-TOF) mass spectrometry; they managed to identify five proteins not previously described in human bile. It has not yet been determined whether these polypeptides play an active role in gallstone formation.

In an effort to investigate the toxic effects of 1,1-dichloroethylene and diclofenac on rat biliary canaliculus, Jones et al. ***(30)*** carried out a comparative proteomic study of bile obtained prior to drug exposure compared with bile obtained during drug treatment in animals. Using 1D gel electrophoresis followed by liquid chromatography-tandem mass spectrometry (LC-MS/MS), the authors identified 23 proteins in bile from nontreated animals ***(30)***. The study also allowed the authors to detect selectively chemical modifications on proteins identified from drug-treated animals.

4.1. Proteomics Applied to Biliary Tract Cancers

Approximately 4500 people (out of 7500 diagnosed with the disease) die from biliary tract carcinomas every year in the United States ***(31)***. Early detection of this cancer is crucial for the outcome, and most patients with a completely resectable cancer achieve a modest 5-yr survival. At present, radiation or conventional chemotherapy does not significantly improve the survival rate, and early detection at a stage at which the cancer is resectable offers the best hope of a cure ***(32)***. However, laboratory tests for detection of biliary tract cancers are usually only modestly sensitive and specific, and distinguishing between benign and malignant causes of biliary tract obstruction based on biopsies is quite challenging. Currently, CA19-9 is the best available fluid-based marker for detection of cholangiocarcinomas and has a sensitivity of only 50 to 60% and a specificity of 80% ***(33,34)***. The use of CA19-9 in early-stage detection of cholangiocarcinomas is even more limited, and better fluid-based markers are required to offer patients a better chance of survival.

In an effort to characterize the human biliary juice proteome, Kristiansen et al. ***(35)*** recently carried out an analysis of bile obtained from patients with biliary tract cancer. Using multiple fractionation techniques and affinity enrichment methods in combination with LC-MS/MS analysis, they identified 87 unique bile-derived proteins. A subset of the proteins identified in the study is known to be secreted by hepatocytes and probably represents a baseline proteome of what would be expected in bile. These included transport proteins (transferrin, albumin, lactoferrin, ceruloplasmin, transthyretin, and α_2-macroglobulin), enzymes (γ-glutamyltransferase and adenosine deaminase), epithelial glycoproteins (CEACAM1) and members of the coagulation cascade (fibrinogen and antithrombin). Also, the data set contained a number of known cancer-associated proteins, which was probably because the sample was derived from a patient with cholangiocarcinoma. Among others, this group of proteins included CA125 ovarian cancer antigen. CA125 is a cell surface marker glycoprotein that is used as a serum tumor marker for gastrointestinal and gynecological cancers ***(36,37)*** but is also known to be highly upregulated in patients with primary sclerosing cholangitis (a chronic inflammation of the biliary tree that predisposes patients to biliary tract cancers) and biliary tract cancers ***(38–40)***.

In addition to these known hepatic and known cancer-associated proteins, several of the proteins identified had not been previously reported in normal or cholangiocarcinoma-associated bile before. In the context of cancer proteomics, three of the proteins were of special interest: lipocalin 2 (also known as oncogene 24p3 or NGAL), deleted in malignant brain tumors 1 (DMBT1), and Mac-2-binding protein (Mac2BP). Lipocalin 2 is overexpressed in a variety of human cancers such as breast cancer, colorectal cancer, and pancreatic carcinomas and has also been suggested as a tumor marker for bladder cancer in urine. DMBT1 is an opsonin receptor that is frequently deleted in gliomas and other malignant brain tumors. Aberrant expression of DMBT1 has been proposed to play a role in the progress of gastric and colorectal cancers. Finally, a peptide screening study identified DMBT1 as a protein specifically secreted by pancreatic adenocarcinoma cells lines but not by cell lines derived from normal pancreatic ductal epithelium. Mac2BP is a secreted galectin-3-binding protein and most likely plays a role in cell surface adhesion. Whereas galectin-3 has been observed to be downregulated during the development of colon cancer, elevated serum levels for Mac2BP have been reported in patients with solid tumors (breast, ovarian, lung, and colorectal cancers).

In a follow-up study to this proteomic analysis, Koopman et al. ***(41)*** evaluated Mac-2-binding protein as a biomarker and showed that it could be used as a diagnostic marker for detection of biliary tract cancers. Screening by immunohistochemistry showed positive staining for Mac2BP in 94% of the biliary tract cancer tissues tested, whereas normal normal biliary tract ductules and gallbladder epithelium were negative for staining. Also, the levels of Mac2BP were elevated by a factor of 3 in bile samples from patients with biliary tract cancer compared with bile obtained from patients with either primary sclerosing cholangitis or benign biliary disorders when enzyme-linked immunosorbent assay (ELISA) was used.

5. Methods

5.1. Sample Collection

Human biliary juice is obtained either during surgery or from endoscopic retrograde cholangiopancreatography (ERCP) examinations of patients. For proteomic purposes, it is desirable to work with ERCP-drained juice compared with juice obtained from surgery. Often, the surgical procedure results in variable amounts and contamination of the sample with serum or pancreatic juice, and thus the quality of the sample will vary greatly. In our experience, ERCP also has contamination issues, but not to the same extent as those of surgery.

1. During an ERCP procedure, a flexible tube containing an endoscope is passed down the throat into the duodenum, from where it is guided either into the common bile duct or into the main pancreatic duct, and the sample is collected. Typical amounts of bile obtained during ERCP range from 5 to 10 mL in volume.
2. After collection, the bile samples should be stored at –80°C, and during any transportation they should be kept in dry ice to prevent degradation.
3. Always thaw the sample on ice and keep it at 4°C while working with it.
4. To remove small impurities and debris introduced during sample collection, centrifuge the sample at 16,000*g* for 5 min at 4°C and transfer the supernatant to a new tube.
5. One can attempt to measure the protein concentration of the sample by Coomassie staining or, Bradford's or Lowry's methods, but owing to bile salts and bilirubin, these results can in some cases be misleading.
6. If lipids interfere with the analysis, it is possible to clean up the sample with Cleanascite™HC (Ligochem, Fairfield, NJ), which forms lipid micelles that subsequently can be removed by centrifugation. Many biliary proteins are embedded in the mixed micelles in biliary juice, so removing the lipids might also remove the protein of interest.

5.2. Fractionation of Biliary Juice

Given the complexity of most body fluids, it is often necessary to fractionate the sample before analysis. Because of the high amounts of bile salts and lipids in biliary juice, using a fractionation technique such as 1-DE is a good choice for unprocessed bile samples.

1. Determination of protein amounts in the bile should be estimated on the gel by loading a standard protein such as BSA.
2. The gel-separated proteins can be visualized by standard techniques such as silver, Coomassie, or colloidal blue staining.
3. For whole-lane analysis of your bile sample by LC-MS/MS, 10 to 20 µg of protein per lane should provide a good starting point. Divide the lane into appropriate sections (for a standard 1D gel dividing the gel into 15–20 pieces will provide decent size of gel slices), and do a standard tryptic in-gel digestion.
4. After tryptic digestion and extraction, dry down the peptides to completion and resuspend in 5 to 10 µL of 5% formic acid.
5. The peptides can be stored indefinitely at –80°C.

5.3. Lectin Affinity Chromatography

Most commercially available lectins can be used for batch purification of biliary juice. The most commonly used lectin is concanavalin A (Con A), which has a preference for high-mannose-type glycans.

1. Mix approximately 500 μg of bile protein (approx 100–300 μL juice) with 2X volume of TBS and 2X volume of 50% slurry of Con A-agarose beads and incubate at 4°C for 4 h under rotation.
2. Wash the beads twice with TBS buffer and elute the bound protein with 2X 100 μL of 100 m*M* methyl α-D-mannopyranoside (elution at room temperature for 10 min).
3. Concentrate the eluate by size-exclusion filtration or column purification. The eluted proteins can be visualized by 1-DE as described for unprocessed bile (*see* **Subheading 5.2.**).

5.4. Liquid Chromatography and Tandem Mass Spectrometry

1. A standard LC-MS/MS setup with a trap column may be used for analysis of tryptic digests of fractionated bile.
2. Load the peptides onto a precolumn packed with C_{18} reverse-phase material, and wash the column with 95% mobile phase A (0.4% acetic acid, 0.005% heptafluorobutyric acid v/v)/5% mobile phase B (90% acetonitrile, 0.4% acetic acid, 0.005% heptafluorobutyric acid v/v). Washing and loading can be done with high flow rates (1–4 μL/min).
3. Elute the peptides onto the separation column (packed with reverse-phase material) and subsequently into the mass spectrometer using a linear gradient of 10 to 40% mobile phase B over 35 to 70 min with a flow rate of 300 nL/min.
4. After the gradient is finished, the columns should be cleaned with a burst of 90% mobile phase B.
5. The performance of LC-MS/MS setups strongly depends on the automated LC system, the type of mass spectrometer, and choice of reverse-phase material. The above conditions can be used as starting parameters, but optimization of conditions is almost always necessary.

6. Perspectives and Future Directions

Thus far, only serum and plasma have been subjected to a coordinated comprehensive proteomic analysis ***(42,43)***. Although a full proteomic catalog of the "normal" constituents of bile is somewhat difficult, defining a subproteome for a specific physiological state using a specific fractionation method can provide a good reference for further exploration. Although further experiments are warranted, Koopman et al. ***(41)*** have demonstrated how one can proceed from such a subproteomic catalog to the evaluation of a diagnostic marker. Quantitative methods, which are applicable to the analysis of body fluids, are becoming more and more widespread. Even though sample variability and availability are obstacles that must be dealt with (especially when the analysis is combined with a fractionation technique), hopefully these methods will provide quantitative data, which will aid in the discovery of both drug targets and diagnostic markers for human diseases associated with the hepatobiliary tree.

Acknowledgments

Anirban Maitra is supported by an AACR-PanCAN Career Development Award, a grant from the Lustgarten Foundation for Pancreatic Cancer Research, NCI 1R01CA113669, and the Michael Rolfe Foundation. Akhilesh Pandey is supported by a pilot project award from the Sol Goldman Trust and a grant (U54 RR020839) from the National Institutes of Health.

References

1. Sherlock S, Dooley J. *Diseases of the Liver and Biliary System*, 11th ed. Malden, MA: Blackwell Science, 2002.
2. Clavien P-A, Baillie J. *Diseases of the Gallbladder and Bile Ducts: Diagnosis and Treatment*. Malden, MA: Blackwell Science, 2001.
3. Guyton AC, Hall JE. *Textbook of Medical Physiology*, 10th ed. Philadelphia: Saunders, 2000.
4. Bittar EE. *The Liver in Biology and Disease (principles of medical biology)*, Amsterdam: Elsevier Science, 2005.
5. Redinger RN. The coming of age of our understanding of the enterohepatic circulation of bile salts. *Am J Surg* 2003;185:168–172.
6. Fuchs M. Bile acid regulation of hepatic physiology: III. Regulation of bile acid synthesis: past progress and future challenges. *Am J Physiol Gastrointest Liver Physiol* 2003;284:G551–G557.
7. Small DM, Dowling RH, Redinger RN. The enterohepatic circulation of bile salts. *Arch Intern Med* 1972;130:552–573.
8. Arrese M, Ananthananarayanan M, Suchy FJ. Hepatobiliary transport: molecular mechanisms of development and cholestasis. *Pediatr Res* 1998;44:141–147.
9. Arrese M, Trauner M. Molecular aspects of bile formation and cholestasis. *Trends Mol Med* 2003;9:558–564.
10. Meier PJ, Stieger B. Bile salt transporters. *Annu Rev Physiol* 2002;64:635–661.
11. Hagenbuch B, Meier PJ. The superfamily of organic anion transporting polypeptides. *Biochim Biophys Acta* 2003;1609:1–18.
12. Borst P, Elferink RO. Mammalian ABC transporters in health and disease. *Annu Rev Biochem* 2002;71:537–592.
13. Mullock BM, Shaw LJ, Fitzharris B, et al. Sources of proteins in human bile. *Gut* 1985;26:500–509.
14. Keulemans YC, Mok KS, de Wit LT, Gouma DJ, Groen AK. Hepatic bile versus gallbladder bile: a comparison of protein and lipid concentration and composition in cholesterol gallstone patients. *Hepatology* 1998;28:11–16.
15. Coleman R. Biochemistry of bile secretion. *Biochem J* 1987;244:249–261.
16. Mullock BM, Dobrota M, Hinton RH. Sources of the proteins of rat bile. *Biochim Biophys Acta* 1978;543:497–507.
17. Reuben A. Biliary proteins. *Hepatology* 1984;4(5 Suppl):46S–50S.
18. Sewell RB, Mao SJ, Kawamoto T, LaRusso NF. Apolipoproteins of high, low, and very low density lipoproteins in human bile. *J Lipid Res* 1983;24:391–401.

19. Albers CJ, Huizenga JR, Krom RA, Vonk RJ, Gips CH. Composition of human hepatic bile. *Ann Clin Biochem* 1985;22:129–132.
20. Gallinger S, Harvey PR, Petrunka CN, Ilson RG, Strasberg SM. Biliary proteins and the nucleation defect in cholesterol cholelithiasis. *Gastroenterology* 1987;92:867–875.
21. Yamazaki K, Powers SP, LaRusso NF. Biliary proteins: assessment of quantitative techniques and comparison in gallstone and nongallstone subjects. *J Lipid Res* 1988;29(8):1055–1063.
22. Paul R, Sreekrishna K. Physicochemical and comparative studies on bile proteins. *Indian J Biochem Biophys* 1979;16:399–402.
23. Osnes T, Sandstad O, Skar V, Osnes M, Kierulf P. Total protein in common duct bile measured by acetonitrile precipitation and a micro bicinchoninic acid (BCA) method. *Scand J Clin Lab Invest* 1993;53:757–763.
24. Reynoso-Paz S, Coppel RL, Mackay IR, Bass NM, Ansari AA, Gershwin ME. The immunobiology of bile and biliary epithelium. *Hepatology* 1999;30:351–357.
25. van Egmond M, Damen CA, van Spriel AB, Vidarsson G, van Garderen E, van de Winkel JG. IgA and the IgA Fc receptor. *Trends Immunol* 2001;22:205–211.
26. Anderson NL, Polanski M, Pieper R, et al. The human plasma proteome: a nonredundant list developed by combination of four separate sources. *Mol Cell Proteomics* 2004;3:311–326.
27. Anderson NL, Anderson NG. The human plasma proteome: history, character, and diagnostic prospects. *Mol Cell Proteomics* 2002;1:845–867.
28. He C, Fischer S, Meyer G, Muller I, Jungst D. Two-dimensional electrophoretic analysis of vesicular and micellar proteins of gallbladder bile. *J Chromatogr A* 1997;776:109–115.
29. Stark M, Jornvall H, Johansson J. Isolation and characterization of hydrophobic polypeptides in human bile. *Eur J Biochem* 1999;266:209–214.
30. Jones JA, Kaphalia L, Treinen-Moslen M, Liebler DC. Proteomic characterization of metabolites, protein adducts, and biliary proteins in rats exposed to 1,1-dichloroethylene or diclofenac. *Chem Res Toxicol* 2003;16:1306–1317.
31. de Groen PC, Gores GJ, LaRusso NF, Gunderson LL, Nagorney DM. Biliary tract cancers. *N Engl J Med* 1999;341:1368–1378.
32. Gores GJ. Early detection and treatment of cholangiocarcinoma. *Liver Transplant* 2000;6(6 Suppl 2):S30–S34.
33. Bjornsson E, Kilander A, Olsson R. CA 19-9 and CEA are unreliable markers for cholangiocarcinoma in patients with primary sclerosing cholangitis. *Liver* 1999; 19:501–508.
34. Patel AH, Harnois DM, Klee GG, LaRusso NF, Gores GJ. The utility of CA 19-9 in the diagnoses of cholangiocarcinoma in patients without primary sclerosing cholangitis. *Am J Gastroenterol* 2000;95:204–207.
35. Kristiansen TZ, Bunkenborg J, Gronborg M, et al. A proteomic analysis of human bile. *Mol Cell Proteomics* 2004;3:715–728.
36. Modugno F. Ovarian cancer and high-risk women—implications for prevention, screening, and early detection. *Gynecol Oncol* 2003;91:15–31.

37. Haga Y, Sakamoto K, Egami H, Yoshimura R, Mori K, Akagi M. Clinical significance of serum CA125 values in patients with cancers of the digestive system. *Am J Med Sci* 1986;292:30–34.
38. Chen CY, Shiesh SC, Tsao HC, Lin XZ. The assessment of biliary CA 125, CA 19-9 and CEA in diagnosing cholangiocarcinoma—the influence of sampling time and hepatolithiasis. *Hepatogastroenterology* 2002;49:616–620.
39. Ker CG, Chen JS, Lee KT, Sheen PC, Wu CC. Assessment of serum and bile levels of CA19-9 and CA125 in cholangitis and bile duct carcinoma. *J Gastroenterol Hepatol* 1991;6:505–508.
40. Brockmann J, Emparan C, Hernandez CA, et al. Gallbladder bile tumor marker quantification for detection of pancreato-biliary malignancies. *Anticancer Res* 2000;20:4941–4947.
41. Koopmann J, Thuluvath PJ, Zahurak ML, et al. Mac-2-binding protein is a diagnostic marker for biliary tract carcinoma. *Cancer* 2004;101:1609–1615.
42. Ping P, Vondriska TM, Creighton CJ, et al. A functional annotation of subproteomes in human plasma. *Proteomics* 2005;5:3506–3519.
43. Omenn GS, States DJ, Adamski M, et al. Overview of the HUPO Plasma Proteome Project: results from the pilot phase with 35 collaborating laboratories and multiple analytical groups, generating a core dataset of 3020 proteins and a publicly available database. *Proteomics* 2005;5:3226–3245.
44. Hay DW, Cahalane MJ, Timofeyeva N, Carey MC. Molecular species of lecithins in human gallbladder bile. *J Lipid Res* 1993;34:759–768.
45. Donovan JM, Jackson AA, Carey MC. Molecular species composition of intermixed micellar/vesicular bile salt concentrations in model bile: dependence upon hydrophilic-hydrophobic balance. *J Lipid Res* 1993;34:1131–1140.

19

Proteomics of Amniotic Fluid

David Crettaz, Lynne Thadikkaran, Denis Gallot, Pierre-Alain Queloz, Vincent Sapin, Joël S. Rossier, Patrick Hohlfeld, and Jean-Daniel Tissot

Summary

Amniotic fluid is fundamental for the development of the fetus. Many proteins detected in the amniotic fluid are already present at a very early stage of gestation, whereas other proteins are detected only at the end of the pregnancy. The concentration of a given protein in amniotic fluid is governed not only by fetal, placental, or maternal synthesis and degradation, but also by exchanges between the mother and the fetus through the placenta. Maternofetal transfer of proteins involves several different mechanisms such as first-order process or active transport. Consequently, the concentration of each amniotic fluid protein results from a balance between opposing dynamic metabolic and physiological processes, which proceed simultaneously. Thus, proteomics that allows simultaneous study of a multitude of proteins may be of importance to gain insight into the physiology of amniotic fluid as well as to identify potential markers of diseases during pregnancy. Here we present a review of proteomic studies of normal amniotic fluid and describe alterations in the amniotic fluid proteome that occur during pregnancy.

Key Words: Amniotic fluid; biochemistry; electrophoresis; premature rupture of fetal membranes; proteomics; review; two-dimensional electrophoresis.

1. Introduction

The presence and integrity of amniotic fluid is fundamental for normal development of the human fetus during pregnancy. Its production rate varies throughout pregnancy and is mainly related to functions of the different fetal and amniotic compartments. Its biochemical composition partly reflects the maturation of the production sites. Thus, amniotic fluid represents an integrative medium for studying the molecular and metabolic events occurring *in utero* during the different developmental stages of mammalians. Quantitative and qualitative properties of proteins found in amniotic fluid summarize the temporal expression

From: *Proteomics of Human Body Fluids: Principles, Methods, and Applications*
Edited by: V. Thongboonkerd © Humana Press Inc., Totowa, NJ

pattern of genes present in the tissues and implicated in the generation of this liquid. Moreover, posttranslational mechanisms as well as other events are able to induce significant changes in the protein profile. Until now, the exploration of complex and dynamic protein patterns in biological samples has been quite difficult to achieve. Standard biochemical approaches have been used to characterize proteins of nonmaternal origin *(1)*, and identification of particular amniotic fluid proteins such as α-fetoprotein has proved useful to develop diagnostic test for premature rupture of the membrane *(2)*.

In preliminary proteomic works published on amniotic fluid, no relevant protein markers have been validated for prenatal diagnosis of fetal or placental pathological conditions. However, recent publications have demonstrated that proteomic tools could be of potential clinical value. Vuadens et al. *(3)* identified a number of peptides that were present in amniotic fluid but not in plasma of the same mother. Two peptides, with molecular masses of about 19 kDa, were fragments of heparan sulfate proteoglycans. They corresponded to COOH-terminal parts of agrin and perlecan, respectively *(3)*. Using electrospray ionization tandem mass spectrometry (ESI-MS/MS), Ramsay et al. *(4)* provided evidence that biomarkers of lysosomal storage diseases are present in amniotic fluid. Finally, Gravett et al. *(5)* as well as Buhimschi et al. *(6)* showed that proteomics allowed identification of novel biomarkers that may prove useful for the diagnosis of intraamniotic infections.

The aim of this review is to present the physiology of amniotic fluid, to give some data on its "normal" proteome, and to report on potential clinical applications of amniotic fluid proteomics.

2. Physiology of Amniotic Fluid

2.1. Overview

The amniotic fluid surrounds the developing fetus and plays a crucial role in normal development (reviewed in **refs.** *7* and *8*). This clear-colored liquid cushions and protects the fetus and provides it with fluids. By the second trimester, the fetus is able to breathe the fluid into the lungs and to swallow it, promoting normal growth and development of the lungs and gastrointestinal system. Amniotic fluid also allows the fetus to move around, which aids in normal development of muscle and bone. The amniotic sac, which contains the embryo, forms at about 12 d after conception. Amniotic fluid immediately begins to fill the sac. In the early weeks of pregnancy, amniotic fluid consists mainly of water supplied by the mother. After about 12 wk, fetal urine makes up most of the fluid. At 15 wk of gestation, the volume of amniotic fluid is about 200 mL, with substantial individual variations. The amount of amniotic fluid is about 800 mL at 24 wk and increases until about 28 to 32 wk of pregnancy. After that time, the level of fluid generally stays about the same until the baby is full term (approx 37–40 wk); thereafter, the level declines ***(9)***.

However, there may be too little or too much amniotic fluid in some pregnancies. These conditions are referred to as oligohydramnios and polyhydramnios, respectively ***(10)***. Various mechanisms are involved in transfer of biological components from the mother to the fetus and several, such as glucose transporters ***(11)*** or thyrotrophin-releasing hormone ***(12)***, have been characterized. By contrast, transport of proteins from the fetus to the amniotic fluid has not been investigated in detail. The transplacental transfer of immunoglobulins has been well studied ***(13,14)***. Several proteins, notably the MHC class I-related receptor (FcRn), are involved in the delivery of maternal IgG to the fetus. With such a transfer across syncytiotrophoblast of the chorionic villi, the concentration of IgG in fetal blood increases from early in pregnancy through term. IgG_1 is the most efficient subclass of IgG transported ***(14)***. Monoclonal IgG produced by the mother can also be transported to the fetus ***(15,16)***. Maternal IgGs are the main immunoglobulins in amniotic fluid, which contains different forms of fetal immunoglobulins ***(17)***. IgM is not detected in amniotic fluid ***(17,18)***.

2.2. Amniotic Fluid Proteome

The concentration of a given protein in amniotic fluid is governed not only by fetal, placental, or maternal synthesis and degradation, but also by exchanges between the mother and the fetus through the placenta. Amniotic membranes certainly play an important role. Maternofetal transfer of proteins involves several different mechanisms such as first-order process or active transport. Consequently, the concentration of each amniotic fluid protein results from a balance between opposing dynamic metabolic and physiological processes, which proceed simultaneously ***(19)***. The concentration of amniotic fluid proteins varies considerably during pregnancy, and mothers giving birth to large-for-gestational-age infants have uniformly lower amniotic fluid protein concentrations at 12 to 20 wk of gestation compared with those of mothers of appropriate-for-gestational-age infants ***(20)***.

There is some indirect evidence demonstrating that fetal defecation is a normal physiological process in the third trimester, and intestinal proteins may also be present in amniotic fluid ***(21)***. The amniotic fluid proteome is therefore composed of urine proteins, intestinal proteins, alveolar fluid proteins, and their degradation products. It is therefore quite important to have data on these different proteomes, and some relevant information is already available ***(22–26)***. In addition, cellular proteins are produced, either by the skin of the fetus or directly by the amnion.

3. Proteomics: A Brief Overview

The genome of an individual determines its potential for protein expression. However, it does not specify which proteins are expressed in the cells. For a long time, complete characterization of the genome of various species has been an aim of the scientific community. The first genome to be sequenced was that of *Haemophilus influenzae* in 1995. Since then, several other genomes have been entirely sequenced. In 2001, human genome sequencing was completed, and this success has made the estimation of the number of encoded genes in humans possible. Because one gene may produce more than one protein, at least 30,000 genes encode for about 100,000 proteins. Furthermore, owing to alternative splicing, different mRNAs are synthesized, leading to the production of different proteins ***(27)***. Therefore, some authors have indicated that the total number of proteins of the human genome is considerable, with estimates of about 500,000 to 1,500,000 distinct molecular species.

In addition, different posttranslational modifications such as phosphorylation or glycosylation can occur in the cells ***(28–32)***. These and other posttranslational modifications are crucial for protein function. They are also of importance for protein stability as well as cellular location. All these changes are not apparent from genomic sequence or mRNA expression data. In addition, during the lifetime of an individual, the synthesis of specific proteins can be activated or suppressed. Consequently, it is important to know which proteins are produced by a given cell at a given moment. To do so, the first step is to study the transcriptome, which is the pool of all transcribed mRNA molecules ***(33,34)***. The transcriptome is translated into proteins, which characterize the status of a cell or an organ ***(35,36)***. The term *proteome* describes the expressed protein complement of a cell or a tissue at a given time ***(30,37,38)***. However, proteomics has to deal with the wide dynamic range of protein expression, with highly abundant proteins such as albumin in plasma or very low protein concentrations produced in body fluids (zeptomole or yactomole per liter) ***(39)***.

The progress made in proteomics over the last few years has been made possible through the developments in mass spectrometry that caused J. B. Fenn and K. Tanaka to be given the 2002 Nobel Prize in chemistry ***(40)***. The ability of mass spectrometry (in association with bioinformatics) to identify small amounts of proteins from complex mixtures is the cornerstone of proteomics ***(41,42)***. The achievements made over the years in proteomic sciences certainly represent a unique opportunity to study various disease processes better in humans ***(43)***, as well as the complex relations between pathogens and the cells/tissues that participate in the pathogenesis of diseases or that are specifically targeted by infectious agents. Impressive developments have been achieved in the past few years in proteomics, allowing the study of proteins in a given cell, protein isoforms, and modifications as well as protein-protein interactions ***(44)***.

4. Methods of Proteomic Analysis of Amniotic Fluid

4.1. Sample Preparation

The proteomics of amniotic fluid is facilitated because most of the fluid's protein content is soluble in water. However, to eliminate blood contaminants, amniotic cells, as well as maternal tissue or fetal tissue fragments, it is essential to obtain amniotic fluid using rigorous preanalytical conditions. This can be done by collecting amniotic fluid from women undergoing cesarean section, after transamniotic puncture or during extramembranous hysterotomy. Amniotic fluid is then immediately placed in plastic tubes and centrifuged and aliquots are stored at –80°C until use. When comparison studies are to be performed with maternal plasma/serum, it is mandatory that blood samples be collected at the same time.

Various conditions have been described to maintain proteins in solution, mainly for isoelectric focusing ***(45–47)***. For the association of ionic and nonionic detergents, chaotropes are mandatory in sample buffers, and reproducible results can be obtained with denaturing solutions containing urea, thiourea, and zwitterionic detergents. To detect low-abundance proteins, removal of major abundant proteins such as albumin may be required ***(48,49)***. Finally, numerous approaches similar to those described for urine or cerebrospinal fluid proteomics can be used to prepare amniotic fluid concentrates ***(25,50)***.

4.2. Protein Separation

There are many different approaches to the separation of proteins in a complex mixture of biological fluids. Electrophoresis is still the cornerstone of protein separation sciences. However, many other different methods have been developed (e.g., capillary zone electrophoresis, size exclusion chromatography, Off-Gel electrophoresis, immunoaffinity subtraction chromatography, liquid chromatography [LC], multidimensional LC, and tryptic digestion followed by multidimensional LC of peptides) that have proved useful to characterize both the plasma/serum ***(51–54)*** and the urine proteomes in depth ***(25)***.

4.2.1. Electrophoresis

Various techniques can separate proteins in solution (reviewed in **ref.** ***55***). Separation of polypeptides by electrophoresis allowed Arne Tiselius to describe protein fractions corresponding to albumin and α-, β-, and γ-globulins in serum. The number of fractions expand into subfractions, identified as α1, α2, β1, β2, γ1, and γ2. These fractions have mobility characteristics and are still used to denote proteins such as α_1-macroglobulin, α_2-antiplasmin, and β_2-microglobulin. Sophisticated new electrophoretic techniques for identifying many proteins simultaneously and relating them to diseases have been developed over the years. Almost all body fluids have been studied by electrophoresis, including serum, urine, and cerebrospinal fluid. Despite the major developments and progress achieved in protein separation sciences, only a restricted number of methods are routinely used in clinical laboratory. Nowadays, 1D polyacrylamide gel electrophoresis in the presence of sodium dodecyl sulfate (SDS-PAGE or 1D-PAGE) is still widely used to separate proteins and is particularly useful to study proteins with low solubility such as membrane proteins. Additionally, 1D-PAGE is particularly useful when associated with affinity isolation techniques ***(56)***.

4.2.2. High-Resolution 2D Polyacrylamide Gel Electrophoresis

Methods used for high-resolution 2D-PAGE have been recently reviewed in detail ***(57)***. Isoelectric focusing can be performed using various types of pH gradients, according to the range of p*I* to be resolved. The protein sample can be loaded onto an immobilized pH gradient (IPG) strip during or after the rehydration step. Isoelectric focusing can be done by progressively increasing the voltage from 300 to 3000 V during the first 3 h, followed by 1 h at 3500 V and finally stabilization at 5000 V for a total of 100 kVh. Before the 2D electrophoresis, proteins on IPG strips have to be equilibrated. The strips are then placed onto the top of homogeneous or gradient polyacrylamide 2D gels, and proteins can be visualized with silver stain, Coomassie Blue, or other stains that are compatible with mass spectrometry ***(58–60)***. Finally, the strategies of combining various stains or sequential protein staining have been proposed, allowing preliminary characterization of the glycosylation and phosphorylation status of protein spots ***(61)***.

4.2.3. Protein Microarrays

Miniaturized ligand binding assays are of great interest because they allow the simultaneous determination of a large number of proteins from a minute amount of sample within a single experiment ***(62)***. Furthermore, protein microarrays are applicable for studying protein-protein interactions, protein-ligand interactions, kinase activity, and posttranslational modifications of proteins ***(63)***. Surface-enhanced laser desorption/ionization time-of-flight mass spectrometry (SELDI-TOF-MS) or ProteinChip technology has emerged as a powerful tool to study complex biological fluids ***(64)***. In theory, the chemical and physical properties of the protein array probe can be used for selective retention of almost all proteins. At the same time, nonspecific binding can also be one of the most difficult tasks to overcome in the development of such arrays, especially when the goal is to develop a broad-range binding surface, as is the case for SELDI targets that employ classical chromatographic moieties.

4.3. Identification of Proteins

Different methods have been used for the identification of proteins *(65)*. Comigration with purified known proteins and Western blotting were employed by the pioneers of the electrophoretic field. Nowadays, the use of specific polyclonal or monoclonal antibodies and the recent developments of antigen-antibody interactions with enhanced chemiluminescence allow detection and identification of traces of proteins. However, monoclonal antibodies may not be able to detect some proteins or peptides. In these cases, microsequencing, amino acid analysis, peptide mass fingerprinting, and/or mass spectrometry are particularly useful. The development of automated, high-throughput technologies for the identification of multiple proteins is progressing rapidly, and automation already exists at several stages of the protein identification process. Finally, bioinformatics allows identification of proteins by mining several databases.

5. Proteomics of Amniotic Fluid

Amniocentesis is frequently used for various diagnostic purposes *(66)*, and several amniotic fluid proteins have been studied over the years *(67–69)*. 2D-PAGE was used more than 20 yr ago to investigate amniotic fluid proteins *(70–74)*, but the first detailed study of the amniotic fluid proteome was published in 1997 by Liberatori et al. *(75)*. Using high-resolution 2D-PAGE, the authors identified more than 100 spots. The methods used at that time were map comparison, N-terminal microsequencing, and immunoblotting.

5.1. 2D Gel-Based Amniotic Fluid Proteomics

Figure 1 shows a 2D proteome map (which is still in construction in our laboratory) of normal amniotic fluid. Spots were identified by comparison either with a reference plasma proteome map (http://www.expasy.org/cgi-bin/map2/def?PLASMA_HUMAN) or with a reference amniotic fluid proteome map (http://www.bio-mol.unisi.it/cgi-bin/map2/def?AF_HUMAN). The latter was based on the work of Liberatori et al. *(75)*. Additionally, several spots were identified by mass spectrometry, as described elsewhere *(3)*. Up to now, at pH 4 to 7, approximately 100 spots corresponding to about 26 gene products have been identified in our 2D proteome map, a number being degradation peptides **(Table 1)**.

Several alterations in the protein spot pattern were evidenced when amniotic fluids from pregnant women at 17 wk of gestation were compared with those of women at term **(Fig. 2)**. The identification of these alterations is of importance because it clearly shows that the proteome of amniotic fluid is dynamic and that adequate reference protein patterns should be used when amniotic fluids from fetuses at different gestational ages and with various diseases are studied.

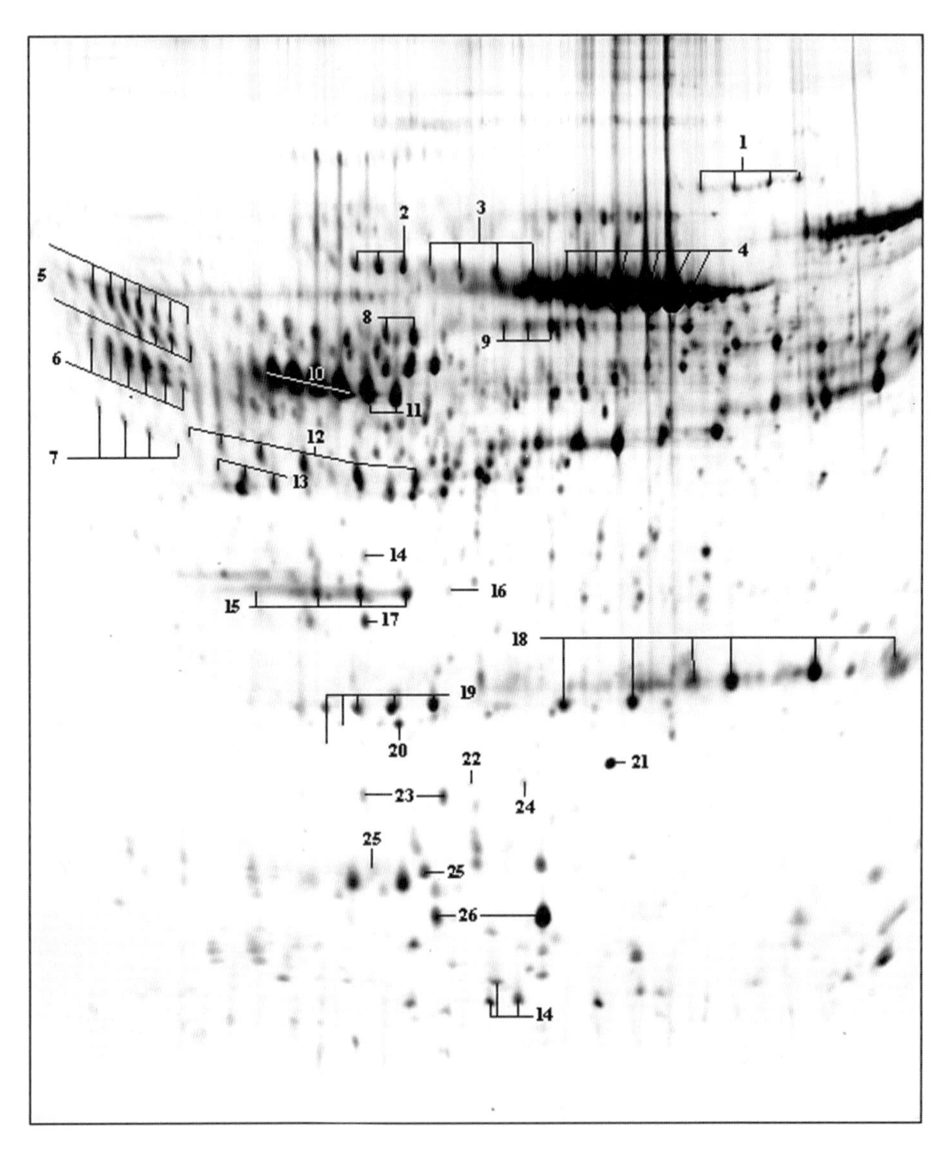

Fig. 1. Silver-stained, high-resolution 2D-PAGE of an amniotic fluid sample at term. First dimension: immobilized 4 to 7 pH gradient; second dimension: 9 to 16% gradient polyacrylamide gel. The spots identified either by map comparison or by mass spectrometry are numbered (*see* **Table 1**).

5.2. Amniotic Fluid Study by LC and Mass Spectrometry

Nilsson et al. *(49)* recently published a study of the amniotic fluid proteome analyzed by LC followed by ESI and Fourier transform ion cyclotron resonance

Table 1
Spots Identified by Mass Spectrometry[a]

Spot		SwissProt		p*I*		Molecular mass (kDa)		
no.	Name	ID	Abbreviation	Measured	Theoretical	Measured	Theoretical	Function
1	Complement factor B	P02751	CFAB_HUMAN	5.98–6.21	6.67	100	85.53	Part of the alternate pathway of the complement system
2	α-1B-glycoprotein	P04217	A1BG_HUMAN	5.05–5.19	5.58	72.8–75.5	54.2	
3	Hemopexin	P02790	HEMO_HUMAN	5.22–5.41	6.55	73.7–76.7	51.7	Binds heme and transports it to the liver
4	Serum albumin	P02768	ALBU_HUMAN	5.5–5.9	5.92	66.3–67.5	69.4	Main plasma protein
5	α_1-antichymotrypsin	P01011	AACT_HUMAN	4.57–4.8	5.33	56.3–64.5	47.6	Inhibits neutrophil cathepsin G and mast cell chymase
6	α-2-HS-glycoprotein	P02765	A2HS_HUMAN	4.56–4.7	5.43	57.4–52.3	39.3	Promotes endocytosis, possesses opsonic properties
7	Leucine-rich α-2-glycoprotein	P02750	A2GL_HUMAN	4.6–4.72	6.45	45–50.3	38.17	Involved in granulocytic differentiation
8	Antithrombin	P01008	ANT3_HUMAN	5.13–5.21	6.32	58.3–58.6	52.6	Serine protease inhibitor in plasma
9	Immunoglobulin chain α			5.3–5.4		62		Structural component of IgA

(Continued)

Table 1 *(Continued)*

Spot no.	Name	SwissProt ID	Abbreviation	p*I* Measured	p*I* Theoretical	Molecular mass (kDa) Measured	Molecular mass (kDa) Theoretical	Function
10	α_1-antitrypsin	P01009	A1AT_HUMAN	4.9–5.15	5.37	53.4–56.2	46.7	Inhibitor of serine proteases
11	Vitamin D-binding protein	P02774	VTDB_HUMAN	5.07–5.16	5.4	53.6	53	Carries the vitamin D sterols
12	Haptoglobin	P00737	HPT_HUMAN	4.74–5.21	6.13	40.9–44.5	45.2	Combines with free plasma hemoglobin
13	Zinc-α-2-glycoprotein	P25311	ZA2G_HUMAN	4.85–4.97	5.57	40.3–41.9	33.8	Stimulates lipid degradation in adipocytes
14	Serum albumin	P02768	ALBU_HUMAN	5.06	5.92	36	69.3	Main structural protein of plasma
15	Insulin-like growth factor binding protein 1	P08833	IBP1_HUMAN	4.9–5.20	5.11	33.2	27.9	Prolongs the half-life of the insulin-like growth factors
16	Collagen α 1(III) chain	P02461	CA13_HUMAN	5.25	6.18	33.2	139	Occurs in most soft connective tissues
17	AMBP protein	P02760	AMBP_HUMAN	5.08	5.95	30	39	Occurs in many physiological fluids
18	Immunoglobulin light chain			5.5–6.4		23–27		Structural component of immunoglobulins
19	Apolipoprotein A-I	P02647	APA1_HUMAN	4.99–5.22	5.56	23	30.7	Participates in the transport of cholesterol from tissues to the liver

20	cAMP-specific 3′,5′-cyclic phosphodiesterase 7B	Q9NP56	CN7B_HUMAN	5.17	6.54	22	51.8	May be involved in the control of cAMP-mediated neural activity and cAMPmetabolism in the brain
21	Perlecan	P98160	PGBM_HUMAN	5.62	6.06	19.6	468.8	Responsible for the fixed negative electrostatic charge
22	Agrin	O00468	AGRN_HUMAN	5.29	6.05	20	214.9	Component of the basal lamina
23	Plasma retinol-binding protein	P02753	RETB_HUMAN	5.06–5.23	5.76	20.2	23	Delivers retinol from the liver stores to the peripheral tissues
24	Chorionic somatomammotropin hormone	P01243	CSH_HUMAN	5.4	5.34	20	25	Plays an important role in growth control
25	Von Ebner's gland protein	P31025	VEGP_HUMAN	5.10–5.21	5.39	15	19.2	Could play a role in taste reception.
26	Transthyretin	P02766	TTHY_HUMAN	5.23–5.44	5.52	13.8	15.9	Thyroid hormone-binding protein

[a]The numbering corresponds to that of **Fig. 1**.

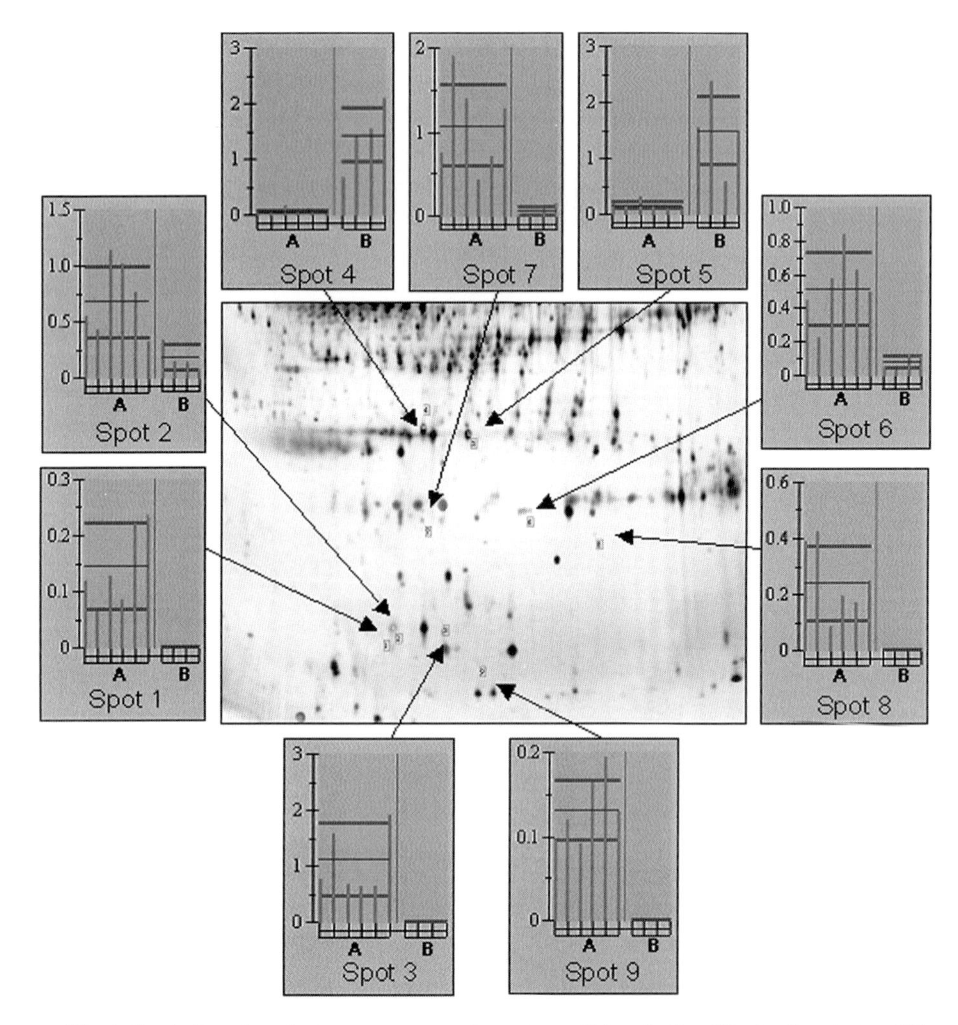

Fig. 2. Spot alterations between pregnancies at term **(A)** and after 17 wk of gestation **(B)**. Quantitative expression of the alterations is presented by histograms, showing the percent intensity volume of the spots in samples A (n = 7) and B (n = 4). Horizontal bars represent mean and SD. The corresponding spots are shown by arrows on a silver-stained 2D-PAGE gel of an amniotic fluid sample, corresponding to p*I* between 4.5 and 7 and to molecular masses between 10,000 and 50,000 Daltons. First dimension: immobilized 4 to 7 pH gradient; second dimension: 9 to 16% gradient polyacrylamide gel. Note that some spots such as 4 and 5 were more abundant at 17 wk of pregnancy, whereas the others such as 1, 2, 3, 6, 7, 8, and 9 were more abundant at term.

(FT-ICR) mass spectrometry. After removal of albumin, tryptic peptides were separated by gradient capillary electrophoresis and identified by high-resolution mass spectrometry. Using this approach, 43 new amniotic fluid proteins were putatively identified. In addition, the study revealed that a combination of different proteomic methods should be used to evaluate the proteome of amniotic fluid globally. Tsangaris et al. *(76)* recently reported on the amniotic fluid cell proteome at the 16th wk of pregnancy analyzed by 2D-PAGE and matrix-assisted laser desorption/ionization time-of-flight mass spectrometry (MALDI-TOF MS). This study resulted in the identification of 432 different gene products, most of which were enzymes and structural proteins.

5.3. Proteomics of Amnion Cells

Because amniotic fluid partially derives from amniotic cells, it is important to mention in this review the studies of Afjehi-Sadat et al. *(77)* and Oh et al. *(78)*, who provided important data on these cells. Using 2D-PAGE analysis of cultivated amniotic cells, Afjehi-Sadat et al. *(77)* identified five "hypothetical" proteins, some of which had putative enzyme activity. Oh et al. *(78)* also used 2D-PAGE to analyze cultivated amniotic cells obtained from controls and patients with Down syndrome. Using mass spectrometry, they identified several proteins that appeared to be potential markers of Down syndrome, particularly enzymes that are involved in several metabolic pathways (purine, carbohydrate, intermediary, and amino acid metabolism).

6. Clinical Applications of Amniotic Fluid Proteomics

6.1. Premature Rupture of the Membranes

Premature rupture of the membranes occurs in about 5% of deliveries with complications such as infection and preterm birth *(79,80)*. For women in labor, when infection or irreversible fetal distress occurs, there is no other option than delivery. For those not in labor, especially at premature gestational ages, the complexity of the decisions (prolonging gestation, reducing complications of prematurity, and choosing the timing and route of delivery) turns management into a difficult task *(81)*. The methods used to diagnose premature rupture of the membranes are variable *(82)*. To identify new potential markers of premature rupture of the membranes, proteomic studies were performed on amniotic fluid and plasma samples collected from women at term, as well as on amniotic fluid samples collected at the 17th wk of gestation *(3)*. Several "amniotic fluid-specific" peptides were identified. Three spots corresponded to fragments of plasma-related proteins (retinol-binding protein, properdin, and apolipoprotein A-1); therefore, they did not appear to be useful potential markers of premature rupture of the membranes.

In contrast, two peptides were fragments of proteins that are not present in normal human plasma. These two peptides were identified as fragments of agrin (SwissProt: O00468) and perlecan (SwissProt: Q9H3V5), respectively, both of which are heparan sulfate proteoglycans *(3)*. Their physiological roles in amniotic fluid remain unknown *(83)*. However, they are thought to mediate the action of growth factors and be involved in developmental processes *(84)*. Agrin has been shown to be highly expressed in basement membranes of adult lung and kidney *(85)*. Perlecan maintains the integrity of cartilage and some basement membranes *(86)* and has recently been reported to be involved in chondrogenesis *(87)* as well as in the development of particular bones *(88)*. In addition, perlecan was identified at the fetal-maternal interface of the placenta *(89)*. Perlecan may have a facilitating role in trophoblast invasion processes, acting together with heparin/heparan sulfate-interacting proteins present at the maternal interface *(89)*.

Recently, important advances have been made by the group of R. V. Iozzo. These investigators have shown that perlecan may have a potential effect on tumor growth through its interaction with progranulin *(90)* as well as an effect on bone formation and angiogenesis through its interactions with extracellular matrix protein 1 *(91)*. Finally, the COOH-terminus of perlecan, named endorepellin, is a novel antiangiogenic factor that may retard neovascularization *(92,93)*. The sequence of the fragment of perlecan observed in amniotic fluid corresponded to that of endorepellin, suggesting a role for this peptide in amniotic fluid. However, this peptide may be a signature of the renal function of the fetus because the COOH-terminal fragment of perlecan, beginning with amino acid residue 4216 to the COOH-terminal, appeared to be present in the urine of patients with end-stage renal failure *(94)*. In addition, fragments of perlecan and agrin, both less than 20 kDa, were identified in normal human urine by Pieper et al. *(25)*.

6.2. Amniotic Fluid Infection

Intraamniotic infection may be linked with preterm birth and adverse neonatal sequelae. By using SELDI-TOF MS, gel electrophoresis, and MS/MS, Gravett et al. *(5)* were able to characterize several amniotic fluid peptides in a model of animal infection. Candidate biomarkers such calgranulin B, azurocidin, vitamin D binding protein, and insulin-like growth factor binding protein 1 were tested in a cohort of 33 women. The results showed that specific biomarkers in amniotic fluid and maternal serum might have application in the early detection of intraamniotic infection. Buhimschi et al. *(6)* also used SELDI-TOF MS to evaluate the presence of biomarkers of intraamniotic inflammation. They studied 104 samples of amniotic fluid and identified four proteins, including neutrophil defensins-1 and -2, calgranulin C, and calgranulin A, as potential markers of inflammation/infection of amniotic fluid.

6.3. Indicators of Lysosomal Storage Diseases

Deficiency of a particular lysosomal protein or a protein involved in lysosomal biogenesis results in lysosomal storage diseases, which are a large group of inherited metabolic disorders *(95)*. Accumulations of either undegraded substrates or catabolic products are observed within the lysosomes. The identification of this group of pathologies is important ***(96)***. Lysosomal storage diseases are usually diagnosed by leukocyte or fibroblast enzyme activities. Recently, Ramsay et al. ***(4)*** showed that analysis of oligosacharides and glycolipids in amniotic fluid by ESI-MS/MS can be used for prenatal diagnosis of various lysosomal storage diseases according to their unique metabolic profile of protein, oligosaccaride, and glycolipid markers.

7. Perspectives: The Proteome of the Human Fetus

Over the last 15 yr, we have extensively analyzed plasma/serum samples from fetuses, newborns, and infants ***(15,97–99)*** and have identified important physiological alterations related to aging. In addition, fetal proteins such as α-fetoprotein and a fetal form of α_1-antitrypsin were identified using 2D-PAGE ***(55)***. The fetal form of α_1-antitrypsin was characterized by an apparent molecular mass of 46 kDa and a p*I* of 5.0. N-terminal microsequencing (25-EDPQ) and immunoblotting using anti-α_1-antitrypsin antibodies revealed that this protein was similar to the fetal form of α_1-antitrypsin that was identified in mouse fetal plasma ***(100)***. The identification of the fetal proteome is important because part of the genome may be used only during embryogenesis and fetal development. The function of many fetal proteins is almost unknown, even with the cumulative knowledge of particular proteins such as α-fetoprotein ***(101)***. With improvements of proteomics and fetal medicine ***(102)***, it will be easier to study the fetal proteome, which represents expression of the genome during development. Recently, there have been several reports on proteomic studies of the amnion ***(77,78)***, placenta ***(103,104)***, human trophoblasts obtained after laser microdissection of placenta ***(105)***, organelles of the placenta such as mitochondria ***(106)***, umbilical vein endothelial cells ***(107,108)***, and human CD34+ stem/progenitor cells isolated from cord blood ***(109)***. These studies clearly represent the first step toward the establishment of the repertoire of the proteins of the human fetus and neonate, and its completion will be a challenge for fetal medicine.

Characterization of the proteome of amniotic fluid and identification of its alterations during pregnancy are in progress. Thus, methods that allow simultaneous study of a multitude of proteins may be of value to gain insight into the physiology of amniotic fluid and to identify potential markers of diseases during pregnancy.

References

1. Drohse H, Christensen H, Myrhoj V, Sorensen S. Characterisation of non-maternal serum proteins in amniotic fluid at weeks 16 to 18 of gestation. *Clin Chim Acta* 1998;276:109–120.

2. Yamada H, Kishida T, Negishi H, et al. Comparison of an improved AFP kit with the intra-amniotic PSP dye-injection method in equivocal cases of preterm premature rupture of the fetal membranes. *J Obstet Gynaecol Res* 1997;23:307–311.
3. Vuadens F, Benay C, Crettaz D, et al. Identification of biologic markers of the premature rupture of fetal membranes: proteomic approach. *Proteomics* 2003;3: 1521–1525.
4. Ramsay SL, Maire I, Bindloss C, et al. Determination of oligosaccharides and glycolipids in amniotic fluid by electrospray ionisation tandem mass spectrometry: in utero indicators of lysosomal storage diseases. *Mol Genet Metab* 2004; 83:231–238.
5. Gravett MG, Novy MJ, Rosenfeld RG, et al. Diagnosis of intra-amniotic infection by proteomic profiling and identification of novel biomarkers. *JAMA* 2004;292:462–469.
6. Buhimschi IA, Christner R, Buhimschi CS. Proteomic biomarker analysis of amniotic fluid for identification of intra-amniotic inflammation. *Br J Obstet Gynaecol* 2005;112:173–181.
7. Brace RA. Physiology of amniotic fluid volume regulation. *Clin Obstet Gynecol* 1997;40:280–289.
8. Sohaey R. Amniotic fluid and the umbilical cord: the fetal milieu and lifeline. *Semin Ultrasound CT MR* 1998;19:355–369.
9. Sherer DM. A review of amniotic fluid dynamics and the enigma of isolated oligohydramnios. *Am J Perinatol* 2002;19:253–266.
10. Hohlfeld P, Marty F, De Grandi P, Tissot JD, Bossart H, Gerber S. Pathologies du liquide amniotique. In *Le Livre de l'Interne Obstétrique.* Paris: Flammarion, 2004:255–258.
11. Illsley NP. Glucose transporters in the human placenta. *Placenta* 2000;21:14–22.
12. Bajoria R, Fisk NM. Maternofetal transfer of thyrotrophin-releasing hormone: effect of concentration and mode of administration. *Pediatr Res* 1997;41: 674–681.
13. Firan M, Bawdon R, Radu C, et al. The MHC class I-related receptor, FcRn, plays an essential role in the maternofetal transfer of gamma-globulin in humans. *Int Immunol* 2001;13:993–1002.
14. Simister NE. Placental transport of immunoglobulin G. *Vaccine* 2003;21: 3365–3369.
15. Tissot JD, Schneider P, Hohlfeld P, Tolsa JF, Calame A, Hochstrasser DF. Monoclonal gammopathy in a 30 weeks old premature infant. *Appl Theor Electroph* 1992;3:67–68.
16. Dolfin T, Pomeranz A, Korzets Z, et al. Acute renal failure in a neonate caused by the transplacental transfer of a nephrotoxic paraprotein: successful resolution by exchange transfusion. *Am J Kidney Dis* 1999;34:1129–1131.
17. Quan CP, Forestier F, Bouvet JP. Immunoglobulins of the human amniotic fluid. *Am J Reprod Immunol* 1999;42:219–225.

18. Jauniaux E, Jurkovic D, Gulbis B, Liesnard C, Lees C, Campbell S. Materno-fetal immunoglobulin transfer and passive immunity during the first trimester of human pregnancy. *Hum Reprod* 1995;10:3297–3300.
19. Malek A, Sager R, Schneider H. Transport of proteins across the human placenta. *Am J Reprod Immunol* 1998;40:347–351.
20. Tisi DK, Emard JJ, Koski KG. Total protein concentration in human amniotic fluid is negatively associated with infant birth weight. *J Nutr* 2004;134: 1754–1758.
21. Kimble RM, Trudenger B, Cass D. Fetal defaecation: is it a normal physiological process? *J Paediatr Child Health* 1999;35:116–119.
22. Magi B, Bini L, Perari MG, et al. Bronchoalveolar lavage fluid protein composition in patients with sarcoidosis and idiopathic pulmonary fibrosis: a two-dimensional electrophoretic study. *Electrophoresis* 2002;23:3434–3444.
23. Sabounchi-Schutt F, Astrom J, Hellman U, Eklund A, Grunewald J. Changes in bronchoalveolar lavage fluid proteins in sarcoidosis: a proteomics approach. *Eur Respir J* 2003;21:414–420.
24. Bai Y, Galetskiy D, Damoc E, et al. High resolution mass spectrometric alveolar proteomics: identification of surfactant protein SP-A and SP-D modifications in proteinosis and cystic fibrosis patients. *Proteomics* 2004;4:2300–2309.
25. Pieper R, Gatlin CL, McGrath AM, et al. Characterization of the human urinary proteome: a method for high-resolution display of urinary proteins on two-dimensional electrophoresis gels with a yield of nearly 1400 distinct protein spots. *Proteomics* 2004;4:1159–1174.
26. Thongboonkerd V, Malasit P. Renal and urinary proteomics: current applications and challenges. *Proteomics* 2005;5:1033–1042.
27. Roberts GC, Smith CW. Alternative splicing: combinatorial output from the genome. *Curr Opin Chem Biol* 2002;6:375–383.
28. Wilkins MR, Gasteiger E, Gooley AA, et al. High-throughput mass spectrometric discovery of protein post-translational modifications. *J Mol Biol* 1999; 289:645–657.
29. Yan JX, Sanchez JC, Binz PA, Williams KL, Hochstrasser DF. Method for identification and quantitative analysis of protein lysine methylation using matrix-assisted laser desorption ionization time-of-flight mass spectrometry and amino acid analysis. *Electrophoresis* 1999;20:749–754.
30. Banks RE, Dunn MJ, Hochstrasser DF, et al. Proteomics: new perspectives, new biomedical opportunities. *Lancet* 2000;356:1749–1756.
31. Sarioglu H, Lottspeich F, Walk T, Jung G, Eckerskorn C. Deamidation as a widespread phenomenon in two-dimensional polyacrylamide gel electrophoresis of human blood plasma proteins. *Electrophoresis* 2000;21:2209–2218.
32. Imam-Sghiouar N, Laude-Lemaire I, Labas V, et al. Subproteomics analysis of phosphorylated proteins: application to the study of B-lymphoblasts from a patient with Scott syndrome. *Proteomics* 2002;2:828–838.

33. Devaux F, Marc P, Jacq C. Transcriptomes, transcription activators and microarrays. *FEBS Lett* 2001;498:140–144.
34. Strausberg RL, Riggins GJ. Navigating the human transcriptome. *Proc Natl Acad Sci U S A* 2001;98:11837–11838.
35. Kettman JR, Frey JR, Lefkovits I. Proteome, transcriptome and genome: top down or bottom up analysis? *Biomol Eng* 2001;18:207–212.
36. Oliver DJ, Nikolau B, Wurtele ES. Functional genomics: high-throughput mRNA, protein, and metabolite analyses. *Metab Eng* 2002;4:98–106.
37. Anderson NL, Anderson NG. Proteome and proteomics: new technologies, new concepts, and new words. *Electrophoresis* 1998;19:1853–1861.
38. Fields S. Proteomics—proteomics in genomeland. *Science* 2001;291:1221–1223.
39. Corthals GL, Wasinger VC, Hochstrasser DF, Sanchez JC. The dynamic range of protein expression: a challenge for proteomic research. *Electrophoresis* 2000;21: 1104–1115.
40. Cho A, Normile D. Nobel Prize in Chemistry. Mastering macromolecules. *Science* 2002;298:527–528.
41. Aebersold R, Mann M. Mass spectrometry-based proteomics. *Nature* 2003;422: 198–207.
42. Boguski MS, McIntosh MW. Biomedical informatics for proteomics. *Nature* 2003; 422:233–237.
43. Hanash S. Disease proteomics. *Nature* 2003;422:226–232.
44. Tyers M, Mann M. From genomics to proteomics. *Nature* 2003;422:193–197.
45. Rabilloud T, Adessi C, Giraudel A, Lunardi J. Improvement of the solubilization of proteins in two-dimensional electrophoresis with immobilized pH gradients. *Electrophoresis* 1997;18:307–316.
46. Hochstrasser DF, Harrington M, Hochstrasser AC, Miller MJ, Merril CR. Methods for increasing the resolution of two-dimensional protein electrophoresis. *Anal Biochem* 1988;173:424–435.
47. Chevallet M, Santoni V, Poinas A, et al. New zwitterionic detergents improve the analysis of membrane proteins by two-dimensional electrophoresis. *Electrophoresis* 1998;19:1901–1909.
48. Ahmed N, Barker G, Oliva K, et al. An approach to remove albumin for the proteomic analysis of low abundance biomarkers in human serum. *Proteomics* 2003;3:1980–1987.
49. Nilsson S, Ramstrom M, Palmblad M, Axelsson O, Bergquist J. Explorative study of the protein composition of amniotic fluid by liquid chromatography electrospray ionization Fourier transform ion cyclotron resonance mass spectrometry. *J Proteome Res* 2004;3:884–889.
50. Burkhard PR, Rodrigo N, May D, et al. Assessing cerebrospinal fluid rhinorrhea: a two-dimensional electrophoresis approach. *Electrophoresis* 2001;22:1826–1833.
51. Anderson NL, Polanski M, Pieper R, et al. The human plasma proteome: a nonredundant list developed by combination of four separate sources. *Mol Cell Proteomics* 2004;3:311–326.

52. Chan KC, Lucas DA, Hise D, et al. Analysis of the human serum proteome. *Clin Proteomics* 2004;1:101–226.
53. Heller M, Michel PE, Crettaz D, et al. Two stage Off-gel™ isoelectricfocusing: protein followed by peptide fractionation and application to proteome analysis of human plasma. *Electrophoresis* 2005;26:1174–1188.
54. Thadikkaran L, Siegenthaler MA, Crettaz D, Queloz PA, Schneider P, Tissot JD. Recent advances in blood-related proteomics. *Proteomics* 2005;5:3019–3034.
55. Tissot JD, Hohlfeld P, Layer A, Forestier F, Schneider P, Henry H. Clinical applications. Gel electrophoresis. In: Wilson I, Adlar TR, Poole CF, Cook M, eds. *Encyclopedia of Separation Science.* London: Academic, 2000:2468–2475.
56. Lee WC, Lee KH. Applications of affinity chromatography in proteomics. *Anal Biochem* 2004;324:1–10.
57. Gorg A, Weiss W, Dunn MJ. Current two-dimensional electrophoresis technology for proteomics. *Proteomics* 2004;4:3665–3685.
58. Lauber WM, Carroll JA, Dufield DR, Kiesel JR, Radabaugh MR, Malone JP. Mass spectrometry compatibility of two-dimensional gel protein stains. *Electrophoresis* 2001;22:906–918.
59. Nesatyy VJ, Dacanay A, Kelly JF, Ross NW. Microwave-assisted protein staining: mass spectrometry compatible methods for rapid protein visualisation. *Rapid Commun Mass Spectrom* 2002;16:272–280.
60. White IR, Pickford R, Wood J, Skehel JM, Gangadharan B, Cutler P. A statistical comparison of silver and SYPRO Ruby staining for proteomic analysis. *Electrophoresis* 2004;25:3048–3054.
61. Wu J, Lenchik NJ, Pabst MJ, Solomon SS, Shull J, Gerling IC. Functional characterization of two-dimensional gel-separated proteins using sequential staining. *Electrophoresis* 2004;26:225–237.
62. Templin MF, Stoll D, Schwenk JM, Potz O, Kramer S, Joos TO. Protein microarrays: promising tools for proteomic research. *Proteomics* 2003;3:2155–2166.
63. Espina V, Woodhouse EC, Wulfkuhle J, Asmussen HD, Petricoin EF, III, Liotta LA. Protein microarray detection strategies: focus on direct detection technologies. *J Immunol Methods* 2004;290:121–133.
64. Tang N, Tornatore P, Weinberger SR. Current developments in SELDI affinity technology. *Mass Spectrom Rev* 2004;23:34–44.
65. Tissot JD, Duchosal MA, Schneider P. Two-dimensional polyacrylamide gel electrophoresis. In: Wilson I, Adlar TR, Poole CF, Cook M, eds. *Encyclopedia of Separation Science.* London: Academic, 2000:1364–1371.
66. Alfirevic Z, Sundberg K, Brigham S. Amniocentesis and chorionic villus sampling for prenatal diagnosis. *Cochrane Database Syst Rev* 2003;CD003252.
67. Sutcliffe RG, Brock DJ, Nicholson LV, Dunn E. Fetal- and uterine-specific antigens in human amniotic fluid. *J Reprod Fertil* 1978;54:85–90.
68. Burnett D, Bradwell AR. The origin of plasma proteins in human amniotic fluid: the significance of alpha 1-antichymotrypsin complexes. *Biol Neonate* 1980;37: 302–307.

69. Prado VF, Reis DD, Pena SD. Biochemical and immunochemical identification of the fetal polypeptides of human amniotic fluid during the second trimester of pregnancy. *Braz J Med Biol Res* 1990;23:121–131.
70. Jones MI, Spragg SP, Webb T. Detection of proteins in human amniotic fluid using two-dimensional gel electrophoresis. *Biol Neonate* 1981;39:171–177.
71. Burdett P, Lizana J, Eneroth P, Bremme K. Proteins of human amniotic fluid. II. Mapping by two-dimensional electrophoresis. *Clin Chem* 1982;28:935–940.
72. Stimson WH, Farquharson DM, Lang GD. Pregnancy-associated alpha 2-macroglobulin—a new serum protein elevated in normal human pregnancy. *J Reprod Immunol* 1983;5:321–327.
73. Kronquist KE, Crandall BF, Cosico LG. Detection of novel fetal polypeptides in human amniotic fluid using two-dimensional gel electrophoresis. *Tumour Biol* 1984;5:15–31.
74. Mackiewicz A, Jakubek P, Sajdak S, Breborowicz J. Microheterogeneity forms of alpha-fetoprotein present in amniotic fluid. *Placenta* 1984;5:373–380.
75. Liberatori S, Bini L, De Felice C, et al. A two-dimensional protein map of human amniotic fluid at 17 weeks' gestation. *Electrophoresis* 1997;18:2816–2822.
76. Tsangaris G, Weitzdorfer R, Pollak D, Lubec G, Fountoulakis M. The amniotic fluid cell proteome. *Electrophoresis* 2005;26:1168–1173.
77. Afjehi-Sadat L, Krapfenbauer K, Slavc I, Fountoulakis M, Lubec G. Hypothetical proteins with putative enzyme activity in human amnion, lymphocyte, bronchial epithelial and kidney cell lines. *Biochim Biophys Acta* 2004;1700:65–74.
78. Oh JE, Fountoulakis M, Juranville JF, Rosner M, Hengstschlager M, Lubec G. Proteomic determination of metabolic enzymes of the amnion cell: basis for a possible diagnostic tool? *Proteomics* 2004;4:1145–1158.
79. Gibbs RS, Blanco JD. Premature rupture of the membranes. *Obstet Gynecol* 1982;60:671–679.
80. Parry S, Strauss JF, III. Premature rupture of the fetal membranes. *N Engl J Med* 1998;338:663–670.
81. Garite TJ. Management of premature rupture of membranes. *Clin Perinatol* 2001;28:837–847.
82. Atterbury JL, Groome LJ, Hoff C. Methods used to diagnose premature rupture of membranes: a national survey of 812 obstetric nurses. *Obstet Gynecol* 1998;92:384–389.
83. Thadikkaran L, Crettaz D, Siegenthaler MA, et al. The role of proteomics in the assessement of premature rupture of fetal membranes. *Clin Chim Acta* 2005;360:27–36.
84. Perrimon N, Bernfield M. Specificities of heparan sulphate proteoglycans in developmental processes. *Nature* 2000;404:725–728.
85. Groffen AJ, Buskens CA, van Kuppevelt TH, Veerkamp JH, Monnens LA, van den Heuvel LP. Primary structure and high expression of human agrin in basement membranes of adult lung and kidney. *Eur J Biochem* 1998;254:123–128.
86. Costell M, Gustafsson E, Aszodi A, et al. Perlecan maintains the integrity of cartilage and some basement membranes. *J Cell Biol* 1999;147:1109–1122.

87. French MM, Smith SE, Akanbi K, et al. Expression of the heparan sulfate proteoglycan, perlecan, during mouse embryogenesis and perlecan chondrogenic activity in vitro. *J Cell Biol* 1999;145:1103–1115.
88. Govindraj P, West L, Koob TJ, Neame P, Doege K, Hassell JR. Isolation and identification of the major heparan sulfate proteoglycans in the developing bovine rib growth plate. *J Biol Chem* 2002;277:19,461–19,469.
89. Rohde LH, Janatpore MJ, McMaster MT, et al. Complementary expression of HIP, a cell-surface heparan sulfate binding protein, and perlecan at the human fetal-maternal interface. *Biol Reprod* 1998;58:1075–1083.
90. Gonzalez EM, Mongiat M, Slater SJ, Baffa R, Iozzo RV. A novel interaction between perlecan protein core and progranulin: potential effects on tumor growth. *J Biol Chem* 2003;278:38,113–38,116.
91. Mongiat M, Fu J, Oldershaw R, Greenhalgh R, Gown AM, Iozzo RV. Perlecan protein core interacts with extracellular matrix protein 1 (ECM1), a glycoprotein involved in bone formation and angiogenesis. *J Biol Chem* 2003;278: 17,491–17,499.
92. Mongiat M, Sweeney C, San Antonio JD, Fu J, Iozzo RV. Endorepellin, a novel inhibitor of angiogenesis dervied from the C terminus of perlecan. *J Biol Chem* 2003;278:4238–4249.
93. Bix G, Fu J, Gonzalez EM, et al. Endorepellin causes endothelial cell disassembly of actin cytoskeleton and focal adhesions through alpha2beta1 integrin. *J Cell Biol* 2004;166:97–109.
94. Oda O, Shinzato T, Ohbayashi K, et al. Purification and characterization of perlecan fragment in urine of end-stage renal failure patients. *Clin Chim Acta* 1996;255:119–132.
95. Meikle PJ, Fietz MJ, Hopwood JJ. Diagnosis of lysosomal storage disorders: current techniques and future directions. *Expert Rev Mol Diagn* 2004;4:677–691.
96. Wilcox WR. Lysosomal storage disorders: the need for better pediatric recognition and comprehensive care. *J Pediatr* 2004;144:S3–S14.
97. Tissot JD, Schneider P, Pelet B, Frei PC, Hochstrasser DF. Mono-oligoclonal production of immunoglobulins in a child with the Wiskott-Aldrich syndrome. *Br J Haematol* 1990;75:436–438.
98. Tissot JD, Hohlfeld P, Hochstrasser DF, Tolsa JF, Calame A, Schneider P. Clonal imbalances of plasma/serum immunoglobulin production in infants. *Electrophoresis* 1993;14:245–247.
99. Tissot JD, Hohlfeld P, Forestier F, et al. Plasma/serum protein patterns in human fetuses and infants: a study by high-resolution two-dimensional polyacrylamide gel electrophoresis. *Appl Theor Electroph* 1993;3:183–190.
100. Nathoo SA, Finlay TH. Fetal-specific forms of alpha 1-protease inhibitors in mouse plasma. *Pediatr Res* 1987;22:1–5.
101. Mizejewski GJ. Biological roles of alpha-fetoprotein during pregnancy and perinatal development. *Exp Biol Med (Maywood)* 2004;229:439–463.
102. Kumar S, O'Brien A. Recent developments in fetal medicine. *BMJ* 2004;328: 1002–1006.

103. Hoang VM, Foulk R, Clauser K, Burlingame A, Gibson BW, Fisher SJ. Functional proteomics: examining the effects of hypoxia on the cytotrophoblast protein repertoire. *Biochemistry* 2001;40:4077–4086.
104. Page NM, Kemp CF, Butlin DJ, Lowry PJ. Placental peptides as markers of gestational disease. *Reproduction* 2002;123:487–495.
105. de Groot CJ, Steegers-Theunissen RP, Guzel C, Steegers EA, Luider TM. Peptide patterns of laser dissected human trophoblasts analyzed by matrix-assisted laser desorption/ionisation-time of flight mass spectrometry. *Proteomics* 2005;5:597–607.
106. Rabilloud T, Kieffer S, Procaccio V, et al. Two-dimensional electrophoresis of human placental mitochondria and protein identification by mass spectrometry: toward a human mitochondrial proteome. *Electrophoresis* 1998;19:1006–1014.
107. Bruneel A, Labas V, Mailloux A, et al. Proteomic study of human umbilical vein endothelial cells in culture. *Proteomics* 2003;3:714–723.
108. Scheurer SB, Rybak JN, Rosli C, Neri D, Elia G. Modulation of gene expression by hypoxia in human umbilical cord vein endothelial cells: a transcriptomic and proteomic study. *Proteomics* 2004;4:1737–1760.
109. Tao W, Wang M, Voss ED, et al. Comparative proteomic analysis of human CD34+ stem/progenitor cells and mature CD15+ myeloid cells. *Stem Cells* 2004;22:1003–1014.

20

Proteomics of Human Milk

Amedeo Conti, Maria Gabriella Giuffrida, and Maria Cavaletto

Summary

Both expression and functional proteomics have been applied to identifying and characterizing human milk proteins. The most extensive expression work, resulting in more than 107 identified proteins reported in an annotated database, was done on the proteins associated with the milk fat globule membranes of human colostrum. Reports on the differences in protein expression between colostrum and mature milk and on the relationship between the mother's diet and protein milk composition may be regarded as functional proteomics studies. Future studies of the human milk proteome will be beneficial for both lactating mothers and nursing children.

Key Words: Human milk; casein; whey proteins; milk fat globule membrane proteins; proteomics.

1. Introduction

Milk proteins have been studied in depth for decades. The proteomic approach to studying milk proteins is different from traditional protein biochemistry methods in its capacity to display highly resolved protein mixtures simultaneously. Like all mammalian species, human mothers provide the specific nutrients needed by newborns by producing milk with a unique composition that mirrors these requirements. Milk samples are complex mixtures of proteins, lipids, lactose, oligosaccharides, and several bioactive factors. Traditionally, milk proteins can be grouped into three main fractions: caseins in colloidal dispersion as micelles, true soluble proteins (whey proteins), and proteins associated with the milk fat globule membranes (MFGMP) **(Fig. 1)**.

The protein concentration of human milk is high during early lactation, mainly owing to the presence of secretory immunoglobulin A (IgA) and lactoferrin; it gradually declines thereafter to the relatively low level of 0.8 to 1.0%

From: *Proteomics of Human Body Fluids: Principles, Methods, and Applications*
Edited by: V. Thongboonkerd © Humana Press Inc., Totowa, NJ

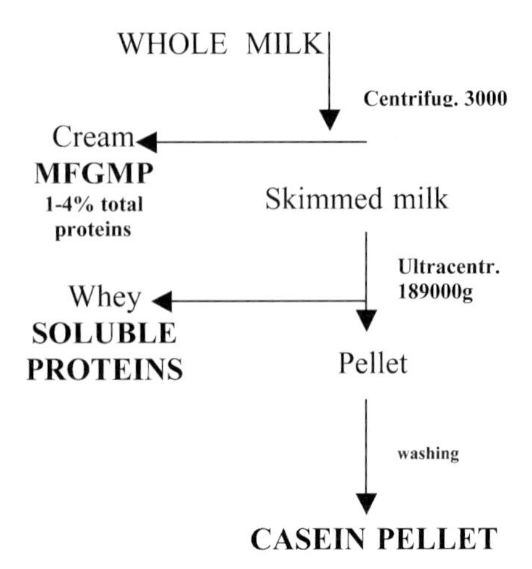

Fig. 1. Outline of fractionation of milk proteins.

Table 1
Variations in Human Milk Protein Concentration During Lactation

	Months post partum	
Elements (g/L)	0–0.5	0.5–1.5
Total nitrogen	3.05 ± 0.59	1.93 ± 0.24
NPN	0.53 ± 0.09	0.46 ± 0.03
True protein[a]	15.80 ± 4.2	9.2 ± 1.8
α-lactalbumin	3.62 ± 0.59	3.26 ± 0.47
Lactoferrin	3.53 ± 0.54	1.94 ± 0.38
Serum albumin	0.39 ± 0.06	0.41 ± 0.07
Secretory IgA	2.0 ± 2.5	1.0 ± 0.3
IgM	0.12 ± 0.03	0.2
IgG	0.34 ± 0.01	0.05 ± 0.03

[a]True protein = (total nitrogen – NPN) × 6.25.
Adapted from **ref.** ***1***.

in mature milk **(Table 1)**. **Table 2** shows the distribution of casein and whey proteins in mature human milk. Many human milk proteins in the casein and whey fractions are involved in facilitating the digestion and uptake of other nutrients in breast milk, as well as providing an important source of amino acids to rapidly growing breastfed infants. For instance, absorption of calcium and iron are assisted by β-casein and lactoferrin, respectively. Human milk proteins from all three fractions also have numerous physiologic activities, including enhancing

Table 2
Distribution of Caseins and Whey Proteins in Human Mature Milk

Protein	%
Casein (3.7 g/kg)	
α-Casein (%)	11.75
β-Casein (%)	64.75
κ-Casein (%)	23.50
True whey proteins (7.6 g/kg)	
α-lactalbumin (%)	42.37
Immunoglobulins (%)	18.15
Serum albumin (%)	7.56
Lactoferrin (%)	30.26
Lysozyme (%)	1.66

Adapted from **ref.** *2*.

the immune function, providing a defense against pathogenic bacteria, viruses, and yeasts, and aiding in the development of the gut and its function *(3)*.

The high nutritional value of human casein and whey proteins depends on the high digestibility of the human casein micelles, owing to their small size, and on the high content in the whey fraction of essential amino acids including lysine, methionine, cysteine, tryptophan, and threonine, which are limited in different dietary protein sources. The frequently cited ratio of 60:40 for whey proteins to casein is an approximation of the ratio during the normal course of lactation, but it can vary from 90:10 in very early lactation to 50:50 in late lactation. Accordingly, the amino acid composition of human milk changes during lactation, probably fulfilling different requirements of the breastfed infant.

Milk contains proteins with native and/or latent biological functionality *(4)*. Activities of the native state are imputed to bioactive molecules from the MFGMP and whey protein fractions. Most of these components exert specific or nonspecific activity against a great variety of pathogenic strains as well as food spoilage microorganisms. Latent biological activity is expressed only upon enzymatic hydrolysis by certain proteolytic digestive enzymes such as chymosin, pepsin, and trypsin. Bioactive peptides have been described and tested for their physiological functionality, which is derived mainly from the hydrolysis of casein fractions. Moreover, breast milk contains many substances including hormones, growth factors, cytokines, and even whole cells that act as mediators between mother and child and establish biochemical or physiological communication. In this sense, breast milk may be considered as a partial and temporary extension of the intrauterine environment to extrauterine life *(5)*.

The milk from each species is dominated by the presence of just a few major proteins. In human milk, these proteins are β-casein, α-lactalbumin, lactoferrin, κ-casein, immunoglobulins, α-casein, and serum albumin. Nevertheless, the milk proteome is still extremely complex as a consequence of posttranslational modifications and the presence of numerous genetic variants of this limited list of proteins. In recent years, proteomic applications of high-resolution 2D gel electrophoresis (2-DE) to milk proteins from a limited number of species have been reported; human and bovine milk dominate these studies, with other limited applications including analysis of goat, wallaby, and mouse milk ***(6)***.

2. Methods

The milk proteome is extremely complex, owing to the presence of genetic variants, posttranslational modifications, and proteolysis processes. For these reasons high-resolution analytical methods are needed; among the methods used for milk protein separation, 2-DE is the most effective and has been used to resolve milk's complexity since 1982. Anderson et al. *(7)* used 2-DE (Iso- and Baso-Dalt and sodium dodecyl sulfate and polyacrylamide gel electrophoresis [SDS-PAGE]) to separate whole milk proteins and fractionate whey proteins. They achieved protein identifications by matching 2-DE maps either with the corresponding cow milk protein profile or with the 2-DE of a single purified protein.

Goldfarb et al. ***(8)*** tried to map skimmed milk proteins by 2-DE (Iso-Dalt followed by 10–20% gradient SDS-PAGE gels) and immunoblotting with specific immunoprobes. The same group applied a similar approach in 1997 and 1999 to identify proteins associated with human MFGM with selected antibodies against the proteins that could be expected in this fraction ***(9)*** and to characterize and quantify the casein content using anticasein monoclonal antibodies and computer imaging ***(10)***. 2-DE followed by immunoblotting with specific antibodies has also been successfully employed to detect some components, like the macrophage migration inhibitory factor, in different human milk fractions ***(11)***.

None of these studies used direct methods to identify proteins; all the identification methods depended on matching 2-DE maps or on selecting suitable antibodies, thus excluding the possibility of identifying new proteins. The first method applied for direct protein identification was N-terminal sequencing, a technique that usually requires electroblotting on a PVDF membrane before sequencing, Coomassie Blue staining, and cutting the band/spot to be sequenced. In 1994, in a comparative study of the casein content of human colostrum and milk, Cavaletto et al. ***(12)*** combined 1-DE or SDS-PAGE separation, immunoblotting, and N-terminal sequencing to characterize all visible electrophoretic bands. This was the first report of direct identification of human milk proteins, even if it employed a less powerful method of protein separation compared with 2-DE. The same group also applied this strategy to identify the major components of human MFGMP ***(13)***.

N-terminal sequencing combined with other techniques was used for protein identification after 2-DE by Murakami et al. in 1998 ***(14)***. This study compared minor components of human whey colostrum with those of mature milk. Owing to the broad dynamic range of distribution of milk proteins, immunoabsorption with the three major whey proteins, α-lactalbumin, lactoferrin, and secretory IgA, was used to highlight the less expressed proteins. Since proteins were identified by N-terminal sequencing, those found blocked at the N-terminal were digested with Lys-C endopeptidase, and the corresponding peptides were separated by C-8 reversed-phase high-performance liquid chromatography (RP-HPLC). Some of the manually collected fractions were sequenced to gain information on the internal amino acid sequence.

In 2001, Quaranta et al. ***(15)*** published a map of colostral MFGMP obtained by a double-extraction protocol using SDS and urea/thiourea/CHAPS. Proteins were identified by both N-terminal sequencing and mass spectrometric analysis. This was the first report of the human milk proteome in which the peptide mass fingerprint with matrix-assisted laser desorption/ionization-time of flight (MALDI-TOF) peptide analysis was used for protein identification. In 2002, the same group applied a zoom-in gel and a combined mass spectrometric approach (MALDI-TOF and electrospray ionization-tandem mass spectrometry [nanoESI-MS/MS]) and evaluated the expression of proteins of the butyrophilin family ***(16)***. At the same time, Charlwood and co-workers ***(17)*** published a proteomic study on MFGMP characterization, in which they confirmed the presence of seven major components and performed the first glycomic study on the major glycosylated human MFGMP. They were also able to analyze the structure/composition of *N*-linked sugars after in-gel enzymatic release of glycans and their derivatization with 3-(acetylamino)-6-aminoacridine using a hybrid mass spectrometer (MALDI-Q-TOF).

The first annotated database of human colostral MFGMP separated by 2-DE was published by Fortunato et al. in 2003 ***(18)***. The list of 107 protein spots is available at http://www.csaapz.to.cnr.it/proteoma/2DE. A different approach to monitoring the presence of potential biologically active peptides from human casein was applied by Ferranti et al. ***(19)***, who used liquid chromatography-mass spectrometry (LC-MS) on both insoluble and soluble 10% trichloroacetic acid (TCA) fractions to separate relatively small peptides that are difficult to separate by 2-DE. Their position in the amino acid sequence was assigned by comparing experimental and theoretical molecular masses deduced from casein sequences. Because they performed LC-ESI-MS using a single quadrupole to achieve peptide structure identification, each manually collected HPLC fraction was submitted to MS/MS using either a MALDI-TOF analysis in PSD mode or an ESI-Q-TOF. Although the LC/LC-MS/MS approach is sometimes proposed as an alternative method to the classical 2-DE-MS procedure, this is still today the only paper reporting on the LC strategy for human milk proteome analysis.

3. Findings and Applications

3.1. Whey Proteins

Anderson et al. ***(7)*** reported on the first separation of human milk proteins by 2-DE with identification (expression proteomics) of the major components: α-lactalbumin, lactoferrin, albumin, and transferrin. They also made a comparison between colostrum and mature milk, thus representing the first attempt at functional proteomics, although the term "proteomics" was only introduced 12 yr later, and no sequencing or MS was used. Using the powerful 2-DE separation method, the same group tried to resolve the complexity of milk samples, but this was still during the initial developmental phase for 2-DE. Goldfarb et al. ***(8)*** produced the 2-DE map of whole human skimmed milk (without separating casein from whey) at 1 mo post partum; 34 proteins were identified. The main spots corresponded to immunoglobulin, caseins, lactoferrin, albumin, α-lactalbumin, and lysozyme. 2-DE separation was applied to 125 samples in an attempt to determine the significance of longitudinal, diurnal, and individual protein pattern variations, and some indication of the poor correlation between the milk protein fraction and the mother's diet emerged. Kim and Jimenez-Flores ***(20)*** tried to obtain comparative 2D patterns of the milk proteins of different mammals in the ambitious framework of the development of transgenic animals with altered milk composition.

Improvements in protein separation technique have led to the feasibility of analyses of a more complex mixture of proteins in biological samples. Only in 1998 did the first high-resolution 2-DE separation of human milk proteins appear, with the introduction of the Immobiline strip gel; thus 2-DE maps with several hundred well-resolved spots were obtained ***(14)***. After removal of the three major proteins (α-lactalbumin, lactoferrin, and secretory IgA), minor whey proteins were analyzed, and about 400 spots were detected. Minor whey proteins were investigated to find possible correlations between the stage of lactation (colostrum and mature milk) and the presence of enzymes, immunomodulators, and growth modulator factors. In an analysis of samples from 22 mothers, no major difference in 2-DE patterns was observed between colostrum and mature milk whey proteins. Despite the dramatic decrease in the major proteins (lactoferrin and secretory IgA) in mature milk compared with colostrum, these results were the first evidence that the concentrations of minor whey proteins remain relatively constant throughout lactation, suggesting that they may play an important role in the health and development of breastfed infants. The study also confirmed that several fragments of milk proteins exist naturally, thus contributing to integrating the biological activity of milk proteins.

3.2. Caseins

Since 1-DE revealed human casein as a heterogeneous group of proteins, the 2-DE approach was employed to improve resolution. A study by Anderson et al. *(7)* showed that human casein has many more components than was previously thought (at least 14 spots, 9 ascribed to posttranslational modifications of β-casein and 5 identified as α-casein). Nevertheless, the posttranslational modifications involved were not identified. In 1999 Goldfarb ***(10)*** described an additional 2-DE pattern of human casein coupled with quantitation by computer imaging. In this case, the human casein profile was completed with the migration position of κ-casein, para-κ-casein, casomorphins, and γ-casein, determined by immunoblotting. Many bioactive milk peptides are derived from casein degradation; among them, casomorphin originates from α-casein and opioid peptides from β-casein. Recently, Ferranti et al. ***(19)*** carried out a structural analysis by LC-ESI-MS to define the pattern of casein fragments present in human milk that are usually detected in both precipitate and soluble forms. The presence of degraded caseins could be an advantage for the immature digestive system of the newborn, in which milk protein adsorption is facilitated by the presence of shorter peptides. They found that the action of a plasmin-like enzyme was the primary step in casein degradation, followed by endopeptidases and/or exopeptidases.

3.3. MFGMP

The 2-DE separation of whole milk generally fails to resolve the membrane proteins of MFGM, since they are only 1 to 4% of total human milk protein. In this case, fractionation of MFGM and the subcellular proteomic approach are recommended ***(21)*** to unravel the protein organization and structure of human milk fat globules. Besides the typical MFGMP detected in classical biochemical studies ***(22)***, which include xanthine oxidase, butyrophilin, and fatty acid-binding protein, the better resolution of 2-DE, as described by Goldfarb ***(9)***, brought about the detection of spots corresponding to IgM heavy chain, IgA heavychain, secretory piece and J chain, actin, albumin, HLA class I heavy chains, α-1-acidic protein, apolipoprotein H, apolipoprotein A-I, and apolipoprotein E. The patterns of apolipoprotein E from different milk samples were more complex than those presented in human plasma, probably reflecting different sources of synthesis (liver, macrophages, smooth muscle cells, brain, and so on.) ***(9)***. Some years later, in a typical expression proteomics study, Quaranta et al. ***(15)*** proposed a new method for extraction of the MFGMP and presented a 2-DE map with approximately 150 spots and a total of 23 proteins identified. The main spots corresponded to lactadherin, adipophilin, butyrophilin, and carbonic anhydrase; the latter has not previously been described in association with MFGM.

Other newly identified MFGMPs were disulphide isomerase and clusterin. Despite their low nutritional value, MFGMPs play important roles in protecting the breastfed infant ***(21)***. Lactadherin inhibits rotavirus binding and infectivity. Clusterin acts as an inhibitor of the cytolytic activity of the complement. Butyrophilin has some immunologic receptorial activity, but its ligand has not yet been discovered. Human butyrophilin expression was evaluated by Cavaletto et al. ***(16)*** using a comparative proteomic approach between colostral and mature milk. They found 13 multiple forms of butyrophilin and one spot corresponding to a new butyrophilin-like protein (BTN2A1). This study employed proteomics to search the protein complement of a genome, in particular, the presence of seven human butyrophilin transcripts known at the mRNA level, two of which were identified at the protein level to be associated with MFGM.

The MFGM glycoproteins maintain their original structure and function even in the acidic environment of the infant's stomach, as a result of their glycol moiety. Charlwood et al. ***(17)*** employed proteomic tools (MALDI-Q-TOF) to confirm the identity of the most abundant proteins and to characterize the sequence of *N*-linked glycans of clusterin, lactoferrin, polymeric Ig receptor, and lactadherin. The presence of multiple fucosylation was evidenced, and this may be a mechanism whereby mothers provide their infants with protective factors during the neonatal stage of development. Magi et al. ***(11)*** evaluated the presence of a cytokine, macrophage migration inhibitory factor (MIF), associated with MFGM and speculated that MIF may affect development of the infant immune system and play a role in preserving functionality of the mammary gland, as a result of its proinflammatory features. Combining data produced by different separation techniques and using specific monoclonal and polyclonal antibodies, this group confirmed the localization of MIF in a gap between the core fat and the surrounding membrane. Entrapment within the fat globules might protect the cytokine from degradation during transit in the gastrointestinal tracts.

The first annotated 2-DE database of human colostral MFGMP appeared in the paper by Fortunato et al. ***(18)*** and can be found at http://www.csaapz.to.cnr.it/proteoma/2DE. **Figure 2** shows the two 2-DE maps (p*I* 3–10 and 4–7) available on the website, and the identified spots can be retrieved by clicking on the marked cross. The 107 proteins annotated are derived from only 39 genes, since many identified spots correspond to multiple forms of the same protein owing to posttranslational modifications. Taking into account that milk is the secretion product of the mammary gland epithelial cells, its protein variety is much less abundant than that of a typical cell. About 60% of the identified proteins are typical MFGMP or mammary gland-secreted proteins, 10% are involved in protein folding and destination, 9% are involved in intracellular transport and receptorial

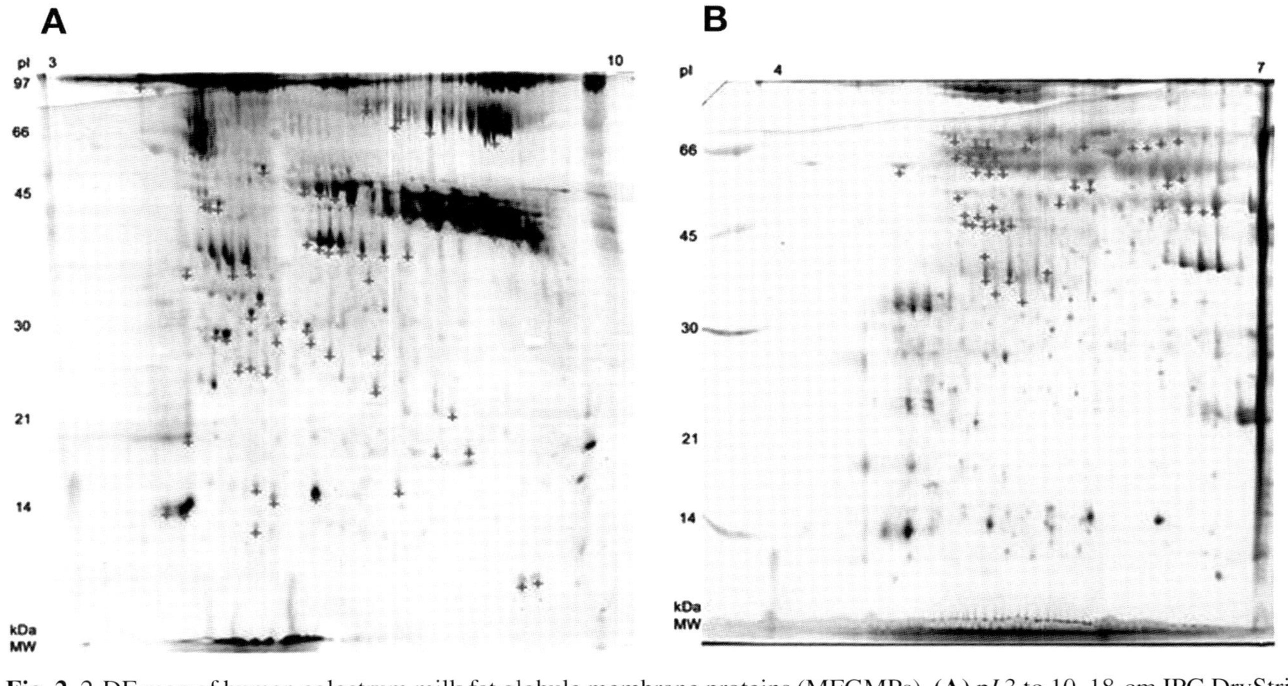

Fig. 2. 2-DE map of human colostrum milk fat globule membrane proteins (MFGMPs). **(A)** p*I* 3 to 10, 18-cm IPG DryStrips (Amersham Biosciences). **(B)** p*I* 4 to 7, 18-cm IPG DryStrips (Amersham Biosciences). 2D gels were both from 11.7% SDS-PAGE.

activities, and the remaining minor proteins belong to signal transduction system, glutathione metabolism, and complement complex. **Table 3** shows the list of all proteins identified in human milk using proteomic tools.

4. Perspectives and Future Directions

The proteomic approach, by combining high-resolution separation techniques and powerful mass spectrometric analysis, allows previously unattainable information to be acquired. Within the major areas of today's proteomics, such as protein detection and identification, studies on posttranslational modifications and expression analysis of the human milk proteome will be beneficial for both the lactating mother and the breastfed child. One problem with mother's milk is that it may contain allergens that have been taken by the mother. Allergens derived from cow milk and/or other allergenic sources may appear in the mother's milk and sensitize her child. The circulation of foreign proteins through a mother's body through the breast into the milk, and hence into the infant's gastrointestinal tracts and body, is a remarkable biological fact. However, this free passage of intact food proteins through many body filters and defense systems is still the subject of debate.

Since food allergens from the mother's diet may appear in her breast milk, the lactating mother may have to modify her diet to protect her infant. Restrictions may include avoidance of milk products and other highly allergenic foods like eggs, peanuts, citrus fruit, chocolate, nuts, and in some cases cereal grains, certain meats, and fish. Lactating mothers should, consequently, follow a diet that is problematic from both the psychological and the physiological standpoint. Clear identification and measurement of heterologous proteins and peptides in breast milk following a controlled maternal diet could indicate which harmful proteins and/or peptides are actually transferred to breast milk, suggesting a less problematic diet for the mother that can still protect the breastfed infant.

The quality of infant formulae is another field that will benefit from proteomic studies on human milk and on some of its possible substitutes such as bovine or equine milk. The debate over breast vs bottle has long been closed. The American Academy of Pediatrics' policy on breastfeeding states that "human milk is the preferred feeding for all infants, including premature and sick newborns, with rare exceptions" *(23)*. The question now becomes "how can we ensure the healthiest start for those infants whose mothers cannot, should not, or will not breastfeed?" Infant formula is an industrially produced milk product designed for infant consumption. Usually based on either cow or soy milk, infant formula strives to duplicate the nutrient contents of human breast milk. Since the exact chemical compositions and properties of breast milk are still unknown, "formulae" are, at present, an imperfect approximation. As scientists learn which bioactive compounds in human milk are responsible

Table 3
Proteins in Human Milk Identified With Proteomic Tools

Protein	Lactation stage	Milk fraction	Identification method and relative reference
α-Lactalbumin	C, M	Whey, skim, C, MFG	EF *(7)*; I *(8)*; S *(12–15)*; PMF *(15,18)*; MS/MS *(17)*
Actin	M	MFG	I *(9)*; PMF *(18)*
Adipophilin	C	MFG	PMF *(15,18)*
Albumin	C, M	Whey, skim, MFG	EF *(7)*; I *(8,9)*; S *(14)*; PMF *(18)*
α-S1 casein	C, M	Whey, C, skim	S *(12,14)*; MS/MS *(19)*
α_1-acid glycoprotein	M	Skim, MFG	I *(8,9)*
α_1-antichymotrypsin	M	Skim	I *(8)*
α_1-antitrypsin	M, C	Skim, whey	I *(8)*; S *(14)*
α_2-HS glycoprotein	M	Skim	I *(8)*
α_2-macroglobulin	M	Skim	I *(8)*
α-casein	C, M	C, skim, MFG	EF *(7)*; I *(8)*; PMF *(15)*
Apolipoprotein A-1	M	Skim, MFG	I *(8,9)*; PMF *(18)*
Apolipoprotein A-2	M	Skim, MFG	I *(8,9)*
Apolipoprotein A-4	C	MFG	PMF *(18)*
Apolipoprotein C1	C	MFG	PMF *(18)*; MS/MS *(18)*
Apolipoprotein E	M	MFG	I *(9)*; PMF *(18)*; MS/MS *(18)*
Apolipoprotein H	M	MFG	I *(9)*
β_2-microglubulin	M, C	Skim, whey	I *(8)*; S *(14)*
β-casein	C, M	C, skim, whey, MFG	EF *(7)*; I *(8,10)*; S *(12,14,15)*; PMF *(18)*; MS/MS *(17–19)*
Breast cancer suppressor 1	C	MFG	PMF *(18)*
Butyrophilin BTN	M, C	MFG	I *(9)*; S *(13,15)*; PMF *(16,18)*; MS/MS *(16,18)*
Butyrophilin BTN2A1	C, M	MFG	PMF *(16,18)*; MS/MS *(16)*
Carbonic anhydrase	C	MFG	PMF *(15,18)*
Cargo protein TIP47	C, M	MFG	PMF *(16,18)*
Casein	M	MFG	I *(9)*

(Continued)

Table 3 *(Continued)*

Protein	Lactation stage	Milk fraction	Identification method and relative reference
Casomorphin	M	Skim	I ***(10)***
CD59 glycoprotein	C	MFG	PMF ***(18)***; MS/MS ***(18)***
Clusterin	C, M	Whey, MFG	S ***(14,15)***; PMF ***(15,17,18)***; MS/MS ***(17)***
Complement C4 γ-chain	C, M	Whey, MFG	S ***(14)***; PMF ***(18)***
CRABP II (P29373)	C	MFG	PMF ***(18)***; MS/MS ***(18)***
Disulfide isomerase	C	MFG	S ***(15)***; PMF ***(18)***
Endoplasmin	C	MFG	PMF ***(18)***
Fatty acid binding protein	C, M	Whey, MFG	S ***(14)***; I ***(9)***; PMF ***(18)***
Fibrinogen	C	MFG	PMF ***(18)***
Folate binding protein	C	MFG	PMF ***(18)***
Fructose-biphosphate aldolase A	C, M	Whey	S ***(14)***
G protein SAR1b (Q9Y6B6)	C	MFG	PMF ***(18)***
γ-casein	M	Skim	I ***(10)***
γ-glutamyl transferase	M, C	Skim, MFG	I ***(8)***; PMF ***(18)***
Gc globulin	M	Skim	I ***(8)***
Growth hormone	M	Skim	I ***(8)***
GRP 78 (P11021)	C	MFG	PMF ***(18)***
GSHH (P36969)	C	MFG	PMF ***(18)***; MS/MS ***(18)***
GTP binding protein	C	MFG	PMF ***(18)***
Heme binding protein	C	MFG	PMF ***(18)***
HLA class I	M	MFG	I ***(9)***
HS7C (P11142)	C	MFG	PMF ***(18)***
Immunoglobulin A	C, M	Milk, skim, MFG	EF ***(7)***; I ***(8,9)***; S ***(14)***
Immunoglobulin D	M	Skim	I ***(8)***
Immunoglobulin E	M	Skim	I ***(8)***

Immunoglobulin G	M	Skim	I *(8)*
Immunoglobulin M	M	Skim, MFG	I *(8,9)*
J chain	M	Skim, MFG	I *(8,9)*
κ-casein	M	C, skim	S *(12)*; I *(10)*; MS/MS *(19)*
Keratin type II	C	MFG	PMF *(18)*
Lactadherin	C, M	MFG	S *(13)*; PMF *(15,17,18)*; MS/MS *(17)*
Lactoferrin	C, M	Whey, skim, C, MFG	EF *(7)*; I *(8)*; S *(12)*; PMF *(17,18)*; MS/MS *(17)*
Lysozyme	M, C	Skim, whey, C, MFG	I *(8)*; S *(12,14)*; MS/MS *(17)*
Migration inhibitor factor (MIF)	M	Milk, MFG	I *(11)*
Para-κ-casein	M	Skim	I *(10)*
Plasminogen	M	Skim	I *(8)*
Poly Ig receptor	C, M	Whey, MFG	S *(14)*; PMF *(16,18)*; MS/MS *(17)*
Prealbumin	M, C	Skim, whey	I *(8)*; S *(14)*
Retinol binding protein	M	Skim	I *(8)*
Rotamase	C	MFG	PMF *(18)*
Secretory piece	M	Skim, MFG	I *(8,9)*
Selenium binding protein	C	MFG	PMF *(18)*
Transferrin	C, M	Milk, skim	EF *(7)*; I *(8)*; S *(14)*
Transforming protein RhoA	C	MFG	PMF *(18)*
UDP-galactosyltransferase	M	Skim	I *(8)*
WNT-2B protein (Q93097)	C	MFG	PMF *(18)*
Xanthine oxidase	M, C	MFG	I *(9)*; S *(13)*
Zn α_2-glycoprotein	M	Skim	I *(8)*

Abbreviations: *Lactation stage:* C, colostrum; M, mature milk; *milk fraction*: milk, whole milk; W, whey; C, casein; skim, skimmed milk; MFG, milk fat globule associated protein; *Identification method*: EM, electrophoretic mobility; I, immunoblotting; S, N-terminal sequencing; PMF, peptide mass fingerprint; MS/MS, tandem mass spectrometry.

for its health benefits, formula companies try to adapt their products accordingly. All cow milk-based formulae contain totally defatted milk, thus losing the fraction of proteins associated with the MFGM. The next step for improving infant formulae will probably be the addition of some of these proteins or of the whole fat fraction from the milk (i.e., equine milk) that is closer to human milk than cow milk. How many breast milk proteins will need to be isolated, purified, sequenced, and completely characterized before infant formula begins to approach human milk in quality? Developing technologies, including specific fluorescent labeling of 2-DE-separated proteins to detect posttranslational modifications like glycosylation and phosphorylation, or the very promising (although still in its infancy) Protein Chip proteomics, could help to answer these questions within a few years.

References

1. Lönnerdal B, Atkinson S. Nitrogenous components of milk. A. Human milk proteins. In: Jensen RG, ed. *Handbook of Milk Composition.* San Diego; Academic, 1995;351–368.
2. Malacarne M, Martuzzi F, Summer A, Mariani P. Protein and fat composition of mare's milk: some nutritional remarks with reference to human and cow's milk. *Int Dairy J* 2002;12:869–877.
3. Lönnerdal B. Nutritional and physiologic significance of human milk proteins. *Am J Clin Nutr* 2003;77(Suppl):1537S–1543S.
4. Hinrichs J. Mediterranean milk and milk products. *Eur J Nutr* 2004;43 (Suppl 1): 1/12–1/17.
5. Bernt KM, Walker WA. Human milk as a carrier of biochemical messages. *Acta Pediatr* 1999;Suppl 430:27–41.
6. O'Donnell R, Holland JW, Deeth HC, Alewood P. Milk proteomics. *Int Dairy J* 2004;14:1013–1023.
7. Anderson NG, Power MT, Tollaksen SL. Proteins of human milk. I. Identification of major components. *Clin Chem* 1982;28:1045–1055.
8. Goldfarb M, Savadove MS, Inman J. Two-dimensional electrophoretic analysis of human milk proteins. *Electrophoresis* 1989;10:67–70.
9. Goldfarb M. Two-dimensional electrophoretic analysis of human milk-fat-globule membrane proteins with attention to apolipoprotein E patterns. *Electrophoresis* 1997;18:511–515.
10. Goldfarb M. Two-dimensional electrophoresis and computer imaging: quantitation of human milk casein. *Electrophoresis* 1999;20:870–874.
11. Magi B, Ietta F, Romagnoli R, et al. Presence of macrophage migration inhibitory factor in human milk: evidence in the acqueous phase and milk fat globules. *Pediatr Res* 2002;51:619–623.
12. Cavaletto M, Cantisani A, Napolitano L, et al. Comparative study of casein content in human colostrum and milk. *Milchwissenschaft* 1994;49:303–305.

13. Cavaletto M, Giuffrida MG, Giunta C, et al. Multiple forms of lactadherin (breast antigen BA46) and butyrophilin are secreted into human milk as major components of milk fat globule membrane. *J Dairy Res* 1999;66:295–301.
14. Murakami K, Lagarde M, Yuki Y. Identification of minor proteins of human colostrum and mature milk by two-dimensional electrophoresis. *Electrophoresis* 1998;19:2521–2525.
15. Quaranta S, Giuffrida MG, Cavaletto M, et al. Human proteome enhancement: high-recovery method and improved two-dimensional map of colostarl fat globules membrane proteins. *Electrophoresis* 2001;22:1810–1818.
16. Cavaletto M, Giuffrida MG, Fortunato D, et al. A proteomic approach to evaluate the butyrophilin gene family expression in human milk fat globule membrane. *Proteomics* 2002;2:850–856.
17. Charlwood J, Hanrahan RT, Langridge J, Dwek M, Camilleri P. Use of proteomic methodology for the characterization of human milk fat globular membrane proteins. *Anal Biochem* 2002;301:314–324.
18. Fortunato D, Giuffrida MG, Cavaletto M, et al. Structural proteome of human colostral fat globule membrane proteins. *Proteomics* 2003;3:897–905.
19. Ferranti P, Traisci MV, Picariello G, et al. Casein proteolysis in human milk: tracing the pattern of casein breakdown and the formation of potential bioactive peptides. *J Dairy Res* 2004;71:74–87.
20. Kim HY, Jemenez-Flores R. Comparison of milk proteins using preparative isoeletric focusing followed by polyacrylamide gel electrophoresis. *J Dairy Sci* 1994;77:2177–2190.
21. Cavaletto M, Giuffrida MG, Conti A. The proteomic approach to analysis of human milk fat globule mambrane. *Clin Chim Acta* 2004;347:41–48.
22. Keenan TW, Patton S. The structure of milk implication for sampling and storage. The milk lipid globule membrane. In: Jensen RG, ed. *Handbook of Milk Composition.* San Diego: Academic, 1995;5–62.
23. AAP Policy. Breastfeeding and the Use of Human Milk (RE9729.) *Pediatrics* 1997;100:1035–1039.

21

Proteomics of Nipple Aspirate Fluid in Nonlactating Women

Edward R. Sauter

Summary

Nipple aspirate fluid (NAF) is collected noninvasively using breast massage and a modified breast pump. The fluid contains concentrated proteins from the ductal and lobular epithelium, the source of 99% of breast cancers. The analysis of NAF has identified a number of proteins associated with the presence of breast cancer, none of which is sufficiently sensitive and specific when analyzed by itself to be clinically useful. The assessment of multiple NAF markers provides great promise in using this body fluid for early breast cancer detection. In this review, the strengths and weakness of various approaches to protein assessment in NAF are discussed.

Key Words: Nipple aspirate fluid; enzyme-linked immunosorbent assay; radioimmunoassay; surface-enhanced laser desorption/ionization time-of-flight mass spectrometry; polyacrylamide gel electrophoresis.

1. Introduction

Proteins in nipple aspirate fluid (NAF) can be analyzed one at a time, in groups of two or more, or on a proteome-wide basis. As with other biological markers of cancer, which might have been anticipated owing to the heterogeneity of breast cancer, we have yet to find a single marker in NAF that is adequately sensitive and/or specific to be clinically useful. As a result, there is significant interest in looking at multiple markers or screening the NAF proteome to determine the optimally sensitive and specific panel of markers to screen for new or recurrent disease.

In the following sections, analyses of NAF combining two or more proteins, as well as proteome-wide approaches, are discussed, with a focus on secreted NAF proteins, since the limited and mixed cellularity of the samples makes

From: *Proteomics of Human Body Fluids: Principles, Methods, and Applications*
Edited by: V. Thongboonkerd © Humana Press Inc., Totowa, NJ

cellular protein analysis unreliable. The methodologies involved as well as the results provide insight into the promise that analysis of this body fluid holds.

2. Methods

2.1. Collection of NAF

One of the common causes of inability to collect NAF is inadequate subject preparation. Fluid is always present in the nipple ducts, but it does not spontaneously drain owing to the nipple sphincter and keratin plugs present in the nipple. If the subject is anxious or in a rush, the sphincter does not relax, with the result that it is less likely that the individual performing nipple aspiration will collect NAF. It is therefore important to put the subject at ease and not attempt to perform the procedure if the subject is pressed for time. The entire procedure takes approximately 30 min, much of which is involved with patient consent, the collection of the subject's health history, changing of clothes, and so on. We warm the breasts prior to aspiration for 2 to 5 min. The aspiration itself takes only seconds.

The nipple is cleansed with alcohol *(1)*. A warm, moist cloth is placed on the breast after the alcohol evaporates. The cloth is removed after 2 min, and the subject massages her breast with both hands while the aspiration device (a 10-mL syringe attached to no. 4 endotracheal tube cap) is withdrawn to the 7-mL level or until she experiences discomfort. Aspiration is repeated on the opposite breast, if present. Fluid in the form of droplets (1–200 μL) is collected in capillary tubes, and the samples are immediately frozen at –80°C. If keratin plugs rather than NAF are obtained after suctioning is completed, the plugs are removed with an alcohol swab and suctioning is repeated. Occasionally, this procedure must be repeated two or three times to remove all the plugs before NAF can be collected.

2.2. Analysis of a Select Number of Proteins

A major strength of selecting one or a few proteins for analysis is the ability to quantitate the amount of each protein present within the sample, which is less feasible when one is doing protein-wide screening. The number of proteins that one can analyze, short of screening the entire NAF proteome, is limited by the amount of NAF sample that one has available. Methodologies to circumvent this limitation involve the development of more sensitive assays, including the use of multiplex analysis. Standard methods of analysis include radioimmunoassays (RIAs) and enzyme-linked immunosorbent assays (ELISAs). Although multiplex protein analysis has been performed using murine serum *(2)* and human bronchoalveolar lavage fluid *(3)*, a recent search of the literature failed to identify a publication using multiplex protein analysis of NAF samples. For this reason, the discussion focuses on RIA and ELISA single-protein analysis methods.

2.2.1. RIA

1. NAF samples are diluted in buffer and incubated with the primary antibody (generally for 2–6 h at room temperature or overnight at 4°C); incubation is terminated by adding dextran-coated charcoal ***(4)***.
2. The samples are centrifuged.
3. The supernatant is dissolved in liquid scintillation fluid, and tritium activity is determined with a liquid scintillation spectrometer. Samples are generally run in duplicate.
4. A linear regression equation is created from standards of known primary antibody concentration; the protein concentrations of unknown samples are fitted to a standard curve regression equation, corrected for aliquot volume.

2.2.2. ELISA

NAF samples are analyzed using a monoclonal or polyclonal antibody to bind competitively with the protein of interest in the standard or sample.

1. Samples are diluted in assay buffer, pipeted into appropriate wells, incubated (generally for 2–6 h at room temperature or overnight at 4°C), and washed.
2. Substrate solution is added followed by incubation.
3. The absorbance is then measured at the appropriate wavelength for detection using a microtiter plate reader.

2.3. Proteome-Wide Approaches

2.3.1. 1- and 2D Polyacrylamide Gel Electrophoresis (1D and 2D-PAGE)

2.3.1.1. Sample Preparation

1. NAF samples for 1D and 2D-PAGE analysis are thawed by breaking the capillary tube directly into denaturing buffer ***(5)***.
2. The proteins are then reduced.
3. Pharmalytes® isoelectric focusing carrier ampholytes pH 3 to 10 (Sigma-Aldrich, St. Louis, MO) are added, the samples are centrifuged, and the supernatants are then kept.

2.3.1.2. Protein Separation

1. The diluted sample (250 or 400 μg) is used to run on 11- or 24-cm strips, respectively ***(5)***.
2. 1D separation can be performed by focusing the proteins at a total of 80,000 v/h, with a 6000-V programmable power supply (Proteome Systems, Boston, MA).
3. 2D separation can be conducted by PAGE on 8 to 18% gradient gel chips (Proteome Systems).
4. Gels are stained overnight using colloidal Coomassie Brilliant Blue, destained in 1% acetic acid, and scanned.

2.3.1.3. Proteomic Analysis

1. Electronic images of 1D and 2D gels can be analyzed using Phoretix Advanced software (Nonlinear Dynamics, Newcastle, UK).
2. Protein spots of interest are excised, destained, and dehydrated.
3. Proteins are digested with trypsin (Promega, Madison, WI). Tryptic peptides are extracted from the gel plugs and concentrated by centrifugal vacuum evaporation.
4. Tryptic peptide samples (0.5 μL) are analyzed by matrix-assisted laser desorption/ionization time-of-flight mass spectrometry (MALDI-TOF-MS), operating in the positive ion delayed extraction reflector mode.
5. The closed external calibration method employing a mixture of standard peptides (Applied Biosystems) provides a mass accuracy of 25 to 50 ppm across the mass range of 600 to 5000 Daltons.
6. Peptide spectra can be automatically processed for baseline correction, noise removal, and peak deisotoping and can be analyzed using Protein Prospector (http://prospector.ucsf.edu). Search criteria are set up to require a match of a minimum number of peptides with a maximum mass error of, for example, less than 50 ppm for a protein assignment. The data are manually reexamined to ensure maximum peptide coverage for the identified proteins.

2.3.2. Surface-Enhanced Laser Desorption/Ionization Time-of-Flight Mass Spectrometry (SELDI-TOF-MS)

2.3.2.1. Sample Preparation

1. The portion of the capillary containing NAF is introduced into a 1.7-mL Eppendorf tube containing 100 μL of 100 m*M* Tris-HCl, pH 8.0.
2. The capillary is then crushed using a glass rod, and the mixture is vortexed to disperse the sample.
3. The mixture is then centrifuged at 14,000*g* for 5 min; the supernatant can be used without further dilution.
4. Protein concentrations are determined on diluted NAF samples using the Pierce BCA (Pierce Chemical, Rockford, IL) or similar kit and then further diluted to a final total protein concentration of 3.6 mg/mL.
5. Samples are coded so that the research assistant performing SELDI-TOF-MS (Ciphergen Biosystems, Fremont, CA) analysis is blinded.

2.3.2.2. Protein Separation

The spectrum of proteins of a given mass/charge ratio can be entered into the software so that only proteins meeting the defined criteria are identified. For NAF, it is not generally possible to screen proteins of all sizes from a single run. The options, therefore, are to select midsize proteins (e.g., 5–60 kDa) or to do two runs, one for smaller proteins (approx 1–40 kDa) and the second for larger proteins (approx 40–250 kDa). A variety of approaches are used to pick the peaks of interest, based on comparing samples from normal subjects with samples from subjects with disease.

A variety of NAF pretreatment procedures have been tested to remove abundant proteins such as albumin, hemoglobin, and immunoglobulins, which tend to mask proteins of lower abundance. Although pretreatment protocols appear to remove the undesirable components, they also remove unknown amounts of protein from the samples ***(6)***. This results in spectra of less than optimal quality. A number of investigators ***(6,7)*** have found that the optimal protocol for SELDI-TOF-MS analysis consists of treating samples with 9.5 *M* urea containing protease inhibitors.

SELDI-TOF-MS employs various chips with different affinities (negatively charged proteins—strong anion exchange or [SAX]; hydrophobic proteins—H4; cationic proteins—WCX, and so forth) to improve the detection of lower abundance proteins. Only 1 μL of NAF sample is sufficient for SELDI-TOF analysis, whereas 3 to 5 μL of NAF is generally required for 2D-PAGE.

2.3.2.3. Proteomic Analysis

A specific protein of known size detected by SELDI-TOF-MS ***(6)*** can be isolated by 1D-PAGE. The Coomassie Blue-stained band is excised from the gel and digested with trypsin using an in-gel digestion procedure. The tryptic peptide masses are then determined by SELDI-TOF, and the protein is identified by comparing these peptide fragments with those present in the virtual tryptic digest prepared from known protein databases and the translated nucleic acid databases using the ProFound program (http://129.85.19.192/profound_bin/WebProFound.exe) ***(8)***.

3. Findings and Applications

Anatomically, the breast is comprised of ducts and lobules, surrounded by supporting adipose and connective tissues. During the immediate postpartum lactation period, the breast glands actively secrete milk into the ducts for the nurture of the newborn infant, but it has long been recognized from histologic studies that the nonpregnant breast also secretes small amounts of fluid containing sloughed epithelial and other cells. The epithelial cells, which line the ducts and lobules, are at risk for malignant degeneration and are the origin of 99% of breast cancers ***(9)***.

Early detection is a major factor contributing to the steady decline in breast cancer death rates, with a 3.2% annual decline over the past 5 yr ***(10)***. Unfortunately, currently available breast cancer screening tools, such as mammography and breast examination, miss up to 40% of early breast cancers and are least effective in detecting cancer in young women, whose tumors are often more aggressive. Thus, there has long been interest in developing a noninvasive method to determine whether a woman has breast cancer.

Proteomic analysis of body fluids holds significant promise to allow noninvasive detection of disease. The focus of this chapter is on nipple aspirate fluid, a fluid present in the breast ducts and lobules, which can be collected using breast massage and a modified breast pump. The major strength of NAF, in addition to the fact that it can be collected noninvasively, is that the proteins secreted into sample are frequently more concentrated compared with blood proteins, making their analysis feasible with as little as 1 μL of NAF.

3.1. Studies Evaluating Proteins in NAF

3.1.1. Single-Protein Analysis

3.1.1.1. Hormones and Growth Factors

A variety of hormones and growth factors have been measured in NAF, including estrogens, androgens, progesterone, dehydroepiandrosterone sulfate, prolactin, growth hormone, epidermal growth factor, transforming growth factor-α, vascular endothelial growth factor, and basic fibroblast growth factor ***(11–14)***. Elevated levels of estrogens, cholesterol, and cholesterol epoxides have been suggested to have etiologic significance in breast disease ***(15)***. Levels of a number of these factors have been compared with disease risk. With the exception of recent parity, no relation has been found between levels of estrogen in NAF and breast cancer risk. Higher levels of estradiol and estrone have been found in the NAF of women with benign breast disease compared with those in normal controls ***(16)***. There is a decrease in estradiol and estrone levels in NAF following pregnancy or lactation that persists for several years before returning to prepregnancy levels ***(17)***. This period of decreased estrogen exposure of the breast epithelium of postpartum women has been suggested to explain partially the breast cancer protective effect of early pregnancy.

Basic fibroblast growth factor (bFGF) is an important angiogenic factor that stimulates tumor growth ***(18,19)***. A preliminary report, which analyzed 10 breast cancer patients and 10 controls, found that bFGF levels in NAF were higher in women with breast cancer than in normal subjects ***(20)***. Leptin is a hormone that plays a central role in food intake and energy expenditure ***(21)***. Systemic levels of leptin are increased in obese individuals and have been found to stimulate the growth of breast cancer cells in vitro. Leptin levels in NAF are more readily measured in postmenopausal than in premenopausal women and are significantly higher in postmenopausal than in premenopausal women with a body mass index (BMI) < 25 ***(22)***. Although NAF leptin levels are not associated with premenopausal or postmenopausal breast cancer, they are associated with premenopausal BMI.

3.1.1.2. Tumor Antigens

A number of proteins present in NAF have previously been associated with cancer using measurements of the marker in the blood. Two of these are prostate-specific antigen (PSA) and carcinoembryonic antigen (CEA). PSA, a chymotrypsin-like protease first found in seminal fluid and associated with prostate cancer ***(23)***, is also found in breast tissue ***(14,24)*** and in NAF. PSA levels in cancerous breast tissue are lower than in benign breast tissue ***(14)***. Most ***(14,25,26)*** but not all ***(27)*** studies indicate that low NAF PSA levels are associated with the presence and progression ***(28)*** of breast cancer. One explanation for the discrepancy in PSA results may be the difference in NAF yield, which was 97% of subjects in the studies finding an association and 34% in the study in which an association between NAF PSA and breast cancer was not found ***(29)***.

Another protein that is concentrated in NAF is CEA. CEA was identified in 1965 as the first human cancer-associated antigen ***(30)***. Serum CEA levels have been used clinically to assess and monitor tumor burden in patients with breast cancer ***(31)***. CEA titers in NAF samples from normal breasts are typically more than 100-fold higher than in corresponding serum ***(32)***. CEA levels in NAF from 388 women, including 44 women with newly diagnosed invasive breast cancer, have been analyzed. CEA levels are significantly higher in breasts with cancer, but the sensitivity of CEA for cancer detection is only 32% ***(27)***.

3.2. Analyzing Multiple Proteins

Although single NAF protein analysis has identified a number of proteins associated with breast cancer, none is sufficiently sensitive and specific for cancer detection in clinical practice. Investigators have, therefore, attempted to analyze two or more NAF proteins, each predictive of cancer, to assess whether multiple protein analysis would improve the sensitivity and/or specificity of predicting breast cancer in women. As already discussed, both bFGF and PSA levels in NAF have been associated with breast cancer. A study that evaluated 143 NAF specimens ***(11)*** found that mean NAF bFGF levels were significantly higher in women with breast cancer than in those without. A logistic regression model including NAF levels of bFGF and clinical variables showed 90% sensitivity and 69% specificity in predicting breast cancer. Adding another biomarker linked to breast cancer, PSA, increased the sensitivity to 91% and the specificity to 83%.

Attempts to improve the predictive ability of single NAF proteins have also been conducted through the analysis of two ***(33)*** and then multiple human kallikreins (hKs) ***(34)***. hKs 2, 3, 6, and 10 are expressed in breast and prostate tissue. hK2 and hK3 (or PSA) are used to screen for prostate cancer. hK6 and hK10 are downregulated in breast cancer compared with normal breast tissue. In the first study ***(33)***, hK2 and PSA were analyzed for their association with breast cancer. Although each was associated with disease in univariate analysis, hK2 did not improve the ability of PSA to predict whether the subject had breast cancer when multivariate analysis was conducted. The second study ***(34)*** came to a similar conclusion for hK2, hK6, and hK10, with only PSA and menopausal status independently contributing to the model.

Insulin-like growth factor (IGF)-1 is an important growth factor for breast cancer cells, and IGF-binding protein-3 (IGFBP-3) is its most prevalent binding protein. PSA enzymatically cleaves IGFBP-3. We observed that IGFBP-3 levels were directly associated with breast cancer, whereas PSA levels were inversely associated ***(26)***. When considered together, PSA, age, menopausal status, and age at menarche (but not IGFBP-3) were significant predictors of breast cancer ***(26)***.

The plasminogen activator (PA) system helps to control the degradation of the extracellular matrix and basement membrane. Levels of three proteins in this family, including urinary PA (uPA), uPA inhibitor (PAI-1), and uPA receptor (uPAR) in breast cancer tissue have been linked to breast cancer prognosis ***(35)***. Logistic regression analysis of these markers in 120 NAF samples found that uPA, PAI-1, and age were independent predictors of breast cancer, suggesting that combining these biological and clinical markers may better predict cancer than using any single marker.

3.3. Proteomic Analysis

Recent advances in comprehensive molecular technologies have allowed the analysis of global gene expression or protein profiles in cancerous vs normal tissues, with the goal of identifying protein markers that are differentially expressed between benign and malignant tissues. One such study ***(36)*** used serial analysis of gene expression to identify molecular alterations involved in breast cancer progression. The authors concluded that many of the highly expressed genes encoded secreted proteins, which in theory would be present in NAF.

Breast tissue contains thousands of intracellular proteins. NAF contains a limited number of cells and extracellular fluid, the composition of which includes a relatively small set of secreted breast-specific proteins. The few cells in NAF can be separated from the extracellular fluid. The remaining proteins are isolated and therefore represent their final processed form, which makes proteomic analyses less ambiguous and can provide clues to the changes in protein translational rates, posttranslational modification, sequestration, and degradation that lead to disease.

3.3.1. 2D-PAGE

3.3.1.1. Candidate Protein Detection

The traditional method of proteomic analysis is 1D or 2D-PAGE. Using 2D rather than 1D-PAGE allows better separation of proteins of equal molecular weight based on charge. Once a protein of interest is found, it can be cut from the gel and identified. 2D-PAGE has been used to screen NAF because it provides a convenient and rapid method for protein identification based on MALDI-TOF-MS. At least two studies have analyzed the NAF proteome. One used liquid chromatography ***(37)***, and the other used 2D-PAGE ***(38)***. Over 60 proteins were identified in the first and 41 in the second study. Many of the proteins were the same, but a significant subset of proteins (35 in the first and 21 in the second) was unique to each study. Both studies should be considered when one is assessing the NAF proteome.

3.3.1.2. Validation

2D-PAGE can serve as a screening platform to identify proteins in NAF that are differentially expressed in cancerous and benign breasts. These proteins can then be validated using one or more high-throughput proteomic approaches. In one study ***(5)***, three protein spots that were detected using 2D-PAGE were upregulated in three or more NAF samples from breasts with cancer. These spots were identified as gross cystic disease fluid protein (GCDFP)-15, apolipoprotein (Apo)D, and α-1 acid glycoprotein (AAG). To validate these three potential biomarkers, 105 samples (53 from benign breasts and 52 from breasts with cancer) were analyzed using ELISA, a high-throughput method of evaluating protein concentration. Considering all subjects, GCDFP-15 levels were significantly lower and AAG levels were significantly higher in breasts with cancer. This was also true in pre- but not postmenopausal women. GCDFP-15 levels were lowest and AAG levels were highest in women with ductal carcinoma *in situ* (DCIS). Menopausal status influenced GCDFP-15 and AAG more in women without than those with breast cancer. ApoD levels did not correlate significantly with breast cancer. Thus the three proteins which appeared to be overexpressed using a proteomic approach were either overexpressed (AAG), underexpressed (GCDFP-15), or not associated with breast cancer (ApoD) in 52 samples from cancer patients using ELISA, a better quantitative method of protein analysis. This confirms the need to validate preliminary findings in small data sets before drawing firm conclusions.

3.3.2. SELDI-TOF-MS

3.3.2.1. Candidate Protein Detection

Although 2D-PAGE is quite powerful, it has limitations in protein separation and sensitivity. The SELDI-TOF technique can be performed with 1 μL of NAF, can detect components in the high femtomole range, and allows the rapid evaluation of 8 to 24 samples simultaneously.

We are aware of three studies ***(6,39,40)*** that have demonstrated the feasibility of SELDI-TOF analysis of NAF and identified one or more protein mass peaks as associated with breast cancer. A potential limitation of these studies is that specific protein identification was performed only for a subset of the protein masses. A limitation of SELDI-TOF-MS is that protein identification of a peak seen with SELDI-TOF is not as simple as that of a spot visualized using 2D-PAGE. Although it has been proposed that this is not necessary ***(41)***, validation studies to confirm that these protein masses are linked to breast cancer are easiest after identification of the specific proteins, eliminating the confounder of multiple proteins of similar mass.

A concern with SELDI-TOF analysis, perhaps more than with other proteomic approaches, is that the criteria used to identify a "breast cancer-associated" protein may influence which proteins are selected. This fact is brought out by the identification of two tumor-associated (TA) peaks (4233 and 9470 *m/z*) in the first report of SELDI-TOF-MS analysis of NAF ***(40)***, of five TA peaks (6500, 8000, 15,940, 28,100, and 31,700 *m/z*) in the second ***(39)***, and of four TA peaks (5200, 11,880, 13,880, and 33,400 *m/z*) in the third ***(6)***. In the third and largest study (114 women), three clinical variables (age, parity, and presence or absence of spontaneous nipple discharge) in addition to the SELDI-identified protein masses contributed to the optimal predictive model. Although the subject populations differed in the three studies, the lack of common breast cancer-associated proteins remains somewhat surprising.

3.3.2.2. Validation

In addition to the importance of the population studied, the criteria used to choose the proteins, sample preparation, and instrument settings, these three studies demonstrate the role of confounding proteins that must be accounted for. In the third report, the authors discussed the protein identification of the most promising (15,940 *m/z*) of the five protein peaks detected in their earlier publication ***(39)***. They focused on the facts that this protein is the β chain of hemoglobin and that the 8000 and 31,770 *m/z* proteins are doubly charged and dimeric forms of the same protein. The many variables, which must be considered, have led to bioinformatics approaches to analyze the data.

3.3.2.3. Data Analysis

A number of approaches have been used to analyze SELDI-TOF-MS data to determine reliability. One of the best known reports demonstrated a high predictive ability of serum analysis using SELDI-TOF-MS to determine which women had ovarian cancer ***(42)***. In this report, the authors document the importance of quality control to obtain reliable results. Additionally, a training set of samples was used to find candidate peaks and then a validation set to confirm that certain peaks were unique to ovarian cancer patients.

Pooled samples of NAF from healthy breasts and breasts with cancer were analyzed using SELDI-TOF-MS on 4 successive days to generate 24 spectra, and then in 36 subsequent experiments using a portion of the same pooled NAF;

the resulting spectra were analyzed to confirm how closely the 36 subsequent experiments agreed with the original 24 spectra ***(43)***. The authors developed algorithms that located peaks and combined peak detection with baseline correction using principal components analysis. They observed that the protein peaks were highly correlated across samples and that 80% of the variance in samples could be explained using six principal components.

The interlaboratory variability in the analysis of serum by SELDI-TOF-MS from subjects with and without prostate cancer was recently reported ***(44)***. Parameters assessed included signal-to-noise ratio, mass accuracy, resolution, and normalized intensity of three *m/z* peaks present in a standard pooled sample. Standard operating procedures were established at all laboratory sites. The across-laboratory measurements found a CV for mass accuracy of 0.1%, variability of signal-to-noise ratio of approx 40%, and variability in normalized intensity of 15 to 36% ***(44)***, which was comparable to the intralaboratory measurements of the same peaks. All six sites achieved perfect blinded classification of subjects with and without prostate cancer using boosting (boosting logistic regression and boosting decision tree analysis).

4. Perspectives and Future Directions

It is likely that a panel of biomarkers will be required to harness optimally the information present in NAF. Preliminary reports suggest that combining protein markers such as bFGF and PSA, or uPA and PAI-1, provides a more predictive model of breast cancer than does either marker alone. Protein-wide assessment of NAF holds great promise but is currently limited by its ability to detect proteins of low abundance and the labor-intensive nature of specific protein identification. Whether protein patterns using SELDI-TOF-MS will prove sufficiently consistent, such that specific protein identification is not required, remains to be demonstrated. To optimize breast cancer detection, both protein and clinical information should be incorporated into predictive models.

NAF is likely to be increasingly useful not only in breast cancer prediction but also in determining response to the ingestion of a food or chemical. Preliminary studies of samples analyzed before and after soy ingestion demonstrate the ability of NAF to assess response to treatment ***(45)***. Further evidence of this comes from the ability to evaluate the effect of prostaglandin E_2 levels in NAF before and after ingestion of celecoxib ***(4)*** and estrogenic markers in NAF before and after taking tamoxifen ***(46)***.

References

1. Sauter ER, Ross E, Daly M, et al. Nipple aspirate fluid: a promising non-invasive method to identify cellular markers of breast cancer risk. *Br J Cancer* 1997;76: 494–501.

2. Khan IH, Kendall LV, Ziman M, et al. Simultaneous serodetection of 10 highly prevalent mouse infectious pathogens in a single reaction by multiplex analysis. *Clin Diagn Lab Immunol* 2005;12:513–519.
3. Hartl D, Griese M, Nicolai T, et al. Pulmonary chemokines and their receptors differentiate children with asthma and chronic cough. *J Allergy Clin Immunol* 2005;115:728–736.
4. Sauter ER, Schlatter L, Hewett J, Koivunen D, Flynn JT. Lack of effect of celecoxib on prostaglandin E2 concentrations in nipple aspirate fluid from women at increased risk of breast cancer. *Cancer Epidemiol Biomarkers Prev* 2004;13:1745–1750.
5. Alexander H, Stegner AL, Wagner-Mann C, Du Bois GC, Alexander S, Sauter ER. Proteomic analysis to identify breast cancer biomarkers in nipple aspirate fluid. *Clin Cancer Res* 2004;10:7500–7510.
6. Sauter ER, Shan S, Hewett JE, Speckman P, Du Bois GC. Proteomic analysis of nipple aspirate fluid using SELDI-TOF-MS. *Int J Cancer* 2005;114:791–796.
7. Cordingley HC, Roberts SL, Tooke P, et al. Multifactorial screening design and analysis of SELDI-TOF ProteinChip array optimization experiments. *Biotechniques* 2003;34:364–365, 368–373.
8. Zhang W, Chait BT. ProFound: an expert system for protein identification using mass spectrometric peptide mapping information. *Anal Chem* 2000;72:2482–2489.
9. Young JL Jr, Ward KC, Wingo PA, Howe HL. The incidence of malignant non-carcinomas of the female breast. *Cancer Causes Control* 2004;15:313–319.
10. Weir HK, Thun MJ, Hankey BF, et al. Annual report to the nation on the status of cancer, 1975–2000, featuring the uses of surveillance data for cancer prevention and control. *J Natl Cancer Inst* 2003;95:1276–1299.
11. Hsiung R, Zhu W, Klein G, et al. High basic fibroblast growth factor levels in nipple aspirate fluid are correlated with breast cancer. *Cancer J* 2002;8:303–310.
12. Chatterton RT Jr, Geiger AS, Khan SA, Helenowski IB, Jovanovic BD, Gann PH. Variation in estradiol, estradiol precursors, and estrogen-related products in nipple aspirate fluid from normal premenopausal women. *Cancer Epidemiol Biomarkers Prev* 2004;13:928–935.
13. Petrakis NL. Oestrogens and other biochemical and cytological components in nipple aspirates of breast fluid: relationship to risk factors for breast cancer. *Proc R Soc Edinburgh* 1989;95B:169–181.
14. Sauter ER, Tichansky DS, Chervoneva I, Diamandis EP. Circulating testosterone and prostate-specific antigen in nipple aspirate fluid and tissue are associated with breast cancer. *Environ Health Perspect* 2002;110:241–246.
15. Petrakis NL. Nipple aspirate fluid in epidemiologic studies of breast disease. *Epidemiol Rev* 1993;15:188–195.
16. Ernster VL, Wrensch MR, Petrakis NL, et al. Benign and malignant breast disease: initial study results of serum and breast fluid analyses of endogenous estrogens. *J Natl Cancer Inst* 1987;79:949–960.
17. Petrakis NL, Wrensch MR, Ernster VL, et al. Influence of pregnancy and lactation on serum and breast fluid estrogen levels: implications for breast cancer risk. *Int J Cancer* 1987;40:587–591.

18. Folkman J, Klagsbrun M. Angiogenic factors. *Science* 1987;235:442–447.
19. Folkman J, Shing Y. Angiogenesis. *J Biol Chem* 1992;267:10,931–10,934.
20. Liu Y, Wang JL, Chang H, Barsky SH, Nguyen M. Breast-cancer diagnosis with nipple fluid bFGF. *Lancet* 2000;356:567.
21. Macajova M, Lamosova D, Zeman M. Role of leptin in farm animals: a review. *J Vet Med A Physiol Pathol Clin Med* 2004;51:157–166.
22. Sauter ER, Garofalo C, Hewett J, Hewett JE, Morelli C, Surmacz E. Leptin expression in breast nipple aspirate fluid (NAF) and serum is influenced by body mass index (BMI) but not by the presence of breast cancer. *Horm Metab Res* 2004;36:336–340.
23. Soderdahl DW, Hernandez J. Prostate cancer screening at an equal access tertiary care center: its impact 10 years after the introduction of PSA. *Prostate Cancer Prostatic Dis* 2002;5:32–35.
24. Howarth DJ, Aronson IB, Diamandis EP. Immunohistochemical localization of prostate-specific antigen in benign and malignant breast tissues. *Br J Cancer* 1997;75:1646–1651.
25. Sauter ER, Daly M, Linahan K, et al. Prostate-specific antigen levels in nipple aspirate fluid correlate with breast cancer risk. *Cancer Epidemiol Biomarkers Prev* 1996;5:967–970.
26. Sauter ER, Chervoneva I, Diamandis A, Khosravi JM, Litwin S, Diamandis EP. Prostate-specific antigen and insulin-like growth factor binding protein-3 in nipple aspirate fluid are associated with breast cancer. *Cancer Detect Prev* 2002;26:149–157.
27. Zhao Y, Verselis SJ, Klar N, et al. Nipple fluid carcinoembryonic antigen and prostate-specific antigen in cancer-bearing and tumor-free breasts. *J Clin Oncol* 2001;19:1462–1467.
28. Sauter ER, Klein G, Wagner-Mann C, Diamandis EP. Prostate-specific antigen expression in nipple aspirate fluid is associated with advanced breast cancer. *Cancer Detect Prev* 2004;28:27–31.
29. Sauter ER, Diamandis EP. Prostate-specific antigen levels in nipple aspirate fluid. *J Clin Oncol* 2001;19:3160.
30. Gold P, Freedman SO. Demonstration of tumor-specific antigens in human colonic carcinomata by immunological tolerance and absorption techniques. *J Exp Med* 1965;121:439–462.
31. Ebeling FG, Stieber P, Untch M, et al. Serum CEA and CA 15-3 as prognostic factors in primary breast cancer. *Br J Cancer* 2002;86:1217–1222.
32. Foretova L, Garber JE, Sadowsky NL, et al. Carcinoembryonic antigen in breast nipple aspirate fluid. *Cancer Epidemiol Biomarkers Prev* 1998;7:195–198.
33. Sauter ER, Welch T, Magklara A, Klein G, Diamandis EP. Ethnic variation in kallikrein expression in nipple aspirate fluid. *Int J Cancer* 2002;100:678–682.
34. Sauter ER, Lininger J, Magklara A, Hewett JE, Diamandis EP. Association of kallikrein expression in nipple aspirate fluid with breast cancer risk. *Int J Cancer* 2004;108:588–591.
35. Qin W, Zhu W, Wagner-Mann C, Folk W, Sauter ER. Association of uPA, PAT-1, and uPAR in nipple aspirate fluid (NAF) with breast cancer. *Cancer J* 2003;9:293–301.

36. Porter DA, Krop IE, Nasser S, et al. A SAGE (serial analysis of gene expression) view of breast tumor progression. *Cancer Res* 2001;61:5697–5702.
37. Varnum SM, Covington CC, Woodbury RL, et al. Proteomic characterization of nipple aspirate fluid: identification of potential biomarkers of breast cancer. *Breast Cancer Res Treat* 2003;80:87–97.
38. Alexander H, Stegner AL, Wagner-Mann C, Du Bois GC, Alexander S, Sauter ER. Identification of breast cancer biomarkers in nipple aspirate fluid using proteomic analysis. *Clin Cancer Res* 2004;10:7500–7510.
39. Sauter ER, Zhu W, Fan XJ, Wassell RP, Chervoneva I, Du Bois GC. Proteomic analysis of nipple aspirate fluid to detect biologic markers of breast cancer. *Br J Cancer* 2002;86:1440–1443.
40. Paweletz CP, Trock B, Pennanen M, et al. Proteomic patterns of nipple aspirate fluids obtained by SELDI-TOF: potential for new biomarkers to aid in the diagnosis of breast cancer. *Dis Markers* 2001;17:301–307.
41. Petricoin EF, Ardekani AM, Hitt BA, et al. Use of proteomic patterns in serum to identify ovarian cancer. *Lancet* 2002;359:572–577.
42. Petricoin EF 3rd, Ardekani AM, Hitt BA, et al. Use of proteomic patterns in serum to identify ovarian cancer. *Lancet* 2002;359:572–577.
43. Coombes KR, Fritsche HA Jr, Clarke C, et al. Quality control and peak finding for proteomics data collected from nipple aspirate fluid by surface-enhanced laser desorption and ionization. *Clin Chem* 2003;49:1615–1623.
44. Semmes OJ, Feng Z, Adam BL, et al. Evaluation of serum protein profiling by surface-enhanced laser desorption/ionization time-of-flight mass spectrometry for the detection of prostate cancer: I. Assessment of platform reproducibility. *Clin Chem* 2005;51:102–112.
45. Hargreaves DF, Potten CS, Harding C, et al. Two-week dietary soy supplementation has an estrogenic effect on normal premenopausal breast. *J Clin Endocrinol Metab* 1999;84:4017–4024.
46. Harding C, Osundeko O, Tetlow L, Faragher EB, Howell A, Bundred NJ. Hormonally-regulated proteins in breast secretions are markers of target organ sensitivity. *Br J Cancer* 2000;82:354–360.

22

Proteomics of Seminal Fluid

Benjamin Solomon and Mark W. Duncan

Summary

Seminal fluid or semen is a complex mixture consisting of spermatozoa suspended within the secretions of male accessory sex glands. Although most studies of seminal fluid have emphasized the importance of spermatozoa morphology, concentration, and motility, and of biochemical constituents such as fructose and zinc, the significance of the protein composition of this biological fluid is receiving increasing attention. As detailed in this chapter, numerous studies have been conducted to investigate the protein composition of spermatozoa, seminal plasma (i.e., the acellular component of seminal fluid), and prostasomes (i.e., membrane-bound vesicles of prostatic origin present in seminal fluid). Most of these investigations were based on 2D gel electrophoresis, but most of the resolved proteins were not identified. Several proteins in seminal fluid have, however, already found potential use as markers for the assessment of reproductive function (e.g., neutral α-glucosidase and semenogelin 1) and as forensic tools for semen detection (e.g., semenogelin 1 and prostate-specific antigen [PSA]). Studies of spermatozoa are providing insights into immune infertility and identifying potential targets for contraceptive vaccines. In addition, because 20 to 30% of the volume of seminal fluid is derived from secretions of the prostate gland, this proximate biological fluid serves as a rich resource for biomarkers of prostatic disease, including cancer. Recent advances in technology for comparative proteomic studies will facilitate future biomarker discovery studies of this important biological fluid.

Key Words: Seminal fluid; seminal vesicles; seminal plasma; spermatozoa; sperm; prostate; prostasomes; proteome; proteomics.

1. Introduction

Seminal fluid or semen refers to the fluid that is expelled from the penis at the time of ejaculation. It consists of spermatozoa suspended within a complex medium composed primarily of the secretions of male accessory sex glands. Although most assessments of human seminal fluid have emphasized the importance of spermatozoa morphology, concentration, and motility, together with a

From: *Proteomics of Human Body Fluids: Principles, Methods, and Applications*
Edited by: V. Thongboonkerd © Humana Press Inc., Totowa, NJ

few biochemical parameters such as fructose, prostaglandins, and zinc, the importance of the protein composition of seminal fluid is being increasingly appreciated. Proteomic studies of seminal fluid can improve our understanding of reproductive biology and have potential clinical applications, not only in the assessment and treatment of infertility, but also for a variety of urological diseases and in forensic medicine.

1.1. Overview

The typical volume of seminal fluid produced at ejaculation is about 3 mL (range 2–6 mL). Reference parameters for seminal fluid are shown in **Table 1** *(1)*. Two major components have generally been recognized in studies of seminal fluid. There is a cellular fraction made up principally of spermatozoa, which accounts for 2 to 5% of the volume of the ejaculate. In addition, nonspermatozoan cellular elements including immature germ cells, leukocytes, and epithelial cells may also be present. The second component, an acellular fraction, also referred to as seminal plasma, accounts for most of the volume of the ejaculate. This component is made up of fluid from the testis, epididymis, and vas deferens, which at the time of ejaculation is mixed with secretions of the accessory sex glands. This fluid is derived predominantly from the seminal vesicles (50–80%) and the prostate gland (20–30%), with minor contributions from the bulbourethral and the periurethral glands **(Fig. 1)** *(2,3)*. It is a biochemically complex fluid rich in fructose, prostaglandins, zinc, citric acid, carnitine, and glycerophosphocholine, and it possesses a characteristic protein composition. This acellular fraction also contains membrane-bound organelles of prostatic origin called prostasomes.

Studies of split-ejaculates, i.e., ejaculates collected in two or more fractions *(4,5)* have shown that sequentially collected fractions of a single ejaculate vary in composition, owing to the organ-specific sequence in which the secretions of the accessory sex glands are mixed. Initially, the bulbourethral and periurethral glands produce alkaline secretions that neutralize and lubricate the urethra. Then, during a process called emission, spermatozoa suspended in fluid from the epididymis and vas deferens pass into the posterior urethra and are mixed, just before ejaculation, first with the prostatic secretions, and finally with seminal vesicle secretions. As a result, the early portions of the ejaculate are rich in spermatozoa and prostatic secretions, and the latter portions contain residual spermatozoa and are enriched with secretions from the seminal vesicles.

Given that seminal fluid represents a mixture of the secretions of several different accessory glands, each with its own unique biochemistry, its composition varies owing to the changing proportions of fluids contributed by the different accessory glands and also as a result of changes in the ratio of spermatozoa to seminal plasma. This may occur because of physiological factors (e.g., the time between ejaculates and the duration of foreplay) as well as pathological conditions

Table 1
Reference Values for Human Seminal Fluid

Parameter	Reference value
Volume	≥2.0 mL
pH	≥7.2
Sperm concentration	≥20 × 10^6 spermatozoa/mL
Total sperm no.	≥40 × 10^6 spermatozoa/ejaculate
Motility	50% or more spermatozoa with forward progression (grades a + b) or 25% or more with rapid forward progression (grade a) within 25 min of ejaculation
Vitality	75% or more alive
White blood cells	<1 × 10^6 /mL
Zinc	≥2.4 μmol per ejaculate
Fructose	≥13 μmol per ejaculate
Neutral α-glucosidase	≥20 mU per ejaculate

From **ref.** ***1***.

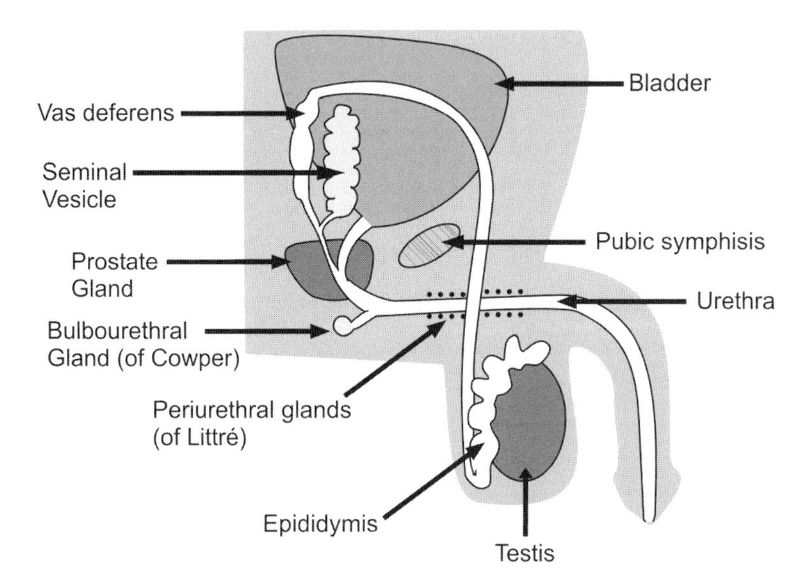

Fig. 1. Diagram demonstrating the position of male accessory sex glands.

affecting the secretory output of any of the glands (e.g., obstruction, infection, or malignancy) ***(6)***.

Moreover, the composition of seminal fluid is dynamic and undergoes rapid changes following ejaculation. Soon after ejaculation, seminal fluid undergoes coagulation, forming a gelatinous mass within which the spermatozoa are immobilized. Thereafter, within a period of about 20 min, proteolytic enzymes derived from prostatic secretions break down the gel-forming proteins within semen coagulum, resulting in its liquefaction and the release of motile spermatozoa.

1.2. Collection and Fractionation of Seminal Fluid

It is important to collect seminal fluid in a standardized manner to minimize variation caused by to the collection method *(1)*. The standard method of collection is to obtain the sample by masturbation into a clean container. Because the time between ejaculations can affect the composition of seminal fluid, samples should be collected after a standardized period of abstinence, e.g., 3 d. (This is the standard practice for samples collected for fertility assessment.) Seminal fluid may also be obtained after vaginal intercourse using a specially designed nonspermicidal Silastic condom. The latter method results in increased volumes and possibly different seminal fluid parameters compared with samples obtained by masturbation.

Following collection, a standardized method of handling is required owing to the changes that occur during the processes of coagulation and liquefaction. Typically, samples are allowed to liquefy at room temperature, are subjected to preliminary analysis, and are then stored at –80°C. Importantly, seminal fluid may contain harmful viruses (e.g., human immunodeficiency virus or hepatitis B virus) and should therefore be handled with appropriate care. Guidelines have been described in Appendix II of the *WHO Laboratory Manual for the Examination of Human Semen and Sperm-Cervical Mucus Interactions* ***(1)***.

1.3. Methods for Separating Components of Seminal Fluid

1.3.1. Acellular Component (Seminal Plasma Including Prostasomes)

In most studies, seminal plasma has been defined as the supernatant remaining after centrifugation and removal of cellular elements from liquefied seminal fluid. Recent work has shown that prostasomes are present, suspended within this fraction. Consequently, the proteins within the acellular component of seminal fluid can be subdivided into a soluble protein fraction, which accounts for at least 85% of the total seminal fluid protein, and a prostasome protein fraction, which accounts for 1 to 3% of the total seminal fluid protein *(7)*. Purified prostasomes may be obtained by further ultracentrifugation of seminal plasma (105,000*g* at 4°C, for 2 h) followed by purification on a 40-mL Sephacryl S-500 HR (or Sephadex G-200) column (as detailed by Renneberg et al. in **ref. *8***).

1.3.2. Cellular Component

The cellular component consists principally of spermatozoa, and these account for about 7% of the total protein content of seminal fluid *(7)*. To obtain the cellular material, liquefied seminal fluid is first centrifuged (e.g., 800*g*, 10 min), and the resulting pellet is then resuspended and washed in an appropriate buffer to remove seminal plasma. Because the cellular pellet may also contain cells other than spermatozoa (collectively referred to as "round cells"), including immature germ cells, leukocytes, and epithelial cells, a purified spermatozoan fraction requires additional workup. Approaches include the use of a discontinuous density gradient (i.e., based on Percoll or alternative products) or the "swim-up method." These procedures are described in detail in the WHO laboratory manual *(1)*.

2. Proteomic Studies of Seminal Fluid: Findings and Applications

2.1. Acellular Component

Most of the volume of seminal fluid consists of the acellular fraction, or seminal plasma. Its protein concentration is approximately 35 to 55 g/L, somewhat lower than that of serum. As noted above, it is composed of the secretions of the accessory sex glands, with major contributions from the seminal vesicles and prostate. Consequently, proteomic studies of this fluid can provide important information about the function of these glands. A minor portion of the protein in seminal plasma (<10%) represents serum-derived protein transudated from intercellular fluids *(9)*. This includes albumin (present at a concentration of 0.6 g/L), transferrin, and some immunoglobulins and complement factors.

Studies of the electrophoretic pattern of human seminal plasma were first conducted in the 1940s by Gray and Huggins *(10)* and Ross et al. *(11,12)*. They described the presence of four or five electrophoretic bands. Subsequent 1D electrophoretic studies demonstrated many more bands that changed temporally and in association with pathological conditions of the male accessory glands. These early studies were reviewed by Mann *(2)*.

The true complexity of the protein composition of seminal plasma has only become evident through the application of 2D gel electrophoresis (2-DE) *(4,13–21)*. These studies established 2D maps of seminal plasma in which several hundred polypeptide spots were resolved, with patterns distinct from those observed in other biological fluids such as serum. Only a handful of studies have gone on to provide identifications of the proteins observed.

Starita-Geribaldi et al. *(20)* performed 2-DE followed by mass spectrometry (MS) and were able to identify 26 spots corresponding to 10 distinct proteins. These included prostate-specific antigen (PSA), prostatic acid phosphatase, Zn-α-2-glycoprotein, glycodelin, and clusterin. All these proteins were identified from multiple spots, attesting to the multiplicity of isoforms and posttranslational modifications present in seminal fluid.

Perhaps the most comprehensive analysis of the protein and peptide composition of human seminal plasma has been performed by Fung et al. *(21)*. These investigators employed several different methods to study pooled seminal fluid from healthy volunteers. In initial work, approximately 100 Coomassie Blue-stained spots were recognized following 2D separation of seminal plasma proteins when pH 3 to 10 immobilized pH gradient (IPG) strips were used for the 1D separation. More spots were resolved with the use of narrow-pH-range strips, i.e., a combination of pH 4 to 7 and pH 6 to 11 IPG strips (about 70 in each pH range), as shown in **Fig. 2**. Seventy-three protein spots corresponding to 20 unique proteins were identified following excision, in-gel digestion with trypsin, matrix-assisted laser desorption/ionization (MALDI)-MS, and peptide mass fingerprinting (PMF) (**Table 2**). Many protein spots corresponded to various modified forms of secreted prostatic proteins including PSA, prostatic acid phosphatase, and Zn-α-2-glycoprotein. Most of the low-molecular-weight (<30 kDa) protein spots were identified as truncated forms of semenogelin I and semenogelin II, together with variants of prolactin-inducible protein.

Given the limitations of 2-DE in resolving several classes of proteins, in particular hydrophobic proteins, high-molecular-weight proteins, and basic proteins, Fung et al. *(21)* also separated seminal plasma proteins by 1-DE. Bands were excised from gels, digested with trypsin, and analyzed by liquid chromatography-tandem MS (LC-MS/MS). Ten to 15 proteins were identified in each fraction analyzed. In total, more unique gene products were found using this approach compared with identification by 2-DE. In all, 42 distinct proteins were identified, including several that were seen in multiple bands (e.g., semenogelin I and semenogelin II). The presence of a single gene product in several distinct bands, as was noted in this study, is indicative of posttranslational modifications (such as proteolytic cleavage in the case of semenogelin).

To evaluate low-molecular-mass proteins and peptides, Fung et al. *(21)* employed MALDI-TOF-MS on unfractionated seminal plasma samples. Thirty peptide components, of between 500 and 10,000 Daltons, were visualized (**Fig. 3**). LC-MS/MS was used to determine the identities of these components, many of which were also found to be proteolytic products of either semenogelin I or semenogelin II.

Kausler and Spiteller *(22)* previously identified several peptides in human seminal fluid by MS, and at least one of these was subsequently shown to be a PSA hydrolysis product of semenogelin 1 *(23)*. Interestingly, three inhibin-like peptides were also identified from seminal plasma *(24)* that also proved to be fragments of semenogelin-1 (α-inhibin-92, -52, and -31, corresponding to residues 45–136, 85–136, and 85–115 *[25]*). Several other peptides have been documented in human seminal plasma, including oxytocin *(26)*, angiotensin II *(27)*, β-endorphin *(28)*, met-enkephalin *(28)*, calcitonin *(29)*, thyrotropin-releasing hormone-like peptides *(30–32)*, prolactin *(33)*, and relaxin *(34)*. Some of these peptides have been found in association with spermatozoa, and it has been suggested that they may play a role in promoting sperm motility *(35)*.

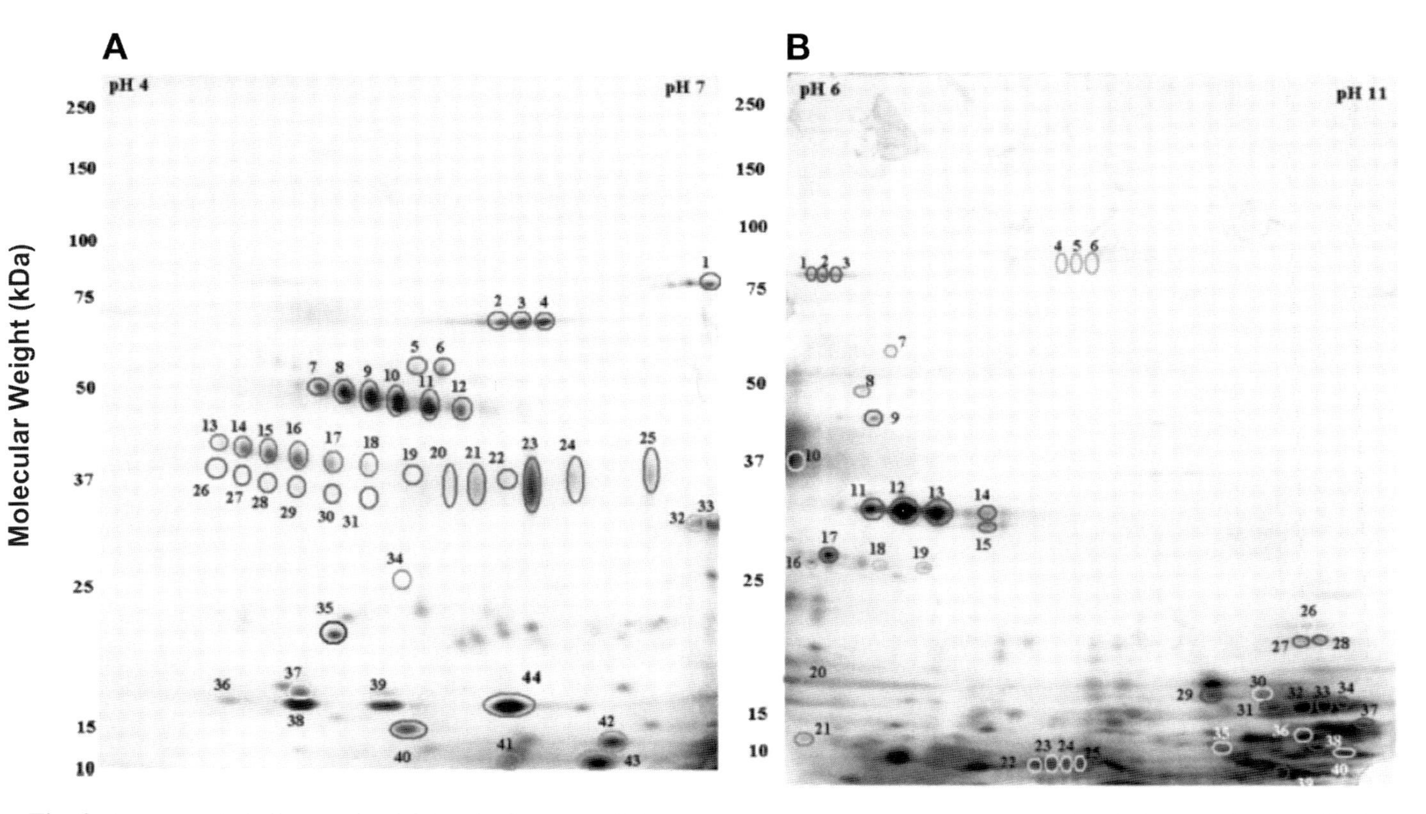

Fig. 2. An annotated silver-stained 2D gel of seminal fluid proteins separated over the p*I* range 4 to 7 **(A)** and 6 to 10 **(B)**. The protein spots that were excised, digested with trypsin, and identified by PMF are circled. Identities for these proteins are provided in **Table 2**. (Reproduced with permission from Fung et al., **ref. *21*.**)

Table 2
Proteins Identified by 2D Gel Electrophoresis of Seminal Fluid[a]

Spot no.	Protein	GI accession no.	Mol. wt./p*I*		Protein coverage (%)
			Theoretical	Experimental	
Atp/*I*4–7					
1	Lactotransferrin	4505043	78/8.5	90/7.0	63
2	Albumin precursor	4502027	69/5.9	70/5.9	43
3				70/5.9	33
4				70/5.9	48
5	α_1-antitrypsin	21361198	47/5.3	60/5.5	20
6				60/5.7	40
7	Prostatic acid phosphatase precursor	6382064	45/5.8	50/5.2	19
8				50/5.3	29
9				50/5.4	25
10				50/5.5	19
11				50/5.6	19
12				50/5.7	30
13	Zn-α-2-glycoprotein	4502337	34/5.7	43/4.8	36
14	Zn-α-2-glycoprotein	4502337	34/5.7	43/4.9	58
15				43/5.0	56
16				43/5.2	56
17				43/5.3	54
18				43/5.4	48
19	Clusterin	4502905	52/5.9	43/5.6	24
20				43/5.8	26
21				43/5.9	28
22				43/6.0	26
23				43/6.1	32
24				43/6.2	29
25				43/6.5	32
26				37/4.7	17
27				37/4.9	21
28				37/5.0	27
29				37/5.2	25
30				37/5.3	18
31				37/5.4	22
32	Prostate-specific antigen	4502173	29/7.6	37/6.9	58
33				37/7.0	57

(Continued)

Table 2 *(Continued)*

Spot no.	Protein	GI accession no.	Mol. wt./p*I*		Protein coverage (%)
			Theoretical	Experimental	
34	Prostaglandin D2 synthase	4506251	21/7.7	27/5.6	25
35	Epididymal secretory protein	5453678	17/7.6	23/5.2	45
36	Prolactin-induced protein	4505821	17/8.3	13/5.5	56
37				17/5.1	66
38				16/5.1	66
39				17/5.5	55
40	β-Microseminoprotein, isoform A, precursor	4557036	13/5.4	16/5.7	28
41	Semenogelin I[b]	4506883	52/9.3	11/5.7	14
42				15/5.9	10
43				13/5.9	14
44	Semenogelin II (LC/MS)	4506885		66/9.1	12
Atp/*I*6–11					
1	Transferrin precursor	4557871	79/6.8	77/6.3	16
2				77/6.4	24
3				77/6.5	31
4	Lactotransferrin	4505043	78/8.5	80/8.3	53
5				80/8.5	31
6				80/8.6	34
7				55/7.1	22
8	α enolase	4503571	47/7.0	48/6.8	47
9	Isocitrate dehydrogenase 1	5174471	47/6.3	42/6.9	25
10	Clusterin	4502905	52/5.9	37/6.1	40
11	Prostate-specific antigen	4502173	29/7.6	30/6.9	42
12				30/7.1	57
13				30/7.4	37
14				30/7.8	25
15				29/7.6	47
16	Albumin precursor	4502027	69/5.9	27/6.3	11
17	Prostate-specific antigen	4502173	29/7.3	26/6.6	35

(Continued)

Table 2 *(Continued)*

Spot no.	Protein	GI accession no.	Mol. wt./p*I* Theoretical	Mol. wt./p*I* Experimental	Protein coverage (%)
18	Albumin precursor	4502027	69/5.9	25/7.0	12
19				25/7.2	16
20	Prolactin-induced protein	4505821	17/8.3	20/6.2	63
21	β_2-microglobulin	4757826	14/6.1	13/6.2	73
21	Semenogelin I[c]	4506883	52/9.3	13/6.2	18
22	Semenogelin I[d]	4506883	52/9.3	10/8.1	14
23				10/8.2	16
24				10/8.4	20
25				10/8.6	19
26				21/10.2	13
27	Semenogelin II[e]	4506885	66/9.0	20/10.1	14
28				20/10.3	17
29				17/9.1	15
30	Semenogelin I[d]	4506883	65/9.0	17/9.6	12
31	Semenogelin II[e]	4506885	65/9.0	16/9.6	17
32				16/9.7	31
33				16/10.0	19
34				16/10.2	18
35	Semenogelin I[d]	4506883	52/9.3	12/9.3	20
36	Semenogelin I[e]	4506883	52/9.3	12/10.0	12
37	Semenogelin II[d]	4506885	65/9.0	15/10.4	13
38	Semenogelin I[e]	4506883	52/9.3	10/9.8	24
39				10/10.3	18
40	Semenogelin II[e]	4506885	65/9.1	10/10.1	17

[a]*See* **Fig. 2**.
[b]C-terminal coverage only for all three forms.
[c]C-terminal coverage only.
[d]Coverage of middle of protein only.
[e]N-terminal coverage.

Efforts have been made to compare the protein profiles of seminal plasma from normal individuals with those of azoospermic or oligospermic individuals ***(14,19,20)***. Starita-Geribaldi et al. ***(19,20)*** compared 2D electrophoretic profiles of seminal plasma from fertile men with seminal plasma from vasectomized or azoospermic men. Differentially expressed spots were noted by gel matching. Although several spots were identified by MS, their biological significance remains to be established ***(20)***. To date, no studies have been reported

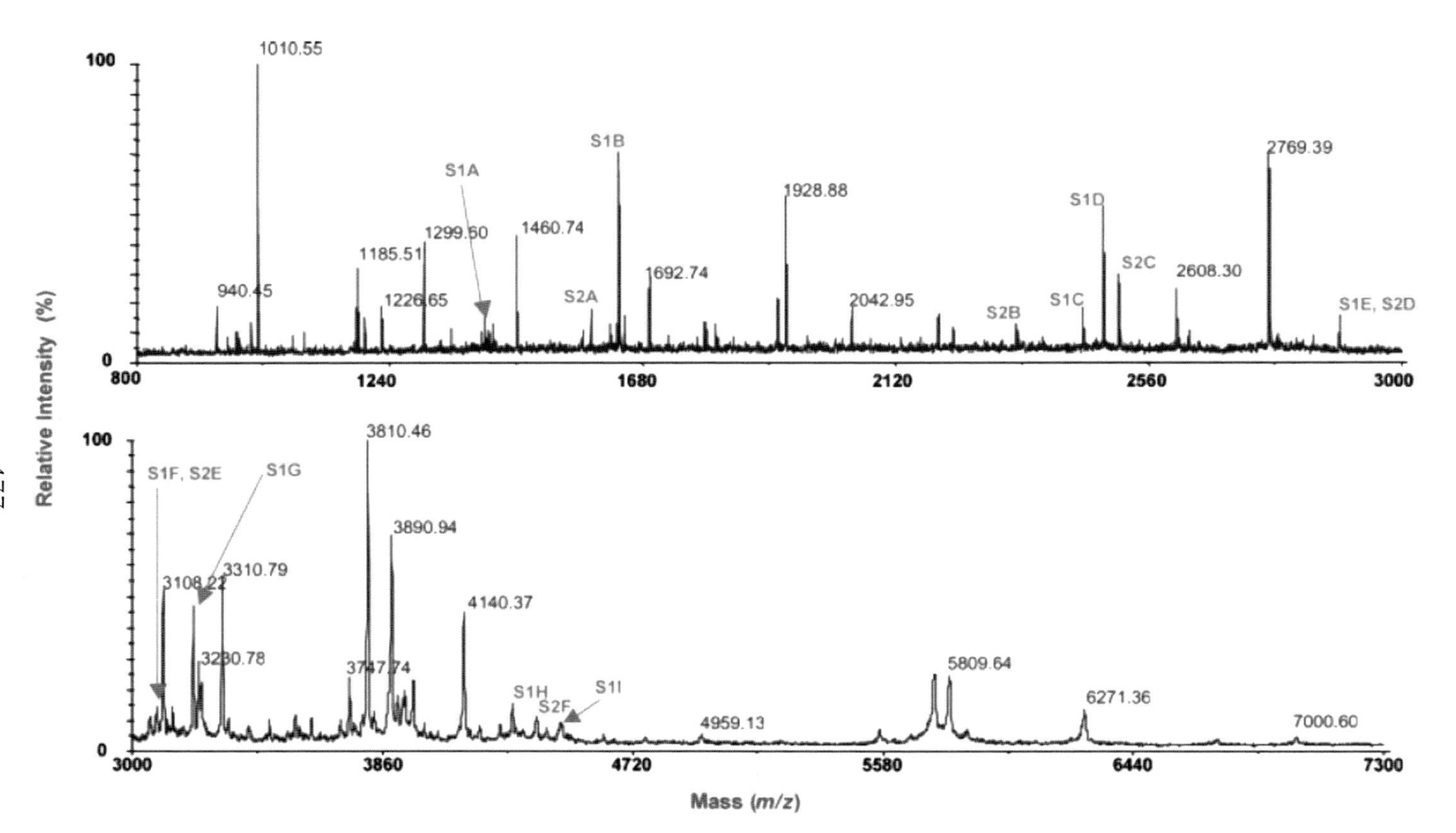

Fig. 3. MALDI mass spectra of the low-molecular-weight components of human seminal fluid. Indicated are some of the peptides that were identified by subsequent ESI-LC-MS/MS. Most of these peptides were identified as fragments of semenogelin I (S1A–S1I) and semenogelin II (S2A–S2F). (Reproduced with permission from Fung et al., **ref. *21*.**)

using modern comparative proteomic methods such as difference gel electrophoresis *(36)* or MS utilizing stable isotope labeling (i.e., ICAT or iTRAQ).

In addition to the studies of seminal plasma detailed above, several investigators have examined the contributions of the individual accessory glands to seminal plasma. Limited studies have been performed in seminal vesicle fluid, but more detailed studies have been conducted in expressed prostatic fluid, as detailed below in **Subheading 2.1.21**. Proteomic studies of prostasomes are also described in the following section.

2.1.1. Contributions of the Seminal Vesicles to Seminal Plasma

The seminal vesicles typically contribute more than half of the seminal volume (i.e., 50–80%) *(2)*. Proteins contained within seminal vesicle secretions are known to play important roles in semen coagulation, sperm motility, and capacitation, as well as suppression of immune activity in the female reproductive tract, and this has been reviewed by Aumuller and Riva *(37)* and by Gonzales *(38)*.

The major protein constituents of seminal vesicle fluid, and indeed in the seminal fluid coagulum, are semenogelin I and semenogelin II *(25,39,40)*. These highly homologous proteins are produced in large quantities by the glandular epithelium of the seminal vesicles. Semenogelin II is produced in smaller amounts by the epididymis *(41)*. The semenogelins undergo rapid proteolytic digestion by PSA, resulting in liquefaction of the semen coagulum and release of motile spermatozoa *(9,23)*. Many of the resulting semenogelin cleavage products have been shown to have biological functions including inhibition of sperm motility, antibacterial activity, and activation of sperm capacitation; others may have inhibin or thyrotropin-releasing hormone (TRH)-like activity. This has been reviewed by Robert and Gagnon *(40)*. Additional seminal vesicle proteins include lactoferrin *(9)*, fibronectin *(9)*, and protein C inhibitor antigen (PCI), which is found at 20-fold higher concentration in the seminal plasma than in blood plasma *(42,43)*.

Only limited proteomic analysis of human seminal vesicular fluid has been performed, primarily because of the difficulties associated with accessing clinical samples. Some 2D gel profiles have been obtained using seminal vesicle fluid aspirated from seminal vesicles removed at the time of operation *(4,16)*. These profiles proved to be distinct from those of prostatic fluid and were characterized by the presence of a group of basic proteins that Lee et al. *(16)* termed seminal vesicle-specific antigens (and that were subsequently shown to be semenogelin). This group of proteins was shown to disappear from seminal fluid after liquefaction *(16)*.

Knowledge of the proteins contributed by the seminal vesicles to seminal plasma has resulted in several clinical applications. For example, levels of specific proteins including semenogelin 1 *(44)* and protein C inhibitor *(45)* have been used as markers of seminal vesicle obstruction or agenesis. Furthermore, semenogelin I has also been used in forensic tests for semen detection ***(46)***.

2.1.2. Contributions of the Prostate to Seminal Plasma

2.1.2.1. Expressed Prostatic Fluid

Secretions from the prostate gland are the major contributors to the initial portion of the ejaculate and represent approximately 20 to 30% of the total volume of the seminal fluid. They are derived from prostatic epithelial cells, are rich in citric acid, zinc, and glyceryl phosphocholine, and contain many proteins including the fibrinolytic enzymes responsible for the liquefaction of coagulated semen. The protein composition of human prostatic fluid has been assessed using expressed prostatic fluid obtained by digital massage of the prostate.

Both 1D ***(47–49)*** and 2D electrophoretic studies ***(16,50–54)*** have been performed on expressed prostatic fluid. A consistent finding in these studies is the preponderance of three proteins: PSA, prostatic acid phosphatase, and prostate-specific protein-94 (PSP-94).

The first of these proteins, PSA, is a chymotrypsin-like serine protease and a member of the tissue kallikrein family of proteases ***(55)***. The primary function of PSA is to liquefy coagulated semen so that the spermatozoa can escape and swim off to fertilize the ovum. It does this by hydrolyzing semenogelin I and semenogelin II, which together with small amounts of fibronectin represent the structural elements of the semen coagulum. PSA is produced, under androgenic control, by secretory epithelial cells located in the acini and ducts of the prostate, and to a lesser extent by the epithelial cells within the periurethral glands of Littré ***(56,57)***. In seminal fluid, PSA exists in various cleavage forms and glycosylation states and is partly bound to the serine protease inhibitor and protein C inhibitor ***(58)***. It is noteworthy that the concentration of PSA is 10^6-fold higher in the seminal fluid (0.5–2.0 mg/mL ***[59–61]***) than in the serum (4 ng/mL in serum ***[62]***). In prostate cancer, however, serum levels of PSA are frequently elevated. In this setting, PSA has been shown to be a useful biomarker for the presence of prostate cancer and for monitoring response to treatment.

Detailed exploration of different molecular forms of PSA has been performed using 2-DE coupled with immunoblot assessment ***(18,63,64)*** or MS ***(64)***. When PSA was chromatographically purified from seminal fluid, Isono et al. ***(64)*** resolved approximately 30 distinct silver-stained protein spots following 2-DE. Of these, 20 spots were shown to have peptide fragments derived from PSA, including some novel peptide forms of PSA. The other silver-stained spots did not contain PSA fragments and may have derived from other proteins that contaminated the material.

The other two major proteins present in prostatic fluid are prostatic acid phosphatase and PSP-94. Prostatic acid phosphatase is a 102-kDa glycoprotein dimer secreted by the prostate under androgenic control ***(65)***. It is found in high concentration in the seminal fluid (0.3–1.0 g/L). Until replaced by PSA, prostatic acid phosphatase was used as a marker for prostate cancer. PSP-94 (also known as β-inhibin and β-microseminoprotein) is a 94-amino acid, cysteine-rich, nonglycosylated protein with a molecular mass of about 16 kDa ***(66,67)***. It acts to inhibit follicle-stimulating hormone (FSH), which appears to play a role as an autocrine or paracrine growth factor in the prostate. Decreased levels of PSP-94 are seen in late-stage prostate cancer and seem to be associated with progression from a hormone-dependent to a hormone-independent state. These findings indicate that PSP-94 may play a role as a biomarker for prostate cancer.

Other proteins present in prostatic secretions include Zn-α_2-glycoprotein ***(15)*** (a 40-kDa glycoprotein), human kallikrein-2 ***(61,68)*** (a serine protease highly homologous to PSA), and progastricsin ***(69)*** (an aspartic protease).

Because prostatic secretions reflect the functional state of the gland, they are potentially fruitful sources of biomarkers of prostatic diseases. Several 2-DE, studies have compared the protein expression profiles of prostatic fluid collected from controls and men with benign prostatic hyperplasia or prostate cancer. Qualitative differences were evident in the profiles ***(51,53,54)***. Studies of purified PSA from prostatic fluid have demonstrated differentially expressed forms (i.e., altered glycosylation and loss of cleaved forms) in benign prostatic hyperplasia and prostate cancer ***(18,64)***. In addition, various cleavage forms of PSA are found to be increased in benign prostatic hyperplasia and decreased in prostate cancer. These include forms of PSA with cleavages between residues 85 and 86, 145 to 146, and 182 to 183—collectively termed benign prostatic hyperplasia-specific antigen (BPSA) ***(70)***. Some of these protein forms have also been detected in the serum. These studies indicate that a panel composed of several distinct proteins, including different isoforms and posttranslational modifications, may prove to be more useful biomarkers for prostate cancer than any single protein.

2.1.2.2. Prostasomes

Prostasomes are membrane-bound organelles of prostatic origin that were first identified in human seminal fluid by Ronquist et al. in 1978 ***(71)***. Structurally, prostasomes are spherical vesicular structures, 40 to 500 nm in diameter, enclosed within a bilaminar or multilaminar membrane. They have a characteristic lipid composition and are unusually rich in cholesterol and sphingomyelin ***(72,73)***. They form in the apical part of the prostatic duct epithelium within large storage vesicles that are secreted either by exocytosis (i.e., the mechanism by which the storage vesicle fuses with the cell membrane and releases its contents into the lumen) or by diacytosis (i.e., the mechanism by which the storage vesicle and its contained prostasomes are translocated through the cell membrane) ***(74)***.

A variety of physiological functions have been attributed to prostasomes, including enhancement of sperm mobility and stabilization of the sperm plasma membrane. They are also involved in the coagulation and liquefaction of semen, regulation of complement, protection of spermatozoa in the acidic environment of the vagina, and immunomodulation. These functions are detailed in several reviews ***(74–79)***. Significantly, prostasomes have been reported to adhere to and even fuse with spermatozoa ***(80–82)***. This allows for the transfer of proteins and lipid directly from prostasomes to spermatozoa ***(83,84)***.

Several investigators have used 2-DE to separate the protein constituents of human prostasomes. Two studies resolved about 80 protein spots in purified prostasomes but did not determine their identities ***(8,85)***. In addition, Carlsson et al. found approximately 200 spots on silver-stained 2D gels of purified prostasomes ***(86)***. Because circulating antisperm antibodies (ASAs) from infertile men and women frequently recognize prostasomes ***(87)***, their gel was also immunoblotted with sera containing ASA. Several immunoreactive spots were subsequently identified by MS as potential prostasomal targets of ASA, including prolactin-inducible protein and clusterin.

Utleg et al. ***(88)*** utilized a different approach, the so-called shotgun or bottom-up proteomic approach, to study the protein composition of human prostasomes. Isolated protasomes were obtained from seminal fluid, tryptically digested, and subjected to single-dimension microcapillary ([μ] LC-MS/MS). One-hundred and thirty-nine putative prostasome proteins were identified, 128 of which had not previously been described in association with prostasomes. Many of the proteins identified were enzymes, but only some were known to be present in prostasomes (e.g., dipeptidyl peptidase IV [CD26] ***[89]***, neprilysin [CD10] ***[90]***, membrane alanyl aminopeptidase ***[90]***, phospholipase A2 ***[85]***, aminopeptidase N [CD13] ***[83]***, and γ-glutamyl transpeptidase ***[91]***). Thirty-seven represented enzymes not known to be associated with prostasomes, including isocitrate dehydrogenase I, sorbitol dehydrogenase, prostate-specific transglutaminase, creatine kinase (b chain), lactoylglutathione lyase, and peroxiredoxin 1. Other proteins identified with this approach included transport proteins (such as annexins A1, A2, A3, A5, A6, and A11), structural proteins (such as actins, tubulins α and β, ezrin, and cofilin) and GTP-binding proteins (including 11 Rab proteins [the regulator Rab-GD11], GTPase activating proteins, Rho-GTPase-activating protein I, and Cdc42). Several of the proteins described are known to be highly expressed in prostatic secretions including PSA and prostatic acid phosphatase. Although this approach provides an extensive list of the proteins present within a sample, no information is available about amounts or the posttranslational modifications present.

2.3. Proteomics of Spermatozoa

Spermatozoa are highly specialized cells that contribute the paternal genetic complement in the reproductive process (*see* reviews by Mann and Lutwak-Mann ***[92]*** and by Millette ***[93]***). The mature sperm cell (spermatozoa) is 0.05 mm long and consists of a head, body, and tail. The head contains a nucleus of dense genetic material from the 23 chromosomes, and the body is rich in the mitochondria necessary to supply the energy required for the sperm's activity. The tail is made of protein fibers that contact, on alternative sides, yielding a wavelike movement that drives the sperm through the seminal fluid. Although they make up only about 5% of the volume, a typical ejaculate contains over 40 million spermatozoa. Because spermatozoa possess limited machinery for transcription and translation of new proteins, they are heavily reliant on their local microenvironment as they undergo the transition from immotile cells with limited capacity for fertilization to mature, motile spermatozoa.

Proteomic studies of spermatozoa are limited, but they clearly have considerable potential to enhance our understanding of reproductive biology and especially clinical infertility. Most reports have employed 2-DE to separate human spermatozoa proteins ***(17,94–109)***. Some of the earliest 2-DE studies employed spermatozoa that were obtained by centrifugation alone, ***(94–98)*** thereby, opening up the potential for contamination with nonspermatozoan cellular elements ***(99)***. More recent studies have obtained purified sperm populations using either the "swim-up method" or Percoll density gradients. Relatively few studies, however, have identified the sperm proteins resolved on these gels ***(17,100,102, 103,108,109)***.

Naaby-Hansen et al. ***(17)*** established a comprehensive 2D map of normal human spermatozoa proteins. When sperm proteins obtained from normal volunteers were purified on a Percoll density gradient, separated in the first dimension by isoelectric focusing (IEF) and nonequilibrium pH gradient electrophoresis (NEPGE), and then separated in the second dimension by polyacrylamide gel electrophoresis, more than 1300 silver-stained spots were resolved. Several of these protein spots, including heat shock protein 70 (HSP70), HSP90, α-tubulin, β-tubulin, calreticulin, PH-20, and gastrin-binding protein were identified, either by MS or by immunoblotting with specific antibodies ***(17,100)***.

Because of their critical role in sperm-oocyte interactions ***(110)***, characterization of the sperm surface membrane proteins is of particular interest in reproductive biology, and a great deal of effort has been directed to this task. Putative sperm surface proteins have been identified using biotinylation or radioiodination to label exposed proteins, presumably on the surface of intact spermatozoa, prior to 2D separation. Naaby-Hansen et al. ***(17)*** found 98 spots on their 2D gel map of over 1300 spermatozoan protein spots that labeled with both techniques, representing strong candidates for surface proteins. This approach has also been employed in several other studies ***(97,100,102)***. Alternative methods used to obtain enriched sperm membrane fractions include nitrogen cavitation of spermatozoa ***(94,99)***, the use of special detergents such as Triton X-115 ***(102,105)***, and a combination of hypoosmotic lysis, homogenization, and ultracentrifugation ***(103,107)***.

Some limited phosphoproteomic studies of human spermatozoa have also been undertaken. Naaby-Hansen et al. found that antiphosphotyrosine antibody bound to five groups of protein spots (a total of 22 spots) on their 2D map ***(17)***. This strategy has also been applied to identify phosphoproteins involved in sperm capacitation, a process known to be associated with increased tyrosine phosphorylation of several proteins. Ficcaro et al. ***(108)*** studied the phosphoproteome of capacitated human sperm using a combination of 2D gels followed by antiphosphotyrosine Western blots and direct MS/MS sequencing of phosphopeptides. The identities of 16 phosphoproteins, including valosin-containing protein (VCP or p97), were determined by excising selected spots from the gels and analyzing them by LC-MS/MS. Immobilized metal affinity chromatography (IMAC) coupled with MS/MS facilitated the mapping of 60 phosphorylation sites.

Many of the reported proteomic studies of spermatozoa aimed to investigate the causes of infertility. The presence of ASAs, which interfere with sperm function, is thought to be responsible for immune infertility in 4 to 8% of infertile men. Because ASAs are not able to penetrate an intact outer cell membrane, the important antigenic targets of ASAs are limited to the outer sperm membrane and the acrosome. The acrosome is included because its inner membrane may become externalized during the acrosome reaction ***(111)***. Bohring et al. ***(101,103)*** separated purified spermatozoan cell membrane proteins by 2-DE and transferred these to a PVDF membrane before probing the resolved spots with ASAs obtained from seminal plasma. Immunoreactive spots were excised and identified by MALDI and PMF. Six proteins were identified, including HSP70, HSP70-2, disulphide isomerase, ER-60, and the caspase 3 and 2 subunits of the proteosome (i.e., component C2 and ζ-chain) ***(103)***.

Similar proteomic studies of spermatozoa have been conducted in an effort to identify candidate proteins for the development of contraceptive vaccines ***(17,100,102,104,105,112)***. Shetty et al. ***(100)*** resolved Percoll density gradient-purified sperm proteins by IEF and NEPGH in the first dimension and polyacrylamide gel electrophoresis in the second dimension and then transferred the proteins onto nitrocellulose membranes. Membranes were subsequently probed with sera containing ASA to identify potential sperm surface antigens. Ninety-eight protein spots were recognized by sera from infertile individuals, but not by control sera from fertile individuals. Six of these spots were identified as potentially relevant sperm surface antigens based on prior labeling of the surface proteins by radioiodination. Shibahara et al. ***(104)*** utilized a similar approach to probe sperm proteins with the sperm-immobilizing antibodies present in the sera of infertile women. Coordinates of the immunoreactive spots were compared with a reference map of sperm proteins ***(17)*** to identify four candidate sperm surface antigens. Notably, these were different from the spots previously identified by Shetty et al. ***(100)***. In the third study, Hao et al. ***(105)*** analyzed 2D gels of sperm extracts enriched for surface proteins using Triton X-114. Of the 12 protein spots analyzed by LC-MS/MS, 4 with isoelectric points (p*I*s) ranging from 4.5 to 5.5 and apparent molecular masses from 32 to 34 kDa were found to contain common peptide sequences. Cloning of the corresponding cDNA revealed that these protein spots were products of a single gene, SAMP32, found to have testis-specific expression and to be localized at the inner acrosomal membrane of mature spermatozoa.

Proteomic studies have also sought the differences between the sperm of infertile and fertile men. This work has utilized both 1D ***(113,114)*** and 2D ***(95,106,109)*** separation strategies. Naz and Leslie ***(113)*** identified a 78-kDa protein band present in the sperm extract of infertile but not fertile men. Rajeev and Reddy ***(114)*** found a 57-kDa band present in normozoospermic individuals but absent or barely detectable in infertile individuals. Studies utilizing 2-DE gel electrophoresis have predictably detected more alterations in protein expression. Morgantaler et al. ***(95)*** noted 36 protein spots in 2-D gels of sperm from 23 normozoospermic individuals that were not consistently seen in the corresponding gels of sperm from 10 infertile oligospermic men. Lefièvre et al. ***(106)*** described approximately 60 spots absent in silver-stained 2D gels of sperm from men with globozoospermia that were present in sperm from normal controls. Perhaps the most detailed study is from Pixton et al. ***(109)***, who compared the 2D gel profiles of sperm from three fertile controls with that of a patient with normal sperm parameters but who experienced fertilization failure at IVF (in vitro fertilization). Twenty different spots were noted (i.e., six spots present in the controls but absent in the patient; three spots present in the patient but absent in the controls; seven spots more intense [more than fourfold] in the patient; and four spots more intense [more than fourfold] in the controls). Identification of the differentially expressed proteins was attempted following manual excision of the spots, ingel tryptic digestion, and LC-MS/MS. However, only two spots, both with increased expression in the patient sample, were identified, including secretory actin-binding protein and an 18-kDa fragment of outer dense fiber protein 2/2.

3. Perspectives and Future Directions

Only limited advantage has been taken of the rapid evolution of proteomic technology for studies of human seminal fluid. Nonetheless, studies of the protein constituents of spermatozoa and the seminal plasma within which they are suspended have made significant contributions to our understanding of reproductive biology in health and in disease. Proteins identified in seminal fluid have already found potential for use as markers for assessment of reproductive function, i.e., neutral α-glucosidase has been used to determine whether azoospermia is caused by epididymal obstruction ***(115)***; semenogelin 1 ***(44)*** and protein C inhibitor ***(45)*** have been used to assess seminal vesicle function; and PSA and prostatic acid phosphatase serve as markers of prostate function. PSA ***(116–118)*** and semenogelin I ***(46)*** have been used in forensic tests for semen detection, even from azoospermic individuals. In addition, proteomic studies of spermatozoa are providing insights into immune infertility and identifying potential targets for contraceptive vaccines.

Future proteomic studies of seminal fluid have the potential to provide new biomarkers of prostate disease, notably prostate cancer. It is instructive to note that the discovery of PSA, arguably the most widely used serum cancer biomarker to date, resulted from studies performed in prostate tissue ***(119,120)*** and seminal fluid ***(59)***. PSA was only found to be a useful circulating marker for the diagnosis and monitoring of prostate cancer after its identification in tissue and the appropriate proximal biological fluid ***(121)***. It is likely that analogous studies will find panels of markers, rather than single analytes, that can be employed for diagnostic purposes. Existing data indicate that candidates for such a biomarker panel include different forms of PSA and other proteins such as kallikrein-2, and PSP-94, as well as various cytokines, growth factors, and cellular adhesion molecules.

New and powerful techniques are available that allow for comparative proteomic studies. One such method is differential gel electrophoresis (DIGE), a technique whereby two or more samples are labeled with spectrally distinct fluorescent dyes and run simultaneously on the same gel. This approach allows precise relative quantification of differences between biological samples and is compatible with subsequent protein identification by MS. DIGE allows rapid and accurate comparisons of two or more samples without the limitations associated with across-gel comparisons. In addition, complementary mass spectrometric techniques using stable isotope labeling (e.g., ICAT or iTRAQ) have the potential to provide similar comparisons of proteins and peptides in complex samples. The use of these techniques in future studies of seminal fluid is likely to provide important insights into the molecular pathobiology of reproductive and urological disease and will likely help in defining clinically useful biomarkers of disease.

References

1. *WHO Laboratory Manual for the Examination of Semen and Sperm-Cervical Mucus interactions*, 4th ed. Cambridge: Cambridge University Press, 1999.
2. Mann T. *The Biochemistry of Semen and of the Male Reproductive Tract*. London: Methuen, 1964.
3. Jenkins AD, Turner TT, Howards SS. Physiology of the male reproductive system. *Urol Clin North Am* 1978;5:437–450.
4. Rui H, Mevag B, Purvis K. Two-dimensional electrophoresis of proteins in various fractions of the human split ejaculate. *Int J Androl* 1984;7:509–520.
5. Amelar RD, Hotchkiss RS. The split ejaculate: its use In the management of male infertility. *Fertil Steril* 1965;16:46–60.
6. Mann T. Biochemistry of semen. In Greep RO, Astwood EB, eds. *Handbook of Physiology: Endocrinology*. American Physiology Society, Washington, 1975.
7. Arienti G, Saccardi C, Carlini E, Verdacchi R, Palmerini CA. Distribution of lipid and protein in human semen fractions. *Clin Chim Acta* 1999;289:111–120.
8. Renneberg H, Konrad L, Dammshauser I, Seitz J, Aumuller G. Immunohistochemistry of prostasomes from human semen. *Prostate* 1997;30:98–106.
9. Lilja H, Oldbring J, Rannevik G, Laurell CB. Seminal vesicle-secreted proteins and their reactions during gelation and liquefaction of human semen. *J Clin Invest* 1987;80:281–285.
10. Gray S, Huggins C. Electrophoretic analysis of human semen. *Proc Soc Exp Biol Med* 1942;50:351.
11. Ross V, Moore DH, Miller EG. Proteins of human seminal plasma. *J Biol Chem* 1942;144:667.
12. Ross V, Miller EG, Moore DH, Sikorski H. Electrophoretic pattern of seminal plasma from some "abnormal" human semens. *Proc Soc Exp Biol Med* 1943;54:179.
13. Edwards JJ, Tollaksen SL, Anderson NG. Proteins of human semen. I. Two-dimensional mapping of human seminal fluid. *Clin Chem* 1981;27:1335–1340.
14. Ayyagari RR, Fazleabas AT, Dawood MY. Seminal plasma proteins of fertile and infertile men analyzed by two-dimensional electrophoresis. *Am J Obstet Gynecol* 1987;157:1528–1533.
15. Frenette G, Dube JY, Lazure C, Paradis G, Chretien M, Tremblay RR. The major 40-kDa glycoprotein in human prostatic fluid is identical to Zn-alpha 2-glycoprotein. *Prostate* 1987;11:257–270.
16. Lee C, Keefer M, Zhao ZW, et al. Demonstration of the role of prostate-specific antigen in semen liquefaction by two-dimensional electrophoresis. *J Androl* 1989;10:432–438.
17. Naaby-Hansen S, Flickinger CJ, Herr JC. Two-dimensional gel electrophoretic analysis of vectorially labeled surface proteins of human spermatozoa. *Biol Reprod* 1997;56:771–787.
18. Charrier JP, Tournel C, Michel S, Dalbon P, Jolivet M. Two-dimensional electrophoresis of prostate-specific antigen in sera of men with prostate cancer or benign prostate hyperplasia. *Electrophoresis* 1999;20:1075–1081.

19. Starita-Geribaldi M, Poggioli S, Zucchini M, et al. Mapping of seminal plasma proteins by two-dimensional gel electrophoresis in men with normal and impaired spermatogenesis. *Mol Hum Reprod* 2001;7:715–722.
20. Starita-Geribaldi M, Roux F, Garin J, Chevallier D, Fenichel P, Pointis G. Development of narrow immobilized pH gradients covering one pH unit for human seminal plasma proteomic analysis. *Proteomics* 2003;3:1611–1619.
21. Fung KY, Glode LM, Green S, Duncan MW. A comprehensive characterization of the peptide and protein constituents of human seminal fluid. *Prostate* 2004;61:171–181.
22. Kausler W, Spiteller G. Analysis of peptides of human seminal plasma by mass spectrometry. *Biol Mass Spectrom* 1992;21:567–575.
23. Robert M, Gibbs BF, Jacobson E, Gagnon C. Characterization of prostate-specific antigen proteolytic activity on its major physiological substrate, the sperm motility inhibitor precursor/semenogelin I. *Biochemistry* 1997;36:3811–3819.
24. Li CH, Hammonds RG Jr, Ramasharma K, Chung D. Human seminal alpha inhibins: isolation, characterization, and structure. *Proc Natl Acad Sci U S A* 1985; 82:4041–4044.
25. Lilja H, Abrahamsson PA, Lundwall A. Semenogelin, the predominant protein in human semen. Primary structure and identification of closely related proteins in the male accessory sex glands and on the spermatozoa. *J Biol Chem* 1989;264: 1894–1900.
26. Goverde HJ, Bisseling JG, Wetzels AM, et al. A neuropeptide in human semen: oxytocin. *Arch Androl* 1998;41:17–22.
27. O'Mahony OA, Djahanbahkch O, Mahmood T, Puddefoot JR, Vinson GP. Angiotensin II in human seminal fluid. *Hum Reprod* 2000;15:1345–1349.
28. Fraioli F, Fabbri A, Gnessi L, et al. Beta-endorphin, Met-enkephalin, and calcitonin in human semen: evidence for a possible role in human sperm motility. *Ann NY Acad Sci* 1984;438:365–370.
29. Foresta C, Caretto A, Indino M, Betterle C, Scandellari C. Calcitonin in human seminal plasma and its localization on human spermatozoa. *Andrologia* 1986;18: 470–473.
30. Cockle SM, Aitken A, Beg F, Morrell JM, Smyth DG. The TRH-related peptide pyroglutamyglutamylprolinamide is present in human semen. *FEBS Lett* 1989; 252:113–117.
31. Cockle SM, Prater GV, Thetford CR, Hamilton C, Malone PR, Mundy AR. Peptides related to thyrotrophin-releasing hormone (TRH) in human prostate and semen. *Biochim Biophys Acta* 1994;1227:60–66.
32. Khan Z, Aitken A, Garcia JR, Smyth DG. Isolation and identification of two neutral thyrotropin releasing hormone-like peptides, pyroglutamylphenylalanineproline amide and pyroglutamylglutamineproline amide, from human seminal fluid. *J Biol Chem* 1992;267:7464–7469.
33. Smith ML, Luqman WA. Prolactin in seminal fluid. *Arch Androl* 1982;9:105–113.
34. Winslow JW, Shih A, Bourell JH, et al. Human seminal relaxin is a product of the same gene as human luteal relaxin. *Endocrinology* 1992;130:2660–2668.

35. Fraser LR, Osiguwa OO. Human sperm responses to calcitonin, angiotensin II and fertilization-promoting peptide in prepared semen samples from normal donors and infertility patients. *Hum Reprod* 2004;19:596–606.
36. Unlu M, Morgan ME, Minden JS. Difference gel electrophoresis: a single gel method for detecting changes in protein extracts. *Electrophoresis* 1997;18: 2071–2077.
37. Aumuller G, Riva A. Morphology and functions of the human seminal vesicle. *Andrologia* 1992;24:183–196.
38. Gonzales GF. Function of seminal vesicles and their role on male fertility. *Asian J Androl* 2001;3:251–258.
39. Lilja H, Laurell CB. The predominant protein in human seminal coagulate. *Scand J Clin Lab Invest* 1985;45:635–641.
40. Robert M, Gagnon C. Semenogelin I: a coagulum forming, multifunctional seminal vesicle protein. *Cell Mol Life Sci* 1999;55:944–960.
41. Bjartell A, Malm J, Moller C, Gunnarsson M, Lundwell A. Distribution and tissue expression of semenogelin I and II in man as demonstrated by in situ hybridization and immunocytochemistry. *J Androl* 1996;17:17–26.
42. Hermans JM, Jones R, Stone SR. Rapid inhibition of the sperm protease acrosin by protein C inhibitor. *Biochemistry* 1994;33:5440–5444.
43. Laurell M, Christensson A, Abrahamsson PA, Stenflo J, Lilja H. Protein C inhibitor in human body fluids. Seminal plasma is rich in inhibitor antigen deriving from cells throughout the male reproductive system. *J Clin Invest* 1992; 89:1094–1101.
44. Calderon I, Barak M, Abramovici H, et al. The use of a seminal vesicle specific protein (MHS-5 antigen) for diagnosis of agenesis of vas deferens and seminal vesicles in azoospermic men. *J Androl* 1994;15:603–607.
45. Kise H, Nishioka J, Satoh K, Okuno T, Kawamura J, Suzuki K. Measurement of protein C inhibitor in seminal plasma is useful for detecting agenesis of seminal vesicles or the vas deferens. *J Androl* 2000;21:207–212.
46. Sato I, Kojima K, Yamasaki T, et al. Rapid detection of semenogelin by one-step immunochromatographic assay for semen identification. *J Immunol Methods* 2004;287:137–145.
47. Nylander G. The electrophoretic pattern of prostatic proteins in normal and pathologic secretion. *Acta Chir Scand* 1955;109:473–482.
48. Atanasov NA, Gikov DG. Disc electrophoresis of prostatic fluid. Relative mobility and molecular weight of some enzymes and proteins. *Clin Chim Acta* 1972;36:213–221.
49. Resnick MI, Stubbs AJ. Age specific electrophoretic patterns of prostatic fluid. *J Surg Res* 1978;24:415–420.
50. Carter DB, Resnick MI. High resolution analysis of human prostatic fluid by two-dimensional electrophoresis. *Prostate* 1982;3:27–33.
51. Tsai YC, Harrison HH, Lee C, Daufeldt JA, Oliver L, Grayhack JT. Systematic characterization of human prostatic fluid proteins with two-dimensional electrophoresis. *Clin Chem* 1984;30:2026–2030.

52. Lee C, Tsai Y, Sensibar J, Oliver L, Grayhack JT. Two-dimensional characterization of prostatic acid phosphatase, prostatic specific antigen and prostate binding protein in expressed prostatic fluid. *Prostate* 1986;9:135–146.
53. Grover PK, Resnick MI. Analysis of prostatic fluid: evidence for the presence of a prospective marker for prostatic cancer. *Prostate* 1995;26:12–18.
54. Guevara J Jr, Herbert BH, Lee C. Two-dimensional electrophoretic analysis of human prostatic fluid proteins. *Cancer Res* 1985;45:1766–1771.
55. Watt KW, Lee PJ, M'Timkulu T, Chan WP, Loor R. Human prostate-specific antigen: structural and functional similarity with serine proteases. *Proc Natl Acad Sci U S A* 1986;83:3166–3170.
56. Iwakiri J, Granbois K, Wehner N, Graves HC, Stamey T. An analysis of urinary prostate specific antigen before and after radical prostatectomy: evidence for secretion of prostate specific antigen by the periurethral glands. *J Urol* 1993;149: 783–786.
57. Breul J, Pickl U, Schaff J. Extraprostatic production of prostate specific antigen is under hormonal control. *J Urol* 1997;157:212–213.
58. Christensson A, Lilja H. Complex formation between protein C inhibitor and prostate-specific antigen in vitro and in human semen. *Eur J Biochem* 1994; 220:45–53.
59. Sensabaugh GF. Isolation and characterization of a semen-specific protein from human seminal plasma: a potential new marker for semen identification. *J Forensic Sci* 1978;23:106–115.
60. Wang MC, Papsidero LD, Kuriyama M, Valenzuela LA, Murphy GP, Chu TM. Prostate antigen: a new potential marker for prostatic cancer. *Prostate* 1981;2:89–96.
61. Lovgren J, Valtonen-Andre C, Marsal K, Lilja H, Lundwall A. Measurement of prostate-specific antigen and human glandular kallikrein 2 in different body fluids. *J Androl* 1999;20:348–355.
62. Catalona WJ, Smith DS, Ratliff TL, et al. Measurement of prostate-specific antigen in serum as a screening test for prostate cancer. *N Engl J Med* 1991;324:1156–1161.
63. Huber PR, Schnell Y, Hering F, Rutishauser G. Prostate specific antigen. Experimental and clinical observations. *Scand J Urol Nephrol Suppl* 1987;104: 33–39.
64. Isono T, Tanaka T, Kageyama S, Yoshiki T. Structural diversity of cancer-related and non-cancer-related prostate-specific antigen. *Clin Chem* 2002;48:2187–2194.
65. Romas NA, Kwan DJ. Prostatic acid phosphatase. Biomolecular features and assays for serum determination. *Urol Clin North Am* 1993;20:581–588.
66. Ulvsback M, Lindstrom C, Weiber H, Abrahamsson PA, Lilja H, Lundwall A. Molecular cloning of a small prostate protein, known as beta-microsemenoprotein, PSP94 or beta-inhibin, and demonstration of transcripts in non-genital tissues. *Biochem Biophys Res Commun* 1989;164:1310–1315.
67. Dube JY, Frenette G, Paquin R, et al. Isolation from human seminal plasma of an abundant 16-kDa protein originating from the prostate, its identification with a 94-residue peptide originally described as beta-inhibin. *J Androl* 1987;8: 182–189.

68. Tremblay RR, Coulombe E, Cloutier S, et al. Assessment of the trypsin-like human prostatic kallikrein, also known as hK2, in the seminal plasma of infertile men: respective contributions of an ELISA procedure and of Western blotting. *J Lab Clin Med* 1998;131:330–335.
69. Szecsi PB, Halgreen H, Wong RN, Kjaer T, Tang J. Cellular origin, complementary deoxyribonucleic acid and N-terminal amino acid sequences of human seminal progastricsin. *Biol Reprod* 1995;53:227–233.
70. Mikolajczyk SD, Millar LS, Marker KM, et al. Seminal plasma contains "BPSA," a molecular form of prostate-specific antigen that is associated with benign prostatic hyperplasia. *Prostate* 2000;45:271–276.
71. Ronquist G, Brody I, Gottfries A, Stegmayr B. An Mg^{2+} and Ca^{2+}-stimulated adenosine triphosphatase in human prostatic fluid—part II. *Andrologia* 1978;10:427–433.
72. Arvidson G, Ronquist G, Wikander G, Ojteg AC. Human prostasome membranes exhibit very high cholesterol/phospholipid ratios yielding high molecular ordering. *Biochim Biophys Acta* 1989;984:167–173.
73. Arienti G, Carlini E, Polci A, Cosmi EV, Palmerini CA. Fatty acid pattern of human prostasome lipid. *Arch Biochem Biophys* 1998;358:391–395.
74. Ronquist G, Brody I. The prostasome: its secretion and function in man. *Biochim Biophys Acta* 1985;822:203–218.
75. Kravets FG, Lee J, Singh B, Trocchia A, Pentyala SN, Khan SA. Prostasomes: current concepts. *Prostate* 2000;43:169–174.
76. Saez F, Frenette G, Sullivan R. Epididymosomes and prostasomes: their roles in posttesticular maturation of the sperm cells. *J Androl* 2003;24:149–154.
77. Arienti G, Carlini E, Saccardi C, Palmerini CA. Role of human prostasomes in the activation of spermatozoa. *J Cell Mol Med* 2004;8:77–84.
78. Ronquist G, Nilsson BO. The Janus-faced nature of prostasomes: their pluripotency favours the normal reproductive process and malignant prostate growth. *Prostate Cancer Prostatic Dis* 2004;71:21–31.
79. Stewart AB, Anderson W, Delves G, Lwaleed BA, Birch B, Cooper A. Prostasomes: a role in prostatic disease? *BJU Int* 2004;94:985–989.
80. Ronquist G, Nilsson BO, Hjerten S. Interaction between prostasomes and spermatozoa from human semen. *Arch Androl* 1990;24:147–157.
81. Carlini E, Palmerini CA, Cosmi EV, Arienti G. Fusion of sperm with prostasomes: effects on membrane fluidity. *Arch Biochem Biophys* 1997;343:6–12.
82. Arienti G, Carlini E, Palmerini CA. Fusion of human sperm to prostasomes at acidic pH. *J Membr Biol* 1997;155:89–94.
83. Arienti G, Carlini E, Verdacchi R, Cosmi EV, Palmerini CA. Prostasome to sperm transfer of CD13/aminopeptidase N (EC 3.4.11.2). *Biochim Biophys Acta* 1997;1336:533–538.
84. Arienti G, Carlini E, Verdacchi R, Palmerini CA. Transfer of aminopeptidase activity from prostasomes to sperm. *Biochim Biophys Acta* 1997;1336:269–274.
85. Lindahl M, Tagesson C, Ronquist G. Phospholipase A2 activity in prostasomes from human seminal plasma. *Urol Int* 1987;42:385–389.

86. Carlsson L, Ronquist G, Nilsson BO, Larsson A. Dominant prostasome immunogens for sperm-agglutinating autoantibodies of infertile men. *J Androl* 2004;25: 699–705.
87. Carlsson L, Nilsson BO, Ronquist G, Lundquist M, Larsson A. A new test for immunological infertility: an ELISA based on prostasomes. *Int J Androl* 2004; 27:130–133.
88. Utleg AG, Yi EC, Xie T, et al. Proteomic analysis of human prostasomes. *Prostate* 2003;56:150–161.
89. Schrimpf SP, Hellman U, Carlsson L, Larsson A, Ronquist G, Nilsson BO. Identification of dipeptidyl peptidase IV as the antigen of a monoclonal anti-prostasome antibody. *Prostate* 1999;38:35–39.
90. Fernandez D, Valdivia A, Irazusta J, Ochoa C, Casis L. Peptidase activities in human semen. *Peptides* 2002;23:461–468.
91. Lilja H, Weiber H. gamma-Glutamyltransferase bound to prostatic subcellular organelles and in free form in human seminal plasma. *Scand J Clin Lab Invest* 1983;43:307–312.
92. Mann T, Lutwak-Mann C. *Male Reproductive Function and Semen*. Berlin: Springer Verlag, 1981.
93. Millette CF. Spermatozoa. In: Knobil E, Neill JD, eds. *Encyclopedia of Reproduction*. Academic Press, San Diego, CA, 1999:586–596.
94. Mack SR, Zaneveld LJ, Peterson RN, Hunt W, Russell LD. Characterization of human sperm plasma membrane: glycolipids and polypeptides. *J Exp Zool* 1987; 243:339–346.
95. Morgentaler A, Schopperle WM, Crocker RH, DeWolf WC. Protein differences between normal and oligospermic human sperm demonstrated by two-dimensional gel electrophoresis. *Fertil Steril* 1990;54:902–905.
96. Naaby-Hansen S. Electrophoretic map of acidic and neutral human spermatozoal proteins. *J Reprod Immunol* 1990;17:167–185.
97. Primakoff P, Lathrop W, Bronson R. Identification of human sperm surface glycoproteins recognized by autoantisera from immune infertile men, women, and vasectomized men. *Biol Reprod* 1990;42:929–942.
98. Kritsas JJ, Schopperle WM, DeWolf WC, Morgentaler A. Rapid high resolution two-dimensional electrophoresis of human sperm proteins. *Electrophoresis* 1992;13:445–449.
99. Xu C, Rigney DR, Anderson DJ. Two-dimensional electrophoretic profile of human sperm membrane proteins. *J Androl* 1994;15:595–602.
100. Shetty J, Naaby-Hansen S, Shibahara H, Bronson R, Flickinger CJ, Herr JC. Human sperm proteome: immunodominant sperm surface antigens identified with sera from infertile men and women. *Biol Reprod* 1999;61:61–69.
101. Bohring C, Krause W. The characterization of human spermatozoa membrane proteins—surface antigens and immunological infertility. *Electrophoresis* 1999; 20:971–976.
102. Shetty J, Diekman AB, Jayes FC, et al. Differential extraction and enrichment of human sperm surface proteins in a proteome: identification of immunocontraceptive candidates. *Electrophoresis* 2001;22:3053–3066.

103. Bohring C, Krause E, Habermann B, Krause W. Isolation and identification of sperm membrane antigens recognized by antisperm antibodies, and their possible role in immunological infertility disease. *Mol Hum Reprod* 2001;7:113–118.
104. Shibahara H, Sato I, Shetty J, et al. Two-dimensional electrophoretic analysis of sperm antigens recognized by sperm immobilizing antibodies detected in infertile women. *J Reprod Immunol* 2002;53:1–12.
105. Hao Z, Wolkowicz MJ, Shetty J, et al. SAMP32, a testis-specific, isoantigenic sperm acrosomal membrane-associated protein. *Biol Reprod* 2002;66:735–744.
106. Lefievre L, Barratt CL, Harper CV, et al. Physiological and proteomic approaches to studying prefertilization events in the human. *Reprod Biomed Online* 2003;7:419–427.
107. Bohring C, Krause W. Characterization of spermatozoa surface antigens by antisperm antibodies and its influence on acrosomal exocytosis. *Am J Reprod Immunol* 2003;50:411–419.
108. Ficarro S, Chertihin O, Westbrook VA, et al. Phosphoproteome analysis of capacitated human sperm. Evidence of tyrosine phosphorylation of a kinase-anchoring protein 3 and valosin-containing protein/p97 during capacitation. *J Biol Chem* 2003;278:11,579–11,589.
109. Pixton KL, Deeks ED, Flesch FM, et al. Sperm proteome mapping of a patient who experienced failed fertilization at IVF reveals altered expression of at least 20 proteins compared with fertile donors: case report. *Hum Reprod* 2004;19:1438–447.
110. Primakoff P, Myles DG. Penetration, adhesion, and fusion in mammalian sperm-egg interaction. *Science* 2002;296:2183–2185.
111. Bohring C, Krause W. Immune infertility: towards a better understanding of sperm (auto)-immunity. The value of proteomic analysis. *Hum Reprod* 2003;18:915–924.
112. Herr JC. Update on the Center for Recombinant Gamete Contraceptive Vaccinogens. *Am J Reprod Immunol* 1996;35:184–189.
113. Naz RK, Leslie MH. Sperm surface protein profiles of fertile and infertile men: search for a diagnostic molecular marker. *Arch Androl* 1999;43:173–181.
114. Rajeev SK, Reddy KVR. Sperm membrane protein profiles of fertile and infertile men: identification and characterization of fertility-associated sperm antigen. *Hum Reprod* 2004;19:234–242.
115. Pena P, Risopatron J, Villegas J, Miska W, Schill WB, Sanchez R. Alpha-glucosidase in the human epididymis: topographic distribution and clinical application. *Andrologia* 2004;36:315–320.
116. Graves HC, Sensabaugh GF, Blake ET. Postcoital detection of a male-specific semen protein. Application to the investigation of rape. *N Engl J Med* 1985;312: 338–343.
117. Hochmeister MN, Budowle B, Rudin O, et al. Evaluation of prostate-specific antigen (PSA) membrane test assays for the forensic identification of seminal fluid. *J Forensic Sci* 1999;44:1057–1060.
118. Simich JP, Morris SL, Klick RL, Rittenhouse-Diakun K. Validation of the use of a commercially available kit for the identification of prostate specific antigen (PSA) in semen stains. *J Forensic Sci* 1999;44:1229–1231.

119. Ablin RJ, Soanes WA, Bronson P, Witebsky E. Precipitating antigens of the normal human prostate. *J Reprod Fertil* 1970;22:573–574.
120. Wang MC, Valenzuela LA, Murphy GP, Chu TM. Purification of a human prostate specific antigen. *Invest Urol* 1979;17:159–163.
121. Papsidero LD, Wang MC, Valenzuela LA, Murphy GP, Chu TM. A prostate antigen in sera of prostatic cancer patients. *Cancer Res* 1980;40:2428–2432.

23

Proteomics of Vitreous Fluid

Atsushi Minamoto, Ken Yamane, and Tomoko Yokoyama

Summary

Vitreous and serum samples were obtained from subjects with diabetic retinopathy (DR; 33 cases) and idiopathic macular hole (MH; 26 cases), at the time of pars plana vitrectomy. The expressed proteins were separated by 2D gel electrophoresis. Separated protein spots were then visualized by silver staining and analyzed by mass spectrometry. For the MH vitreous samples, more than 400 spots were detected on 2D gels, of which 78 spots were identified as 18 unique proteins, including pigment epithelium-derived factor (PEDF), prostaglandin-D2 synthase, plasma glutathione peroxidase, and interphotoreceptor retinoid-binding protein (IRBP), which were not identified in the corresponding serum samples. For the DR vitreous samples, more than 600 spots were detected on gels, and 141 spots were identified as 38 unique proteins, some of which were derived from serum. Enolase and catalase were identified among four detected spots; neither of them was found in MH vitreous or DR serum samples. The increased protein expression observed in DR vitreous samples may be due to barrier dysfunction and/or production in the eye.

Key Words: Diabetic retinopathy; macular hole; oxidative stress; PEDF; proteome; vitrectomy; vitreous.

1. Introduction

The intraocular space between the lens, zonules, ciliary body, and retina is filled with a transparent gel or liquid that is termed the vitreous. The vitreous contains more than 99% water; the rest is composed of solids. The two major structural components are collagen and hyaluronic acid. The former confers a gel-like property, whereas the latter provides a viscoelastic consistency that resists compression and also stabilizes the collagen network. Also contained in the vitreous are soluble proteins. In addition to its space-filling function, the vitreous provides mechanical support to the surrounding ocular tissues and, because of its viscoelastic property, serves as a shock absorber against mechanical impact ***(1,2)***.

From: *Proteomics of Human Body Fluids: Principles, Methods, and Applications*
Edited by: V. Thongboonkerd © Humana Press Inc., Totowa, NJ

In the normal eye, the vitreous appears to be a quiescent compartment. However, the juxtaposition of the retina, a metabolically active tissue, with the vitreous suggests that changes in the retina would affect the vitreous. Retinal vascular occlusive diseases such as diabetic retinopathy and retinal vein occlusion are commonly associated with significant vitreous change ***(3)***. In diabetic retinopathy (DR), the neovascularization arises from the retina and is attached to the posterior vitreous surface. Fibrous tissue that accompanies new vessels may lie on the surface of the retina, forming epiretinal membrane, or may extend forward onto the posterior surface of the vitreous. It is most developed at the sites where the vitreous is attached firmly along the vascular arcades surrounding the central retina (macula). Fibrous tissue sheets in front of the retina cause visual loss. Additionally, fibrous bands connecting fibrous tissue along superior and inferior vessels may result in traction retinal detachment of the macula, resulting in profound visual loss. Moreover, DR is the leading cause of visual loss in working, aged people in the industrialized world.

Currently, intense interest exists in vasoproliferative factors released by the retina itself, retinal vessels, and the retinal pigment epithelium, which are thought to induce neovascularization and subsequent fibrovascular proliferation of the retina and vitreous in diabetic eyes. The blood-retinal barrier (BRB) is a blood barrier to control the permeability of proteins and water-soluble substances to prevent retinal edema and interference with neurotransmission. A breakdown of the BRB is caused by an intraocular increase in vascular endothelial growth factor (VEGF) ***(4–10)***, interleukin-6, angiotensin II, and many other cytokines and/or growth factors in DR. New vessel formation is a complex multistep process and is regulated by many proteins including cytokines and/or growth factors. In addition to the factors mentioned, basic fibroblast growth factor (bFGF) ***(11,12)***, insulin-like growth factor-1 ***(7)***, hepatocyte growth factor (HGF) ***(9,10)***, and others are known to be involved during the destructive process of the endogenous ocular tissue. Changes in expression levels of these factors have been described by the measurement of these substances using enzyme-linked immunosorbent assay (ELISA) in the aqueous and vitreous of eyes with DR. In these studies, however, substances for measurement were targeted in advance, and the targets were limited because of the small amount of available sample material.

To examine systematically and exhaustively which factors are increased and what the causes of those increases are, a methodology to examine without preconceptions is needed. We have adopted 2D gel-based proteome analysis ***(13)*** to investigate protein profiles of human vitreous with DR compared with idiopathic macular hole (MH) as the control. MH formation is caused by a spontaneous, usually abrupt, focal contraction of the vitreous cortex, which elevates the retina at the macula ***(14)***, and thus MH is considered to be a silent and localized retinal lesion.

2. Methods

Proteomic analysis was performed as described previously ***(15)***.

2.1. 2D Polyacrylamide Gel Electrophoresis (2D-PAGE)

1. Vitreous sample (40 μL) and serum sample (3 μL) were diluted in a lysis buffer containing 7 *M* urea, 2 *M* thiourea, 4% 3-[(3-cholamidopropyl)dimethylammonio]-l-propanesulflonate (CHAPS), 2% ampholytes, pH 3.5 to 10 (Amersham Pharmacia Biotech, Uppsala, Sweden), and 1% dithiothreitol to a final volume of 400 μL.
2. Protein samples were applied overnight to Immobiline Dry Strips (Amersham Pharmacia Biotech; pH 4–7, 3–10, 18 cm) by in-gel rehydration ***(16,17)***.
3. For the first dimension, a Multiphor II electrophoresis chamber (Amersham Pharmacia Biotech) was used. Isoelectric focusing (IEF) was performed with the following voltage program: 500 V for 2 min, 3500 V for 1.5 h, and then 3500 V for 6 h.
4. Immobiline Dry Strips were stored at –80°C until the 2D electrophoresis was carried out.
5. 2D separation was performed in 9 to 18% acrylamide gradient gels (20 × 20 cm) using the Iso-Dalt system (Amersham Pharmacia Biotech).
6. The protein spots were visualized by silver staining ***(18)***, and the 2D gels were scanned on an Epson ES 80000 scanner (Seiko Epson, Suwa, Japan).
7. After scanning, the 2D gels were sandwiched between two cellophane sheets and dried.
8. Image analysis and 2D gel proteome database management were done using the Melanie II 2D-PAGE software package (Bio-Rad).

2.2. In-Gel Tryptic Digestion

1. Protein spots were excised from the dried silver-stained gels by a gel cutter and rehydrated in 100 m*M* ammonium carbonate.
2. The gel pieces were destained with 30 m*M* potassium ferricyanide and 100 m*M* sodium thiosulfate and then rinsed a few times with Milli-Q (Millipore, Billerica, MA) water and once in 100 m*M* ammonium carbonate.
3. Dehydration was performed using acetonitrile until the gel pieces turned opaque white; they were subsequently dried in vacuum centrifuge.
4. The gel pieces were then rehydrated in a dilution buffer containing trypsin, and proteins underwent in-gel digestion overnight at 37°C.
5. The digestion was stopped by covering the gel pieces with 5% trifluoroacetic acid, and the peptides were extracted three times with 5% trifluoroacetic acid in 50% acetonitrile.
6. The extracted peptides were pooled and dried in a vacuum centrifuge.

2.3. Electrospray Ionization (ESI) Mass Spectrometry (MS) and Protein Identification

1. The peptides were resuspended with 1% formic acid in 4% methanol and loaded onto an OLIGO R3 column (PerSeptive Biosystems, Framingham, MA).
2. After the column was washed with 1% formic acid, the peptides were eluted with 1% formic acid in 70% methanol and subjected to mass analysis.
3. The eluted peptides were loaded into Au/Pd-coated nanoES spray capillaries (Protana, Odense, Denmark).
4. The capillaries were inserted into the nano-flow Z-spray source of a quadrupole time-of-flight (Q-TOF) mass spectrometer (Micromass, Manchester, UK).
5. Instrument operation, data acquisition, and analysis were performed by MassLynx/Biolynx 3.2 software (Micromass).
6. The Q-TOF was operated in two modes: MS and MS/MS.
7. The proteins were identified by matching the obtained amino acid sequences against the Swiss-Prot and Genbank databases using the GenomeNet Internet server of Kyoto University, Japan (www.fasta.genome.ad.jp/).

2.4. Matrix-Assisted Laser Desorption/Ionization-Mass Spectrometry (MALDI-MS) and Protein Identification

1. The peptides were dissolved in 5 μL of 0.1% trifluoroacetic acid and 50% acetonitrile. The matrix was α-cyano-4-hydroxy cinnamic acid (CHCA; 10 mg) in a 1-mL mixture of 0.1% trifluoroacetic acid and 50% acetonitrile.
2. The peptide mixture (1 μL) was deposited on the sample plate, and the solvents were removed by air-drying at room temperature.
3. The matrix mixture (1 μL) was then deposited on the sample plate, on top of the peptide mixture.
4. MALDI-MS was carried out on a Voyager-DE STR (PerSeptive Biosystems) in the reflector mode.
5. The laser wavelength was 337 nm, and the laser repetition rate was 3 Hz.
6. The MALDI spectra were averaged at 500 laser pulses.
7. Calibration was performed with four peptides: des-Arg1-bradykinin, angiotensin, Glu1-fibrinopeptide B, and adrenocorticotropic hormone (ACTH) ***(18–39)*** with monoisotopic $(M+H)^+$ at *m/z* 904.4681, 1296.6853, 1570.6774, and 2465.1989, respectively.
8. Peak lists were searched against the NCBInr protein sequence database using the MS-Fit search tool (MS tolerance 0.3 Da) to identify the proteins.

3. Findings and Applications

3.1. Patients and Clinical Samples

Undiluted vitreous samples and corresponding serum samples were obtained from 59 eyes (51 patients) with DR (n = 33) and MH (n = 26) at the time of vitreous surgery, after securing written permission from all patients. All patients were informed of the purpose of the study and the nature and potential adverse effects of sampling procedures. The Institutional Review Board of Hiroshima University approved the protocol for collection of samples.

1. Three-port pars plana vitrectomy was carried out in all subjects.
2. A slit-like sclerotomy was made at the lower temporal quadrant, 3.5 mm from the limbus.
3. An infusion cannula with a 4-mm shaft was then inserted through the sclerotomy.
4. Sclerotomy for insertion of the light pipe and vitrectomy probe was made in the same manner as that for the infusion cannula but was located 3.5 mm from the limbus in superior quadrants.
5. After 0.3 to 0.8 mL undiluted vitreous samples were obtained, the infusion line was opened for the maintenance of intraocular pressure.
6. Harvested vitreous and blood samples were collected in tubes, placed immediately on ice, centrifuged for 15 min to separate the cell contents, and stored at –80°C until use.
7. The protein concentration of each sample was measured on a V-1500 spectrophotometer (Hitachi, Tokyo, Japan) using a Bio-Rad Protein Assay Kit (Bio-Rad, Hercules, CA). The average protein concentration of DR vitreous was 4.13 μg/μL (range: 1.28–7.26 μg/μL), whereas that of MH vitreous was 0.47 μg/μL (range: 0.10–1.00 μg/μL).

3.2. 2D-PAGE of the Vitreous From Patients With DR and MH

Totals of 511 to 785 spots (average, 617 spots) and 558 to 965 spots (average, 779 spots) were detected on silver-stained 2D gels of DR vitreous proteins using immobilized pH gradient (IPG) 4 to 7 (28 of 33 eyes) and 3 to 10 (12 of 33 eyes), respectively; 371 to 519 spots (average, 437 spots) and 359 to 580 spots (average, 463 spots) were detected in vitreous samples from eyes with MH using IPG 4 to 7 (23 of 26 eyes) and 3 to 10 (10 of 26 eyes), respectively (**Figs. 1** and **2**). The pattern of protein expression of each disorder was reproducible among vitreous samples, regardless of the patient's age and gender.

3.3. Comparison Between Vitreous and Serum Samples (Vitreous-Specific Proteins)

A total of 18 proteins were identified from 78 spots in MH, and most of them corresponded to serum proteins. These included transferrin, albumin, α_1-antitrypsin, α_1-antichymotrypsin, α2-HG-glycoprotein, antithrombin III, hemopexin, fibrinogen γ-chain, haptoglobin-1, apolipoprotein J, apolipoprotein A-1, IgG heavy chain, IgG light chain, and transthyretin. However, 26 spots of vitreous samples were not identified in the corresponding serum samples. Of these, 16 spots were identified as polypeptide fragments of pigment epithelium-derived factor (PEDF), prostaglandin-D2 (PGD2) synthase, plasma glutathione peroxidase, and interphotoreceptor retinoid-binding protein (IRBP) **(Fig. 1)**.

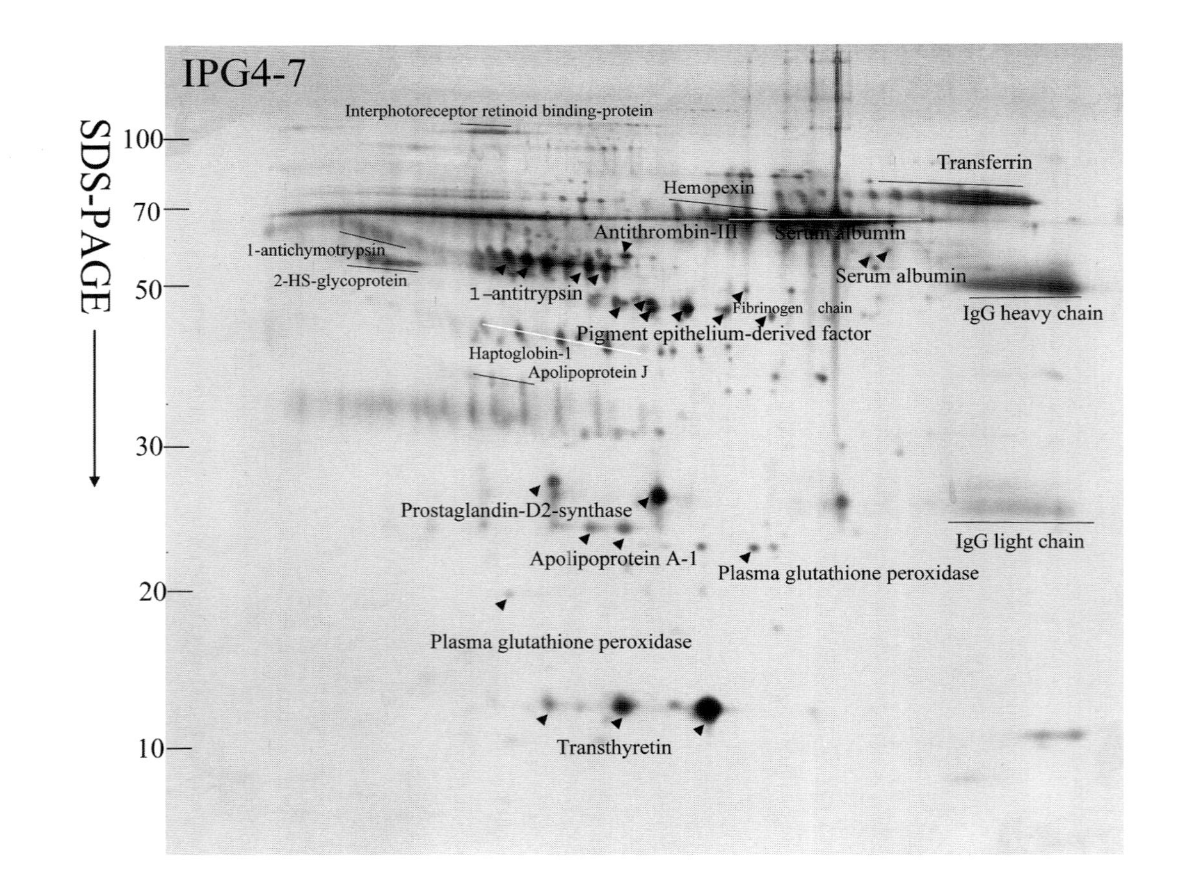

Fig. 1. Silver-stained 2D gel of vitreous proteins obtained from eyes with MH. IEF was performed using IPG 4-7. A total of 18 proteins were identified from 78 spots.

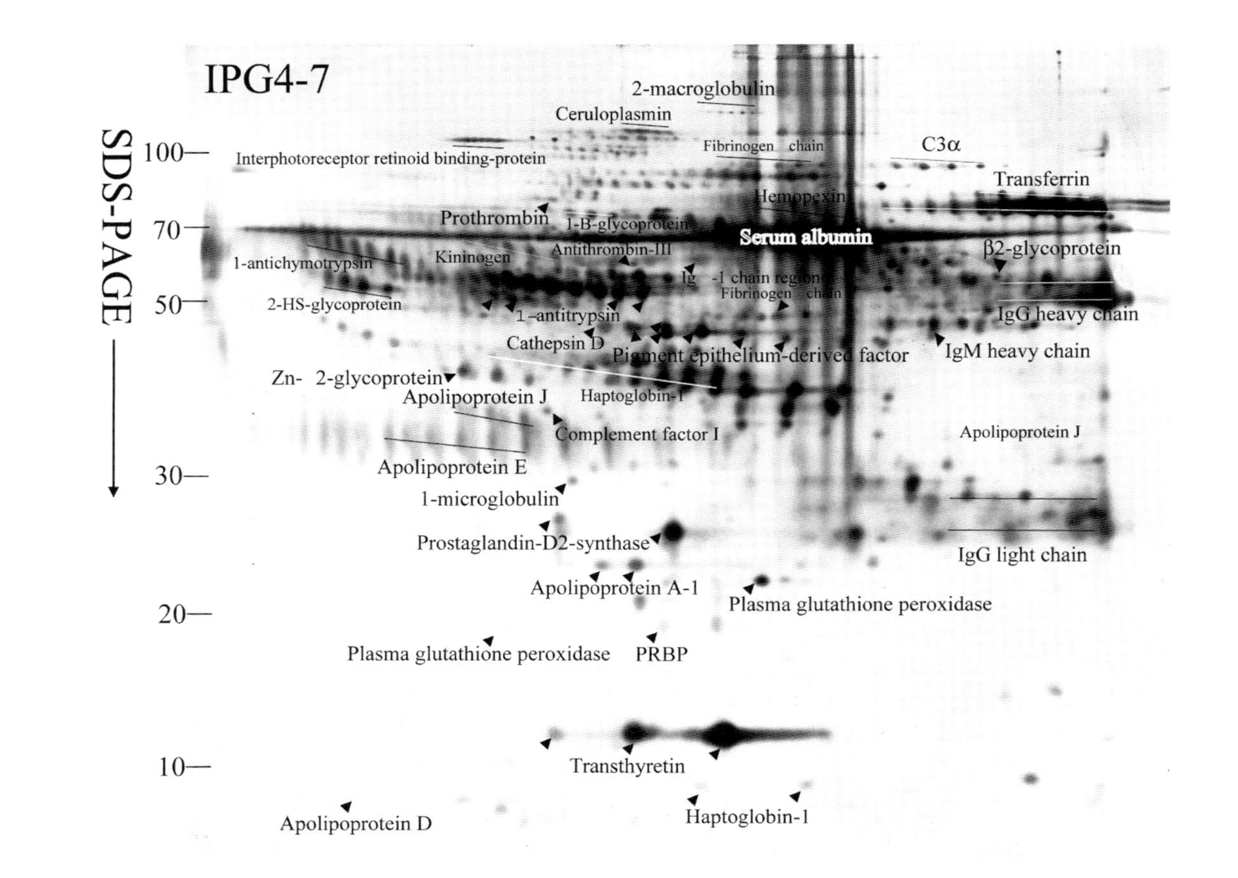

Fig. 2. Silver-stained 2D gel of vitreous proteins obtained from eyes with DR. IEF was performed using IPG 4-7. A total of 36 proteins were identified from 136 spots.

PGD2 synthase, a 26-kDa glycoprotein, is a member of the lipocalin family of secretory proteins and hydrophobic molecule transporters such as β-lactoglobulin and retinol-binding protein *(19)*. PGD2 is the major prostaglandin formed in the eye *(20)*. PEDF is present in aqueous humor, interphotoreceptor matrix, and vitreous *(21)* and has been reported to play a role as a potent inhibitor of angiogenesis *(22–24)*. However, in our study, there was no distinct tendency toward difference in PEDF spot intensity volume between DR and MH samples. Therefore, we supposed that the balance between angiogenic stimulators and inhibitors should be chiefly regulated by the increase or decrease of angiogenic stimulators such as VEGF and that PEDF, as an angiogenic inhibitor, should support the effect of increased angiogenic stimulation to a certain level. The PEDF level in the vitreous has been reported to vary far less than that of VEGF between DR and MH *(25)*.

3.4. Comparison Between Vitreous Samples of DR and MH

A total of 36 proteins were identified from 136 spots in DR. The spots identified in MH vitreous samples were also identified in DR vitreous samples, and these common spots with increased volume were all identified in serum samples as well. However, a total of 18 proteins were present in vitreous samples of DR and in sera but were not identified in MH vitreous samples. These 18 proteins included α_2-macroglobulin, Ig α-1 chain C region, α1-B-glycoprotein, α_1-microglobulin, C3α, ceruloplasmins, apolipoprotein E, complement C4, complement factor D, PRBP, complement factor I, Zn-α2-glycoprotein, kininogen, apolipoprotein D, cathepsin D, and prothrombin. The presence of these proteins in vitreous considered to be the result of the breakdown of the BRB in DR (**Fig. 2**). Cathepsin D, one of the components of the proteolytic machinery for misfolded proteins, has been known to be upregulated in VEGF-activated human endothelial cells *(26)*. This seems to suggest the presence of inflammation and decomposition of proteins *(27)*, consistent with new evidence indicating that DR may be an inflammatory disease *(28)*.

The spots among the area of higher p*I* range (p*I* > 6) with 20 to 70 kDa molecular mass were overshadowed by polypeptide fragments of IgG, which were abundant and of great diversity. Therefore, as a separate series of analyses, we adopted IgG removal procedures in the sample preparation prior to 2D-PAGE with IPG 3 to 10, with samples of MH (n = 3) and DR (n = 5). Briefly:

1. Larger amounts of vitreous (200 μL) and serum (15 μL) samples were shaken overnight with 15 μL Protein A Sepharose 4 Fast Flow (Amersham Pharmacia Biotech).
2. The Protein A beads were precipitated by centrifugation at 5200*g*, and then clear supernatant liquid was collected.

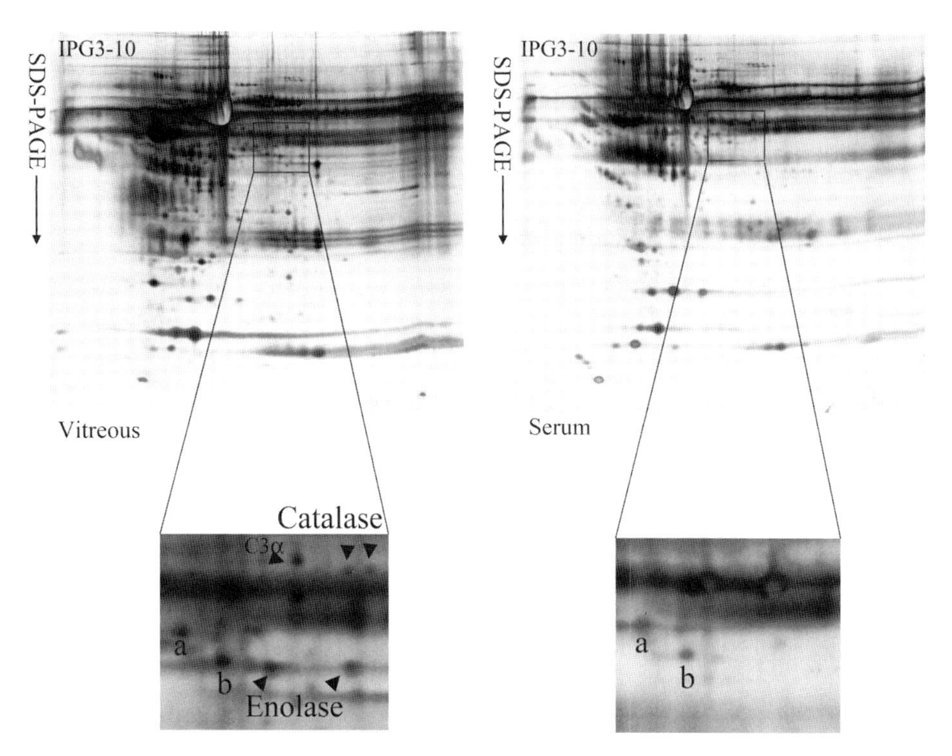

Fig. 3. Comparison between vitreous and corresponding serum sample of DR. IgG removal procedures were adopted in the sample preparation prior to 2D PAGE with IPG 3-10. The clear up-regulation of enolase and catalase was seen in the vitreous sample.

3. After the Protein A bead was washed five times with 500 mL PBS, clear supernatant liquid was collected.
4. An UltraFree column (Millipore) was used to concentrate sample fluids.

With this modification of sample preparation, we could recognize five spots in DR vitreous samples that were found neither in MH vitreous samples nor in DR serum samples. Among these five spots, four were identified as enolase and catalase **(Fig. 3)**. The presence of these two proteins in DR vitreous samples, which were undetected in corresponding serum samples and MH vitreous samples, may suggest their roles in the development and exacerbation of vitreoretinal disorders in DR, or they may be related to the damage of retinal tissue.

Enolase, which was confirmed to be neuron-specific enolase (NSE) by Western blotting ***(13)***, may be used as a marker of acute neuronal damage in humans with neurological disorders. Cerebral ischemia in rats and humans induces the release of NSE to the extracellular matrix ***(29,30)***. Retinal neuron injury in retinal detachment releases sufficient NSE to be detected in subretinal fluid, aqueous humor, and even serum ***(31)***.

It has been suggested that, in diabetes, oxidative stress plays a key role in the pathogenesis of vascular complications such as DR *(32)*. Catalase, an antioxidant enzyme, is possibly serving as a protective substance within the diseased eye. Its presence, therefore, implies a role in acute ischemic disease of the retina, such as DR *(27,33,34)*.

4. Perspectives and Future Directions

At present, completing the reference library of proteins expressed in the vitreous is difficult because sample preparation of vitreous fluid for proteomic analysis presents several problems. It has been reported that albumin and immunoglobulin account for over 80% of whole-vitreous protein. Because the large area of multiple spots of albumin and immunoglobulin obscure small other spots, low-abundant proteins might be unrecognized. Moreover, proteins are lost during 2D-PAGE and extraction from stained gels. Poor solubility of some proteins and charge heterogeneity are often refractory to 2D-PAGE *(35)*. It has been reported that more proteins, including angiogenic and antiangiogenic factors, could be identified from vitreous samples in diabetes by using 1D-PAGE, blind cutting of gels, and extraction from unstained gels, compared with identification using 2D-gel-based proteomics *(36)*.

Further improvements in gel electrophoretic technique, staining methods, and MS technologies should contribute to a more detailed analysis of vitreous protein profiles, thereby deepening our knowledge of the pathological features of sight-threatening vitreoretinal disorders.

Acknowledgments

This work was supported by a Health Science Research Grant, no. 12120101, from the Japanese Ministry of Health and Welfare for Research on Eye and Ear Sciences, Immunology, Allergy, and Organ Transplantation. The authors thank Prof. Hiromu K. Mishima and Prof. Hidetoshi Yamashita for their support.

References

1. Berman ER. Vitreous. In: *Biochemistry of the Eye*. New York: Plenum, 1991:291–307.
2. Lund-Andersen H, Sander B. The vitreous. In: *Adler's Physiology of the Eye*. St. Louis: Mosby, 2003:293–316.
3. McMeel JW, Jalkh AE. Vitreous changes in vascular diseases. In: Schepens CL and Neetens A, eds. *The Vitreous and Vitreoretinal Interface*. New York: Springer-Verlag, 1987:133–153.
4. Hemandez C, Burgos R, Canton A, Garcia-Arumi J, Segura RM, Simo R. Vitreous levels of vascular cell adhesion molecule and vascular endothelial growth factor in patients with proliferative diabetic retinopathy: a case-control study. *Diabetes Care* 2001;24:516–521.

5. Funatsu H, Yamashita H, Nakanishi Y, Hori S. Angiotensin II and vascular endothelial growth factor in the vitreous fluid of patients with proliferative diabetic retinopathy. *Br J Ophthalmol* 2002;86:311–315.
6. Hemandez C, Lecube A, Segura RM, Sararols L, Simo R. Nitric oxide and vascular endothelial growth factor concentrations are increased but not related in vitreous fluid of patients with proliferative diabetic retinopathy. *Diabet Med* 2002; 19:655–660.
7. Simo R, Lecube A, Segura RM, Garcia-Arumi J, Hernandez C. Free insulin growth factor-1 and vascular endothelial growth factor in the vitreous fluid of patients with proliferative diabetic retinopathy. *Am J Ophthalmol* 2002;134: 376–382.
8. Ambati J, Chalam KV, Chawla DK, et al. Elevated gamma-aminobutyric acid, glutamate, and vascular endothelial growth factor levels in the vitreous of patients with proliferative diabetic retinopathy. *Arch Ophthalmol* 1997;115: 1161–1166.
9. Katsura Y, Okano T, Noritake M, et al. Hepatocyte growth factor in vitreous fluid of patients with proliferative diabetic retinopathy and other retinal disorders. *Diabetes Care* 1998;21:1759–1763.
10. Shinoda K, Ishida S, Kawashima S, et al. Clinical factors related to the aqueous levels of vascular endothelial growth factor and hepatocyte growth factor in proliferative diabetic retinopathy. *Curr Eye Res* 2000;21:655–661.
11. La Heij EC, Van de Waarenburg MP, Blaauwgeers HG, et al. Levels of basic fibroblast growth factor, glutamine synthetase, and interleukin-6 in subretinal fluid from patients with retinal detachment. *Am J Ophthalmol* 2001;132: 544–550.
12. Cassidy L, Barry P, Shaw C, Duffy J, Kennedy S. Platelet derived growth factor and fibroblast growth factor basic levels in the vitreous of patients with vitreoretinal disorders. *Br J Ophthalmol* 1998;82:181–185.
13. Yamane K, Minamoto A, Yamashita H, et al. Proteome analysis of human vitreous proteins. *Mol Cell Proteomics* 2003;2:1177–1187.
14. Duker JS. Macular hole. In Yanoff M and Duker JS, eds. *Ophthalmology*, 2nd ed. St. Louis: Mosby, 2004:942–946.
15. Kristensen DB, Imamura K, Miyamoto Y, Yoshizato K. Mass spectrometric approaches for the characterization of proteins on a hybrid quadrupole time-of-flight (Q-TOF) mass spectrometer. *Electrophoresis* 2000;21:430–439.
16. Rabilloud T, Valette C, Lawrence JJ. Sample application by in-gel rehydration improves the resolution of two-dimensional electrophoresis with immobilized pH gradients. *Electrophoresis* 1994;15:1552–1558.
17. Sanchez JC, Rouge V, Pisteur M, et al. Improved and simplified in-gel sample application using reswelling of dry immobilized pH gradients. *Electrophoresis* 1997;18:324–327.
18. Shevchenko A, Wilm M, Vorm O, Mann M. Mass spectrometric sequencing of proteins silver-stained polyacrylamide gels. *Anal Chem* 1996;68:850–858.

19. Beuckmann CT, Gordon WC, Kanaoka Y, et al. Lipocalin-type prostaglandin D synthase (β-trace) is located in pigment epithelial cells of rat retina and accumulates within interphotoreceptor matrix. *J Neurosci* 1996;16:6119–6124.
20. Goh Y, Urade Y, Fujimoto N, Hayaishi O. Content and formation of prostaglandins and distribution of prostaglandin-related enzyme activities in the rat ocular system. *Biochim Biophys Acta* 1987;921:302–311.
21. Wu YQ, Becerra SP. Proteolytic activity directed toward pigment epithelium-derived factor in vitreous of bovine eyes. Implications of proteolytic processing. *Invest Ophthalmol Vis Sci* 1996;37:1984–1993.
22. Dawson DW, Volpert OV, Gillis P, et al. Pigment epithelium-derived factor: a potent inhibitor of angiogenesis. *Science* 1999;285:245–248.
23. Ogata N, Tombran-Tink J, Nishikawa M, et al. Pigment epithelium-derived factor in the vitreous is low in diabetic retinopathy and high in rhegmatogenous retinal detachment. *Am J Ophthalmol* 2001;132:378–382.
24. Matsuoka M, Ogata N, Otsuji T, Nishimura T, Takahashi K, Matsumura M. Expression of pigment epithelium derived factor and vascular endothelial growth factor in choroidal neovascular membranes and polypoidal choroidal vasculopathy. *Br J Ophthalmol* 2004;88:809–815.
25. Ogata N, Nishikawa M, Nishimura T, Mitsuma Y, Matsumura M. Unbalanced vitreous levels of pigment epithelium-derived factor in diabetic retinopathy. *Am J Ophthalmol* 2002;134:348–353.
26. Pawlowska Z, Baranska P, Jerczynska H, Koziolkiewicz W, Cierniewski CS. Heat shock proteins and other components of cellular machinery for protein synthesis are up-regulated in vascular endothelial growth factor-activated human endothelial cells. *Proteomics* 2005;5:1217–1227.
27. Yamashita H. Pathogenesis of diabetic retinopathy and strategy to develop new therapeutic modalities. *Ophthalmologica* 2004;218(Suppl 1):19–28.
28. Adamis AP. Is diabetic retinopathy an inflammatory disease? *Br J Ophthalmol* 2002;86:363–365.
29. Horn M, Seger F, Schlote W. Neuron-specific enolase in gerbil brain and serum after transient cerebral ischemia. *Stroke* 1995;26:290–296.
30. Barone FC, Clark RK, Price WJ, et al. Neuron-specific enolase increases in cerebral and systemic circulation following focal ischemia. *Brain Res* 1993;623:77–82.
31. Dunker S, Sadun AA, Sebag J. Neuron specific enolase in retinal detachment. *Curr Eye Res* 2001;23:382–385.
32. Giugliano D, Ceriello A, Paolisso G. Oxidative stress and diabetic vascular complications. *Diabetes Care* 1996;19:257–267.
33. Nayak MS, Kita M, Marmor MF. Protection of rabbit retina from ischemic injury by superoxide dismutase and catalase. *Invest Ophthalmol Vis Sci* 1993;34:2018–2022.
34. Ceriello A. New insights on oxidative stress and diabetic complications may lead to a "causal" antioxidant therapy. *Diabetes Care* 2003;26:1589–1596.

35. Simpson RJ, Connolly LM, Eddes JS, Pereira JJ, Moritz RL, Reid GE. Proteomic analysis of the human colon carcinoma cell line (LIM 1215): development of a membrane protein database. *Electrophoresis* 2000;21:1707–1732.
36. Koyama R, Nakanishi T, Ikeda T, Shimizu A. Catalogue of soluble proteins in human vitreous humor by one-dimensional sodium dodecyl sulfate-polyacrylamide gel electrophoresis and electrospray ionization mass spectrometry including seven angiogenesis-regulating factors. *J Chromatogr B* 2003;792:5–21.

24

Proteomics of Human Dialysate and Ultrafiltrate Fluids Yielded by Renal Replacement Therapy

Michael Walden, Stefan Wittke, Harald Mischak, and Raymond C. Vanholder, for the European Uremic Toxin Work Group (EUTox)

Summary

This chapter discusses different methods for determining proteins and peptides present in human dialysate and ultrafiltrate fluids. The main focus is on separation-coupled MS techniques such as gel electrophoresis-MS, LC-MS, and CE-MS. The increasing number of patients on renal replacement therapy shows the need for techniques that may improve the dialysis process and consequently the quality of life as well as life expectancy of these patients. Dialysate and ultrafiltrate fluids are a rich source of information on relevant peptides and proteins, especially the uremic toxins. Furthermore, these liquids are available in high amounts compared with serum or plasma. The different proteomic methods such as HPLC separation of samples and subsequent MALDI-TOF-MS and 2D-PAGE followed by MALDI-TOF-MS or LC-MS, in combination with MS/MS methods, as well as CE-MS are reviewed, and some of the most recent papers are summarized. The knowledge obtained by proteome analysis holds great promise to allow improvements in current dialysis techniques.

Key Words: Dialysate; ultrafiltrate; renal replacement therapy; dialysis; uremia; uremic toxins; renal failure; capillary electrophoresis; proteome.

1. Introduction

The use of proteomics in the analysis of complex biological fluids has grown rapidly during the last few years, as seen by the review of Thongboonkerd in 2004 *(1)*. In particular, the high sensitivity, speed, and reproducibility of mass spectrometry (MS) have boosted its application in all aspects of protein analysis, including discovery, identification (i.e., peptide mapping, sequencing), and structural characterization. Proteome analysis of human dialysate, hemofiltrate (HF), and ultrafiltrate (UF) fluids can roughly be divided into two major types: (1) methods used to investigate single molecules with techniques like Western

From: *Proteomics of Human Body Fluids: Principles, Methods, and Applications*
Edited by: V. Thongboonkerd © Humana Press Inc., Totowa, NJ

blotting, enzyme-linked immunosorbent assay (ELISA), different chromatographic steps, and/or MS; and (2) "shotgun" methods, which are used to identify as many individual polypeptides as possible using coupling techniques, e.g., gel electrophoresis-MS, liquid chromatography (LC)-MS or capillary electrophoresis (CE)-MS.

We will focus on the second type, or "shotgun" methods, which are well known and well suited for the investigation of the proteome of plasma or urine. This is reflected by some 300+ manuscripts currently present in Medline when one searches "proteomics AND serum/plasma" and about 60+ manuscripts for "proteomics AND urine." In the field of analysis of dialysate and UF fluids, only a few reports are currently available. The adoption of these techniques is driven by the need to identify proteins and polypeptides responsible for several severe complications in renal failure patients or during renal replacement therapy, mostly owing to so-called uremic retention molecules.

Peptides constitute a heterogeneous group of uremic retention molecules. Under normal conditions, they are excreted by glomerular filtration and degraded by renal tubules. These two elimination pathways are hampered once kidney function starts to fail. Many peptidic compounds show an increased concentration in uremia. Most of these molecules have a molecular weight exceeding 500 Daltons, which corresponds to the characteristics of a classical subgroup of uremic retention compounds, the so-called middle molecules. These are difficult to remove and can be eliminated efficiently from the body of uremic patients only by advanced convective dialysis strategies ***(2,3)***.

Several peptides have been shown to interfere with biochemical/biological functions, which are part of the uremic syndrome, a term that covers the symptoms and clinical complications in patients with renal failure. Subsequently, their identification and removal might be germane in combating the morbidity and mortality of uremic patients. This might be especially relevant for the reduction of cardiovascular damage ***(4)***, which is a major problem affecting a substantial number of patients with renal failure, even in the predialysis stage. In the next paragraphs, current knowledge about some of the known peptidic uremic solutes will be summarized.

β_2-microglobulin (β_2-M) is a 12-kDa component of the major histocompatibility antigen. Uremia-related amyloidosis is a disease affecting mainly patients who have been dialyzed for many years. This disease essentially affects bone, joints, and tendons, causing incapacitating symptons, and is, to a large extent, related to β_2-M ***(5)***. The inflammatory characteristics of the disease are most likely related to oxidative modifications of β_2-M, which in part have been identified as advanced glycation end products (AGEs) ***(6)***.

Parathyroid hormone (PTH) is a 9-kDa hormone mainly attributable to parathyroid glandular secretion, in response to hypocalcemia, hyperphosphatemia, and hypovitaminosis D *(7)*, rather than to a direct effect of reduced renal function. PTH acts as a calcium ionophore and provokes cellular calcium influx in many cell systems. This in turn results in cellular activation, e.g., of immune cells, with a proinflammatory impact *(8)*. In addition, PTH mobilizes calcium out of bone and causes calcium deposits in the tissues, e.g., in the vessel walls, potentially contributing to uremic vascular disease *(9)*. In addition, PTH fragments are retained in uremia *(10)*, and these might have their own toxic effects.

Granulocyte inhibiting protein I (GIP-I) is a 28-kDa structural analog of κ-light chains with proinflammatory properties *(11)*. Atrial natriuretic peptide (ANP) (3.1 kDa) and endothelin (3.5 kDa) play a role in the hemodynamics and regulation of blood pressure. Endothelin is a potential contributor to uremic hypertension *(12)* as well as to insulin resistance *(13)*. Neuropeptide Y (NPY) is a 4.3-kDa peptide with vasoconstrictive properties that indicates cardiovascular complications in dialysis patients *(14)*. Leptin, a 16-kDa plasma protein, suppresses appetite, induces weight loss in mice *(15)*, and is related to weight loss in peritoneal dialysis patients *(16)*. It also has been associated with the malnutrition that is a major problem in a substantial number of uremic patients.

Cytokines, such as interleukin-1 (IL-1), IL-6, IL-18, and tumor necrosis factor-α (TNF-α), are proinflammatory compounds that are retained in renal failure, in part because they are inadequately excreted by the failing kidneys. They may play a prominent role in the inflammatory status, which affects a substantial part of the uremic population *(17)*. Inflammation, in turn, is related to atherogenesis *(18)*. Complement factor D is part of the chain of compounds responsible for complement activation. It is retained in uremia and considered as another proinflammatory agent *(19)*.

In summary, a host of peptidic compounds are retained in renal failure and evidently play a role in the biochemical, biological, and clinical alterations in patients with renal failure. The retention of peptidic compounds is closely related to one of the most important problems of nephrology today, the dramatically increased cardiovascular risk. Their clinical potential is underscored by a number of clinical studies, suggesting better cardiovascular outcome if dialysis membranes that improve peptidic removal are used *(20)*.

2. Methods, Findings, and Applications

Hemodialysis fluids are known to serve as a favorable source for proteomic analysis owing to their low content of albumin and other interfering large proteins. Compared with plasma or serum, the concentration of albumin in HF is 1500-fold less, whereas concentrations of polypeptides in the range of 1 to 30 kDa remain nearly unaffected *(21)*. One of the first attempts to analyze the so-called middle-molecules from HF of patients with chronic uremia was described by Brunner et al. in 1978 *(22)*. Lack of suitable techniques for rapid identification of polypeptides in those days restricted the methods to Western blotting, ELISA, radioimmunoassay (RIA), and so on to identify single proteins, one at a time.

2.1. Hemofiltrate as the Source of Bioactive Peptides

An advanced approach to identify polypeptides from HF was initiated by Forssmann et al. around 1995. This work resulted in the characterization of HF as a source of circulating bioactive peptides *(21)*. In addition, a "peptide bank" containing up to 300 different chromatographic fractions generated from up to 10,000 human HFs was established *(23)*. Starting from this peptide bank, bioactive peptides were isolated. The first proteomic approach using LC-MS-guided purification was the isolation of the human peptide hormone guanylin from HF *(24)*. Using LC-MS-guided purification, additional peptides with various biochemical functions, e.g., endostatin, resitin, angiogenesis inhibitors, and a proopiomelanocortin-derived peptide with lipolytic activity, could be isolated from HF *(25,26)*. Additional improvements to the LC-MS technique were made by Wagner et al. *(27)* with the development of an automated 2D high-performance liquid chromatography (HPLC) system with integrated sample preparation and matrix-assisted laser desorption/ionization time-of-flight mass spectrometry (MALDI-TOF-MS) for the analysis of proteins and peptides less than 20 kDa.

2.2. Gel Electrophoresis–Mass Spectrometry

A recent proteomic approach using 2D polyacrylamide gel electrophoresis (2D-PAGE) and MALDI-TOF-MS to identify uremic toxins from UF was presented by Ward and Brinkley in 2004 *(28)*. Briefly:

1. The sample was prepared by a 50-fold concentration step using a 1-kDa cutoff membrane, followed by a desalting step using a dialysis cassette.
2. Gel electrophoresis was carried out using a 22 × 22-cm Duracryl gel after isoelectric focusing (IEF) on a 17-cm immobilized linear pH gradient (IPG) strip.
3. Protein spots of interest were excised and digested by trypsin (for details *see* **ref.** *28*), followed by mass spectrometric analysis, which was performed on a MALDI-TOF-MS (Micromass Tof-Spec 2E; Micromass, Manchester, UK) by peptide mass fingerprinting using Mascot and ProFound search engines to identify the proteins.

Using this approach, Ward and Brinkley were able to identify 21 spots from the proteome map representing six proteins harboring several posttranslational modifications (PTMs), which resulted in multiple spots of the same protein. The identified proteins were β_2-M, one of the major uremic toxins *(5)*, as well as α_1-antitrypsin, albumin (mature and complexed forms), complement factor D, cystatin C, and retinol-binding protein. The results indicate that small proteins like β2-M, with concentrations of 1 to 2 mg/L in normal volunteers and 30 to

50 mg/L in hemodialysis patients, have higher concentrations in patients, whereas the concentration of larger molecules like albumin is not increased.

Likewise, in 2004, Lefler et al. presented in a combination of reversed-phase (RP) chromatography, 2D-PAGE, and MALDI-TOF/TOF tandem MS for the identification of proteins in ultrafiltrate ***(29)***.

1. The first sample preparation step was an RP C-4 HPLC.
2. UF was loaded onto the column and eluted stepwise with 10% (fraction 1), 25% (fraction 2), and 50% buffer B (fraction 3). Buffer A was water, and buffer B was acetonitrile.
3. After lyophilization, each fraction was applied to 2D gel electrophoresis, IEF followed by separation on an 8 to 16% gradient gel.
4. Staining was performed with SYPRO Ruby, and images were analyzed with PDQuest software.
5. For identification of protein spots, peptide mass fingerprinting onto a Micromass MALDI-TOF instrument was performed.
6. For the identification of spots that resulted in low Mascot scores, additional MS/MS experiments were made using a MALDI-TOF/TOF instrument (Applied Biosystems 4700 Proteomics Analyzer).

With this approach, Lefler et al. were able to identify 47 protein spots representing 10 different proteins. The most identified protein was albumin (nine spots), followed by transferrin (eight spots). β_2-M was also identified in this study. Evidently, the addition of RP chromatography as a third dimension of separation to the established 2D gel MS resulted in a higher number of visible protein spots.

In a very recent and detailed study, Molina et al. presented a proteome analysis of human hemodialysis fluid using gel electrophoresis (1D) in combination with LC-MS/MS ***(30)***.

1. During sample preparation, the hemodialysis fluid was first desalted and concentrated using a 3-kDa cutoff filter.
2. Subsequently, the desalted sample was resolved by sodium dodecyl sulfate (SDS)-PAGE, and silver-stained; the bands were then excised, in-gel-digested with trypsin, and analyzed by LC-MS/MS.
3. Nanoflow RP C-18 chromatography coupled via nanoelectrospray sources to either a quadrupole time-of-flight MS (Q-Tof API-US, Micromass) or an ion trap MS (LC/MSD Trap XCT, Agilent Technologies, Palo Alto, CA) was used.

With this approach, 292 different proteins from hemodialysis fluid were identified; 205 of them had not previously been reported in serum or plasma. Additional Western blot analysis of a subset of these proteins revealed their presence in normal serum, indicating that the sensitivity of detection might be the major reason why most of these proteins had not been identified previously in serum or plasma. The authors conclude that this might mainly be owing to the greater dynamic range of protein concentration in serum/plasma samples and the enrichment of the lower molecular weight proteins in the hemodialysis fluid. A further outcome of this proteomic analysis was that, similar to those reported by Ward and Brinkley, proteins in hemodialysis fluid often harbor PTMs, which makes identification by MS/MS difficult when one is searching databases with standard settings only. Most PTMs reported by Molina et al. were oxidization at methionine or tryptophan residues, pyroglutamine formation, N-terminal acetylation, *N*-glycosylation of peptides, and proline hydroxylation.

In conclusion, all these gel electrophoresis-based proteomic techniques showed differences in protein expression within a mass range of more than 10 kDa, and many of these proteins could be identified. The results of these studies are of great relevance in the evaluation of uremic toxins. However, all these techniques lack identification of uremic retention molecules in the lower molecular range, from 1 kDa up to 10 kDa ("middle molecules"), owing to methodological restriction to the analysis of proteins with higher molecular masses (>10 kDa). In addition, these approaches are most likely not applicable to a larger number of individual patient samples. In the next section we describe CE-MS as an approach to assess the proteome of dialysate and as a method especially suited to fill the gaps in the other techniques.

2.3. CE-MS

Two- or multidimensional approaches are labor intensive and time consuming and hence are suited for the analysis of a limited number of different samples. An alternative, which permits the analysis of hundreds of samples in a timely fashion, is CE-MS. We have successfully used this technology in the analysis of different body fluids, including, dialysate and UF ***(31,32)***. This technology combines the high-resolution separation capability of capillary electrophoresis (CE) with the high-resolution detection capability of electrospray time-of-flight mass spectrometry (ESI-TOF-MS). CE-MS permits the analysis of up to 2000 polypeptides within 45 to 60 min, in a small volume and with a high sensitivity.

In the first approach, we examined the effect of different dialysis membranes (low-flux vs high-flux) on the number of polypeptides in the dialysate ***(31)***. The sample preparation used for this study was anion-exchange chromatography with DEAE-Sepharose to remove interfering salts and uncharged elements and to concentrate the final sample, followed by lyophilization. Because of the relative insensitivity of CE toward salts, this fast and simple procedure has resulted in a reproducible sample matrix.

1. CE-MS analysis was performed on a Beckman P/ACE MDQ CE coupled via a CE-ESI-MS sprayer kit from Agilent to an Applied Biosystems Mariner ESI-TOF-MS.

2. The sample was injected hydrodynamically (1 psi, 20 s) on an untreated silica capillary (i.d. 75 and o.d. 360 μm, length 90 cm).
3. When a separation buffer composed of 30% methanol and 0.5% formic acid in water (pH 2.4) is used, the electrophoretic run at 30 kV is completed within 45 to 60 min.
4. After each run, the capillary was rinsed for 5 min with 0.1 *M* NaOH, followed by a 5-min rinse with water and another 5-min rinse with running buffer.
5. The capillary temperature was held constant at 35°C.
6. The sheath flow was applied at a rate of 5 μL/min coaxial to the capillary, and the sheath liquid was identical to the separation buffer.

This setup resulted in a stable and reproducible CE-MS method. The huge amount of data generated from each single run was evaluated with a specialized software package, MosaiquesVisu (available online at www.proteomiques.com), which is described in detail elsewhere ***(33,34)***. In first experiments, more than 600 polypeptides could be analyzed in a single sample. As shown in **Fig. 1**, larger polypeptides (>10 kDa) were only present in the dialysates from high-flux membranes **(Fig. 1A)**, whereas most of those in dialysates from low-flux membranes **(Fig. 1B)** were smaller than 10 kDa.

In a further study, the potential of CE-MS followed by CE-MS/MS to identify uremic retention molecules in dialysis fluids from low-flux and high-flux membranes was examined ***(32)***. To obtain further insight into uremic toxins within a mass range of 800 up to 15,000 Daltons, the same CE-MS setup as described above was used, combined with a different sample preparation procedure.

1. The dialysates were applied onto a Merck LiChrospher RP C-18 ADS column with a flow rate of 0.8 mL/min.
2. After washing, the polypeptide fraction was eluted with a step gradient of 80% methanol and 20% water with a flow rate of 0.8 mL/min.
3. The elution profile was monitored by UV detection at 200 nm.
4. Approximately 6 mL of eluate was collected from each sample, frozen, and lyophilized.
5. Shortly before use, samples were resuspended in 20 μL of HPLC-grade water, yielding a 300-fold enrichment of polypeptides present in each sample.

The results of these experiments are shown in **Fig. 2**. Although the data from the two different membranes appeared quite similar at first sight, several differences became obvious upon closer examination. When we compared the different CE-MS runs, the signal intensity was always higher in the effluent of high-flux membranes. The signal intensity is shown as a color code (0–10,000 MS counts) for both the raw data plot and the protein plot. In addition, more polypeptides were detectable in the dialysate from the high-flux membrane. In all, 1394 different polypeptides were detected using the high-flux membrane, whereas only 1046 polypeptides were recovered in the dialysate of the same

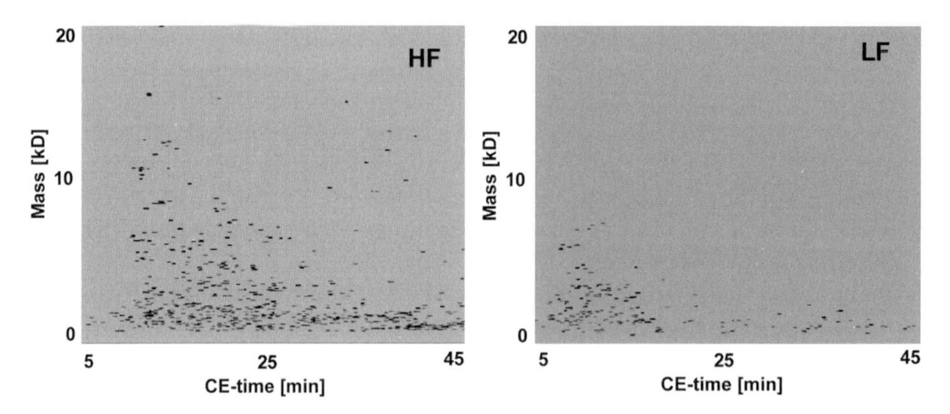

Fig. 1. Comparison of the abundance of polypeptides detected in high-flux (HF) and low-flux (LF) dialysates defined by their molecular mass (in kDa) and migration time (in min). Although the distribution, but not the absolute number, is quite similar for both types of membranes in the low-molecular-mass range (<10 kDa), an increase in the number of polypeptides can be observed in the high-molecular-mass range of the high-flux dialysates.

patient obtained from the low-flux membrane, indicating a more efficient removal of polypeptides by the high-flux membrane. Furthermore, the mass distribution of the removed molecules was different; more molecules with molecular masses above 5 kDa were present in the UF from high-flux membranes. To identify peptides contained in the UF, CE-MS/MS analysis was performed. A complete CE run was spotted onto a MALDI target plate (one spot for every 15 s) and examined subsequently in MS mode on a MALDI-TOF/TOF instrument (Bruker Daltonics). Polypeptides of interest were fragmented in MS/MS mode, and their sequences were identified with a Mascot search against the Swiss-Prot database.

Taken together, the results demonstrate that CE-MS allows fast analysis of large numbers of individual compounds; up to 1400 compounds with a molecular mass of more than 800 Daltons could be recognized. These findings demonstrate the potential of the CE-MS application for proteomics and the identification of yet unknown uremic retention molecules.

3. Perspectives and Future Directions

Modern technologies for proteome analysis are well suited for the identification of polypeptides and proteins in ultrafiltrate, hemofiltrate, and dialysate fluids. The results also strongly suggest that, with respect to uremic retention molecules, only the tip of the iceberg is known. Currently, only a few of these molecules are defined ***(35)***, but it is conceivable that a much larger number is

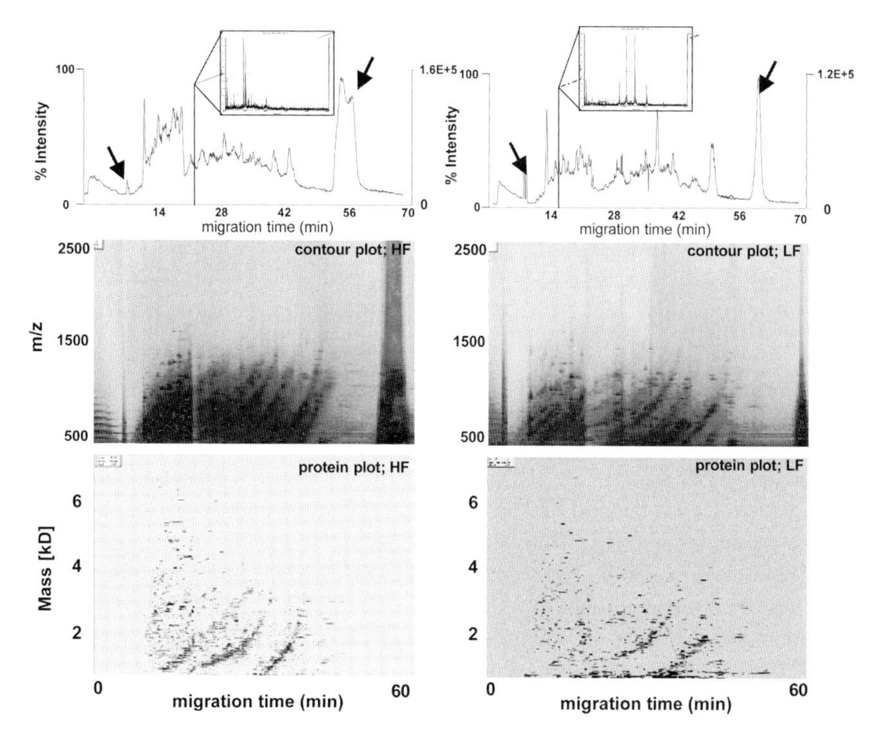

Fig. 2. Comparison of proteomic analysis of ultrafiltrate obtained from high-flux (left) and low-flux (right) polysulfone membranes. The total ion chromatograms (TICs; upper graphs) were obtained after CE-MS analysis. The inserts show individual spectra, obtained every 3 s, yielding the total ion chromatograms. These data have been converted to a 3D contour plot (middle panels). Mass per charge is shown on the *y*-axis against the migration time (in min) on the *x*-axis. Even before the polypeptides, an array of highly mobile ions like formiates appear, whereas the end of the spectrum is marked by the appearance of organic polymers, most likely a result of the sample preparation. The lower panels show a graphic depiction of the peak list, as calculated actual mass plotted against migration time. This peak list contains more than 1000 different molecules. Evidently, both the number of individual compounds and the intensity are higher for the high-flux membrane.

retained and many compounds that play a role in uremia remain unrecognized. Evidently, the problem of uremic retention molecules is substantially more complex than initially anticipated. In uremia, all substances that are excreted by the healthy kidneys under normal conditions are potential candidates for retention and may be of pathophysiologic importance.

The results also indicate that proteome analysis allows examination of the dialysis process and especially the efficiency of the removal of polypeptides by different dialysis membranes. A comparison of polypeptides in plasma and dialysis fluid of patients with those present in the plasma of healthy individuals should allow the identification of polypeptides that are increased in dialysis patients and are consequently the first-line uremic toxin candidates. This knowledge and the ability to evaluate the dialysis process on a molecular level with respect to the polypeptides present should allow substantial improvement in dialysis protocols. Future research will concentrate on the identification of new, as yet unknown uremic retention molecules. In this context, proteomic analysis might be germane in the identification of key molecules responsible for some of the principal complications of uremia, such as cardiovascular disease, malnutrition, or inflammation.

References

1. Thongboonkerd V. Proteomics in nephrology: current status and future directions. *Am J Nephrol* 2004;24:360–378.
2. Locatelli F, Mastrangelo F, Redaelli B, et al. Effects of different membranes and dialysis technologies on patient treatment tolerance and nutritional parameters. The Italian Cooperative Dialysis Study Group. *Kidney Int* 1996;50:1293–1302.
3. Raj DS, Ouwendyk M, Francoeur R, Pierratos A. Beta(2)-microglobulin kinetics in nocturnal haemodialysis. *Nephrol Dial Transplant* 2000;15:58–64.
4. Vanholder R, Argiles A, Baurmeister U, et al. Uremic toxicity: present state of the art. *Int J Artif Organs* 2001;24:695–725.
5. Gejyo F, Yamada T, Odani S, et al. A new form of amyloid protein associated with chronic hemodialysis was identified as beta 2-microglobulin. *Biochem Biophys Res Commun* 1985;129:701–706.
6. Miyata T, Oda O, Inagi R, et al. Beta 2-microglobulin modified with advanced glycation end products is a major component of hemodialysis-associated amyloidosis. *J Clin Invest* 1993;92:1243–1252.
7. Rodriguez M, Canalejo A, Garfia B, Aguilera E, Almaden Y. Pathogenesis of refractory secondary hyperparathyroidism. *Kidney Int Suppl* 2002;155–160.
8. Massry SG, Smogorzewski M. Mechanisms through which parathyroid hormone mediates its deleterious effects on organ function in uremia. *Semin Nephrol* 1994; 14:219–231.
9. London GM. Cardiovascular calcifications in uremic patients: clinical impact on cardiovascular function. *J Am Soc Nephrol* 2003;14(9 Suppl 4):S305–S309.
10. Gao P, Scheibel S, D'Amour P, et al. Development of a novel immunoradiometric assay exclusively for biologically active whole parathyroid hormone 1-84: implications for improvement of accurate assessment of parathyroid function. *J Bone Miner Res* 2001;16:605–614.
11. Cohen G, Rudnicki M, Horl WH. Uremic toxins modulate the spontaneous apoptotic cell death and essential functions of neutrophils. *Kidney Int Suppl* 2001;78:S48–S52.
12. Brochu E, Lacasse S, Moreau C, et al. Endothelin ET(A) receptor blockade prevents the progression of renal failure and hypertension in uraemic rats. *Nephrol Dial Transplant* 1999;14:1881–1888.

13. Ottosson-Seeberger A, Lundberg JM, Alvestrand A, Ahlborg G. Exogenous endothelin-1 causes peripheral insulin resistance in healthy humans. *Acta Physiol Scand* 1997;161:211–220.
14. Zoccali C, Mallamaci F, Tripepi G, et al. Prospective study of neuropeptide Y as an adverse cardiovascular risk factor in end-stage renal disease. *J Am Soc Nephrol* 2003;14:2611–2617.
15. Stephens TW, Basinski M, Bristow PK, et al. The role of neuropeptide Y in the antiobesity action of the obese gene product. *Nature* 1995;377:530–532.
16. Stenvinkel P, Lindholm B, Lonnqvist F, Katzarski K, Heimburger O. Increases in serum leptin levels during peritoneal dialysis are associated with inflammation and a decrease in lean body mass. *J Am Soc Nephrol* 2000;11:1303–1309.
17. Yeun JY, Levine RA, Mantadilok V, Kaysen GA. C-Reactive protein predicts all-cause and cardiovascular mortality in hemodialysis patients. *Am J Kidney Dis* 2000;35:469–476.
18. Ross R. Atherosclerosis—an inflammatory disease. *N Engl J Med* 1999;340:115–126.
19. Deppisch RM, Beck W, Goehl H, Ritz E. Complement components as uremic toxins and their potential role as mediators of microinflammation. *Kidney Int Suppl* 2001;78:S271–S277.
20. Eknoyan G, Beck GJ, Cheung AK, et al. Effect of dialysis dose and membrane flux in maintenance hemodialysis. *N Engl J Med* 2002;347:2010–2019.
21. Schepky AG, Bensch KW, Schulz-Knappe P, Forssmann WG. Human hemofiltrate as a source of circulating bioactive peptides: determination of amino acids, peptides and proteins. *Biomed Chromatogr* 1994;8:90–94.
22. Brunner H, Mann H, Essers U, Schultheis R, Byrne T, Heintz R. Preparative isolation of middle molecular weight fractions from the hemofiltrate of patients with chronic uremia. *Artif Organs* 1978;2:375–377.
23. Schulz-Knappe P, Schrader M, Standker L, et al. Peptide bank generated by large-scale preparation of circulating human peptides. *J Chromatogr A* 1997;776:125–132.
24. Schrader M, Jurgens M, Hess R, Schulz-Knappe P, Raida M, Forssmann WG. Matrix-assisted laser desorption/ionisation mass spectrometry guided purification of human guanylin from blood ultrafiltrate. *J Chromatogr A* 1997;776:139–145.
25. John H, Radtke K, Standker L, Forssmann WG. Identification and characterization of novel endogenous proteolytic forms of the human angiogenesis inhibitors restin and endostatin. *Biochim Biophys Acta* 2005;1747:161–170.
26. Fricke K, Schulz A, John H, Forssmann WG, Maronde E. Isolation and characterization of a novel proopiomelanocortin-derived peptide from hemofiltrate of chronic renal failure patients. *Endocrinology* 2005;146:2060–2068.
27. Wagner K, Miliotis T, Marko-Varga G, Bischoff R, Unger KK. An automated on-line multidimensional HPLC system for protein and peptide mapping with integrated sample preparation. *Anal Chem* 2002;74:809–820.
28. Ward RA, Brinkley KA. A proteomic analysis of proteins removed by ultrafiltration during extracorporeal renal replacement therapy. *Contrib Nephrol* 2004;141: 280–291.

29. Lefler DM, Pafford RG, Black NA, Raymond JR, Arthur JM. Identification of proteins in slow continuous ultrafiltrate by reversed-phase chromatography and proteomics. *J Proteome Res* 2004;3:1254–1260.
30. Molina H, Bunkenborg J, Reddy GH, Muthusami B, Scheel PJ, Pandey A. A proteomic analysis of human hemodialysis fluid. *Mol Cell Proteomics* 2005;4: 637–650.
31. Kaiser T, Hermann A, Kielstein JT, et al. Capillary electrophoresis coupled to mass spectrometry to establish polypeptide patterns in dialysis fluids. *J Chromatogr A* 2003;1013:157–171.
32. Weissinger EM, Kaiser T, Meert N, et al. Proteomics: a novel tool to unravel the patho-physiology of uraemia. *Nephrol Dial Transplant* 2004;19:3068–3077.
33. Neuhoff N, Kaiser T, Wittke S, et al. Mass spectrometry for the detection of differentially expressed proteins: a comparison of surface-enhanced laser desorption/ionization and capillary electrophoresis/mass spectrometry. *Rapid Commun Mass Spectrom* 2004;18:149–156.
34. Wittke S, Mischak H, Walden M, Kolch W, Radler T, Wiedemann K. Discovery of biomarkers in human urine and cerebrospinal fluid by capillary electrophoresis coupled to mass spectrometry: towards new diagnostic and therapeutic approaches. *Electrophoresis* 2005;26:1476–1487.
35. Vanholder R, De Smet R, Glorieux G, et al. Review on uremic toxins: classification, concentration, and interindividual variability. *Kidney Int* 2003;63:1934–1943.

Index

A

ABPP, *see* Activity-based protein profiling
Acinar cell,
  pancreas, 379
  salivary secretions, 348, 349
Activity-based protein profiling (ABPP), principles, 46–49, 95, 96
Albumin, removal before proteomic analysis, 41, 42, 44, 45, 314
Allergic asthma, bronchoalveolar lavage fluid and sputum proteomics, 319
Alzheimer's disease, biomarker discovery, 190
Amniotic fluid,
  protein sources, 417
  proteomics
    amnion cells, 427
    clinical applications,
      fetal proteins, 429
      infection, 428
      lysosomal storage disease, 428, 429
      premature rupture of membranes, 427, 428
      prospects, 415, 416
    liquid chromatography/mass spectrometry, 422, 426, 427
    overview, 418
    polyacrylamide gel electrophoresis, 419, 420
    protein microarray, 420
    protein types, 423–425
    sample preparation, 419
    two-dimensional gel electrophoresis, 420, 421
  synthesis, 416, 417
Apolipoprotein A1,
  biomarker prospects, 304, 305
  studies in pleural effusion, 300–304
Ascites, CA125 levels in ovarian cancer, 6, 7

B

BALF, *see* Bronchoalveolar lavage fluid
Basic fibroblast growth factor (bFGF), breast cancer marker in nipple aspirate fluid, 458
Bernard, Claude, 3
bFGF, *see* Basic fibroblast growth factor
Bile,
  biliary tree anatomy, 399–401
  composition, 401–403
  protein sources, 405, 406
  proteomics,
    biliary tract cancer studies, 408, 409
    fractionation, 410
    lectin affinity chromatography, 410, 411
    liquid chromatography/mass spectrometry, 411
    overview of studies, 406–408
    prospects, 411
    sample collection, 409, 410
  salts,
    composition, 401, 403
    transport, 403–405
  secretion, 399
Bioinformatics,
  biomarker discovery, see Biomarker discovery

omic technology integration and analysis, 183, 184
Biomarker discovery,
Alzheimer's disease, 190
bioinformatics,
biological variation and power analysis, 162–165
completeness of analysis, 158, 159
data normalization and transformation, 159, 160
dynamic range, linearity, and detection limits, 155, 156, 158
experimental design principles, 155
randomization to decrease bias, 155
reproducibility and variance, 160, 161
statistical analysis, 165–169
strategy, 154
biomarker types,
differentially expressed proteins, 151
differentially posttranslationally modified proteins, 151, 153
isoforms, 153
overview, 151, 152
primary versus secondary, 153, 154
protein fragments, 153
cancer biomarkers,
breast cancer, 186, 187
head and neck cancer, 178
lung cancer, 187, 188
ovarian cancer, 186
pancreatic cancer, 189, 386–391
prostate cancer, 188
cerebrospinal fluid proteomics, 279, 280
comparison of proteomics technologies, 148–151
drug discovery,
applications, 190
challenges, 185
hepatitis, 189
lung allograft rejection, 320, 321
lymph advantages, 6
plasma or serum samples, 215, 216, 218, 219
plasma versus serum for industrial scale proteomic studies, 184, 185
rheumatoid arthritis, 189
Blood, *see also* Plasma; Serum,
coagulation, 9, 32–34, 199
collection, 32, 40
Breast cancer,
biomarker discovery, 186, 187
nipple aspirate fluid proteomics, *see* Nipple aspirate fluid
Bronchoalveolar lavage fluid (BALF),
protein sources, 309, 310
proteome features, 315
sample collection, 310, 311
sample preparation, 53, 313
two-dimensional gel electrophoresis,
differential display proteomics,
allergic asthma, 319
chronic lung allograft rejection biomarker discovery, 320, 321
interstitial lung disease, 317–319
overview, 311, 312
prospects, 322
sample preparation,
enrichment of low-abundance proteins, 314, 315
microdialysis, 313
protein precipitation, 312, 313
ultrafiltration, 313

C

CA125, body fluid levels in ovarian cancer, 6, 7
Capillary electrophoresis/mass spectrometry,
hemofiltrate, ultrafiltrate fluid proteomics, 514–516

urinary proteomics,
applications, 242, 245, 246
coupling, 237–239
data analysis, 239, 240
sample preparation, 239
Carbonylation, detection of proteins, 93
Carcinoembryonic antigen (CEA),
breast cancer marker in nipple aspirate fluid, 459
gel loading and levels, 10–11
Caseins, milk protein proteomics, 439, 443
CCD camera, fluorescent dye detection in gels, 97
CEA, *see* Carcinoembryonic antigen
Cell culture, suspension antibody microarray sample preparation, 77
Cerebrospinal fluid (CSF),
function, 269
protein abundance, 269–271
proteomics,
biomarker discovery, 279, 280
liquid chromatography/mass spectrometry, 274, 276, 279
prospects, 280, 281
sample collection,
lumbar puncture, 271
ventricular cerebrospinal fluid, 271
two-dimensional gel electrophoresis, 271–274
synthesis, 269
Chemoproteomics, overview, 176
Citrate, anticoagulant in blood sample preparation, 9, 40, 199
Coefficient of variation, calculation, 161
CSF, *see* Cerebrospinal fluid
Cytokines, dialysate, hemofiltrate, ultrafiltrate fluid proteomics, 511

D

Diabetic retinopathy,
vision loss, 496
blood-retinal barrier breakdown, 496
vitreous fluid proteomics, 499, 502–504
Dialysate, hemofiltrate, ultrafiltrate fluids
proteins of interest in disease, 510, 511
proteomics,
capillary electrophoresis/mass spectrometry, 514—516
hemofiltrate bioactive peptides, 512
overview, 509, 510
prospects, 516–518
rationale, 510
two-dimensional gel electrophoresis, 512–514
Differential gel electrophoresis (DIGE),
fluorescent dyes, 88
pleural fluid analysis in pleural effusion
applications in pleural effusion,
apolipoprotein A1 as biomarker, 300–304
diagnosis, 294
lung cancer, 296
mesothelioma, 299, 300
serum comparison, 297, 298
automatic spot picking, 291
gel electrophoresis and staining, 290
mass spectrometry and database searching, 293, 296
sample collection and preparation, 289
tryptic digestion in-gel, 291, 293
two-dimensional differential gel electrophoresis, 290, 291
Western blot, 293
principles, 18, 288, 289
seminal fluid proteomics, 485
DIGE, *see* Differential gel electrophoresis
Dimethylbenzylamine (DMBA), protein response in nasal lavage fluid, 340, 341

DMBA, *see* Dimethylbenzylamine
DMBT1, biliary tract cancer marker, 409

E

EDTA, anticoagulant in blood sample preparation, 9, 40, 199
ELISA, *see* Enzyme-linked immunosorbent assay
Endoglycosidase D, deglycosylation, 122, 123, 125
Endoglycosidase H, deglycosylation, 112, 122, 123, 125
Enzyme-linked immunosorbent assay (ELISA),
  nipple aspirate fluid, 454, 455
  protein capture, 200, 201
Expression proteomics, definition, 7
Extracellular fluid, volume, 3, 4

F

FFE, *see* Free-flow electrophoresis
Fluorescent dyes, gel staining,
  glycoproteins, 91, 92
  histidine-tagged proteins, 94, 95
  imaging instrumentation, 96–98
  oxidative posttranslational modification detection, 92, 93
  phosphoproteins, 91
  prefractionation strategies, 99, 100
  prospects, 101
  proteins, 84, 85, 87, 88
  trifunctional probe, 100
Free-flow electrophoresis (FFE), prefractionation, 16, 17

G

GIP-I, *see* Granulocyte inhibiting protein I
GlycoProfile III, glycoprotein staining, 93
Glycoproteins,
  capture for proteomic analysis, 14, 15, 46, 109–111, 121
  CD59 complexity, 108
  deglycosylation, 112, 113, 121–123, 125
  fluorescent dyes, 91, 92
  hydrophilic interaction chromatography, 111, 121, 122, 125
  linkage types, 108
  mass spectrometry of deglycosylated peptides,
    liquid chromatography/mass spectrometry, 124, 125
    matrix-assisted laser desorption mass spectrometry, 123, 125
    overview, 114–116
  materials for analysis, 116, 119, 120
Glycoproteomics, overview, 176
Granulocyte inhibiting protein I (GIP-I), dialysate, hemofiltrate, ultrafiltrate fluid proteomics, 511

H

Head and neck cancer, biomarker discovery, 178
Hemofiltrate fluid, *see* Dialysate, hemofiltrate, ultrafiltrate fluids
Heparin, anticoagulant in blood sample preparation, 9, 40, 199
Hepatitis, biomarker discovery, 189
High-performance liquid chromatography (HPLC), *see also* Liquid chromatography/mass spectrometry,
  nanoflow liquid chromatography, 182
  prefractionation, 19, 20, 38, 39
Histidine-tagged proteins, detection, 94, 95
HPLC, *see* High-performance liquid chromatography
HPPP, *see* Human Plasma Proteome Project
Human Plasma Proteome Project (HPPP),
  access, 201, 202

aims, 205, 206
data management, 207
database, *see* Plasma Protein Database
error rate estimation, 210
Gene Ontology annotation, 210–212
organization, 130
protein abundance, 207, 208
protein diversity in individual specimens, 210, 212–213
protein identification list
comparison with other databases, 208, 209
rationale, 129, 130
sample collection, 8
Hydrophilic interaction chromatography, glycoproteins, 111, 121, 122, 125

I

IEF, *see* Isoelectric focusing
IGFBP-3, *see* Insulin-like growth factor-binding protein-3
Immunoglobulin A, nasal lavage fluid, 338
Immunoglobulin G, removal before proteomic analysis, 41, 42, 44, 45, 314
Insulin-like growth factor-binding protein-3 (IGFBP-3), breast cancer marker in nipple aspirate fluid, 460
Integrative systems biology, 24
Interstitial lung disease, bronchoalveolar lavage fluid and sputum proteomics, 317–319
Intracellular fluid, volume, 3, 4
Isoelectric focusing (IEF), *see also* Two-dimensional gel electrophoresis,
free-flow electrophoresis and chromatofocusing, 16, 17
narrow-range gradients, 15, 16, 55

K

Kallikreins,
cancer markers, 459, 460
seminal fluid proteomics, 480

L

LC-MS, *see* Liquid chromatography/mass spectrometry
Lectins,
affinity chromatography of bile, 410, 411
glycoprotein capture for proteomic analysis, 46, 109–111, 121
Lipocalins,
biliary tract cancer marker, 409
nasal lavage fluid, 338, 339
Liquid chromatography/mass spectrometry (LC-MS),
amniotic fluid proteomics, 422, 426, 427
bile proteomics, 411
cerebrospinal fluid proteomics, 274, 276, 279
deglycosylated peptides, 124, 125
milk protein proteomics, 441, 442
nasal lavage fluid proteomics,
mass spectrometry, 329
sample collection, 328, 329
sample preparation, 329
pancreatic juice proteomics, 392
peptidomics, 56
saliva peptides, 360–363
seminal fluid proteomics, 472, 477
shotgun proteomics, 38, 57, 59, 60
two-dimensional liquid chromatography, *see* Shotgun proteomics
urinary proteomics, 229–233
Lung allograft rejection, biomarker discovery, 320, 321
Lung cancer, biomarker discovery, 187, 188
Lymph, advantages in biomarker discovery, 6

Lysosomal storage disease, amniotic fluid proteomics, 428, 429

M

Mac-2-binding protein, biliary tract cancer marker, 409
Macular hole,
formation, 496, 497
vitreous fluid proteomics, 499, 502–504
Magnetic beads, sample preparation, 50, 51
Mass spectrometric immunoassay (MSIA), urinary proteomics,
β_2-microglobulin quantification, 248–250, 252, 253
inter-individual protein analysis, 253, 256–258
principles, 246–248
Mass spectrometry (MS), *see also* Capillary electrophoresis/mass spectrometry; Liquid chromatography/mass spectrometry; Mass spectrometric immunoassay; Surface-enhanced laser desorption/ionization mass spectrometry,
deglycosylated peptides,
liquid chromatography/mass spectrometry, 124, 125
matrix-assisted laser desorption mass spectrometry, 123, 125
overview, 114–116
pattern recognition, 201
protein chip technology, 49, 50
protein identification,
false-positives, 198–199
tandem mass spectrometry, 21, 22
vitreous fluid proteomics, 498
quantitative protein analysis, 22, 23, 200
MDLC, *see* Multidimensional high-performance liquid chromatography
Metabolomics,
overview, 177
technologies, 178–180
Metabonomics,
overview, 176, 177
technologies, 180
Methyltetrahydrophthalic anhydride (MHHPA), protein response in nasal lavage fluid, 341
MHHPA, *see* Methyltetrahydrophthalic anhydride
Microfluidics, technologies, 180, 181
β_2-Microglobulin, dialysate, hemofiltrate, ultrafiltrate fluid proteomics, 510
Milk,
proteins,
casein and whey protein types, 439
classification, 437
major proteins, 440
variation during human lactation, 437, 438
proteomics,
caseins, 443
liquid chromatography/mass spectrometry, 441, 442
milk fat globule membrane proteins, 443, 444, 446
overview, 440–442
prospects, 446, 450
protein types identified, 447–449
two-dimensional gel electrophoresis, 440, 441
whey proteins, 442, 443
MS, *see* Mass spectrometry
MSIA, *see* Mass spectrometric immunoassay
Mucins, saliva analysis, 363–365
Mucous cell, nasal lavage fluid proteins, 339
MudPIT, *see* Shotgun proteomics
Multidimensional high-performance liquid chromatography (MDLC), prefractionation, 19, 20

Multiplexed immunoassay, *see* Suspension antibody microarray

N

NAF, *see* Nipple aspirate fluid
Nanofluidics, technologies, 181, 182
Nanotechnology, planar waveguides, 182, 183
Nasal lavage fluid (NLF),
cytokeratins, 339, 340
immunoglobulin A, 338
lipocalin proteins, 338, 339
liquid chromatography/mass spectrometry,
mass spectrometry, 329
sample collection, 328, 329
sample preparation, 329
mucous cell proteins, 339
nasal provocation,
dimethylbenzylamine, 340, 341
hypertonic saline, 340
methyltetrahydrophthalic anhydride, 341
smoking, 341
plasma exudation, 339
protein exocytosis, 339
protein sources, 331, 337
proteomics,
approaches and rationale, 327, 328
prospects for study, 343
serous cell proteins, 337, 338
sinusitis proteomics,
acute sinusitis findings, 341, 342
normal versus sinusitis protein findings, 331–336
posttreatment findings, 342, 343
two-dimensional gel electrophoresis, 330, 331
Neuron-specific enolase (NSE), vitreous fluid proteomics, 503
Nile Red, gel staining, 87
Nipple aspirate fluid (NAF),
collection, 454
enzyme-linked immunosorbent assay, 454, 455
proteomics,
breast cancer prognostic markers, 460
hormones and growth factors in breast disease, 458, 460
kallikreins as cancer markers, 459, 460
prospects, 463
surface-enhanced laser desorption/ionization mass spectrometry,
candidate protein detection, 461, 462
data analysis, 462, 463
protein identification, 457
protein separation, 456, 457
sample preparation, 456
validation, 462
tumor antigens, 458, 459
two-dimensional gel electrophoresis,
candidate protein detection, 460, 461
electrophoresis, 455, 456
mass spectrometry identification of proteins, 456
sample preparation, 455
validation, 461
radioimmunoassay, 454, 455
S-Nitrosylation, detection of proteins, 92, 93
NLF, *see* Nasal lavage fluid
NSE, *see* Neuron-specific enolase

O

Ovarian cancer,
biomarker discovery, 186
CA125 levels, 6, 7

P

PAGE, *see* Polyacrylamide gel electrophoresis

PAI-1, breast cancer prognostic marker in nipple aspirate fluid, 460
Pancreas,
  anatomy, 379
  endocrine secretion, 380
  exocrine secretion, 379, 380
Pancreatic cancer,
  biomarker discovery, 189
  epidemiology, 386
Pancreatic juice,
  collection, 381, 391
  components, 379, 380
  production, 377, 379
  proteomics,
    liquid chromatography/mass spectrometry, 392
    pancreatic cancer biomarkers
      discovery and validation, 388–391
      types, 386–388
    prospects, 392, 393
    protein types, 382–385
    two-dimensional gel electrophoresis, 391, 392
Parathyroid hormone (PTH), dialysate, hemofiltrate, ultrafiltrate fluid proteomics, 510, 511
PEDF, *see* Pigment epithelium-derived factor
Phosphatidylinositol phospholipase C (PI-PLC), GPI anchor removal, 113, 114
Phosphoproteins, detection in gels, 90, 91, 98, 99
Photomultiplier tube (PMT), fluorescent dye detection in gels, 97
PI-PLC, *see* Phosphatidylinositol phospholipase C
Pigment epithelium-derived factor (PEDF), vitreous fluid proteomics, 499, 502
Planar waveguide (PWG), nanotechnology, 182, 183
Plasma,
  complexity of proteins, 198
  detectable protein abundance, 201–205
  disease protein markers, 130, 131
  dynamic range of proteins, 196–198
  industrial scale biomarker discovery, 184, 185
  prospects for proteomics, 219, 220
  protein classification, 129
  proteome advantages and disadvantages, 195, 196
  range of proteins, 36
  relationship with other body fluids, 5
  sample preparation, 9, 10, 39–41
  serum proteomics comparison, 213, 215
  storage, 10
  suspension antibody microarray sample preparation, 76, 77
Plasma Protein Database (PPD),
  annotation, 132, 135
  availability, 130
  biomarker discovery, 215, 216, 218, 219
  overview, 130, 132
Pleural fluid,
  differential gel electrophoresis
    applications in pleural effusion,
      apolipoprotein A1 as biomarker, 300–304
      diagnosis, 294
      lung cancer, 296
      mesothelioma, 299, 300
      serum comparison, 297, 298
    automatic spot picking, 291
    gel electrophoresis and staining, 290
    mass spectrometry and database searching, 293, 296
    principles, 288, 289
    sample collection and preparation, 289
    tryptic digestion in-gel, 291, 293
    two-dimensional differential gel electrophoresis, 290, 291

Western blot, 293
pleural effusion,
diagnosis, 286, 287
pathophysiology, 285, 286
protein complexity, 287, 288, 296
suspension antibody microarray sample preparation, 77
PMT, *see* Photomultiplier tube
PNGase F, deglycosylation, 112, 122, 125
Polyacrylamide gel electrophoresis (PAGE),
one-dimensional,
amniotic fluid proteins, 419, 420
saliva proteins, 352, 353
two-dimensional, *see* Two-dimensional gel electrophoresis
PPD, *see* Plasma Protein Database
Prefractionation,
depletion of high-abundance proteins, 13, 14, 39, 41, 42, 44, 45, 55
fluorescent dye detection strategies, 99, 100
glycoprotein capture, 14, 15, 46, 109–111, 121
immunoprecipitation, 14
isoelectric focusing,
prefractionation for narrow-range gradients, 15, 16, 55
free-flow electrophoresis and chromatofocusing, 16, 17
liquid chromatography, 19, 20, 38, 39
magnetic beads, 50, 51
peptide ligand chromatography, 14
protein chip technology, 49, 50
rationale, 13
restricted access materials, 38
ultrafiltration, 37, 38
Premature rupture of membranes, amniotic fluid proteomics, 427, 428
Pro-Q Diamond, phosphoprotein staining, 91, 98, 99
Pro-Q Emerald 300, glycoprotein staining, 91, 92
Pro-Q Sapphire, histidine-tagged protein detection, 94, 95
Prostaglandin-D2 synthase, vitreous fluid proteomics, 499, 502
Prostasome,
functions, 481
seminal fluid proteomics, 480–482
Prostate cancer, biomarker discovery, 188
Prostate-specific antigen (PSA),
breast cancer marker in nipple aspirate fluid, 458, 460
seminal fluid proteomics, 479
Prostatic acid phosphatase, seminal fluid proteomics, 480
Protein chip technology, overview, 49, 50, 150, 181
Protein microarray, *see also* Suspension antibody microarray,
amniotic fluid analysis, 420
principles, 150, 151
Proteome, definition, 418
ProXPRESS®, 2D Proteomic Imaging System, 79, 98
PSA, *see* Prostate-specific antigen
PSP-94, seminal fluid proteomics, 480
PTH, *see* Parathyroid hormone
PWG, *see* Planar waveguide

R

Radioimmunoassay (RIA), nipple aspirate fluid, 454, 455
RAMs, *see* Restricted access materials
Restricted access materials (RAMs), prefractionation, 38
Rheumatoid arthritis, biomarker discovery, 189
RIA, *see* Radioimmunoassay
RuBPSA, *see* Ruthenium tris (bathophenanthroline disulfonate)

Ruthenium tris (bathophenanthroline disulfonate) (RuBPSA), gel staining, 85, 87

S

Saliva,
polyacrylamide gel electrophoresis, 352, 353
production, 348, 349
protein sources, 348, 349
proteomics,
clinical applications, 365, 366
liquid chromatography/mass spectrometry of peptides, 360–363
mucin analysis, 363–365
MudPIT analysis, 358–360
overview of approaches, 349, 350
prospects, 366
sample collection, 350, 352
two-dimensional gel electrophoresis,
advantages and limitations, 353, 354
electrophoresis, 354, 355
mass spectrometry identification of proteins, 357, 358
proline-rich proteins, 356
staining, 355, 356
Western blot, 357
SELDI mass spectrometry, *see* Surface-enhanced laser desorption/ionization mass spectrometry
Semenogelins, seminal fluid proteomics, 478
Seminal fluid,
collection and storage, 470
composition, 467, 468
dynamics of composition, 468, 469
fractionation,
acellular component, 470
cellular component, 470, 471
proteomics,
acellular component, 471, 472, 476, 478
liquid chromatography/mass spectrometry, 472, 477
prospects, 485
prostate proteins,
prostasome proteins, 480–482
prostate-specific antigen, 479
prostatic acid phosphatase, 480
PSP-94, 480
protein types, 474–476
seminal vesicle proteins, 478, 479
spermatozoa proteomics, 482–484
two-dimensional gel electrophoresis, 471, 472, 476
reference values for humans, 468, 469
Serous cell, nasal lavage fluid proteins, 337, 338
Serum,
biomarker discovery, 215, 216, 218, 219
detectable protein abundance, 201–205
industrial scale biomarker discovery, 184, 185
plasma proteomics comparison, 213, 215
sample preparation, 9, 32, 39–41
suspension antibody microarray sample preparation, 76, 77
Shotgun proteomics,
principles, 38, 57, 59, 60, 148, 150
salivary protein MudPIT analysis, 358–360
seminal fluid proteomics, 481
Sinusitis proteomics,
acute sinusitis findings, 341, 342
approaches and rationale, 327, 328
liquid chromatography/mass spectrometry,
mass spectrometry, 329
sample collection, 328, 329
sample preparation, 329

normal versus sinusitis protein findings, 331–336
post-treatment findings, 342, 343
prospects for study, 343
two-dimensional gel electrophoresis, 330, 331

Smoking, protein response in nasal lavage fluid, 341

Spermatozoa, proteomics, 482–484

Sputum,
protein sources, 309, 310
proteome features, 315
sample collection, 310, 311
two-dimensional gel electrophoresis,
differential display proteomics,
allergic asthma, 319
chronic lung allograft rejection biomarker discovery, 320, 321
interstitial lung disease, 317–319
overview, 311, 312
prospects, 322
sample preparation
enrichment of low-abundance proteins, 314, 315
microdialysis, 313
protein precipitation, 312, 313
ultrafiltration, 313

Surface-enhanced laser desorption/ ionization (SELDI) mass spectrometry,
nipple aspirate fluid proteomics,
candidate protein detection, 461, 462
data analysis, 462, 463
protein identification, 457
protein separation, 456, 457
sample preparation, 456
validation, 462
pattern recognition, 201
urinary proteomics,
applications, 235–237
principles, 233–235

Suspension antibody microarray,
antibody coupling to carboxylated microspheres,
bead activation, 73, 75, 81
bead counting, 75
coupling reaction, 75
washing and storage, 75
detection antibody concentration, 78
materials, 72, 73
overview, 71, 72
principles, 73, 74
processing of beads, 78, 79, 81
sample preparation,
cell culture, 77
diluent, 77, 78
plasma, 76, 77
pleural fluid, 77
serum, 76, 77
synovial fluid, 77
tears, 77
validation,
accuracy, 79
linearity, 80
precision, 80
range, 81
repeatability, 80
reproducibility, 80
robustness, 81
selectivity, 79
sensitivity, 80, 81
specificity, 79, 80

SWISS-2DPAGE Database,
annotation, 138, 140
availability, 140, 141
usage, 141, 142

Synovial fluid, suspension antibody microarray sample preparation, 77

SYPRO® Ruby, gel staining, 84, 85, 87, 88

T

Targeted proteomics, definition, 8

Tears, suspension antibody microarray sample preparation, 77
Transmembrane domains, detection of proteins, 94
Two-dimensional gel electrophoresis,
amniotic fluid proteomics, 420, 421
bronchoalveolar lavage fluid and sputum samples,
differential display proteomics,
allergic asthma, 319
chronic lung allograft rejection biomarker discovery, 320, 321
interstitial lung disease, 317–319
overview, 311, 312
prospects, 322
sample preparation,
enrichment of low-abundance proteins, 314, 315
microdialysis, 313
protein precipitation, 312, 313
ultrafiltration, 313
cerebrospinal fluid proteomics, 271–274
databases,
overview, 137–138
SWISS-2DPAGE Database,
annotation, 138, 140
availability, 140, 141
usage, 141, 142
types, 139, 143, 144
difference gel electrophoresis, *see* Differential gel electrophoresis
hemofiltrate, ultrafiltrate fluid proteomics, 512–514
milk protein proteomics, 440, 441
nasal lavage fluid proteomics, 330, 331
nipple aspirate fluid proteomics,
candidate protein detection, 460, 461
electrophoresis, 455, 456
mass spectrometry identification of proteins, 456
sample preparation, 455
validation, 461
pancreatic juice proteomics, 391, 392
principles, 17, 18
resolution, 12, 148
saliva proteomics,
advantages and limitations, 353, 354
electrophoresis, 354, 355
mass spectrometry identification of proteins, 357, 358
proline-rich proteins, 356
staining, 355, 356
sample loading quantitative factors and normalization, 10–12, 53, 55
seminal fluid proteomics, 471, 472, 476
sensitivity, 18
staining, 18, 84, 85, 87, 88
urinary proteomics, 228, 229
vitreous fluid proteomics, 497, 499
Tyrosyl radicals, detection, 93

U

Ultrafiltrate fluid, *see* Dialysate, hemofiltrate, ultrafiltrate fluids
Ultrafiltration, prefractionation, 37, 38
uPA, breast cancer prognostic marker in nipple aspirate fluid, 460
Urine,
protein sources, 225, 226
proteomics analysis approaches,
capillary electrophoresis/mass spectrometry
applications, 242, 245, 246
coupling, 237–239
data analysis, 239, 240
sample preparation, 239
liquid chromatography/mass spectrometry, 229–233
mass spectrometric immunoassay,
inter-individual protein analysis, 253, 256–258

β_2-microglobulin
quantification, 248–250, 252, 253
principles, 246–248
overview, 226—228
prospects, 258, 259
surface-enhanced laser desorption/ionization mass spectrometry,
applications, 235–237
principles, 233–235
two-dimensional gel electrophoresis, 228, 229
sample preparation, 51, 52

V

Vascular endothelial growth factor (VEGF), vitreous fluid proteomics, 502
VEGF, *see* Vascular endothelial growth factor
Vitreous fluid,
composition, 495
pathology, 496
proteomics,
diabetic retinopathy, 499, 502–504
macular hole, 499, 502–504
mass spectrometry for protein identification, 498
prospects, 504
serum sample comparison, 499, 502
tryptic digestion in-gel, 497, 498
two-dimensional gel electrophoresis, 497, 499

W

Western blot,
differential gels in pleural effusion, 293
saliva proteomics, 357
Whey proteins, milk protein proteomics, 439, 442, 443

If he had known his child to be alive, if no deceit had been ever practiced and he had grown up beneath his eye, he might have been a careless, indifferent, rough, harsh father – like enough – he felt that; but the thought would come that he might have been otherwise, and that his son might have been a comfort to him and they two happy together. He began to think now, that his [Smike's] supposed death and his wife's flight had had some share in making him the morose, hard man he was. He seemed to remember a time when he was not quite so rough and obdurate, and almost thought that he had first hated Nicholas because he was young and gallant, and perhaps like the stripling who had brought dishonour and loss of fortune on his head. (p. 904)

The richness of Dickens' Ralph Nickleby is largely due to the narrative voice enveloping him. Narrative reflection of this or any sort is impossible – or at least extremely difficult to pull off – in melodrama, even for the rare playwright interested in psychological realism. And in spite of the novel's transforming narrative, readers of *Nickleby* must have seen in Ralph not only Stirling's character but the stock melodrama capitalist, willing victimizer of honest young men and women for the sake of financial gain. Doggrass, a character in Douglas Jerrold's popular *Black Ey'd Susan* (1829), which Dickens and a large number of his readers undoubtedly saw on the stage during one of its many runs, resembles Ralph in certain important ways. As a popular theatrical type, he was undoubtedly present in the novelist's and his readers' imaginations.

DOGGRASS. Now, Susan, you know my business – I say, you know my business. I come for money.
SUSAN. I have none, Sir.
DOGGRASS. A pretty answer, truly. Are people to let their houses to beggars?
SUSAN. Beggars! Sir, I am your brother's orphan child.[44]

One might easily substitute Ralph and Kate Nickleby here for Doggrass and Susan. (Ralph, when he learns that his brother is dead, his brother's wife and children living, confounds the survivors for seeking his assistance in London and exploits them for his own gain.) The same potential for conflation exists in the sea of Dickens adaptations, like the Stirling play quoted above. Whatever the relative merits of these texts, plays and the novel, the fact remains that they spoke to each other in compelling ways, with the plays responding to Dickens' text, and audience/readers drawn in, engaged as it were in that space I have called the imaginary text.

The Victorian reader's imaginary text was heavily overdetermined, layer upon layer of voice, image, phrase, rhythm. Its layers included novels and their specific adaptations, as well as the entire theatrical and literary repertoires, works of social criticism, politics, journalism – all of the written and performance-based genres which played the same figures and voices – the same social stereotypes – against a backdrop of normalcy, of standard speech, dress, behavior. If we peel back the layers of *Nicholas Nickleby*, we find more of the same beneath – more plays and more prose, a litany of Nicholases, Ralphs, Mrs. Nicklebys. The core of *Nickleby,* which to us seems easily identified, the hard kernel of Dickens' novel, becomes elusive, provisional, when we place the text in its contemporary contexts. As I have suggested, the original readers of the novel, those consuming the monthly numbers in 1838/9, received the text as part of a larger *Nickleby* experience, a novel plus adaptations. Because there were at least twenty-five *Nickleby* plays produced before the novel's completion,[45] Dickens' characters and episodes were circulating outside of his novel, a part of this metatext, this popular discussion in which the public could participate regardless of whether or not they were reading the novel. Even those who recognized the inauthenticity of the adaptations and distinguished a proper "Dickensian" text from among the substantial body of related materials generally read the novels on that plane of shared assumptions and social agreements created by the current field of popular entertainments; we can see this in their responses to Dickens, which accept, without comment, those elements of his fiction which grew out of this theatre of popular assumptions. A writer for the *Edinburgh Review* (1838) remarks,

> We think him a very original writer – well entitled to his popularity – and not likely to lose it – and the truest and most spirited delineator of English life, amongst the middle and lower classes, since the days of Smollett and Fielding . . . We would compare him . . . with the painter Hogarth. What Hogarth was in painting, such very nearly is Mr. Dickens in prose fiction . . . Like Hogarth he takes a keen and practical view of life – is an able satirist – very successful in depicting the ludicrous side of human nature . . . The reader is led through scenes of poverty and crime, and all the characters are made to discourse in the appropriate language of their respective classes . . . His vicious characters are just what experience shows the average to be . . . we find no monsters . . . no creatures blending with their crimes the most incongruous and romantic virtues . . . In short, he has eschewed that vulgar and theatrical device for producing effect – the representation of human beings as they are likely *not* to be.[46]

What is most interesting about this review (of Dickens' first four novels) is its assumptions about human nature. Dickens and Hogarth stand out as great artists of the human condition: their characters are true to class, with the appropriate lineaments, voices, gestures. But we have begun to see that – in spite of this critic's assertion to the contrary – Dickens' characters are ultimately theatrical products, and we shall soon see that the languages the reviewer finds so realistic are somewhat more sophisticated versions of standard theatrical idioms. Like Dickens, Hogarth was indebted to the theatre for many of his subjects and compositions, which would partly explain the conjunction of the two in this review. That he applauds Dickens and Hogarth for the *truth* of their representations may seem curious to twentieth-century readers. The engravings to which he refers are brilliant, detailed, satirical caricatures, pictures of human beings with somewhat grotesque features, engaged in somewhat grotesque behaviors. The comparison itself is quite appropriate, although for different reasons than the *Edinburgh* critic imagines: both Hogarth and Dickens perceived their subjects in theatrical space, their creative pulse moving to theatrical rhythms, and they both exaggerated, using conventional types (elevated in their hands to unusual complexity) to convey a message. What looks to us like a serious oversight in the review suggests that its writer observed Hogarth, and read Dickens, from a completely different place than we do. He is not unsophisticated: he recognizes that Dickens transcends the "vulgar" oversimplifications of the theatre, that his satire resembles Hogarth's in some ways. But his culture, generally speaking, had accepted as authentic the characters and actions which we can only receive as caricatures, or in Dickens' case, as exaggerated or sentimental; the popular culture had embraced and disseminated these types so thoroughly that they rang true.

Of course, not every reader found absolute truth in Dickens' characters. Harriet Martineau, as ever rigorously logical, wrote: "While he tells us a world of things that are natural and even true, his personages are generally, as I suppose is undeniable, profoundly unreal."[47] For that matter, Dickens himself disparaged broad caricature or conventionality in characterization; in response to a poor production of Bulwer Lytton's *Not So Bad As We Seem,* he wrote,

> A miserable thing of no note enacts Colonel Flint. Nothing is changed for the better in the drunken scene of Tonson's part, which I am told is very humorously done by Buckstone. If you had seen the dense conventionality of some of them . . . Mr. Stuart, who does the Duke, actually steamed with conventionality. I saw it passing off from the pores of his skin . . . [all of the staging was done] according to theatrical precedent.[48]

But his friend Macready praises the "force and precision" of Dickens' characters – a concession, one would assume, to their verisimilitude. He then, somewhat unexpectedly, likens them to familiar portraits: "*Nickleby* is much superior to *Pickwick* . . . in the force and precision of its characters – and already includes a gallery of faces as familiar to us as our own."[49] The portrait metaphor strongly suggests that these are not faces we know from our strolls along the Strand, or from our clubs, but are rather the collective images of the popular culture. Curiously, Macready sees no conflict in insisting upon both their universal prior existence and their faithfulness to real life; a year after he published the review quoted above he wrote,

> [Dickens] seizes the eager attention of his readers by the strong power of reality. He thoroughly individualizes what he takes in hand . . . In everything of that kind that he presents to us, there is, in his manner of doing it . . . the truth of life as it is . . . Now by all the various readers of this tale of *Nicholas Nickleby* – and they separate themselves into classes most widely apart from each other – these various qualities can be in an almost equal degree appreciated and felt.[50]

Macready was correct in his assessment of Dickens' equal appeal and relevance to people of every class; the sheer volume of novel sales indicates it. And his attribution of this popularity to the characters' realism is also correct. Clearly, Victorian "reality" – like any industrial or postindustrial culture's reality – merged at some level with the images of the popular culture; how else do we explain the fact that even those Dickensian characters who seem most exaggerated to us were routinely accepted as true to life?

Of course, not all of Dickens' characters strike us as exaggerations: a few of them, particularly from the profound later novels, achieve such a height of emotional and psychological complexity that one can imagine even Martineau converted in the end. Yet all of the novels, even the late ones, rely on melodramatic codes. The stories he tells, from Oliver Twist's to Sidney Carton's, feature such elements as obscure births, long-lost relations, unexpected fortunes, desolate orphan-heroes, and threatened orphaness-heroines – all standard melodrama devices. Even the psychologically richer characters in the late novels can be recognized as more sophisticated versions of theatrical types. Edith Dombey, for example, whose motivations and desires are complex, who suffers acute mental anguish over what she sees as her transgressions, occupies a highly theatrical space in *Dombey and Son*. Beautiful, haunted, emotionally charged, her words and movements enact a melodrama of danger and betrayal, and of the struggle between good and evil forces, as the novel darkens with its burden of moral decay.

"Dear Mama, what have you done to your hand?" said Florence. Edith drew it suddenly away, and, for a moment, looked at her with the same strange dread (there was a sort of wild avoidance in it) as before; but presently she said, "Nothing, nothing, a blow." And then she said, "My Florence!" and then her bosom heaved, and she was weeping passionately.

Strange dreads and wild passions are the stuff of Victorian theatre, as is the language of extremes, the eruption of suppressed emotion in apostrophe, which typify Edith's discourse:

"I have dreamed," she said, "of such indifference and callousness, arising from this self-contempt; this wretched, inefficient, miserable pride; that it has gone on with listless steps even to the altar, yielding to the old, familiar, beckoning finger, – oh mother, oh mother! – while it spurned it; and willing to be hateful to itself for once and for all, rather than be stung daily in some new form. Mean, poor thing!"[51]

Like Edith Dombey, Bradley Headstone, appropriately considered one of the great psychological studies of nineteenth-century fiction, a mass of paranoias and psychoses who is assumed to have inspired no less a creation than Raskolnikov, can be fairly easily traced to his melodramatic roots. Not only do specific scenes in *Our Mutual Friend* indulge in the emotional extremes and the heavy, heady atmosphere of staged melodrama, but many of Headstone's signifying properties – his voice and dialogue, his gesture, for instance – have theatrical referents. His introduction into the novel emphasizes certain gestural peculiarities – a "stiffness of manner," "a kind of settled trouble in the face"[52] – in a way that establishes a theatrical frame of reference for Headstone, so that by the time he proposes to Lizzie in the churchyard, his readers see him as one of a class of violent, unlucky lovers they know from the theatre. His genealogy is confirmed by his melodramatic idiom and the narrative's emphasis on his wild emotion.

"whatever I say to you seems, even in my own ears, below what I want to say, and different from what I want to say. I can't help it. So it is. You are the ruin of me."

She started at the passionate sound of the last words, and at the passionate action of his hands [which was like flinging his heart's blood down before her in drops upon the pavement-stones] with which they were accompanied.

"Yes! you are the ruin – the ruin – the ruin – of me. I have no resources in myself, I have no confidence in myself, I have no government of myself when you are near me or in my thoughts . . . I have never been quit of you since I first saw you. Oh, that was a wretched day for me! That was a wretched, miserable day!"

A few pages later, Bradley responds to Lizzie's refusal with passionate violence:

> "Then," he said, suddenly changing his tone and turning to her, and bringing his clenched hand down upon the stone with a force that laid the knuckles raw and bleeding; "then I hope that I may never kill him!" The dark look of hatred and revenge with which the words broke from his livid lips, and with which he stood holding out his smeared hand . . . made her so afraid of him that she turned to run away.[53]

While these are clearly the words and actions of a psychologically troubled man, they are also conventional signifiers of jealousy, rage, and passion, suggesting that Victorian understandings of the psyche were heavily dependent on externalized signs. In other words, the Victorians acknowledged the inner self and even devoted time to its study, but as in the novels which increasingly explored human psychology, the internal was framed by certain popular assumptions – in particular, gestural and vocal assumptions – which made their way into the collective imagination via the stage. Thus, Bradley's tone changes significantly with his change in emotion, and his dialogue is embellished with the heavy-handed repetition of key words that was typical of the melodrama: "you are the ruin – the ruin – the ruin – of me" (read pauses, saturated with emotion, in the dashes). The compulsive gesture, "like flinging his heart's blood down before her in drops upon the pavement-stones," which accompanies Bradley's passionate utterances serves the melodramatic function of classifying him as a certain type – as someone with an ugly secret, with guilt to express. And the dark look of hatred and revenge, the livid lips and bloody hand were visual images readers could identify with; there were gestures, facial and bodily, used on the stage to indicate these emotions, and there was makeup to simulate blood and livid tissue.

One sees a similar scene in C. H. Hazlewood's adaptation of *Lady Audley's Secret* (1863). In the following passage the haunted George Talboys describes his reaction to the news of his wife's death, an event which provokes emotions resembling Bradley's:

GEORGE. I read the words, "Died in London, May 2nd, 1860, Helen, wife of George Talboys, aged 20." Oh, Bob, what a blow that was to me. I was toiling – saving for her – her who was my life – my soul – my joy! and woke from my dream of hope to know my darling wife was dead – dead – dead!

ROBERT. My poor friend, it must, indeed, have been a shock to you.

GEORGE. It crushed me for a time; but I was obliged to fulfill my duties, or sacrifice my appointment. But during all that time, Bob, the scorching sun

> of India was nothing to the fire that was raging here – here – here! [*Pressing his hand to his forehead*][54]

The shared semiotic patterns in the two scenes – dramatic repetition, significant physical gesture – accompany some more subtle points of connection. Both characters are strangely, and emasculatingly, dependent on the objects of their fixation. Both bow rather mechanically to the pressures of "duty" at the expense of emotional and sexual fulfillment. Each man, in other words, represents a certain kind of failure – failure to attain the material and social capital, for example, that would enable him to establish a conventional middle-class domestic life; failure to master, sexually and otherwise, the woman he desires. But for Victorian readers of Dickens, this sort of critical probing would have been unnecessary; Headstone and Talboys were merely accepted as versions of the same character. They habitually found themselves in similar straits, used the same species of utterance, wore similar clothes, and met the same kind of lonely, often violent end. In the melodramatic universe, they figure as thwarted lovers, vaguely tragic, vaguely sinister figures who suffer from unrequited love and passionate dispositions. It was possible for readers to fully understand Headstone because they had had contact with Talboys and other characters like him.[55]

And this returns us to the difference between nineteenth- and twentieth-century reading. With so many theatrical sources and adaptations in circulation, even the Arthur Clennams of Dickens' world must have turned conventional, losing something of their sharpness and singularity to the swell of images and voices. Even limiting that swell to specific adaptations results in some compromise, some loosening of plot and character. Most Victorian adaptations are fairly homogeneous, the major differences residing in general plot details and in the scenarios actually dramatized. But they existed independently of the novels, whispering amongst themselves, jockeying for position on the boards. Stirling's Nickleby spoke to its fellow dramatizations, and they to it, as loudly as any of them spoke to the novel.

For example, W. T. Moncrieff's *Nicholas Nickleby* (1838) pointedly dramatizes scenes which Stirling's omits, and while both end with Smike the heir to a large and unexpected fortune, Moncrieff devises an elaborate twist of circumstance (and in fact, rewrites the relationships between some major characters) to justify taking that leap. To some extent, of course, the plays are rivals, competing for audience and profits; hence the dramatizations of alternate scenes, and Moncrieff's outrageously complicated ending, an act of one-upmanship. At the same time, they rely on the same dramatic conventions, with characters who speak in standard stage idioms and behave according to generic expectations. In Moncrieff's version, Ralph Nickleby is once again the melodramatic villain, actively and explicitly engaged in plots to injure Smike, Nicholas, and Kate. Nicholas, whose moral and dramatic opposition to his uncle is seated in his diction, speaks the language of sentimental heroism – not as Dickens represents it in the novel, although his Nicholas is certainly a "sentimental hero," but as it was currently represented on the stage. Waiting in his cold, dark room at Dotheboys Hall, he soliloquizes, "What am I to think of all this – what horrid place [*sic*] – what horrid people am I leagued with. Can my uncle have been aware – I would fain hope not – may he not consign poor Kate to equal misery – horrible thought – no, 'tis I alone am the object of his hatred." These are cold and chaste words, apparently unlike anything spoken by the "real" Nicholas. But in fact they reverberated in the public's ears, obscuring or at least complicating the very identity of the character whom we would call the "real" Nicholas Nickleby.

Variations on those words of Moncrieff's Nicholas could be heard on any Victorian stage. The hundreds of melodramas in repertoire over the course of the century each had their Nicholas – one could choose almost at random among them and find young men who looked and talked like the young hero. In Isaac Pocock's *The Miller and His Men* – which Schlicke claims the juvenile Dickens directed in a toy theatre[56] – his name is Lothair and he is a peasant, but he is obviously the standard type to which these Dickens dramatists matched their heroes: an idealized, brave, and for a peasant, surprisingly well-spoken youth. (For that matter, George Eliot used the type in *Felix Holt*, and a dialect version of it in *Adam Bede*.) *The Miller and His Men* was first produced at Covent Garden in 1813, where it played fifty-one times, and was periodically revived in nearly all of the legitimate theatres throughout the 1860s.[57] In June of 1837, while Dickens' novel was in progress, it was revived at Sadlers Wells Theatre; given his childhood attachment to the play, it is not unlikely that he went to see it at this time. A Gothic melodrama set in Bohemia, *Miller* centers on the trials of two poor lovers, Claudine, daughter of the bankrupt Kelmar, and Lothair. Claudine is pursued by Grindoff, a rich miller who is also a *banditto* in disguise; he and his men plan to abduct Claudine and murder a variety of people, including Lothair. The play ends happily, although not without some spectacular scenes of violence. Its plot bears no resemblance to *Nickleby,* but the languages spoken by its hero and villain reverberate in those later plays. Notice the similarities between Lothair's dialogue and that of Moncrieff's Nicholas :

LOTHAIR. A sudden exclamation burst from my lips, and arrested their intent; they turned to seek me, and with dreadful imprecations vowed death to the intruder. Stretched beneath a bush of holly, I lay concealed; they passed within my reach. I scarcely breathed, while I observed them to be ruffians, uncouth and savage – they were banditti.[58]

Here, again, is Moncrieff's Nicholas upon his arrival at Dotheboys Hall:

NIC. "What am I to think of all this – what horrid place [*sic*] – what horrid people am I leagued with. Can my uncle have been aware – I would fain hope not – may he not consign poor Kate to equal misery – horrible thought – no, 'tis I alone am the object of his hatred."

These passages resemble each other in several ways. Stilted and archaic, they do not reflect the authentic spoken language of any class. In fact, the language spoken by Lothair and Nicholas is in one sense classless – spoken indiscriminately, in a way that must have satisfied audiences, by the son of a country gentleman and a peasant! It represented an ideal rather than any verbal or social reality – a unanimously accepted signifier of young and honest masculinity. If the idiom and its particular speakers in some ways evaded class specificity, the ideal did not – he may have occasionally sported peasant clothes and worked out of doors, but he stood for middle-class values: hence his verbal fluency. Furthermore, the tendency of both characters to exaggerate their difference from a degraded other group – "horrid people," "ruffians, uncouth and savage" – suggests that both plays presume the same class politics, the same set of social codes.

Similar dialogue may be sampled from almost any melodrama of the early to mid-century. At the end of Jerrold's *Black-Ey'd Susan,* Susan's sailor-husband William, about to be hanged for a crime he did not commit, responds to the drowned body of his wife's uncle, the capitalist Doggrass, in similarly "melodramatic" language. "What, Susan's Uncle! villain, may the greatest – [bell tolls] – no, no, – I shall soon be like him; why should the dying triumph over the dead? [after a moment] I forgive him."[59] Conventionally differentiating the villainous from the good, he is interrupted, like the melodrama heroes before and after him, by his own surging emotions – here, cleverly materialized in the bell – which induce in him those typical heavy pauses. Melodrama always establishes clear boundaries between the good and the bad, its moral delineation often serving as a metaphor for class hierarchy. Even in Dickens, who devised substantially more complicated plots, and wrote more realistic and complex dialogue than may be found in the melodramas, the boundaries between good and bad generally adhere, and are dramatized in characters' voices.

Having now established the existence of "real" voices in Dickens' novels (a later chapter on Dickens and Charles Mathews will invoke some actual prototypes), and having identified a highly specific context for Dickens and his characters, placing them in the center of an enormous theatrical event, I wish to raise certain questions about the place of vocal performance in English culture, about the historical ties between reading and voice, about acting, singing, and storytelling. To confine this study to the relatively tiny moment and place of nineteenth-century London entertainments would be insufficient, a begging of important questions. Much of this historical–theoretical work is undertaken in the next chapter, but it will be useful to prepare for that discussion by introducing here the voice which lies at the center of this study, and which was popularly used to dramatize social marginality in the nineteenth century. I am referring to the *pattered* voice, a verbal gesture historically linked to performance and subversion.

Patter was a common theatrical trope in the eighteenth and nineteenth centuries. In the nineteenth century it appeared with increasing frequency in narrative as well, and therefore my attention to voice in the novel necessarily includes observations on the written word, the inscriptions of variant speech in English prose. In fact, representations of patter always presume the word as a kind of literal or essential presence, because patter is *about* words, it assumes an excess of them, oddly arranged or pronounced. In this sense it presupposes textuality even when it is purely performative. This confluence of the written and the spoken, the read and the acted, inspires this project throughout. I am equally interested in the transformation of text into theatre, the insistence of the theatrical on text, and the implications of performance in the act of reading. This book, then, is as much about playing Dickens as it is about reading him, because in the process of reading we in varying degrees perform. In any case, I have written with a conviction of the potential for gesture or activity in scripted language, and I wish my reader to believe it too – to feel the energy in words which turns certain types of narrative into potential or actual theatre.

This study originated with a set of specific questions. Why did

Dickens use such excessive, extravagantly textured and rhythmed (i.e. pattered) dialogue for his middle-aged, single, or widowed female characters? What was this dialogue's relation to the nonmusical theatre and to the *opera buffa* patter aria, which it in some ways resembled? And most importantly, what did his readers make of it? To me it seemed obviously derisive – try reading one of Flora Finching's or Mrs. Nickleby's passages without feeling annoyed or giving up in disgust! But what did the Victorian reading public think? The answer to this last question in particular surprised me. I am led to believe that they did not think about the pattered voice at all – at least, not in the self-conscious way that I have done here. Rather, patter was a vocal sound and social condition so discreetly integrated into the popular imagination, so long in circulation by the mid-nineteenth century, that people heard it or read it as a social fact, a real voice with real significance in everyday life. They did not think about it so much as experience it. Thus, two witnesses of Dickens as Gabblewig in *Mr. Nightingale's Diary* were able, at least fifteen years after the performance in question, to faithfully describe the structures, rhythms, and cadences of the patter with which his character impersonates an old Gamp-like woman.[60] This suggests that patter was a part of the Victorian vernacular, familiar enough to two writers, in any case, to enable them to reproduce it accurately in writing from memory. The 1855 *Nightingale* performance at Tavistock house[61] was not, obviously, their only exposure to the idiom. That they, and a large segment of the Victorian public, had internalized patter, and understood it, as we shall see, to exist off as well as on the stage, indicates that the period's collective imagination was essentially theatrical: people observed their world through a theatrical frame, often drawing on conventional stage gestures, like patter, to order their experience. In other words, patter, like the larger theatrical apparatus, was implicated in the formation of a Victorian social consciousness.

Of the strains of voice Victorian readers heard in Dickens' texts, one of the most prominent was this pattered voice, which accompanied, in a sort of comic *continuo*, the *obbligato* of the standard idiom. Patter, in Dickens and for that matter in the entire field of contemporary popular entertainments, exists as a sort of background interference, a buzz of potential disruption which occasionally flashes forth in discordant solo. It is always there as a reminder of social difference as well as a generic tag. Its universe is comedic. Patter figures as one of the earliest eccentric voices on historical record; from the Middle Ages through the end of the nineteenth century, unlicensed speech was frequently identified as a form of patter. The fourth chapter of this study traces the declension of patter from its beginnings in medieval religious discourse to its secular life in eighteenth- and nineteenth-century English popular cultures. Here I will not devote as much space to definition. Suffice it to say that patter confounds the conventional structures of verbal discourse, violating the standard idiom with its eccentric lexicons and/or its verbal excess, logical discontinuities, and rapid, repetitive delivery. It was prominent among the strains of unlicensed speech and song at play in the English imagination, and it always existed as a kind of theatricality. As far as it is possible to determine, patter first appeared in the medieval European church, the play-Latin of illiterates who could only mimic the rhythms and forms of the paternoster. Around this fact evolved a set of ideas about speech and social place: pattering the paternoster established a potential for thinking heresy, which rendered suspect a large universe of illiterate commoners. This identification of the lower orders with subversive patter continued into the modern age, and as Western culture became secular, patter was increasingly identified with the street: beggars, thieves, low-brow professionals. In an indirect fashion, then, by way of medieval church and Renaissance minstrel, Italian opera and Victorian street culture, we come to one of the most popular and meaningful comic voices of the nineteenth century, a voice sung on the stage and on the page, one which everyone knew to represent certain social conditions. People may have heard it as a reference to the stage delivered in the manner of their favorite comedian but they did hear it, and possibly nowhere as clearly as in the pages of Dickens.

It is impossible to discuss Victorian patter without reference to gender and class, because it materialized in class-specific "male" and "female" forms in the nineteenth century. "Female" patter, believed to be spoken primarily by middle- or lower middle-class spinsters and widows, is essentially comic, with its high-speed, run-on sentences, sparce punctuation, and logical circularity. It typically has only contextual, and not inherent, meaning – and in this it differs from earlier forms of comic female speech, which, like Mrs. Malaprop's or Mrs. Slipslop's, require that the reader/observer absorb and evaluate every word if the joke is to be effective.[62] "Male" patter lived on the streets with the lower portion of the working class; less amusing than its feminine counterpart, it reinvents standard English, infusing the language with new words, new syntactic patterns, new (and generally subversive) ideas. Its meanings are inherent but obscure, and hence sinister.[63] One may occasionally locate characters who speak the "wrong" patter, but such idiomatic boundary crossings generally accompany crossings of gender or class boundaries by their speakers, so that a lower-class male using "female" patter may be received as slipping into the proscribed realms of femininity or, less frequently, bourgeois culture.[64] These "slippings" represent a certain threat: hence the linguistic insult. In other words, it is the character, and not the patter, who loses his or her gender or class integrity when these crossings occur.

Any attempts to contextualize Dickens' use of patter will lead directly to these gender and class margins. Patter was used, in theatre and prose, to publicize and politicize social differences, to render a negative physical and verbal presence, and lend a sense of danger or impropriety to social groups who lay outside of mainstream (bourgeois) social life. The gesture derives its power from a suspicion of double or frivolous languages still in force throughout most of the nineteenth century; direct associations of eccentric speech with heresy and lunacy obtained at least until the last quarter of the century. The end of the nineteenth century saw an increased interest, even a delight, in slippery language. Gilbert and Sullivan, for example, constructed their dramatic universe out of verbal jokes and deceptions, and the very patter which had recently caused so much anxiety.[65] But the *fin de siècle* was a time of social and aesthetic re-visioning, precipitated by a perceived need to cast off the old millstone of mid-Victorian morality. During the first two-thirds of the century, the rough time period with which this study concerns itself, eccentric speech almost always signified moral and social malaise.

I do not mean to suggest that sensibilities changed abruptly around the year 1870, or that the possibility of eccentric idioms as pleasurable or interesting never arose until Gilbert and Sullivan or some of their contemporaries found them to be so. Eccentric languages like patter typically attended comic performances and were generally intended to serve as instruments of social control – reminders of the "unitary" language which they violated, and indeed, proved illusory – although subversions of the "standard" writerly idiom (if in fact one may isolate a standard prose style) emerge here and there without the usual negative attachments. For example, what is Thomas Carlyle's idiom if not a sort of second cousin to patter, an excessive, recursive explosion of sounds signifying in unconventional ways? *Sartor Resartus* never spares an extra syllable or word when it can help it, and its circular tale unwinds in mysterious ways, saving itself, I suppose, from the ignominy of verbal frivolity or extravagance by its firm moral grounding, and perhaps as well by the reputation of its author. It opens with a circumlocution on a grand scale:

> Considering our present state of culture, and how the Torch of Science has now been brandished and borne about, with more or less effect, for five thousand years and upwards; how, in these times, especially, not only the Torch still burns, and perhaps more fiercely than ever, but innumerable Rushlights, and Sulpher-matches, kindled threat, are also glancing in every direction, so that not the smallest cranny or doghole in Nature or Art can remain unilluminated, – it might strike the reflective mind with some surprise that hitherto little or nothing of a fundamental character, whether in the way of Philosophy or History, has been written on the subject of Clothes.[66]

Notwithstanding some obvious comic elements, *Sartor Resartus* is a serious work, and Teufelsdröckh, despite his name, a serious character. The rhythms and syntax of Carlyle's text may recall us to examples of comic patter – it shares with patter an overabundance of words, if not a lack of intelligibility (it is difficult, not impossible) – but, as the contemporary reviews indicate, his readers – even those who found his prose bizarre – did not make this association.[67] The same may be said for the tortured narrative voice Dickens uses at times.[68] Take the opening passage of *Bleak House,* for example:

> London. Michaelmas term lately over, and the Lord Chancellor sitting in Lincoln's Inn Hall. Implacable November weather. As much mud in the street, as if the waters had but newly retired from the face of the earth, and it would not be wonderful to meet a Megalosaurus, forty feet long or so, waddling like an elephantine lizard up Holborne Hill . . . Foot passengers, jostling one another's umbrellas, in a general infection of ill temper, and losing their foot-hold at street-corners, where tens of thousands of other foot passengers have been slipping and sliding since the day broke (if ever this day broke), adding new deposits to the crust upon crust of mud, sticking at those points tenaciously to the pavement, and accumulating at compound interest.[69]

This passage lumbers like the forty-foot elephantine lizard it describes, and its general lengthiness and curliness encourages suggestions of words like "serpentine" and "alexandrine," in Dickens' "elephantine." In fact, one cannot help but recall the "needless alexandrine" in Pope's "Essay on Criticism," although this lacks Pope's ever-tight verbal and metrical control, and of course does not set out to satirize itself.[70] And this raises an important question: why is the excessive voice not self-satirical or parodic when it appears to issue from the author's own lips? Dickens often forces his readers to endure taxing syntactic excursions. These may, like the cited passage, serve the serious purpose of dramatizing unacceptable social conditions, but they do bear an ironic resemblance to the comic patter he used in an effort to contain his redundant women – although, like Carlyle's prose, they stop well short of patter's typical opacity.

Miss Flyte, for example, delivers a speech in *Bleak House* which is no less lengthy, curly, and fragmented than the opening narrative quoted above – just a bit more obscure, thematically if not linguistically.

> Most extraordinary . . . You never heard of such a thing, my dear! Every Saturday, Conversation Kenge, or Guppy (clerk to Conversation K.) places in my hand a paper of shillings. Shillings. I assure you! Always the same number in the paper. Always one for every day in the week. Now you know, really! So well-timed, is it not? Ye-es. From whence do those papers come, you say? That is the great question. Naturally. Shall I tell you what I think?[71]

Why should readers have accepted in the authorial voice a stylistic device which they found comic or pathetic in the speech of Dickens' female patterers, or for that matter which they found threatening when it surfaced on the streets? Why, in this specific instance, is Miss Flyte's speech less serious than Dickens'? Partly because of its lighter subject matter, its repetitions and extraneous matter. The narrator's passage, with its heavy social commentary, its heightened language, its attention to the mud and its preponderance of labials, is grounded by its own impressive weight, while Miss Flyte's nonmetaphoric, crisp-consonanted prose feels light and giddy – flighty, if you will. Her language is not only nonmetaphoric, but aggressively prosaic, while the narrator's opening passage might almost be refigured as a slightly irregular blank verse: London. Michaelmas term lately over, / And the Lord Chancellor sitting in Lincoln's Inn Hall. / Implacable November weather. / As much mud in the street, as if the waters / Had but newly retired from the face / Of the earth . . .

These textual differences may have influenced readers, but it seems that the seriousness of that first voice is inextricably tied to the larger issues surrounding intellectual currency and authorship. First of all, educated middle-class men, like the century's leading acknowledged intellectuals, *did not* patter; by collective agreement, that trope belonged to other social types. Furthermore, authorship, in an age increasingly dominated by the press, represented a peculiar kind of social and financial power, an absolute authority. Writers like Dickens and Carlyle achieved the status of cultural heroes or myths. Both were subjects of public discussion and speculation, creatures of the public sphere. When a Carlyle, or a Dickens, speaks in his own voice or in the privileged voice of "narrator," he maintains his dignity no matter how eccentric the content or style of his dialogue – this was, in a sense, in keeping with the terms of his agreement with the public. When Dickens broke the agreement, by theatrically abandoning and humiliating his wife, his stock dropped considerably in some quarters. But certainly through the early and middle years of his career he was virtually universally recognized as the century's greatest novelist: any voice associated with him, even the comic ones, must have been heard as "serious," in the sense of authoritative. He himself took his comic voice quite seriously, as an important dramatic tool, as rhetorical marks in the margins of his prompt copies indicate.[72] Patter was a verbal trope of such importance, then, because the voice itself was a highly privileged social construction.

We may assume that the dramatic voice was a privileged Victorian entity from the preponderance of staged readings in the middle of the nineteenth century, but the idea of a reading that explores all of the dramatic possibilities in language is not exclusively Victorian. Contemporary theatre places increasing value on it. Let us dwell for a moment in the present, with some comments of Royal Shakespeare Company voice coach Cicely Berry, under the assumption that an understanding of current theatrical practice will enable us to conjecture about practices in the past. Berry has written several books on the role of voice in theatre, and she works actively on things like poetry reading and vocal exercise with the company. Her insistence on the primacy of the voice in acting confirms an impulse one may notice in Victorian theatre criticism, to search an actor's diction, tones, and inflection for the crux of dramatic meaning.[73] At a certain level, then, one might argue that the English theatrical experience has primarily to do with vocality – the accents, idioms, and cadences of the voice. As Berry asserts,

> The voice is the means by which, in everyday life, you communicate with other people, and though, of course, how you present yourself – your posture, movement, dress and involuntary gesture – gives an impression of your personality, it is through the speaking voice that you convey your precise thoughts and feelings.

She uses poetry to teach a sensitivity to vocal rhythms and resonance:

> [Poetry's] value to the actor , whether or not you are going to do Shakespeare or other poetic drama, is that it increases this sensibility to words, and rhythms and meanings which come to you from sound – meanings which cannot necessarily be explained and which go deeper than our conscious logical mind.[74]

This, of course, conflates the textual and performative in suggestive ways, and is relevant to the present study in that this sort of commerce between text and performance was so central to Victorian experience. The emphasis on the role of the voice by the central British dramatic organization parallels a larger, cultural focus on accent and diction: the English, as I have already argued, have always identified social groups on the basis of their respective speech types. This basic attention to patterns of voice has interesting implications for the act of reading, drawing it closer to the act of performance.

So how does one "perform" readings of performative texts? Dramatic voices may be easily identified and sounded in certain – usually "musical" – types of text. One text well-equipped to demonstrate the workings of voice in drama is by American playwright Wallace Shawn. I shall use this play expressly to demonstrate the possibilities of aural reading, so that we may later approach some very different texts with an ear to their peculiar voices. Shawn's *Marie and Bruce,* a sort of perverse, 1980s comedy of manners, underscores the music in dramatic verbal expression, sings a complicated, syncopated duet in which semantic meaning is complemented by vocal gesture. It gives a clear sense of two (sung) vocal lines, rhythmically and morally in conflict with each other. The play opens with a barrage of verbal abuse: Marie, loud, extravagant, compulsively wordy, has decided to walk out on Bruce.

> MARIE [*to audience*]. Let me tell you something. I find my husband so God-damned irritating that I'm planning to leave him. And that's a fact. [*To* BRUCE] Yes! I'm sick of you! Do you get it? You're driving me insane! I can't stand living with you for one more minute! I'm sick of it! I hate it! I hate my life with you! Do you hear me? I hate it!
>
> BRUCE. Oh – hello, darling. Is it time to get up?
>
> MARIE. No! No! God damn you, it's the middle of the night! Now go back to sleep – please!
>
> BRUCE. Well, don't be irritable, darling –
>
> MARIE. Irritable? Irritable? You call me *irritable*? God damn you, I've had about *enough* of your disgusting *insults*, you God-damned cheap God-damned idiotic pig, you shit! Now go back to sleep!
>
> BRUCE. Well – all right, darling – [75]

The baroque structure of the text – Marie's rhythmic, staccato repetitions in counterpoint with Bruce's banal, legato responses – yields music. What we hear – and we *do* hear, even as silent readers – are the patterns of their verbal parries. Marie's lines grow more thickly repetitive as she gains momentum, Bruce's remain maddeningly smooth (although he too occasionally falls into repetition), and out of their mutual play is generated the drama's meaning: that words are themselves a kind of action, and that in the bland easiness of ingratiating speech (Bruce's) lurks a grotesque insensibility. We may turn to some very different sorts of text and experience a similar musicodramatic movement. W. S. Gilbert, one of the most successful theatrical writers of the nineteenth century, wrote a type of drama that feels surprisingly similar to Shawn's, despite their tremendous thematic and idiomatic differences. Like Shawn's, Gilbert's characters speak in tightly choreographed patterns – although what might be described as a *pas de deux* in *Marie and Bruce* expands, in Gilbert, into something larger and more complex, a reel or contradance, perhaps. Take *Ruddigore*, for example. Here, the difficulty and delicacy of social relationships is dramatized in the give and take of syncopated dialogue, in strains of contrapuntal voice, sounded and silenced to the rhythms of dramatic prose.

MARGARET. Shall I tell you one of Mad Margaret's odd thoughts? Well, then, when I am lying awake at night, and the pale moonlight streams through the latticed casement, strange fancies crowd upon my poor mad brain, and I sometimes think if we could hit upon some word for you to use whenever I am about to relapse – some word that teems with hidden meaning – like "Basingstoke" it might recall me to my saner self. For, after all, I am only Mad Margaret! Daft Meg! Poor Meg! He! he! he!
DESPARD. Poor child, she wanders. But soft – someone comes. Margaret, pray recollect yourself – Basingstoke, I beg! Margaret, if you don't Basingstoke at once, I shall be seriously angry.
MARGARET. [*recovering herself*]. Basingstoke it is!
DESPARD. Then make it so.
[*Enter* ROBIN. *He starts on seeing them.*]
ROBIN. Despard! And his young wife! This visit is unexpected.
MARGARET. Shall I fly at him? Shall I tear him limb from limb? Shall I render him asunder? Say but the word, and –
DESPARD. Basingstoke!
MARGARET. [*suddenly demure*]. Basingstoke it is!
DESPARD. [*aside*]. Then make it so.

The tightness of repetition, the deftness of repartee, make this prose sing, as they do Shawn's. Margaret's "Shall I fly at him? / Shall I tear him limb from limb? / Shall I render him asunder?" is really highly rhythmic poetry, with the metronomal regularity typical of doggerel verse. (One is reminded by Mad Meg's verse of the children's rhyme that begins "Do your ears hang low? / Do they wobble to and fro?") These voices alternate regularly and rapidly, without losing the rhythm of the piece, which is generally sustained throughout. We do not need to know the music to hear the vocal–verbal relationships Gilbert has used to metaphorically suggest the social relationships between his characters. Mad Meg's verbal flights, her linguistic excesses, match her emotional and mental instability, her excess of feeling, and they are punctuated by Despard's rational, economical commands; Despard's speech loses its sharpness and brevity when the power relationships change and he finds himself in conversation with his brother. Perhaps *Ruddigore*'s whole verbal structure fits together so tightly and "naturally," adjusting to its own rules and rhythms, because its universe is so outrageously eccentric: these voices are in sync with each other precisely because they are out of sync with the real world. In any case, understanding the ways in which such voices may be heard (or visually imagined: my mixing of musical and dance metaphors is deliberate) even in silent reading, feeling the peculiar rhythms and tones inscribed in these explicitly vocal–musical texts, should prepare us for hearing and feeling the voices in Dickens. This is why I have chosen to discuss them here. *Ruddigore* and *Marie and Bruce* graphically illustrate that quality of voice (and gesture) which is so important in Dickens.

Using musical metaphors for Dickens' prose allows one to emphasize the real performativity, the genuine vocal presence, that typical analyses tend to ignore. One way to recover Dickens' voices is by tapping into his theatrical sources – a task which the fifth chapter of this book devotes itself to undertaking. Here, let us return to the Dickensian imaginary text. So much of Dickens criticism has focused on text as written word, literary artifact, and has neglected the layers of extratextual movement in his novels. But biographer Peter Ackroyd has celebrated these complications in Dickens: "So strong was Dickens' imaginative hold upon his readers, in fact, that it is also entirely probable that people began to behave in a 'Dickensian' fashion when they were in his presence; in other words they unconsciously exaggerated their own mannerisms and behaviour in order to conform to the types which he had already created."[76] Ackroyd's recognition of the slipping of generic boundaries which seems always to mark Dickens' work and life – his self-consciously theatrical self, his tendency to infuse other people, as well as any prose he set his hand to, with theatricality – results, rather disappointingly, in the conclusion that Dickens himself was the ultimate source of his fictional–theatrical types, that he single-handedly spawned a set of characters, voices, and gestures and then sent them forth into society. This is certainly not the case: brilliant as he was, he fed from a common cultural plate. What Ackroyd has grasped, however, is the way in which dramatic voices resounded in the Victorian imagination, and in Dickens' work.

In Dickens, voices speak in counterpoint, delivering lines with clear rhythms and tones, as well as general and specific theatrical referents. Take, for example, the Crummles episodes in *Nicholas Nickleby,* Dickens' most explicitly theatrical novel. (As Ackroyd suggests, "*Nicholas Nickleby* is written by someone whose understanding of appearance, of gesture, of speech and of character has been very strongly influenced by his experience of acting" [pp. 283–284].) Here the strains of dialogue (including that of the narrator) are carefully choreographed, much like the novel's mock sword fight on the road to Portsmouth – verbal thrusts and parries as different in character as the dueling Crummles brothers are in size. Each character owns a specific voice, and these voices strike and retreat in cadenced steps, generating meaning in the patterns and rhythms, not the words, of conversation. One might read the staged fight, described almost entirely in the voice of the narrator, as a sort of physical metaphor for the verbal politicking that takes place in the theatre itself.

> "There's a picture," said Mr. Crummles, motioning Nicholas not to advance and spoil it. "The little 'un has him; if the big 'un doesn't knock under in three seconds he's a dead man. Do that again, boys."
>
> The two combatants went to work afresh, and chopped away until the swords emitted a great shower of sparks, to the great satisfaction of Mr. Crummles . . . The engagement commenced with about two hundred chops administered by the short sailor and the tall sailor alternately, without producing any particular result until the short sailor was chopped down on one knee, but this was nothing to him, for he worked himself about on one knee with the assistance of his left hand, and fought most desperately until the tall sailor chopped his sword out of his grasp. Now the inference was, that the short sailor, reduced to this extremity, would give in at once and cry quarter, but instead of that he all of a sudden drew a large pistol from his belt and presented it at the face of the tall sailor, who was so overcome at this (not expecting it) that he let the short sailor pick up his sword and begin again. Then the chopping recommenced.[77]

This passage describes a dance, actors in pantomime combat, but the narrative, with its sympathetic diction, its reliance on quick, explosive words like "chop" and "legs" and "short," its repetitions and its bracing pace, dances as well. The narrator shifts back and forth between combatants, often in mid-sentence, signaling the contextual changes with tag words like "but" and "now." What I am describing is a kind of dialogism, to borrow Bakhtin's term, a plurality of voice within the ostensible monovocality of third-person narrative.[78] There is always this sort of vocal resonance in Dickens, and although here I have described it as dancing, it also exists as a kind of scored music, structured and resonant. Typical of Dickens is a sensitivity to the rhythms and syntax of human speech and thought, a willingness, on the part of the author, to hear, at a level deeper than that of semantics, the voices of his characters.

Let us try to complicate our experience of the Crummles episode by introducing another version of it, an 1838 dramatic adaptation of *Nickleby* called *The Infant Phenomenon; Or, A Rehearsal Rehearsed,*[79] by playing one against the other and locating the third text, the imaginary text, which is generated in the process. This piece takes as its entire subject the scenes which occur in the Crummles' theatre. The swordfight, which occupies a place of such descriptive importance in the novel, can only be performed here, and exists in the text as a stage direction ("a couple of boys, one tall, the other short, dressed as sailors, with pigtails and buckles complete, fighting a theatrical combat"). The locus of dramatic meaning has been shifted, from the elaborate, prosy, descriptive passages in which it primarily resides in Dickens' novel, to spoken dialogue. Apart from the necessary plot contractions, and the omission of that descriptive narrative which constitutes the real flesh of a Dickens novel, the playwright, H. Horncastle, has made some specific adjustments to the text. I use "adjustments" to suggest a conscious adaptation to his medium. Most of these changes lie in the diction of the two main characters, Nicholas and Mr. Crummles, and they serve the purpose of drawing both characters closer to the stock theatrical types which originally inspired them, an ironic artistic regression. Thus, Crummles uses a more vulgarly comic idiom than we see in Dickens, drawing heavily, in fact, upon current lower-class slang. An exchange which, in the novel, reads

> "Excuse me saying so," said the manager, leaning over to Nicholas and sinking his voice, "but – what a capital countenance your friend has got!"
>
> "Poor fellow!" said Nicholas, with a half smile, "I wish it were a little more plump and less haggard."
>
> "Plump!" exclaimed the manager, quite horrified, "you'd spoil it forever."[80]

is rewritten, in the play, to read as follows:

CRUMMLES. You will excuse me saying so, but what a capital *mug* your friend has.

NICHOLAS. Poor fellow – I wish his face were a little more plump and less haggard.

CRUMMLES. Plump? Would you spoil it forever – if the young gentleman was a *plumper*, you'd spoil him forever. (My italics.)

Horncastle's Crummles often substitutes slang terms for the "legitimate" English of Dickens' character, and his brand of comedy verges on the slapstick, as when he runs through a litany of homonyms each time he refers to Smike. "Stike, Dyke, Tyke or Mike, or what you please" is used repeatedly, a tag line for Crummles, as if audiences needed the cue, and the gag was too good to drop. To us he is obviously a caricature of the other Crummles, although Victorian audiences, steeped in his sort of comic gesture, brought up to expect characters of his cast to "really" talk like that, would not have seen him as such, but would have seen something of him in Dickens' character, just as the *Edinburgh Review* critic saw the subjects of Hogarth's engravings walking the streets of London.

If Crummles' dialogue transforms him into the conventional "low comedic" figure of the day, Nicholas' idiom turns him into a standard theatrical hero. It is a language, like Moncrieff's, far more chaste and formal than the one Dickens had him speak. In the novel, Nicholas responds to the theatre manager's job offer ambivalently: he is not entirely adverse to trying his hand at the "genteel comedy" and "juvenile tragedy" so suited to his handsome face and figure.

"I don't know anything about it," rejoined Nicholas, whose breath had been almost taken away by this sudden proposal. "I never acted a part in my life, except at school."

"There's genteel comedy in your walk and manner, juvenile tragedy in your eye, and touch-and-go farce in your laugh," said Mr. Vincent Crummles. "You'll do as well as if you had thought of nothing else but the lamps from your birth downwards."

Nicholas thought of the small amount of change there would remain in his pocket after paying the tavern bill and he hesitated.

"You can be useful to us in a hundred ways," said Mr. Crummles. "Think what capital bills a man of your education could write for the shop-windows."

"Well, I think I could manage that department," said Nicholas.[81]

Horncastle's Nicholas is less attracted to the offer, at first demurring for the sake of his friend Smike: "You flatter me, sir, I can only speak for myself, and to speak truthfully, I should not like to part from poor Smike." Dickens' Nicholas *is* flattered, tempted by what looks like an exciting way to earn some cash, and his speech – casual, hesitant –

shows it. His theatrical double, however, shows more restraint; one can imagine him drawing himself up to accuse Crummles of flattery, and delivering his rather formal response from a place of moral, and possibly physical height, depending on the staging.

And there were other versions of the Crummles scene in circulation at the time. Dickens himself was responsible for some of them. His performances in *Mr. Nightingale's Diary* included the role of "'Sam Weller,' an actor turned waiter, whom Lemon . . . recognizes as a fellow member of a provincial acting company in years gone by. They recall a 'dreadful combat together in a wood.'" R. H. Horne described the action:

> First they prowled round and round each other – now darting in, very nearly, and as suddenly starting back; next, a passing cut is exchanged, then two or three cuts, the swords emitting sparks . . . At last they close, and strike and parry to a regular measured time very like the one known as *Lodoiska* in the Lancer Quadrilles.[82]

This is Dickens performing, as he would continue to do till the end of his life, a scene from one of his novels. While it seems clear that he and his audience were indulging, here and in the *Nickleby* scene, in a bit of melodrama parody, it seems equally clear that the parodied conventions held substantial cultural currency: why parody them if no one cared about them?

An equally parodic prose version of the scene by the young Boz, a sketch called "Private Theatres," anticipated *Nickleby's* mock sword fight by a few years but sounds quite similar to it; the novel's scene was hardly original, to Dickens or the corpus of early to mid-century popular writings.

> Then the love scene with Lady Ann, and the bustle of the fourth act can't be dear at ten shillings more – that's only one pound ten, including the "off with his head!" – which is sure to bring down the applause, and it is very easy to do – "Orf with his ed" (very quick and loud; – then slow and sneeringly) – "So much for Bu-u-u-uckingham!" Lay the emphasis on the "uck"; get yourself gradually into a corner, and work with your right hand, while you're saying it, as if you were feeling your way, and it's sure to do. The tent scene is confessedly worth half a sovereign, and so you have the fight in, gratis, and everybody knows what an effect may be produced by a good combat. One – two – three – four – over; then one – two – three – four – under; then thrust; then dodge and slide about; then fall down on one knee; then fight upon it; then get up again and stagger.[83]

Ostensibly a satire of the "private theatre," an institution which encouraged anyone with a few spare pounds to buy himself a leading dramatic role (all roles were for sale), the sketch is itself an exercise in theatricality, in the skipping and starting and thrusting of dramatic action. The broken narrative shifts from one context to another with a sort of ventriloquistic movement, a sleight-of-hand which keeps one looking – or reading – back and forth, stage left to downstage center and back, as if following a cast of characters through an intricately choreographed scene. These choppy little segments of narrative, attentive to their own dramatic rhythms and pieced together into a frame, make the theatre in this prose piece. It is choreographed like a dance, syncopated like music, and if we let ourselves be drawn into its movements we experience the private theatre live, as real drama as opposed to prose description. And thus this parody of melodrama generates real theatricality, as if the form refuses to relinquish its primacy. Dickens cannot escape it even when he wants to.

So far I have emphasized the physical gesture of staged combat, the piecing of text into dance patterns, over the strains of voice in these Dickensian scenes. But voices make theatre as well, darting, dashing, clashing, and in *Nicholas Nickleby,* often fighting to monopolize our ears. In fact, the scene of Nicholas' introduction into the Crummles' acting company echoes the sword fights above. Mrs. Crummles, a very grand tragedian, greets Nicholas thus: "I am glad to see you, sir . . . I am very glad to see you, and still more happy to hail you as a promising member of our corps."[84] As if the explicit stage directions for Mrs. Crummles ("in a sepulchral voice," "crossing to Smike, as tragic actresses cross when they obey a stage direction") were not enough, Dickens provides a significant number of identifying markers in her speech. "Each word tastes of the context and contexts in which it has lived its socially charged life," to borrow from Bakhtin.[85] Mrs. Crummles' dialogue is a semiotic goldmine, a rich field of social and theatrical signs. Her fairly regular trochaic meter, her fondness for broad vowels and words that linger in the mouth, her stately repetition, all conspire to suggest artificiality, obsolescence of method, even largeness of girth. She is powerful and innocuous; great and silly. Even her dinner-table conversation quivers with dignity, each word an act in itself, clearly enunciated and wholly aware of its antecedents.

"We have but a shoulder of mutton with onion sauce," said Mrs. Crummles, in the same charnel-house voice; "but such as our dinner is, we beg you to partake of it."

"You are very good," replied Nicholas, "I shall do it ample justice."

"Vincent," said Mrs. Crummles, "what is the hour?"

"Five minutes past dinner-time," said Mr. Crummles.

Mrs. Crummles rang the bell. "Let the mutton and onion sauce appear."[86]

Mrs. Crummles is on the stage. Her formal diction and syntax, the ceremony of the dinner ritual – asking the hour, ringing the bell – gleam with paint and gaslight, theatrical tricks. Each step must proceed in order because it is required by the script. Her *coup de grâce,* "Let the mutton and onion sauce appear," evokes Roman games and conjuring, ancient and modern spectacles. She is, in fact, theatre itself, large, real and unreal, self-referential.

Mrs. Crummles' dialogue sits in contrast with her husband's shorter, clipped, unresonant speech, which is suggestive of his whitey-brown paper snuff bag, his distraction, and his sloppy exterior; and also of the current "low comic" idiom. Her voice seems always to challenge his to assume greater dimension, more tragic feeling, and between them they cultivate some extremely funny moments. For example, her broad, unbroken "They are going through the Indian Savage and the Maiden" is followed by his glottal, highly punctuated "Oh! the little ballet interlude. Very good, go on. A little this way, if you please, Mr. Johnson. That'll do. Now!" (p. 364). Thrust and parry, glide and chop: they alternate lines as deftly as duelists.

And they are not the only ones to do so. Mr. Folair, the company's pantomimist, and Mr. Lenville, the first tragedian, pronounce their offstage lines in character, so to speak, with Folair adopting an exaggerated verbal flourish, and his friend punctuating his sentences with phrases from the stock belonging to his type, such as "What ho! Within there!" (instead of "Are you at home, and may we come in?"), and "Gadzooks! You astonish me!" Each one supports the other with voice and body, as if, as pantomimist and first tragedian, they serve complementary roles, exist as parts rather than wholes and must share the stage, gracefully retreating in time when their lines have been uttered. The whole company operates in this way, emerging and retreating with perfect rhythmic integrity.

"Well, Tommy," said [the tragedian], making a thrust at his friend, who parried it dexterously with his slipper, "what's the news?"

"A new appearance, that's all," replied Mr. Folair, looking at Nicholas.

"Do the honours, Tommy, do the honours," said the other gentleman, tapping him reproachfully on the crown of the hat with his stick.

"This is Mr. Lenville, who does our first tragedy, Mr. Johnson," said the pantomimist.

"Except when old bricks and mortar takes it into his head to do it himself, you should add, Tommy," remarked Mr. Lenville. "You know who bricks and mortar is, I suppose, sir?"

"I do not, indeed," replied Nicholas.

"We call Crummles that, because his style of acting is rather in the heavy and ponderous way."[87]

But if they complement each other, these characters also compete with each other, for parts, applause, remuneration – the petty and important details which reduce all of us, actors and entrepreneurs and academics alike, to politicians – and they live almost continually in a kind of verbal sparring match, a pitting of style against style, voice against voice. It seems evident that this playing of the regularity and artifice of professional acting against the volatility of human emotions, breaks some the barriers between staged theatre and "real life," confuses the acts of performance and living by mixing them up in suggestive ways. If the tragedian never falls out of verbal character, he does occasionally act in ways unbefitting his role (though perfectly appropriate to a narrow-minded, unexceptional man) and in doing so muddles the self he has created with the self that he is. In some ways, this is precisely what we, as readers of Dickens, constantly do: we create, we perform, we hybridize our reading selves by becoming active participants – actors, property managers, directors – in the Dickens theatre, to borrow a phrase from Robert Garis.[88]

I am proposing that we read Dickens in this way, with an ear to verbal patterns and relationships, and with other voices and stories ringing in our ears, because I am convinced that he *was* read in this way, and ought still to be read thus. While Victorian readers could identify characters with real London or provincial actors, we may also, with a little imagination, make the leap to performance, to theatre, that Dickens' readers made. I am helped to make that leap by my early experience as a listener to the novels, and also by the Royal Shakespeare Company's production of *Nickleby,* which displays the theatre of voices in the novel, from Mrs. Crummles' quavery dramatic contralto, to Smike's haunted, halting tones, to Mrs. Nickleby's extravagant, high-pitched patter.

In order to locate the theatrical Dickens, the RSC spoke him, and that reading aloud yielded a tremendous theatrical performance. This makes sense, given the strong historical relationships between reading and performance, explored in detail in the next chapter, and the Victorian institutions of public and family reading. Under the proper conditions, reading is theatre, a concert of verbal and physical dramatic gestures gathered up into the frame of text. Victorian England boasted the proper conditions: a strong, class-inclusive theatre industry; a strong printing industry; a vital street culture. If theatrical culture infused the streets and entered bourgeois homes, then it surely influenced the act of reading, with its tropes and myths the stuff of everyday life, evidenced in the mere act of strolling through Regent Street or the Seven Dials. And this is the condition I have tried to recover in writing this book, which imagines theatre as an actual and metaphorical influence on Victorian readings and writings. Throughout the whole, I have sustained a commitment to listening – to voices, stories, theatrical performances, and written words.

CHAPTER 3

Theatrical attitudes: performance and the English imagination

~

It was a marvellous sight: a mighty revelation.
It was a spectacle low, horrible, immoral.[1]

In spite of the fact that theatricality flowed through the veins of Victorian England, with the producers and consumers of popular culture equally indebted to the theatre, bourgeois culture was deeply ambivalent about performance, assuming play acting to be connected more or less directly to the streets. That connection was embodied in the costermonger, the strolling actor, the prostitutes who staked out theatres and music halls. If performance yielded pleasure it also created guilt, even in its most prestigious venues, the legitimate theatres. This chapter will chart a history of English responses to performance, identifying some of the imaginary texts which shaped public opinion about the profession, and broadening my earlier discussion of the role of the theatre in mediating Victorian acts of reading. At its center lies a curious contradiction: the swell of antitheatrical sentiment in a culture which thoroughly embraced the tropes of performance, incorporating them into its conceptual framework.

Victorian propriety may have shuddered at the thought of theatrical entertainment, but a single ethos, essentializing and hierarchical, governed the dominant social discourse and the dominant theatrical mode, so that efforts to exorcise the theatrical from society attempted the impossible, an act of self-extinguishment. These attempts were made constantly, though, by the arbiters of taste and morality: bourgeois legislators, writers, journalists, and social scientists, who recognized latent currents of theatricality beneath the surface of their culture, and – paradoxically, but appropriately – used the language of the stage to articulate their social experience, describing scenes of London life as variations on standard melodrama plots, complete with standard characters. Often, attacks on the theatre were themselves acts of "theatrical" display, which suggests that the institutions implicitly or explicitly pitted against performance, like family, church, and government, depended on a healthy theatre industry to provide them with voices and to secure their own identities. Certainly they used its tropes and paradigms to construct their rhetorical and political positions. And despite their campaigns against it, English theatre flourished in the nineteenth century, because it *was*, in a sense, the nineteenth century, and for the more practical reason that people, notwithstanding their ambivalence, supported it. These are all facts we need to know if we want to understand the cultural significance of dramatic voices and gestures in Victorian England, if we are curious about the codes which allowed the period's novelists to use theatrical tropes like patter in a kind of shorthand commentary, confident that their readers would construe a set of social facts from them.

One might say that it is a mechanism of bourgeois self-preservation, the enlistment of the disenfranchised – women and a portion of the working class – in their own cultural devaluation, which is the true subject of this study. Certainly, the end to which the "theatrical" social and literary structures defined here aspired – even if they often failed to attain it – is a state of social regulation, in which the marginal and mainstream are delineated. The theatre of voices in Dickens, the imaginary texts which enabled Victorian novel-reading, the pattered diction which wrote itself into the margins of text and society with wild pen-strokes, these all served a common deity: the god of social regulation, of power delegation, of labor division and cash disbursement. This book's work is to peel back layers of cultural sediment, of normalizing pictures and homogenizing stories, and expose the machinery beneath them. But it is also to linger in the beauties and carnival pleasures of Victorian life – in other words, to successfully negotiate the often treacherous passage between the aesthetic and political, the mystical and mystified, the entertaining and the oppressive. Sometimes the passage swells, sometimes it disappears. It is always on the verge of slipping out of one's grasp.

Implicit in any study of cultural production and reception lies a series of negotiations like this, negotiations to be made by the writer, and to be revealed in the culture which is her subject. The issues we have begun to explore, those concerning modes of novel reading, the integrity of genre, relationships between high and popular art, lean towards these sites of negotiation, these dialogic spaces in which art and identity are made possible, with a certain inevitability, as if the teleology of cultural studies "naturally" follows this course. This tendency suggests that social or aesthetic entities, like Dickens, his novels, his dramatists, their adaptations, the Victorian reading public, the playgoing public, and an epic catalogue of entities besides are generated or at least granted identity in the fractures of discourse, in the fray of a cultural argument, as it were, about class and taste and power. We discover them, when we resist the impulse to totalize them, in the slippery passages between art and politics, between the classes, between readers and publishers and love and money – meaningful, with an abundance of meaning, in all ways to all people. It takes some effort to recover them in this spirit, but it is well worth it. This may involve some degree of loss, like the loss Matthew Arnold felt upon reading Dowden's *Life of Shelley*, which, in exploring the details of the poet's "private life" violated the "charming picture" evoked by his poetry,[2] but it brings with it an exhilarating kind of freedom, the combined pleasure and disappointment in abandoning, say, formal portraits of Mozart for his racy, moving letters. We ought to leave behind as well the beautiful portrait of 1839 of Dickens by Daniel Maclise, and only return to it as a reference, or for the pleasure of it. After all, what was the Victorian Dickens if not a shifting, contextual social idea: a good investment, cheap entertainment, a social reformer, a social reactionary, a bad husband, a fine performer, a hack writer, and a brilliant artist?

We can, of course, assign a certain static identity to Dickens, the mid-Victorian novelist and journalist, the young man half-turned dreamily from his desk in Maclise's portrait, the object of boundless scholarly attention. But to most Victorian consumers his identity, even at his professional apex, was somewhat less fixed; he was the product of an expanding and diversifying entertainment industry, a writer whose voice reverberated among the genres, slipping fluidly in and out of journalism, fiction, theatre, ballad opera, even poetry, projecting its peculiar rhythms and tones among the entire field of current popular entertainments. If there was a common aspect of the Dickens experience it was established by the imaginary texts enveloping the novelist and his work; Victorian readers of any class could read the novels in theatrical context, participants in an elaborate cultural–aesthetic compact, an all-inclusive "inside joke," because people of all classes enjoyed virtually the same theatrical diet in the mid-nineteenth century.[3] If segments of the audience processed the experience differently from each other, they at least presumed the same semiotic language in assimilating what they saw on the stage.

If we ourselves are to participate in the compact, to discern the imaginary texts generated among popular sources for, and adaptations of, Dickens' novels, we must also understand the general significance of theatre in Victorian lives and imaginations. We may achieve this by identifying some of the theatre-related imaginary texts circulating among the public. The fact that the theatre mediated readings of text and culture does not prove that the institution enjoyed a privileged place in Victorian moral discourse, although as the century progressed middle-class opinion did improve. But the ambivalence persisted. While huge numbers of people attended the theatres and music halls and purchased theatrical paraphernalia, hostility towards the theatre was expressed in other segments of the popular culture, particularly in the print industry, which may well have perceived the theatre as unwanted competition. Victorian novels, even, incredibly, some of Dickens' novels, often reject the theatre as an unwholesome type of entertainment, and portions of the nonfiction print industry – magazine and newspaper journalism, texts explicitly moral or conduct-related – frequently attacked actors and performance. A *London Times* critic summed up the situation: "There are so many persons in this country who run after everything theatrical, save a theatre itself."[4] William Hazlitt, writing early in the century, blamed the profession's notoriously loose morals on bourgeois public opinion: "If there is any tendency to dissipation in the profession of a player, it is owing to the prejudices entertained against them."[5]

Notwithstanding the nineteenth century's powerful antitheatrical sentiment, performative tropes and gestures were received as examples of realism, and the social world borrowed its paradigms from the stage.[6] We have seen this in critical responses to Dickens' novels, which most readers found compelling and truthful – populated with people one might meet at one's club, or on the street – yet which were heavily indebted to the contemporary melodrama. We shall see it again in social commentaries which read like theatre reviews, and in a wholesale subscription to the veracity of theatrical voices. Why, then, if the business of Victorian life was recorded, by novelists, journalists, and sociologists, in such a way to suggest its affinity to staged performance, did people feel so ambivalent about staged performance? This paradox is not easily solved, but its very existence may explain certain things about the relationship between Victorian class and gender politics and popular culture, work undertaken by the fourth and sixth chapters of this book. Here we shall explore the paradox itself – an essentially theatrical culture promoting antitheatrical ideology – and its role in the formation of English identity.[7] Set in the context of this deep cultural ambivalence about theatricality, the simultaneous attraction and repulsion of patter may be properly understood.

Some interesting aesthetic, political, and even legal texts were born of this ambivalence. One of its most significant by-products was the evolution of a dramatic reading industry, exemplified by Dickens and his extraordinary reading tours, but actively engaged in by many of the period's reputable (and indifferent) writers.[8] The increasing popularity of this sort of performance, a form of drama which eschews the proscenium theatre and dispenses with standard theatrical formalities like ensemble performers, costumes, and sets, suggests that a significant portion of the population sought a sanitized alternative to the morally dubious conventional theatre. Public reading represented a legitimation of performance, drawing it closer to the more "serious," the purer, literary genres; such modified performance privileged text and voice but rejected the extravagances, the baroque complexities of detail, typical of standard theatrical entertainment. It also reproduced in large the Victorian ideal of reading around the hearth, a father with his family clustered round him as he read from Mr. Dickens' latest novel or Miss Barrett's newest book of poems. As Philip Collins suggests, "Dickens' public-readings developed . . . from the way in which his novels were often in fact read aloud in private households."[9] But his novels were equally influenced by the theatrical modes demonstrated, in modified form, in dramatic declamation, so that his late-career public readings, in spirit at least, anticipated even the earliest novels. If, as I shall argue, the theatrical spirit implicit in certain kinds of literary and cultural text could erupt suddenly and unexpectedly, with spectacular force, there was a peculiar kind of safety in the domestic and public readings of those texts: the father could be expected to expurgate and annotate, to contain and regulate, as necessary.

Dramatic reading represented a significant branch of Victorian theatre, and indeed, constituted a sort of social comment on the profession. Collins has located a flourishing reading industry in mid-century London, with literary figures rushing to join ranks with actors such as Charles and Frances Kemble, and Mrs. Siddons. If such activity represented a concession by the theatre industry (or its representatives) to certain bourgeois objections to the stage, and a mercenary interest on the part of writers eager to enlarge their markets, it may also indicate some genuine anxieties on all sides about the undefinable and potentially dangerous power of theatricality. It is common knowledge, for example, that Fanny Kemble hated the stage, its personalities, its politics, and perhaps most of all its reputation; when she took to the platform with Shakespeare in hand, it was partly because she needed the money, and partly because she loved Shakespeare and wanted a "better" way to perform him. Dickens' friend, the distinguished actor William Macready, disliked his profession as well; he wrote:

> My experience has taught me that whilst the law, the church, the army and navy give a man the rank of gentleman, on the stage that designation must be obtained in society (though the law and the Court decline to recognize it) by the individual bearing . . . I was not aware, in taking it, that this step in life was a descent from the equality in which I had felt myself to stand with those . . . whom our education had made my companion.[10]

These anxieties had, of course, a history; Barish finds a dominant strain of sixteenth-century thought to promote reading over acting, and cites texts such as Stephen Gosson's *Plays Confuted,* which contends that "whatsoever such Playes as conteine good matter . . . may be read with profite, but cannot be played, without a manifest breach of God's commaundement."[11] In other words, reading was thought to obviate the dangerous distractions and seductions inherent in "playing." Dickens himself harbored some of these negative feelings about play-acting, as we shall see. Surely his own public readings were prompted at least in part by a desire to control the flow of indiscriminate dramatizations of his novels; or, to put it differently, to harness the powerful theatrical impulses in his texts which, if unpoliced, could erupt into dubious adaptations by the likes of Edward Stirling and W. T. Moncrieff – and more seriously, perhaps, into acts of social disruption.

One sees evidence of such "harnessing" in Dickens' prompt copies, privately printed portions of novels and stories – those he read from in public – which bear the author's notes and cues, usually marked in blue ink and serving the purpose of containing the text and directing its oral delivery. For example, the prompt copies of "The Poor Traveller" and "Mrs. Gamp" are heavily marked up with single and double underlines, indicating, apparently, modes of enunciation, levels of stress or volume, and demonstrating to the reader the best way to exploit and control the dramatic rhythms in the text. These underscores are more interesting than the standard stage directions occasionally scrawled in a margin, because they suggest a kind of struggle, or negotiation, between Dickens and his text. One can almost sense danger beneath those crosses and bars, as if a story unmarked and undirected could erupt into the unexpected, the theatre of voices and rhythms inside of it surging out of control.

To read early modern and modern English cultures as if they were themselves Dickensian prompt books, seeking the marks and scores designed to order an array of cultural signs, is to disclose similar acts of policing on a larger scale,[12] attempted containments of the theatrical impulses embedded in social communities as inevitably as class hierarchies, with the theatrical occupying, as Richard Schechner puts it, "a 'cultural' as well as 'natural' space."[13] These acts of repression resemble, in their own way, the short, broad strokes of Dickens' pen, his notes to himself about when to ride the dramatic crest of his own words and when to quiet or enclose them. They include a body of official proclamations and laws designed to defuse theatrical activity over the centuries, and a set of popular depictions, in writing and art, of performers and acts of performance. The history of theatre in England is a history of preemptive legislative measures, a history of dread and danger, but of pleasure as well; English culture, particularly in the industrial nineteenth century, with its mass promotion of popular entertainments, was dangerously and irrepressibly dialogic. Popular assumptions about the theatre were mediated by this history, and by the imaginary texts it generated.

While novel readers' imaginary texts drew them deeply into the structures of theatre by confirming a "reality" established by the theatre, imaginary texts forming and formed by other products of Victorian culture repelled the theatre, constructing theatricality as "the ultimate, deceitful mobility."[14] The novel itself often deliberately fans the antitheatrical fire, its conscious agenda in direct conflict with its subconscious. Attacks on the theatre flourish among nineteenth-century novels, even in the works of novelists who were friendly to performance.[15] The list is long, but I will name a few: *Mansfield Park*, with its disastrous private theatrical of the *Lover's Vows*; *Daniel Deronda*, with its slurs on the theatre and theatrical characters, like Lapidoth; *Little Dorrit,* which peers into the wings and the orchestra pit at the opera house and finds some grim material there; *Vanity Fair,* whose voracious and dangerous Becky Sharp is born to a French opera-girl; *Jane Eyre,* whose Adèle is born to a French actress-courtesan; and *Villette,* whose renowned actress "Vashti" (i.e. Rachel Felix) is a French Jew, and appears to start a fire in the theatre with her passionate performance.[16] These novels were born of a culture which still actively legislated against performance, and they mediated the public's relationship to the theatre, just as public relationships to the novel were mediated by the gestures and voices of the theatre.

This is a complicated situation, with theatre participating in novel-writing and novel-reading, lending its voices, its gestures, its ethos, to texts and imaginations, while being simultaneously repulsed by some of the novels and readers it had in effect created. Reading, in this context, becomes a sort of dialectical struggle. One wonders how readers of, say, *Villette*, Charlotte Brontë's most theatrical novel, reconciled the novel's fear and disapprobation of the theatre with its reluctant admiration of the actress Vashti, and with its heavy reliance on melodramatic effects – if indeed such reconciliation was necessary for a reading subject whose "reality" was created on the stage – including spectacular storms, dangerous villainesses, mental illnesses, haunted nuns, shipwrecks, and a language often extravagant and emotionally heightened. For example, the Vashti scene quivers with theatricality, using the actress's own dramatic technique, the gestures of which Lucy Snowe cannot quite approve, to make its point. In fact, Lucy's description becomes one with the sexually charged actress, exhibiting the same conflict between substance and ellipsis, the same indulgence in erotic power, the same richness of gesture:

> She rose at nine that December night: above the horizon I saw her come. She could shine with pale grandeur and steady might; but that star verged already on its judgment day. Seen near, it was a chaos – hollow, half-consumed: an orb perished or perishing – half lava, half glow.[17]

Vashti's body is provisional, as is the highly metaphorical language, which both fails and succeeds in describing it. Neither woman nor man ("in each of her eyes sat a devil"[18]) she comes to represent disembodied appetite and is vanquished to the stage, meaningful only as she takes on, chameleon-like, the characteristics of passions which she does not "naturally" feel, passions which lie dormant, perhaps, but potentially responsive, inside of her audience. "To her, what hurts becomes immediately embodied: she looks on it as a thing that can be attacked, worried down, torn in shreds. Scarcely a substance herself, she grapples to conflict with abstractions."[19]

Here, the inversion of abstraction and substance suggests Lucy's awareness of the strangeness and danger of women like Vashti and the work they do. (It is noteworthy that Forster celebrates in Dickens' dramatic reading the same quality Lucy finds so unsettling in Vashti's acting, the "power of projecting himself into shapes and suggestions of his fancy . . . What he desired to express he became." If acting represented a "descent from equality" for men, for women it constituted a plummet.[20]) Yet she lingers thirstily on her descriptions of the actress, playing with metaphor, searching for the appropriate language. Her description of the woman as a "sunflower, turned from the south to a fierce light, not solar – a rushing, red, cometary light – hot on vision and to sensation"[21] draws attention to what must be an erotic specular image; her language itself is an erotic event. Lucy's struggle is with her strong identification with the actress and the ominous warnings her socially attuned self keeps sending her: acting is immoral, actresses embody appetite. Hazlitt's comments on actors may help to contextualize Lucy's ambivalence: "[Players] are the only honest hypocrites. Their life is like a voluntary dream; a studied madness. The height of their ambition is to be beside themselves. To-day Kings, to-morrow beggars, it is only when they are themselves that they are nothing".[22]

Villette explores the sexual politics of the theatre as frequently as it invokes melodramatic spectacular and verbal gestures. For example, Madame Beck's birthday *fête* features a planned theatrical performance as well as another, more complex form of theatre, a ball. The headmistress has imported boys, brothers of the female pupils, for the occasion, but these unfortunates are carefully cordoned off from the ballroom, where the girls dance with each other or M. Paul, the only unmarried man granted entry. "Others there were admitted as spectators – with (seeming) reluctance, through prayers, by influence, under restriction . . . [and] kept far aloof at the remotest, drearest, darkest side of the carre."[23] The teasing and denial at play produces tremendous sexual energy, so that what appears an aggressively antisexual space becomes in fact an erotic hothouse, a theatre of passions, with female sexuality (insufficiently) contained and displayed to a male viewing audience. This conflation of sexual fever and theatricality echoes the Vashti episode, as well as similar gestures in a large body of English and European texts.

Perhaps Brontë had read one of the most explicit and articulate of these, Rousseau's "Letter to M. D'Alembert on the Theatre." If not, she was surely familiar with writings like it. In the "Letter," his response to D'Alembert's proposal of a Genevan theatre in the *Encyclopédie*, Rousseau employs the standard set of stereotypes about actors and acting. The piece is frequently persuasive, although at times it borders on the hysterical as it warns of the corruption worked by theatres.

How many questions I find to discuss in what you appear to have settled! Whether the theatre is good or bad in itself? Whether it can be united with morals [manners]? . . . Whether it ought to be tolerated in a little city? Whether the actor's profession can be a decent one? Whether actresses can be as well behaved as other women? . . . At first glance given to these institutions I see immediately that the theatre is a form of amusement; and if it is true that amusements are necessary to man, you will at least admit that they are only permissible insofar as they are necessary . . . People think they come together in the theatre, and it is there that they are isolated. It is there that they go to forget their friends, neighbors, and relations in order to concern themselves with fables, in order to cry for the misfortunes of the dead, or to laugh at the expense of the living . . . The stage is, in general, a painting of the human passions, the original of which is in every heart. But if the painter neglected to flatter these passions, the spectators would soon be repelled and would not want to see themselves in a light that made them despise themselves . . . Hence the author, in this respect, only follows public sentiment . . . It is only reason that is good for nothing on the stage.[24]

Rousseau objects most strongly to the display of false sentiment on the stage, to a manipulation of the passions which flatters the admirable, deplores the vicious, and ignores the most common human vices. He was not the first to express such concerns. His arguments echo early modern attacks on performance, appeals for social order and for truth in self-representation: acting was assumed to promote anarchy or insurrection, as we shall see. They also anticipate Brontë's anxieties over Vashti's flexibility of attitude, her ability to embody abstract emotions in a way that seems to promote erotic awareness among her audience, to awaken their repressed theatrical impulses. Rousseau argues that playwrights and actors portray those behaviors which society has agreed to recognize, using comedy and tragedy as vehicles for promoting, or at least codifying, the generally accepted set of values – which may not, he suggests, be the best possible values for that society. What he describes as his commitment to the moral health of the society at large moves him to reject D'Alembert's proposal: however problematic his argument might be, Rousseau was clear-sighted in his recognition of the potential for widespread influence in acts of public performance.

I have quoted Rousseau for several reasons. First, to demonstrate the gradual evolution of ideas over the centuries. The nineteenth century had obvious cultural antecedents. This is not to say that one may speak of early modern and modern social discourses as identical, or even consistent. But the industrial imagination shared certain values with the past, some of which were preserved in collective memory, some of which were revived in theatres and in books. To limit a discussion of Victorian cultural values to the nineteenth century would be misleading, because cultural values, like novels, do not spontaneously generate: hence, this study pays brief visits to medieval and early modern England and continental Europe, as well as to various nonliterary aesthetic disciplines, on its way to a highly specific Victorian phenomenon. One of those visits is with Rousseau, whose voice was influential in the eighteenth and nineteenth centuries, and who spoke for many in his distaste for theatrical amusements. Rousseau's attention to the moral health of theatre audiences concedes either to an inherent theatricality in nontheatrical social bodies, or, more probably, to the infective power of performance, both of which assumptions informed Victorian antitheatre discourse.[25] And his insistence on social responsibility reverberates in Victorian literature and social criticism: the postromantic ethos rejected solipsism for a commitment to the social organism. Rousseau's objections to performance provide a moral–philosophical context for the antitheatrical imaginary texts current in the middle of the nineteenth century.

Those "texts," and the social climate which produced them, had powerful antecedents. A proliferation of legislative acts, designed to control actors and acts of performance over about four centuries, offer a commentary on the reception of the profession in early modern and modern Europe. This library of "nondramatic" theatrical documents may serve as a context for theatrical activity during this period of time,[26] and their interest is enhanced by the fact that they generated imaginary texts among the literate and nonliterate public. They include any number of sixteenth- and seventeenth-century proclamations and acts of Privy Council, like the Elizabethan Vagrancy Acts; the 1737 Theatres Act, or Act Against Strolling Players; and the 1843 Act for Regulating Theatres.[27] Historically, the power of Privy Council in regulating performance was absolute and potentially devastating. For instance, in 1556, Queen Mary's Privy Council ordered the suppression of a theatrical troupe who had "wandered abowt the North partes, and represented certaine playes and enterludes conteyning very naughty and seditious matter touching the King and Quene . . . and to the slaunder of Christe's true and Catholik religion, contrary to all good ordre."[28] This very public flexing of the state muscle served as warning and instruction to the average citizen: acts of public performance hovered dangerously close to acts of subversion, even treason, especially when such performance traveled about the country.

The state's anxiety over these activities makes sense, given their potential for disruption, and so does its campaign to denormalize performance. Legislative actions of this sort were not confined to a feudal, parochial England; as late as 1810 the Privy Council convened to determine the fate of the two London patent theatres.[29] Their decision to sustain the Covent Garden–Drury Lane monopoly by declining to recognize a third theatre did as much, in its own way, to control the spread of theatrical activities as did the more aggressive act of suppression by Queen Mary's Privy Council. At any historical moment, such antitheatrical acts served the purpose of defining social regularity or normalcy by proscribing the illicit, identifying and containing the dangerous. More often than not they focus on class status and property ownership in differentiating between the legitimate citizen and the actor: "normalcy" is hence strictly connected to structures of social, political, and economic hegemony. For example, the "Minutes of the Committee of the Council on Education Reports by Her Majesty's Inspector of Schools 1852–3" cites the innate vulgarity of "lower" tastes as a justification for curbing low entertainments and even eradicating the study of music from working-class education.

> It is known to your Lordships, and Mr. Mayhew's painfully interesting work on London Labour and London Poor has made it known to all readers of that book, that it is an habitual practice in public houses of an inferior stamp, all over the country, to have frequent musical entertainments of a low and immoral character, for which performers are in great demand. A certain cultivation of the voice and a due knowledge of exciting and profligate songs, in the best of which "sentiment" is allied with indelicacy, and humour is depraved into "slang," constitute the requisite qualifications; and it is worth reflecting whether that semi-professional skill which is the ordinary method of teaching music in schools is calculated to impart . . . may not possibly become a fatal gift to those who acquire it.[30]

Because they were publicized by way of proclamations, newspaper and journal articles, and parliamentary reports, these antitheatrical gestures and the social politics they promoted entered the field of popular ideas. People read about them and talked about them, and some responded to them by avoiding theatres and traveling or street performance in its various incarnations. In this way, laws designed to control performance formed the bases for imaginary texts, cultivating, with their symbolic as well as actual power, a set of popular attitudes toward the theatre. The significance of these cultural signs exceeds their literal or textual forms, which are useful to the extent that they enable later generations to recover patterns of Victorian popular thought, but limited by their tendency to display themselves, whether proclamations, paintings, song-sheets, or campaign slogans, as static cultural artifacts. More important to us are their symbolic value, their metonymic and metaphoric relationships to dominant social ideologies, and the depth of their integration into the cultural mechanism.

Imaginary texts' sources are usually obscured during their periods of cultural currency; the attitudes they promote and express seem "natural." This explains their powerful hegemonic force: works of Victorian popular culture tend to reify the ideas and ideologies they espouse. These ideas are usually conservative; many products of Victorian culture, even those which are informed by "radical" or reformist principles (for example, the "liberal" Matthew Arnold's *Culture and Anarchy,* and the "radical" Felix Holt's "Address to Working Men") ultimately privilege the bourgeois values they would appear to question. Thus, the theatrical myths and anecdotes, the visual and textual images in circulation, generally attempt to conserve middle-class values; they define and delineate society along class lines, with actors and acting overwhelmingly associated with either marginalized classes or marginalized behavior because of their association with carnival or ludic activities, with play in a work-driven economy.

Early in the eighteenth century, William Hogarth did much to popularize the association of theatrical and low cultures, and his influence obtained through the nineteenth century as his engravings continued to be circulated, displayed, and discussed, and his legacy supplied contemporary illustrators, writers, and as Meisel has argued, even theatre directors with rich materials. Hogarth appreciated the theatre, and indeed, visualized his drawings in terms of theatrical space and dramatic gesture, placing his subjects within proscenium frames and drawing heavily on stock theatrical postures and expressions, although his engravings on the subject suggest its degradation. If this tendency expresses Hogarth's personal nostalgia for a lost classical theatre, it also expresses some collective cultural anxieties about what takes place on the stage, and to whom dramatic spectacle appeals. *A Just View of the British Stage* (1724) illustrates the aesthetic and dramatic dissolutions of the post-Restoration theatre, and suggests an increasing shift in the tastes and values of British audiences. The engraving represents Drury Lane managers Robert Wilks, Colley Cibber, and Barton Booth as actors in

a new Farce that will Include ye two famous Entertainments *Dr. Faustus & Harlequin Shepherd* to Wch will be added *Scaramouch Jack Hall the Chimney Sweeper's*

Plate 4. William Hogarth, *A Just View of the British Stage*. Engraving (1724).

Escape from Newgate through ye Privy, with ye comical Humours of *Ben Johnson's* [*sic*] *Ghost,* Concluding Wth the Hay-Dance Perform'd in ye Air . . . Assisted by Ropes from ye *Muses,* Note, there are no Conjurors concern'd in it as ye ignorant imagine. The Bricks, Rubbish, &c. will be real, but the Excrements upon *Jack Hall* will be made of Chew'd Gingerbread to prevent Offence.[31]

The stage is in chaos. The three actor-managers, working puppets at a precarious table, recite fractured, unrelated lines while slightly beneath them and to their right, Ben Jonson's ghost rises through a trapdoor in the stage floor. Broken mannequins and rubbish litter the stage, slum-like, and a row of privies, presumably the same through which Jack Hall will make his escape from prison, sits to the actors' left. The scatological humor and the general lowness of the scene – with its broken people and props, its dirt and disarray – links the degradation of the theatre to the fallenness of human society, to poverty and depravity. Implicit in the engraving is an alternative condition: a bourgeois rejection of feces-covered criminals, and on a larger scale, of theatrical extravaganzas which muddle serious drama with spectacular hay-dances and low farce.

The history of performance is inseparable from such impulses toward economic and social definition. Indeed, one might argue that early modern and modern English cultures defined themselves partly in relation to performance, placing moral value on "truth" in language, place, property, and social definition – all of which theatrical entertainments were perceived to obscure or foil. But, paradoxically, English performance was intimately linked to an economics which would eventually be called capitalism, to the ideologies of profit and property ownership; in this sense the institution, the larger profession, did in fact participate in the accumulation of wealth and property which was the prerogative of the powered class. Even pre-industrial theatre was a money-making enterprise, although the average actor may not have personally prospered. According to popular accounts, actors also did not, at least through the end of the eighteenth century, typically own property, although that claim has been disputed by a few contemporary critics.[32] In 1817, Hazlitt wrote "they live from hand to mouth; they plunge from want into luxury; they have no means of making money breed."[33] Early modern depictions of acting, in art, literary, and social documents, even in drama, represent actors as indigent; shifty creatures whose lack of property and social identity – the latter defect magnified by their habitual assumption of character – made them difficult to understand. These factors also made them difficult to locate, and potentially, to prosecute.

This lack of place was a fact of life for many actors through at least the beginning of the eighteenth century, not merely the invention of an anxious propertied class.[34] Perhaps the most famous description of the condition is Shakespeare's. *Hamlet* features a troupe of players who arrive in Elsinore seeking work, forced out of their old venue by the "late innovation" of church-sponsored boy actors.

HAMLET. How chances it they travel? Their residence, both in reputation and profit, was better both ways.
GUILDENSTERN. I think their inhibition comes by the means of the late innovation.
HAMLET. Do they hold the same estimation they did when I was in the city? Are they so followed?
ROSENCRANTZ. No indeed, they are not.
HAMLET. How comes it? Do they grow rusty?

ROSENCRANTZ. Nay, their endeavour keeps in the wonted pace, but there is, sir, an eyrie of children, little eyases, that cry out on top of the question and are most tyrannically clapped for it. They are now the fashion. (II. ii. 323–334)[35]

Changes in "fashion," in political climate, in theatre management, could force actors to travel. And for some companies "strolling" was a regular part of the job: they sought their audiences. Popular sketches of the strolling actor were plentiful, even after the tradition had died out and "legitimate" performance was contained, with the help of eighteenth- and nineteenth-century licensing acts, in the patent theatres.

Hogarth captured the harshness of the stroller's life in an engraving entitled *Strolling Actresses Dressing in a Barn* (1738). The picture shows actresses in various states of dress, readying themselves to take the stage as Diana, Juno, Aurora, and a host of similarly exalted characters. But the scene is anything but exalted. The barn floor is cluttered with bizarre props, clothes, farming implements. Laundry – including underwear – hangs from ropes and rafters. Beer mugs and pipes suggest the offstage coarseness of the actresses, and a few children dressed as devils and grotesques lend the picture an air of lewdness and degradation. Hogarth indicates a political context for the engraving in the lower left-hand corner: a partially unrolled playbill states "By a Company of Comedians from the Theatres at London at the GEORGE INN This Present Evening will be Presented THE DEVIL to Pay in HEAVEN Being the last time of *Acting* Before Ye *Act* Commences" (my emphasis). In spite of the suggestive repetition of "act" here, which tends to emphasize the connections between the performance in question and the legislation designed to control it, and in spite of the various potential alternative readings which may suggest themselves to today's Hogarth enthusiast, the engraving is essentially conservative. It asks us to focus on the strollers' *dishabille* and moral transgressions as opposed to the enforced marginalization which has likely created the conditions under which they work. The act in question, one 1737 "Act against Strolling Players," purported to "[reduce] the laws relating to rogues, vagabonds, sturdy beggars and vagrants, into one act of parliament; and for the more effectual punishing such rogues . . . sending them whither they ought to be sent, *as relates to common players of interludes.*"[36] The language of the proposal suggests, with its echoes of Elizabethan proclamations against "rogues, vagabonds, and masterless men," that attitudes about actors changed very little over two centuries.

Plate 5. William Hogarth, *Strolling Actresses Dressing in a Barn*. Engraving (1738).

The fears expressed in the 1737 Act had been cultivated for centuries. In the late Middle Ages and early Renaissance English anxieties over performance were so great that even vagrants who were not, technically speaking, actors, were suspected of violating the "natural" social order by acting destitute in order to receive alms. Thus in 1545, Henry VIII issued a warrant " 'for the punishment of vagabonds, Ruffins, and Idle psons' . . . [and] employed on galleys and other vessels, for service in his wars, 'all such Ruffyns, Vagabonds, Masterles men, Comon players, and euil disposed psons.' "[37] And in 1566, Thomas Harman lamented "the abominable, wicked and detestable behaviour of all these rowsey, ragged rabblement of rakehells that – under the pretense of great misery, diseases, and other innumerable calamities that they feign – through great hypocrisy do win and gain great alms."[38] The difficulty was in distinguishing the real beggars from the fakes. The vagrant maintains his pretense "by three principal modes of semiotic larceny: disguise, or the wearing of false apparel; the carrying of forged letters and patents; and the use of an impenetrable criminal dialect, or, in

Harman's phrase, an 'unlawful language.'"[39] Harman's "unlawful language" will be termed "patter" by the eighteenth century, but it will still be located in the streets, the slang-talk of beggars, thieves, and mountebanks.

As Harman points out, the language of the street, of the disenfranchised and derelict, rebels against the standards of linguistic normalcy, to which it neither has, nor perhaps desires to have, access. This is true today in western countries, with disempowered groups often linguistically and topographically segregated from the powered classes, devising their own slang in what may appear to be pointed defiance of the standard lexicon. The perceived danger in such eccentric idioms lies in their tendency towards display. Their semantic, syntactic, rhythmic, inflective differences may beguile the ear as they draw attention to themselves; this, in fact, is precisely how Henry Mayhew describes the costermonger's patter, as an enticing, entrapping barrage of words.[40] In other words, there is a theatricality about certain types of "low" language which metonymically connects performance to the streets, and nowhere is this evidenced as clearly as in works of Victorian literature and popular culture. While the theatre of the street may be colorful and exciting – particularly when the bourgeois subject, with his or her controlling gaze, plays the role of audience, as Victorian journalists and social scientists so often did – it may also, under different conditions, surge out of control, a theatre of gibbering voices and low, violent passions; a writhing, seething animus uncontained by any frame.

This sort of primal theatre is what Dickens saw at the public hanging of the infamous murderers, the Mannings – a mob/audience which spilled into the streets, panting, leering, swearing, with a fevered, almost sexual collective energy in anticipation of the morning's entertainment. He described the spectacle in an impassioned letter to the editor of the *Times*.

SIR, – I was a witness of the execution at Horsemonger-lane on Tuesday morning. I went there with the intention of observing the crowd gathered to behold it . . .

I do not wish to address you on the subject, with any intention of discussing the abstract question of capital punishment . . . I simply wish to [suggest] . . . that the government might be induced to give its support to a measure, making the infliction of capital punishment a *private solemnity within the prison walls* . . .

I believe that a sight so inconceivably awful as the wickedness and levity of the immense crowd, collected at that execution . . . could be imagined by no man . . . The horrors of the gibbet, and of the crime which had brought the wretched murderers to it, faded in my mind before the *atrocious bearing, looks, language of the assembled spectators. When I came upon the scene at midnight, the shrillness of the cries and howls that were raised from time to time* . . . made my blood run cold. As the night went on, *screeching, and yelling in strong chorus of parodies of Negro melodies with substitutions of "Mrs. Manning" for "Susannah,"* and the like, were added to these. When the day dawned, *thieves, low prostitutes, ruffians and vagabonds of every kind, flocked to the ground, with every variety of offensive and foul behaviour. Fightings, faintings, whistlings, imitations of Punch, brutal jokes, tumultuous demonstrations of indecent delight, when swooning women were dragged out of the crowd by the police, with their dresses disordered, gave new zest to the entertainment.* (my italics)[41]

The letter describes a sort of carnival gone bad: a discord of shrill voices and vicious behaviors. To extend the theatre metaphor, this audience is the gallery audience, frequenters of bad shows and patrons of cheap seats. Dickens often described their bad manners in the theatre in similar, though of course less impassioned, language, and in all cases he finds their collective voice, its harsh tones and vulgar lingo, disturbing and offensive. For example, an article in *Household Words* (1852), entitled "Shakespeare and Newgate," posits a causal relationship between low entertainments and debased audiences, using the same moral and rhetorical strategies as the capital punishment letter had a few years earlier.

Seven or eight years ago, Sadler's Wells Theatre, in London, was in the condition of being entirely delivered over to as ruffianly an audience as London could shake together. Without, the Theatre, by night, was like the worst part of the worst Fair in the worst kind of town. Within, it was a bear-garden, resounding with foul language, oaths, catcalls, shrieks, yells, blasphemy, obscenity – a truly diabolical clamour . . . The audience were of course directly addressed in the entertainments. An improving melo-drama, called *Barrington the Pickpocket,* being then extremely popular at another similar Theatre, a powerful counter-attraction, happily entitled *Jack Ketch,* was produced here, and received with great approbation. It was in the contemplation of the management to add the physical stimulus of a pint of porter to the moral refreshments offered to every purchaser of a pit ticket, when the management collapsed and the Theatre shut up. At this crisis of the career of Mr. Ketch and his pupils, [the tragedian] Mr. Phelps . . . conceived the desperate idea of changing the character of the dramatic entertainments presented at this den, from the lowest to the highest, and of utterly changing with it the character of the audience.[42]

In both cases, the merest contact with the violent or degraded infects the lower-class audience, which explodes into a venal frenzy. What happens in such explosions is a spectacular transformation: the spectators themselves become actors in a peripheral drama – notice that Dickens' purpose in attending the Manning executions was to play audience to the crowd – by displacing the spectacle from scaffold to the streets, where people occupy themselves as they wait for the criminals' arrival with singing, fighting, play-acting.

Foucault, of course, has written about the rather complex theatricality of the public execution,[43] but his reconstruction of the institution places much greater faith in its efficacy as an instrument of instruction and control than Dickens and others who lived in its midst were willing to concede it. His representation of public execution as representation assumes that the theatre of capital punishment terrifies its audience, whose very presence is a necessary component of the punishment-drama, into conformity with law and order, whereas we have every reason to believe that, like the violent deaths of every description which are a staple of late twentieth-century film and television, the performance is not didactic, serving primarily to entertain. The executions committed during the reign of terror in France, for example, drew wildly excited crowds, whose interaction with the condemned and their dispatchers resembled the audience participation in a working-class London theatre;[44] and at public hangings in London the site was often so packed with spectators that people in the crowd suffered serious injuries. What Foucault ignores is the element of choice: no one, in early modern or modern Europe, was forced to attend a public execution by state or royal coercion. People turned up because they wanted to watch death occur. The uneasy resemblance of capital punishment, in any public form, to theatrical entertainment was not lost on Dickens, who went on to write a second letter condemning the culture of low amusement, violence, and sexuality promoted by such proceedings.

The *Times* letter raises some questions about the role of the audience in modern England. Descriptions of theatre and opera audiences in the eighteenth and nineteenth centuries indicate some powerful anxieties about maintaining the distance between stage and spectators, as we saw in Rousseau's "Letter to M. D'Alembert." This is virtually impossible to do, and therefore most representations of theatre audiences locate a version of the stage performance among the spectators, who refuse to play the role of passive viewing body, generating their own dramatic stories, watching each other as eagerly as – sometimes more eagerly than – they watch the stage. This gives us a peculiar context for the flashes of dramatic display among the Mannings mob. In the theatre, such transgression was not limited to the gallery and pit; as Frances Burney has shown us, the elite opera audience engaged in its own form of play during the performance – playing cards, dining, watching the social and domestic dramas unfolding in adjacent boxes, and only occasionally directing their attention to the stage, when a favorite castrato or diva came on to sing.

This situation was mirrored on the continent, as the young Mozart learned when

> two leading actresses [of the Mannheim Company] have already been hissed off the stage [in Munich], and there was such an uproar that the Elector himself leaned out of his box and said, "Sh!" When not a soul took any notice he sent someone down to put a stop to it, but when Count Seau asked certain officers to make less noise, the answer he got was that they paid for their seats and were answerable to no one![45]

Mozart's letter is really a complaint about the public performance hall, which commodified performance, reducing it to the value of a seat in the auditorium – hence the officers' legitimate claim to have bought their evening's amusement – and which in one sense democratized audiences by making seats available for purchase. He might have written it in England, for it incisively captures the contemporaneous mood there. Implicit in the problem he describes is a discord of voices as well as a conflict of behaviors. The officers speak, laugh, no doubt curse in a social space reserved for other sounds: music, refined murmuring, *bravos*, or whatever was encompassed in the governing codes for the classical opera audience. Their difference, a class difference, is inscribed in their voices, which distract and unsettle both the "legitimate" audience and the performers. Dickens describes a similar situation in the execution letter, with the shrill, raucous collective voice of the underclass – the people who turn out in droves for such events – resounding painfully in his ears, signifying their poor education, bad morals, and worse behavior. Both commentaries suggest that disruptive theatrical impulses inhere in social communities, waiting to be released by contact with staged spectacle.

Marc Baer's interesting work on the "Old Price Riots"[46] demonstrates the lengths to which audience disruption could go in the early nineteenth century, although Baer argues that theatre – meaning both the architectural and performative phenomena – actually defuses serious social disruption by "confirm[ing] and sustain[ing] disorder, but [doing] so theatrically, i.e. fostering non-revolutionary discontent" (p. 14). In other words, the audience's "subversive" behavior is really play-acting – fictitious, and not seriously dangerous. While I would challenge Baer's somewhat facile assumptions about the distinction between fiction or play-acting and reality – and while the actors and managers who suffered physical and financial injury as a result of the riots might disagree, if they had the chance, with his assessment of these events as ultimately benign – his treatment of the whole situation in its complexity, a lower- to middle-class segment of the audience demanding the "old prices" in the wake of certain management decisions, including a hike in the cost of admission, and the resultant prejudices against both the gallery and the theatre in general, is strongly suggestive. Even if the "rioters" were merely playing their rage, the public response to the unrest was fairly extreme, as was usually the case with visible working-class agitation, such as the later Chartist actions. But the "old price" violence was relatively contained, rarely spreading physically or in principle beyond the theatre. Something about the nature of audience itself, then, must have fed the public's fears.

The audience is a fairly complex social body, governed by a politics of hierarchy and behavior which in some ways differs little from that of the world outside. But audiences also maintain an exclusive organizing principle: there are certain activities whose legitimacy begins and ends within the boundaries of a theatre. For example, the prologue to Dryden's *Marriage à la Mode* describes the social and sexual politics of the Restoration playhouse with a cutting sarcasm which implies that such bad behavior was reserved solely for that space:

> Lord, how reformed and quiet we are grown,
> Since all our braves and wits are gone.
> Fop corner now is free from civil war:
> White wig and vizard make no longer jar.
> France, and the fleet, have swept the town so clear,
> That we can act in peace, and you can hear.
> 'Twas a sad sight, before they marched from home,
> To see our warriors, in red waistcoats come
> With hair tucked up into our tiring room.
> But 'twas more sad to hear their last adieu;
> The women sobbed, and swore they would be true.
> And so they were, so long as e'er they could:
> But powerful guinea cannot be with stood,
> And they were made of playhouse flesh and blood.[47]

Posturing, disruptiveness, and even infidelity count among the theatregoer's bad habits. Dryden neutralizes his flattery of the present audience (who can "hear" now that the fop-corner regulars are off at war) when he accuses them, later in the prologue, of running after the newest special effects at the Duke's house (his company's rival), and promises not "to come behind in courtesy, / [But] follow the new mode which they begin, / And treat 'em with a room and couch within. / For that's one way, howe'er the play fall short, / T'oblige the town, the city, and the court."[48] Clearly, the theatre space promoted the kind of overtly sexual behavior which was forbidden in other seventeenth-century public spaces, and audiences, their playhouse blood pumping, responded on cue.

In the eighteenth century, the theatre's codes of behavior permitted – and its architecture accommodated – a different kind of erotic behavior: overt surveillance of other people. The occupants of boxes in the newly transformed proscenium patent theatres frequently turned their glasses on other boxes, which suggests that social performances competed successfully with what was happening on the stage. (In the 1780s, during one stage of Drury Lane's architectural history, the boxes were subdivided and enclosed, reportedly providing working quarters for prostitutes.[49]) And in the nineteenth century, actors and writers expected active participation from the audience, making their artistic decisions with such participation in mind. Here, individual theatregoers had "the courage to express themselves theatrically,"[50] even if their lines were stylized, predetermined by the play's context. Theatre audiences, in other words, violated some of the culture's most important rules, because theatres encouraged erotic and social acts which were unacceptable outside of their walls.[51] Dickens shows us such transgressions in *Great Expectations*, where the audience to Mr. Wopsle's *Hamlet* responds insolently to the action, mocking prince and grave-digger with equal energy and enjoyment, orchestrating their moves into a situation ostensibly closed to them – violating the strictures of social decorum with the relish of individuals who rarely receive the privilege of sneering at a prince.

> Whenever that undecided Prince had to ask a question, the public helped him out with it. As for example; on the question whether 'twas nobler in the mind to suffer, some roared yes, and some no, and some inclining to both opinions said "toss up for it"; and quite a Debating Society arose . . . When he appeared with his stocking disordered . . . a conversation took place in the gallery respecting the paleness of his leg and whether it was occasioned by the turn the ghost had given him. On his taking the recorders . . . he was called upon unanimously for Rule Britannia. When he recommended the player not to saw the air thus, the

sulky man [in the audience] said, "And don't *you* do it, neither; you're a deal worse than *him*!" And I grieve to add that peals of laughter greeted Mr. Wopsle on every one of these occasions.[52]

The greatest threat in performance lay, then, in the susceptibility of any contiguous social body to theatrical infection, to an irresistible desire to reenact the performed stories and attitudes, or to create their own stories. Fears about catching something in a theatre – the plague, moral laxity, venereal disease? – dominate nineteenth-century attitudes about the profession. That theatre audiences could themselves be infected, and enter into acts of performance, has interesting implications for novel readers as well, certainly for readers of Dickens. Dickens' readers actively participated in a creative–dramatic process, with the reading public itself a fertile field in which novel and theatre merged and became indistinguishable, thoroughly integrated into a larger fabric of ideas and desires. The public as vanishing point is a suggestive metaphor. If, to borrow a construction from Herbert Blau, "theatre is a model of life that takes its cues from the theatre [and hence] has dominated the activity of perception,"[53] then in one sense the consumer of plays and novels was both writer and actor in the theatre of Victorian social–aesthetic life. The reading experience depended on an actively creative reading subject. Of course, one might argue (and many have) that all acts of reading involve acts of "writing," and some, like Barthes, find that certain texts encourage such reading more than others.[54] But what I am suggesting here is specifically a theatrical reader response, a reflexive, subconscious turning to the tropes of the stage. These tropes, especially the verbal ones, were essential to Victorian ideology. Ideas about how certain social groups talked informed trends in education, for example with middle-class schoolboys cultivating a standard verbal affect; in social-reform journalism, as we see in Henry Mayhew's street interviews; and of course in popular entertainments. One's voice told elaborate stories about one's background, geographical origins, and values; and truisms about accent and dialect reverberated on the streets and echoed through bourgeois drawing-rooms.

What is this condition if not itself a form of theatre, a verbal enactment of the perceived social order, a public agreement about the verbal signs of social difference? If audiences could be "read" or viewed like performers on a stage, so too could people in other social communities. Writing on a similar phenomenon in pre-industrial English society, Harold Perkin explains that eighteenth-century social consciousness was formalized, and behavior stylized, citing Lord Chesterfield's advice to his son, who was to request that his dancing-master "teach you every genteel attitude that the human body can be put into: let him make you go in and out of his room frequently, and present yourself to him, as if he were by turns different persons; such as a (royal) minister, a lady, a superior, an equal, an inferior, etc."[55] Lord Chesterfield has described a form of theatre, a semiotics of dramatic posture, of the body. Patterns of physical gesture and feature were deeply integrated into popular conceptions of social place in the eighteenth century. We see an example of this in the use, among the upper class, of patches on the face and breasts, signifiers of the pox early in the century, later, marks of the beauty that comes with wealth and power. Fiction and drama of the period also play on the importance of physical marks in determining social status. In Mozart's *Le Nozze di Figaro*, for example, a "spatule" on Figaro's arm confirms his "illustrious" birth: "L'oro, le gemme, e i ricamati panni, che ne' più teneri anni mi ritrovaro addosso i masnadieri, sono gl'indizi veri di mia nascita illustre: e sopra tutto questo al mio braccio impresso geroglifico."[56] ["The gold, jewels, and embroidered clothes in which the robbers found me when I was a baby, are proof of my illustrious birth, as is this mark upon my arm."] And in Richardson's popular novel of the 1740s, Clarissa's failure to correctly read the physiognomy of the widow Sinclair's face – a face read by Lovelace as animalistic, "eyebrows erect, like the bristles upon a hog's back . . . [which] more than half-hid her ferret eyes" – ends in a series of assaults on the heroine's body.[57] Hence, while the moment of Figaro's identification, and the social training that Chesterfield desires for his son, may seem grossly artificial to the modern sensibility – indeed, Chesterfield's advice does suggest the highly stylized painting and performance aesthetics current at the time – such posturing was meaningful to the contemporary observer.[58]

Compare Chesterfield's letter with another eighteenth-century testament to the power of marks and postures: Diderot's description of Rameau's nephew:

> Nothing is less like him than himself. At times he is thin and gaunt like somebody in the last stages of consumption . . . A month later he is sleek and plump as though he had never left some millionaire's table . . . Today, in dirty linen . . . Tomorrow, powdered, well-shod, hair curled . . . He lives for the day . . . according to circumstances.[59]

At first glance, Rameau's nephew and Chesterfield's son appear to be poles apart: the former seems to represent a kind of social fluidity, slipping in and out of roles, free-thinking and free-living, while the young aristocrat at his social best exemplifies controlled decorum. But both paradigms of behavior, with their spectacular, theatrical elements, their presumption of a viewing audience, suggest a denial of subjectivity in that, ultimately, they merely *signify* to other people. If Rameau's nephew is completely "unlike himself," he is less the self-fashioning character than he imagines himself to be. Both he and Chesterfield's son are signs, their bodies exist to be interpreted. In both cases the signifying bodies function within a specific set of signifying possibilities. Writing on the late seventeenth- and early eighteenth-century German stage, Erika Fischer-Lichte argues that "the meanings for which theatre in the baroque era tried to find suitable representations in the form of theatrical signs were preordained by a code external to theatre . . . Models for the form such gestures should take on the stage . . . could only be gleaned from a study of social reality."[60]

While the eighteenth century organized body parts and coverings into a social language, the nineteenth century's dominant "language" was a vocal one. This shift attended industrialization. Suddenly, in the noon of industrial life, wealth surpassed birth as the *passe-partout* to high society, as Dickens so often reminds us; and as capital and power settled in the mid-section of society, affording the nameless and landless stunning opportunities for material acquisition, the upper and middle classes looked increasingly alike, drove equally splendid carriages, entered the doors of the same houses and clubs.

As the external marks of nobility were embraced by a middle class with the taste – and capital – for living stylishly, the voice was increasingly perceived as a more reliable social register, one which could function in much the same way as Chesterfield's social postures, and which eventually dominated the field of popular stereotypes, blessed with the ability to embody truth. A successful Victorian merchant could readily buy clothes in imitation of a peer, and engage for his children the peer's dancing-master; he could not as easily purchase the peer's education and social experience. His voice – accent, idiom, and topic of conversation – and not his figure, would now indicate his class and profession. The Victorian social, literary, and theatrical establishments actively engaged in a semiotics of the voice, locating the signifiers of social place in speech. Dickens continually demonstrates this procedure: think of Mr. Merdle, whose lack of appropriate conversation marks his difference from the society in which he travels, or Mr. Dombey, whose obsession with business and the making of money dominates his discourse and belies the gentility of his establishment. We know them immediately by what they say.

We have begun to see how the socially inscribed voice was woven into antitheatrical discourse, with working-class voices raising respectable blood pressures.[61] Early modern and modern European statements about performance frequently centered on the insolent or subversive voice, as Dickens', Harman's, and Mozart's did. Speech acts in general held a kind of sinister power, with words, clad in their own particular rhythms and tones, potentially revolutionary. Actors, of course, live on words, troping their way through forgeries of social situations. In a universe where social relationships are no more nor less than semiotic relationships, consisting of assumed names, feigned marriages, bogus fights, and rehearsed dialogue, social order – the balance of the earth, the harmony of the spheres – is perpetually disrupted. These words have reference only to themselves, and with this autonomy comes an exquisite kind of creative power, particularly when the operative cultural assumption places words in direct, moral contact with the actual, the real. That these "forgeries" were often thought empowered to affect the social world is demonstrated in *Hamlet*, where the players' representation of a real murder in act III precipitates a series of new events which lead to mass death and the prospect of reversion to an archaic social order under Fortinbras. "False speaking" is empowered with the potential to destroy old worlds and create new ones – or to destroy new ones and reestablish old ones. It is as if dramatic voices hold the essence of human behavior inside of them, germs of being and action, and threaten to loose the dangerous, the eccentric, the deviant of these into the world.

Victorian responses to performance, and to the "dramas" of real life as well, confirm this belief, seeking and locating signs of truth or potentiality in the voice. This is not to deny the importance of the body, its physical gestures and postures in nineteenth-century social and theatrical discourses; the stage was from the beginning a forum for a semiotic display of bodies, and certainly Victorian dramaturgy played on collective social assumptions about body signs. So did the novel; one need only examine Hablôt Browne's illustrations of novels like *David Copperfield* to see how physical dramatic effect was integrated into fictional "realism," how the gestures of melodrama must have satisfied a novel-reading public in search of social verisimilitude. (For example, in a plate entitled *Mr. Peggotty's Dream Comes True* Browne has drawn a melodrama tableau, with a wild-eyed Mr. Peggotty clutching a fainting Little Em'ly, front left, and David in the background with clasped hands and eyes turned up to Heaven.[62]) I want merely to suggest here that the voice, overwhelmingly neglected by studies of Victorian literature and culture, increasingly exceeded the physical gesture in semiotic value over the course of the nineteenth century.

We may confirm this privileging of the voice by reading theatre criticism and journalistic prose as well as novels. Works of popular and high culture fetishized voices, receiving them as literal manifestations of a "natural" social order. Nineteenth-century theatre reviews typically illustrate this belief. Take, for example, *Punch*'s comments on *The Knights of the Round Table*, an adaptation by Planché from a French novel called *Les Chevaliers du Lansquenet*:

> BUCKSTONE was ripe as a ripe peach in *Tom*; turning just as sunny a face on all things. His words, too, came with a flavour of richness that, at times, had withall a rare delicacy . . . MR. VANDENHOFF'S *Captain Cozens* is all to nothing his best effort in the modern drama. Cold, subtle, venomous, he seems as though he lived on snakes, a swindler whose syllables are drops of poison.[63]

Tom's identity is established in the resonance of Buckstone's voice; the same principle distills Captain Cozens' evil into the syllabic droplets issued from Vandenhoff's mouth. Of course, to those of us unacquainted with nineteenth-century theatrical customs and the general context in which this play and its performance lived, these voices – even if we could hear them – are obviously constructed rather than the strains of pure meaning, the passions and personalities distilled into a stream of breath and vocal sound, that *Punch* supposes them to be. Instead, we recognize the highly artificial and politically conservative ethos of melodrama in the pure goodness in Tom's character, the drops of poison from his nemesis.

Interestingly, the same assumptions about the purity or actuality of the voice, and the same melodramatic ethos, often motivate sociological writings of the period. For example, the author of an *All the Year Round* article on "South Wales Colliers" qualifies the social attainments of certain "iron kings" – factory owners, self-made men – with the inevitable reference to "Their manners [which] were rough, their speech, rougher; and what with frequently enforcing their orders with a volley of strong language, and sometimes a blow, it is hard to say which was the most civilized, the master or the man."[64] If Welsh inferiority to the English may be discerned in the low speech of their moneyed class, a more generic inferiority is expressed in their song, which, although "hearty and refreshing," and capable of great technical feats, is marked

Plate 6. Hablôt K. Browne (Phiz), *Mr. Peggotty's Dream Comes True*. Original illustration for Charles Dickens, *David Copperfield* (1849–50).

by "very defective" intonation. With cultural and national identities at stake in this paternalistic, journalistic exploration – how different is it, after all, from Mayhew's sociology of the London slums? – the essentialized voice takes on tremendous importance. The writer locates in it the indelible marks of cultural hierarchy. The semiotics of his approach is indistinguishable from the current semiotics of theatre criticism – both listen for signs of sociodramatic truth in the voices of their subjects. And both readings imply a potential for action, for the actual and the active, in these voices.

The fact that theatrical discourse provided a framework for social "reality" may help us to account for the tremendous importance ascribed to the voice, to vocal gesture, and pattern in, say, bourgeois accounts of working-class or female social communities. If the crux of dramatic meaning was thought to be located in a performer's voice, then, using the theatrical paradigm an observer might isolate the truth of social experience in the collective voices of social groups, or the single voices of their representatives, as did the reporter for *All the Year Round*. More accurately, truth tends to accumulate in the spaces between voices, in the negotiations between the pitches and patterns and accents of voice which so often manifest English social identities. This, of course, is my reading of Victorian vocal politics; the voice itself, and not the resonance generated between voices, is usually represented as the seat of meaning in Victorian writings.

For example, George Eliot's description of the election mechanism in chapter 30 of *Felix Holt* relies on the truth distilled in voices to establish the basic relationships between class, politics, and education in the novel. A unionist supporter of Harold Transome, a man whose muscular appearance initially attracts Felix, ascends the block to demand "universal suffrage, and annual parliaments, and the vote by ballot, and electoral districts" in a "voice . . . high and not strong, but [in which] Felix recognized the fluency and the method of a habitual preacher or lecturer."[65] Here, the voice belies the body: "fluency" in this case suggests insincerity, the insolent ease with which a con man or costermonger makes his pitch. Felix's "sonorous" voice sits in contrast with the speaker's – it has its very life, I would argue, not in its sonority but in the field of contrast, the dialogic space, as I described it earlier, between entities, finding its meaning less in itself than in the difference between two pitches, two *tessituras*. As he stepped up to the platform,

> The effect of his figure in relief against the stone background was unlike that of the previous speaker. He was considerably taller, his head and neck were more massive, and the expression of his mouth and eyes was something very different from the mere acuteness and rather hard-lipped antagonism of the trades union man . . . When he began to speak, the contrast of voice was still stronger than that of appearance. The man in the flannel shirt had not been heard – had probably not cared to be heard – beyond the immediate group of listeners. But Felix at once drew the attention of persons comparatively at a distance.[66]

Felix's voice is a promissory note, a signifier of moral rectitude and good behavior, in large part because it is not the high-pitched sales pitch which immediately precedes it in Eliot's story. It is the voice, of course, of the middle class, a voice strikingly similar, in what it speaks if not in tone, to Matthew Arnold's voice in *Culture and Anarchy*. If his antagonist's high, thin, fluent voice, an instrument of mercenary ends, fades and dies a short distance from the lips, his own instrument bears extraordinary physical and moral power, projecting itself beyond the normal range, drawing listeners on the strength of its conviction and truth. When later, the mob stages its insurrection at the Treby election, *its* voice is clearly a variant of that of the trades unionist (and, not surprisingly, of Dickens' execution crowd):

Cheers, sarcasms, and oaths, which seemed to have a flavour of wit for many hearers, were beginning to be reinforced by more practical demonstrations, dubiously jocose . . . [A voter] was cut off from his companions and hemmed in; asked, by voices with hot breath close to his ear, how many horses he had, how many cows, how many fat pigs; then jostled from one to another, who made trumpets with their hands and deafened him by telling him to vote for Debarry.[67]

These are voices used in coercion and loose-talking, pregnant with seeds of dangerous action – unmanly voices which utter words capable of becoming wildly disruptive deeds. Eliot inscribed volumes of story in the sounds which emanate from the mob's mouth, assured that her readers possessed the skills to read them, to apprehend and assimilate the set of social facts implied in a way of speaking, a tonal quality, an inflection or accent.

Perhaps contributing to the perceived power – subversive, infective, or constructive – of the spoken voice was its proximity to song, assumed to be a beguiling or deranging force in both mythology and real life, as recourse to Homer and some more pedestrian testimonies, like the previously cited evidence from Her Majesty's Inspector of Schools, will confirm. Popular assumptions about song certainly included a belief in its natural affinity to low culture, to the pubs and music halls where revelers were known to break into spontaneous choruses, and where porter was a natural companion to singing. Often the voice of social difference, the pattered or accented or generally strange language of otherness, is sung: music, usually of the lower sort, was frequently linked to "unlawful" language in popular English discourse, and strollers were often, in fact and in fiction, a combination of musician, actor, and costermonger. Sometimes, in a romantic gesture, the minstrel and highwayman are conflated: we see this in the proliferation of English ballads about highwaymen (as if danger and violence are the natural companions of story-singing), and even in real life, as in the fifteenth-century French poet-criminal François Villon.[68]

A humorous article in Dickens' journal *All the Year Round* explores the politics of music-making, insinuating the danger and violence that lie barely suppressed beneath the surface of street music. It also articulates a set of fairly typical middle-class complaints about the strolling instrumentalists who graced London's streets, small, colorful knots of unwelcome music:

in the case of music, we find ourselves altogether in a different position. While, as I have pointed out, literature and art both wait till we seek them out, and let us alone if we let *them* alone, music is altogether of a less retiring character, comes to us often uninvited, often continues with us unsolicited, and sometimes even refuses to withdraw its beneficent influences when directly requested to do so . . . To what purpose is it that one musician should be removed from before my house, when in the course of a few minutes his place is filled by another? Consider, too, the loss of temper that ensues after a row with one of these men – and they will seldom go without a row . . . There are some *quiet* streets in London where ten or twelve of these musicians will turn up in the course of a single day; why, one need keep a servant (and a man-servant, too) on purpose to drive them away . . . Why should we have street musicians at all? Why should not a clean sweep be made of the whole organ and hurdy-gurdy tribe . . . When the professional poet comes in between Brutus and Cassius, at the end of the celebrated quarrel scene between them, Cassius asks with pardonable irritability, "What do the wars do with these jiggling fools?" Substituting "the streets" for "the wars" in the above quotation, may we not make the same inquiry with regard to our street musicians? They do us no good, they give us no pleasure, they interfere with our occupation, they chafe our nerves; what do we want with them?[69]

Dickens himself is behind this piece, his approbation if not his pen.[70] But Dickens loved popular music – why should he object so strongly to hearing it on the streets? Largely, because in this case he has no control over it – the music comes unbidden, and it, like the Welsh colliers' voices, publishes the details of its social status in its very tones, in the jangles and jingles of strange, hybrid instruments, street organs and hurdy-gurdies. Obviously, the infringement described is the insistent trespassing of the itinerant poor on "quiet" London streets. But there is another sort of trespass implied here as well: an assumption of the privilege of musical interpretation by uneducated musicians on eccentric instruments. The confrontation, then, is between those with property (owned or rented, material or intellectual) and those without, and what motivates the attack is its writer's belief in the primacy of certain types of music or voice over others.

Hogarth's *The Enraged Musician* (1741) anticipated "An Unreported Speech" by over 100 years, pitting the same low street music against educated ears, in this case the ears of a periwigged court violinist, who glares out of his window at the noisy street commerce outside.[71] To the left of the musician's window, a playbill announces forthcoming attractions at the Theatre Royal: "The Sixty Second Day . . . Comedians . . . The Beggar's Opera" – metonymically connecting theatrical entertainment with the gutter. While Hogarth does not sympathize unequivocally with the representative of refined taste, whose place in the frame is eccentric, and whose affect is feminine, neither does he celebrate the mob, which urinates on the street, raises a maelstrom of dust and noise, and, typically, leaves rubbish in its wake. As Dickens would play out a version of the scenario in *Our Mutual Friend,* "musical" street culture, there in the form of ballad-monger Silas Wegg, insidiously takes over, creeping into respectable neighborhoods, taking possession of house fronts with the confidence of squatters, just as London street musicians take entire neighborhoods hostage to their music by refusing to leave until they are paid off or arrested. Music, in other words, essentialized and empowered, carries the germs of lower-class culture into bourgeois neighborhoods, drawing-rooms, and imaginations, awakening the theatre already latent in them.

In 1834, Henry Mayhew wrote a play which illustrates this very condition. *The Wandering Minstrel,* a one-act farce which played over seventy times in London, illustrates how street culture may insinuate itself into "decent" society with the assistance of deceptive song or speech. Mr. and Mrs. Crincum, a wealthy, status-conscious bourgeois couple, are supervising their niece Julia's entrance into the marriage market. When Mrs. Crincum reads a notice in the paper about "The Wandering Minstrel," her romanticism – and snobbery – gets the best of her.

> We understand a bet is on the *tapis,* between two persons of distinction, that a well-known musical nobleman, will collect a certain sum of money by traveling through the country, under the disguise of a Wandering Minstrel. The titled votary is now on his tour, and invariably experiences the kindest receptions from the gentry . . . it being easy to perceive from his noble air and courtly demeanor that his character is assumed.[72]

When Jem Bags, a genuine street singer without a trace of gentility arrives at her door, Mrs. Crincum happily invites him in, much to the dismay of her husband, who recognizes his gamey accent for what it is ("What – make my house a home for such a vagabond as that!"). But even vagrancy, when sweetened with a title, can be fashionable. Mrs. Crincum persists until the end, when Julia's intended spouse disguises *himself* as the Wandering Minstrel, and proceeds to outsing, "outnoble," and disgrace the interloper. What makes this play so interesting is its assumptions about minstrelsy, which had lost its romantic sheen in the industrial nineteenth century and evolved into disruptive, annoying street commerce. The jongleur's song, once a source of real information, was now superfluous, vulgar entertainment, its rhythms and tones harsh on the ears. Hence, when called upon to perform at the evening concert, Jem can only sing the patterer's tune, a litany of cheap ballads. "Marm vot vill you have? (As if crying ballads) 'Ere you has 'em here' – 'Nancy Dawson' – 'I met her at the Fancy Fair' – 'My love is like the Red Red Rose' – . . . and all for the small charge of one penny." While I do not mean to suggest that bourgeois Victorian society eschewed popular songs – Dickens doted on them, and indeed the song publishing business thrived throughout the entire century – Mayhew's affiliation of the day's ballads with the patterer degrades them. Perhaps the working-class mouth articulates them differently than a Mayhew or a Dickens would have done, sings them as threats or insults or, in accent and cadences unfamiliar to the bourgeois balladeer, as the private property of the underclass.

If Jem's language is ugly and irrelevant, it is also insistently – and delightfully – slippery. One of the more amusing sequences in the play is a sort of verbal minuet, in which Jem and Tweedle, a band master engaged by the Crincums, pun their way through a conversation in which no intellectual contact is made.

TWEEDLE. I am delighted, Sir, to have the honour of meeting a gentleman whose musical talents promise so much.

JEM. [*aside*]. They may promise a great deal – but hang me! if they don't perform very little.

TWEEDLE. Respecting the selection of music for this evening['s concert] – what school do you prefer?

JEM. What school! [*aside*] Blow me, if I was ever even inside on one! but I must not let him know nothing about that. Vhy, I thinks as how, the Parish School is a pretty tidy un.

TWEEDLE. The Paris school – aye, Sir, France, certainly contains some very excellent masters.

Of the two voices in this dialogue, Tweedle's is the "normal" one, his idiom familiar to a middle-class auditor. This is somewhat surprising, given his name and occupation, neither of which is elevated – the name, in particular, recalling Fielding's graveyard "warrior," the fiddler Jemmy Tweedle.[73] In fact, the undermining of Tweedle's standard English with the derisive reference suggests something attractive or interesting about Jem Bags' alternative speech, and this inconsistency is typical of ostensibly negative representations of patter, which often succumb to the pleasures of the idiom they seek to contain. At any rate, their mutual misreadings dramatize the potential danger, from the hegemonic perspective, in nonstandard verbal forms (although as entertainment the scene is spectacular): here, a conversation proceeds, a compact is made, with a total – and admittedly comic – misunderstanding on each side. On one level, Mayhew's play is entirely about misreadings: Mrs. Crincum's, Tweedle's, even Jem Bags'. This comes as no surprise, as the minstrel figure historically represents opacity as well as clarity, destructive as well as instructive stories, magic as well as reality. I want to emphasize here the rich register of social meanings in the Victorian voice, susceptible, given the malleability of voices, to misreading, but implicated in an elaborate social compact, an agreement between people of all classes about the realities of their world. If eccentric idioms level a certain threat, it is because the dissimulation which in this case fails (the level-headed characters do *not* mistake Bags for a nobleman) always carries the potential for success, for class disruption; dialogic voices may just as easily deconstruct as fortify the desired social order. Even in this case the potential remains, for Bags vows he will return every day for a week, to annoy Tweedle.

Vagrants like Jem Bags were traditionally thought to have a special relationship to madness as well as music. Madness was itself a form of theatre until the nineteenth century, according to Foucault,[74] and I would argue, counter to Foucault, that madness *continued* an object of ritual and display throughout the nineteenth century. Its murky languages and gestures inspired fear and admiration, and literalized the differences between sanity and lunacy. Frances Burney demonstrates this in *Camilla*, whose heroine's loving father arranges for his daughters, the beautiful one and the hunchback, a lesson in right thinking and humility when he takes them to see a "beautiful idiot," histrionic in her craziness, and available for viewing;[75] but so does Dickens, in novels as diverse as *David Copperfield* (Mr. Dick), *Bleak House* (Miss Flyte) and *Nicholas Nickleby* (the gentleman next door). Here, the reader occupies the audience position, deriving both amusement and instruction from madness, but the situation is the same as it had been in the previous century for Burney: cultural values are "performed" in very public venues, using the externalized signs of a heteroglossic and highly diverse society. The introduction of performance into the "private" sphere in the form of reading aloud, or its reintroduction into the public sphere in the controlled form of dramatic reading, represented an attempted containment of the dangerous, the infectious or lunatic, in acting, just as actual lunacy itself was to be enclosed and quieted in Victorian asylums. But it also contributed to a diffusion of the tropes and gestures of performance (and madness) among the public, admitting them into the vernacular.

Such diffusion meant that Victorians read in their world the same signs they read at the theatre. Their "reading" skills were cultivated in the theatre, the place where words spawned worlds of action, where voices could inspire sexual fever and acts of insurrection. But why did theatrical voices carry such weight, given the general disapprobation of the theatre? If mainstream culture questioned the legitimacy of the theatre, and if even some of the century's brightest performers publicly disparaged their profession, why did people fear the potential of theatrical entertainments? Why did readers fail to recognize novel characters as adaptations of theatrical types, accepting, for example, Eliot's representations of Felix Holt and the mob as true to life if representations of that sort were standard fare at any theatre, from the legitimate Drury Lane to the questionable Royal Victoria? And why should the theatre have provoked such anger and anxiety when its characters, plots, and voices turned up without comment in novels like *Felix Holt*?

The answer to these questions lies in the perceived power of spoken dramatic words, in the resonance of the human voice. Although theatrical voices could certainly threaten from the pages of novels, as I suggested earlier in my discussion of Dickens' "preemptive" public readings, of the control which reading his own text aloud gave him over that text, the power of a convincing actress, a Vashti, over her audience, was perceived to be absolute and transforming. Actors and actresses not only embodied the passions, but invoked them in their audience, "exciting the histrionic sensibility of their audience."[76] In other words, while Eliot's narrative representations of violence and venality in the mob are relatively controlled – signifying diachronically, not in the moment but in their metonymic and thematic relationships to other representations – a dramatic rendition of the scene would bring the audience into direct contact with the very thing, increasing the possibility of corruption, of imitation, of a real mob exploding into being. Dickens proved as much in "Shakespeare and Newgate": plays about pickpockets breed pickpockets; elevated entertainments attract – and even create – the refined audience. It is also likely, as I suggested earlier, that because theatrical structures, gestures, and postures were so finely integrated into the cultural mechanism, people saw them as socially authentic; enacting the immoral on the stage meant doing something immoral, realizing a dubious or antisocial act in the same form it took on the streets.

Whether "social reality" was inscribed on the body or in the voice, the dominant interpretive paradigms used in processing that meaning came from the theatre. Assumptions about class and gender were often supported by identifying and interpreting signs, in the speech or dress or behavior of a group, which closely resembled the signs used on the stage to differentiate between character types. The redundant speech of widows, the slick and deceptive patter of the street sellers, the vulgar and explosive cries of the mob – these all "proved" that the characteristics assigned those groups by the popular culture were natural and justified, that single women were silly and superfluous, the working class dangerous and degraded, although in fact their wide dissemination as social ideas is more probably owed to the theatre than to any other cultural form. We have already seen evidence to suggest that this is so, that these ideas were first articulated on, and disseminated by way of, the stage. But of course one must avoid oversimplifying: the epistemology of the stereotype is complicated.

I opened this chapter by introducing an inscrutible Victorian paradox: the simultaneous embrace and rejection of the theatrical in nineteenth-century England. Indeed, it wanders into my next chapter, a rich and necessary enigma, and so I have a strong incentive to allow the paradox its play even if I could easily strip it of all of its resonant contradictions. While this tactic leaves certain questions unanswered, it also lays open a great field of possibility. This is fitting, because Victorian culture was anything but a closed signifying unit, a linear story with beginning, middle and end in proper sequence. It was, in fact, something of an unchoreographed dance, an undirected – or overdetermined – performance, in spite of the strong directorial presence created by the likes of Dickens. If the act required hero, or pedant, or vicar, or mother, to cry out against the theatre's immorality, it also required an appreciative gallery, a community of playwrights, and a shared understanding of the language in which the entertainment itself was conducted. That shared understanding lies at the heart of this project; it lay at the center of every nineteenth-century novel writer's and novel reader's consciousness.

CHAPTER 4

Patter and the politics of standard speech in Victorian England

This particularly rapid, unintelligible patter
Isn't generally heard, and if it is, it doesn't matter!

Gilbert and Sullivan, *Ruddigore*[1]

London is a large village on the Thames where the principle industries carried on are music halls and the confidence trick.

Dan Leno[2]

Ruddigore, like all of the Gilbert and Sullivan collaborations, with its vocally aware universe, its tightly rhythmed and choreographed mode of expression, reflects the underlying theatricality in nineteenth-century English culture. To make this claim is to assume a certain risk; after all, the world of *Ruddigore* is slightly mad, and certainly not mimetically descriptive of the "real" world. In fact, one might argue that its various boundaries – the beginning and end of its plot, the rectangular containment of its staged performance – and the hysterics and histrionics enclosed within them, mimic the architecture and business of the lunatic asylum. If this is so then the suggestion that *Ruddigore,* a tale of supernatural visitation, sexual intrigue, and disguise, expresses the truth of the outside world is a suggestion that the outside world is both theatrical and mad. If we take "lunacy" to represent something larger and more diffuse than clinical insanity, if we link it to certain symptoms of the carnival – eccentricities, strange vocalities, social disruptions; irregularities, not severe enough to require confinement, of the odd but relatively harmless sort we see in Miss Flyte of *Bleak House* – then this is of course true, and *Ruddigore,* both theatrical and mad, a fair representative of Victorian culture. We can confirm this truth in the patter-trio quoted above, which makes oblique reference to the theatre of voices in its own internal economy and that of the outside world: its "rapid, unintelligible" verbal gesture implies a better way of speaking, a moderately paced, transparent idiom against which patter – or theatricality, or madness – plays its eccentric music.

One may locate the dominant paradigms of national and class identity in the counterpoint of these competing voices. In fact, it becomes apparent, in examining representations of voice in popular entertainments, that normative English culture not only positioned itself in relation to the theatrical, but to the lunatic as well – not so much in the inward, panoptical mode in which Foucault figures the lunatic/carceral/theatrical in the nineteenth century, as in an externalized, semiotic way, with the processes of containment and regulation taking eminently readable, spectacular, or aural forms. English popular entertainments dramatized the discursive order in a series of very public "dialogues" between normalcy and eccentricity; the comic–dramatic idiom known as "patter" was typically used in this fashion, to underscore the clarity and transparency of standard English speech and the superiority of those who spoke it.

While we may, with a certain critical distance from our subjects, reason away the inevitability of binarisms like "standard" and "eccentric" speech, the binary paradigm was invested with tremendous power in the nineteenth century, and so perhaps the best way to approach such constructions is with an awareness of their simultaneous strength and vulnerability. Certainly, the current discussion assumes at all times the operation of certain structures of social regulation, the maintenance of a standard discourse which continually regroups and redefines itself in relation to shifting pockets of nonstandard culture and speech, like patter, but which believes in its integrity despite the movement. In fact, the very word *patter*, with its derivation from "paternoster,"[3] reflects the shifty nature of standardness: its impressive etymology failed from the start to protect it from an overwhelmingly negative cultural value. In medieval religious discourse, the word was used to describe rote delivery of the paternoster, a hypocritical devotion.[4] When I use the term *patter*, I intend to suggest the secular meaning that had evolved by the early eighteenth century: a frivolous language, a language of form rather than content, in which words are deceptive because they deviate from conventional (elite) semantics, rhythms, and syntax; a language which means meaninglessness, and therefore means something desperately significant.[5] Secular patter characterizes, from an elite point of view, the languages specific to socially marginal groups, groups which may change at different historical moments (the eighteenth century counted servants, beggars, and thieves among its patterers; the nineteenth century found patterers to be prevalent among street folk and redundant women). Hence its importance in questions of social identity. Implicit in this trope is a decentering of English words, which, in the act of pattering, have no moral authority but merely slip in and out of social and linguistic contexts. This is why patter has always such an urgency about it: presumably, patterers could promote revolution or anarchy by banding together and speaking a language inaccessible to middle-class men, a language as empty and inspired as lunacy. Some of them actually did cluster fiercely together on the borders of respectability, like the London street-sellers or "patterers" described by Henry Mayhew, a social group with a trade slang of their own which played on the standard idiom.[6]

Likewise, this chapter's purpose is to play patter against the standard English that defined it, so that its political and linguistic offenses against that Behemoth are fully realized. It is also to imagine the reception of patterers in middle-class drawing-rooms and theatres. If patter is hard for the reader or listener to grasp, it's also hard to define: while I suggest in this book that it belongs first to the stage, and then to the novel, patter's history is more complex than that, encompassing sacred and secular cultural ideas as well. Even its forms are shifty; while one may – and should – differentiate between certain species of patter, at times such classification robs the trope of its full resonance. But if patter's philology is opaque, its sociopolitical implications are transparent. Those who would not or could not speak the language properly occupied an unambiguously marginal place in the popular imagination.

Patter was affiliated, in the nineteenth century, with street culture and the condition of "redundancy"; this chapter explores both affiliations but grants slightly more attention to the former, while chapter 6 is devoted entirely to redundant women and spinster patter. The Victorians did not differentiate between the types of patter spoken by spinsters or widows and costermongers, although I shall adopt here the designations *street patter* and *spinster patter* when such classification is necessary to prevent any confusion of the two versions of the idiom (sometimes the generic *patter* is preferable). That the nineteenth-century public understood the languages imagined to belong to these very different social groups as identical, or at least sufficiently similar to obviate a classifying nomenclature, makes sense; the two idioms exhibit a like slipperiness and surreality. And redundant women and street-sellers shared certain attributes in the popular imagination. These points of connection may have been unarticulated, or circumlocuted, but the dominant discourses on both groups indicate that they were perceived to violate the "natural" order of industrial capitalism, creating, on the one hand, an artificial demand for a supply of bogus goods; disrupting, on the other, the well-regulated domestic apparatus which, among other things, ensured a male monopoly on the marketplace.

By insisting upon this un-Victorian differentiation between the two popular forms of patter, I mean not to diminish these points of contact and linguistic similarities, but rather to emphasize the fact that Victorian class and gender politics were neither identical nor equal social forces. Women waited much longer than the working-class men who were the primary speakers of street patter for their various rights, and the effect of their representations in the popular culture was more devastating, I would argue, than those of these men. In other words, the two forms of patter were effectually different. The men's slangy, sinister street argot fed an undercurrent of dread and thus generated a kind of respect among the middle class, even as it made them smile; whereas the repetitive, illogical, under-punctuated, and redundant monologues delivered by widows and spinsters in novels, in journals, and on the stage, inspired mainly jollity and contempt. Ironically, this suggests a greater danger on the part of redundant women, who threaten not only the country's politico-economic structure but its very moral fiber as well – an argument I set forth in the sixth chapter of this book.

Other derisive speech patterns coexisted with patter, some quite similar to it although less ubiquitous. "Malapropism," for example, turns up frequently in English plays, and later in novels, from the Renaissance through the nineteenth century. Malapropism consists of semantic substitution: the audience expects a word and the comic character substitutes another, usually a near-homonym, for it. This trope is derisive – it draws laughs at the expense of the speaker – but it depends ultimately on its comprehensibility for its humor. In other words, it only works if the audience hears and understands every word. Its difference from patter lies, then, in the attention it demands to be given: this is speech that is listened to and understood, not ignored or dismissed out of hand like spinster speech, or listened to and misunderstood like coster speech. It absorbs meaning instead of repelling it. Compare, for example, Constable Dogberry's comic idiom with Sairey Gamp's:

DOGBERRY. One word, sir. Our watch, sir, have indeed comprehended two aspicious persons, and we would have them this morning examined before your worship.[7]

GAMP. A thing . . . as hardly ever, Mrs. Mould, occurs with me unless it is when I am indispoged, and find my half a pint of porter settling heavy on the chest. Mrs. Harris often and often says to me, "Sairey Gamp," she says, "you raley do amaze me!" "Mrs. Harris," I says to her, "why so? Give it a name, I beg." "Telling the truth then, ma'am," says Mrs. Harris, "and shaming him as shall be nameless betwixt you and me, never did I think till I know'd you, as any woman could sick nurse and monthly likeways, on the little that you takes to drink."[8]

The difference between these idioms should be obvious: Dogberry – and the same can be said for his linguistic sisters and daughters – uses measured English cadences. His pace is moderate and his diction, even when he "malaprops," is transparent. We know as well what "comprehended" and "aspicious" signify as if he had used the proper words. Mrs. Gamp, on the other hand, speaks in a way that confounds English diction and cadences. Crowded, rushed, with two voices crammed into the narrative space conventionally designated for one, her speech threatens to take over. It is as if all of her potential fecundity were misdirected and channeled into her speech, which grows wildly out of control and fills the available space around it. Both passages are meant to be funny, but the second, far more eccentric than the first, is much less good-natured. It is, I would argue, downright hostile.

One finds another relatively good-natured comic verbal trope in dialect. Dialect differs from patter, like malapropism, in its basic clarity; it honors most of the rules of English speech, except the pronunciative ones. Dickens counted on his readers' ability to process Sam Weller's speech as surely as he intended the more serious Stephen Blackpool to be understood. That their voices may be difficult for us to understand does not suggest that Victorian readers had the same trouble; on the contrary, even upper middle-class readers encountered these voices frequently, in novels and plays if not on the streets.

Sometimes, male speakers of dialect in Dickens appear to appropriate spinster patter: they produce larger blocks of dialogue than the typical Victorian hero does, and in contrast appear as verbally redundant as his Saireys and Floras. Alfred Jingle, Sam Weller, and Toby Magsman come to mind, as do the later Joe Gargery and Magwitch. Weller, for example, treats the reader to so much of his characteristic dialect in a sitting that one is tempted to call it patter. (Really, it lies closer to a diluted street patter than spinster patter, with its occasional slang.) His sentences, though, are short and relatively tight. "They puts things into old gen'lm'ns heads as they never dreamed of. My father, sir, vos a coachman . . . His missus dies, and leaves him four hundred pound. Down he goes to the Commons, to see the lawyer and raw the blunt – wery smart – top boots on – nose-gay in his button-hole – broad-brimmed tile . . . "[9] Even the dashes serve the purpose of control here; they allow the speaker to paint a deliberate portrait, the rests they establish between images emphasizing each part of the visual whole. Dickens employs the strategic dash in Alfred Jingle's monologues as well; Jingle strings nouns and verbs in a row like Christmas popcorn, separating them with commas and dashes but denying them the connective tissue usually provided by other parts of speech. So that his retort to Rachael's accusation that he "runs on" – "Run on – nothing to the hours, days, weeks, months, years, when we're united – run on – they'll fly on – bolt – mizzle – steam-engine – thousand horse-power – nothing to it" (p. 141) – is elliptical, and of necessity moderately paced. The opposite, in other words, of spinster patter, even if it is rather silly. There are later examples of such nonpattered eccentric male speech patterns as well: Joe Gargery's long-winded discourses, say, are relatively self-contained, well-punctuated with short sentences – reasonable if ungrammatical communications. Dickens may have found something amusing about *petit bourgeois* male speech, but as social entities, lower middle-class men were clearly innocuous. He felt no need to choke them with a rush of wild words and phrases.

Patter is persistently defined, in eighteenth- and nineteenth-century English letters and thought, as a language of social undesirability, and its definition as such indicates anxiety on the part of the desirables. When Moll Flanders recounts her term in Newgate, she describes her prayer for deliverance as one that emphasizes word forms over meaning – in other words, as a kind of pattering.

> truly I may well call it *saying my prayers*; for I was in such a confusion, and had such horror upon my mind, that tho' I cry'd, and repeated several times the ordinary expression of, *Lord, have mercy upon me,* I never brought myself to any sense of being a miserable sinner, as indeed I was, and of confessing my sins to God.[10]

Defoe uses a conventional stereotype of Catholic liturgy, in which sincerity is compromised by the rapid-gunfire delivery of hypocritical, bored, or merely illiterate congregants; the emphasis on form rather than substance that necessarily characterizes spoken language learned by rote implies a social threat, in that it permits the speaker to say one thing and think another simultaneously, to patter with the lips but not pray from the heart. Moll herself underscores the role of convention in robbing language of truth, "repeat[ing] several times the ordinary expression of, *Lord, have mercy upon me,*" without internalizing its sense. This disjunction between word and meaning suggests the possibility of verbal deception, as well as a frivolity in which words exist merely as words and not as signs of some moral or true condition. There is a truth, of course, in the frivolity of such speech – in that it is thought to be truly or actually superfluous – but this is not the representational truth which readers, particularly Christian readers, had been taught to value. Defoe's association of Moll the criminal with Catholicism illuminates patter's cultural history, which encompasses both Catholicism and criminality.[11] The connection between liturgical patter and street patter is obvious when we consider the negative values – deception, frivolity, hypocrisy – perceived as characteristic of each. In emphasizing the gulf between Moll's words and religious truth, Defoe suggests that "good people" are morally obliged to use properly meaningful language, and he suggests it despite his obvious admiration for his "bad" heroine, perhaps the novel's own *raison d'être.* In doing so he utters a cultural idea centuries in the making.

That the Victorian middle-class imagination excluded the languages of beggars and thieves, of street vendors, widows, and spinsters from its sympathy, is as much attributable to the stage as it is to novels like *Moll Flanders.* The theatre did much to shape Victorian ideas concerning patter, informing even "neutral" journalistic discussions such as Henry Mayhew's *London Labour and the London Poor.* Beginning in the eighteenth century, patter found its place in the dictionary and on the stage, a situation which codified its form and its function. Of course, the word *patter* had been employed in writing and conversation for centuries. One might have used it to describe one's servants' speech, or the talk around Covent Garden; and a good Protestant might have linked the term to some dark devotion involving staccato "Hail Marys." But now one might associate a literal sound and look – and for operatic and music-hall audiences a very specific sort of song – with patter. At the opera, a newly evolved class of character, the *buffo,* sang breathlessly fast and repetitive arias, virtuosic for the performer but derisive of his or her character.[12] At the theatre, widows and old maids spoke rapid, illogical dialogue meant to provoke contemptuous laughter – a laughter touched with uneasiness over the problem of controlling these silly, superfluous folk who spilled out of the proscenium frame and into real life. And if one wanted an "authoritative" definition of the concept now frequently dramatized on the stage, one might consult a dictionary. Dr. Johnson's *opus* does not define *patter* as either street slang or Latin prayer,[13] but Captain Grose's 1785 "vulgar" dictionary does, and so likewise does *The Century Dictionary,* published a century later.[14] Their definitions imply immorality or unrespectability and they describe a language of alien forms and obscure meanings.

The formation of social identity always involves differentiation, the construction of binarisms like "we" and "they"; in modern English culture that differentiation usually takes linguistic form, as it does in *Ruddigore.* There, patter takes on the value of otherness, possibly madness, but it also delights, being both absurdly silly and absurdly difficult to perform. We saw this overdetermination of meaning in this chapter's epigraph, which in one small sentence demonstrates patter's marginality as a social idiom, describes and embodies its typical form, and celebrates its unexpected pleasures, as well as harboring implications of normalcy, performativity, and madness in its structure and content. Patter is typically multiplicitous in this way; its inherent slipperiness makes it a curious choice for an instrument of containment, although it was typically employed in that service by English popular culture, as we shall see. Yet, patter refuses to be mastered. Its richness of possibility has subversive potential; it writes its own stories in the margins of other people's stories, ever threatening to take over the main body of text. But people used patter in their stories anyway, because in some sense it embodied what it described, and because, in a culture which believes its social idioms are invested with authority, there are fewer choices than we might imagine in matters of verbal representation.

That one small sentence, the line from Gilbert's patter trio, should contain within its parts so much of the information about social identity and performance extrapolated, in chapters 2 and 3, and at the cost of some effort, from a much larger field of signs, suggests that linguistic patterns incorporate the dominant structures of social discourse, or that social discourse is somehow patterned on linguistic structures. Either way, it is with a strict attention to the parts and particles of text that one may begin to explore the uses of patter in English entertainments. Of course, the vocality and theatricality recovered in the earlier chapters of this study are always implicit in this chapter's close textual readings, in which the spoken voice inspires or haunts the written. We may, if we try, acquire the necessary skills for excavating this buried vocality, witnessing the transformation of written script into living dialect, and for imagining the power for social disruption contained in such transformations. Once we do so, we enter into what was described earlier as a Victorian cultural–aesthetic compact, into the field of shared assumptions which produced the Victorian reader.

One finds stereotyped comic characters, speakers of eccentric English, inhabiting mid-Victorian drama, journalism, fiction, and social anecdote, and their language is almost always framed by the same set of class- or gender-based assumptions, suggesting a sort of infinitely mirrored verbal reality, in which strains of speech refer only to other strains of speech, the whole social utterance orchestrated in the popular imagination as it might have been – indeed, as it was – on the stage. Written or spoken, their voices resonate with theatricality, with real timbre and movement; as we have seen, Victorian readers recognized authentic sociodramatic power in them. We have also seen how eccentric voices were typically employed, in Victorian popular culture, to contrast with "proper" speech, to befuddle and disrupt, although rarely to overthrow, the conventional structures of verbal discourse. But sometimes this strategy of verbal policing failed – if, for example, the alternative idiom was too comic or interesting or flamboyant to be effectively suppressed. Most of Dickens' fiction demonstrates this employment of verbal difference to reinforce standard speech, as well as the propensity of the eccentric idiom to subvert the very social and verbal structures it is meant to uphold. Consider the widowed Flora Finching, who uses a culturally significant kind of speech, one which allowed Dickens to assert a set of social facts about her:

> One last remark . . . I wish to make, one last explanation I wish to offer, there *was* a time ere Mr. F first paid attentions incapable of being mistaken, but that is past and was not to be, dear Mr. Clennam you no longer wear a golden chain you are free I trust you may be happy, here is Papa who is always tiresome and putting in his nose everywhere where he is not wanted.[15]

This passage is typical of Flora's dialogue, which evades the responsibilities of conventional verbal discourse, sowing confusion rather than reason, and is in equal parts tedious and funny. Dickens' audience undoubtedly recognized Flora's redundant, circuitous speech as spinster patter. Such patter was easy to recognize. It was regularly employed by Victorian novelists, including women; *Mary Barton*'s Alice Wilson is a spinster patterer, for example, and Mrs. Gaskell no doubt benefited from Jane Austen's demonstration of the idiom a few decades earlier in the mouth of *Emma*'s Miss Bates. In such characters, spinster patter signifies redundancy: as single women in a culture that sanctifies marriage and reproduction, they have no clear social place. Their very presence threatens the structures of social hierarchy, particularly in the case of widows, who are presumably financially independent, legally and socially autonomous – in theory, if not in practice, impervious to a father's or husband's control.[16] Redundancy was a social condition, in the nineteenth century, with deep political implications and a strong hold on the popular imagination; the problem of redundancy in Victorian society and popular culture is explored in depth in the sixth chapter of this book. Here, I am concerned mainly with the imagined connections between redundant speech and single women, the shared assumptions that patter was the natural idiom of such people, and that redundancy was one of a very few social conditions that patter was peculiarly suited to express. Clearly, spinster patter provided Victorian readers with a shorthand key to the redundant woman's character: a speaker of frivolous, meaningless words is herself frivolous and unintelligent, even if, as in Flora's and Alice's cases, basically well-intentioned. More seriously, a character like Flora poses a potential threat to social propriety with her extravagant and grotesque sexual flirtation, her widow's shameless groping for marriage and sexual or financial gratification. Spinster patter enacts grotesqueness and extravagance, and was itself often perceived to be as threatening as the characters who spoke it.

Thus, in constructing the speech of spinsters and widows as uncontained, breathless, and confused, Victorian novelists attempted to contain them by establishing their collective voice as a sign of social eccentricity.[17] While most readers possessed the key to this linguistic shorthand, understanding the trope's broadest implications and hearing in it certain theatrical allusions, such attempts at defusing the single woman's speech founder because patter always draws attention to its transgressive qualities, placing those who speak it in an aggressively dramatic and public space. While it may succeed in degrading characters like Flora and Mrs. Holt, their language also raises interesting questions about them, rescuing them, at least temporarily, from the margins of social respectability and narrative interest.

The regulatory work undertaken in Victorian popular culture – identifying and suppressing socioverbal eccentricities – suggests that bourgeois industrial ideology depended on linguistic controls in order to maintain social "balance," to dramatize the difference between those with social power and those with none. Some interesting work has been done on the ideology of standard speech in western culture.[18] Mikhail Bakhtin argues that the process of linguistic regulation, the construction of a "correct," homogeneous verbal discourse in literature and popular culture, is constantly undermined by alternative voices which refuse to be muffled.[19] These voices may play like comic variations against the standard language. What Bakhtin formulates as a general rule of western culture can be applied with particular success to the Victorian period. As D. A. Miller describes it, the voice of hegemony, in the nineteenth century, "continually needs to confirm its authority by qualifying, canceling, endorsing, subsuming all the other voices it lets speak."[20] I would be more cautious than Miller about believing in the dominant ideology's unconditional authority, vocal or otherwise; its "unity" is somewhat compromised by its constant strategic, contextual shifts. In Victorian letters and popular culture we see explicit homages to what Bakhtin terms "unitary language," with marginal verbal modes, like patter, used strategically to reinforce the unitary language by chafing against it – forcing it to reinstate itself, as in Miller's reading, although failing, I would argue, to confirm its authority – with the insistence of irritating foreign matter. But we also see these highly controlled voices slip their contextual harnesses, pulling us from "correct" speech and articulating narratives of working-class or female subversion and middle-class anxiety. When this happens, the heavily patrolled discourse of verbal normalcy falls victim to its own strategies for maintaining power.

In his writings on speech politics, Foucault points out that nonstandard idioms have traditionally been associated with the obscure and the dangerous.[21] Nonreferential speech is supposed mad speech in a culture invested in the meaningfulness of its language, and mad speech itself takes on a sort of double life: not only is it deceptive, meaning what it does not mean, or locating its meaning outside of the standard register in some other sociolinguistic plane, but it is perceived as both empty and inspired, innocuous and darkly menacing. This is especially true of patter, which, throughout its philological and cultural history was figured as a comic voice of irrelevancy, but which inspired a kind of dread in those who spoke the standard idiom. Foucault's critical imagination accommodates the eccentric discourses that mainstream culture tries to suppress, creating what is in fact a somewhat un-Foucauldian resonance in the play of the nonstandard against the standard, a play which seriously undermines the latter, and perhaps even compromises his own argument for the efficacy of verbal controls. He unleashes these voices of difference, and they howl and swirl about him, pattering, mad, dangerous. Their richness proves the vulnerability of standard discourse, suspended as it is within the limited area of cultural–historical possibility. If, as Foucault argues, there is a basic difference between "speaking truth" and speaking "*dans le vrai*," or speaking within the boundaries of what is already accepted as true,[22] then the very possibility of speaking false vanishes, except as it is constructed "*dans le vrai.*" This unsettling of established meaning opens up more problems than can possibly be explored here. But I do want to destabilize the categories of "standard" and "eccentric" so that my use of them will be met with a proper skepticism.

The *opera buffa* was among the first artistic genres to dramatize patter, and although such dramatizations evolved significantly over the next century, the patter in eighteenth-century operas and plays (its presence in the Augustan fiction is negligible) was recognizable and meaningful to Victorian audiences.[23] In fact, the frequent invocation of eighteenth-century comic opera in English theatre and literature suggests that the form influenced Victorian popular entertainments.[24] Mid- to late eighteenth-century composers like Haydn, Paisiello, and Mozart used patter as a comic vocal trope, and inasmuch as it resembles the structure-over-content emphasis and rapid delivery peculiar to the paternoster, the spinster's overflow, and the cheap-jack's pitch, we can presume a connection between operatic patter and its other manifestations. It seems reasonable to assume that Mozart, who composed *Le Nozze di Figaro* in 1785, understood patter as Captain Grose defined it in his 1785 dictionary – the concept sustained a cross-cultural negativity – and used it to demarcate and control his socially eccentric characters.[25] Mozart's *Figaro* and *Don Giovanni* were venerated by Victorian operagoers (they thought Mozart himself a high priest of high culture); a discussion of his use of patter in these favorite operas will shed some light on the gesture's function in English popular entertainments. It will also underscore the inherent musicality in patter, which, even in its textual embodiments, sings with a kind of rhythmic and tonal honesty, generating its meanings in the play of these nonsignifying parts of speech.

Several characters from *Le Nozze di Figaro* sing in patter, but the most explicit and interesting use of the idiom occurs in Dr. Bartolo's act I aria, "La Vendetta." "Tutta Siviglia conosce Bartolo," booms the foolish man in his aria of revenge against Figaro, and the orchestra accompanies him with a playful staccato, a sort of chortling, undermining acquiescence. Typical of this aria is the disjunction between Bartolo's words and his music; the grand seriousness of his vow for revenge – a literate audience would associate revenge with characters of epic or tragic proportion, like Achilles or Hamlet – is constantly subverted by the playfulness of Mozart's score. While the actual pattering begins later in the aria, Mozart presents Bartolo as a dealer in deceptive language from the start. Of course, music always qualifies language in *Figaro* – it is the privileged medium – but Mozart seems to be interested here in conflating what Henry Mayhew would later call "oral puffery" with social and dramatic eccentricity.

The patter in "La Vendetta" aids the audience in evaluating Bartolo's character; it undermines his seriousness and disqualifies his speech, in much the same way that Flora Finching's or Mrs. Holt's patter, as I suggested earlier, indicates their silliness and attempts to deflect their subversive potential. Bartolo narrates a grand romance of revenge, while his music reminds us of his origins in *commedia dell'arte* farce. A sort of stout, middle-class Hamlet, he will act in the interests of decency and Marcellina's loan. "La vendetta, O la vendetta!" he bellows, and the orchestra speaks with him in one assertive voice, qualified only by the grace note with which the flutes, oboes, violins, and violas compromise the *det* of the first "vendetta." But after the dignified arpeggio which precedes his next words, the music changes in nature; light and sweeping, with the second violins playing sixteenth notes in ascending and descending scales, it belies the pompous gravity in Bartolo's words. There is a futility in those second violins, which, in the next eight measures, run continuously up and down, mostly confined to the middle register, and which undercut the strong vocal melody. Bartolo's repetitions blunt the force of his words, rendering them nonsensical; the dogged repetition furthermore suggests a bar to the singer's desired object, words being substituted for action. He begins to regret his bluff halfway through the aria. "Il fatto è serio, ma credete si farà" ["The situation is serious, but trust me, I'll take care of it"] he assures Marcellina, recovering his typically phlegmatic air, but the music prevents him from backing down. It starts to precede him, the first violins commencing into triplets (m. 56), and forces him into finishing what he started. In order to catch up to the music, Bartolo breaks into patter; the meaning in his words is entirely obscured by the verbal/musical gesture which becomes in itself the locus of dramatic value. To the extent that it controls rhythm and tempo, music, here in the service of conservative values, determines character.

One necessary condition of musical patter is speed; this sort of comic sung-speech is only powerful insofar as language spills into a space insufficient to contain it, as Bartolo's does when he is forced to sing twice as many notes as he was earlier, but in the same amount of time. (We can hear the same dynamic in textual patter, particularly spinster patter, which implies the same spatial–temporal imbalance.) I have described music as basically conservative in *Figaro* because it is used by Mozart in the service of hegemonic values: the marginal character is forced into patter and despite our enjoyment of his aria, our appreciation of his virtuosic skill, he is subjected to comic denigration. Bartolo may be relatively innocuous, but the opera's social universe strives to keep him in his place – a silly old professional, when all is said and done. In its pattering, his voice is meant to assume a merely conventional meaning; despite the lack of any significant staging of patter in the past, a contemporary audience would have recognized this comic device as a version of the traditional comic speech forms, like those used by Shakespeare, for instance, that signified social inferiority.[26] On the other hand, of course, patter makes Bartolo's ordinary *basso* voice interesting; it suggests that this is a character to watch, one who may do something funny. Operatic patter illustrates the doubleness, the emptiness and power, of an eccentric idiom particularly well, partly because a certain excitement at witnessing such extraordinary vocal control will almost certainly attend any incomprehension occasioned by the patter, and partly because, serious or comic, there is always pleasure to be had in the music itself if the story fails to amuse.

A proliferation of opera reviews and references in Victorian periodicals suggests that operatic influence on English popular culture was strong; certainly the *opera buffa* articulated some of the same ideas about verbal excess, in its patter arias, that were circulating in eighteenth- and nineteenth-century England. Music criticism lacked its own theoretical language until Robert Schumann redefined the practice in his periodical *Neue Zeitschrift für Musik*;[27] one may witness the halting half-steps of a new discipline in English music criticism of the first half of the nineteenth century. It is interesting to note that, despite the lack of a common vocabulary and critical purpose, mid-Victorian music reviewers typically expected the musical genres to be bound to the same sort of faithful or truthful signification – the same standard of communication – as were nonmusical entertainments. So, for example, a *Punch* reviewer of 1842, describing one production of *Figaro*, levels his criticism not at the singing, the acting, or the artistic direction, but rather at an improperly balanced relationship between vocal and instrumental "dialogues":

The band is always too loud in its accompaniment . . . This is very unfair play to the enchanting instrumentation of our beloved Wolfgang – more so than to the music of any of his rivals. The charming playfulness with which he works his subjects about in almost every part of his scores – *the beautiful fitness of each passage to the capabilities and genius of the instrument to which it is allotted,* and the consummate skill with which his melodies are developed, not by solos of voice or instrument, but by the combined effects of the whole – cause an undue prominency of the orchestra to be instantly detected and painfully felt by the least tutored ear. (My emphasis.)[28]

This reviewer tenders a prescription for operatic excellence: never violate the truth – the "beautiful fitness" – of music's natural relationship to instrument. For every musical idea there exists a correlative instrument; the expectation, that instruments function as signifiers of musical possibility, attached ultimately to appropriate sounds or passages, is analogous to the current expectations that language, particularly spoken language, contains within its parts pearls of truth. (Those expectations also, of course, informed theatrical discourse, which assumed dramatic gestures and voices to be anchored to the passions.) In each case signification represents a kind of compact, and when signifier and signified, or form and substance, fail to generate "beautifully fit" music ("truth"), or when form, as is the case with patter, makes an eccentric music of its own, something of balance or fitness is lost. This desire for "balance," for a standard of social and aesthetic conduct, was deeply etched in the English consciousness, representing a much larger discursive value than the reviews suggest, or than the reviewers themselves may have realized. These are principles underlying capitalism and industrialism – both of which are perceived as driven by internal rhythms, checks and balances – and were fully integrated into late eighteenth- and nineteenth-century English imaginations. Charles Burney articulated them in a letter of 1771, to Denis Diderot:

the poets who wrote for [Lully and Rameau] knew but little of true Lyric poetry – & crouded the Airs with so *many words* & *ideas,* & in *measures* so incommensurate, with respect to *Melody,* that no symmetry or connection could be preserved in the Musical phrases . . . our ears have been too long accustomed to [Modern Melody] to allow us to return back to the ancient Simplicity of Elementary Sounds, of only one note to one Syllable.[29]

Whether the imbalance is due to an excess of syllables per note or an instrumental voice that fails to adequately express its correlative musical idea, one trend in modern English music criticism was to eschew redundancy or irrelevancy (arguably an "industrial" attitude), and to privilege a kind of wholeness or purity of musical expression, where parts fit seamlessly together and yield a recognizable tonal idea. Both Dr. Burney's and *Punch*'s musical examples serve as metaphors for verbal patter, in which linguistic structure deviates from the expected – is redundant or alien – and "meaning" is thus compromised, the whole thrown off balance. In fact, we might reverse the metaphor and read patter as a kind of syncopation or variation against the standard (middle-class) verbal gesture, one which creates movement and heterogeneity but produces threatening sounds of disruption and difference. Yet unlike musical syncopation, which renders what the listener perceives as meaning and balance, patter's violation of standard semantics and syntax represents a violation of balance: the balance of form and meaning typical of "proper" speech; the social "balance" (really an imbalance) in conventional class power relations, which eccentric speech may subvert by emphasizing alternative verbal and social structures.

Opera criticism was a regular feature of Victorian periodicals, and reviewers usually invoked that properly motivated relationship between structure and meaning, text and music, dramatic vehicle and truth, as their primary critical criterion. For example, a critic hostile to Mercadante delivers a flippant and satirical condemnation of *Elena Uberti*: "A 'grand tragic opera' – all Italian except the words, and all music except the monotonous recitatives and the bad singing of the inferior performers"[30] is not what it pretends to be. This writer complains that the title "grand tragic opera" is deceptive: "To the learned in the works of the best operatic writers, a 'grand tragic opera' by Mercadante . . . must suggest the idea of a symphony of Beethoven played upon a German flute" – rendered, that is, by an unfit instrument. (This recalls the sort of failure, noted by *Punch*, in the aforementioned production of *Figaro*, and by Burney in the poetic texts set by Lully and Rameau.) The criticisms here have generally to do with the unsuitedness of Mercadante's light and mechanically pretty score to the depths of emotion suggested by the plot. This writer locates human truth in the story but finds the dramatic vehicle, music, deficient. ("The music is no[t] . . . in accordance with the story.") But the libretto also falls short of meaningful in this English translation:

Boemondo and *Ubaldo* open the second act, in order that *Elena* might come on to plead to them for her father's life. This they agree to spare if she will marry the

Count, and after some hesitation *Miss Uberti* consents. *Ubaldo* is in raptures, and raves as follows: –

> Ah, more warm – more wild each day,
> Shall my passion burn before thee;
> Fair as saint of poet's lay,
> How shall lover blest adore thee?
> &c. &c.

Trying, as one would suppose, to increase the contempt she already feels for him, by singing the most silly nonsense it would be possible for a man without brains to utter.[31]

Ubaldo "raves," he speaks "silly nonsense," earning the contempt of his beloved and, we are to presume, his audience. It is not merely that Ubaldo's poetry is bad, although no reader or listener of taste could endure a line like "Fair as saint of poet's lay." Far worse than its affectedness is the passage's failure to effectively represent the drama of the situation. Count Ubaldo is a sinister criminal; having abducted Elena so she cannot marry his best friend, Count Guido, he forces her to accept *him* by threatening to kill her father. When she complies, he kills her father anyway, and overwhelmed by the murder, and by her own impending rape, Elena stabs herself with Ubaldo's dagger. An opera heavy with melodramatic tragedy deserves a villain who disdains to utter pastoral verse; no doubt the critic's disappointment was enhanced by his participation in a melodramatic theatrical tradition of properly sinister villains, who just did not speak in this Count Ubaldo's silly manner. But he also voices a conventional demand for truth in language, verbal or musical. This opera's failure to generate truth in the interaction between music and libretto renders it worthy of harsh derision. While a "grand tragic opera" like *Elena Uberti* would not use patter (a comic trope) to thematize the debasement of its Ubaldos, *Punch*'s reviewer finds a patter-like disjunction of structure and content in its verbal and musical languages, and reacts with an anger which seems disproportionate to the opera's failures – its superficial score and poorly translated libretto. Really, he reacts to the sociomoral implications of a loose or deceptive medium of communication, a language whose forms and content are ill-matched and therefore confound conventional structures of meaning.

While a form of patter was used by *opera buffa* composers and invoked by opera critics, it was also articulated in the eighteenth-century English comic drama, in which we see a form of dialogue that is excessively wordy and fragmented, or else fraught with slang, used to represent socially or morally eccentric characters. (This patter is Dickensian patter, or the patter performed by Victorian actors like Charles Mathews, in embryo – not yet fully developed but bearing unmistakable elements of its later manifestation.) *The Old Maid* (1761), an invective against spinsterhood, illustrates both the promise and the potential difficulties in using patter as an instrument of sociolinguistic control.[32] In this cruel comedy, women speak less truthfully than men, and spinsters speak even less truthfully than wives. The story begins with a mistaken identity, and the subsequent blind courtship of Miss Harlow (the old maid), who is assumed by her "lover" – a man who knows "Miss Harlow" only by reputation – to be her young sister-in-law Mrs. Harlow, results eventually in the soured spinster's humiliation. The women's dialogue, especially Miss Harlow's, is breathlessly wordy and broken.

MRS. HARLOW. And to be at this work of sour grapes, till one is turned of three and forty –

MISS HARLOW. Three and forty, ma'am! – I desire, sister – I desire, ma'am – Three and Forty, ma'am –

MRS. HARLOW. Nay – Nay – Nay – don't be angry – don't blame me – blame your husband; he is your own brother, and he knows your age – he told me so.

MISS HARLOW. Oh! Ma'am, I feel your drift – but you need not give yourself those airs, ma'am – years, indeed! Three and forty, truly! – I'll assure you – upon my word – hah! Very fine – But I see plainly ma'am, what you are at – Mr. Clerimont, madam! – Mr. Clerimont, sister! That's what frets you – a young husband, ma'am – younger than your husband, ma'am – Mr. Clerimont, let me tell you, ma'am – .

The excessive use of dashes in Miss Harlow's speech fractures it, gives it a flighty, frivolous air; these are not the same dashes later used by Dickens to season Alfred Jingle's speech.[33] She takes forever to get to the point. Her repetition of unimportant words and phrases calls attention to the potential emptiness – and deceptiveness – of an old maid's speech; one suspects that the contemporary audience felt both contemptuous and amused at her hysterical patter. Of course, a sympathetic listener might identify in her dialogue the sort of anguish which robs speech of its order and efficacy, but that response is clearly not what this playwright intended. Pregnant or empty, Miss Harlow's repetitions suggest, as Dr. Bartolo's do in "La Vendetta," a kind of ineffectuality, an earnest against her taking serious or decisive action. She can but protest, after all – and that only to her sister-in-law. In contrast to Miss Harlow's flighty speech, her brother utters compact, almost epigrammatic lines. "An old maid in the house is the devil," he laments. "Her peevish humours, and her maiden temper are becoming unsupportable." The difference in their speech symbolizes the inequity of their positions: the old maid is superfluous, desired in no man's home yet denied her own home, while her brother is comfortably situated with home and wife and graced with social respectability. Furthermore, her aggressive sexuality, her uncontrolled appetite, threatens to undo the cultural constructions of masculinity and femininity, and her dialogue enacts these dangerous excesses of behavior by undoing the cultural construction of "proper" speech.

If Miss Harlow's dialogue represents an attempt by the author to contain her dangerous excesses, it also registers her subversive potential by failing, ultimately, to neutralize the threat of alternative idioms or eccentric social positions. Like any comic speaker, Miss Harlow draws attention to herself; her speech may confound common sense, but it is far more interesting than the male dialogue in the play. One knows it is meant to be scorned or dismissed and is nonetheless drawn to it, listens to it, laughs at it. Its magnetic power may have had something to do with the architecture of the theatre, which, in the eighteenth century, was enlarged and given a proscenium stage, requiring an exaggeration of tone and diction, the actress losing herself – and holding her audience – in the swell of her voice. Perhaps it had to do as well with the customs of performance: after Garrick, for example, the audience, which had once the privilege of a certain number of seats in the fray of performance, was denied its place on the stage. The resultant distance between audience and actors may have imposed a psychological as well as a physical barrier, which the voice had to penetrate if performers wished to captivate their public.

The Old Maid was popular enough to play several times at the Theatre Royal, Brighton, in 1803, where it was revived "for the first time in six years." At the time there was no dearth of plays to replace it if necessary. One must assume that something about the play touched its audiences, and that the disruptive possibilities in Miss Harlow's speech and position seemed to them effectively circumscribed by the play's conservative machinery. Certainly they approved of its explicitly derisive treatment of Miss Harlow, and understood her patter to be a reason for, as well as an instrument of, that derision. If she cannot harness her own speech to truth – and if her very name is so incongruous with the "truth" of her body as to deceive young Clerimont – then she clearly represents a threat of social disruption, of the sexual, the carnival, the lunatic; that threat was presumably deflected by the exaggerated ineffectuality of her speech, although the very existence of so many Harlow-like patterers on the English stage suggests that this general strategy of deflection failed.

Street patter appeared as frequently as its feminine counterpart on a stage increasingly interested in the criminal and the deviant, an interest confirmed by the immense popularity of pieces like Gay's eighteenth-century sensation *The Beggar's Opera*, and a century later, W. T. Moncrieff's *Tom and Jerry; or, Life in London*. Another play of this sort is Isaac Pocock's popular farce of 1810 *Hit or Miss*. Like *The Old Maid, Hit or Miss* demonstrates the simultaneously conservative and subversive powers of patter.[34] *Hit or Miss* is, typically, about the trials and ultimate union of two young lovers. The tortuous plot ends with Clara Stirling and Janus Jumble united, but not without first teasing the audience with the prospect of a match between Clara and a slang-talking attorney named Dick Cypher. Cypher is a gamester; he speaks the cant of the racetrack and the prize-ring, and smacks of a dangerous sort of vulgarity. His name is nicely evocative, for it suggests "zero" as well as linguistic obscurity, and Dick Cypher is both fond of slang and socially undesirable if not insignificant (his historical model was a young man of family taken to low pursuits). Anne Jackson Mathews, whose husband Charles originated the role of Cypher, wrote the following comments on *Hit or Miss:*

> a farce called "Hit or Miss" was produced from the pen of Mr. Pocock, with a mere outline (as it often happened) for Mr. Mathews to fill up. His character was of course the one that touched upon the peculiarities of the [Four-in-Hand] club, and he presented a faithful copy of its dress, using all the slang of that day, which I fear was too often employed, at the time, by those not "unto the manner born." It was, however, very amusing in itself, and the character of *Dick Cypher* was a faithful copy of a young man of good family (then in the law) who contrived to mix up this jargon with the most gentlemanlike manner and character.[35]

The "Four-in-Hand Club," to which Mrs. Mathews refers, had in fact been established, shortly before the publication of Pocock's satire, by a group of wealthy young men-about-town who exhibited their driving skills – and their affluence – by dashing about in expensive equipages. This feat was generally performed with a coachman seated in splendid livery to the left of the driver.[36] Mathews' pun on "manner/manor" emphasizes the social and linguistic ambiguities engendered by this practice: does she mean to say that the patter was unfortunately employed by those not born unto that degraded *manner* of speaking – i.e. educated men – or does she refer to the lower orders, those "not unto the *manor* born"?

Hit or Miss may be, as Mrs. Mathews suggests, a satire against one specific social club, but it also speaks to general concerns, among the privileged classes, about the insidious powers of subversive speech, and the confusion of identity that could result from the "master" language failing to maintain its proper relation to other languages. As a caricature of the Four-in-Handers, Dick Cypher plies a mild rebuke on the thoughtlessness and exuberance of wealthy young men, but as a figure of real social ambiguity – as one who might just as easily be a costermonger with genteel aspirations as a young gentleman with a penchant for slang and fast driving – he represents the possibility of working-class bodies, along with working-class speech patterns, infiltrating, infecting, the upper classes.

Cypher's first entrance, in act II, establishes his participation in the confusion of "manner" and "manor." Bursting into Jumble's office, he treats the clerk, Quill, to a vivid description of his drive hence, during which "If I hadn't turned the leader neatly over the old woman I should have dash'd neck and crop into the china shop." Quill quite reasonably assumes that Cypher is a coachman – he has, after all, been driving – and neither his language nor his clothes nor his potential violence indicate otherwise. Cypher's colorful, enthusiastic response to the clerk's error establishes a positive social and moral fact about him: he has been keeping the wrong company.

CYPHER. Discharged? and fellow servant? Why spoonies – sawnies – clods, have you the superlative ignorance and impudence to mistake Richard Cypher, Esq., attorney and solicitor, for a servant?

JERRY. To be sure. Didn't you sit cheek by jowl, and take all the trouble, while he [the real coachman] sat at his ease and chatter'd to you as if you were his groom?

CYPHER. What the devil then! – do you suppose I let my coachman drive me?

QUILL. If not, why did you hire him?

CYPHER. Hey! That's a poser, a proper setter – d–n me if I know, though I've studied the law.

I would define Cypher's speech as patter because it flouts the standard lexicon, calling attention to the parts of which it is composed rather than quietly reflecting "truth," as standard speech was supposed to do. Note its structural similarity to the patter – about a hat – in a popular 1860s patter song, "Cheap John:" "A tile – a castor – a skull case – a nutshell, or whatever you please to call it."[37]

One can imagine Mathews, who was celebrated for his prodigious verbal talents, rattling off Cypher's lines with beautiful fluency. Yet if Cypher's dialogue indicates that he is not a gentleman, it also proves he is not merely a working-class brute; the inclusion of phrases like "superlative ignorance and impudence" in his dialogue, and the "Esq." after his name, promise a certain level of education to the audience, who would have found little to amuse them in the tale of a boxer, or a farmer, with pretensions to the hand of a wealthy young lady like Clara Stirling. The fact that he owns a carriage and maintains idle servants places him at least ostensibly among the moneyed order. Better yet, he has lost his entire fortune to gaming, and lives as fashionably and irresponsibly under prodigious debt as a young lord. These recognizable habits, mingled with the eccentric conversation which marks his proximity to the lower classes, might have eased an educated audience's anxieties over Cypher, although they also confirm his social ambiguity. Like Miss Harlow, he has no explicit social place but teases from his margin, a constant threat of disruption or revolution. (Is he one of us or one of them? Born to the manor or to that manner? A person might well ask.) He may have been funny, but Dick Cypher represented an unhinging of fact and meaning. His verbal mode reflects the uneasy combination of the comic and the terrible in him; it sets him apart from mainstream society and at the same time allows him to steal the show. In the opinion of Drury Lane's house prompter, William Powell, it was the brilliance of Mathews' improvised patter which caused such a tremendous stir when the play opened: "The farce which I have sent was published last night & . . . is the most popular piece we have produced for many years . . . The speaking at the end of each verse of Cypher's song must depend upon the genius of [Mr. Mathews]."[38] Cypher's dialogue, especially when rendered by a comedic actor of Mathews' caliber, makes him the center of attention, the main attraction; it posits itself in irrevocable opposition to standard speech, refusing to be hushed, and forces the possibility of verbal difference as lively or interesting rather than unequivocally negative.

This subversive pulse must have made some playgoers unhappy. Rendered more subversive, more explicit, as it was in a revised version of *Hit or Miss,* Cypher's patter provoked strong objections. One reviewer found the speech of the new play's comic hero insulting and immoral.[39]

The musical farce of *Hit or Miss,* with "a new feature," was last night revived here [the English Opera House], when Mr. Mathews appeared in the character of Dick Cypher, which was originally performed by him. The new feature strikes us as being exceedingly objectionable; so much so, indeed, that we think it ought not to be tolerated by a genteel audience. When the piece was first produced . . . nothing of an offensive nature was contained either in [Dick Cypher's] song or in his dialogue, though both were foolish enough. But as the farce has been remodelled, the character abounds in that nauseous slang so liberally made use of by the persons who report the proceedings of the prize ring, and in defense of which neither humour nor originality can be pleaded. *Cypher* now appears before the audience an accomplished master of every species of blackguard amusement. As one of "the fancy" he is initiated into all the mysteries of boxing, dog-fighting, badger-baiting, *et omni genus ludorum,* and his conversation was, of course, of the same delicate nature as his acquirements. The first song is so descriptive of pugilistic contest, and of the language and manners of those by whom such infamous exhibitions are usually attended. We cannot conceive what pleasure . . . the audience could derive from the speaking part of this miserable song, which is, from beginning to end, a repetition of the low cant of professional bruisers and their admirers.[40]

The reviewer's palpable disgust at Cypher's "nauseous slang," a language unrelieved by any "proper" diction, reducing him to the level of "professional bruisers and their admirers," derives at least in part from his inability to locate semantic meaning in it, although one can be sure it was, in another sense, emphatically meaningful to him. These are words unlinked to middle-class experience – specific, in fact, to lower-class recreation – and they represent a sort of dark region, mysterious and menacing to "respectable" society. Cypher's patter shares with the street talk described later in the century by Mayhew not only enthusiastic transgressions of mainstream speech, but implications of working-class union against the upper classes. Here there appear to be few familiarities, no "Esquire"s or "superlative"s to reassure the speakers of standard English; the Four-in-Hand Club had long since disbanded, and new audiences would have discerned no traces of gentility in Cypher. Almost defensively, the reviewer takes care to circumscribe his own language within the bounds of decency: the sterilized "pugilistic contest," for example. He intends his correct and elegant words to foster meaning rather than obscure it – that is, to do what Dick Cypher's idiom refuses to do.

Tom Taylor's 1863 drama *The Ticket of Leave Man* suggests that the street patter Isaac Pocock found to be an effective dramatic tool was still, fifty years later, meaningful to the Victorian theatre public. The piece opens in a Vauxhall-inspired tea garden, the likes of which were frequented by increasingly lower-class visitors as the century progressed. The present visitors include James Dalton ("alias Downey, alias the Tiger") and Melter Moss, a couple of confidence men on the make, who speak very much the same idiom which Henry Mayhew describes in *London Labour and the London Poor,* or one like Dick Cypher's, for that matter, except rather oriented toward the thieving than the sporting subculture.

MOSS [*stirring and sipping his brandy and peppermint*]. Warm and comfortable. Tiger ought to be here before this. [*As he stirs, his eye falls upon the spoon, he takes it up; weighs it in his fingers.*] Uncommon neat article – might take in a good many people – plated, though, plated.

[*While* MOSS *is looking at the spoon,* DALTON *takes his seat at* MOSS's *table, unobserved by him.*]

DALTON. Not worth flimping, eh?

MOSS. [*right of table, starting, but not recognizing him*]. Eh, did you speak to me, sir?

DALTON. What? Don't twig me? Then it is a good get up.[41]

Here, as in *Hit or Miss,* street patter violates middle-class semantics, although the words in question – and there are relatively few of these – can be fairly easily deciphered. The characters immediately locate themselves within the street culture that audiences had learned to identifiy with their kind of speech, a world of tricks and trade. Tiger/Dalton/Downey's ability to shed identities like dirty clothes reflects a similar quality in his words, which obscure conventional meanings in strange packages. His most interesting – and frightening – feature is this linguistic and social suppleness.

Characters like the revised Dick Cypher and James Dalton existed as a kind of sore on their culture's tongue, irritating and persistently asserting their presence in the standard utterances which in fact guaranteed their existence even as they articulated a desire to do them in. The dominant language *needed* them, even as the high moral discourse needed its theatre. Mozart dramatized this paradox in his beloved *Don Giovanni*, an opera which anyone who was anyone, and quite a few who were no one at all, knew. (One can imagine its appeal for Victorians – it even inspired adaptations like Madam Vestris' *Don Giovanni in London.*) *Don Giovanni* orchestrates its vocal universe in such a way that patter is simultaneously marginalized and crucial to the opera's politico-aesthetic whole. The world, it seems to say, must have its Leporellos.

The opera establishes in its first scene that the servant's speech and music differ materially from those of his betters. In the emotional trio

"Non sperar, se non m'uccidi," Leporello sings patter to Don Giovanni's and Donna Anna's serious and highly charged parts (the latter two have emerged, struggling, from an attempted rape while Leporello has stood watch) a gesture which disarms the servant by reducing him to a comic foil. We can be sure from the start that Leporello poses no actual threat to anyone – although, as we know, the plebeian was eminently threatening in late eighteenth-century Europe – because his words are sung fast and staccato, inappropriate to the scene's heavy dramatic weight (he is, among other things, a bad reader of outward social signs); his vocal part represents an imbalance similar to that denounced by Victorian music critics, in which music and story or text are unsuited to each other, and hence fail to yield appropriate meaning.

While most operagoers do not, of course, analyze scenes like this while they watch them, rather abandoning themselves to the pleasure of the encounter, their experience will probably confirm this imbalance: Leporello exists as a droning peripheral distraction from the "real" action. (Thoughtful productions often stage the scene so that one must look aside to watch him at that moment.) But to watch him, to tune him in, is to realize just how important he is to the integrity of the scene, and to revel in his comic difference. Don Giovanni's brutality, Donna Anna's anguish, are registered against the banality of the comic, and we suddenly realize how the whole scene recreates the human condition in miniature. We dance, in life, with the atrocious and the mundane. Each may be profound to us, and each unreal; the one creates the other's meaning. But western culture favors the brutal over the mundane, and its aesthetic productions tend to denigrate the latter (note how even "realist" works from the nineteenth century tend to heighten the painful, the gruesome, the unlikely, the ravishing). In eighteenth- and nineteenth-century England, the existence of inappropriate lexicons and modes of delivery, like Leporello's patter, formed a topic of discussion which helped to preserve social divisions by constantly dramatizing them; compared with Don Giovanni's "serious" musical diction – a privilege which is only revoked when he enters the peasants' recreation and adopts their idiom – Leporello's voice sounds disadvantageously rustic, silly. However attractive he might appear – and however popular the singers who made *buffo* characters their specialty – there is always the power and beauty of his master's voice to put him in his place. In works like *Don Giovanni*, the Victorian public was provided with descriptions of the way "they" talk, the poor, the servile, the negligible, and was thus reminded of the basic, even biological, differences between the classes.

This was Henry Mayhew's project eighty years later in *London Labour and the London Poor,* his ambitious study of life on the London streets. *London Labour* offers one of the most explicit dramatizations of street patter in English cultural history. I use "dramatization" with the consciousness of Mayhew's connection to the stage (he wrote and co-wrote several plays, which were subsequently produced), and with the intention of emphasizing his reliance on theatrical conventions in even this most serious nonfiction prose work. *London Labour* muddles the difference between fiction or drama and "real life"; its "realizations" of life on the street, to borrow Martin Meisel's terminology, partake generously of dramatic and literary stereotypes.[42] That Mayhew was commissioned by the *Morning Chronicle* to undertake the study indicates a market of middle-class readers hungry for details of lower-class lives and habits; that he violates what twentieth-century readers might consider the integrity of genre in this profound work suggests that his readers were accustomed to reading plays in their novels, novels in their nonfiction prose, and so on, without necessarily recognizing the generic slippage as anything of the sort. And Mayhew was only one of many exploring these themes through a medium of shared representational structures. For example, Victorian interest in the moral implications of idleness or "inappropriate" forms of work ran high, and such situations were frequently explored, in nonfiction prose as well as novels, plays, poems, and visual arts. Meisel notes Victorian theatrical realizations of Hogarth's engravings, particularly of *Industry and Idleness,* such as "a new grand Melo-Dramatic and moral Ilustration [*sic*] founded on Hogarth's Apprentices."[43] These clearly grew out of the same soil that produced *London Labour,* with which they share a thematic interest as well as an organic, intergeneric structure, using and creating, both, the imaginary texts which focused Victorian eyes on the problem of the working class.

One of Mayhew's primary foci is on verbal manifestations of class difference; he devotes, in fact, a significant amount of attention to street patter and the people who spoke it. In a manner typical of middle-class representations of working-class life, Mayhew's long section on patterers obscures the patterers' own perspective on their work and social identities, constructing instead an elitist drama of life on the street. If Mayhew's audience did not share his knowledge of street vocabulary, they shared his general assumptions about the working class. His definition of patter resonates with those from the eighteenth century and earlier, cited previously, assuming that "moral" speech – that is, the Queen's English – is transparent and motivated, and that eccentric locutions are at least a sign of bad faith, if not outright immorality.

> We now come to a class of street-folk wholly distinct from any before treated of . . . The street-sellers of stationery, literature and the fine arts . . . constitute principally the class of street-orators known in these days as "patterers," and formerly termed "mountebanks" – people who, in the words of Strutt, strive to "help off their wares by pompous speeches, in which little regard is paid to truth or propriety." To patter, is a slang term, meaning to speak.[44]

Regardless of the neutral definition with which the cited passage concludes, "patter" means social and moral deviance, a *manipulation* of language which results in increased trade. The suggestion here is that pattering violates the "natural" rhythm of supply and demand by fabricating or coercing a state of demand for ballads or pictures or poems. It seems unlikely that the group engaged in pattering would identify themselves in such negative terms; they probably perceived themselves as savvy men and women of business rather than the unethical subversives they appear to be here. In other words, patter in Mayhew's version of the street serves middle-class values: it represents deviance from the "proper" methods of speech and commerce, and locates itself among working-class transients who violate capitalist industry, engaging in illegitimate forms of work.

If Mayhew did not himself choose the term *patter* to describe street selling, he surely appropriated it in his own interests. He intends it to hold the same conservative value he himself would have recognized in its manifestations in fiction and on the stage. But here it represents as well a genuine sociopolitical threat, a language – and social group – well beyond middle-class control. This patter was used by tacit or explicit agreement to dupe middle-class consumers; uncontained within a proscenium or narrative frame, it teased and insulted real people. Mayhew's implicit task was to domesticate these street-folk for the drawing-room, to preserve their color while disabling them, much as a naturalist would pin insects to cards so they could be examined dispassionately by ladies and gentlemen alike.[45] To do this he relied upon the theatrical tradition of invoking social differences in comic eccentric speech.

Mayhew separates his patterers into several classes: "running patterers," "standing patterers," "long song-sellers," the "publishers and authors of street-literature," and others. Each type of hawker engages in a different form of "oral puffery"; the running patterers, for example, line each side of the street and make a prodigious racket, enveloping the quieter streets of London with their sales pitches:

> a "mob" or "school" of running patterers (for both those words are used) . . . consists of two, three or four men. All these men state that the greater the noise they make, the better is the chance of sale, and better still when the noise is on each side of a street, for it appears as if the vendors were proclaiming such interesting or important intelligence, that they were vieing with one another who should supply the demand which must ensue.[46]

Since "it is not possible to ascertain with any certitude *what* the patterers are so anxious to sell, for only a few leading words are audible,"[47] the whole operation smacks of deception, of illegitimacy. Again Mayhew describes the cultivation of an artificial demand for a superfluous supply of goods – fictions, poems, romantic embellishments of historical events. That these folk seem to compete with one another over the privilege of "supply[ing] the demand which must ensue" suggests that they can somehow – with the help of their deceptive speech – alter the "natural" laws of supply and demand. Furthermore, the designation of two or three patterers as a "mob" affirms a conventional middle-class stereotype of the working class, and suggests that in a commercial as well as a verbal or a political sense, working-class solidarity represented dangerous subversion. With "mob," as with "patter," Mayhew permits some ambiguity of origin (by whom, and how, are the words used?) but in the context of his other remarks on street-hawking the word exerts a profoundly negative value, and would signal to Mayhew's middle-class readers that this group violates the set of acceptable social and political values that constitute the dominant culture. In this respect *London Labour and the London Poor* engages in an influential kind of social policing; Mayhew could expect that his audience shared assumptions about the sound and look of degradation or marginality. The composers and playwrights who introduced patter to the stage were partly responsible for this.

London Labour's chapter on patterers consists of lengthy segments of testimonial from the patterers themselves, laced with Mayhew's commentary. The piece presents itself as factual and accurate, a pure example of investigative journalism, but the expansiveness of his subjects' monologues indicates that Mayhew must have written their speeches himself, honoring, no doubt, the spirit but not the letter of their words. Without the help of a tape recorder, he could not have preserved such lengthy blocks of speech.[48] The logistical necessity of Mayhew's interventions forces us to question the truth of his representations. Particularly, we must consider what assumptions he made in reconstructing these characters for publication. Clearly, his reconstructions privilege conventional middle-class stereotypes about the working class. On the other hand, at a time when education came with a high price tag and strict social criteria (formal schooling became available to the working classes during the second half of the century), lower-class people did indeed sound "different," and the written English of those home-taught individuals who could write often reads like patter. For example, note the unchecked verbal flow of this subversive letter, penned in 1762 and addressed to James Bailey, a JP:

This his to asquaint you that We poor of Rosendale Rochdale Oldham Saddleworth Ashton have all mutaly and firmly agreed to Word and Covinent and oath to Fight and Stand by Each Other as long as Life doth last for We may as well all be hanged as starved to Death and to see ower Children weep for Bread and none to give Them nor no liklyness of ever mending wile You all take part with Brommal and Markits drops at all the principle Markits elcewhere but take this for a shure Maxon, That if you dont put those good Laws in Executions against all Those Canables or Men Slayers That have the Curse of God and all honest Men both by Gods Laws and Mens Laws so take Notice . . .[49]

Such examples of lower-class writing only further complicate the "truth" of Mayhew's dramatizations, which may not be completely accurate but fall well short of unjustified or libelous. In any case, they suggest some potential real-life sources for his patter.

Whether the real world of nineteenth-century London or contemporary popular entertainments (or a mixture of both) served as his model, Mayhew invests his characters with a fully evolved social idiom, one which accommodates personal and professional situations. The patterers' conversational speech is excessive and illogical; their professional speech rather tighter, and deceptive. Under the "Experience of a Running Patterer," Mayhew illustrates both types of discourse. Here, the patterer chats casually with Mayhew about his work:

"Well, sir," he said, "I think, take them altogether, things hasn't been so good this year as the year before. But the Pope, God bless him! he's the best friend I've had since Rush, but Rush licked his Holiness. You see, the Pope and Cardinal Wiseman is a one-sided affair; of course the Catholics won't buy anything against the Pope, but *all* religions could go for Rush. Our mob once thought of starting a cardinal's dress, and I thought of wearing a red hat myself.

I did wear a shovel hat when the Bishop of London was our racket; but I thought the hat began to feel too hot, so I shovelled it off. There was plenty of paper that would have suited to work with a cardinal's hat. There was one – "Cardinal Wiseman's Lament," – and it was giving his own words, like, and a red hat would have capped it . . . They shod me, sir. *Who's* they? Why, the Pope and Cardinal Wiseman. I call my clothes after them I earn money by to buy them with. My shoes I call Pope Pius; my trowsers and braces, Calcraft [the hangman "up for starving his mother"]; my waistcoat and shirt, Jael Denny [a murderess]; and my coat, Love Letters. A man must show a sense of gratitude in the best way he can. But I didn't start the cardinal's hat; I thought it might prove disagreeable to Sir Robert Peele's dress lodgers . . . There was very little doing," he continued, "for some time after I gave you an account before; hardly a slum worth a crust and a pipe of tobacco to us. A slum's a paper fake, – make a footnote of that, sir." (*London Labour*, vol. I, p. 224)

The patterer's excessive wordiness and lapses in logic (he moves from one subject to another without breaking stride, noting with innocent surprise his interlocutor's confusion) rather unexpectedly resemble dramatic renditions of spinster patter; Mrs. Nickleby's speech, for example, is merely an exaggerated version of the street hawker's conversational dialogue. And one could meet his verbal type in plays like Moncrieff's hugely popular *Tom and Jerry*, whose streetwise Bob Logic explains the meaning of "blunt" – like "slum," a bit of the slang patter:

Blunt, my dear boy, is – in short, what is it not? It's everything now o'days – to be able to flash the screens – sport the rhino – shew the needful – post the pony . . . [it] is to be at once good, great, handsome, accomplished, and every thing that's desirable – money, money, is your universal good, – only get into Tip Street, Jerry.[50]

It was not necessary for a middle-class audience to understand the slang in any literal sense; it sets up its own highly logical context, signifying, as patter so often does, in nonrepresentational ways. Like the dissipated Logic, Mayhew's patterer gets caught up in language, explaining how he names his clothes, how he almost adopted papal dress for pattering about papal scandal. He obscures his point (Mayhew has asked about the state of the trade) with a rush of words. Of course, from Mayhew's perspective, that *is* the point – the patterer's speech manifests the duplicity which characterizes one in his social position.

The running patterer also speaks a professional cant influenced, at least as it appears in Mayhew's text, by popular entertainments. Oddly enough, his "pitch" is more tightly knit than his conversation – indeed, comparatively almost poetic.

"Here's a bit of the patter, now:
'Let us look at William Calcraft,' says the eminent author, 'in his earliest days. He was born about the year 1801, of humble but industrious parents, at a little village in Essex. His infant ears often listened to the children belonging to the Sunday schools of his native place, singing the well known words of Watt's beautiful hymn,

When e'er I take my walks abroad,
How many poor I see, &c.

'But alas for the poor farmer's boy, he never had the opportunity of going to that school to be taught how to shun "the broad way leading to destruction." To seek a chance fortune he travelled up to London where his ignorance and forlorn condition shortly enabled that fell demon, which ever haunts the footsteps of the wretched, to mark him for her own.'
'Isn't that stunning, sir? Here it is in print for you. "Mark him for her own!" . . . You may believe me, and I can prove the fact – the author of that beautiful writing ain't in parliament! Think of the mental flame, sir! O, dear.'" (*London Labour*, vol. 1, p. 225)

The written script avoids breathless wordiness, slang, and breaches in logic, but it shares with the patterer's everyday speech that element of deception which offended middle-class Victorian values. Calcraft's biography is certainly forged; the street author who composed the piece could hardly have known what sort of strains entered the hangman's "infant ears." Furthermore, the story draws upon the melodramatic material which filled Victorian theatres and novels. ("Think of the mental flame, sir! O, dear!" sounds startlingly Dickensian; in fact, as John Rosenberg notes in the introduction to the Dover edition of *London Labour*, "reading Dickens persuades us that Mayhew is not merely a fine reporter but also a superb artist" – and vice versa.[51]) The tender young country-bred child, who, through poverty and ill-chance, embraces London and the devil, lived, along with the fallen woman and the aristocratic seducer, in the center of Victorian consciousness. The speech with which the hawker sells his Calcraft narrative, borrowed from the paper itself, is only more elegant in its inventiveness than the more typical low cant generally associated with social and moral degradation. Both the personal and professional versions of his dialogue hint at economic or political subversion ("the author of that beautiful writing ain't in parliament!") and demonstrate that disjunction of word and truth which suggested subversion or anarchy to the middle-class imagination.

Mayhew's remarks on street pattering define patter as an unethical form of speech that emphasizes verbal delivery over content. In street patter and in the other types of patter I have discussed, meaning generally resides in rhythm, tempo, word shapes, and patterns – in the style and method, that is, rather than in the semantics of speech. Mayhew posits the patterers' speech as inseparable from their identities – as indeed, they seem to encourage him to do – and he makes no effort to conceal his distaste for both their slang and their industry. His philanthropic interest in the poor was certainly genuine, but he could not observe complacently and without disgust what he perceived as a flagrant rejection of "truthful" speech and commerce, a violation of bourgeois values. Mayhew's genuine philanthropism, informed as it was by conventional dramas of lower-class culture, resembles that of his friend and colleague Dickens, whose very earnest public works helped to underscore social divisions, to confine the recipients of his largess within a controlled social space.[52] For his rendition of street vending, Dickens undoubtedly referred to Mayhew's work. Nearly fifteen years after the publication of *London Labour and the London Poor* in 1851, he created Silas Wegg, whose "stock in trade" includes a "few small lots of fruits and sweets . . . [and] a choice collection of halfpenny ballads."[53] Wegg is one of the few costermongers to appear in a Victorian novel. His produce is unhealthy ("It gave you the face-ache to look at his apples, the stomach-ache to look at his oranges, the tooth-ache to look at his nuts"), and his ballads almost all suffer from misquotation or improvisation ("and you needn't Mr. Venus be your black bottle, / For surely I'll be mine, / And we'll take a glass with a slice of lemon in it to which you're partial, / For auld lang syne" [*Our Mutual Friend*, p. 539]).[54] The comic possibility in this character is tempered by his sinister potential for disrupting the benign order of the Boffin household, his avarice and envy and innate violence. While Dickens uses comedy and punishment to deflect Wegg's destructive power, he never seems quite easy with his character. Himself a devotee of the popular music that decorates Wegg's stall, perhaps he understood that these songs – many of which bear the patterer's personal mark – infiltrated the English home, and more importantly, the English imagination.

Despite his affiliation with what Dickens clearly perceived, at its best, as "wholesome" popular entertainment, Wegg is decidedly unwholesome. As a forger of his own literacy, he persistently fails to anchor language to "true" meaning, so that his seriously rendered ballads read like parodies of their originals, and his readings of books like *The Fall of the Roman Empire* pervert historical "truth" (in "taking all the hard words

. . . [he] get[s] rather shaken by Hadrian and Trajan, and the Antonines; stumbling at Polybius," and so on [p. 103]). Interestingly, his idiom, in spite of its violations of truth and decorum, is not really a species of patter; like malapropism, or the verbose working-class male dialogue generally favored by Dickens, it restrains itself from excess. Wegg merely uses a more or less cockney dialect, occasionally substituting one word for another in a way that maintains the integrity of the correct word. While one might protest that his verbal improvisations indicate creativity rather than error, the narrative affiliates him with gross physical and mercenary appetite, placing sinister constructions on his reading, writing, and actions. (In contrast, Noddy Boffin, who is also illiterate, represents the humane and often playful imaginative side of life in Dickens' social vision.) And like Mayhew before him, Dickens conflates verbal eccentricity with commercial fraudulence: Wegg's ambition throughout most of the novel is to appropriate wealth by deception and extortion – a violation of middle-class industrial ideology, which assumes that fortunes are made by participating honestly in the capitalist process. Wegg's deviation from "normal" speech and labor would have been a familiar story to Dickens' audience; a middle-class reader familiar with *London Labour,* London theatre, or London street hawking, as many of them were, would naturally assume Wegg's dishonesty in language and in commerce.

As a purveyor of song, Silas Wegg is much more powerful than even he supposes himself to be; as I suggested above, Dickens' ruthless treatment of him indicates his profound anxiety over Wegg's possible social influence. Wegg is potentially powerful because he chooses – and often rewrites – the ballads he sells; assuming a substantial market (and while Wegg's market is small, Mayhew indicates that the real ballad sellers did a brisk business),[55] he would certainly have contributed to the formation of an industrial social consciousness.

The songs one could buy on the street from vendors like Wegg often came from the music hall, and included patter songs.[56] The patter song was a type of comic song typical of music-hall entertainment. It consists of sung portions interspersed with spoken dialogue. The spoken dialogue, or patter, is generally wordy and unwieldy, with lapses in grammar and logic that make it somewhat difficult to follow. The whole would be delivered by a costermonger or an aspirant clerk, a Dick Cypher type whose cockney or northern dialect contributed to the derisive effect of the patter. J. A. Hardwick, a prolific writer of comic music-hall songs, used patter effectively to indicate the class and professions – both low – of his *personae*. For example, the character (and singer) of "Cheap John" is a patterer of Mayhew's sort, a hawker who dazzles his credulous customer with a blast of words.[57]

> Cheap John's on his travels and calls upon you,
> Ladies and Gentlemen, stand round and view.
> Sacrifices alarming I am bent upon;
> Shut up your fly-traps and list to Cheap John.
> CHORUS
> Making a living, my own hook upon,
> Things away giving – original Cheap John.

PATTER

In the first place, my noble swells, here's a bran-new ventilating hat, to begin with. A tile – a castor – a skull case – a nutshell, or whatever you please to call it, my pretty little dears. Look at this 'ere hat – there's a brimmer! With a bit let in at the back – it will fit everybody. Just examine this hat, while I twirl it round on a stick. Ain't it a non-such, and a bit over? Don't *I* look well in it? [*puts it on his head*] And if such a plain fellow as me looks well in it, what would *your* handsome, intelligent, phizzes be? . . . You may sport a cap on your knowledge box, but if it's ever so good it don't get the respect a tile does . . . Now it's a common saying – "get out of that hat"; but I want somebody to get *into* this hat, and hand me over the "sugar" – alias the cash – for it. Now who'll stand three half-crowns for this hat? Every inch of it beaver, where there's not a foot of silk. To cut pattering about it, I'll come down at once to half a crown, and if I can't sell it at that, why, I'll give it away, like most of the world, to them as don't want it. Now, why is this hat like the French emperor? Because it's *Nappy*. Talking of him, why is the conclusion of the Italian war like an unsuccessful drama? Well, because it's a *bad peace* (piece). And why is Louis Napoleon like Iago in "Othello?" Why, because he's the *villain of the peace* (piece).

Chorus, etc.

Hardwick's song, like the previous examples of patter, depends upon the audience recognizing that its words are duplicitous, its intentions dubious. The feigned street hawker represents his hat as stylish and luxurious; in truth it is seedy and torn – hence "ventilating hat." His speech is peppered with professional jargon, estranging words to a middle-class audience; the extravagant wordplay ("a tile – a castor – a skull case – a nutshell . . . a brimmer") suggests his delight in the polyvalence of language. He is clearly a scoundrel, justifying Mayhew's insulting descriptions of his trade by trying to divert the potential "buyer" (in this case the audience) with equivocation and jokes – and in case there is any doubt about his intentions he explicitly identifies his verbal gesture as patter. Here is more evidence that the term *patter* and the verbal and commercial gestures associated with it were peculiarly meaningful to the Victorian public. Cheap John tells the crowd with a wink that he'll "cut pattering about it," and the crowd, whether or not they believe him, knows exactly what he means. *Now I'll play straight – the hat's worthless, and if I can't sell it to some dupe I'll give it away.* The fact of patter's prominence on the music-hall stage illustrates my point about its resistance to containment: it was a delectable comic device, better suited to center stage than the wings of society.

Late in the century, patter singers like Dan Leno were still drawing crowds. Leno's "The Swimming Master" plays on the same assumptions about patterers' deceptiveness as "Cheap Jack" and so many of the others did.

> When the water is wet and the air is dry
> A beautiful sight you may then espy,
> On the pier in the summer-time there am I
> Teaching the ladies to swim.
> Though frightened at first of the water they be,
> Their confidence soon will return, don't you see,
> When they have feasted their eyes upon me,
> And noticed my figure so trim.

PATTER

You didn't notice my figure when I first appeared – I came on you to suddenly. You weren't able to grasp me altogether, as it were – I'll go off and come on again. [*retires off and re-enters*] There! Now you can notice me properly. You see you've got a north-east view of me. It is really remarkable the effect I leave on people who see me for the first time.[58]

The song includes three choruses and three pattered interludes. Each of the latter opens with a suggestion of subterfuge. The swimming master himself confuses people – his body, wraith-like, appears only to disappear. It takes two entrances for the singer to establish himself as a corporeal presence; the audience cannot "grasp him" the first time. The second burst of patter undermines the singer's veracity: "I could tell you things you'd hardly believe – in fact, I could tell you things I don't believe myself." And finally, "You wouldn't believe how strong you get having so much to do with water."

Patter is the mundane tweaked to strangeness. Nowhere was there an idiom more familiar, and more strange, to the bourgeois imagination. It amused and unnerved. Patter songs monopolized much of the space afforded to nonlegitimate dramatic forms during the second half of the nineteenth century; they were written, published, and performed with unflagging enthusiasm. They were also indelibly impressed upon the century's collective imagination; people knew them and sang them.[59]

The image of people singing patter songs, humming operatic tunes, invoking dramatic dialogues, or relating comic stories, is an appropriate one with which to close this chapter on the sounds and shapes of verbal difference. I have primarily examined written texts here, although they are texts haunted by living idioms, intimately connected to songs and plays which shared – and staged – their assumptions about speech in all of its social forms. That patter was used so frequently in eighteenth- and nineteenth-century entertainments to indicate social eccentricity confirms that it was a resonant social gesture, deeply meaningful to a large segment of the English public. But if Victorian dramas of lower class or female life, of eccentricity, redundancy, and subversion, tell stories of dangerous street industrialists and crazy spinsters, they also tell stories about language – its rhythms and tones, and its surprising social power.

CHAPTER 5

Charles Mathews, Charles Dickens, and the comic female voice

In his biography of Dickens, Peter Ackroyd, with one finger always on the novelist's theatrical pulse, describes the first performance of "a new farce, by Mark Lemon and Dickens himself. . . entitled *Mr. Nightingale's Diary*." The piece allowed Dickens to undertake a variety of characters, "in manner, if not in method, so similar to the impersonations of Charles Mathews so many years before. In rapid succession and with equally rapid 'patter' Dickens took the shape of a deaf sexton and an invalid, as well as characters not a million miles removed from his own Sam Weller and Mrs. Gamp."[1] Ackroyd is not, of course, the first Dickens biographer to note the influence of Mr. Mathews on the blossoming novelist-performer; many, starting with Forster, have found that relationship worth mentioning. Mathews taught Dickens much about the comic voice, so much, in fact, that it seems reasonable to give him some credit for the Panckses and Boffins and Gamps for which the world is so much richer. It is almost certainly the case that Dickens borrowed freely from Mathews' *dramatis personae*, for I have found prototypes for some of his most famous characters in the comedian's famous one-man shows, "At Home With Charles Mathews." If he studied with particular attention a specific voice from the actor, it was the pattered voice: the "At Homes" included pattering women who appear to have served as structural models for Dickensian widows and spinsters, like Flora Finching, Mrs. Nickleby, Miss Knag, and Miss Flyte.

Understanding Mathews' peculiar interest for Dickens, locating his mark on the novelist's work, involves more than tracing a direct line of influence from one to the other, although this chapter's teleology pulls towards an ultimate reading of Mathews in Dickens. But our first concern is to isolate the vehicle of influence – that is, the voice – and explore its parts. We have seen many instances of a Victorian faith in the efficacy of voice, and of the ways in which patter and other idioms expressed or embodied social truths. The fact that the nineteenth century received dramatic and literary voices as authoritative, or authentic, raises certain questions about those voices themselves. What did they sound like? Whence did they derive their force? Because this chapter largely concerns itself with voices, those belonging to Mathews, Dickens, and others, both real and imaginary, it seems appropriate to begin by exploring the vocal organ, imagining it, hearing it, as the Victorians themselves did, and then moving on to Mathews, Dickens, and pattering women.

One might well begin by claiming that the voice is seated, not in the physical body, diaphragm, or mouth, but in the social body. We have seen how the nineteenth-century English social body is imagined in linguistic terms, with class specified primarily in speech patterns, and how the verbal differences attendant upon social marginality may, if unpoliced, violate as well as confirm discursive order. Dramatizations of verbal differences are, of course, an integral element, a "natural" component, of aesthetic works and popular entertainments. A play or a novel without heteroglot dialogue, an opera without heterogeneous recitative or song, is unthinkable, as are any dramas of social relationship which do not, at least implicitly, express discursive patterns in the play of voices. In the nineteenth century, these genres, even the word-bound novel, were vocal genres, as we have seen, and the significance of their characters' voices derived in part from contemporary popular assumptions about speech, and from the strains of dialogue which ran in Victorian heads – from the current imaginary texts, in other words, which mediated acts of novel reading and writing.

Knowing what those imagined voices sounded like, or even that they existed at all, is of course problematic, and involves at best a level of speculation. But we can start with the following piece of analysis, an obituary of Charles James Mathews, Victorian actor, playwright, and theatre manager:

> His utterance was clearness itself, and with his winning charm of manner was allied extraordinary incisiveness in speaking. Every word was finished, and every sentence was delivered with a crispness that brought its full meaning home at once to the mind of the listener . . . No one who knew him in the old days can forget his pleasant glibness, his extraordinary volubility and his most remarkable distinctness of articulation . . . Rapidity of delivery or pronunciation was, in his case, allied with a distinctness that has been very rarely equalled, and certainly not in our time.[2]

C. J. Mathews was the son – and pupil – of Charles Mathews. A light comedian in his own right, Mathews the younger wrote and played roles that depended for their humor on glib, rapid articulation, on the ability to patter at high speed. His obituary is not merely descriptive, although that is what it seems to intend. It also articulates a set of assumptions about comic voice; its writer knew, most likely before he ever saw Mathews on the stage, exactly what a light comedian's speech should sound like. That is because dramatic forms and voices were so seamlessly integrated into the urban popular imagination, that an average Londoner would know the difference, say, between light-comic and low-farcical idioms – even if he or she could not articulate that difference, or the dramatic–linguistic rules governing each mode of expression – as readily as most people today can distinguish operatic from pop tones. This knowledge of voice is more visceral than intellectual. One may make the latter discernment without ever having sat through *Aida*, because the distinction between operatic and popular singing is deeply inscribed in contemporary popular culture, played out in all of our popular venues: commercials, athletic events ("it ain't over till the fat lady sings"), books, dramas, etc. When the baseball announcer makes his fat-lady joke, the crowd sees and hears her, the stereotyped overweight diva.

Victorian popular stereotypes were disseminated as thoroughly and effectively as our "fat lady" is today, so that in a culture which *believed* in its voices, the light-comic, or the sentimental–heroic idiom, with or without the label, was a "real" idiom, one invested with social as well as dramatic significance, a voice heard on and off the stage. Thus, when C. J. Mathews' obituary writer sounds the deceased actor's specific comic voice, he speaks to an audience who has already heard it sung. From the description it sounds clear and quick, glib and smooth and perhaps middle-registered – not too distinguished-deep. The fact that he knows it with such confidence suggests that the light comedian's idiom was common knowledge, circulating with other Victorian stereotypes and myths. People would even have "heard" it in books, staged and recited in their own minds as they read written dialogue.

Ostensibly, the obituary notice praises Mathews' comic voice because that voice worked precisely as his professional roles required it to. His vocal dimensions perfectly matched the current model of "light comedian." But apart from the technical merit indicated here, one senses that Mathews had a certain theatrical something – charm? comic timing? magnetism? – which touched his audiences, and which was understood to manifest itself in his voice. Most studies of nineteenth-century theatre, however, stress the semiotic dominance of the body in performance. Joseph Donohue historicizes the rise of the theatrical body, pinpointing its ascension in "the years following the transformation, in the 1790s, of Covent Garden and Drury Lane into unprecedentedly large structures [which promoted] an increasing tendency to forsake the demands of speech for the pleasures of the eye."[3] While Donohue's implication that the pre-1790s theatre did not stage "pleasures for the eye" is troubling, his assumptions about the effect of increased theatrical space on performance would seem logical, and indeed, are consonant with the general scholarly opinion. But nineteenth-century theatre reviews indicate a different reality: that dramatic experience was thought to be seated primarily in the voice, which was perceived as more important than other theatrical signifiers, like physical gesture and costume. This is not to suggest that physical gesture was unimportant; on the contrary, bodily movements and postures were meaningful elements of the current theatrical "language." But the voice dominated. Consider, for example, John Coleman's review of Fanny Kemble in J. S. Knowles' *The Hunchback*:

> Tortured, despairing, maddened, she sprang to her feet erect and terrible. With fiery eyes and dilated form she turned at bay, even as a wounded hind might turn upon the hunter's spear, then with quivering lips she commenced the famous speech, extending over some thirty lines. As it proceeded her voice gained strength, changing from the flute to the bell – from the bell to the clarion. Then upon a rising *sostenuto* of concentrated agony and defiance, she smote and stabbed Walter with that awful "Do it! Nor leave the task to me!" Even as the last word left her lips, she strode down to the right-hand corner, returned to the centre, and then came to anchor, her right hand clutched on the back of the great oaken chair, her left thrown out towards Walter, her blazing eyes fixed on him in an attitude of denunciation and defiance. Then it was, and not till then, that the breathless and enthralled auditors rose in such an outburst of wild enthusiasm as I have never heard equalled before or since.[4]

The review starts with the actress's extravagant physical gesture, which, though powerful itself, serves mainly as a prologue to her "famous speech." The writer's description increases in intensity with Kemble's escalating verbal performance, with her rising *crescendo* and *sostenuto* of voice. Her voice is supremely flexible, capable of music and murder: trilling, tolling, smiting, and stabbing. Her audience finds *its* voice only after Kemble's has ceased, subsided back into physical gesture. While the dramatic descriptions of Kemble's movements confirm their importance, the power of her voice was clearly perceived as the greater, perhaps because of its mysterious hold on the imagination. Actions may almost always be described literally, while vocal gesture – by this I refer not to representational language, but the voice which clothes it – lingers in the anteroom to realism, the semiotic to the symbolic, to borrow the language of Julia Kristeva's feminist psychoanalysis, which posits a psychological and also poetic plane of "maternal" rhythms and tonality, a level of signification beneath discursive language which in certain types of literary expression surges up into linear discourse, disruptive, resonant bursts of the carnivalesque.[5] Kemble's voice seems to have functioned in the same way, for Coleman, as Kristeva's semiotic "voice" does in entering the realm of the symbolic: it thrills the standard discourse with alternative possibilities, with hints of nonrepresentational signification, with its proximity to madness. It is both haunted and familiar, and in that paradox lies perhaps its greatest power.

George Henry Lewes found similar resonances in the actress Rachel's voice, her greatest dramatic asset:

> Hermione, in "Andromaque," was also another very fine part of hers, especially in the two great scenes with Pyrrhus. In the first her withering sarcasm, calm, polished, implacable, was beyond description; in the second she displayed her manifold resources in expressing rage, scorn, grief and defiance. In her eyes charged with lightning, in her thin, convulsive frame, in the spasms of her voice, changing from melodious clearness to a hoarseness that made us shudder, the demoniac element was felt . . . In describing how she will avenge the insult to her beauty by slaying Pyrrhus . . . her wail was so piercing and so musical that the whole audience rose in a transport to applaud her.[6]

The voice receives by far the most attention from Lewes, a vehicle of indescribable but potent powers; gifted, siren-like, with the ability to inspire physical responses from an audience. One of the author's stranger essays, on "Shakespeare as Actor and Critic," performs an inverse reading, searching the bard's voice and finding it barren, locating no evidence to support the possibility that he could act. The lack of reference to him in any of his acting roles confirms, argues Lewes, a basic weakness of his vocal organ.

> I dare say he declaimed finely, as far as rhythmic cadence and a nice accentuation went. But his non-success implies that his voice was intractable, or limited in its range. Without a sympathetic voice, no declamation can be effective. The tones which stir us need not be musical, need not be pleasant, even, but they must have a penetrating, vibrating quality. Had Shakespeare possessed such a voice he would have been famous as an actor. Without it all his other gifts were as nothing on the stage.[7]

The idea of a sympathetic voice is deeply suggestive, with "sympathetic" signifying in its conventional sense, an identification with character, but also in a more archaic – and musical – sense, as in the resonant bodily thrill seen in sympathetic vibration, a physical phenomenon found in certain Baroque stringed instruments like the viola d'amore. There, a peripheral string vibrates in sympathy with a primary string sounding a particular pitch, lending another song, somewhat eerie but very rich, to the musical whole and making it fuller. A sympathetic voice, then, would sing with the passions established by a dramatic situation, perfectly flexible and responsive. Shakespeare indeed possessed a sympathetic pen, but his voice, argues Lewes, failed him.

Dickens, apparently, was better equipped. A description of his public reading skills suggests that his voice, like Kemble's, conveyed a sense of mystery and musicality, but it also notes with approbation that the performer maintained control over an instrument which could, if unmoderated, flow into excess, extravagance, lunacy.

> The reports of Mr. Dickens's success in the provinces as a reader, which at the time seemed exaggerated, scarcely did justice to his peculiar power; his oral interpretation . . . being admirable. In the first place, Mr. Dickens's voice, naturally powerful and expressive, and specially rich in its lower tones, is completely under his control, and he modulates it with . . . perfect ease.[8]

It seems that Dickens' performance was safer, if not more successful, than Kemble's, with the performer policing his dramatic delivery much as he scored and barred his prompt copies to contain the theatrical impulses beneath their surface. But despite his self-control, Dickens' voice charmed, carried its auditors away on its dramatic swells and valleys, affecting at the most basic level of reception. The fact of a dramatic attraction like Dickens' or Kemble's is difficult to explain: today, one might try to describe what "it is" about a Derek Jacobi, or on the operatic stage, a Thomas Allen, and resort to comments about resonance of tone, integrity, sensitivity, or intelligence without ever touching the more essential quality. The same must be said for Victorian accounts of actors and actresses who "had it" (whatever "it" may have been), including both of the Charles Mathews, elder and younger. They dance, these commentaries, around some inexpressible quality – a quality of voice, and, to a lesser extent, of gesture or bodily presence.

I expect to be dancing around the question of dramatic voice myself. Part of the problem is that dramatic scholarship has traditionally bound itself to the written word, the literary play. Perhaps the relative neglect of the theatrical personalities (apart from the Garricks, Kembles, and

Macreadys), who kept theatre alive in the eighteenth and nineteenth centuries, has something to do with the ephemeral nature of the majority of their roles: most of the popular English plays from those periods have not weathered well. It seems also that literary criticism has failed to hear the music in these people's voices, the rhythms and cadences of dramatic speech which open it to a form of musicoliterary inquiry. But whatever the reason for their suppression, I seek to re-sound them here, because they lay at the center of English popular culture, their voices – or the types of voice they dramatized – articulating the boundaries that circumscribed the Victorian imagination, playing all of the possible parts in the sociopolitical drama that was London in the 1800s. Certainly the actors and idioms of the nineteenth-century London theatre, like Mathews and his pattering females, populate Dickens' novels, playing his characters, singing his dialogue.

Several critics have addressed the uses of eccentric voices, like dialect and idiolect, in Dickens,[9] but little has been written on other forms of verbal stereotyping in his work, including patter,[10] and no study really addresses these larger questions about theatrical or dramatic voice, which seem germane to a study of Dickensian speech. Peter Ackroyd has perhaps come closest to approaching the theatrical, in the sense in which I have defined the term, in Dickens. He notes of the early writings, that

> he is able to bring the light and exaggeration and animation of the theatre to the streets of London . . . he can, as it were, jump on the stage and do all the voices . . . But what his contemporaries heard *were* the voices, and the prevailing admiration for Dickens' early work came from what was described as the "vivid" and "graphic" way in which he had described the speech of London.[11]

Ackroyd's comments implicitly collapse the lines between staged and social speech, suggesting that, even if Dickens turned to theatrical idioms for his journalistic renditions of London life, his readers heard those idioms as authentic rather than dramatic. We have always to contend with this uncertainty of linguistic origins in Dickens, and in Victorian popular culture as a whole. It is difficult, if not impossible, to separate the theatrical from the social strands in what is left of Victorian voices.

Dickens himself acknowledged this uncertainty, writing of Flora Finching, the garrulous widow in *Little Dorrit*, "It is a wonderful gratification to find that everyone knows her."[12] His comment also, intentionally or not, bows to the scores of nameless hacks and the handful of recognizable figures whose works, previous to *Little Dorrit*, had already published a set of verbal eccentricities which included patter. Dickens attributes the general acquaintance with Flora Finching to an actual, collective experience – "We have all had our Floras"[13] – but his statement, irresponsible and exaggerated (is it possible that everyone knew someone just like her?) really suggests that we have had our Floras on the stage, we have found them, florid and loose-tongued, between the papers that bind periodicals, the stiff vellum covers of books. If we meet them on the street, we only recognize them because we have seen them, displayed, many times previously, as Macready had when he read *Nickleby* and remembered all of the characters as if they were old acquaintances: "a gallery of faces familiar to us as our own."[14] How else does one explain the universality of the experience?

We would probably call this a confusion of theatrical characters and real people, although clearly Macready and Dickens would not. Explicable or not, it was a "confusion" which occurred frequently, despite the fact that recent scholarship has largely either neglected or dismissed it. This may be due to our own narrative orientation: contemporary western culture is hardly theatrical in the Victorian way. Of the handful of critics who have explored the commerce between fiction and theatre in the nineteenth century, most have understood the theatrical more or less novelistically, reading schematically, and with a certain linearity, the relations between the genres, or telling the stories of Victorian novelists' theatrical adventures. These approaches are of course valuable, but they tacitly accept the premise that the nineteenth century was the age of the novel, and indeed owe their own structures to the post-theatrical – filmic or television-based – discourse in which we live now.

Joseph Litvak's work on theatricality and the nineteenth-century English novel invokes the Foucauldian paradigm of privatization and surveillance to suggest a sort of inverse theatricality, in which "[the novels'] implication in a widespread social network of vigilance and visibility – of looking and of being looked at – renders them inherently, if covertly, theatrical."[15] Litvak's argument, while suggestive, is somewhat diminished by its insistence on the covert, the interior, nature of literary theatricality in the nineteenth-century novel; theatricality, as we have seen, was worn externally, a kind of semiotic clothing, publicized and shared even when it was perceived as something shameful, something to be hidden. His argument about the anxieties in these texts' treatments of the theatre is in itself a reasonable reading of theatricality in the novel –

Victorian novels do indeed express ambivalence about the theatre – but he tends to ignore, or to read in a literary, or a historico-narrative manner following Foucault, the latent theatricality in the novels themselves, their hopeless entanglement in the culture of the stage. The paradox of a simultaneous assumption and rejection of theatricality in the nineteenth century is in itself worthy of study;[16] it poses a strange doubleness which we must not "correct" into a unitary or totalizing discourse. Theatre was both interesting and threatening to the middle class (Litvak points to a rhetoric of infection in accounts of the stage), who fetishized actresses and singers, attracted to and repulsed by the eroticism that seemed inseparable from acting.[17] Whether or not they explicitly endorsed it – and a large proportion of upright middle-class Londoners did not – theatre, as we have seen, shaped their imaginations, providing in large part their stereotypes, their bigotries, their cultural mythologies. The theatricality in Victorian literature was extravagantly public rather than privatized and voyeuristic, despite the novel's frequent privatization of performance, because theatre was a shared experience, "the chief medium of popular entertainment and popular understanding,"[18] and strongly formative of public opinion. It was also a maker of subjects and objects, heavily implicated in contemporary notions of self and other.

Litvak constructs a theatricality based on Foucault's "panopticism," a model of the theatre drawn on patently nontheatrical plans, a theatre of penal discipline which shares a vague architectural anatomy with the proscenium stage, and depends upon a situation of surveillance and display which imitates the positions and roles of actors and audience. In one sense, then, the theatricality he describes is a metaphor for other, larger and more sinister things, for the relentless performance of discipline. My intention, when I use "theatricality," is also to suggest strategies of containment or "discipline" but also to demonstrate the outwardness of the disciplinary apparatus, which was integrated into theatre itself, and to invoke as well strains of the actual Victorian theatre in Victorian life. After all, it was the existence of these strains of the carnival in nineteenth-century London which gave Dickens' comic characters their cultural significance, and which drew the novelist so insistently toward the bright, gaslit theatre, one of his homes. It also led him to adopt Mathews' comic voice as a model of certain social idioms.

A theatrical culture, a theatrical literature, is populated by dialogic voices, often, as in Dickens, borrowed from the stage; also, as in Dickens, lent to the stage. Theatricality must be heard and seen rather than read or logically deduced: it is not, ultimately, reducible to a discursive paradigm. Mikhail Bakhtin has described a state of polyglossic flux which may be seen as theatricality; he speaks, for example, of "centrifugal" forces of decentralization, which work "alongside verbal-ideological centralization and unification" to form a body of social utterances, standard and eccentric, competing to be heard.[19] Always, in Bakhtin, the voice is implicated in the making of discourse, and in the constructions of "normalcy" and "deviance." One should, for obvious reasons, affiliate the vocal with the theatrical, and the theatrical with the carnival; I would add to those associations a linking of the carnival, with its disruptive voices and rhythms, with music. Think of Robert Schumann's *Carnaval,* which turns the fair into music, and Robert Browning's *Fifine at the Fair,* which confirms Schumann's impulse.[20] I want to invoke all of these forces – voice, music, carnival – in the concept of theatricality; to speak of the theatrical in Dickens or in Victorian culture is to speak of these, as well as the actual people who inhabited the voices found in books, on streets and stages.

Assuming Victorian theatricality to be a live, and lively, sort of condition, vocal, physical, and fluid, presents some baffling problems to the scholar bent on recovering that theatricality. For example, how does one unstop the long-dead theatre of voices, so that we may hear them as Dickens did, and likewise so we may "hear" Dickens' novels as Victorian readers heard them? One way is to read Dickens aloud, which I suggested we do at the outset of this study; we may, with practice, learn the inflections and cadences of his typical idioms. Victorian theatrical and literary reviews may also tune our ears to this music. Of course, certain aspects of the voices themselves died with the last people to attend a Victorian staged entertainment; as we shall see when we come to Charles Mathews, there is something lost – something to do with audience and gesture as well as voice – which textual transcription cannot recover. Lost also are the authentic social, political, and comic contexts of these characters and acts, contexts which we cannot adequately recover through scholarship. (This problem is more or less endemic to theatrical entertainment; the same contextual loss prevents me from appreciating the World War II era comedians who continue to touch my parents.) But we can, I think, recover the dramatic spirit, the peculiar sensibility, which belonged to Victorian readers. This would involve attempting to read with the ears and the eyes, rather than merely the brain: to notice vocal patterns and rhythms and tones; to visualize the physical gesture Dickens so brilliantly conveys.

Bakhtin manages this neatly, although in a narrativistic, slightly tone-deaf sort of way; his "theatrical" reading of *Little Dorrit* attends to the function of spoken language in written text, the jostlings and clashes, the politics, one might say, of different modes of expression vying for place in a single narrative episode. He describes a layering of heterogeneous voices in narrative, a "comic-parodic re-processing of almost all the levels of literary language, both conversational and written, that were current at the time."[21] Bakhtin locates the voices of eccentricity, professionalism, rhetorical genre in shifting relation to the "common language," the "neutral" or "normal" idiom used by Dickens to represent consensus and regularity. His analysis of speech in *Little Dorrit* examines the often seamless changes in voice which indicate changes in the author's relation to his narrative, a relationship constantly in flux. Ceremonial language in a description of the still-ascendent Mr. Merdle; high epic style and chorus in a description of Merdle's dinner party – these subtle changes in tone and style serve satiric purposes. He is most interested in the structural machinery in such vocal play, in the ways in which, for example,

> an utterance [may belong], by its grammatical (syntactic) and compositional markers, to a single speaker, but which actually contains mixed within it two utterances, two speech manners, two styles, two "languages," two semantic and axiological belief systems . . . the division of voices and languages takes place within the limits of a single syntactic whole, often within the limits of a single sentence.[22]

There are limits to Bakhtin's reading of *Little Dorrit*, particularly in his subordination of cultural–political implications to structural concerns. He presents the heteroglossia in Dickens as fact, a matter of linguistic interest, something that just occurs, and in doing so he begs some important questions. For example, why does Dickens employ alternative verbal modes to disrupt the flow of "common" speech, sculpting a sort of narrative frieze, synchronic and diachronic, which generates meaning in the raising of certain figures to aural–spectacular emphasis? And what do the "raised figures," the prominently eccentric voices, mean; how are we to read them?

We might begin to answer these questions by reading Forster, who notes Dickens' own entrance into the thriving dramatic culture of his day, his introduction, at "the very tenderest age," to the stage:

> he was at any rate old enough to recollect how . . . his young heart leapt with terror as the wicked King Richard, struggling for life against the virtuous

Richmond, backed up and bumped against the box in which he was; and subsequent visits to the same sanctuary . . . revealed to him many wondrous secrets, "of which not the least terrific were, that the witches in *Macbeth* bore an awful resemblance to the Thanes and other proper inhabitants of Scotland."[23]

While the reaction to *Macbeth* may represent the workings of a heated childish imagination, it also demonstrates an intimacy between theatre and social stereotyping. Shakespeare's intention was not to affiliate the entire Scottish people with grotesque witches, but the adult Dickens nevertheless renders this association in reconstructing his immature reaction to the play. He knows the reaction is immature, he treats it with an air of fond condescension, but even so he propagates it. The memory suggests that, from a very early age, his consciousness was shaped by the representations of social life he saw on the stage; his literary construction of it suggests an adult sensibility prone to similar influence.

Dickens frequently mixed his personal, social, and theatrical observations in this way, producing a creative genre which rarely confines itself to either the fictive or the "real," and which actively participated in disseminating and codifying conservative social values like linguistic standardization. Ackroyd has also described Dickens' work in these terms, remarking on the thick layering of sources beneath the fiction – for example, his comments in a speech given at the first annual banquet of the Metropolitan Sanitary Association, which state, in short, one of the central tenets of *Bleak House*, the novel Dickens was about to commence: "certain it is, that the air from Gin Lane will be carried when the wind is Easterly, into Mayfair, and that if you once have a vigorous pestilence raging furiously in Saint Giles's, no mortal list of Lady Patronesses can keep it out of Almack's."[24] It was less a propensity on Dickens' part to look for appropriate fictional images in real life, and vice versa, than the general "confusion," in Victorian culture, of the theatrical or fictive and the "real," which allowed him to draw indiscriminately – because discrimination was unnecessary – on both. This possibility seems peculiarly Victorian: in what other historical moment could one see a "realist" novel performed on the stage before consuming the novel's last chapters – indeed, before the novel itself was fully written? And where, other than Victorian London, could one have "already seen" the characters in a novel walking along the street? This, of course, is a simplistic line of reasoning, and ultimately vulnerable, but it does, I think, express the inadequacy of concepts like "real" and "fictive" for describing the Victorian experience.

The *Sketches by Boz* (1836), a collection of hybrid fictional–journalistic forays into the heart and mind of industrial London, reflect this epistemological confusion. The sketches often collapse the boundaries between "real life" and the stage, using theatrical metaphors to describe ordinary people going about their daily business, as well as illustrating the importance of popular entertainments to middle- and lower-class family life in London. "Seven Dials," a vivid description of one of London's most degraded – and interesting – quarters, a neighborhood affiliated with the street life and low entertainment later described by Mayhew in *London Labour and the London Poor,* demonstrates how fiction or myth may enter the realm of the actual. The sketch purports to offer an honest view of the neighborhood, and indeed, there is no reason to doubt that "Boz" saw what he described, a "region of song and poetry – first effusions, and last dying speeches: hallowed by the names of Catnac and of Pitts – names that will entwine themselves with costermongers, and barrel-organs, when penny-magazines shall have superseded penny yard of song."[25] The Seven Dials was a place of legendary bad character, the infamous subject of rumor, story, even drama, and this problematizes contemporary accounts of the area. In this very sketch, in fact, sits some of the same "entwining" of truth and fiction that Dickens represents as a quasi-negative characteristic of the Seven Dials itself. Of course, to some extent he celebrated such communion, and freely admitted to shoring up "fact" with fiction: that, as he told his friend Forster, is one of the novelist's jobs. "It does not seem to me to be enough to say of any description that it is the exact truth. The exact truth must be there; but the merit or art in the narrator, is in the manner of stating the truth. As to which thing in literature, it always seems to me there is a world to be done." But despite this open concession to the demands of creativity, one frequently finds quiet suppressions of his "narrator's art" in Dickens' texts, implicit claims to factuality.

A moment ago I suggested that "Seven Dials" illustrates that suppression of fictional origin which turns "art" into "life." One of the events its narrator records witnessing, for example, looks decidedly familiar:

> On one side, a little crowd has collected round a couple of ladies, who . . . have at length differed on some point of domestic arrangement, and are on the eve of settling the quarrel by an appeal to blows, greatly to the interest of other ladies who live in the same house . . .
>
> "Vy don't you pitch into her, Sarah?" exclaims one half-dressed matron, by way of encouragement. "Vy don't you? if *my* 'usband had treated her with a drain last night, unbeknown to me, I'd tear her precious eyes out – a wixen!"

"What's the matter, ma'am?" inquires another old woman, who has just bustled up to the spot.

"Matter!" replies the first speaker . . . "Here's poor dear Mrs. Sulliwin . . . can't go out a charing for one afternoon, but what hussies must be a comin', and 'ticing avay her oun' 'usband . . . "

"What do you mean by hussies?" interrupts a champion of the other party, who has evinced a strong inclination throughout to get up a branch fight on her own account . . .

"Niver mind," replies the opposition, expressively, "niver mind; *you* go home, and, ven you're quite sober, mend your stockings."

This somewhat personal allusion . . . rouses her utmost ire, and she accordingly complies with the urgent request of the bystanders to "pitch in," with considerable alacrity. The scuffle became general, and terminates, in minor play-bill phraseology, with "arrival of the policemen, interior of the station-house, and impressive *dénouement*."[26]

This passage is quite forthright about its "theatricality," using "minor play-bill phraseology" to achieve comic effect, but it is silent on the matter of its debt to *Tom Jones*, whose mock-heroic churchyard battle pits a "half-dressed" Molly Seagrim against other women and men of her class, who resent her sexual and material commerce with Tom.[27] The "Seven Dials" sketch invokes Fielding's action and even his dialogue, although it stops short of his epic pretense. In Fielding's version,

so roared forth the Somerset mob an hallalloo . . . some were inspired by rage, others alarmed by fear . . . but chiefly Envy, the sister of Satan and his constant companion, rushed among the crowd and blew up the fury of the women; who no sooner came up to Molly than they pelted her with dirt and rubbish.

Molly, having endeavoured in vain to make a handsome retreat, faced about; and laying hold of ragged Bess . . . she at one blow felled her to the ground."[28]

While Fielding narrates Molly's battle scene without dialogue, the rural dialect used throughout *Tom Jones* matches that employed by Dickens above, especially the substitution of "v" for "w" or "f," and vice versa, which constitutes a stock dialectical gesture for both Fielding and Dickens (a rustic's comment on the battle – "an't please your honour, here hath been a vight" – demonstrates this usage in *Tom Jones*). Both episodes end with invocations of male authority – in the forms of Tom and the police – as well as the restoration of order in highly dramatic dénouements. One interesting difference should be noted: the "mob" in "Seven Dials" is exclusively female, whereas in Fielding it consists of working-class people of both sexes led by a few domineering women. This shift from the working class to a female "class" as locus of socioverbal deviance occurs as well on a larger scale in the nineteenth century, as eccentric idioms like patter and gossip were increasingly associated with women. Dickens participated in the working of this change: his patter-speakers are nearly always female.[29] But apart from its political revision of Fielding's scene, Dickens' reference to *Tom Jones* is significant in that it poses as original, as a spontaneous experience in London, 1836. It is highly probable that the young journalist witnessed the event – he could have, at any rate, and his readers would certainly have thought it credible, just as contemporary readers of the *New Yorker* will assume that its writers traverse the city culling material for the "Talk of the Town" feature, the heir to *Sketches by Boz* – but his presentation of it incorporates fictional elements, cultural stereotypes, and dramatic gestures. My point here is to illustrate the ease and inevitability with which fictitious or dramatic episodes may enter the field of "actual" experience, and contribute to the formation of collective social opinions. "Seven Dials" is heavily textual: one peels back layers of fiction to find a dubious truth at the center. This illustrates how the very discourses of normalcy, which are themselves part and parcel of the distribution of power, the promotion of order, are woven in complex patterns into the artifacts of popular culture, sometimes producing, sometimes being born to, a body of creative texts.

Dickens' literary sources, like *Tom Jones*, may be fairly explicit, but his theatrical influences often elude positive identification and systematic study, simply because their remains (where there are any) are skeletal, bits of articulated plot and dialogue stripped of the flesh of gesture, inflection, audience. Some of the popular entertainments of the early nineteenth century have been granted scholarly attention; fortunately for Dickens scholars these include the phenomenally successful "At Home with Charles Mathews" productions. According to Forster, Dickens was fascinated by Mathews, a famous comic actor whose one-man shows captured public attention with their caricatures and impersonations.[30] "[Dickens] went to theatres almost every night for a long time; studied and practiced himself in parts; was so much attracted by the 'At Homes' of the elder Mathews, that he resolved to make his first plunge in a similar direction; and finally wrote off to make offer of himself at Covent Garden."[31] Ackroyd cites sources which attest to Dickens' debt to the actor,[32] and Robert Golding has pointed out that one of Mathews' characters, Mrs. Neverend, may have served as a structural model for the novelist's voluble female characters. I have located others of equal influence.[33]

We know a fair amount about the structure and content of Mathews' "At Homes," thanks in large part to Richard Klepac's monograph on the subject.[34] Klepac has reconstructed the format of the entertainments, and tried as well to imagine – with the help of letters, reviews, et cetera – all that is unrecordable about them, like Mathews' physical and vocal gestures, his comic timing, his relations with his audience.[35] Anne Mathews' posthumous "autobiography" of her husband sheds a larger if hazier light on the nature of his comic vision, on the dramatic sensibility which apparently moved Dickens to "make offer of himself at Covent Garden." The text begins, in Mathews' own voice, at the beginning, and it reads remarkably like the opening of a Dickens novel – *David Copperfield* or *Great Expectations*, for example.

I was born on the 20th of June, 1776, at half-past two o'clock "and a cloudy morning" at No. 18 Strand, London . . . The agreeable twist of my would-be features, was occasioned . . . – indeed, I have heard my mother with great tenderness and delicacy confirm it, – by a species of hysteric fits to which I was subject in infancy, one of which distorted my mouth and eyebrows to such a degree as to render me almost hideous for a time . . . the "off-side" of my mouth, as a coachman would say, took such an affection for my ear, that it seemed to make a perpetual struggle to form a closer communication with it; and one eyebrow became fixed as a rusty weathercock.[36]

Mathews' self-portrait resembles some of his own stage *personae*, drawn, in the published editions of his scripts, by illustrators like George Cruikshank (another tie to Dickens) with mouth and eyebrow askew – variously to the left and right – and may as well recall other popular male characters who dominated the English stage during the actor's life. This raises an interesting question: was Mathews' perception of his disfigurement colored by the comic visages he saw on the stage, and were illustrators drawing his features to a conventional comic model? (The latter possibility is raised by the inconsistent location, in illustrations, of the "off-side" of his mouth.) Here is a problem of origin like the one I noted earlier in Dickens' "factual" writings. How can we distinguish the real from the fictive if "actual" events or characters are anticipated on the page or on the stage, as was the case with "Seven Dials"? The Mathews quotation underscores the problem further as it anticipates in style and tone, another, yet unwritten, autobiographical narrative: his entrance into the world is described with the same extravagance of detail, the same slant of humor, as David Copperfield's beginnings would be some eleven years later. This is not to suggest that Dickens borrowed narrative material directly from the Mathews biography – although one must assume that he read it – but rather to illustrate the intimacy between "realist" and theatrical or fictional forms, as well as the likeness of sensibility that attracted the era's most popular novelist to its most popular comedian. Given their similar narrative styles, their taste for verbal-centered humor and like comic types, the suggestion that Mathews influenced Dickens is hardly debatable. Assuming that relationship, we must try to hear the tones and rhythms of his comic voice as Dickens might have heard them.

Mathews' "biography," partly self-written, emphasizes the actor's acute ear for sound as well as his eye for the ridiculous. Among his favorite comic butts, as a child, were the clergy; his favorite preacher was a "huge-wigged old devotee whom we called Daddy Berridge."[37] His impersonations of Daddy Berridge, he explains, always focused on the old man's oratory, his brimstone-and-fire verbal flourishes. The other preachers at Daddy Berridge's church failed to realize the building's structural flaw, and consequently neglected to touch the poorest part of the congregation – whose cheap seats in "the oven," a recessed area in the floor, lay out of the speaker's range. But

> Daddy Berridge was a regular old-stager. He was well aware of the select portion confined in the black hole . . . He, therefore, when he had any choice bit of consolation for his flock, encored himself in his most eloquent passages. Turning his body entirely round in the pulpit, exhibiting his lank desponding visage to those of the gallery, who were delighted with this indulgence . . . he would roar out the repetition of his last sentence, which frequently reaching their ears without the context, could not, I fear, have tended much to their spiritual comfort . . . he would elevate his gutteral voice to a ludicrous pitch, peep down to the half-stifled wretches underneath, and cry, "You will all be damned, – do ye hear below?" This being all they heard of the sentence, they might very naturally have asked, "For what?" . . . here again is the lack of manner and tone of voice, but I pledge myself to the truth of my description.[38]

Mathews gives us the church as theatre, complete with galleries, pit, stage, and actor. This, of course, is not an original conceit, but that he chooses to use it suggests that he experienced his social world theatrically. In Mathews' reenactment, the dramatic vehicle is Daddy Berridge's voice, which literally conveys or withholds grace to the unelect in "the oven," and which rasps and roars them into submission. Mathews' sense that writing fails to evoke the scene with fidelity to its vocal and dramatic resonances accords with the larger cultural idea about the powers of the voice. He writes in like spirit about a musical episode: "I wish it were in the power of my pen to give effect to this scene; it requires the aid of a practical and vocal elucidation to convey it with full force .'"[39] *I* wish that he had lived to read the mature Dickens, and had left us record of the experience; his feeble confidence in the dramatic efficacy of the written word might have been bolstered.

Mathews accumulated an extensive theatrical record by the end of his life, playing virtually the entire late eighteenth- and early nineteenth-century repertoires of secondary male roles, apart from the "At Homes" which would ultimately confirm his transatlantic reputation. In a letter of December 28, 1794 to his friend John Litchfield, Mathews catalogued his dramatic roles, which included parts in obscure and recognizable plays.[40] While he does not appear to have been cast as leading man (hardly a surprise, given his comedic thrust), his appearances – and the subsequent reviews of them – were frequent enough to have insured him public recognition. He was particularly acclaimed for his Dick Cypher, the slang-pattering "gentleman" of dubious character in Isaac Pocock's *Hit or Miss*.[41]

Before turning to his sketches, let us try to imagine Mathews' own dramatic voice, the physical qualities of his comic idiom. Contemporary reviews indicate that the actor's voice was chameleon-like, "negatively capable" of losing itself in character. In an 1812 production of *Ways and Means*,

> His Sir David Dunder took a strong and immediate hold of the audience by its truth and originality . . . Before he had been five minutes on the stage, a thorough understanding was established betwixt him and the audience. But it is not in this species of regular delineation that the powers of this singular performer are to be seen to full advantage. – The part, or rather parts, of Dick Buskin [in *Killing No Murder*, performed on the same bill] unfold them completely; certainly a more whimsical example of versatility was never exhibited in the annals of acting. – We have heard of Foote and Wilkinson in the arts of personation and mimicry, but it is impossible to conceive that they could excel Mathews. In several of his changes, the individuality of the performer was totally lost; voice, features and deportment were so radically altered, that a new man was brought before the audience on each appearance.[42]

The words "truth" and "originality," the suggestion of a magical metamorphosis in the actor's "radical alterations," attest to a mystification of the acting process: when Mathews does it he does it right, strikes truth, *becomes* rather than *acts*. Theatre criticism lacked – and still lacks – the words to describe what happens on the stage, and so relied on conventional metaphors, or the often vague assumptions, about truth, inspiration, and so on, shared by contemporary theatre audiences; this piece resembles the review of Fanny Kemble in that respect. But if we still have difficulty in speaking of the dramatic process of *becoming*, we can, at least, examine the critical lexicon that envelops it, the words like "truth" and "inspiration" which we take to mean good acting.

In the nineteenth century "good acting" meant something different than it does today; it involved taking on the external characteristics assumed to represent certain conditions, and the very possibility of a "successful performance" presumed, as it does in the present, that those meaningful characteristics, whatever they might be, already existed in prototype, circulating in the popular imagination as the necessary components of successful drama. On the most obvious level, one might point to the Edinburgh critic's familiarity with Dick Buskin and Sir David Dunder as evidence of this: he knows these characters, has seen them played before, and holds Mathews to an amalgam of performances, written texts, and printed reviews which have collected in his imagination over the plays' lifetimes. Add to these presuppositions a set of similar comic performances of similar comic stage characters, and one begins to see how heavily overdetermined theatrical experience is. But despite this rich store of actual sources which might reveal Mathews to us, the scripts, reviews, letters, and journals of the Victorian theatre, we actually know very little, only that his performance compared favorably with other performances of his day; that his voice matched the prototypical comic idiom.

Studying systematically all of the late eighteenth- and early nineteenth-century comic roles which might have relevance to the roles Mathews played would be impossible here, and probably of limited helpfulness anyway. Focusing instead on his own dramatic creations – particularly the female ones, given this study's ultimate interest in gendered speech in Dickens – makes more sense. But first an important question remains to be answered: how did he speak his patter, what were his rhythms and inflections like? What features of his delivery found such approbation and delight in his audience, critic and layperson alike? With Mathews in particular, we must focus on spoken voice rather than written word because his scripts were often so very bad – a fact acknowledged, incidentally, by some of his most faithful admirers. It is not sufficient to assume, as one might be inclined to, that early nineteenth-century theatre audiences were less discerning or more naïve than we are; we must rather assume that Mathews' scripts succeeded despite their serious failings, that something magnificent happened between the written word and the staged character.

I take this magnificent "happening" to have been a vocal flexibility which allowed Mathews to mimic specific sounds, and more importantly, an attention to the rhythms of speech, which may change with changes in social setting and status, and which evoke specific species of responses – a grasping, not so much of character in the psychological sense, but of the verbal forms assumed by character. This is the "sympathy" intended by Lewes in his piece on Shakespeare's acting. If one examines the current stock of dramatic idioms – the comic, tragic, melodramatic theatrical voices – one finds metrical patterns, even in prose, which are endemic to each type. Thus, the tragic or sentimental hero usually speaks fairly regular "metrical" lines, bound to the iambus or the trochee if not the decasyllabic verse line;[43] while one must allow for some metrical irregularities in dramatic speech, odd fluctuations which often enter during the metamorphosis from written to spoken language, this character's idiom is steady in beat, refined. He adds dramatic force to his speech by way of inflection, emotion (perhaps vibratoed tones?), and of course physical gesture. Absent is the sort of rhythmical chaos that typifies comic speech. Conversely, Victorian comic characters, particularly, though not exclusively, female ones, speak in eccentric meters. The comic voice pushes too many feet into a line; too many words into a sentence; too many sentences in a dialogue. Juxtaposed against the "regularity" of serious speech, this voice sounds disruptive, hurried, undisciplined. When Mathews used it, it must have been both funny and unsettling, in the way that I described Dick Cypher's speech to be in the previous chapter. There I argued that Cypher's comic speech unsettled because of its alien diction, the slang patter of low sporting and gambling. Here I want to go further, to suggest that patter subverts the standard idiom not only in its rapid delivery and its semantic and syntactic eccentricities, but also in its disruptive rhythms, as we shall see.

Mathews' "At Homes" are difficult to reconstruct; although most seem to have been published shortly after they appeared on the stage, in both English and American editions, the extant scripts are plagued with inconsistencies.[44] It is clear from the available texts that a typical evening "At Home" included short character sketches, in which Mathews exercised his considerable talent for impersonation; comic patter songs; and a final "monopolylogue," in which the actor portrayed each of several characters sharing a single physical space, like a boat or a hot-air balloon. We have already seen how the Victorian patter song dramatized redundant speech as aberrant, enforcing cultural prejudices against the gesture and the people who were thought to use it: comic songs like those written and performed by Mathews certainly helped to shape Dickens' dramatic voice. But Mathews' nonmusical character impersonations also provided the novelist with a set of verbal and social stereotypes. For example, a character sketch printed in *Mr. Mathews' Memorandum Book, of Peculiarities, Characters, and Manners*[45] introduces Mrs. Chyle, a precursor of Dickens' Mrs. Jellyby. Whether or not Mrs. Chyle was a direct source for Mrs. Jellyby, we can at any rate assume that the female reformer cum domestic failure had made her way into the Victorian social imagination before Dickens embodied her in *Bleak House*, and that Mathews played some role in defining and standardizing her character type. With his "accurate description of, and Critical Remarks on, Mr. Mathews' Memorandum Book," the compiler – not, apparently, the actor himself – of *Mr. Mathews' Memorandum Book* combines critical opinion with paraphrase and direct quotation to describe and reenact for his readers Mathews' performance:

Mr. Mathews commenced his Annual Lecture on Thursday, March 10. Immediately on the opening of the doors, the house was crowded to an overflow in all parts . . . That one man should excite so considerable an interest, is indeed extraordinary . . . [In one of his acts] Mr. Mathews introduces the domestic miseries of Mr. Chyle, which are humorously portrayed.

How do you do? ah! I have long wished to tell you something of my wife; you seem to have a comfortable home, I wish I had; it is all owing to the uncommon *benevolence* of Mrs. Chyle: from the moment she became a good woman, I have been the most miserable man in existence; – she lets the children run about with hardly any dress upon them, that she may make more linen up for the Lying-in Charity, than any of her neighbours; she employs all the servants in making soup for the poor, while I am obliged to go out to get a dinner.[46]

Compare Mr. Chyle's complaint of his wife's domestic mismanagement with Esther Summerson's narration of her first visit to Mrs. Jellyby.

Nobody had appeared belonging to the house, except a person in pattens . . . I therefore assumed that Mrs. Jellyby was not at home; and was quite surprised when the person appeared in the passage without the pattens, and going up to the back room on the first floor before Ada and me, announced us as, "Them two young ladies, Missis Jellyby!" We passed several more children on the way up, whom it was difficult to avoid treading on in the dark; and as we came into Mrs. Jellyby's presence, one of the poor little things fell downstairs . . .

Mrs. Jellyby, whose face reflected none of the uneasiness which we could not help showing in our own faces, as the dear child's head recorded its passage with a bump on every stair . . . received us with perfect equanimity. We . . . sat down behind the door where there was a lame invalid of a sofa. Mrs. Jellyby had very good hair, but was too much occupied with her African duties to brush it. The shawl in which she had been loosely muffled, dropped onto her chair when she advanced to us; and as she turned to resume her seat, we could not help noticing that her dress didn't nearly meet up the back, and that the open space was railed across with a lattice-work of stay-lace – like a summerhouse . . . "you find me, my dears, as usual, very busy . . . The African project at present employs my whole time. It involves me in correspondence with public bodies, and with private individuals anxious for the welfare of their species all over the country. I am happy to say it is advancing."[47]

These are two versions of the same character. The irony, and the humor, in Mrs. Jellyby's effectual "correspondence with public bodies" when her own very public body, and those of her children, are grossly neglected, give this moment a great comic–dramatic power which seems lacking from Mathews' script, although the thematic similarities between the two suggest that Dickens found something quite interesting in the latter. Of course, there are several pronounced differences between the two scenes. Most obviously, Dickens' character is developed in minute detail: we are forced even to notice her dirty hair and the breadth of her excessive body, which bulges through the gap in her too-small dress. More importantly, while Mathews tells his story from the injured husband's perspective, Dickens' Mr. Jellyby has no voice at all. Variously described as a "non-entity," "Merged – in the more shining qualities of his wife"; figured as a gloomy corporeal presence at the dinner table, Mr. Jellyby seems to inflict a harsher censure on the undomestic wife than does his predecessor, Mr. Chyle, although it is impossible to gauge the power of Mathews' staged sketch. But if either attempt was successful, its author owed his success partly to the way in which his characters' voices harmonized with the voices already sounding in the collective Victorian imagination.

The "Mrs. Chyle" sketch articulated popular sentiments about woman's place, and presented the English public with a comic rendition of antidomesticity which Dickens would later develop. Mathews also dramatized the pattering redundant woman in a manner which Dickens later adopted for such characters as *Nickleby*'s Mrs. Nickleby and Miss Knag; *Little Dorrit*'s Flora Finching; Miss Flyte, of *Bleak House*; and Mrs. Lirriper, of the *Christmas Tales*. *Mr. Mathews' Comic Annual for 1831*[48] offers the monologue of Miss Euphemia Blight, whom Mathews portrayed, in drag, as a gossipy, malicious spinster.

Were any of you ever troubled with a *kantankerous* aunt? I find in my next look into the volume [of personalities] a mention of Miss Euphemia Blight, a maid, rather on the wrong side of youth. She was the very quintessence of politesse; yet, though she spoke fairly of every person she met, she would always find some method of picking a hole in people's coats when she *got behind their backs*; and her manner of expressing her opinion . . . was as novel as it was ingenious . . . "I must say, I enjoy myself as a woman of fashion should do; and though some people might be spiteful enough to call my reputation into question, because I may use a little extra innocent freedom – yet I am thoroughly convinced, that where a woman acts conscientiously within herself, comparatively speaking, and as the word goes, she ought not to care what is said. I may be wrong, certainly – yet that, comparatively speaking, is my fixed opinion. There's that Miss Primblossom – she's a sweet girl! understands music and poetry. She's a very clever girl, but what a pity it was of her father committing that error; it went hard with him – great pity to be sure; it will always follow his children – but then she hadn't ought to be a sufferer for her father's failings; she couldn't help it. Why should she care what people said, providing she does right herself? Where a girl was never known to do wrong – comparatively speaking, and taken in every sense of the word, and as people's opinions go – she has no cause to mind what they think . . . Mr. Dashly is a very good-hearted, affable, polite young man – but what a pity his father cut his throat; and there is the insanity always attends the mother's side. Yet he can't help that – he shouldn't be the sufferer; yet in a measure, though people know the facts of the case, yet there are some that will talk – but it is my decided opinion, and my thorough belief, taken on an average, that he has no conscious reason to mind. I may, certainly, be in error, but that is my rooted opinion."

Miss Euphemia's monologue is interesting for several reasons. First, it propagates some strongly negative sentiments about redundant women, drawing on the set of derisive female stereotypes in currency at the time.[49] (This subject is taken up fully in the sixth chapter of this book.) She talks too much, and talks about all the wrong things; the undertones of financial and sexual misconduct in the Primblossom and Dashly stories implicate Euphemia in two areas to which she should have no access, two types of experience of which she should have no knowledge. Of course, the association of redundant women – widows as well as spinsters – with sexual and financial greed was already firmly established in the English cultural consciousness by this time; Mathews engages in comfortable, conservative satire here, plucking a resonant string of associations which would have been shared by most of his audience.

But he also attempts a less obvious kind of derision in Euphemia's speech. Generally speaking, gossip is a language empowered with real social effect, representing the possibility of verbal and social subversions, but Euphemia's gossip is a study in futility, undoing every story it constructs, succumbing to language itself. She uses a form of patter, an extravagance of words with a paucity of ideas; a thick knot of clauses and hyphens and various parts of speech. In the last chapter, I quoted a passage from *Felix Holt*, arguing that Eliot conflates gossip and patter, and in so doing creates a doubly threatening idiom; conversely, Mathews' pattered gossip defuses its own destructive potential. The inconsistency in my readings of these two examples of pattered gossip underscores the doubleness of patter itself, which is, as I also argued in the last chapter, simultaneously an agent of subversion and control; interesting and tedious; pregnant and empty. If Eliot insists on its sinister powers, Mathews is fairly optimistic about patter, confident that it contains the seeds of its own destruction. Every certainty becomes a relativity in Miss Euphemia's patter; she erases whatever fact thoroughly convinces her by revoking its status as fact. Hence, female virtue, "comparatively speaking, and taken in every sense of the word, and as people's opinions go" is undefinable, "virtue" itself having lost its authority. This, as any proper middle-class Victorian knew, was impossible, nonsensical – and to suggest it committed egregious violence to the most sacred of cultural values. Even so, Mathews made his spinster say it, and no matter how comical her delivery, Miss Euphemia cannot have helped to improve the social standing of single women. But she certainly contributed to the codification of patter as a metaphor of redundancy.

Given Dickens' faithful attention to Mathews' London career, it is inevitable that he learned something about what I have called "redundant speech" from the actor, who by the 1830s had assumed considerable authority as a writer and comedian. Of course, patter had already been dramatized on the stage for at least half a century by the time Mathews entered the scene, and the genealogy of Dickens' sources should be understood as complicated at best. But their explicit similarities, especially in the rendition of verbal eccentricity, suggests the strength of Mathews' influence. Consider, for example, the speech of Miss Knag, the spinster milliner's assistant in *Nicholas Nickleby*, whose volubility and strange verbal tics – repetitions, echolalias, jerky stops and starts – accompany an obsequious self-righteousness. Her tone differs from Miss Euphemia Blight's, but the structure and spirit of the spinsters' idioms is similar enough to suggest that they are two versions of a single type.

"Well, now, indeed, Madame Mantalini," said Miss Knag, as Kate was taking her weary way homewards . . . "that Miss Nickleby is a very creditable young person – a very creditable young person indeed – hem – upon my word, Madame Mantalini, it does very extraordinary credit even to your discrimination that you should have found such a very excellent, very well-behaved, very – hem – very unassuming young woman to assist in the fitting on. I have seen some young women when they had the opportunity of displaying before their betters, behave in such a – oh dear – well – but you're always right, Madame Mantalini, always; and as I very often tell the young ladies, how you do contrive to be always right, when so many people are so often wrong, is to me a mystery indeed."[50]

Miss Knag's patter depends less upon obvious paradoxes and logical transgressions than on a larger psychosocial context, provided by the narrator, for its humor. We already know enough about her personality to recognize the hypocrisy in her words: she resembles Uriah Heep, in fact, submissive yet slippery, capable of serious subversion. Mathews, of course, had to expose his spinster's whole character in her dialogue – hence her explicit stupidity and maliciousness, which take verbal form in her self-contradictions and her moral relativism – whereas, if one abstracts Miss Knag's speech from the surrounding narrative, it is notable only for its excessiveness, its odd stammers and half-halts, and of course its shameless toadyism.[51] Dickens' greater subtlety is as much attributable to the nature of his genre, which accommodates interiority, narrative reflection, as it must be to his greater writerly genius; it might also be that with the voice of Mathews' spinster resounding in the public ear, he could leave Miss Knag's more exaggerated spinsterly noises to be imagined, and formulate a complex social drama rather than a comic caricature.

Nicholas Nickleby was published seven years after Miss Euphemia Blight appeared on the London stage in 1831. If Forster speaks truly, Dickens was almost certainly in the audience for Euphemia's debut. This may account for the idiomatic similarities to the spinster's monologue in Miss Knag's dialogue. If so, then it also accounts for the strains of Mathews' comic female voice in Mrs. Nickleby, who descends from Miss Euphemia's line as well. That Miss Knag and Mrs. Nickleby occupy the same marginalized social corner is indicated in this text by their continual verbal droning, two voices which at once chafe against each other, different, and meld together into an impossible crescendo.

After concluding this effort of invention without being interrupted, Miss Knag fell into many more recollections, no less interesting than true, the full tide of which Mrs. Nickleby in vain attempting to stem, at length sailed smoothly down, by adding an undercurrent of her own recollections; and so both ladies went on talking together in perfect contentment; the only difference between them being, that whereas Miss Knag addressed herself to Kate, and talked very loud, Mrs. Nickleby kept on in one unbroken monotonous flow, perfectly satisfied to be talking, and caring very little whether anyone listened or not.[52]

This is an extraordinarily funny, and brilliantly conceived, verbal duel – perhaps a rowing match would be the more appropriate metaphor – but unlike its masculine prototype, it produces no victor. Both participants fade into a sort of vocal obscurity, their words melting into sameness. One is forced to feel disgust at their lack of control, and at the insistent buzz of their frivolous stories. They hover, gnats (and nags) about the reader's ears, impossible, unshakable, and incredibly irritating. In some ways this extravagance of words is a strategy of verbal suppression: one may find oneself skimming, even skipping, the dialogue of characters like these because one knows in advance what it will sound like, what it will (or will not) say, how annoying it will be.

If Mrs. Nickleby's idiom shares with the others a verbal superfluity and a poverty of important facts or ideas, it lacks their sharp edge: unlike Miss Euphemia and Miss Knag, who speak with the intention of self-aggrandizement, or if that fails, then injuring other people, Mrs. Nickleby's words are flabby, innocuous. A reader would hear her intoning her rapid, lazy verbal circles, hear the pitter-patter of words that are pure rhythm or form, beating time against nothing, and probably feel more amused than explicitly threatened – although a sense of their subversive power certainly lies behind such derisive renditions of redundant women, as I shall argue in the next chapter. I use "rhythm" here with the intention of provoking some interesting comparisons. It is no coincidence that the word *patter* is defined in reference to the proverbial "little feet" in that most canonical of dictionaries, Dr. Johnson's.[53] Patter is as much about about rhythm as it is about too many words. It is also about music. There is always a music in patter – even in its nonoperatic forms. When the emphasis of speech falls on its tones, rhythmic patterns, morphological and syntactic structures, that speech sings, leaving one undistracted by "what the words mean." (One sometimes hears people claim that they prefer opera sung in foreign languages because the words do not get in the way; pattered speech can work in the same way.) One might turn to Julia Kristeva at this point; she writes of a "carnivalesque" idiom, polyvocal, antirational, subversive, which resembles Dickens' female patter, and which "produce[s] a more flagrant dialogism than any other discourse." For Kristeva, "the scene of the carnival is proffered as the only space in which language escapes linearity (law) to live as drama in three dimensions."[54] Dialogism, then, sits in opposition to social and linguistic authority; both "music" and "drama," in Kristeva's reading, are implicated in this protest against the standard discourse, and so, of course, is a certain type of novel. Certainly, a language laden with musicodramatic nuances, with the unrepresentable, the vocal, the corporeal, the colorful, finds itself at odds with "transparent" prose – and there is a certain charm in that. We have seen that patter, for all of its negativity, has a power of attraction which undermines the generally conservative agendas of the authors who used it. This attractive power is found in what one might describe as its "carnival," its explicit, distracting rhythmicality and vocality; in the almost physical, rather than cerebral, response it provokes.

But this again raises the inevitability of patter's inconsistencies, its concurrent power and irrelevance. If an emphasis on the rhythms of speech make speech sing, an extravagantly rhythmed idiom may also distract and annoy. Either way, patter affects at the level of emotion, evoking feelings of impatience, annoyance, dread. Comparing Mrs. Nickleby's comic idiom with Nicholas' serious idiom will illuminate the difference between speech calculated to provoke "cerebral" and "physical" responses, as well as the equal possibility that patter will alienate or move us. When Nicholas speaks, we always understand his accents, his diction. Furthermore, his fairly regular iambic and trochaic feet (used, as this is prose, more interchangeably than would be permissible in verse), never distract us from his meaning. Instead, they blend invisibly into the whole fabric of his speech: this kind of dialogue is meant to promote a standard of speaking, a transparent idiom.

"You have disregarded all my quiet interference in the miserable lad's behalf," said Nicholas; "returned no answer to the letter in which I begged forgiveness for him, and offered to be responsible that he would remain quietly here. Don't blame me for this public interference. You have brought it upon yourself; not I."[55]

If one abstracts this passage from its narrative context, bleeds it of its dramatic rhythms, one finds a largely regular trochaic meter, stresses falling consistently on the first syllables of two- and three-syllable feet, disrupted only by occasional reversals of stress pattern into the iambic. Add the rise and fall of emotional pitch, the occasional surfacing of suppressed emotion manifested in rhythmic irregularity, a loud and strong gesture out of place, and Nicholas' speech will sound as it ought to: human, moved, dramatic. But it will not – could not – stray too far outside of metrical and idiomatic regularity, no matter how much anguish or anger one pours into the reading. That is because it is neutral speech, neutral in the sense that it fulfills the requirements imposed on the standard idiom, is "correct," nonviolating, unremarkable. Its power lies in its logical, syntactical, semantic clarity, in its appeal to the reason, and to speak it is a privilege conferred upon educated male characters, usually of the middle class.

If Nicholas speaks the language of reason and good-breeding, his mother's speech violates both of those principles. She shares her idiom with a large body of spinsters and widows who filled the pages of books and graced – or disgraced – the English stage. So far my discussions of female patter have focused on its syntax, diction, punctuation, and speed of delivery, but its rhythms are also eccentric, contributing to its distracting, unsettling and humorous effects. Consider, for example, the following passage:

> "Recommend!" cried Mrs Nickleby. "Isn't it obvious, my dear, that of all occupations in this world for a young lady situated as you are, that of companion to some amiable lady is the very thing for which your education, and manners, and personal appearance, and everything else, exactly qualify you? Did you never hear your poor dear papa speak of the young lady who boarded in the same house that he boarded in once, when he was a bachelor – what was her name again? I know it began with a B, and ended with a g, but whether it was Waters or – no, it couldn't have been that, either; but whatever her name was, don't you know that that young lady went as companion to a married lady who died soon afterwards, and that she married the husband, and had one of the finest little boys the medical man had ever seen – all within eighteen months?"[56]

This particular dialogue is notable for its excess of unstressed syllables – usually three in a row, as in the phrase "who boarded in the same house that he boarded in once, when he was a bachelor." This compromises its clarity, and brings parts of it down to the pitches and rhythms of muttering, of madness. An extravagance of syllables and words – she uses about twice as many words per sentence, and juxtaposes more polysyllabic words than Nicholas does – Mrs. Nickleby's speech assaults her reader, who emerges breathless from any attempt at processing it. Her language does not encourage sympathy from the reader, who, as I have already suggested, may walk away from it rather than struggle with it. These gestures, the preponderance of unstressed syllables and the excess of words, occur frequently enough in Mrs. Nickleby's dialogue for us to conclude that it characterizes her peculiar idiom.

Other redundant female characters overuse the unstressed syllable in their patter; there are a large cluster of them in the literary canon, for our purposes the most notable starting with Austen's Miss Bates and persisting at least through Eliot, in Mrs. Holt and Mrs. Tulliver.[57] In Dickens alone we find several – and we can only imagine the number of theatrical daughters and granddaughters of Miss Euphemia Blight. This suggests that when people heard unassertive speech, words that recoiled from semantic or social authority, nouns and verbs struck with the low and rapid beats of unimportant parts of speech-like articles, they heard disenfranchised female speech. (I use "disenfranchised" not in its literal sense here – "disenfranchised female" would be, appropriately, redundant – but to suggest a female situation outside of mainstream Victorian situations, one which does not include, for one of several social reasons, marriage and reproduction.) That Victorian readers and audiences heard this sort of female language as excessive, unstressed, unauthoritative, even mad, is indicated by the constant play of dialogue with those qualities in the mouths of redundant women, in popular entertainments like the "At Homes" of Charles Mathews and the novels of Charles Dickens.

If Mathews inspired in Dickens a feeling for the comic voice, he did the same for his adoring domestic audience – as did Dickens after him. With theatre and fiction as two of the chief sources of Victorian entertainment, people were quite familiar with the conventional strains of dialogue, which crossed generic boundaries and tied written and oral genres to a common dramatic base. I want to stress the conventionality of these dialogues: even a writer of Dickens' stature is surrounded by an ever-evolving body of texts, sometimes of indifferent merit but nevertheless read, and therefore he writes for readers who already expect and understand certain dramatic cues. The more distinguished purveyors of art and entertainment, like Mathews and Dickens, were of course able to introduce the unfamiliar as they replayed the usual. But nineteenth-century English entertainments, reflecting the popular taste, were politically and aesthetically conservative; one finds more of the old than the new, even, I think, in the hyper-creative Dickens. This contention may draw disapproval and displeasure, but the risk seems well taken. How else, than with reference to his conservative appeal, might one explain the tremendous public trust in this man, maintained without a flicker of serious doubt until the details of his sordid personal affairs began to emerge? The following letter to Dickens from C. J. Mathews confirms the novelist's considerable public following, and implies the magnitude of his writerly influence as well.

My dear Mr. Dickens,

I conclude that you have lived too long in this world to be easily surprised, and yet I fancy I am presumptuous enough to fancy I can make you open your eyes when I tell you that I am off to Australia! . . .

On Tuesday morning January 4th my farewell benefit comes out at Covent Garden and will be composed of contributions from all the London Theatres, with a little assemblage of friends of the stage *à la française*, and I should feel more pride than I can express if I could see Mr. Charles Dickens among them, if only as a "visitor" and a few words spoken by him by way of "God speed" . . .

If you object to words, a shake of the hand will be as much esteemed as an hour's speech but I confess I should like [at least?] to be honoured by your presence.[58]

The letter seems unnecessarily groveling, coming as it does from a powerful man in the theatre industry, and one well known to Dickens, but it was a matter of great importance – financial and personal importance – to Mathews that his benefit come off successfully. Clearly, Dickens was a man whose publicized presence meant big money for any theatre; if he wanted to speak – anywhere, not just on the stage – he was always assured of a voice, and of a public ear greedy for its barest whispers. I do not know whether he obliged the son of his old idol, although he might have returned a debt to the father in doing so, to the man who gave him a verbal and vocal model for some of his most memorable characters.

This chapter has attempted to convey to its reader the sounds of certain comic-dramatic voices and their existence in Dickens' dialogic novels. Reading such musical, multivocal prose as Dickens writes involves, for the reader, a loss of self as he or she enters the fairground, an abandonment of reason to the novel's enveloping tones and rhythms. This is a loss which threatens. Entering a text which drenches the senses with its music, one may drown. This, I think, is the key to understanding the force of eccentric idioms like patter. One must try to hear the chaos under the joke, a strong and sinister undertow which threatens to sweep reason out from under the world, to carry one away on its slippery back. Dickens heard it, surely, as he sat "At Home" in the theatre with Charles Mathews.

CHAPTER 6

Patter and the problem of redundancy: odd women and "Little Dorrit"

> But half the sorrows of women would be averted if they could repress the speech they know to be useless.
>
> George Eliot, *Felix Holt*[1]

This book has concerned itself, to this point, with the construction of a Victorian reading subject, a reader, of Dickens and others, primarily constituted in the dialogic spaces between theatre and novel. We have seen how the "public" constitution of the reader called into social play the gestures of acting, lifting the contextual signs of the theatre out of their staged universe and into the realms of social commentary and literary realism – freezing them, I would like to suggest now, into the status of mythical or universal ideas. A novel reader situated outside of linear narrativity and subjectivity, and inside the very public and expressive machinery of the theatre, creates a somewhat different kind of Victorian novel than we are accustomed to imagining. It also cultivated a peculiar sort of political consciousness, one which framed the mid-century debate over redundant or "odd" women.

Let us begin this discussion with a slight digression, a little story about truth, drama, and performance which is in fact relevant to the larger concerns of the chapter. When William Macready reread *Strafford*, the tragedy he requested from Robert Browning in 1836 ("Write a play, Browning, and keep me from going to America!"), he found that in his first perusal "I had been too much carried away by the truth of character to observe the meanness of plot and occasional obscurity."[2] This is a curious comment, an implicit faulting of the "truth" in dialogue that we have seen celebrated in so much criticism of the day. As if the truth of Browning's characters were too great to be borne, a scorching scirocco of meaning liable to carry one away, Macready turned his face to the softer wind of its weak plot and syntactic tangles, focusing almost exclusively on them in all of his subsequent comments on the play. He ultimately determined *Strafford* unplayable: Victorian audiences demanded plot movement, coherency and causality in their drama, none of which this play provided. *Strafford* was finally produced, though not with much success, and the process chafed every temper involved with it.

It may seem an irrelevance, even an irreverence, to introduce a chapter on odd women in *Little Dorrit* with an anecdote about Macready and Browning, but in fact that episode might have been written expressly for this purpose. It demonstrates an important point about the Victorian voice: although, as this study has argued extensively, dramatic voices held within them seeds of truth and identity, their concentrated meaning was not always perceived as a constructive interpretive tool. If truth inhered in social speech – even, paradoxically, in deceptive or frivolous speech – then the spoken word could be literally dangerous. We saw this, for example, in the critical reaction to Isaac Pocock's *Hit or Miss*, and, indeed, in the whole history of responses to English performance. We see it as well in Macready's anxieties about the strength of Browning's dramatic voices: whatever it was that he heard in them made him uneasy, and he deflected it with compaints about the play's dramatic action.

Scrolls of socioaesthetic history lie coiled in Macready's comments, in his mere desire for a certain kind of verisimilitude in drama, one that attends both character and plot and which does not unsettle the balance between them. Those scrolls unfurl when we contextualize the comments. For example, while most students of Crowning, even his greatest lovers, will cheerfully admit his few plays, like *Strafford* and *A Blot on the 'Scutcheon*, to be dramatic failures, they would also conclude the same about J. S. Knowles' *Virginius*, a play which Macready much admired, with a leading role which made his reputation. To the late twentieth-century reader, both plays read uncomfortably, and neither would appear to offer a mimetic reproduction of nineteenth-century society, speech, or manners. But *Strafford* does its own thing, its blank verse thick and prosy and fractured in Browning's peculiar style, its story – a story about character – blithely unaware of its theatrical antecedents. Hence, critics found the poetry difficult to follow, and the plot feeble. In fact, critics who lacked Macready's intelligence and sensibility often failed to hear the truth in *Strafford*'s voices at all; a reviewer for the *Athanaeum* (May 1837) complained that "the speeches generally contain so many broken sentences, that they become quite unintelligible," and *La Belle Assemblée* found the actors, with the exception of Macready, unequal to the performance of such lines.[3] In contrast, *Virginius* follows the melodrama formula, with its conservative emphasis on virtue and the forces which threaten it, and "remained for many years the most popular tragedy of the century."[4]

If Macready found particles of truth in Browning's resonant verse, it was a synchronic kind of truth, universal "facts" about character, and not the truth of sequential events, or of human behaviors in changing conditions and moving landscapes. In *Virginius* he seems to have found a balance of both. Implied in Macready's attack on *Strafford* is a confirmation of plays like *Virginius*, standard historical melodramas with linear, causal plots and regular if artificial verse or prose dialogue. For reasons with which we are now familiar, the structures and languages of such melodramas were received as realistic by the nineteenth century, a fact which the Macready anecdote bears out.

It is significant to us that the ethos and structures of melodrama filled the Victorian imagination in the ways illustrated above; we may use the interpretive paradigm demonstrated in the Macready–Browning anecdote, the seeking for "truth" in a voice, in examining the cultural problem of redundant women and its relation to *Little Dorrit*. The story that anecdote tells about "truth" in language, particularly spoken language, is one of *Little Dorrit*'s stories as well – not one of its plots, but a driving force behind its treatment of character. Nothing could illustrate the Victorian faith in vocal-linguistic power more effectively than Macready's description of Browning's verse, whose truth is so concentrated as to become dangerous, beguiling or distracting. My argument about "odd women" depends upon this fact, upon the implicit meaning in the pattered dialogue written by novelists like Dickens for their widows and spinsters, speech considered at once to be frivolous and true, a literal embodiment of redundancy, and a motivated sign of certain kinds of social marginality. While I do not mean to suggest, in juxtaposing these two examples of truth in speech, that *Strafford*'s dialogue is structured like comic patter, they do share certain qualities – an excess of words and vocal patterns in relation to action, and the power of beguilement.

There is also an interesting parallel in the failures of Browning and Dickens – both splendidly theatrical writers – at writing for the theatre, although each failed for different reasons. Dickens' problems with the genre are more difficult to isolate than Browning's, and may have had as much to do with complications in production and direction as flaws in the writing. But we do know that he mastered, in prose, the popular voices of his day, adding resonance and depth to them while maintaining their conventionality, and without obscuring the "truth" in his complex but generally causal plots. Among his best efforts was the pattered voice, used almost exclusively, by him, for widow speech; a voice featured prominently in *Little Dorrit*.

Male characters, of course, speak in comic tongues as well in Dickens; these figures are abundant and wonderful, but they use various comic dialects instead of patter, modes of speech which may be derisive but are always – unlike the speech of "odd" women – understood. I have not treated them at length in this study because they and their originals fit much more neatly into the Dickensian and Victorian social universes – nobody talked about transporting working-class men merely because they *were* working-class men – and because their voices, at least as they are sung in the novels, were far less disruptive and carnivalesque than those of the widows. Even Silas Wegg, who, as a ballad monger is a genuine "patterer" in Mayhew's sense, articulates his cockneyed sentences intelligibly. Of course, Dickens is merely reproducing the current social perceptions in his gendered speech patterns. Lower-class males were not understood to be superfluous; they were, however, perceived as less intelligent, less worldly, and less attractive than their betters, and their typical speech patterns, in novels and other popular entertainments, reflect these prejudices. Strong cockney or northern dialect characterizes lower male speech in Dickens, and while it performs comic variations on the standard idiom, it never exceeds the bourgeois grasp, although we know that authentic London street patter did. That novelists like Dickens refuse to capitulate to the incomprehensible working-class male by representing him as incomprehensible suggests their strong anxieties on the subject. They were anxious, too, about redundant women, and found some relief in proposing that such women are incapable of controlled thought and speech – that all of their energy is expended in pointless talk.

Although Dickens used the sketches of Charles Mathews as a model for his pattering women, he was also reading, simultaneously, a much larger and more diffuse cultural text on single women. So was his audience, who found a complex social narrative, an imaginary text, about single women, economics, and politics between the lines of novels like *Little Dorrit*. In using this verbal trope, with its social and theatrical histories imprinted in the collective imagination, Dickens was able to take a quiet but explicit position in the Victorian debate over redundant women, a sociopolitical argument structured upon the same myths about female speech, sexuality, and power that informed the use of pattered female speech in popular entertainments.

If Victorian England was able to generate and sustain a collective knowledge of theatrical voices like patter, it was largely because the various tones, rhythms, and textures of the voice, the sounds that filled the air, were encoded, gathered into small pockets of meaning and arranged, phonemically, syntactically, semantically, into imaginary texts. Popular entertainments provided people with the interpretive tools necessary for reading these texts. While every individual is born into a discourse in which signs are already named and valued, or are involved in a slow and imperceptible evolution towards new social meanings, one still has to learn to read those signs, a process which begins in the earliest part of domestic socialization, and which is confirmed daily, particularly in industrial societies, in the products of popular culture. Julia Kristeva defines the act of literary reading as a submission to, and subsequent ordering of, the "music" in text, a definition which might relate to acts of cultural reading as well. "The identity of the reading *I* loses itself there, atomizes itself; it is a time of jouissance, where one discovers one text under another . . . At the same time, *already,* a regularity comes forth to gather these atoms: a grid lays out jouissance . . . A harmony organizes sounds around us . . . We must find a way to communicate this music by finding a code."[5] I have invoked Kristeva's reading paradigm, the swoon under a rush of sensory stimuli which ends in recovery, an almost retaliatory gesture of control, because it parallels, on a local level, the acts of theatrical policing I have described in English cultural history. It also invites discussion of the carnival, pockets of festive, spectacular, or musical disruption; literal or figurative fairgrounds on the bourgeois landscape. In Kristeva's reading, the carnival, the rush of impulse, must be contained or coded in order for the reading subject to materialize. There is something of the carnivalesque in patter – an explosive, disjointed burst of unformed energy, as one might describe it – which helps to explain its negative social value and its perceived potential for social disruption.

That Victorian culture placed women among the primary speakers of patter (the others being the costermongers discussed above in chapter 4) suggests some interesting relationships between femaleness and the carnival. Peter Stallybrass and Allon White have written about the carnivalesque and female hysteria in Freud, noting that "the broken fragments of carnival, terrifying and disconnected, glide through the discourse of the hysteric" and concluding that "bourgeois society problematized its own relationship to the power of the 'low', enclosing itself, indeed, often defining itself, by its suppression of the 'base' languages of carnival."[6] We have seen how bourgeois culture defined itself by this kind of linguistic suppression, and also how patter, carnivalesque danger, and madness are conflated in bourgeois English culture. All three of these conditions were imagined to congregate in certain forms of female eccentricity. By encoding the redundant woman's speech as patter, an idiom with a long and familiar sociodramatic history, Dickens and other Victorian writers controlled her, recovering their reading/writing subjectivities by dispersing hers – into the phonemic, rhythmic particles of narrative, to borrow Kristeva's metaphor once again; and then constructing, out of those particles, an object, a signifying unit.

If we recognize the pulses of madness and disruption in carnivalesque languages like patter, we are forced to reconsider the standard readings of fictional patterers like Flora Finching and Mrs. Nickleby, who have been described by various critics as speakers of idiolect, as fictionalizations of historical women, like Mrs. Dickens and Maria Beadnell Winters, and as transparent symptoms of Dickens' misogyny,[7] but never as popular or mythical figures with complex cultural–theatrical genealogies. Nor have they been integrated into the cultural–political present of Victorian England, situated in the center of social or political debate, but in fact their very presence invokes histories and narratives that reach far beyond their textual lives, into the distant past and the nonliterary Victorian present. The best way to understand them, to recover their generally suppressed stories, is to remythologize these women, to reimagine their largest and most diffuse significance, and then, moving from the general to the specific, to place them in the center of a highly topical Victorian debate, reading the patterer as redundant woman by invoking her mythical identity – "mythical," in the sense of archetypal, general, and culturally resonant.[8]

Let us begin this process by defining the "specific" to which this discussion will tend, the mid-century debate over redundant women, and in doing so to set the stage for a later reading of *Little Dorrit*. The theatrical metaphor, "setting the stage," is strategically placed here – irredeemable cliché as it is – to collapse the boundaries between the worlds of staged entertainment and social or political action: each fed the other in Victorian England, as the present study to this point illustrates. The "debate over redundancy," clearly a symptom of intense cultural anxiety, spanned practically the entire second half of the century; the topic, in full public discussion in the 1840s, was still current thirty years later. In mid-Victorian political rhetoric the phrase "redundant woman" was used synonymously with "spinster," although the period's popular entertainments conflate spinsters with widows, and I have assumed that this conflation accurately reflects the general attitude toward these two social types – unpaired, presumably celibate, and, potentially, economically independent women.[9] Mainstream public opinion shunted these women to the furthest margins of "normal" society; articles, novels, and theatrical spectacles either erased them from public sight, debased them with low comic insult, or – less often – emphasized the pathos of their situation with melodramatic flourish. Dickens shared the popular sentiments about redundancy – aversion and anxiety, sweetened perhaps with a touch of pity – and his writings position him on the conservative side of the debate, the side which saw redundant women as odd, worthy of marginalization: erasure, silencing, humiliation, even, in the extreme, transportation.

This position was carefully articulated in William Rathbone Greg's influential article of 1862, "Why are Women Redundant?"[10] Greg rejects spinsterhood as an insidious form of idleness; in a culture which supposed the natural "work" of middle-class women to be the production and moral maintenance of family, spinsters were useless, and indeed, unnatural. Greg was not the only one who thought so. John Stuart Mill complained about a generalized attitude towards single women which denied them the most basic social function or sanction.

> A single woman . . . is felt both by herself and others as a kind of excrescence on the surface of society, having no use or function or office there. She is not indeed precluded from useful and honorable exertion of various kinds: but a married woman is *presumed* to be a useful member of society . . . a single woman must establish what very few either women or men do establish, an *individual* claim.[11]

In other words, single women confounded the accepted hierarchy of social purposes, the standard arrangement of middle-class social and sexual roles. Greg proposed forced emigration of redundant women: they should be used, he and his supporters argued, to balance the disproportionately male colonial populations. The idea that at least some of Britain's colonies housed criminals and other social outcasts would not have been lost on Greg's public: spinsterhood allied a woman with the social deviants who were thought to represent her best possibility for marriage.

The institution of several agencies which promoted female emigration, and, implicitly, marriage, in the middle of the century, including the Society for the Employment of Women and the Female Middle-Class Emigration Society[12] indicates how deeply the Victorian middle class believed in the sort of "domestic capitalism" which relegated to women the labors of hearth and cradle. The Victorian domestic picture – a prescriptive rather than faithfully mimetic version of bourgeois society, propagated by the popular culture – showed women in various states of dependence upon husbands, fathers, or brothers. Scenes of domestic pleasure, in prose, in painting, on the stage, featured young brides or mothers, nursing, sewing, making trifling music on the fortepiano, adoring their husbands and children. That this domestic scene was represented in a large portion of the nineteenth-century's cultural–aesthetic productions (Helene Roberts notes that "the *Spectator* in 1852 added the category 'Domestic Picture' in its review column to cope with the increasing numbers"[13]) suggests that the bourgeois ranks of contemporary artists felt anxious about it, as a vision less often realized than might now seem to be the case. And women, like Charlotte Brontë, sometimes felt anxious about it as well, although for different reasons. When Brontë's Lucy Snowe attends an art exhibition, she reacts with disgust to two conventional representations of women: a fleshy odalisque, the "Cleopatra," and a series of didactic paintings entitled *La vie d'une femme.* Turning her face from the four pious women of *La vie*, she exclaims "What women to live with! insincere, ill-humoured, bloodless, brainless non-entities! As bad in their way as the indolent gipsy-giantess, the Cleopatra, in hers."[14]

Of course, Lucy's feeling of exclusion is hardly surprising. As a spinster she has no place in conventional domestic arrangements, and is painted out of the picture.[15] Spinsters spoiled symmetry and perspective in the Victorian domestic landscape, and their numbers increased alarmingly over the course of the century. "Do you know that there are half a million more women than men in this happy country of ours?" asks Gissing's Rhoda Nunn. "So many *odd* women – no making a pair with them. The pessimists call them useless, lost, futile lives."[16] Widows appeared useless and futile to the pessimists as well, many of whom were women. For example, Anne Brontë describes the widowed Mrs. Wilson, in *The Tenant of Wildfell Hall*, as "a narrow-minded, tattling old gossip, whose character is not worth describing" (although she describes it in full detail); Eliza Millward, from the same novel, isolates old maids and cats as the two creatures gentlemen hate most, suggesting, perhaps, a basic connection between spinsters and witches.[17]

The debate over single women was taken up in the popular press, which devoted serious journalistic articles to "the woman question" in all of its variations: the "new" woman question, the prostitute question, the enfranchised woman question, and particularly the redundant woman question.[18] I emphasize the seriousness of these articles on single women because elsewhere – in fiction, theatrical display, song – the very presence of spinsters and widows instigated verbal parody. One rarely reads a story about old maids, or a satiric article about widows, without being forced to endure these poor women's endless locutions. The seriousness of articles devoted specifically to redundancy and its socioeconomic implications plays in counterpoint against these endless comic treatments of redundant women, which both preceded and outnumbered the serious articles themselves. The music metaphor should promote a musical consciousness of these dialogic voices, these contrapuntal strands which blended into a single if differentiated sensory experience. In other words, the fact that some writers could employ serious professional language in their writings on the subject did not promote the status of redundant women: in the cultural orchestration, people heard the comic, derisive treatment as the dominant melody.

This is partly because articles like "Why are Women Redundant," and its companions, like Eliza Lynn Linton's expansive corpus of misogynistic essays, were brow-furrowing, heavy-going affairs, humorless, depressing – even, I should think, for people who agreed with their politics – and they and their serious liberal counterparts had to compete against more colorful tidbits, biting journalistic squibs which shopped, unashamedly, for their juiciest parts in the popular market. Pieces of this type appeared often in *Punch*, a magazine notoriously hostile to women. For example, a satiric "letter to the editor" appeared in a *Punch* article entitled "Politics and Petticoats" in January 1860, to demonstrate the social, political, and verbal frivolity of women, particularly spinsters.

My Dear Mr. Punch,

You so very often ridicule us *poor weak* women, and more especially the *stronger-minded* of the sex, that I declare I'm half afraid of writing to you seriously, for fear you'll print my letter for the sake of making fun of me . . . although *of course* you know that it's written to you *privately* . . . However, I *must* write, whatever *mean* advantage you may take of my so doing. I can't let that dear duck and darling of an Empress be laughed at by you men for her Crinoline *absurdities*, as you are pleased to term them, without calling your attention to a most convincing proof that she devotes herself to *far more serious* pursuits, and is a *great stateswoman* as well as a good dresser. If you doubt me, read this passage from the *Illustrated News*, where it recently appeared with the account of a new bonnet, and other highly interesting and *most important* French intelligence; –

"The Empress Eugenie has assisted for the last few days at the Council of Ministers presided over by the Emperor."

There now, *Mr. Punch*, what say you to *that*, Sir! Only think, that sweet Eugenie assisting at a council, not of milliners and bonnet-makers, but of veritable councillors and ministers of state. "*Assisting*," you observe, Sir! . . . I would not encroach, Sir, on your *valuable* space, but I cannot help just saying, that it would in my opinion be a good thing for the country, if *our* Ministers would take example by the French . . . At least thinks one whose name until, to aid him in his councils some *stupid husband* changes it, is Xantippe Rose Sophia Sophonisba Smith.

P.S. That darling, Mr. Roebuck, I remember, once confessed that he felt perfectly *convinced* that if Woman had her *rights* she ought to have a vote. If I were either of the Ladies Palmerston or Punch, I would not let my husband rest till he had promised he would get a law made that should give her one.

P.P.S. Do you know – I ask in confidence; *is* Mr. Roebuck married? If not, *will* you tell me; *has* he got red hair? And *would* you call his nose a classically chiselled one?[19]

Few Victorian readers would have to wait for the end of the piece to learn that Xantippe is a spinster. A study in extravagance and silliness, her letter demonstrates her redundancy in various ways: its verbal excess; its inability to end itself, bleeding on into two postscripts; the sprinkling of italics throughout, which suggest a species of ornamental spoken emphasis; the childlike, coquettish tone, which indicates that this stong-minded woman is making a pathetic attempt at flirtation and that she wants to land a husband more than she wants a vote. Even her name is rococo, redundant. All of the middle-class journals of the mid-nineteenth century, with the exception of feminist mouthpieces like the *Victoria Magazine*, published articles of this sort, along with cartoons, poems, and songs to the same effect. It is no wonder that those who advocated training and education for women as a solution to the redundancy problem were ridiculed – "evidence" showed that spinsters were incapable of intellectual exertion, and preferred marriage to independence in any case.

Despite – perhaps because of – their unpopularity, redundant women held a particular fascination for the bourgeois community; so did overtly different or eccentric women, whose stories were often told in popular writings and dramas, two obvious examples of this being the dwarves Miss Mowcher (*David Copperfield*) and Mme. Walravens (*Villette*). A little digging uncovers a whole body of literature on the subject of "odd women," much of it now obscure, moldering between the leaves of unread journals from the seventeenth, eighteenth and nineteenth centuries. Some of these take the form of scientific inquiry – for example, the papers published by the Royal Society – but most, anecdotal histories of the physically or socially unfortunate, read like literary circus-sideshows, textual embodiments – and containments – of the carnivalesque.

For instance, in November of 1874, four years after Dickens' death, the magazine he had edited and published in his later years ran the first part of a serial about female "freaks of nature," women with grotesque physical deformities, entitled "Odd Women." This was a species of article which frequently graced the pages of *All the Year Round* during Dickens' tenure as editor,[20] and it exposes a particularly cruel vein of Victorian misogyny, with its perverse interest in the dangerous or deforming carnival impulse in women. That George Gissing published a novel by the same name in 1893, but one explicitly about spinsterhood and the sociopolitical implications of female celibacy, suggests that the nineteenth century saw certain connections between physical oddity and the "odd" or leftover women who represented "as many as 30 percent of all English women between the ages of twenty and forty" by the middle of the century.[21] It also suggests the height and duration of public interest in the problem of single women.[22] The popular notion of the English spinster or widow was of a woman with a spectacular verbal deformity, a soul sister to the dwarf, the bearded lady, the human pig. Women in both groups were thought to be "odd" in the word's two senses: strange and leftover, unpaired.

The "Odd Women" article confirms the intimacy between these two kinds of oddness by foregrounding, as it tells its tales, the "natural" relationship between physical or social normalcy and marriageability. It begins with an act of definition:

> There are, for instance, the women who, through some freak of nature, are compelled to work their way in life without the advantages which come to human beings generally. The blind, the deaf, the dumb, the idiot, are too mournful to be called odd; and the anecdotes referring to them are so well known that they need not be touched on here.[23]

The anonymous author goes on to describe the kinds of physical difference that *are* narratable: women born without arms, who engage in painting and other kinds of fancy-work by using their toes; "Bearded women [who] are more odd than loveable," and so on. Their stories read like a perverse variation on the marriage-plot novel, centering on a search – here unsuccessful – for husbands. For example, the author recounts some remarkable stories about pig-faced women, forced to resort at last to the most denigrating measures, like bribery or advertising, in the hopes of avoiding spinsterhood:

[One] story is that of Janakin Skinker, born in Rhenish Holland in 1618, well proportioned in form else, but pig-faced, and having no other power of language than a grunt. She, or her parents, offered forty thousand pounds to any gentleman who would marry her; many gallants came, but one and all begged to decline when they had seen her.[24]

Another version ends happily, after the woman's face recovers its human shape:

Equally veracious, we suppose, was the story of a Belgian gentleman who renounced the church and embraced Judaism; the first child born to him afterwards had a pig's face; but in later years, when the father recanted and the daughter was baptised, the face miraculously changed to human form.[25]

The explicitly misogynistic, anti-Semitic, and xenophobic rhetoric offers a disturbing view of mid-Victorian patriarchal bigotry. It also expresses the widely held belief that female celibacy constituted a disgusting form of abnormalcy. One is tempted to read this cultural bias through the lens of capitalism, with the single middle-class woman a slub in the otherwise silky fabric of bourgeois economic discourse, an irregularity that displays itself with a kind of quiet insistence. In other words, these are women who, "through some freak of nature, are compelled to work their way in life" and they disrupt the balance of labor and gender roles, in that, despite outward evidence of elevated social class (the armless Miss Hawlin, for example, wore a powdered wig and "frilled trowsers") they are compelled to support themselves because they are unmarriageable. Capitalism does indeed play an important role in the formulation of redundancy as a social problem; its doctrine pervades the work of conservative writers on the subject like Eliza Lynn Linton, who presumes a division of middle-class male and female labor into separate camps, marketplace and home, and warns that "the woman who is 'too gifted, too intelligent, to find scope for her mind and heart' in domestic duties 'is simply not a woman. She is a natural blunder, a mere unfinished sketch.'"[26]

But, if the language and principles of capitalism informed bourgeois anxieties about single women, the languages and gestures of social definition, the signs which populated the theatre and the collective imagination, were antecedent and necessary to the current understanding of socioeconomic patterns and rhythms. Worker–boss relationships were imagined as general class relations were imagined, in terms of the finite set of voices, statures, and plots established in popular entertainments of both radical and conservative politics. One sees this in industrial novels like *Mary Barton*, and industrial melodramas like John Walker's *The Factory Lad* and J. B. Johnstone's *How We live in London*, all of which use verbal differences to manifest social or moral facts, and one sees it in the public acceptance of such gestures as authentic. An eruption of strange figures and voices on the scene, pigs and dwarves, clowns and madwomen, was as threatening as any disruption in the division of labor, especially since these pieces of the carnival, bursting out of their cultural repression, leveled a subrational threat, an assault on the machinery of bourgeois regulation by the unstructured, the licentious, the mad, a guerilla attack not easily deflected and difficult to contain.

The madness and spectacle dormant in bourgeois industrial culture had its corollary in bourgeois women, from whom the carnivalesque sometimes issued in the form of verbal display, the mad, redundant patter of widows and spinsters. It is hardly surprising to see single women implicitly or explicitly associated with crazy language and deformed or half-formed bodies in a culture which regularly associated women and female physiological functions with lunacy or perpetual infancy. According to a dominant strain of Victorian thought, lunacy or other forms of mental deficiency were shared by all women, married or single, and were thought to be related to menses; the suggestions of lunar influence in the monthly wax and wane of menstrual fluid might indeed encourage those so inclined to deduce lunacy as a "natural" female condition.[27] Mental deficiency was a "natural" condition of femaleness, and verbal or physical deficiencies were presented, in the popular culture, as companions to redundancy, variant or secondary symptoms of the same basic problem.

Patter represented verbal deformity, a motivated sign of abnormality in the nineteenth century, in the same way that physical malformation is always in itself a literal sign of difference, with literal and figurative social meanings attached to it. The excessive, recursive, estranging speech of single women was read as a sign of oddity, in the word's dual sense of strangeness and unmarriageability. This belief in the literality of physical or verbal signs was widespread, as one constantly sees in the day's popular journals, and it extended far beyond the aesthetic–political contexts we are concerned with here. For example, an article on nutrition in *All the Year Round* posits a direct relationship between diet and racial characteristics, with the "almost exclusively animal diet . . . [of] the Indians of North America . . . [producing] A considerable degree of physical development, of grace, vigour, and even of beauty," attainments exceeded only, according to this writer, by those of "an ordinary white man of active habits," and set in direct contrast with the disadvantages of "the hideous Rootdigger, [with] his stunted growth and anxious face."[28] This suggests that one might read the human body as a kind of ethnographic map, locating the details of geographic location, race, diet, and physical activity in such variables as length of limb, cleanness of gait, facial expression, and so on. The voice, as we have seen, worked in much the same way, offering, in its rhythms, accents, and inflections crucial pieces of information, facts upon which society could be organized. I have already recovered some of those facts; here I shall invoke some of the odd woman's popular theatrical and literary sources before finally turning to *Little Dorrit*, so that we too may have the advantage of this larger context as we approach the novel.

The pattering spinster is as least as old as the middle of the eighteenth century, when she began to appear on the stage in such plays as Murphy's *The Old Maid*.[29] A sister, in spirit, of female characters like Mrs. Malaprop and Mrs. Slipslop, the pattering spinster provokes laughter with her sexual and financial greed, and her verbal eccentricity. The "malapropists" of the English stage, who are not exclusively female but include characters like Shakespeare's Constable Dogberry, speak clearly and in normal syntactic cadences; the humor in their speech lies in semantic error, as in Dogberry's injunction to the watch that they "shall also make no noise in the streets; for, for the watch to babble and to talk, is most tolerable, and not to be endur'd."[30] But patterers, as we have seen, use either slang or "standard" words in strange rhythmic patterns and syntactic cadences; the humor – and danger – in their idiom lies in its baffling difference from standard English, its rich and beguiling dialogism.

The Lord Chamberlain's files contain an enormous number of eighteenth- and nineteenth-century English plays with spinsters, old maids, and widows in their titles; one also finds wives, daughters, lovers in abundance, as in the innumerable "Schools for" all of the preceding. Trailing Murphy's *The Old Maid* by about a century, J. S. Knowles' *The Old Maids* continued its predecessor's hegemonizing work in more sophisticated and subtle form. This piece, by a reputable and successful playwright, was licensed for Covent Garden in 1841, and seems to have enjoyed a successful run. Its rather complicated plot centers on two sworn spinsters, the Ladies Anne and Blanche, and the gradual dissolution of their resolve as two lovers, Sir Philip Brilliant and Tom Blunt, enter the scene. The play's deconstruction of the women's alternative domestic plan is interesting if predictable, with its introduction of a kind of class cross-dressing (the Countess Anne spends much of her time dressed as a yeoman's daughter) and its highly unusual rejection of the urgency of class divisions by marrying Anne to Tom Blunt, an army colonel and goldsmith's son.

But its greater interest lies in its dialogue, which contains seeds of character, distilled truth, in its verbal patterns. While the entire play is written in more or less regular blank verse, the women's dialogue undoes itself, undermining the control suggested in the very choice of blank verse with its repetitions, confusions, recursiveness. Lady Anne, for example, whose antipathy to marriage runs deep ("Oh! how I doat upon a staunch old maid! I'll die one!"), speaks in rushed, highly hyphenated blocks of dialogue, reminiscent of her predecessor Miss Harlow's fractured, doubled speech.

> Well! Things will jump into one's memory,
> When we least look for them – why do you laugh?
> Don't laugh, dear Bess, and I will tell you more –
> I took the goldsmith to my milliner's one day,
> When he would perforce see me home.
> A yeoman's daughter could not well, you know,
> O'er rule a goldsmith's son. Well, at the door
> In vain I dropped him curtsey after curtsey,
> In linsey-woolsey mode! he would not go –
> He must have speech with me a minute – Nay!
> Indeed, he must – then said I Nay, again –
> He must, in pity – Still did I say nay –
> But what's the use of nay, said fifty times,
> If yes at last will come – and come it did –
> He might have speech a minute – what's a minute?
> A portion of an hour – A portion gone,
> The hour is broken! What's the value of
> A broken thing? As well he have the hour!
> The hour he had – the goldsmith's son was smitten –
> Love at first sight! – the arrow in the core!
> Whereat the maid answered – it may be, pleased;
> Touch'd, will you have it so – well, she was touch'd.[31]

Were her dialogue written in prose, the typical form taken by patter, its excesses would be far more extreme than these. Even so, we can easily recognize the linguistic strategies and the general spirit behind Lady Anne's dialogue. With its breathiness, its unfocused narration, its confusion of speaking subjects, her speech suggests a degree of distraction or excitement on Anne's part; in this sense its structure can be linked to the dramatic moment. Indeed, these are the words of a specific character caught up in a particular sequence of events, and we must read them as such. But they also suggest certain social and even physiological facts about their speaker. They assert, for example, with their peculiar theatrical resonance, her sorority with the standard English spinster (for example Miss Harlow and Miss Euphemia Blight); and they may imply as well that her female physiology has ravaged her intellect and logic, with her broken, unfocused idiom the "natural" result. But they are also the words of a storyteller, lingering in small detail, speaking every voice and touching the emotions. Here, as always, patter is a gift as well as a curse; the irony in such derisive treatments of it lies in its potential suitability for the novelist's or playwright's own work, and in their typical dependence on it for narrating certain kinds of stories.

As Anne begins to relinquish her notions of celibacy, her speech tightens up, becomes more organized, although it loses, at the same time, its intensity of feeling, its hints of the carnival, of pleasure and pain. If this is merely a shift from one conventionality to another, from the spinster's patter to the wife's standard English, it is a shift to a safer conventionality, to a voice contained in the rhythms of regularity, a voice of clarity but one without the power to touch the emotions. This voice is exemplified in the speech of her future husband, whose moral fiber seeps into his very words – firm, unembellished little packages like "To say the least [gambling] never can consist / With proper manhood, to enjoy the thing / Was not one's own an hour ago, and now / One's merit has not won him!"

The Old Maids conveys a sense of disaster averted, a budding Amazon society thwarted. It was not the only mid-nineteenth-century contemplation of such a society; Tennyson imagined one in *The Princess* (1847) and Elizabeth Gaskell did as well, but much more sympathetically, in *Cranford.* The idea of female community suggests the possibility of female sexual autonomy: perhaps a fear of lesbianism partly motivated literary and theatrical attacks on redundant women. Implicit in Rhoda Nunn's claim that there is "no making a pair with them" lie possibilities of alternative pairings. Gissing's novel certainly hints at lesbian relationships between unmarried or unmarriageable women, who share small living spaces, beds, and housekeeping arrangements. That women could only be safely and morally employed in marriage and mothering suggests a real fear of their sexuality, as if it must be harnessed and directed to prevent its flowing into dark territories, into the unthinkable, the immoral, the lunatic. Of course, there has been much done on this subject (less, perhaps, on lesbian sexuality than female sexuality and lunacy) and I shall only touch lightly on it here. But spinster- or widowhood may over-sex as well as de-sex, just as patter, the voice of celibacy, may harness or amplify a character's speech and physical presence. This very richness of possibility intensified the redundant woman's threat, confirming her moral and potentially physical difference from the Victorian feminine ideal.

Whether the heightened sexuality sometimes imagined to attend female celibacy was homosexual or heterosexual, it was generally manifested in patter, which, with typical doubleness, suggested either potency or redundancy. Henry and Augustus Mayhew imagined it to signify both. Their exploration of spinsterhood, courtship, and marriage, *Whom to Marry and How to Get Married, or the Adventures of a Lady in Search of a Good Husband*,[32] charts the progress of a young woman who fields several offers of marriage, is married and widowed several times, and decides to commit her experience to paper. Most of the narrative details Charlotte de Roos' spinsterhood and widowhood – as if these are the only instances of her authentic self – leaving the scenes of her married life to be imagined or ignored, unnarratable stories, to borrow an idea from D. A. Miller.[33] Her first-person narrative swells and sinks to its own eccentric rhythms. Like most specimens of "redundant" speech it rushes forth, neglectful of punctuation and the rules of logic, fractured by hyphens and generally overcome by its own verbal bounty. But Charlotte's storytelling also teases the reader by deferring "consummation," or the yielding of certain key elements of the tale, and by offering itself as kittenish and voluptuous, the voice of a child and a *femme fatale.*

> Offer the First . . . Which certainly was not what – with my improved views of life – I should now designate an excellent offer, for I really don't believe that the poor, poor, wretch of a man has sixpence in the world, except what he got by the perspiration of his brow; still, as he was my first love, and certainly remarkably fond of me, perhaps it is but right that I should begin this *petite histoire* with some account of that insinuating *vaut-rien*; especially as it will show the gentle reader what a silly, silly, truthful disinterested little puss I was at "sweet sixteen."

At this point the gentle reader has already recognized the signs of redundancy in Charlotte's narrative: serpentine sentences of enormous length; repetitions; a heightened interest in the erotic and pecuniary; general frivolity. What is most significant about this book is that the idiom in which it is written so faithfully embodies the truth of its protagonist that her married life is elided, an irrelevancy, an unnatural condition for one of her verbal cast. One is led to suspect that Charlotte will continue a spinster regardless of her marital status, as if the condition were biologically, not sociologically, determined.

William Greg and his supporters thought that this was true, classing spinsters into two groups, the natural and unnatural. "Natural" spinsters included those "who seem utterly devoid of the *fibre feminin* . . . [those] too passionately fond of a wild existence . . . [those] who seem made for charitable uses . . . ideal old maids . . . [and those] in whom the spiritual [or the masculine] predominates."[34] These represent pure old maidism, a kind of genetic deformation which results in the redundant woman. All others, argues Greg, are spinsters by accident, socialization, or circumstance; women who would marry if they could. However, if the Mayhews implicitly endorsed Greg's paradigm in their book, others refuted it. *The Truth About Man, by A Spinster*, published anonymously at the turn of the century, responds to mid-Victorian assumptions about spinsterhood by proposing a spinster of another kind, one "who regards marriage not as a prize, but as a snare to be cleverly avoided, while she sports round the rim of it. Believing that to travel hopefully is better than to arrive, she looks upon such arrival at the altar as a stern conclusion to a delightful frolic."[35] Whether or not the work of a spinster (and I suspect, having tasted its acid, that it was), this book defies the conventional discourse, not only in its celebration of female independence and play, but in its clear, often epigrammatic diction. Its later date may partly explain its liberalism; by 1905, feminism had established itself as a strong countervoice in Britain.

Clearly, the conflation of female celibacy with physical or verbal grotesqueness is a common device in Victorian literature and journalism, and it reflects a cultural investment in preserving domestic balance. When Dickens made gestures of antagonism towards spinsters or widows, he did so with this idea of violated domestic balance in mind. In *David Copperfield*, for example, a spinster inserted into the domestic machinery precipitates a colossal breakdown. While the problem is somewhat larger and more complex than the mere presence of an old maid – this one being the sister and pawn of David's vicious stepfather, the real villain of the story – Miss Murdstone does stand for a set of unnatural and "un-English" values. Steely, flint-like, and antimaternal, she is incapable of fulfilling the "natural" female roles of wife and mother which she usurps from her sister-in-law. She disciplines David with the hard hand of a lower sort of schoolmaster rather than the gentleness of a mother, and her maternal inadequacy is inscribed upon her body:

> a gloomy-looking lady [Miss Murdstone] was; dark, like her brother, whom she greatly resembled in face and voice; and with very heavy eyebrows, nearly meeting over her large nose, as if, being disabled by the wrongs of her sex from wearing whiskers, she had carried them to that account.[36]

Miss Murdstone's femininity is compromised by "whiskers," large nose, and masculine deportment. Dickens insinuates that she is not only incapable of mothering, but incapable of marrying as well. And this poses a frustrating paradox: to marry the confirmed spinster would be like marrying her brother, a violence to the institution of marriage, yet to leave her unmarried does the same thing – violates the social "balance" sustained by marriage.

Miss Murdstone is basically a comic character, despite her cruelty and the pathos in David's subjection to her mothering. Redundant female characters are almost always meant to be comic, even if hints of darkness and danger occasionally emanate from their verbal or bodily display. Let us linger for a moment on this fact, and explore some larger questions about comedy before moving on to *Little Dorrit.* We have already seen how patter may be simultaneously funny and threatening; the doubleness of a trope like patter inheres not only in its forms and its reception, but in its generic affiliations as well. If patter belongs to the comic universe, it at least occasionally pays a visit to tragedy, rendering a difficulty in definition not at all unusual in the aesthetic world, a problem often solved by imagining a hybrid category as Mozart did for *Don Giovanni,* which is neither *opera buffa* nor *opera seria,* but a *dramma giocosa,* and as Polonius does in introducing in court "The best actors in the world, either for tragedy, comedy, history, pastoral, pastoral-comical, historical-pastoral, tragical-historical, tragical-comical-historical-pastoral."[37] Dickens himself acknowledges this difficulty, in the complexity of feeling evoked by Flora Finching in Arthur Clennam, whose "sense of the sorrowful and his sense of the comical were curiously blended" as he observed the changes in his old sweetheart.[38] We might take their example, and define patter as comi-tragic or comi-tragipathetic or comi-patheti-ethical, or we might, as I have chosen to do, stick with "comic" but understand the word to accommodate other impulses, to cover a larger field of significance than, say, Aristotle, or Henry Fielding, would allow.

Fielding, despite the rigidity of his definitions, writes about comedy in ways which I find compelling, and which provide an interesting contrast to the use of comedy in nineteenth-century treatments of female eccentricity. I wish to explore for a moment his theory of comedy, specifically as it relates to the writing and reading of romance. Fielding introduces his novel *Joseph Andrews* with some basic instruction on reading novels, including descriptions of the various genres and the appropriate reader responses to different types of writing.

> [The comic romance] differs from the serious romance in its fable and action, in that as in the one of these are grave and solemn, so in the other are light and ridiculous: it differs in its characters by introducing persons of inferior rank, and consequently of inferior manners . . . lastly, in its sentiments and diction, by preserving the ludicrous instead of the sublime. In the diction, I think, burlesque itself may sometimes be admitted . . . The only source of the true ridiculous (as it appears to me) is affectation . . . from affectation only, the misfortunes and calamities of life, or the imperfections of nature, may become the objects of ridicule.[39]

Fielding's argument constructs a reader who expects a certain type of truthfulness in social interactions; when that truth is violated, for example by social pretense, the violator is treated to ridicule or satire. This anxiety about social definition is clearly related to earlier anxieties about rogues, vagabonds, and masterless men, vagrants who, in the sixteenth and seventeenth centuries, were thought to disrupt the social order by taking on and discarding identities at whim, performers *par excellence.* The passage I have quoted also suggests, implicitly, that what issues from the author's pen signifies a kind of literal or biological truth, with affectation the "source of the true ridiculous," and the language of affectation "naturally" ludicrous. We have seen these ideas promoted in Victorian letters and criticism; much of Fielding's ideology remained in currency throughout the following century.

For Fielding, social climbing represented a threat of anarchy and as such, demanded the dramatic sort of social policing that he subjected it to himself in *Joseph Andrews* and other works. His cruel derision of Mrs. Slipslop, for example, saps her of any potential power for social disruption because it renders her completely ridiculous. Not only a spinster, but a boastful and affected servant, Slipslop embodies vulgarity and insignificance, and Fielding, in a manner typical of the eighteenth century, presents her as "a mighty affecter of hard words,"[40] a woman of appetite, and extremely ugly.

She was a maiden gentlewoman of about forty-five years of age, who, having made a small slip in her youth, had continued a good maid ever since. She was not at this time remarkably handsome; being very short, and rather too corpulent in body, and somewhat red, with the addition of pimples in the face.[41]

But the fact of her spinsterhood seems less important than her misplaced social (and amatory) ambition when one compares her with similar characters in Sheridan, Smollett, and even Burney, and finds vulgar social ambition the common trait among them.[42] As Fielding himself asserted, eighteenth-century fiction isolates and punishes transgressions of manners and social place by ridiculing the guilty parties, emphasizing the grotesqueness of their bodies and their speech.

In contrast to the Augustan attention to behavior (eighteenth-century attacks on spinsterhood are cloaked, at least, as attacks on violations of certain rules of social conduct), Victorian popular literature and entertainments tend to construct the facts of marriage or of redundancy as biologically determined traits, and worthy on their own of approbation or reprobation. If Victorian spinsters misbehave, it is because they are inherently flawed, unnatural women. We saw this Victorian focus on the literal self in the Mayhews' Charlotte De Roos, whose very soul is bared in her words, whose spinsterhood is represented as pure and irrevocable as a shriveled limb. Charlotte's narrative suggests that her actions are symptomatic of a biological condition, and therefore secondary to a larger transgression. Patricia Ingham describes this condition as an excess of femaleness, or femaleness unschooled and uncontained,[43] although Victorian social critics insisted that it marked a deficiency of proper female impulses. Either way, it is a condition inseparable from the darkest regions of body and mind, inherent in certain women in the same way that truth was inherent in the spoken voice. Indeed, redundant women did serve as semiotic markers in Victorian popular culture, as figures encoded with certain meanings. They, like the patter that so often issued from their lips, contained deep within them seeds of disruption, ugly and infectious truths which demanded constant nipping in the bud in order to prevent their flowering into full chaos.

And that note on *flora* and chaos brings us neatly to *Little Dorrit* and the extravagant Flora Finching. Here is a novel which seems to concern itself with almost every contemporary social issue except the problem of redundant women. Money, poverty, ignorance, bureaucracy, prostitution, child abuse, all enter explicitly into the novel's moral agenda, but nowhere, despite the fact that the single woman problem was bandied about while this novel was in progress, does Dickens explicitly engage with the debate over redundancy, although there are several redundant women in *Little Dorrit.* Flora, Miss Wade, Mrs. Clennam, Mrs. Tickit, Mr. F.'s Aunt, and Mrs. General constitute the sum, and they differ greatly in character, voice, shape, and social stature. While Flora is ridiculously loquacious, fat and florid and good-hearted (an unnipped bud), her companion and aunt by marriage is almost hidden, a shriveled body under voluminous clothes, speaking, only occasionally, in eery *non sequiturs,* small verbal eruptions of psychosis. Mrs. General and Mrs. Clennam, both tight and wary widows, police their own speech, the former for lapses in the "prunes-and-prisms" of propriety, the latter for religious sacrilege. Mrs. Tickit patters, though with somewhat less force than Flora, and Miss Wade is dark and foreign-flavored, cold, beautiful, and dangerous.

While the characterological variations in this novel's "odd women" suggests some looseness, or richness, in the popular conception of redundancy (in fact they match fairly accurately Greg's redundant "types") they share among them, with the exception of Miss Wade, several significant traits. Each is identified by or with her speech, whether it be uncontained or overcontrolled, and each, with the possible exception of Mrs. Tickit, is after sex or money. Flora represents the odd woman in her most common form, and she will figure at the center of this discussion. But Miss Wade, in some ways the oddest of them all, is an enigma, and I cannot proceed with my discussion of the novel without acknowledging her and the difficulties she strews in my path, as she does to so many of her fellow characters in Dickens' novel. It may be that Dickens did not perceive her as redundant, but as her age, which would assign to her either the role of girl or spinster if we knew it, is withheld from the narrative, it is difficult to place her at all. Described as a "handsome young Englishwoman" with a gloomy beauty, she is obviously still young enough to expect offers of marriage: this is the language conventionally used to describe heroines, not spinsters, in Victorian literature. Even so, everything about her suggests the dark sort of wisdom which comes with experience, and the insistent "Miss" attached to her name constantly reminds us that she is single. The complications in her social status – not quite definitively young or mature, English or foreign, attractive or repulsive, and *apropos,* marriageable or headed for spinsterhood – suggest that she represents a kind of femaleness which Dickens had not learned how to contain, which had no prototype on the stage or in the imagination, and hence which defies categorization, darkly generating its dark stories in the margins of discourse, a malignancy feeding on the depths of female consciousness. Her kind of oddness, one which hints of lesbianism, sexual and financial victimization, sadomasochism, and madness, is less easily condensed into a verbal tic or physical deformity; it seems that Dickens hardly knew what to do with her, beyond letting her tell her own story in the interpolated "History of a Self Tormentor."

I feel something of the same bewilderment, and for the very same reasons: Miss Wade has no obvious antecedents, no line of women characters before her, to whom we might refer. She is one of Dickens' few truly psychological creations, and the reading public's relation to her is much more difficult to imagine than their relation to, say, Flora Finching. If Miss Wade defies our attempts at placing her, Flora Finching sits squarely within the lines that demarcated redundancy in English popular culture. Clearly on the far slope of middle age, physically and verbally gross, sexually frustrated, hers is a familiar figure. We have seen her grandmothers and great-aunts Tabby Bramble, Mrs. Slipslop, and Miss Harlow; her sisters Mrs. Nickleby, Miss Knag, Miss Euphemia Blight. Victorian readers received her in this context, recognizing her mythical dimensions, her theatrical and literary antecedents, and, as I shall argue, her significance in the dispute over what to do with redundant women.

Flora enters the text in a chapter appropriately titled "Patriarchal." The Patriarch is Flora's father, Christopher Casby, who disarms everyone except Arthur Clennam and the narrator with his shining knobs of forehead, his long gray ruff of hair, and his benign expression. "Patriarch was the name which many people delighted to give him. Various old ladies in the neighborhood spoke of him as The Last of the Patriarchs. So grey, so slow, so quiet, so impassionate, so very bumpy in the head, Patriarch was the word for him."[44] A forgery of paternal benevolence, bourgeois to his very bones and without a hint of the *noblesse*, he might almost be mistaken for a feudal lord if Bleeding Heart Yard were not located in the shabby middle of Victorian London. If a feudal lord exacts high rents and social obedience, and in doing so preserves the gulf which lies between himself and his dependents, then Casby is like him, but if he offers in return a form of protection, a guarantee of resources and employment, then here the two ways part. Casby exacts high rents without pledging anything in return. A confirmed humbug by the novel's close, he demonstrates the limits of paternalism in industrial London.

The chapter entitled "Patriarchal" also demonstrates the limits of paternalism in ways which its author probably did not intend. Dickens was nothing if not a champion of paternalism, with his public works of philanthropy and his fictional visions of marriage as a union between father- and daughter-figures, perhaps nowhere as evident as in *Little Dorrit.* Still, as if she bucks his authority and flings herself bodily and emotionally to the very edges of his narrative control, Dickens' own Flora insists that we notice the father's failures. This has partly to do with her idiom; patter, as I have already argued, undoes its own work of containment by flaming forth, noisy, spectacular, demanding of attention. Thus, when Flora enters the chapter named for her father, she fills the frame with her full-blown body and speech, neither of which are meant to "work," the one unsuited for catching a husband, the other for capturing and delivering an idea. Despite her good nature, Flora tends to disgust; Arthur Clennam, reacting as Dickens is supposed to have reacted to the grown-up Maria Beadnell, recoils from her overwhelming physical–verbal presence. Even so, she takes over, an enthralling presence, far more interesting than the pale men around her.

A close reading of *Little Dorrit* shows just how closely its author imagined language, written and spoken, phoneme, semanteme, phrase, and sentence, to be related to individual and social meanings. His is a semiotic universe, where people are signs – the best of them true, connected, they and their speech, to some external reality; the worst of them disconnected, deceptive like mountebanks, *poseurs*, performers. Christopher Casby is one of the latter, "a mere Inn signpost without any Inn – an invitation to rest and be thankful, when there was no place to put up at, and nothing whatever to be thankful for."[45] He deceives by saying nothing, having discovered that "to get through life with ease and credit, he had but to hold his tongue, keep the bald part of his head well polished, and leave his hair alone";[46] when he does speak, he utters gentle, meaningless repetitions, not patter but flacid, empty speech. Casby's meaning resides in certain parts of his body, which, curiously enough, promote a "true" false impression about him – to some extent he is what he appears to be – and in this sense the Patriarch's body resembles the stocks, bonds, and notes which signify another patriarch's wealth to the world, even though that man, Mr. Merdle, is as bereft of actual wealth as Casby is of paternal benevolence.

In a universe where semiotic systems regularly fail to expose the truth, constructing instead alternative, morally deficient universes, the very act of reading is burdened with a certain kind of ethical responsibility; the stakes go up, as it were, and the reader requires special equipment to successfully negotiate the passage. Of course, this all takes place under Dickens' controlling eye; he insures that his readers do indeed possess the ability to discern the fakes from the real thing, and, like any novelist, metes out parcels of information in such a way as to sustain a level of suspense, giving us only the most obscure clues to the truth about Merdle, for example, until the moments before his suicide and the stupendous ruin that follows it. Reading, in *Little Dorrit*, takes place on two levels: the internal acts of social interaction and negotiation, in which characters try – and often fail – to read each other, and the external, in which the novel reader reads the characters' reading attempts, and assigns meaning to the narrative as a whole. Thus, while Arthur Clennam analyzes Flora in the parlor of her father's home, the Victorian reader analyzes that analysis, and draws a conclusion based as much on Flora's relationship to the imaginary text on redundancy as on her relationship to Clennam and the other characters in the novel.

That imaginary text, as we have seen, constructed redundancy as biological as well as social; as a violation of domestic and economic balance; as a sign to be read on the bodies and in the speech of spinsters and widows. Of course, there were plenty of dissenting voices participating in the debate over what to do with redundant women: the Mills, Mary Taylor, Jessie Boucherett and others published rebuttals to the conservative position – the biologically centered arguments, the calls for moderate persecution (Greg suggested that spinsters' lives be made as uncomfortable as possible, to incline them towards marriage) – in such journals as the *Englishwoman's Review* and *Victoria Magazine*, but they were in fact fighting the entire popular culture machine, the circulating constructions – theatrical, literary, musical, and visual – of female normalcy and eccentricity, and not merely the handful of journalists who wrote explicitly about the problem. No matter how compelling and eloquent the attack, it competes, perhaps without knowing it, against novels like *Little Dorrit*, novels often not even interested in determining "what to do" with redundant women, but contributing to the debate, nonetheless, with their reliance on old and powerful stereotypes like patter.

Little Dorrit certainly participates in the debate, sending up Flora Finching as its delegate, a living sign of the disruptions caused by such women, although for every step towards the conservative end (eradication) she retreats a step, existing simultaneously as proof that single women are superfluous and silly, more trouble than they are worth, and victims, symptoms of the cancer in the paternalist system. Both signs can be detected in her behavior, which, if gauche and oversexual, tends towards acts of great kindness; and more importantly, in her speech, which, patently redundant, demonstrates her redundancy, but at the same time engages in the histrionic fiction-making of a fertile if unschooled mind. Any reading of Flora must center on her speech. Most Victorian readers would have assumed this, because, as we have seen, they "naturally" turned to the voice for information, and also because, as Patricia Ingham explains, "Conversation . . . [was] frequently seen as women's sphere and a testing ground for their progress towards the ideal. [Social critics had] many recommendations to make about it, always on the assumption that, left to themselves, women produce undesirable kinds of talk."[47]

Given his culture's vested interest in vocal and linguistic patterns, it is only natural that Arthur Clennam, after a brief inspection of Flora's changed face and figure, turns to her dialogue for information about her. Once he enters its labyrinthine corridors, he remains trapped inside of it for the rest of their acquaintance, unable to trace its winding ways to any real conclusion; this perhaps newly acquired verbal excessiveness disturbs him more than her ruddy corpulence.

> Flora, always tall, had grown to be very broad too, and short of breath; but that was not much. Flora, whom he had left a lily, had become a peony; but that was not much. Flora, who had seemed enchanting in all she said and thought, was diffuse and silly. That was much. Flora, who had been spoiled and artless long ago, was determined to be spoiled and artless now. That was a fatal blow.[48]

Flora's "progress towards the ideal" has failed. If her overblown physical and verbal presence is in some sense a rebuke to the patriarchy, a taking by force of the stage, a spectacular alternative to both kinds of paternalism – the false, embodied in her father, and the true, exemplified by Clennam – it also plays into the father's hand, confounding order and clarity, posing an alternative to "natural" romance, which in Dickens often occurs between very young women and middle-aged men, as in Clennam and Amy Dorrit, but never between such men and fat, florid, oversexed widows. Indeed, all of Clennam's energy is expended in the interpretation of Flora's speech – one suspects that, had he wanted anything more, he would have been too exhausted to claim it. "The inconsistent and profoundly unreasonable way in which she instantly went on, nevertheless, to interweave their long-abandoned boy and girl relations with their present interview, made Clennam feel as if he were lightheaded."[49] As one of many semiotic systems in the novel which yields only circularity and opacity, that confounds the reason and clarity which do exist in the world for those who know where to seek them, Flora is a constant drain on Clennam's – and the narrative's – energy, pulling both, with an insistent drag, towards a black, sticky hole of moral, economic, and epistemological uncertainty.

Flora's real-life counterparts were thought to do the same, disrupting the well-oiled works of the bourgeois economic and domestic systems, pointing insistently, even sternly, and perhaps with the same effect that the Ghost of Christmas Yet to Come had on Ebenezer Scrooge, to the inevitable conclusions of those systems. As feminist writers claimed, the exclusion of women from the training and education which would enable them to support themselves, turned an innocuous social fact into a serious economic problem, with single women, unequipped to provide for themselves, imposing a drain on other people's resources. Their very presence served as a kind of rebuke, a bony, accusatory finger; hence, perhaps, the serious proposals of forced emigration. Flora's verbal excesses are themselves a sort of forced emigration, a relegation of her voice to the furthest margins of intelligibility, an overload of sense that stops the works altogether and forces the reader to move on to more reasonable ground.

These excesses, as I have suggested, constitute a sapping of narrative energy, a derailment of logical and chronological sequence into small swamps of indeterminacy. When Flora opens her mouth, the story sinks into inertia, its sharpness and logic congealed, if you will, into a thick soup, fluid but formless, ostensibly going nowhere. The same might be said for the novel's other redundants, whose speech almost always disrupts the plot's progress, as in Mrs. Clennam's and Miss Wade's sinister deflections of certain facts, acts of verbal bad faith that impede the novel's amateur investigators, or in the self-contained dialogue spoken by Mr. F.'s Aunt and Mrs. General, which refers to some internal system of meaning rather than the events of the real world, and which Dickens holds up to view as articles of strangeness and humor rather than the seamless dialogue which participates in social and narrative progress. Most of the novel's redundant speech shares a quality of illogicality, or an alternative (and inferior) "female" logic, and in this it merely confirms the standard Victorian assumptions about female intelligence.

Punch often published articles about this feminine deficiency, as it did in 1860 with a piece entitled "Logic For Ladies, By One Of Them." This "one of them" is a spinster by the looks of the caricature in the margin, an ugly, large-headed, small-bodied, middle-aged woman – apparently, with her withered lower body and over-developed cranium, ill-equipped for childbearing. The article itself confirms this suspicion, as it ends with a pointed reference to nieces and nephews rather than her own children. Not unlike the "letter to the editor" I quoted earlier, this piece proves that spinsters cannot sustain serious thought; that any high-minded discourse on politics, logic, money, must inevitably droop, like a millinery feather, into the lower regions of fashion, society, and marriage. "Ladies ought to be fine logicians . . ." this exemplary specimen of the sex informs us. Ought to be, perhaps, but cannot be – even the most pedantic among them.

> Logic teaches us to train our mental faculties, so that we may firmly hold the thread of our discourse, and prevent it getting into a tangle. Under its guidance alone we draw from sound premises a safe conclusion. If people's premises are untenable, they must necessarily break down, just as if an Alderman and his august consort were to dance a *pas de fascination* in one of our modern compound villas; assuming (as we may) the premises to be unsound, one can easily predicate with what a disaster such an imprudent step must conclude.
>
> Of mental operations there are these three: simple apprehension, judgement, and discourse or reasoning. Thus, if a lady's hoop should entrap a gentleman's hat and carry it out of a church pew, it is simple apprehension; but it requires judgement to drop it gracefully at the porch; and logical *acumen* to prove that such an abduction is sanctioned by the law of licensed carriers.[50]

This logician, having loosened her grasp on the thread of discourse, ends up tangled in technical language she does not really understand, and undone, as well, by her "natural" and irrepressible interest in crinolines and hoops. In this sense her idiom resembles Flora's, although its confusions and excesses are less ferocious. Filled with irrelevancies, much longer than it needs to be, and patently illogical (as in her syllogism "All men are heartless / A Parrot is not a Man / A Parrot, therefore, is not heartless"), this treatise on logic constitutes an assault on its own subject, a parody of "serious" philosophy. It takes its place in the larger cultural subtext on redundant women, a text which we know Dickens consulted as he wrote his novels, and which the public consulted as they read them.

While Flora Finching speaks a language marked by its violations of the logic of standard English, it is possible to read in that language a deferral to the requirements of a different species of logic, or of the same species of logic operating under different conditions and in a different time zone. For instance, while Flora's locutions seem unconnected to the conversations and events around her, they do sustain a kind of sequence and logic among themselves, so that one may, by extracting them from the surrounding text and piecing them together, discover a continuous train of thought and a degree of linguistic consistency in them. Almost all of her conversation attends to the changes wrought upon her appearance and her relationship with Clennam; she generally picks up the abandoned threads of one dialogue in the next, no matter how much later it occurs. Furthermore, the majority of her verbal explosions contain the following phrase: "One last remark I might wish to make one last explanation I might wish to offer" – a statement referring to the dissolution of the love affair by her father and Clennam's mother, and Flora's subsequent, and presumably willing, marriage to Mr. F. The sum of Flora's dialogue represents an attempt to explain an affair of the past: she is stuck in a verbal hole, groping for the best way to justify her past behavior without damaging her future prospects. The love affair itself might be a fiction – even Clennam is unsure – and Flora's insistence on reclaiming it, refiguring and reenacting it in the harsh light of the present involves her in an act of supreme fiction-making. The patterer may not be a classical logician, but she is, despite the novelist's best efforts to conceal it, a spinner (spinster) of stories, a speaker of strange and resonant tongues, disruptive, exciting, disconcerting.

My point, in juxtaposing Flora with the popular culture-text on redundant women, has been to imagine Dickens' own assumptions as he wrote the novel, to locate his voice among the many strands of voice competing to be heard on the problem of unpaired women, and to disclose some of the popular ideas which fed the debate over redundancy. That debate was not, of course, based on pure sociopolitical facts or needs, if indeed such purity ever exists. The bourgeois anxieties and desires expressed in the debate – the need for social balance, the belief in a feminine ideal, the assumptions of biological urgency in gender delineations – were articulated in the popular culture, in plays and novels and paintings. The crisis came when the reality of industrial England no longer matched popular assumptions, when marriage and motherhood eluded more and more middle-class women, and the derisive treatments of female celibacy in circulation failed to contain the redundant women whose numbers increased as the century wore on.

Dickens' reliance on conventional theatrical tropes in characterizing his spinsters and widows suggests not only his own debt to theatrical culture, but the tangling of popular entertainments and social politics as well. *Little Dorrit* fed the debate over redundancy, with its troupe of bizarre and generally distasteful single women, just as the actual debate, playing out in the periodical press, influenced the novel, and a great number of mid-Victorian popular entertainments besides.

This study has explored these tangles, the communions of theatrical and fictional, imaginative and real, aesthetic and political, which dominated Victorian socioaesthetic experience. If there is a doubleness in patter, the voice with which I have spent so much of my time, lately, there is a doubleness – or more accurately, a polyvalence – in almost everything Victorian, certainly in the novels and novel readers of the period, and particularly in Dickens. Thus, if *Little Dorrit* contributes to the conservative side of the redundancy debate, it also, with typical resonance and flexibility, opens itself to something better, yields ever so slightly to the carnival impulse which runs deep inside of it, bends to the wild *flora* which bloom so brightly and insistently that they threaten to overrun the orderly domestic garden.

Little Dorrit ends in marriage, a union between Amy Dorrit and Clennam which deprives Flora Finching of the closure she has desired for her narrative. This denial of marital closure is, of course, essential to the redundant woman's condition. But widows and old maids demonstrate other kinds of open-endedness in Victorian popular culture – linguistic diffusion; the possibility of alternative sociosexual and economic arrangements; demographical imbalance, the oddness of the unpaired. The novel itself posits this openness against the closure of the wedding, a buzz of noise against the quiet regularity of bourgeois married life. After Flora treats Amy to a kidney pie at the bakery shop, a gesture of selflessness and self-indulgence which the younger woman understands though unable "closely to follow Mrs. Finching through [her verbal] Labyrinth," she attends the wedding, the event which seals her own eccentricity, which leaves her in the marginal place which she prefers after all.

> "The withered chaplet my dear," said Flora, with great enjoyment, "is then perished the column is crumbled and the pyramid is standing upside down upon what's-his-name call it not giddiness call it not weakness call it not folly I must now retire into privacy and look upon the ashes of departed joys no more but taking the further liberty of paying for the pastry which has formed the humble pretext for our interview will forever say Adieu!"[51]

Far from the proposed retirement into privacy, Flora will live out her days as the public, spectacular, histrionic widow, the popular myth; inhabitant of Drury Lane and the drawing-rooms of London, of journals and pictures and popular songs as well as the pages of novels like *Little Dorrit.* Her voice will be intractable; it will motivate generations of politicians and writers against her, and it will prove its fascination in the prominence of its role in English popular entertainments. Little Dorrit and Arthur Clennam may retreat into silence, "inseparable and blessed," but all around them the likes of Flora Finching, "the noisy and the eager, the arrogant and the froward and the vain, fretted and chafed and made their usual uproar."[52] Theirs – hers – are the voices we remember.

CHAPTER 7

Conclusion

A person once told me (a lover of Shakespeare's plays, who thought Victorian novels inferior beings) that Dickens was a businessman first and an artist last: if the novels showed rare flashes of brilliance they were due perhaps to a sparkle of pecuniary anticipation that sometimes brightened his eyes. I cannot remember the eyes of the speaker as he delivered this comment, but his voice I remember as highly serious. I think about his words every time I pick up *David Copperfield* or *Our Mutual Friend*, every time I marvel at the richness and resonance of Dickens' prose, and the peculiarity – in both senses of the word – of his vision.

"Dickens," the man said, "was in it for the money; his purpose was utilitarian, to bleed as much prose out of his brain as possible so that he could buy his great house and live at the height of fashion."

Dickens was, certainly, "in it for the money"; there is no question about that. But if he wrote – and performed – to make piles of cash, if a kind of brittle utility browns the pages of his books and his life, then he is in good company indeed. Mozart rarely wrote a page of music, as far as we know, which someone had not bought from him. Beethoven made his living at composition; Schubert did not, and as the story goes, died young and hungry.

Now when I think of this incident, and how I might respond to it today, I think, curiously enough, about snow, and about a large illustrated book on snow crystals in my bookcase. Snow is unimaginative, the product of certain atmospheric conditions, and, one might argue, utilitarian in that its larger purpose is to replenish water in the ground and in the rivers. But each snow crystal is a gratuitous work of art, beautiful, cleanly executed, original. A Dickens novel may lack the precision and self-confidence of a snow crystal, but if the whole of Dickens' professional effort served the larger purpose of replenishing money depleted from the family bank – and served, or attempted to serve, on a cultural level, the function of discursive regulation – each individual piece lives its own aesthetic life, vivid, complex, and true to itself, while participating in the larger process. This "larger process" is one I have described in this book, of individual characters, or voices, fitting into a larger signifying apparatus, the whole unified into a dominant theatricality. But the single voice may retain its singularity, its distinct pattern, so that it lies slightly in relief upon the normalized background, a snow crystal on a pile of undifferentiated white snow.

That some voices will stand out, retaining a certain carnival power, is confirmed by the perennial interest, among actors and actresses, in exploring those voices. Over the past several years some very distinguished dramatic explorations of Dickensian eccentrics have entered circulation, including Patrick Stewart's "A Christmas Carol" and Miriam Margolyes' "Dickens' Women."[1] The success of these performances is due to their performers' understanding that the characters they dramatize are ultimately theatrical beings; that a Dickens novel may best be understood as an amalgam of dramatic voices and rhythms, a medium more for the voice and the ear than the eye. In fact, the *Carol* has captured the imaginations of the nineteenth and twentieth centuries precisely because it gives itself so fully to theatrical adaptation: many people today know the story of Scrooge and the three Christmas spirits – or think they do – without ever having read Dickens' text.[2] But they have heard the tale told in many voices, some awful but most participating insistently in an overdetermined *Carol* experience; they have seen it in various forms, performed by Muppets, cartoons, third-rate comedians, and even, if they are lucky, by the likes of the intelligent and gifted Stewart.

Even today, then, something of the Victorian reading experience lingers, at least in the rare instances that a literary work has attained such a level of cultural resonance – and relevance – that some people have "read" it without reading it, and many cannot read it without hearing the dialogue delivered in a certain way, without imagining the characters to have certain faces, without hearing strains of conventional music complement the dramatic situation. Perhaps *A Christmas Carol* is the only one of these works left; in any case, it demonstrates, more or less practically, the complex layering of nineteenth-century literary experience. Imagine a situation in which novels are always complicated in this way, in which readings are bound by a set of very specific common experiences, in this case theatrical experiences. Of course, all readings, textual and cultural, depend on a specific and mutual semiotic network, but the type of reading this study has explored shares with other, contemporaneous, readings a circumscribed imaginative universe, in which novel characters are virtually always old friends, with little variation in the set of dramatic gestures available to them – that is, available to the writer creating them, and the reader processing them. While this situation may limit readers' creative agency, it also enriches their general experience. Again, the *Carol* will demonstrate this. If we assume that a substantial proportion of the English-speaking public who fall into the appropriate age category, own a television and/or a videocassette recorder, and occasionally go to the movies, have seen at least one of several adaptations of the story produced over the past twenty years or so, we may also assume that they will remember parts of those adaptations if and when they read (or reread) the original. The popular renditions will affect readings of Dickens' story, adding a certain kind of resonance whether or not the renditions were "good" ones, or received as such by the viewer-reader.

Perhaps this stretches too far into the hypothetical – there are many "ifs" in the scenario – but it is the reading experience I have striven to recover here. If we lack hard evidence that *A Christmas Carol* functions thus today, we do have evidence that it and other narratives, particularly but not exclusively Dickensian narratives, resonated in the Victorian popular imagination, that nineteenth-century novels and their readers embraced the stage in ways that we can only with difficulty imagine. Disputing the currently dominant historical teleology, questioning the inevitability of the nineteenth century's interiorized mechanisms of control, may help us to make that leap, to imagine the novel reader not as isolated, curled up on a window seat in the privacy of the sitting room, as D. A. Miller has suggested, but sitting in a theatre, the imagination's playhouse, with a book in one hand and an eye and ear on the stage.

Notes

I INTRODUCTION

1 Michel Foucault, *Madness and Civilization: A History of Insanity in the Age of Reason*, trans. Richard Howard (New York: Vintage Books, 1973); *Discipline and Punish: The Birth of the Prison*, trans. Alan Sheridan (New York: Vintage Books, 1979).

2 D. A. Miller, *The Novel and the Police* (Berkeley: University of California Press, 1988), p. 82.

3 "Popular theatre" generally refers to the melodrama here, as the melodrama genre dominated the nineteenth-century stage, and as the performance styles which we tend to associate with melodrama were, in fact, the standard in all kinds of Victorian plays. See chapter 2, note 3 for a more detailed explanation.

4 Miller, *The Novel and the Police*; D. A. Miller, *Narrative and its Discontents: Problems of Closure in the Traditional Novel* (Princeton University Press, 1981).

5 Martin Meisel, *Realizations* (Princeton University Press, 1983).

6 Joseph Litvak, *Caught in the Act: Theatricality in the Nineteenth-Century English Novel* (Berkeley: University of California Press, 1992).

7 George Taylor, *Players and Performances in the Victorian Theatre* (Manchester University Press, 1989); Joseph Roach, *The Player's Passion* (Ann Arbor: University of Michigan Press, 1993).

8 See, for example, Nina Auerbach's *Private Theatricals* (Cambridge, Mass.: Harvard University Press, 1990); Philip Collins' *Charles Dickens: The Public Readings* (Oxford: Clarendon Press, 1975) and *Reading Aloud: A Victorian Metier* (London: Tennyson Society, 1972); Michael Booth's *English Plays of the Nineteenth Century*, vols. I–IV (Oxford: Clarendon Press, 1969–76) and *Theatre in the Victorian Age* (Cambridge University Press, 1991); George Rowell's *The Victorian Theatre: A Survey* (Oxford: Clarendon Press, 1967); Edwin Eigner's *The Dickens Pantomime* (Berkeley: University of California Press, 1989); and Robert Garis' *The Dickens Theatre* (Oxford: Clarendon Press, 1965).

9 Paul Davis, *The Lives and Times of Ebenezer Scrooge* (New Haven: Yale University Press, 1990).

2 DICKENS AND THE "IMAGINARY TEXT"

1 This is not to suggest that Victorianists have not recognized and analyzed the public sphere, or the shared assumptions of the culture. Some excellent books have recently explored publicity and the nineteenth-century popular imagination, including Elaine Hadley's *Melodramatic Tactics* (Stanford University Press, 1995), Richard Altick's *Writers, Readers and Occasions* (Columbus: Ohio State University Press, 1989), and Linda Hughes and Michael Lund's *The Victorian Serial* (Charlottesville and London: University Press of Virginia, 1991). I am referring specifically to the presumption, which governs so many works of scholarship, of a privatized Victorian subject, and a Victorian novel which is itself "privatized" – a discrete aesthetic unit, fixed and unresponsive to the conditions around it. And a large number of representative critical studies focus on such matters as identity, sexuality, domesticity, and death, as if private practices are the only legitimately Victorian ones.

2 Miller, *The Novel and the Police*, p. 82.

3 I typically use "popular theatre" in this study to refer to the melodrama, partly because melodrama dominated the Victorian theatre scene, and partly because most Victorian productions assumed what one might describe as a "semiotics of melodrama." In other words, for Victorian actors and directors, the theatrical signs that we receive as quintessentially "melodramatic" today were not specific to that genre. They were simply the elements of good theatre and employed in all genres of plays. In a sense, then, "melodrama" takes on a much more expansive significance than its generic one. This study imagines it as this larger phenomenon, a concept that shaped all Victorian theatrical performances. In some cases a specific theatrical genre is specified, such as comedy or farce in my discussions of patter in performance, but even then one should assume that the spirit of melodrama, at least pertaining to acting style, dominates.

4 Miller, like many of the scholars whose work has contributed so richly to the field of nineteenth-century studies, embraces too uncritically Foucault's historical paradigm, assuming the nineteenth century to be an age of interiority. Even some major writings on the Victorian theatre and theatricality have been shaped by Foucault's privatization of the nineteenth century. These include two significant books: Joseph Litvak's *Caught in the Act*, and Nina Auerbach's *Private Theatricals*, which is ultimately concerned, as its title suggests, with theatricality of a private rather than public sort – performances of self-actualization, surveillance, spirituality, and death.

5 For example by Collins (*The Public Readings*), Paul Schlicke (*Dickens and Popular Entertainment* [London: Unwin Hyman, 1985]), Garis (*The Dickens Theatre*), and Meisel (*Realizations*). Taylor (*Players and Performances*) suggests Frederick Robson's dramatic characters as sources for some of Dickens' creations, and several writers, starting with Macready, have found models in the "At Homes" of Charles Mathews. Indispensable is Philip Bolton's work on Dickensian adaptations, *Dickens Dramatized* (London: Mansell, 1987).

6 Herbert Blau, *The Audience* (Baltimore and London: Johns Hopkins University Press, 1990), p. 8.

7 Schlicke, *Dickens and Popular Entertainment.*

8 While Hughes and Lund's interesting work, *The Victorian Serial*, asserts that serial publication "required readers to stay with a story a long time and to postpone learning a story's outcome" (p. 4), the appearance of novels on the stage before the end of their serialization in fact compromised the deferral of gratification that Hughes and Lund cite as one of the cultural values driving serialization, even if the endings contrived by playwrights did not resemble the conclusions eventually written by the novelists.

9 I have focused almost exclusively on theatrical sources, although I do not mean to dismiss the textual influences on Victorian novels; clearly writers like Dickens inherited literary forms and conventions from their predecessors. But the novels of the mid-nineteenth century also bear explicit marks of early and mid-nineteenth-century theatre. One finds countless examples of this in the major novelists alone, most of whom relied on character types which had been popularized in the melodrama, farce, comedy, and pantomime: the sentimental or Gothic hero (e.g. Nicholas Nickleby, Charles Darnay, Felix Holt, Mr. Rochester, Daniel Deronda, Adam Bede); the comic loquacious woman (e.g. Flora Finching, Mrs. Nickleby, Mrs. Tulliver, Mrs. Holt, Miss Clack); the "villain" (e.g. Ralph Nickleby, Mr. Murdstone, Mr. Dombey, Count Fosco, Sir Pitt Crawley); the *buffo* (e.g. Mark Tapley, Mr. Pancks). All of these characters are drawn to type – they speak or act like characters currently walking the boards. And Dickens' novels were not the only ones to be adapted: popular sensation novels like *East Lynne* and *Lady Audley's Secret* practically begged to be dramatized, and even "serious" fiction like *Jane Eyre* and *Adam Bede* found its way on to the stage. Dramatist did not stop at fiction, either: Henry Mayhew's muck-raking *London Labour and the London Poor*, an unlikely candidate for adaptation, turned up on the stage in dramas such as *Want and Vice, or, London Labour and the London Poor*, and *How We Live in London.*

10 Several commentators, from John Forster (*The Life of Charles Dickens*, ed. J. W. T. Ley [London, 1928]) to Earle Davis (*The Flint and the Flame* [Columbia: University of Missouri Press, 1963]) and Robert Golding (*Idiolects in Dickens* [New York: St. Martins, 1985]) have noted Dickens' debt to the early nineteenth-century comedian Charles Mathews.

11 Mary Poovey, in *Uneven Developments* (University of Chicago Press, 1988), briefly describes this saturation of "real life" with the structures and characters of the melodrama in her discussion of Caroline Norton and the Matrimonial Causes Act.

12 *Edinburgh Review*, 68, October 1838: p. 84.

13 Paul N. Campbell, "Communication aesthetics," *Today's Speech*, 19, 1971: p. 9.

14 See Leon Rubin, *The "Nicholas Nickleby" Story* (Harmondsworth: Penguin Books, 1981). I am also indebted to Stephen Rashbrook, an actor from the

company, who discussed the process with me at some length.

15 David Edgar, "The Life and Adventures of Nicholas Nickleby," in *Plays from the Contemporary British Theater*, ed. Brooks McNamara (New York: Mentor Books, 1982), pp. 329–330.

16 Rubin, *The "Nicholas Nickleby" Story*, p. 79.

17 For a look at nineteenth-century theatrical gesture, see Henry Siddons, *Practical Illustrations of Rhetorical Gesture and Action* (New York: Benjamin Blom, 1968).

18 Charles Dickens, *"The Amusements of the People" and Other Papers: Reports, Essays and Reviews, 1834–51*, ed. Michael Slater (Columbus: Ohio State University Press, 1996), pp. 57–58.

19 Peter Ackroyd, *Dickens* (New York: Harper Perennial, 1992), p. 268.

20 One might also borrow Paul Davis' term "culture text" to describe this overdetermined literary–cultural experience, although the culture text, as Davis constructs it, is perhaps less shifty and permissive than what I have in mind here – a fluid, multifaceted experience in which the fictive and the real are continually confused or conflated, and readings are performed through a filter of eclectic popular ideas with various etiologies and affiliations. See *The Lives and Times of Ebenezer Scrooge*.

21 Tracy Davis, "The Actress in Victorian Pornography," in *Victorian Scandals*, ed. Kristine Ottesen Garrigan (Athens: Ohio University Press, 1992), pp. 99–133.

22 Turner defines the liminal and liminoid in *From Ritual to Theatre* (New York: Performing Arts Journal, 1982).

23 Jürgen Habermas, *The Structural Transformation of the Public Sphere* (Cambridge, Mass. MIT Press, 1991), p. 31.

24 Siddons, *Practical Illustrations*, pp. 2–3.

25 Ibid., p. 27.

26 It was Descartes, of course, who published the definitive treatise on the affections and their physical manifestations, *Les Passions de l'ame*. Reprinted in *The Philosophical Writing of Descartes*, vol. 1 (Cambridge University Press, 1985).

27 See Taylor, *Players and Performances*, for a discussion of the passions in eighteenth- and nineteenth-century English theatre.

28 Meisel, *Realizations*, p. 7.

29 Charles Dickens, *Nicholas Nickleby* (Harmondsworth: Penguin Books, 1986), pp. 83–84; my italics. All parenthetical text references are to this edition.

30 George Eliot, *Felix Holt, the Radical* (Harmondsworth: Penguin Books, 1987), p. 398.

31 Charlotte Brontë, *Jane Eyre* (New York: St. Martins Press, 1996), pp. 281 and 339.

32 Michael Booth, *English Melodrama* (London: Herbert Jenkins, 1965), p. 192.

33 Hadley, *Melodramatic Tactics*, p. 4.

34 Charles Darwin, *On the Expression of Emotions in Man and Animals* (New York: D. Appleton and Co., 1899), p. 239.

35 On page 323 of the Penguin edition.

36 Edmund Shaftesbury, *Lessons in the Art of Acting: A Practical and Thorough Work for All Persons Who Aim to Become Professional Actors* (Washington, D.C.: Webster Edgerly, 1889).

37 Dickens, *Nicholas Nickleby*, p. 662.

38 Tracy Davis explores the gendered signifying properties of costume in "The Actress in Victorian Pornography."

39 Erika Fischer-Lichte, *The Semiotics of Theater*, trans. Jeremy Gaines and Doris Jones (Bloomington and Indianapolis: Indiana University Press, 1992), pp. 137–138.

40 Herbert Blau, *To All Appearances: Ideology and Performance* (New York and London: Routledge, 1992), p. 54.

41 That this interdependence should be perceived as "peculiarly Victorian," and not, say, as typical also of the eighteenth century, which saw the early developmental stages of the English novel and sustained a strong theatre industry, is confirmed by the forms taken by eighteenth-century novels themselves. The period's strong interest in the epistolary and the picaresque – neither of which opens itself easily to theatrical influence – suggests that the novel, in its infancy, sought to differentiate itself from other forms of public entertainment or instruction. Even Frances Burney, who grew up in the dazzling glare of operatic and theatrical personalities and performances, wrote novels more heavily dependent on discursivity, on metonymic relationship – "novelistic" rather than "theatrical" qualities – than those of her Victorian successors.

42 Edward Stirling, *Nicholas Nickleby, a Burletta in Two Acts.* First performed November 15, 1838 (British Library, Lord Chamberlain Collection, Add. MS 42949, fos. 636–672b). Other adaptations of 1838 include the ones discussed in this chapter by W. T. Moncrieff (Add. MS 42951, fos. 594–684) and H. Horncastle (*Dicks' Standard Plays*, no. 572 [London: J. Dicks, 1882]). See Bolton, *Dickens Dramatized*, for the entire list. Meisel has argued that Stirling's staged work "contributed pictorially to the memorable realization of the novel" (*Realizations*, p. 258).

43 Taylor, *Players and Performances*, p. 143.

44 George Rowell, ed., *Nineteenth-Century Plays* (Oxford University Press, 1972), p. 10.

45 Bolton, *Dickens Dramatized*, p. 154.

46 "Dickens's Tales," *Edinburgh Review*, 68 (1838/9): pp. 76–78.

47 Harriet Martineau, *Autobiography* (Boston: J. R. Osgood, 1877), pp. 378–379.

48 Letter to Richard Henry Horne, March 2, 1853, printed in *An Account of the Performance of Lytton's Comedy "Not So Bad As We Seem," with Other Matters of Interest by Charles Dickens* (London, printed for private circulation by Richard Clay and Sons, 1919).

49 Review in the *Examiner*, September 23, 1838: p. 595.

50 *Examiner*, October 27, 1839: pp. 677–678.

51 Charles Dickens, *Dombey and Son* (Harmondsworth: Penguin Books, 1970), p. 700.
52 Charles Dickens, *Our Mutual Friend* (Harmondsworth: Penguin Books, 1971), pp. 266–267.
53 Ibid., pp. 452 and 456.
54 Rowell, ed., *Nineteenth-Century Plays*, p. 243.
55 Of course, *Lady Audley's Secret* was a novel – a highly theatrical one – before it became a play, but given the generous dialogism between the two genres it seems appropriate to treat the adaptation as hardly less authoritative than the original, to do the Victorian thing and talk about the play first. Braddon herself drew quite self-consciously on the theatrical gestures and scenarios in *Lady Audley*, as even a cursory look into the novel reveals; she wrote a story her public could not fail to understand.
56 Schlicke, *Dickens and Popular Entertainment*, p. 3.
57 Michael Booth, ed., *English Plays of the Nineteenth Century*, vol. I (Oxford: Clarendon Press, 1969), p. 34.
58 Isaac Pocock, *The Miller and His Men*, reprinted in Booth, ed., *English Plays*, pp. 37–38.
59 Douglas Jerrold, *Black-Ey'd Susan*, reprinted in Rowell, ed., *Nineteenth-Century Plays*, p. 42.
60 Leona Weaver Fisher, *Lemon, Dickens, and "Mr. Nightingale's Diary": A Victorian Farce* (University of Victoria Press, 1988), pp. 22ff.
61 Ibid., p. 23.
62 Both Malaprop and Slipslop speak in a comic idiom which is – has to be – eminently comprehensible: measured and precise in their speech, they substitute, at the crucial moment, one word for a like word, so that the audience recognizes the slip. Thus, Mrs. Slipslop, when asked to acknowledge an old acquaintance, replies "I think I *reflect* [recollect] something of her." While her idiom seems, at least in the eighteenth century, to be a marginal feminine one, it differs from spinster patter in too many ways to be related, except, of course socially – they do the same sort of work.
63 Even though women participated fully in urban street life, as Mayhew has shown us, street patter is represented, in middle-class cultural productions, as predominantly male, which makes sense given its affiliation with commerce.
64 An early nineteenth-century example would be Isaac Pocock's Dick Cypher (see chapter 4 of this book), a slang-pattering attorney with upper-class pretensions. A later example, from Dickens himself, is Toby Magsman, whose speech encompasses the entire narrative of "Going into Society," and who, as a gossipy, idle man appropriates a conventionally "female" role.
65 Although they were aware of its negative history; *Ruddigore*, for example, satirizes its own patter with the lines "This particularly rapid, unintelligible patter / Isn't generally heard, and if it is, it doesn't matter" (Alan Jefferson, ed., *The Complete Gilbert and Sullivan Opera Guide* [Exeter: Webb and Bower,

1984], p. 229). On the other hand, the gleeful exploration of the aftermath of verbal error in *Pirates of Penzance* – the misconstruction of "pilot" as "pirate" – is really unimaginable earlier in the century, when one suspects people would have been properly appalled to see an innocent child sent off to a life of piracy because his deaf nurse "misheard" her instructions.

66 Thomas Carlyle, *Sartor Resartus* (New York: A. L. Burt Co., 1910), p. 5.

67 George Eliot's review of *Sartor Resartus* reflects the general attitude towards Carlyle and his prose: "For there is hardly a superior or active mind of this generation that has not been modified by Carlyle's writing; there has hardly been an English book written for the last ten or twelve years that would not have been different if Carlyle had not lived. The character of his influence is best seen in the fact that many of the men who have the least agreement with his opinions are those to whom the reading of *Sartor Resartus* was an epoch in the history of their minds" (quoted in *Essays of George Eliot*, ed. Thomas Pinney [London: Routledge and Kegan Paul, 1963], pp. 213–214). See also J. P. Seigel, ed., *Thomas Carlyle: The Critical Heritage* (New York: Barnes and Noble, 1971).

68 Other canonical novelists use it as well. Charlotte Brontë, for example, wrote some fabulously tangled and circular prose, particularly in describing moments of anguish or stress.

69 Charles Dickens, *Bleak House* (New York: Bantam Books, 1980), p. 1.

70 Alexander Pope, "An Essay on Criticism," in *The Poems of Alexander Pope*, ed. John Butt (New Haven: Yale University Press, 1963), lines 356–357.

71 Dickens, *Bleak House*, p. 184.

72 Many of Dickens' prompt copies are available at the Berg Collection of the New York Public Library. See chapter 3 below for more on Dickens' prompt copies.

73 See the discussions of Victorian theatre and theatre criticism in chapter 5 below.

74 Cicely Berry, *Voice and the Actor* (New York: Collier Books, 1973), pp. 7, 101.

75 Wallace Shawn, *Marie and Bruce* (New York: Grove Press, 1980), pp. 1–2.

76 Ackroyd, *Dickens*, p. 260.

77 Dickens, *Nicholas Nickleby*, p. 354.

78 Bakhtin introduces this concept in *The Dialogic Imagination*, trans. and ed. Caryl Emerson and Michael Holquist (Austin: University of Texas Press, 1981).

79 H. Horncastle, *The Infant Phenomenon; Or, A Rehearsal Rehearsed.* Originally produced on July 8, 1838, at the Strand Theatre. Printed in *Dicks' Standard Plays*, no. 572.

80 Dickens, *Nicholas Nickleby*, p. 356.

81 Ibid., p. 359.

82 Quoted in Fisher, *Lemon, Dickens, and "Mr. Nightingale's Diary,"* p. 16.

83 Charles Dickens, *Sketches by Boz* (London: Oxford University Press, 1957), pp. 119–120.

84 Dickens, *Nicholas Nickleby*, p. 363.

85 Bakhtin, *Dialogic Imagination*, p. 293.
86 Dickens, *Nicholas Nickleby*, p. 372.
87 Ibid., p. 367.
88 Garis, *Dickens Theatre*.

3 THEATRICAL ATTITUDES: PERFORMANCE AND THE ENGLISH IMAGINATION

1 Charlotte Brontë, *Villette* (Harmondsworth: Penguin Books, 1979), p. 339.
2 Matthew Arnold, "Shelley," in *The Poetry and Criticism of Matthew Arnold*, ed. A. Dwight Culler (Boston: Houghton Mifflin Co., 1961), p. 365.
3 There are a variety of good books on the Victorian theatre, its repertoires, audiences, players, and so on. Michael Booth has written extensively on the subject. See his *Theatre in the Victorian Age* (Cambridge University Press, 1991) for a discussion of theatre audiences in the nineteenth century.
4 *London Times*, February 7, 1870, p. 11.
5 William Hazlitt, "On Actors and Acting" (1817), in *Criticisms and Dramatic Essays of the English Stage* (London, 1854), p. 13.
6 Herbert Blau discerns an organic connection between a culture's "reality" – "a newer realism," he calls it – and its "possibilities and strategies of performance." *To All Appearances*, p. 54.
7 Jonas Barish has devoted an entire book (*The Anti-Theatrical Prejudice* [Berkeley: University of California Press, 1981]) to the subject of antitheatricality, essential reading for anyone interested in the subject; William Worthen's *The Idea of the Actor* (Princeton University Press, 1984) explores the projection of these anxieties on to the actor.
8 See Collins, *Reading Aloud* and *The Public Readings* for discussion of Victorian dramatic readings.
9 Collins, *Reading Aloud*, p. 11.
10 *The Journal of William Charles Macready*, ed. J. C. Trewin (London: Longmans, 1967), pp. xv–xvi.
11 Barish, *The Anti-Theatrical Prejudice*, pp. 324–325.
12 My construction of cultural self-regulation as "policing" is obviously indebted to Miller's *The Novel and the Police*.
13 Richard Schechner, *Performance Theory* (New York and London: Routledge, 1988), p. 158.
14 Auerbach, *Private Theatricals*, p. 4.
15 See Litvak's discussions of these in *Caught in the Act*.
16 The association of theatre with French morals is not uncommon, and was perhaps encouraged by the French dominance of the European theatrical scene by actresses like Rachel Felix and Sarah Bernhardt.
17 Brontë, *Villette*, p. 339.
18 Ibid.
19 Ibid., p. 340.
20 John Forster, *The Life of Charles Dickens*, ed. J. W. T. Ley (London, 1928), pp. 380, 381.

21 Brontë, *Villette*, p. 341.
22 Hazlitt, "On Actors," p. 1.
23 Brontë, *Villette*, p. 213.
24 Jean-Jaques Rousseau, "Letter to M. D'Alembert on the Theatre," in *Politics and the Arts*, ed. and trans. Allan Bloom (Ithaca: Cornell University Press, 1960), pp. 15–18.
25 See Litvak's *Caught in the Act* on the infectious in performance.
26 Virginia Guildersleeve's *Government Regulation of the Elizabethan Drama* (New York: Columbia University Press, 1908) offers a comprehensive study of theatrical controls during the Renaissance.
27 See Frank Fowell and Frank Palmer, *Censorship in England* (New York and London: Benjamin Blom, 1913) for details of these acts.
28 Guildersleeve, *Government Regulation*, p. 11.
29 See Watson Nicholson, *The Struggle for a Free London Stage* (New York: Benjamin Blom, 1966) for an outline of the proceedings of the 1810 Privy Council on this matter.
30 Quoted in D. F. Cheshire, *Music Hall in Britain* (Rutherford, N.J.: Fairleigh Dickinson University Press, 1974), p. 24.
31 Printed in *Engravings by Hogarth*, ed. Sean Shesgreen (New York: Dover Publications Inc., 1973), plate 4.
32 See, for example, William Ingram's *The Business of Playing: The Beginnings of the Adult Professional Theater in Elizabethan London* (Ithaca, N.Y.: Cornell University Press, 1992).
33 Hazlitt, "On Actors," p. 11.
34 This is not to suggest that all actors were penniless and vagrant. A certain type of professional, the actor-manager, owned or leased the theatre with which a nonstrolling troup was affiliated. The Burbages are among the most famous of these; others include David Garrick in the eighteenth century, Charles and C. J. Mathews and the celebrated Madame Vestris in the nineteenth century.
35 William Shakespeare, *Hamlet* (Harmondsworth: Penguin Books, 1970).
36 Fowell and Palmer, *Censorship in England*, p. 368. This act established the office of Lord Chamberlain, and restricted "legitimate" drama to two licensed theatres, Drury Lane and Covent Garden. Foucault has written about the containment of vagrants as a discursive strategy in *Madness and Civilization.*
37 Guildersleeve, *Government Regulation*, p. 25.
38 Quoted in Barry Taylor, *Vagrant Writing: Social and Semiotic Disorders in the English Renaissance* (University of Toronto Press, 1991), p. 2.
39 Ibid., p. 3.
40 Henry Mayhew, *London Labour and the London Poor*, 2 vols. (New York: Dover Publications Inc., 1968) vol. 1.
41 Charles Dickens, "Two Letters on Public Executions, by Charles Dickens, Esq., Published with the Sanction of the Author" (letter to *The Times* 1849).
42 *Household Words*, 4, 1852.

43 Foucault, *Discipline and Punish.*

44 See *The Reign of Terror: A Collection of Authentic Narratives of the Horrors Committed by the Revolutionary Government of France Under Marat and Robespierre* (London: Leonard Smithers and Co., 1899) for eyewitness accounts of mob performances during the French Revolution.

45 Mozart to his father, November 12, 1778, in *The Letters of Wolfgang Amadeus Mozart*, ed. Hans Mersmann, trans. M. M. Bozman (New York: Dover Publications Inc., 1972), p. 133.

46 Marc Baer, *Theatre and Disorder in Late Georgian London* (Oxford: Clarendon Press, 1992).

47 John Dryden, *Marriage à la Mode* (Lincoln: University of Nebraska Press, 1981), p. 7.

48 Ibid., p. 8.

49 See Victor Glasstone, *Victorian and Edwardian Theatres* (Cambridge, Mass.: Harvard University Press, 1975), p. 12.

50 Baer, *Theatre and Disorder*, p. 175.

51 Andrew Davies describes theatrical "transgressions" in his discussion of the "penny gaffs" in *Other Theatres* (Totowa, N.J.: Barnes and Noble, 1987).

52 Charles Dickens, *Great Expectations* (Harmondsworth: Penguin Books, 1965), p. 275.

53 Blau, *To All Appearances*, p. 52.

54 Roland Barthes, *S/Z* (New York: Hill and Wang, 1974), defines the "writerly" text as one which is constantly being written, "productive" rather than representative.

55 Harold Perkin, *The Origins of Modern English Society, 1780–1880* (University of Toronto Press, 1969), p. 25.

56 Wolfgang Amadeus Mozart and Lorenzo DaPonte, *Le Nozze di Figaro* (New York: Dover Publications Inc., 1979), p. 272.

57 Samuel Richardson, *Clarissa* (Harmondsworth: Penguin Books, 1985), p. 883.

58 The most thoughtful contemporary stagings of eighteenth-century literature and opera explore this semiotics of marks and postures. Stanley Kubric's gorgeous 1975 film adaptation of *Barry Lyndon*, for example, tells its story through a series of painterly quotations – one recognizes real Gainsboroughs and Hogarths in Kubric's gallery – and relies much more on visual combinations of static cultural signs, like periwigs, makeup, patches, and brocades, than on plot movement. And Jean-Pierre Ponnelle's 1980 video direction of Mozart's *La Clemenza di Tito*, filmed at the Baths of Caracalla in Rome, is instructive in its insistence on its own artifice, both as an *opera seria*, an artificial genre even in its Baroque moment, and as an eighteenth-century cultural artifact. Here, the acting (stagy and stylized), the editing (choppy, abrupt), the costumes (self-conscious mixtures of eighteenth-century court dress and stereotypical Roman costume, a sort of imagined Augustan imagining of Roman culture), construct a world which may be read by piecing together a field of displayed signs.

59 Denis Diderot, *Rameau's Nephew*, trans. Leonard Tancock (Harmondsworth: Penguin Books, 1966), p. 34.
60 Fischer-Lichte, *The Semiotics of Theatre*, pp. 151, 160.
61 Chapter 5 of this study demonstrates the uses of voice in Victorian performance, focusing on its more constructive qualities.
62 Reproduced in *Phiz: Illustrations from the Novels of Charles Dickens*, ed. Albert Johannsen (University of Chicago Press, 1956), plate 32. See Meisel's *Realizations* for more on connections between book illustrations and theatrical tableaux.
63 *Punch*, 26, 1854: p. 225.
64 *All the Year Round*, April 17, 1875.
65 George Eliot, *Felix Holt*, p. 395.
66 Ibid., pp. 398–399.
67 Ibid., pp. 411–412.
68 Appropriately, a play about Villon (*Villon, Poet and Cut Throat*, by S. C. Courte) was produced at the Royalty Theatre in 1894.
69 "An Unreported Speech," *All the Year Round*, 6, 1861: pp. 179–180.
70 Ella Ann Oppenlander ascribes it to Charles Allston Collins. *Dickens' "All the Year Round" Descriptive Index Contributor List* (Troy, N.Y.: Whitston, 1984).
71 Shesgreen, ed., *Engravings by Hogarth*, plate 47.
72 Henry Mayhew, *The Wandering Minstrel*, reprinted in *English and American Drama of the Nineteenth Century*, ed. Allardyce Nicoll and George Freedley (New York: Readex Microprint, 1969).
73 Henry Fielding, *Tom Jones* (New York: W. W. Norton, 1973), pp. 134–138. "First *Jemmy Tweedle* felt on his hinder Head the direful Bone. Him the pleasant Banks of sweetly winding *Stour* had nourished, where he first learnt the vocal Art, with which . . . he cheered the rural Nymphs and Swains, when upon the Green they interweav'd the sprightly Dance; while he himself stood fiddling and jumping to his own Music" (pp. 135–136).
74 Foucault, *Madness and Civilization*.
75 Frances Burney, *Camilla, or, A Picture of Youth*, book IV, chapter 6 (London: Oxford University Press, 1972).
76 Worthen, *The Idea of the Actor*, p. 3.

4 PATTER AND THE POLITICS OF STANDARD SPEECH IN VICTORIAN ENGLAND

1 Jefferson, ed., *The Complete Gilbert and Sullivan*, pp. 228–229.
2 Peter Davison, ed., *Songs of the British Music Hall* (New York: Oak Publications, 1971).
3 *Oxford Dictionary of English Etymology* (Oxford University Press, 1966), p. 269.
4 The *Chronological English Dictionary* (eds. Thomas Finkenstaedt, Ernst Leisi, and Wolff [Heidelberg: C. Winter, 1970]) estimates that the first use of *patter* occurred between 1350 and 1450. The *Oxford English Dictionary*, 2nd edn (Oxford: Clarendon Press, 1989) and Captain Grose's *Dictionary of the Vulgar Tongue* of 1785 (ed. Robert Cromie [Chicago: Follett Publishing Co., 1971])

provide historical examples: "I patter with the lyppes, as one dothe that maketh as though he prayed and dothe nat, *je papelarde*." Grose's definitions of "Pattering" include the "Pattering of prayers; the confused sound of a number of persons praying together."

5 The *OED* offers several pertinent eighteenth-century usages. For instance, "The master who teaches them [young thieves] should be . . . well versed in the cant language commonly called the slang patter" (1758). Grose defines "pattering" as "The maundering or pert replies of servants," while "To Patter" means "to speak . . . the language used by thieves."

6 Henry Mayhew quotes "an educated gentleman . . . [who] had been driven to live among the classes he described"; this gentleman describes the initiation of a friend into the pattering brotherhood:

> "I had lived . . . more than a year among the tradesmen and tramps, who herd promiscuously together in low lodging-houses. One afternoon I was taking tea at the same table with a brace of patterers . . . determined to know their proceedings, I launched out the only cant word I had then learned. They spoke of going to Chatham. Of course, I knew the place, and asked them, 'Where do you stall to in the huey?' which fairly translated means, 'Where do you lodge in the town?' Convinced that I was 'fly,' one of them said, 'We drop the main toper (go off the main road) and slink into the crib (house) in the back drum (street).'" (*London Labour and the London Poor*, vol. 1, p. 217.)

Mayhew's informant's friend goes on to provide a brief dictionary of patter.

7 William Shakespeare, *Much Ado About Nothing* (New York: Signet, 1964), p. 93.

8 Charles Dickens, *Martin Chuzzlewit* (Oxford University Press, 1982), p. 349.

9 Charles Dickens, *The Pickwick Papers* (Oxford University Press, 1986), p. 139.

10 Daniel Defoe, *Moll Flanders* (Boston: Houghton Mifflin, 1959), p. 247.

11 It is interesting to note that Moll does not generally speak the "slang patter,' the street language of London thieves; Defoe's decision not to write a more authentic street narrative might have been partly due to the admiration he had for his heroine. Moll's liturgical pattering is a gentler reference to her immorality, less derisive than vulgar slang, from a dramatic point of view, and probably less alienating to a middle-class audience, many of whom, even as Protestants, would have been guilty themselves of "pattering with the lips."

12 See the *New Grove Dictionary* (ed. Stanley Sadie [20 vols.; London: Macmillan, 1980]) for discussions of *buffo* characters and their patter arias.

13 Samuel Johnson offers the more "decent" definition: "To make a noise like quick steps of many feet." *A Dictionary of the English Language*, vol. II (New York: AMS Press, 1967).

14 *The Century Dictionary*, ed. William D. Whitney. (New York: Century Co., 1889).

15 Charles Dickens, *Little Dorrit* (Oxford University Press, 1982), p. 129.

16 This autonomy was articulated in biblical law as well as English culture. See, for example, Richardson's reference in *Clarissa* (p. 361 footnote a) to the special status of widows in relation to social or legal obligation; the Book of

Numbers denotes a widow's word as fully binding, like any man's. This helps to explain the perception of widows as potential usurpers of male power, as socially and sexually dangerous, which continued into the nineteenth century.

17 In Dickens, early and late "redundant" female characters – Mrs. Nickleby, Mrs. Gamp, Miss Flyte, Flora Finching – are united by their common mode of speaking, even though the earlier and later novels differ tremendously in letter and spirit. In each case their speech enacts their superfluity, suggesting a gender politics that is consistent with the social attitudes about redundancy which dominated from fairly early in the century until near its end.

18 I am thinking in particular of Michel Foucault's "The Discourse on Language" (in *The Archaeology of Knowledge* [New York: Pantheon Books, 1971]) and Bakhtin's *The Dialogic Imagination.* Tony Crowley's *Standard English and the Politics of Language* (Urbana and Chicago: University of Illinois Press, 1989) also engages with these issues.

19 Bakhtin, *The Dialogic Imagination.*

20 Miller, *The Novel and the Police*, p. 25.

21 See, for example, "The Discourse on Language."

22 Ibid., p. 224.

23 While patter had existed as an inherently performative concept since the Middle Ages (it presumes a displaying object and an evaluating audience), it was not, as far as I can tell, actually performed much prior to the eighteenth century, during and after which its theatrical and novelistic forms evolved significantly.

24 *Punch* and other Victorian journals regularly reviewed operatic productions, and an evening at the music hall or even a licensed theatre might include selections from some of the more popular operas.

25 It is possible to read *Figaro* as liberal if not revolutionary. Certainly, the opera is politically charged, not merely interested in love and human foibles. But I maintain that *Figaro* is basically conservative; while it critiques European social discourse it ultimately preserves the oppressive feudal system by effecting a closure which marries servant to servant and reinforces social distances between the classes. Mozart's position on class is uneasy and complicated, but he clearly stops well short of promoting democracy, or even of rejecting the aristocracy.

26 For example, the semantic misapplication of Constable Dogberry (*Much Ado About Nothing*) or the Nurse (*Romeo and Juliet*).

27 *Neue Zeitschrift für Musik* was edited by Robert Schumann from 1834 to 1844.

28 *Punch*, 2, 1842: p. 133.

29 Charles Burney, *The Letters of Charles Burney*, vol. 1 (1751–84), ed. Alvaro Ribeiro (Oxford: Clarendon Press, 1991), pp. 99–100.

30 *Punch*, 2, 1842: p. 42.

31 Ibid.

32 *The Old Maid* manuscript is ascribed to "Mr Murphy." It is available from the Huntington Library's Larpent Collection, a collection of manuscripts

submitted to the Lord Chamberlain prior to 1834, and on microcard at the British Library in London. Also reprinted in *"The Way to Keep Him" and Five other Plays by Arthur Murphy*, ed. John P. Emery (New York: New York University Press, 1956).

33 The author uses the idiosyncratic broken dash, as above, in a way that suggests diffusion rather than ellipsis. He also uses the conventional dash.

34 *Hit or Miss* and the other nineteenth-century plays discussed in this chapter are located in the British Library's Lord Chamberlain Collection (Department of Manuscripts), unless otherwise noted.

35 Anne Jackson Mathews, *The Memoirs of Charles Mathews, Comedian* (London: R. Bentley, 1838–9), vol. II, pp. 73–74.

36 Ibid., p. 73.

37 J. A. Hardwick, "Cheap John," *Music Hall Song Book* (London, 1862), p. 7.

38 Letter of March 6, 1810, from William Powell (prompter, Drury Lane Company) to John Wilkinson. Harvard Theatre Collection, bMS Thr 32.

39 I was unable to obtain the revised play, and can only assume that it erases every trace of gentility from Dick's speech, so that he is, after all, indistinguishable from the boxers in whose cant he delights.

40 *The London Times*, 1823.

41 Tom Taylor, *The Ticket of Leave Man*, in Rowell, ed., *Nineteenth-Century Plays*, p. 272.

42 In fact, Mayhew's text was itself dramatized. Plays based on *London Labour* include *How We Live in London*, by J. B. Johnstone, and *London Labour and the London Poor; or Want and Vice*, by J. Elphinstone.

43 Meisel, *Realizations*, p. 116.

44 Mayhew, *London Labour*, vol. I, p. 213.

45 I do not mean to undermine Mayhew's philanthropic purpose: he obviously undertook this work to expose the dire conditions in which London's lower classes lived. In order to do so, he had to bring these people to the attention of the middle class, make their stories as readable and unalienating as possible. He also had to overcome his own very middle-class prejudices against them, and he only partly succeeded in doing so. It is this failure in sympathy which interests me most, because it testifies to the social and psychic power of verbal/social stereotypes like patter.

46 Mayhew, *London Labour*, vol. I, pp. 221–222.

47 Ibid., p. 222.

48 See Anne Humpherys, *Henry Mayhew* (Boston: Twayne Publishers, 1984) for a discussion of Mayhew's methodology.

49 Quoted in Roy Porter, *English Society in the Eighteenth Century* (London: Penguin Books, 1990), pp. 102–103.

50 W. T. Moncrieff, *Tom and Jerry; or, Life in London* (London: Richardson, 1828), p. 19.

51 John D. Rosenberg, introduction to Mayhew, *London Labour*, p. vii.

52 Consider his establishment of services for reformed prostitutes, for example, which included obtaining for them passage to the colonies.

53 Dickens, *Our Mutual Friend*, p. 87.

54 Wilfred Dvorak argues that the average Victorian reader would have known all of the songs referred to in *Our Mutual Friend*, and would hence have understood the ballad revisions as explicit commentaries on Wegg's character. "Dickens and Popular Culture: Silas Wegg's Ballads in *Our Mutual Friend*," *The Dickensian*, 86, 3, 1990: pp. 142–157.

55 On the "curious and important theme" of "out street and public-house literature," Mayhew asserts that "the street-ballad, and the street-narrative, like all popular things, have their influence on masses of people" (*London Labour*, p. 220).

56 Wegg's ballads, for example, represent part of the mid-Victorian music hall repertoire. Some of them come from popular ballad operas like *Clari; or The Maid of Milan* and *The Beggar's Opera*; songs from these operas were performed on the music-hall stage and in the pleasure gardens, as well as in the theatre. See the notes to the Penguin edition of *Our Mutual Friend* for specific song identifications.

57 *Music Hall Song Book*, p. 7.

58 Herbert Darnley, "The Swimming Master," sung by Dan Leno. Davison, ed., *Songs of the British Music Hall*, pp. 68–70.

59 A *Catalogue of English Song Books, Forming a Portion of the Library of Sir John Stainer* was privately printed in England in 1891 so that "it may prove a useful instalment towards a Bibliography of Song Books." This text illustrates the tremendous activity of songwriters and publishers in the eighteenth and nineteenth centuries. It lists hundreds of songbooks, many of which include comic music-hall and theatre songs. For example, "*The Apollo*; a collection of all the modern songs which have been sung at the theatres, Vauxhall, the Circus, and all public places" was published in 1810. The periodical "*Yankee Smith's London Comic Songster*, a selection of the best comic songs and parodies," was issued in penny monthly numbers, and at that price would have been accessible to a large segment of the English public. The catalogue represents a large number of popular song collections from the theatre, the pleasure gardens, the opera, and the music hall; clearly, these songs entered people's homes and their imaginations. Jane Stedman indicates as much in *Gilbert Before Sullivan* (University of Chicago Press, 1967): "Drowning out the older folk ballad and operatic borrowings . . . music-hall melodies had so dominated the sixties that Gilbert's burlesque of *The Princess* (1870) was heavily praised for drawing its music from Offenbach and Herve" (p. 11).

5 CHARLES MATHEWS, CHARLES DICKENS, AND THE COMIC FEMALE VOICE

1 Ackroyd, *Dickens*, p. 631.

2 *Morning Advertiser*, June 29, 1878, quoted in Booth, ed., *English Plays of the Nineteenth Century*, vol. IV, p. 147.

3 Joseph Donohue, *Theatre in the Age of Kean* (Totowa, N.J.: Rowman and Littlefield, 1975), pp. 50–51.

4 Ibid., p. 21.
5 Julia Kristeva, *Desire in Language*, ed. Leon Roudiez (New York: Columbia University Press, 1980).
6 George Henry Lewes, *On Actors and the Art of Acting* (New York: Grove Press, 1957), p. 35.
7 Ibid., p. 85.
8 Quoted in Collins, *The Public Readings*, p. 3.
9 For example, Norman Page's *Speech in the English Novel* (London: Longman, 1973); Golding, *Idiolects in Dickens*.
10 Golding describes the gesture the Victorians knew as "patter" as idiolectic, an assertion based on his reading of Mrs. Nickleby – and her sister patterers, one would presume – as drawn from real life, in this case, the novelist's mother. In fact, patter was too widely in currency to be considered idiolectic in Dickens; it was used extensively as a metaphor of social degradation or redundancy, in eighteenth- and nineteenth-century popular cultures. Despite Dickens' own comments linking Mrs. Nickleby to his mother, there were plenty of literary and theatrical models for the character who are, at the very least, equally credible sources.
11 Ackroyd, *Dickens*, p. 167.
12 Letter of July 5, 1856, cited in the introduction to *Little Dorrit*, p. x.
13 Ibid.
14 *Examiner*, 23 September, 1838, p. 595.
15 Litvak, *Caught in the Act*, p. x.
16 This work is undertaken in chapter 3, above.
17 Consider the Vashti episode in *Villette*, which plays out this conflict with great force. The historical fascination with castrati, some of whom were still living in the mid-nineteenth century, also underscores the attraction to explicit but tabooed sexuality. Fanny Burney, in her journals, records her culture's obsession with the castrato.
18 Ackroyd, *Dickens*, p. 153.
19 Bakhtin, *The Dialogic Imagination*, p. 272.
20 See my "Music and Dramatic Voice in Robert Browning and Robert Schumann," *Victorian Poetry*, 29, 3, 1991.
21 Bakhtin, *The Dialogic Imagination*, p. 301.
22 Ibid., pp. 304–305.
23 Forster, *Dickens*, p. 7.
24 Charles Dickens, *"The Amusements of the People" and Other Papers: Reports, Essays and Reviews, 1834–51*, ed. Michael Slater (Columbus: Ohio State University Press, 1996), pp. 640–642.
25 Dickens, *Sketches by Boz*, p. 65.
26 Ibid., p. 66.
27 Ackroyd claims that the "Seven Dials" dialogue derives from actual lower-class speech, citing an 1883 court transcript, printed in the *London Times*, which reads as follows: "'I'se quite hinnocent, your vorship. I was valking along, and I see these here trousers a hanging, vich I vishes I'd a never seed

at all.' It is the same vernacular, the same idiom, the same cadence" (*Dickens*, pp. 166–167). This may indeed be the same language we find in Dickens' sketch, but it also strongly resembles the lower-class idiom in *Tom Jones*. Ackroyd acknowledges Dickens' debt to actual London speech, but fails to consider the inverse possibility: that London speech, particularly as transcribed in the *London Times*, may partly derive from popular entertainments, novels like *Tom Jones*, sketches like "Seven Dials." I want to emphasize the reciprocity of influence here, the possibility that "real life" is already staged or fictionalized. The "Seven Dials" sketch, with its magic-box deflection of origins, suggests that this is true.

28 Fielding, *Tom Jones*, p. 135.

29 Chapter 6 below engages explicitly with this issue.

30 If Mathews the elder inspired the young novelist with a fervent theatre-lust, Charles James Mathews, his son, was Dickens' contemporary and colleague. As a long-time theatre manager, actor, and playwright, C. J. Mathews had some claim to Dickens' notice. Incidentally, his popular farce, *Patter v. Clatter* (which played 116 times during his sixteen years at Covent Garden and the Lyceum) dramatized – and named – the idiom used by his father for Mrs. Neverend and Miss Blight, although in this case it marks the speech of a male *buffo* character.

31 Forster, *Dickens*, p. 59.

32 Ackroyd, *Dickens*, pp. 138–139.

33 Robert Golding suggests that Mrs. Nickleby, Mrs. Lirriper, and Flora Finching all speak like a Mathews character called Mrs. Neverend, although he posits the latter as an ultimate source, an original "loquacious" speaker rather than one of many dramatic examples of a cultural idea (*Idiolects in Dickens* [London: Macmillan, 1985], p. 21).

34 Richard Klepac, *Mr. Mathews At Home* (London: Society for Theatre Research, 1979).

35 One of the more ephemeral elements of nineteenth-century theatre, which would of course elude Klepac or any scholar, is the ad-libbing which formed an integral part of comic acting. I have no doubt that Mathews changed bits of his act every time he walked on to the stage, and am indebted to Mr. John Culme for initially suggesting this possibility to me. A theatre expert from whose considerable knowledge of the Victorian and Edwardian stage this study has benefitted, Mr. Culme pointed out that in Edwardian music-hall recordings, song lyrics frequently differ from their printed versions, and they differ in such a way as to suggest ad-libbing, not error. Inconsistencies of the same sort occur in the various published editions of Mathews' shows, and indicate variations in actual performances as well as the inevitable transcription errors.

36 Anne Jackson Mathews, *The Memoirs of Charles Mathews, Comedian*, vol. 1, pp. 16–18. The text combines autobiographic sketches by Mathews himself, and – especially in the second volume – sections of biography written by Anne Jackson Mathews.

37 Ibid., p. 23.
38 Ibid., pp. 23–24.
39 Ibid., p. 26.
40 They include waiters and officers in plays like *The Jew* and *The Maid of Normandy*; Guildenstern in *Hamlet*; Trip in the *School for Scandal*. See ibid., vol. 1, pp. 82–86.
41 Chapter 4 above discusses Cypher and *Hit or Miss*.
42 Edinburgh, Monday April 6, 1912. This and other reviews, unless explicitly noted, are taken from *Scrapbook of Criticisms, etc.*, vol. 1 (1794–1817), compiler Anne Jackson Mathews, Harvard Theatre Collection, Pusey Library, Harvard University.
43 See the introduction to Booth's *English Nineteenth-Century Plays*, vol. 1, for a discussion of the "serious" drama, its theory and practice, in nineteenth-century England.
44 Richard Klepac (*Mr. Mathews at Home*) and John A. Degen ("Charles Mathews' 'At Homes': The Textual Morass," *Theatre Survey*, 28, 2, 1987: pp. 75–88) both point to textual discrepancies among Mathews' written and published materials.
45 *Mr. Mathews' Memorandum Book* (London: Duncombe, [1830?]). This book and a large number of other Mathews materials can be found in the Harvard Theatre Collection, Pusey Library, Harvard University.
46 Ibid., p. 19.
47 Dickens, *Bleak House*, pp. 34–35.
48 *Mr. Mathews' Comic Annual for 1831* (London: Duncombe), Harvard Theatre Collection, Pusey Library, Harvard University.
49 The sketch also plays on certain cultural anxieties about aunts, who could actually usurp a dead mother's place under auspicious conditions, and were perceived by some as a threat to domesticity. Legislation was even enacted, in the form of the Deceased Wife's Sister Act (1835–1907), to defuse an aunt's power. This anxiety has persisted into the twentieth century; consider Bertie Wooster's collection of nephew-bashing aunts!
50 Dickens, *Nicholas Nickleby*, p. 286.
51 As a matter of fact, Miss Knag's idiom closely resembles that of the spinster Miss Harlow (except that it is more calculated); see my discussion of Miss Harlow in *The Old Maid* (1761) in chapter 4 of this volume.
52 Dickens, *Nicholas Nickleby*, p. 291.
53 "To make a noise like the quick steps of many feet." Johnson, *A Dictionary of the English Language*, vol. II.
54 Kristeva, *Desire in Language*, p. 79.
55 Dickens, *Nicholas Nickleby*, p. 222.
56 Ibid., pp. 337–338.
57 Mrs. Tulliver is not a redundant women in the usual sense, being neither single nor widowed, but she does share with the others her eccentric position in relation to her culture's "reproductive economy" – she is, in other words, middle-aged, probably post-menopausal, and unattractive – as

well as their form of speech.

58 Harvard Theatre Collection, bMS Thr. 357 (105).

6 PATTER AND THE PROBLEM OF REDUNDANCY: ODD WOMEN AND *LITTLE DORRIT*

1 Eliot, *Felix Holt*, p. 117.

2 J. C. Trewin, *The Journal of William Charles Macready* (London: Longmans, 1967), p. 84.

3 Reviews quoted in Boyd Litzinger and Donald Smalley, eds., *Robert Browning: The Critical Heritage* (New York: Barnes and Noble, 1970), pp. 53, 56.

4 Booth, ed., *English Plays of the Nineteenth Century*, vol. 1, pp. 76–77.

5 Kristeva, *Desire in Language*, pp. 119–120.

6 Peter Stallybrass and Allon White, "Bourgeois Hysteria and the Carnivalesque," in *The Politics and Poetics of Transgression* (Ithaca, N.Y.: Cornell University Press, 1986).

7 By Norman Page (*Speech in the English Novel* [London: Longman, 1973]), Golding (*Idiolects in Dickens*); Michael Slater (*Dickens and Women* [Stanford University Press, 1983]); by Forster (*Dickens*), and a host of critics and biographers in his wake. Patricia Ingham's useful book *Dickens, Women and Language* (University of Toronto Press, 1992) explores the "excessive" woman in some depth and, while it fails to make some of the larger cultural connections one would expect, and makes unsatisfactory assumptions about linguistic idiosyncrasy, its reading of female speech are insightful and suggestive.

8 For a reading of spinsterhood that is opposed to mine, see Auerbach's *Woman and the Demon* (Cambridge, Mass.: Harvard University Press, 1982), which mythologizes the redundant woman as a kind of epic-heroic figure. Auerbach traces "the outline of a general Victorian myth of the spinster as hero" in Charlotte Brontë's private letters.

> Like all heroic myths, this one is fraught with unspoken doubts and terrors, and thus is repeatedly undermined by diffident self-qualification. But if we piece it together, often from little-known sources of women's biography, autobiography, and fiction, we recover glimpses of an animating myth that worked beneath the surface of its age and has not entirely faded in our own, in which the old maid transcends the laughter and tears with which cultural complacency endows her, to "establish her own landmarks" with the audacity and aplomb of an authentic hero. (p. 112)

My objection to Auerbach's position is that it grants a current of feeling located in the obscure and private writings of a relatively small portion of women the status of cultural myth, when in fact these "heroic" sentiments were far too localized – and private – to have taken on the cultural resonance of myth. Auerbach is over-optimistic about the cultural value of spinsterhood, both then and now, although she is accurate in assessing the implicit power and danger in these women.

9 As we shall see, spinsters and widows are usually dramatized with the same comic attributes despite their potentially different social statuses. Widow-

hood implies inheritance or settlement – the possibility of financial independence made widows peculiarly threatening to the patriarchy – while spinsters, unless born into considerable wealth, remained dependent on male family members or employers. This is a substantial difference, and the conflation of the two figures in popular culture is curious. See my note on widowhood and *Clarissa* in chapter 4 above. I use the phrase "redundant women" to signify both social groups, unless one in particular is specified.

10 W. R. Greg, "Why are Women Redundant?," *National Review*, 15, 1862. This essay is excerpted in Janet Horowitz Murray, ed., *Strong-Minded Women and Other Lost Voices of the Nineteenth Century* (New York: Pantheon Books, 1982), pp. 50–54.

11 John Stuart Mill and Harriet Taylor Mill, *Essays on Sex Equality*, ed. Alice S. Rossi (University of Chicago Press, 1970), p. 73.

12 See M. Jeanne Peterson, "The Victorian Governess: Status Incongruence in Family and Society," in *Suffer and Be Still*, ed. Martha Vicinus (Bloomington: Indiana University Press, 1973), pp. 16–17, for a discussion of emigration societies and unmarried middle-class women in Victorian society.

13 See Helene Roberts, "Marriage, Redundancy or Sin: The Painter's View of Women in the First Half of Victoria's Reign," in Vicinus, ed., *Suffer and Be Still*, for a discussion of Victorian paintings of women.

14 Brontë, *Villette*, p. 278.

15 Spinsters did sometimes make compelling subjects for painters. George Redgrave was one Victorian artist who painted them, with sympathy, as governesses, seamstresses, and the like. The silent, suffering gentlewoman, reduced to earning her bread and literally dying of the work that saps her health and spirits, competed with the other more popular stereotype of redundant woman – the patterer – but failed to amuse the middle-class public. Perhaps because they couldn't separate themselves with laughter or scorn from the thin, silent spinster, people preferred her *alter ego*, the one who alienated them with her breadth of body and extravagance of talk. We should not be surprised that Redgrave "suffered criticism for his choice of subjects . . . [including] accusations of 'theatrical,' 'commonplace,' and 'trite.'" (Roberts, "Marriages, Redundancy," p. 60).

16 George Gissing, *The Odd Women* (New York: W. W. Norton, 1977), p. 37.

17 Anne Brontë, *The Tenant of Wildfell Hall* (Harmondsworth: Penguin Books, 1979), pp. 43, 49.

18 Murray's *Strong-Minded Women* explores the "woman questions" as they were formulated in journalism, diaries and letters, and parliamentary reports.

19 *Punch*, January 28, 1860, p. 33.

20 Another example of such "sociological" writing in *All the Year Round* is "A Prodigy Hunter" (December 1861), which reviews a seventeenth-century text by one James Paris Du Plessis entitled "A Short History of Human Prodigious and Monstrous Births of Dwarfs, Sleepers, Giants, Strong Men, Hermaphrodites, Numerous Births, and Extreme Old Age, &c." This

article differs from "Odd Women" primarily in its gender inclusiveness.

21 Murray, *Strong-Minded Women*, p. 48.

22 But Gissing, at the end of the century, was able to imagine single life as a real option for "normal" women, in a way that Dickens could not. The differences in their positions on spinsterhood are various and important (remember that Gissing wrote during a period of feminist activism), although not the central subject of this discussion.

23 "Odd Women," *All the Year Round*, November 14, 1874: pp. 113–117.

24 Ibid., p. 114.

25 Ibid.

26 Nancy Fix Anderson, *Women Against Women in Victorian England: A Life of Eliza Lynn Linton* (Bloomington and Indianapolis: Indiana University Press, 1987), p. 71.

27 Of course, not everyone was so inclined. R. M. Pankhurst, in an article entitled "The Right of Women to Vote Under the Reform Act, 1867" (*Fortnightly Review*, 10, 1868, pp. 250–254) critiques the conservative position, performing a close reading of "The Representation of the People Act" to prove that women's suffrage is guaranteed under the language of the bill. He also challenges mainstream assumptions, represented by parliament and legal court, about women's intellect.

> If it is maintained that by the law of England women are incompetent to vote in parliamentary election, it . . . can be maintained on only one ground – the same and only ground that disqualifies the insane and infants – mental imbecility . . . That this really the position to which the opponents of women's suffrage are reduced is indisputable. In Olive v. Ingram, the Chief Justice cited strong authority as to the right of women to vote . . . One of the judges, however, said, "In the election of members of Parliament women are not now admitted, whatever they were formerly. They (women) are not allowed to vote . . . because of the judgement required in it." And further: – "This (the decision as to the office of sexton, which, it was held, a woman might fill, it being a 'servile ministerial office, requiring neither skill nor understanding') cannot determine that women may vote for members of Parliament, as the choice requires an improved understanding, which women are not supposed to have . . . infants cannot vote, and women are perpetual infants." Therefore, to hold that women are incompetent to vote . . . is to hold them to be the subjects of absolute and incurable mental defect, and, as it were, to sign against the duly qualified women of England a certificate of perpetual lunacy. (p. 253)

28 "Scientific Aspects of Nutrition," *All the Year Round*, February 13, 1875: p. 414.

29 See chapter 4 above for a discussion of this and other plays about pattering women.

30 Shakespeare, *Much Ado About Nothing*, III.iii. 34–36.

31 J. S. Knowles, *The Old Maids* (1841), British Library, Department of Manuscripts, Add. MS 42960 fos. 189–246b.

32 The brothers Mayhew, *Whom to Marry and How to Get Married, or the Adventures of a Lady in Search of a Good Husband* (London, n.d.).

33 Miller, *Narrative and its Discontents.*
34 Murray, *Strong-Minded Women*, p. 51.
35 *The Truth About Man, by A Spinster* (London, 1905).
36 Charles Dickens, *David Copperfield* (Harmondsworth: Penguin Books, 1966), p. 97.
37 Shakespeare, *Hamlet*, II.ii. 395–400.
38 Dickens, *Little Dorrit*, p. 129.
39 Henry Fielding, *Joseph Andrews* (Boston: Houghton Mifflin, 1961), pp. 8–11.
40 Ibid., p. 19.
41 Ibid., p. 25.
42 Sheridan's Mrs. Malaprop (*The Rivals*), Smollett's Tabitha Bramble (*Humphry Clinker*) and Burney's Mrs. Mittin (*Camilla*) all share with Mrs. Slipslop socio-amatory ambitions and affectations, although their marital backgrounds differ. They demonstrate their ignorance and vulgarity in their speech, which is generally excessive, uncontrolled, and incorrect.
43 Ingham, *Dickens, Women, and Language*, pp. 68–69.
44 Dickens, *Little Dorrit*, p. 121.
45 Ibid., p. 124.
46 Ibid.
47 Ingham, *Dickens, Women, and Language*, p. 70.
48 Dickens, *Little Dorrit*, p. 125.
49 Ibid., p. 128.
50 *Punch*, September 22, 1860, p. 117.
51 Dickens, *Little Dorrit*, p. 683.
52 Ibid., p. 688.

7 CONCLUSION

1 These have both been released in audio recordings, a testament to the force of the nonrepresentational signification, the play of tones, rhythms, vocal textures, which figures so strongly in the Dickens experience.
2 Mr. Stewart has told me that he himself "knew" the *Carol* before reading it for the first time as an adult; while he subsequently became intimate with Dickens' text, and discarded some old associations collected over a lifetime of contact with popular renditions of the story, he clearly established a relationship with the tale which acknowledged its extraordinary extratextual life, its larger cultural value. For a large-scale study of *A Christmas Carol*'s extratextual life, its cultural legacy, see Davis, *The Lives and Times of Ebenezer Scrooge.*

Bibliography

Ackroyd, Peter. *Dickens* (New York: Harper Perennial, 1992).

Altick, Richard. *Writers, Readers and Occasions* (Columbus: Ohio State University Press, 1989).

Anderson, Nancy Fix. *Women Against Women in Victorian England: A Life of Eliza Lynn Linton* (Bloomington and Indianapolis: Indiana University Press, 1987).

Anon., *The Truth About Man, by A Spinster* (London, 1905). British Library, Department of Printed Books.

Arnold, Matthew. "The Function of Criticism at the Present Time." In *The Poetry and Criticism of Matthew Arnold*, ed. A. Dwight Culler (Boston: Houghton Mifflin Co., 1961).

"Shelley." In *The Poetry and Criticism of Matthew Arnold*, ed. A. Dwight Culler (Boston: Houghton Mifflin Co., 1961).

Auerbach, Nina. *Private Theatricals* (Cambridge, Mass.: Harvard University Press, 1990).

Woman and the Demon (Cambridge, Mass.: Harvard University Press, 1982).

Baer, Marc. *Theatre and Disorder in Late Georgian London* (Oxford: Clarendon Press, 1992).

Barish, Jonas. *The Anti-Theatrical Prejudice* (Berkeley: University of California Press, 1981).

Bakhtin, M. M. *The Dialogic Imagination*, trans. and ed. Caryl Emerson and Michael Holquist (Austin: University of Texas Press, 1981).

Barthes, Roland. *S/Z* (New York: Hill and Wang, 1974).

Berry, Cicely. *Voice and the Actor* (New York: Collier Books, 1973).

Blau, Herbert. *The Audience* (Baltimore and London: Johns Hopkins University Press, 1990).

To all Appearances: Ideology and Performance (New York and London: Routledge, 1992).

Bolton, Philip. *Dickens Dramatized* (London: Mansell, 1987).

Booth, Michael. *English Melodrama* (London: Herbert Jenkins, 1965).

Theatre in the Victorian Age (Cambridge University Press, 1991).

Booth, Michael, ed., *English Plays of the Nineteenth Century*, vols. I–IV (Oxford: Clarendon Press, 1969–76).

Brontë, Anne. *The Tenant of Wildfell Hall* (Harmondsworth: Penguin Books, 1979).

Brontë, Charlotte. *Jane Eyre* (New York: St. Martins Press, 1996).

Villette (Harmondsworth: Penguin Books, 1979).

Browning, Robert. *Strafford* (1836). Reprinted in *The Complete Works of Robert Browning*, vol. II (New York: Thomas Y. Crowell, 1898).

Burney, Charles. *The Letters of Charles Burney*, vol. I (1751–84), ed. Alvaro Ribeiro (Oxford: Clarendon Press, 1991).

Burney, Frances. *Camilla, or, A Picture of Youth* (London: Oxford University Press, 1972).

Campbell, Paul N. "Communication aesthetics," *Today's Speech*, 19, 1971.

Carlyle, Thomas. *Sartor Resartus* (New York: A. L. Burt Co., 1910).

Catalogue of English Song Books, Forming a Portion of the Library of Sir John Stainer (1891).

Century Dictionary, ed. William D. Whitney (New York: Century Co., 1889).

Cheshire, D. F. *Music Hall in Britain* (Rutherford, N.J.: Fairleigh Dickinson University Press, 1974).

Chronological English Dictionary, ed. Thomas Finkenstaedt (Heidelberg: C. Winter, 1970).

Collins, Philip. *Charles Dickens: The Public Readings* (Oxford: Clarendon Press, 1975).

Reading Aloud: A Victorian Metier (London: Tennyson Society, 1972).

Cromie, Robert, ed. *1811 Dictionary of the Vulgar Tongue: A Dictionary of Buckish Slang, University Wit, and Pickpocket Eloquence* (Chicago: Follett Publishing Co., 1971).

Crowley, Tony. *Standard English and the Politics of Language* (Urbana and Chicago: University of Illinois Press, 1989).

Darwin, Charles. *On the Expression of Emotions in Man and Animals* (New York: D. Appleton and Co., 1899).

Davies, Andrew. *Other Theatres* (Totowa, N.J.: Barnes and Noble, 1987).

Davis, Paul. *The Lives and Times of Ebenezer Scrooge* (New Haven: Yale University Press, 1990).

Davis, Tracy. "The Actress in Victorian Pornography." In *Victorian Scandals*, ed. Kristine Ottesen Garrigan (Athens: Ohio University Press, 1992).

Davison, Peter, ed. *Songs of the British Music Hall* (New York: Oak Publications, 1971).

Defoe, Daniel. *Moll Flanders* (Boston: Houghton Mifflin, 1959).

Degen, John A. "Charles Mathews' 'At Homes': The Textual Morass." *Theatre Survey*, 28, 2, 1987.

Descartes, René. *Les Passions de l'âme*. In *The Philosophical Writings of Descartes*, vol. I (Cambridge University Press, 1985).

Dickens, Charles. *"The Amusements of the People" and Other Papers: Reports, Essays, and Reviews, 1834–51*, ed. Michael Slater (Colombus: Ohio State University Press, 1996).

An Account of the Performance of Lytton's Comedy "Not So Bad As We Seem," with Other

Matters of Interest by Charles Dickens (London: printed for private circulation by Richard Clay and Sons, 1919).
Bleak House (New York: Bantam Books, 1983).
David Copperfield (Harmondsworth: Penguin Books, 1966).
Dombey and Son (Harmondsworth: Penguin Books, 1970).
Great Expectations (Harmondsworth: Penguin Books, 1965).
Little Dorrit (Oxford University Press, 1982).
Martin Chuzzlewit (Oxford University Press, 1982).
Nicholas Nickleby (Harmondsworth: Penguin Books, 1986).
Our Mutual Friend (Harmondsworth: Penguin Books, 1971).
The Pickwick Papers (Oxford University Press, 1986).
Sketches by Boz (London: Oxford University Press, 1957).
"Two Letters on Public Executions, by Charles Dickens, Esq., Published with the Sanction of the Author" (London: Dyson, E. Wilson, 1849). New York Public Library, Berg Collection.

Diderot, Denis. *Rameau's Nephew*, trans. Leonard Tancock (Harmondsworth: Penguin Books, 1966).

Donohue, Joseph. *Theatre in the Age of Kean* (Totowa, N.J.: Rowman and Littlefield, 1975).

Dryden, John. *Marriage à la Mode* (Lincoln: University of Nebraska Press, 1981).

Dvorak, Wilfred. "Dickens and Popular Culture: Silas Wegg's Ballads in *Our Mutual Friend*." *The Dickensian*, 86, 3, 1990.

Edgar, David. "The Life and Adventures of Nicholas Nickleby." In *Plays from the Contemporary British Theater*, ed. Brooks McNamara (New York: Mentor Books, 1982).

Eigner, Edwin. *The Dickens Pantomime* (Berkeley: University of California Press, 1989).

Eliot, George. *Essays of George Eliot*, ed. Thomas Pinney (London: Routledge and Kegan Paul, 1963).
Felix Holt, the Radical (Harmondsworth: Penguin Books, 1987).

Elphinstone, J. *London Labour and the London Poor; or Want and Vice* (London, 1854).

Fielding, Henry. *Joseph Andrews* (Boston: Houghton Mifflin, 1961).
Tom Jones. (New York: W. W. Norton, 1973).

Fischer-Lichte, Erika. *The Semiotics of Theater*, trans. Jeremy Gaines and Doris Jones (Bloomington and Indianapolis: Indiana University Press, 1992).

Fisher, Leona Weaver. *Lemon, Dickens, and "Mr. Nightingale's Diary": A Victorian Farce* (University of Victoria Press, 1988).

Forster, John. *The Life of Charles Dickens*, ed. J. W. T. Ley (London: Palmer, 1928).

Foucault, Michel. *Discipline and Punish: The Birth of the Prison*, trans. Alan Sheridan (New York: Vintage Books, 1979).
Madness and Civilization: A History of Insanity in the Age of Reason, trans. Richard Howard (New York: Vintage Books, 1973).
"The Discourse on Language." In *The Archaeology of Knowledge* (New York: Pantheon Books, 1971).

Fowell, Frank and Frank Palmer. *Censorship in England* (New York and London: Benjamin Blom, 1913).

Garis, Robert. *The Dickens Theatre* (Oxford: Clarendon Press, 1965).

Gilbert, W. S. and Arthur Sullivan. *Ruddigore.* In *The Complete Gilbert and Sullivan Opera Guide*, ed. Alan Jefferson (Devon: Webb and Bower, 1984).

Gissing, George. *The Odd Women* (New York and London: W. W. Norton, 1977).

Glasstone, Victor. *Victorian and Edwardian Theatres* (Cambridge, Mass.: Harvard University Press, 1975).

Golding, Robert. *Idiolects in Dickens* (London: Macmillan, 1985).

Greg, W. R. "Why are Women Redundant?" *National Review*, 15, 1862.

Guildersleeve, Virginia. *Government Regulation of the Elizabethan Drama* (New York: Columbia University Press, 1908).

Habermas, Jürgen. *The Structural Transformation of the Public Sphere* (Cambridge, Mass.: MIT Press, 1991).

"The Public Sphere: An Encyclopedia Article." In S. E. Bronner and D. M. Kellner, eds., *Critical Theory and Society: A Reader* (New York: Routledge, 1989).

Hadley, Elaine. *Melodramatic Tactics* (Stanford University Press, 1995).

Hazlitt, William. "On Actors and Acting" (1817). In *Criticisms and Dramatic Essays of the English Stage* (London: G. Routledge, 1854).

Hogarth, William. *Engravings by Hogarth*, ed. Sean Shesgreen (New York: Dover Publications Inc., 1973).

Horncastle, H. *The Infant Phenomenon; Or, A Rehearsal Rehearsed* (1838). Printed in *Dicks' Standard Plays*, no. 572 (London: J. Dicks, 1882).

Hughes, Linda and Michael Lund. *The Victorian Serial* (Charlottesville: University Press of Virginia, 1991).

Humpherys, Anne. *Henry Mayhew* (Boston: Twayne Publishers, 1984).

Ingham, Patricia. *Dickens, Women, and Language* (University of Toronto Press, 1992).

Ingram, William. *The Business of Playing: The Beginnings of the Adult Professional Theater in Elizabethan London* (Ithaca, N.Y.: Cornell University Press, 1992).

Jefferson, Alan, ed. *The Complete Gilbert and Sullivan Opera Guide* (Exeter: Webb and Bower, 1984).

Jerrold, Douglas. *Black Ey'd Susan.* In *Nineteenth-Century Plays*, ed. George Rowell (Oxford University Press, 1972).

Mr. Paul Pry, or, I Hope I Don't Intrude. In *English Nineteenth-Century Plays*, vol. IV, ed. Michael Booth (Oxford: Clarendon Press, 1973).

Johannsen, Albert, ed. *Phiz: Illustrations from the Novels of Charles Dickens* (University of Chicago Press, 1956).

Johnson, Samuel. *A Dictionary of the English Language*, 2 vols. (New York: AMS Press, 1967).

Johnstone, J. B. *How We Live in London* (1856).

Klepac, Richard. *Mr. Mathews At Home* (London: Society for Theatre Research, 1979).

Knowles, J. S. *The Old Maids* (1841). British Library, Lord Chamberlain Collection, add. MS 42960.

Virginius (1820). Reprinted in *Representative British Dramas*, ed. Montrose Moses (Boston: Little, Brown, 1931).

Kristeva, Julia. *Desire in Language*, ed. Leon Roudiez (New York: Columbia University Press, 1980).

Lewes, George Henry. *On Actors and the Art of Acting* (New York: Grove Press, 1957).

Litvak, Joseph. *Caught in the Act: Theatricality in the Nineteenth-Century English Novel* (Berkeley: University of California Press, 1992).

Litzinger, Boyd and Donald Smalley, eds. *Robert Browning: The Critical Heritage* (New York: Barnes and Noble, 1970).

Macready, W. C. *The Journal of William Charles Macready, 1832–1857*, ed. J. C. Trewin (London: Longmans, 1967).

Martineau, Harriet. *Autobiography* (Boston: J. R. Osgood, 1877).

Mathews, Anne Jackson. *The Memoirs of Charles Mathews, Comedian*, 4 vols. (London: Bentley, 1838–9). Harvard Theatre Collection, Pusey Library, Harvard University.

Scrapbook of Criticisms, etc. vol. I (1794–1817). Harvard Theatre Collection, Pusey Library, Harvard University.

Mathews, Charles. *Mr. Mathews' Comic Annual for 1831* (London: Duncombe, n.d.). Harvard Theatre Collection, Pusey Library, Harvard University.

Mr. Mathews' Memorandum Book (London: Duncombe, [1830?]). Harvard Theatre Collection, Pusey Library, Harvard University.

Mayhew, brothers (Henry and Augustus). *Whom to Marry and How to Get Married, or the Adventures of a Lady in Search of a Good Husband* (London, n.d.). British Library, Department of Printed Books.

Mayhew, Henry. *London Labour and the London Poor*, vols. I–IV (New York: Dover Publications Inc., 1968).

The Wandering Minstrel. In *English and American Drama of the Nineteenth Century*, ed. Allardyce Nicoll and George Freedley (New York: Readex Microprint, 1969).

Meisel, Martin. *Realizations* (Princeton University Press, 1983).

Mill, John Stuart and Harriet Taylor. *Essays on Sex Equality*, ed. Alice S. Rossi (University of Chicago Press, 1970).

Miller, D. A. *Narrative and its Discontents: Problems of Closure in the Traditional Novel* (Princeton University Press, 1981).

The Novel and The Police (Berkeley: University of California Press, 1988).

Moncrieff, W. T. *Nicholas Nickleby* (1838). British Library, Lord Chamberlain Collection, add. MS 42951.

Tom and Jerry; or, Life in London (London: Richardson, 1828).

Mozart, W. A. *The Letters of Wolfgang Amadeus Mozart*, ed. Hans Mersmann, trans. M. M. Bozman (New York: Dover Publications Inc., 1972) .

Mozart, W. A. and Lorenzo DaPonte. *Don Giovanni* (New York: Dover Publications Inc., 1974).

Le Nozze di Figaro (New York: Dover Publications Inc., 1979).

Murphy, Arthur. *The Old Maid* (1761). Huntington Library, Larpent Collection. Reprinted in *"The Way to Keep Him" and Five Other Plays by Arthur Murphy*, ed. John P. Emery (New York: New York University Press, 1956).
Murray, Janet Horowitz, ed. *Strong-Minded Women and Other Lost Voices of the Nineteenth Century* (New York: Pantheon Books, 1982).
Music Hall Song Book (London, 1862). British Library, Department of Printed Books.
New Grove Dictionary of Music, ed. Stanley Sadie, 20 vols. (London: Macmillan, 1980).
Nicholson, Watson. *The Struggle for a Free London Stage* (New York: Benjamin Blom, 1966).
Oppenlander, Ella Ann. *Dickens' "All the Year Round" Descriptive Index Contributor List* (Troy, N.Y.: Whitston, 1984).
Page, Norman. *Speech in the English Novel* (London: Longman, 1973).
Oxford Dictionary of English Etymology (Oxford University Press, 1966).
Oxford English Dictionary, 2nd edn, 20 vols. (Oxford: Clarendon Press, 1989).
Pankhurst, R. M. "The Right of Women to Vote Under the Reform Act, 1867." *Fortnightly Review*, 10, 1868.
Perkin, Harold. *The Origins of Modern English Society, 1780–1880* (University of Toronto Press, 1969).
Pocock, Isaac. *Hit or Miss* (London: W. H. Wyatt, 1810). British Library, Lord Chamberlain Collection.
The Miller and His Men. In *English Plays of the Nineteenth Century*, vol. I, ed. Michael Booth (Oxford: Clarendon Press, 1979).
Poovey, Mary. *Uneven Developments* (University of Chicago Press, 1988).
Pope, Alexander. "An Essay on Criticism." In *The Poems of Alexander Pope*, ed. John Butt (New Haven: Yale University Press, 1963).
Porter, Roy, ed. *English Society in the Eighteenth Century* (Harmondsworth: Penguin Books, 1990).
The Reign of Terror: A Collection of Authentic Narratives of the Horrors Committed by the Revolutionary Government of France Under Marat and Robespierre (London: Leonard Smithers and Co., 1899).
Richardson, Samuel. *Clarissa* (Harmondsworth: Penguin Books, 1985).
Roach, Joseph. *The Player's Passion.* (Ann Arbor: University of Michigan Press, 1993).
Rousseau, Jean-Jacques. "Letter to M. D'Alembert on the Theatre." In *Politics and the Arts*, ed. and trans. Allan Bloom (Ithaca, N.Y.: Cornell University Press, 1960).
Rowell, George, ed. *Nineteenth-Century Plays* (Oxford University Press, 1972).
The Victorian Theatre: A Survey (Oxford: Clarendon Press, 1967).
Rubin, Leon. *The "Nicholas Nickleby" Story* (Harmondsworth: Penguin Books, 1981).
Schechner, Richard. *Performance Theory* (New York and London: Routledge, 1988).
Schlicke, Paul. *Dickens and Popular Entertainment* (London: Unwin Hyman, 1985).

Seigel, J. P., ed. *Thomas Carlyle: The Critical Heritage* (New York: Barnes and Noble, 1971).
Shaftesbury, Edmund. *Lessons in the Art of Acting: A Practical and Thorough Work For All Persons Who Aim to Become Professional Actors* (Washington, D.C.: Webster Edgerly, 1889).
Shakespeare, William. *Hamlet* (Harmondsworth: Penguin Books, 1970).
Much Ado About Nothing (New York: Signet, 1964).
Shawn, Wallace. *Marie and Bruce* (New York: Grove Press, 1980).
Siddons, Henry. *Practical Illustrations of Rhetorical Gesture and Action* (New York: Benjamin Blom, 1968).
Slater, Michael. *Dickens and Women* (Stanford University Press, 1983).
Spacks, Patricia Meyer. *Gossip* (New York: Alfred A. Knopf, 1985).
Stallybrass, Peter and Allon White. *The Politics and Poetics of Transgression* (Ithaca, N.Y.: Cornell University Press, 1986).
Stedman, Jane. *Gilbert Before Sullivan* (University of Chicago Press, 1967).
Stirling, Edward. *Nicholas Nickleby, a Burletta in Two Acts* (1838). British Library, Lord Chamberlain Collection, add. MS 42949.
Taylor, Barry. *Vagrant Writing: Social and Semiotic Disorders in the English Renaissance* (University of Toronto Press, 1991).
Taylor, George. *Players and Performances in the Victorian Theatre* (Manchester University Press, 1989).
Trewin, John Courtenay, ed. *The Journal of William Charles Macready, 1832–1857* (London: Longmans, 1967).
Turner, Victor. *From Ritual to Theatre* (New York: Performing Arts Journal, 1982).
Vicinus, Martha, ed. *Suffer and Be Still* (Bloomington: Indiana University Press, 1973).
Vlock-Keyes, Deborah. "Music and Dramatic Voice in Robert Browning and Robert Schumann." *Victorian Poetry* 29, 3, 1991.
Worthen, William. *The Idea of the Actor* (Princeton University Press, 1984).

ARCHIVES CONSULTED

British Library, Department of Manuscripts for the Lord Chamberlain's collection of plays, and for letters; Department of Printed Books for rare books, playbills, reviews, and posters.
Harvard Theatre Collection for printed and manuscript materials relating to Charles Mathews, elder and Dickens.
Royal Opera House for materials relating to Charles Mathews, elder and younger.

AUDIO RECORDINGS

Margolyes, Miriam. *Dickens' Women* (London: BBC Enterprises, 1993).
Stewart, Patrick. *A Christmas Carol* (Camm Lane, Inc., 1991).

FILMS

Mozart, W. A. *La Clemenza di Tito*. Directed by Jean-Pierre Ponnelle (Deutsche Grammophon Video, 1980).

Thackeray, W. M. *Barry Lyndon*. Directed by Stanley Kubric (Warner Brothers, 1975).

VICTORIAN PERIODICALS

Athanaeum
All the Year Round
Edinburgh Review
Englishwoman's Review
Examiner
Fortnightly Review
Household Words
La Belle Assemblée
London Times
Punch
Victoria Magazine

Index

Ackroyd, Peter, 129, 135, 140, 143
acting manuals, 20–21
actors, 56, 59
affections, 20, 21, 22, 65
antitheatricality, 56, 59
  legislation, 66–67, 71–72
  and the novel, 62
  political effects of, 60, 62
audience, 9, 15, 50, 74–77, 91
Auerbach, Nina, 5
Austen, Jane
  Emma, 101
  Mansfirld Park, 62

Baer, Marc, 76–77
Bakhtin, Mikhail Mikhailovitch, 3, 52
  and dialogism, 3, 102–103, 139
Barish, Jonas, 61
Barthes, Roland, 79
Berry, Cicely, 44–45
Blau, Herbert, 9, 23, 79
Booth, Michael, 5, 21
Brontë, Anne
  The Tenant of Wildfell Hall, 166
Brontë, Charlotte, 21
  Jane Eyre, 21, 62
  Villette, 62–65, 166
Brown, Hablôt ("Phiz"), 22, 82
Browning, Elizabeth Barrett, 10
Browning, Robert, 10, 161
  Fifine at the fair, 138
  Strafford, 159–161
Burney, Frances
  Camilla, 90

Campanella, Tommaso, 3
Campbell, Paul, 13
Carlyle, Thomas, 41
carnival, 74, 93, 137–138, 154–155, 163, 169
character, 12, 13, 18, 21
Chesterfield, Lord, 80–81
class
  in Victorian culture, 142–143
  and the Victorian novel, 23, 58
  and Victorian theatre, 37, 58
Collins, Philip, 5, 60
comedy, 177–179
Cruikshank, George, 144

Darwin, Charles, 21
Davis, Paul, 6
Davis, Tracy, 19
Defoe, Daniel
  Moll Flanders, 98–99
Descartes, René, 20, 22
dialect, 97–98, 142, 162
dialogism, 13, 49, 138, 154–155
Dickens, Charles
  adaptations, 13–16, 27, 30, 35–36, 49, 61, 191–192
  Bleak House, 42–43, 90, 93, 149
  A Christmas Carol, 191–192
  David Copperfield, 23, 82–83, 90, 176–177
  Dombey and Son, 32–33
  Great Expectations, 78–79
  Little Dorrit, 62, 101, 135–136, 139, 179–189
  Martin Chuzzlewit, 97
  Nicholas Nickleby, 13, 16, 20–21, 22, 27, 29, 48–54, 90, 152–154
  Oliver Twist, 23
  Our Mutual Friend, 33–34, 88, 124–125
  The Pickwick Papers, 97–98
  "Private Theatres," 51
  prompt copies, 61
  public readings, 60, 135
  "Seven Dials," 141–143
  "Shakespeare and Newgate," 74
Diderot, Denis
  Rameau's Nephew, 80–81
domestic sphere, 4, 8
Donohue, Joseph, 132

Dryden, John
Marriage à la Mode, 77–78

Edgar, David, 15–17
Eigner, Edwin, 5
Eliot, George, 21
Adam Bede, 21, 36
Felix Holt, 21, 36, 85–86, 152

Felix, Rachel, 62, 133
Fielding, Henry, 16, 89
and comedy, 178–179
Joseph Andrews, 178–179
Tom Jones, 142–143
Fischer-Lichte, Erika, 23
Forster, John, 63–64, 139–140, 143
Foucault, Michel, 1–3, 5, 8, 75, 103–104, 136–137

Garis, Robert, 5
Gaskell, Elizabeth
Cranford, 174
Mary Barton, 101
generic slippage, 118, 140, 177
genre, 10, 18–19, 28, 47
Gilbert and Sullivan, 41
Ruddigore, 46–47, 93, 100
Gissing, George
The Odd Women, 166, 174–175
Golding, Robert, 143
gossip, 151–152
Greg, William Rathbone, 165–166, 176, 183

Habermas, Jürgen, 19
Hardwick, J. A., 125–126
"Hauteur," 24
Hazlitt, William, 59, 64, 70
heteroglossia
see dialogism
Hogarth, William, 30–31, 68–70, 71–72, 87–88
Horncastle, H., 49–50

illustration
see painting
imaginary text, 6, 11, 15, 18, 19, 29, 58, 59, 67–68, 183
imagination, 6
popular, 8, 11–12, 19–20, 137
Ingham, Patricia, 179, 184

Jerrold, Douglas
Black Ey'd Susan, 29, 37
Johnson, Samuel, 100, 154
"A Just View of the British Stage" 69

Kemble, Frances, 61, 132
Klepac, Richard, 144
Knowles, John Sheridan, 132
The Old Maids, 173–174
Kristeva, Julia, 133, 154–155, 163

Lemon, Mark
Mr. Nightingale's Diary, 39, 51
Leno, Dan, 127–128
Lewes, George Henry, 133–134
Linton, Eliza Lynn, 167, 170
Litvak, Joseph, 4, 136–137

Macready, William Charles, 17–18, 31–32, 61, 159–161
madness, 90
and the carnivalesque, 93, 171
and eccentric speech, 90, 93, 95, 103, 171
malapropism, 96–97, 172
Margolyes, Miriam, 191
Mathews, Anne Jackson, 112–113, 144
Mathews, Charles, 112–114, 129, 143
"At Home" with, 129, 143–144, 148–152, 157
Autobiography, 144–145
"Comic Annual for 1831," 150–152
and comic voice, 146–148
"Memorandum Book," 149
Mathews, Charles James, 130–131, 158
Mayhew, Brothers (Henry and Augustus), 175–176
Mayhew, Henry
London Labour and the London Poor, 10, 118–124
The Wandering Minstrel, 88–90
Meisel, Martin, 4, 6, 19, 20, 118
melodrama, 3, 27–29, 33, 34, 37, 161
Mercadante, Saverio
Elena Uberti, 108–109
meter,
148, 155–157, 173–174
Miller, D. A., 2, 4, 8–9, 103, 192
Moncrieff, W. T., 4
Nicholas Nickleby, 35–37
Tom and Jerry, 122
Mozart, W. A., 58, 76
Don Giovanni, 194, 116–117
Le Nozze di Figaro, 80, 104–106
"Mr. Peggotty's Dream Comes True," 84
"Mr. Ralph Nickleby's 'Honest Composure,'" 26
Murphy, Arthur
The Old Maid, 110–112
music
and class, 87–88
influence on reading/writing of literature, 6, 15, 45–46, 47, 49, 52, 163, 167

and theatricality, 138

narrative/narrativity, 5, 7, 18, 19, 28, 136, 139
New Yorker, 143

odd women
see redundant women
old maids
see redundant women
old price riots, 76–77
opera buffa, 104

"Painful Recollection," 25
painting
influence on reading/writing of literature, 6, 19, 118
passions
see affections
paternalism, 181–182, 184
paternoster, 94, 99
patter, 12
and Catholicism, 98–99
and class, 40, 119
and creativity, 174, 184, 187
definition, 40, 94–95, 99–100
and gender, 40, 43
and hegemony, 94
history of, 38–40, 94–95
and logic, 185–187
and the novel, 41, 153–154, 156, 185
in *opera buffa*, 99, 104–106
place in English culture, 39, 41, 99
in plays, 41, 89, 99, 113–114
spinster, 95–97, 101–102, 151–153, 156–157, 167–168, 172–174, 185–189
street, 49, 89, 95–96, 112–116, 118–124
patter songs, 125–128, 148
performance
of everyday life, 6, 12, 19
and infection, 66, 79, 137
in reading, 12, 45, 54, 79
theatrical, 13–17, 18, 19, 20, 45
physical gesture, 20, 21, 33
physiognomy, 21
Pocock, Isaac,
Hit or Miss, 112–115
The Miller and His Men, 36–37
popular entertainments, 10, 12, 20, 58, 163
private subject, 1, 4, 8
proscenium theatre, 78
public execution, 73, 75
public sphere
Habermas and, 19–20
Victorian, 1

reading subject, 1, 4, 6, 9, 12, 17, 19, 20, 159
realism
literary, 5, 18, 19, 29, 140, 159
social, 59, 63, 140
redundancy, 95, 101–102, 152, 161
and capitalism, 170–171
debate over, 164, 167, 187–188
and lesbianism, 174
redundant women, 102, 110–112, 151–152, 157, 165–180
"natural" v. "accidental," 176
and transportation, 165
rhetorical gesture
see physical gesture
rhythm, 6, 12, 15, 148, 154, 156
Richardson, Samuel
Clarissa, 80
Roach, Joseph, 5
Rousseau, Jean Jacques, 64–66
Rowell, George, 5
Royal Shakespeare Company, 13, 17, 54

Schechner, Richard, 62
Schlicke, Paul, 10, 36
Schumann, Robert
"Carnaval," 138
Neue Zeitschrift für Musik, 106
Scott, Sir Walter, 16
semiotics, 11, 19, 22, 23
of the body, 80–82
of music, 107
and reading, 191
social, 80, 91, 94, 182–183
of the theatre, 5, 6, 9, 16, 21, 23, 159
of the voice, 81–82, 132–134
Shaftesbury, Edmund, 22
Shakespeare, William
Hamlet, 70–71, 82
Macbeth, 140
Much Ado About Nothing, 17–18, 96
Shawn, Wallace, 45–46
Siddons, Henry, 20
slang
see street patter
song
and class, 88–90
and subversion, 67, 86, 88–90
speech
and class, 162
in Dickens, 16, 23
eccentric, 16, 18, 23, 41, 73, 90, 94, 104, 119, 139, 155–157
standard, 16, 18, 23, 94, 104, 119, 139, 155–157
in Victorian literature, 16, 37

spectacle, 1–2, 4–5
spinsters
  see redundant women
stereotypes
  physical, 21
  verbal, 12, 18, 101
Sterne, Laurence, 16
Stewart, Patrick
  A Christmas Carol, 191
"Strolling Actresses Dressing in a Barn," 72
Stirling Edward
  Nicholas Nickleby, 4, 27

tableaux, 15, 19, 22
Taylor, George, 5, 28
Taylor, Tom
  The Ticket of Leave Man, 115–116
Tennyson, Alfred Lord
  The Princess, 174
Thackeray, William
  Vanity Fair, 62
Theatre
  and class, 58, 68
  influence on novel reading, 6, 9, 10–11, 18–19, 21, 33, 63, 159
  influence on novel writing, 6, 9–10, 22, 33, 188
  influence on social life, 19, 21, 23, 26, 66, 79, 92, 140, 159, 188
  versus drama, 4, 5
  Victorian, 4, 5, 15
theatricality, 136–138
  and infection, 137
  and semiotics, 136–137
Turner, Victor, 19

Victorian novel, 3, 4
  reading aloud, 6, 9, 12–13, 90–91, 138
  reading of, 1, 4, 6, 9, 17, 20, 183
  writing of, 1, 4, 6
voice, 129–130
  and class, 76, 81–82, 84
  comic, 12, 44, 53, 130–131, 146–148
  and identity, 83–84, 94, 160
  in literature, 85–86, 100
  and subversion, 91
  and society, 90
  theatrical, 6–8, 10, 12–13, 15–16, 18, 21, 34, 44–47, 52, 129–135, 160

widows
  see redundant women

CAMBRIDGE STUDIES IN NINETEENTH-CENTURY LITERATURE AND CULTURE

General editor

GILLIAN BEER, *University of Cambridge*

Titles published

1. The Sickroom in Victorian Fiction: The Art of Being Ill
by MIRIAM BAILIN, *Washington University*

2. Muscular Christianity: Embodying the Victorian Age
edited by DONALD E. HALL, *California State University, Northridge*

3. Victorian Masculinities: Manhood and Masculine Poetics in early Victorian Literature and Art
by HERBERT SUSSMAN, *Northeastern University*

4. Byron and the Victorians
by ANDREW ELFENBEIN, *University of Minnesota*

5. Literature in the Marketplace: Nineteenth-century British Publishing and the Circulation of Books
edited by JOHN O. JORDAN, *University of California, Santa Cruz*
and ROBERT L. PATTEN, *Rice University*

6. Victorian Photography, Painting and Poetry
The Enigma of Visibility in Ruskin, Morris and the Pre-Raphaelites
by LINDSAY SMITH, *University of Sussex*

7. Charlotte Brontë and Victorian Psychology
by SALLY SHUTTLEWORTH, *University of Sheffield*

8. The Gothic Body
Sexuality, Materialism, and Degeneration at the *Fin de Siècle*
by KELLY HURLEY, *University of Colorado at Boulder*

9. Rereading Walter Pater
by WILLIAM F. SHUTER, *Eastern Michigan University*

10. Remaking Queen Victoria
edited by MARGARET HOMANS, *Yale University*
and ADRIENNE MUNICH, *State University of New York, Stony Brook*

11. Disease, Desire, and the Body in Victorian Women's Popular Novels
by PAMELA K. GILBERT, *University of Florida*

12. Realism, Representation, and the Arts in Nineteenth-century Literature
by ALISON BYERLY, *Middlebury College*

13. Literary Culture and the Pacific: Nineteenth-century Textual Encounters
by VANESSA SMITH, *King's College, Cambridge*

14. Professional Domesticity in the Victorian Novel: Women, Work and Home
by MONICA F. COHEN

15. Victorian Renovations of the Novel
Narrative Annexes and the Boundaries of Representation
by SUZANNE KEEN, *Washington and Lee University*

16. Actresses on the Victorian Stage
Feminine Performance and the Galatea Myth
by GAIL MARSHALL, *University of Leeds*

17. Death and the Mother from Dickens to Freud
Victorian Fiction and the Anxiety of Origins
by CAROLYN DEVER, *New York University*

18. Ancestry and Narrative in Nineteenth-Century British Literature
Blood Relations from Edgeworth to Hardy
by SOPHIE GILMARTIN, *University of London*

19. Dickens, Novel Reading, and the Victorian Popular Theatre
by DEBORAH VLOCK